W9-CAX-834

Visit the Freeman and Herron Web Companion at
www.prenhall.com/freeman

EVOLUTIONARY ANALYSIS

EVOLUTIONARY ANALYSIS

SECOND EDITION

Scott Freeman
University of Washington

Jon C. Herron
University of Washington

PRENTICE HALL
Upper Saddle River, NJ 07458

Library of Congress Cataloging-in-Publication Data

FREEMAN, SCOTT,

    Evolutionary analysis / Scott Freeman, Jon C. Herron.—2nd ed.

    Includes bibliographical references

    1. Evolution (Biology)   2. Evolution (Biology)—Research.   I. Herron, Jon C.   II. Title

QH366.2.F73 2001   576.8—dc21   00-034663

ISBN 0-13-017291-X

Editor in Chief, Biology: Sheri L. Snavely
Production Editor: Debra A. Wechsler
Editorial Director: Paul F. Corey
Executive Managing Editor: Kathleen Schiaparelli
Assitant Managing Editor: Beth Sturla
Assitant Managing Editor, Science Media: Alison Lorber
Project Manager: Karen Horton
Media Editor: Andrew Stull
Assistant Vice President of Production and Manufacturing: David W. Riccardi
Marketing Manager: Jennifer Welchans
Director of Marketing, ESM: John Tweeddale
Manufacturing Buyer: Michael Bell
Manufacturing Manager: Trudy Pisciotti
Art Director: Anne France/Joseph Sengotta/Jon Boylan
Assistant to Art Director: John Christiana
Director of Creative Services: Paul Belfanti
Art Manager: Gus Vibal
Art Editor: Karen Branson
Interior Designer: Ximena Tamvakopoulos
Cover Designer: Joseph Sengotta
Photo Research: Stuart Kenter Associates
Photo Research Administrator: Beth Boyd
Illustrators: Jon C. Herron; Robin S. Manasse; Imagineering
    Scientific and Technical Artworks, Inc.
Editorial Assistant: Lisa Tarabokjia

Cover photos (from left to right): Two color morphs of
    Northern Water Snake, *Nerodia sipedon*, in cypress
    swamp near New Orleans, Louisiana, May, 1988 by
    Photo Researchers, Inc.; cast of trilobite fossil,
    Cryptolithoides ulrichi, Viola Limestone, Ordovician,
    Murray County, Oklahoma, May, 1998 by A.
    Graffham/Visuals Unlimited; four-winged fly by
    Edward B. Lewis; bottlenose dolphins, *Tursiops*
    truncatus, captive, Waikaloa Hyatt, Hawaii by Flip
    Nicklin/Minden Pictures; *Begonia involucrata*, two
    inflorescences, male & female flowers by Douglas W.
    Schemske; finch (*Geospiza fortis*) by T.J. Ulrich,
    Academy of Natural Sciences of Philadelphia.
Cover illustrations: Jon C. Herron; Robin S. Manasse

Prentice Hall

©2001, 1998 by Scott Freeman and Jon C. Herron
Published by Prentice-Hall, Inc.
Upper Saddle River, New Jersey 07458

Printed in the United States of America
10 9 8 7 6 5 4 3 2 1

ISBN 0-13-017291-X

Prentice-Hall International (UK) Limited, *London*
Prentice-Hall of Australia Pty. Limited, *Sydney*
Prentice-Hall Canada, Inc., *Toronto*
Prentice-Hall Hispanoamericana, S.A., *Mexico*
Prentice-Hall of India Private Limited, *New Delhi*
Prentice-Hall of Japan, Inc., *Tokyo*
Pearson Education Asia Pte. Ltd.
Editora Prentice-Hall do Brasil, Ltda., *Rio de Janeiro*

Brief Contents

Contents

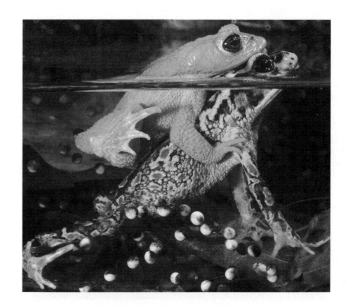

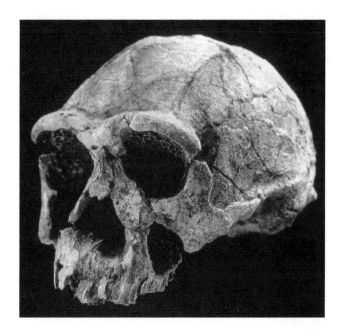

Preface

The aims and audience of *Evolutionary Analysis* have not changed from the first edition to the second. Our goal is still to help students learn how to think like evolutionary biologists. The presentation is intended for undergraduates who are majoring in the biological sciences in preparation for careers in medicine, conservation, education, science journalism, or research. We assume that our readers have finished their introductory coursework and are ready to explore how a course in evolutionary biology can enrich their personal and professional lives.

Our approach and philosophy are also unchanged. Our tack is to present the topics that form the core of evolutionary biology in the same spirit of inquiry that drives research. Wherever possible, we motivate the material with the types of questions that evolutionary biologists ask. Are humans more closely related to chimpanzees or gorillas? If people with the *CCR5-Δ32* mutation are resistant to infection by HIV, will this allele increase in frequency in populations afflicted by the AIDS epidemic? Why did the dinosaurs suddenly go extinct, after dominating the land vertebrates for over 150 million years? Often a theoretical treatment will help to focus these questions, generate hypotheses, and make predictions that can be tested. After introducing the experiments and observations that biologists have used to test competing hypotheses, we analyze the data that resulted and consider what work remains to be done. Throughout the book, our objective is to present evolutionary biology as a dynamic and increasingly interdisciplinary enterprise.

Although the fundamental premise and approach of the book have not changed, its organization has. To align the sequence of chapters more closely with the way that most professors teach the course, we have reorganized the chapters into five units:

- **Part I, Introduction**, demonstrates why evolution is relevant to real-world problems, establishes the fact of evolution, and presents natural selection as an observable process.

- **Part II, Mechanisms of Evolutionary Change**, develops the theoretical underpinnings of the Modern Synthesis by exploring how mutation, selection, migration, and drift produce evolutionary change. The population genetics coverage is dramatically expanded from the first edition, but simplified by the placement of most algebraic treatments in boxes. These chapters have also been enriched by an increased focus on how population and quantitative genetic models can be applied to real-life problems in medicine and conservation.

- **Part III, Adaptation**, is a new unit that begins by introducing methods for studying adaptation, and follows up by offering detailed investigations into sexual selection, kin selection, and selection on life history characters.

- **Part IV, The History of Life**, starts with an analysis of speciation and phylogeny inference methods. Subsequent chapters focus on Precambrian evolution, the Phanerozoic, and human evolution.
- **Part V, Current Research—A Sampler**, includes a chapter treating classical and recent topics in molecular evolution. The unit also contains two new chapters. One of these focuses on evolutionary insights that have emerged from advances in developmental genetics; the other explores applications of evolutionary biology in epidemiology, medical physiology, human behavior, and public health.

As in the first edition, most chapters include boxes that cover special topics or methods, provide more detailed analyses, or offer derivations of equations. All chapters end with a set of questions that encourage students to review the material, apply concepts to new issues, and explore the primary literature.

Website and Transparencies

The companion website for Evolutionary Analysis has been revised and expanded. Each unit now includes two case studies. These tutorials challenge students to pose questions, formulate hypotheses, design experiments, analyze data, and draw conclusions. A tutorial for population genetics features problems students can solve using a downloadable simulation. The website also provides answers to selected end-of-chapter questions, guides to exploring the literature, links to other evolution-related sites, and an opportunity to email us with suggestions and comments.

The website for Evolutionary Analysis is accessible through the book's homepage at

http://www.prenhall.com/freeman

Prentice Hall's commitment to a four-color format for this edition of Evolutionary Analysis has enabled us to make the diagrams, data graphics, and photographs easier to interpret and the overall presentation brighter and more accessible. In response to requests from professors using the first edition, a set of 100 full-color overhead transparencies has been developed for the second edition. All transparencies are labeled with large, boldface type for easy reading in the classroom. Professors can get the transparency set by contacting either their local Prentice Hall representative, or Prentice Hall faculty services at (800) 526-0485.

Acknowledgments

We'd like to extend our heartfelt thanks to all of the colleagues who have helped this book serve students better by contributing published reviews, commissioned reviews, emailed suggestions, packets of reprints, and conversations with us at meetings. In particular, the additions and revisions in this edition were inspired by a series of thoughtful and constructive critiques provided by the following colleagues:

Leticia Aviles, *University of Arizona*
Loren E. Babcock, *The Ohio State University*
David Begun, *University of Toronto*
Andrew Berry, *Harvard University*
Keith A. Crandall, *Brigham Young University*
W. Ford Doolittle, *Dalhousie University*
Jerry F. Downhower, *The Ohio State University*
Scott Edwards, *University of Washington*
Gary M. Fortier, *Delaware Valley College*
Glenys Gibson, *Acadia University*
Lonnie J. Guralnick, *Western Oregon University*
Werner G. Heim, *Colorado College*

Joel Kingsolver, *University of Washington*
Andrew H. Knoll, *Harvard University*
Robert A. Krebs, *Cleveland State University*
Catherine S. McFadden, *Harvey Mudd College*
Andrew Martin, *University of Colorado*
Arnold I. Miller, *University of Cincinnati*
Nancy A. Moran, *University of Arizona*
Stephen W. Schaeffer, *Pennsylvania State University*
Robert Sinsabaugh, *University of Toronto*
Maureen Stanton, *University of California, Davis*
Daniel B. Thompson, *University of Nevada, Las Vegas*
Sara E. Via, *University of Maryland, College Park*

It would be difficult to overemphasize how much this edition has benefited from the care and creativity of these reviewers. In a very real sense, this book is a product of a community that is passionate about the study of evolution and devoted to teaching.

In addition, we'd like to extend our sincere thanks to students who emailed suggestions for improvement. Especially helpful were the formal written critiques contributed by Dr. Peter Wimberger's class at the University of Puget Sound, and many comments and suggestions from our own students at the University of Washington.

We owe a special debt to three colleagues who contributed revisions of key chapters and to Dr. Kathleen Hunt, who contributed revised versions of the end-of-chapter questions. We are tremendously fortunate to have had such talented contributors involved in the preparation of this edition. Specifically, Dr. Michael Hart of Dalhousie University revised the chapters on life history, phylogeny inference, and the Phanerozoic and originated the chapter on evolution and development. His command of the literature and enthusiasm for evolutionary biology were a tremendous asset during the revision process, and his ingenuity and excitement about science made him an absolute joy to work with. Dr. Niles Lehman of the State University of New York, Albany, revised the chapter on the origin of life and Precambrian evolution. His creativity, passion, knowledge, and rapid work were vital in bringing students a fresh and accurate read on this dynamic topic. Dr. David Begun of the University of Toronto updated the chapter on human evolution. His expertise was an invaluable contribution to our coverage of this fast-moving field, and his willingness to work to a tight deadline was greatly appreciated.

This book would not exist without the skill, professionalism, and dedication of the editorial and production team at Prentice Hall. Paul Corey's leadership and commitment as Editorial Director of Science and Mathematics have been instrumental. Senior Production Editor Debra Wechsler did a superb job managing the thousands of details required to make a book as free of errors and omissions as is humanly possible. Art Editor Karen Branson assembled and managed a talented group of artists to create the four-color figure program. Robin S. Manasse did most of the new biological illustrations, including the animals on the cover. Art Directors Anne France, Joe Sengotta, and Jon Boylan created an attractive design and cover for the book. Andrew Stull was responsible for expanding and improving the website. Executive Marketing Manager Jennifer Welchans is supplying the drive and tools necessary to ensure that professors throughout the world have an opportunity to consider the book for their course.

Finally and most, we thank Editor in Chief of Biology Sheri Snavely, whose vision and energy are the wellspring of this book. Her passion for publishing, infallible instincts, humor, and devotion are a constant source of inspiration.

Scott Freeman
Jon C. Herron
Seattle, Washington

The Galápagos islands have served as a laboratory for evolutionary studies for over 160 years. (Frans Lanting/Minden Pictures)

INTRODUCTION

AMONG THE ATTRACTIVE ASPECTS OF EVOLUTIONARY BIOLOGY IS THAT IT IS ROOTED in a single, organizing mechanism. This mechanism is natural selection. Gaining a thorough understanding of natural selection is thus the first challenge for anyone studying evolution. Natural selection is simple in concept, but widely misunderstood. Understanding how the process works requires moving beyond slogans like "survival of the fittest."

Our primary goal in Part I (Chapters 1–3) is to introduce natural selection as an agent of evolutionary change. Our exploration of the process begins with a chapter devoted to understanding the human immunodeficiency virus (HIV) and the AIDS epidemic. Chapter 2 expands on this presentation by reviewing evidence that all organisms have changed through time, or evolved, throughout the history of life—much as HIV has evolved over the past half century. In Chapter 3 we explore how natural selection works in more detail. This sequence of chapters lays the groundwork for exploring the other mechanisms that cause change through time, in Part II.

Our secondary objective in these chapters is to introduce the experimental and analytical methods used by researchers in evolutionary science. These methods are a prominent theme throughout the text. We emphasize them to help readers learn how to ask questions, design experiments, analyze data, and critically review scientific papers, in addition to mastering facts. The detailed examples we present not only make the general concepts of evolutionary biology clear but also provide insight into how we know what we know.

CHAPTER 1

A Case for Evolutionary Thinking: Understanding HIV

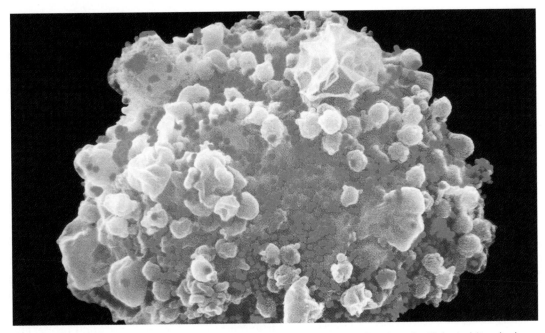

The red particles emerging from this human T cell are HIV virions. (National Institute for Biological Standards and Control, England/Science Photo Library/Photo Researchers, Inc.)

A T THE START OF A COURSE, IT CAN BE USEFUL TO STEP BACK AND ASK TWO questions: What sort of content will be covered? and How will this information help me in my professional and everyday life? To help answer these questions, we explore the evolution of human immunodeficiency virus (HIV). This is the pathogen that causes the acquired immune deficiency syndrome (AIDS).

In this chapter, we introduce the scope of evolutionary analysis through an in-depth look at a prominent contemporary issue. Our goals are to illustrate the kinds of questions that evolutionary biologists investigate, demonstrate how an evolutionary perspective can inform research throughout the biological sciences, and introduce concepts that will be explored in detail later in the text.

HIV is a compelling case study because it raises issues that are almost certain to influence the professional and personal lives of every one of our readers. As an emerging virus that rapidly evolves drug resistance, HIV exemplifies two pressing public health issues. AIDS already qualifies as one of the most devastating epidemics ever experienced by our species.

Here are the questions we address:

- Why have promising AIDS treatments, like the drug azidothymidine (AZT), proven ineffective in the long run?
- Why are some people resistant to becoming infected or to progressing to disease once they are infected?
- Why does HIV kill people?
- Could a vaccine provide protection from the diversity of HIV strains found today?

As a case study, HIV will demonstrate how evolutionary biologists study adaptation and diversity.

These questions may not sound as if they have anything to do with evolutionary biology, but evolutionary biology is the science devoted to understanding two things: (1) how populations change through time in response to modifications in their environment, and (2) how new species come into being. More formally, evolutionary biologists study adaptation and diversity. These are exactly the issues targeted by our questions about AIDS and HIV. Before we tackle them, however, we need to delve into some background biology.

1.1 The Natural History of the Epidemic

In December, 1999 the United Nations' AIDS program estimated that about 40 million people worldwide were infected with HIV (Piot 1998; Figure 1.1). The epidemic has already been responsible for more deaths than the black plague that devastated Europe in the 14th century. By the end of 2001, AIDS is almost certain to overtake the "Spanish" influenza outbreak of 1918 as the epidemic responsible for the most deaths in all of human history.

Most HIV infections result from two related but distinct epidemics that occurred during the 1980s and 1990s: one among heterosexual men and women in sub-Saharan Africa and south and southeast Asia, and the other among homosexual men and intravenous drug users in the United States and Europe. The dual epidemics are distinguished by the number of people involved and the disease's mode of transmission.

In sub-Saharan Africa, the number of AIDS cases is almost overwhelming (see Mann and Tarantola 1998; Piot 1998). Epidemiologists estimate that over 23 million Africans are currently infected with HIV-1. In most sub-Saharan cities, over 75% of

Figure 1.1   Distribution of HIV infections   This map shows the geographic distribution of HIV-1 infections. The most dramatic conclusion to be drawn from this figure is that AIDS is primarily a disease of developing nations. It is estimated that over 90% of HIV-infected people live in the poor countries of the southern hemisphere. Two-thirds of these are in sub-Saharan Africa alone.

920,000

520,000

360,000

530,000

220,000

6 million

1.7 million

23.3 million

12,000

all adult deaths are now due to AIDS; if current projections hold, one-third of the workforce in some countries may eventually die of the disease. In this epidemic—and in the outbreak now underway in India and China—the virus is usually transmitted to new hosts through heterosexual intercourse.

In contrast, the AIDS epidemic in North America and Europe is primarily due to transmission via homosexual intercourse and needle sharing among intravenous drug users. The frequency of infection in the general population is low in the developed nations—about 0.56% in North America versus 10% or greater in many African countries. Further, Figure 1.2 shows that, while the number of people infected continues to increase in Asia and Africa, efforts by educators and researchers are paying off in North America and Europe: In the industrialized nations, the number of new HIV infections is declining.

What is HIV?

Like all viruses, HIV is an obligate, intracellular parasite. It is incapable of an independent life and is highly specific in the cell types it afflicts. HIV parasitizes components of the human immune system called macrophages and T cells. It uses the enzymatic machinery and energy found in these cells to make copies of itself, killing the host cell in the process.

HIV is a parasite that afflicts cells in the human immune system.

Figure 1.3 outlines HIV's life cycle in more detail. Step 1 shows the extracellular or infectious phase, when the virus can be transmitted from host to host. In this form, a virus is called a virion or particle. Steps 2–8 describe the intracellular or parasitic phase, when the virus replicates.

In HIV's case, the replication process begins when a virion attaches to a specific protein, called CD4, found on the surface of certain macrophages and T cells. Binding to the host cell surface is completed when HIV attaches to a second protein, called a coreceptor, on the cell surface. When binding occurs, the envelope around the virion and the cell membrane fuse (steps 2 and 3) and the contents of the virion enter the cell (step 3). These contents include the virus' diploid genome (in the form of two identical RNA molecules) and a protein, called reverse transcriptase, that transcribes this genome.

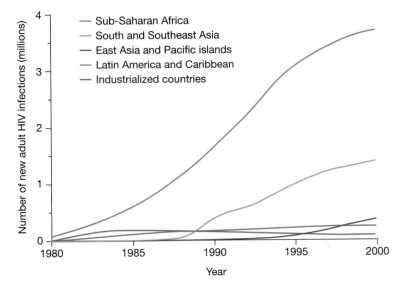

Figure 1.2   **HIV infection rates by geographic area**   This graph shows the estimated number of new HIV infections, annually, for five broad geographic areas. The number of new infections is dropping in the North America and Europe, but increasing steadily in Africa and Asia.

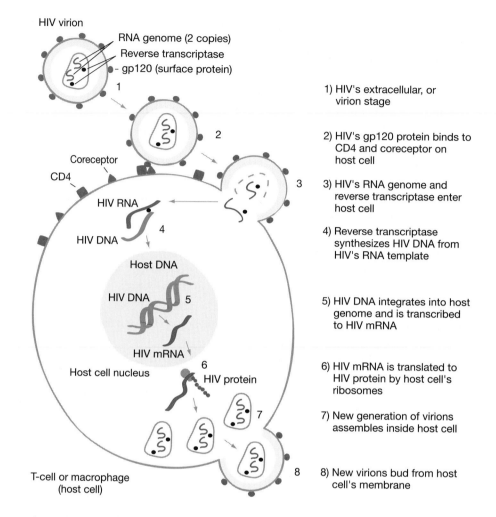

Figure 1.3   The life cycle of HIV

Labels within figure:

HIV virion
RNA genome (2 copies)
Reverse transcriptase
gp120 (surface protein)

Coreceptor
CD4
HIV RNA
HIV DNA
Host DNA
HIV DNA
HIV mRNA
Host cell nucleus
HIV protein
T-cell or macrophage
(host cell)

1) HIV's extracellular, or virion stage

2) HIV's gp120 protein binds to CD4 and coreceptor on host cell

3) HIV's RNA genome and reverse transcriptase enter host cell

4) Reverse transcriptase synthesizes HIV DNA from HIV's RNA template

5) HIV DNA integrates into host genome and is transcribed to HIV mRNA

6) HIV mRNA is translated to HIV protein by host cell's ribosomes

7) New generation of virions assembles inside host cell

8) New virions bud from host cell's membrane

The next step in the life cycle is illustrated in step 4. Reverse transcriptase synthesizes viral DNA in the host cell cytoplasm, using ATP and nucleotides provided by the host cell. The viral DNA then enters the nucleus, integrates into the host's genome, and is transcribed by the host cell's DNA polymerase II (step 5). The viral mRNAs that result are translated into proteins by the host cell's ribosomes (step 6).

In HIV and other "retroviruses," then, the flow of genetic information is different than it is in cells and in the many viruses with DNA genomes. In retroviruses like HIV, genetic information does not follow the familiar route of DNA to mRNA to proteins. Instead, information flow is from RNA to DNA to mRNA to proteins. It is that first step, featuring a backward flow of information, that inspired the "retro" in retrovirus and the "reverse" in reverse transcriptase.

After viral proteins and copies of the RNA genome have been synthesized, a new generation of virions assembles in the host cell cytoplasm (step 7). The life cycle concludes when virions bud off of the cell membrane to the exterior, as shown in step 8. These virions then begin floating in the bloodstream. If they encounter another cell in the same host containing a CD4 protein on its surface, the life cycle begins anew. Or, virions can be transmitted to a new host individual through a blood transfusion, needle sharing, or sexual intercourse.

The important thing to note about this life cycle is that the virus uses the host cell's own enzymatic machinery—its polymerases, ribosomes, and tRNAs—in almost every step. This is why viral diseases are so difficult to treat. Drugs that interrupt the virus life cycle are almost certain to interfere with the host cell's enzymatic functions as well and cause debilitating side effects.

How Does HIV Cause AIDS?

The human body responds to HIV infection in two ways: by destroying virions floating in the bloodstream, and by killing its own infected cells before new virions are assembled and released. The cells infected by HIV—macrophages and helper T cells—are crucial to both aspects of the immune response. Because HIV infection kills these cells, advanced HIV infections begin to undermine the immune response. The eventual collapse of the immune system leads to the condition known as AIDS. The syndrome is characterized by opportunistic infections with bacterial and fungal pathogens that rarely cause problems for people with robust immune systems. Once an HIV-infected individual begins showing symptoms of AIDS, death usually occurs within 2 years.

AIDS begins when HIV infection has progressed to a point where the immune system does not function properly.

With the preceding information as background, we are ready to explore questions about the evolution of HIV. The first issue has frustrated everyone involved in fighting the epidemic: Why has it been so hard to develop drugs capable of combating HIV? It is certainly not for lack of trying; governments and private companies have committed hundreds of millions of dollars to AIDS research and drug development. The story of AZT, one of the first anti-AIDS drugs, has turned out to be typical. Early on, AZT looked promising, but ultimately it proved a disappointment. To explain why, we need to introduce the principle of evolution by natural selection.

1.2 Why Does AZT Work in the Short Run, But Fail in the Long Run?

To combat viral infections, researchers search for drugs that are capable of inhibiting virus-specific enzymes. For example, a drug that interrupts reverse transcription should effectively and specifically kill retroviruses with minimal side effects. This is exactly the rationale behind azidothymidine, or AZT.

Note the *thymidine* in AZT's name: AZT is a nucleotide analog, similar in structure to normal thymidine, that "fools" reverse transcriptase. When AZT is present in a cell, reverse transcriptase mistakenly adds it to the growing DNA strand where a thymidine 5′-triphosphate should go. This misstep terminates reverse transcription, because AZT will not accept the addition of the next nucleotide to the growing chain. AZT thus interrupts the pathway to new viral proteins and new virions, and stops the infection.

When reverse transcriptase adds AZT instead of thymidine to a growing copy of HIV's genome, synthesis stops. In this way, AZT can slow or stop the spread of an infection.

In early tests, AZT worked. It effectively halted the loss of macrophages and T cells in AIDS patients. Because it also sometimes fools DNA polymerase and interrupts DNA synthesis in host cells, AZT produced serious side effects. But it did promise to inhibit or at least slow the progression of the disease. By 1989, however, after only a few years of use, patients stopped responding to treatment and their CD4 cell counts again began to decline. Why?

To answer this question, consider a thought experiment. If we wanted to genetically engineer an HIV virion capable of replicating in the presence of AZT, what would we do? The simplest answer might be to change the active site in the reverse transcriptase enzyme, making it less likely to mistake AZT for the normal nucleotide. In practice, we could use a mutagenic chemical or ionizing radiation to generate strains of HIV with altered nucleotide sequences in their genomes, and thus altered amino acid sequences in their proteins. If many mutants were generated, at least a few would have a change in the part of the reverse transcriptase molecule that recognizes and binds to normal thymidine (Figure 1.4a). If one of these altered sequences were less likely to mistake AZT for the normal nucleotide, then the mutant variant of HIV would be able to continue replicating in the presence of the drug. In populations of HIV virions treated with AZT, strains unable to replicate in the presence of AZT would decline in numbers, and the new form would come to dominate the HIV populations.

The steps involved in this thought experiment have actually occurred inside the bodies of individual HIV patients. How do we know? Researchers took repeated samples of the HIV virions from patients receiving AZT throughout the course of their treatment. In each sample of the virus, the researchers sequenced the reverse transcriptase gene. They found that viral strains present late in treatment were genetically different from viral strains present before treatment *in the same host individuals.* The viral populations had become resistant to AZT. The mutations associated with resistance were often the same from patient to patient (St. Clair et al. 1991; Mohri et al. 1993; Shirasaka et al. 1993), and were located in the active site of reverse transcriptase (Figure 1.4b). Researchers have directly observed the evolution of AZT resistance in dozens of AIDS patients. In each individual, mutations in the HIV genome led to specific amino acid substitutions in the active site of reverse transcriptase. These genetic changes allowed the mutant strains of the virus to replicate in the presence of AZT. Unlike the situation in our thought experiment, however, no conscious manipulation took place. How, then, did the change in the viral strains occur?

The key is twofold: Reverse transcriptase is error-prone, and HIV's genome has no instructions for making error-correcting enzymes. (In this regard HIV is like most retroviruses, but unlike DNA-based cellular organisms such as ourselves.) As a result, over half of the DNA transcripts produced by reverse transcriptase contain at least one mistake or mutation (Hübner et al. 1992; Wain-Hobson 1993). HIV has the highest mutation rate of any virus or organism observed to date. Be-

Some mutations in the active site of reverse transcriptase make the enzyme less likely to add AZT instead of thymidine.

Figure 1.4 Computer-generated images of reverse transcriptase (a) This image shows the large groove in the reverse transcriptase enzyme where the substrate (RNA) binds. (Thomas A. Steitz, Yale University) (b) The red spheres on this image indicate the locations of amino acid substitutions correlated with resistance to AZT. Note that they are localized in the enzyme's groove, or active site. (Lori Kohlstaedt from Jon Cohen, "AIDS Research: The Mood is Uncertain," *Science*, Vol. 260, May 28, 1993.)

(a)
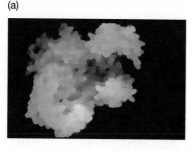

(b)
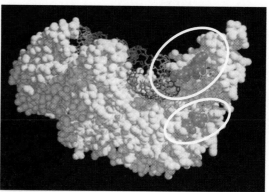

cause thousands of generations of HIV replication take place within each patient during the course of an infection, a single strain of HIV can produce hundreds of different reverse transcriptase variants over time.

Simply because of their numbers, it is a virtual certainty that one or more of these variants contains an amino acid substitution that lessens the reverse transcriptase molecule's affinity for AZT. If the patient takes AZT, the replication of unaltered HIV variants is suppressed, but the resistant mutants will still be able to synthesize some DNA and produce new virions. As the resistant virions reproduce and the nonresistant virions fail, the fraction of the virions in the patient's body that are resistant to AZT increases over time. Furthermore, each new generation of virions is likely to contain virions with additional new mutations. Some of these additional mutations may further enhance the ability of reverse transcriptase to function in the presence of AZT. Because they reproduce faster, the virions that carry these new mutations will also increase in frequency at the expense of their less resistant ancestors.

This process of change over time in the composition of the viral population is called evolution by natural selection. It has occurred so consistently in patients taking AZT that the use of AZT alone has been abandoned as an AIDS therapy. In addition, when other nucleotide analogs, like ddI or ddC, have been used by themselves or in conjunction with AZT, HIV populations have evolved multiple-drug resistance (Larder et al. 1993; Shirasaka et al. 1993; Mohri et al. 1993). Resistance to the class of drugs called protease inhibitors evolved within two years of their introduction (Ala et al. 1997; Deeks et al. 1997).

Changes in the genetic make-up of HIV populations over time have led to increased drug resistance. This is an example of evolution by natural selection.

Now consider a slightly different question. We have been following the fate of virions carrying different versions of the reverse transcriptase gene when AZT is present in the environment (that is, in host cells). Are the mutant HIV strains also more efficient at reproducing when host cells do not contain AZT? The answer is no: When AZT therapy has been started and then stopped, the proportion of AZT-resistant virions in the virus population has fallen over time, back toward what it was before AZT treatment began. Back-mutations that restored reverse transcriptase's amino acid sequence to its original configuration were favored by natural selection (St. Clair et al. 1991). Notice how dynamic natural selection is: Without AZT present, natural selection favors nonmutant virions; with AZT present, natural selection favors mutant virions. Is evolution by natural selection unidirectional and irreversible? The answer, emphatically, is no.

Note that the process we have just reviewed involves four steps:

1. Transcription errors made by reverse transcriptase led to mutations in the reverse transcriptase gene.
2. These mutations produced variability in enzyme function among virions.
3. Some virions were better able to survive and reproduce in an environment containing AZT than others, because of the functional properties of their mutant reverse transcriptase enzymes.
4. These mutations were passed on to the offspring of AZT-resistant virions.

The outcome of this process is that new virion forms came to dominate HIV populations within hosts. The genetic composition of the HIV population at the end of the process was different from what it was at the start. This is evolution by natural selection.

1.3 Why is HIV Fatal?

One of the keys to becoming an evolutionary biologist is to learn how to think like an organism. This means adopting what biologists call "selection thinking." For example, from HIV's point of view, the tendency to cause disease in a host—the trait called **virulence**—is largely a function of its growth rate. Illness in the host is a side effect of high growth rates. Extremely high growth rates can result in the death of the host. According to selection thinking, the key to understanding why HIV is fatal is to understand why it is advantageous for virions to replicate as quickly as they do. If HIV can evolve so readily in response to drug therapies, why has it not evolved to have a more benign impact on its host?

Sometimes, the answer to questions like this is because it cannot. That is, perhaps competing HIV strains would be more successful—in the sense of infecting more people—if they grew more slowly and did not kill their hosts, but are not able to because of some unchangeable property of reverse transcriptase, or because targeting CD4 cells inevitably makes infections fatal. It is important to recognize that organisms are constrained in a variety of ways. Natural selection cannot optimize every aspect of a life cycle.

The evidence, however, suggests that constraints are not the cause of HIV's high virulence. Several nonlethal diseases afflict CD4 cells, including human herpes virus 6, which appears to cause only a mild rash similar to the childhood affliction called roseola (see Culliton 1990), and a virus called HIV-2, which may often be nonlethal (Ewald 1994; Marlink et al. 1994). These observations suggest that diseases of CD4 cells are not highly virulent by definition.

Another explanation for HIV's virulence is that evolution to a benign state has not occurred, simply due to a lack of variation in the degree of virulence. If there are no mutations that alter the level of virulence, then virulence cannot evolve by natural selection. Three lines of evidence argue against this hypothesis, however. First, HIV-1 strains that dominate late in a given infection, when the patient is symptomatic, grow much faster in culture than strains present earlier in the infection, suggesting that they are more virulent (see Goldsmith 1990; Ewald 1994). Second, specific base substitutions have been identified that are associated with increased virulence (these base substitutions are found in the gene coding for the gp120 protein found on the surface of HIV virions; see Groenink et al. 1993). Third, specific genetic changes have been identified that are associated with decreased virulence (these base substitutions are found in a gene coding for one of HIV's regulator proteins; see Deacon et al. 1995).

A third hypothesis is that natural selection has favored highly virulent strains of HIV-1. This is the claim that Paul Ewald makes with his transmission rate hypothesis (Ewald 1994) illustrated in Figure 1.5. The hypothesis predicts that if transmission of sexually transmitted diseases from an existing host to new hosts is frequent, then natural selection will favor increased virulence. But if transmission to new hosts is infrequent, then selection will favor more benign strains.

Several alternative hypotheses exist to explain why HIV is fatal. HIV's virulence could be (1) an inevitable outcome of infecting immune system cells, (2) due to a lack of genetic variation, or (3) a trait that allows certain HIV strains to thrive in particular environments.

The key to the transmission rate hypothesis is the concept of trade-offs. Evolutionary biologists analyze trade-offs in terms of the costs and benefits of competing strategies. In this case, the strategies are growing slowly or growing rapidly. For a virion, the benefit of rapid growth is increasing its prevalence in the bloodstream of its host, and consequently increasing its chance of becoming transmit-

a) Assumption 1: More virulent strains maintain a higher concentration of HIV virions in the patient's blood, and kill the patient faster.

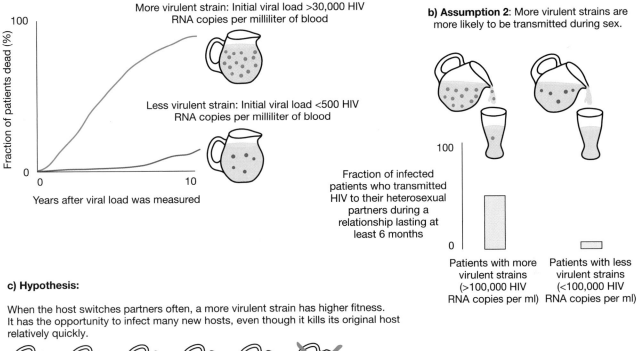

More virulent strain: Initial viral load >30,000 HIV RNA copies per milliliter of blood

Less virulent strain: Initial viral load <500 HIV RNA copies per milliliter of blood

Fraction of patients dead (%)

100

0

0   10

Years after viral load was measured

b) Assumption 2: More virulent strains are more likely to be transmitted during sex.

100

0

Fraction of infected patients who transmitted HIV to their heterosexual partners during a relationship lasting at least 6 months

Patients with more virulent strains (>100,000 HIV RNA copies per ml)

Patients with less virulent strains (<100,000 HIV RNA copies per ml)

c) Hypothesis:

When the host switches partners often, a more virulent strain has higher fitness. It has the opportunity to infect many new hosts, even though it kills its original host relatively quickly.

When the host switches partners rarely, a less virulent strain has higher fitness. It allows its host to live long enough to give it the opportunity to infect more than one new host.

Figure 1.5   The transmission rate hypothesis.   The diagrams at the top illustrate two important assumptions that underlie the hypothesis. The data in part (a) are from J. W. Mellors, 1998. *Scientific American* 279 (July): 90; the data in (b) are from M. V. Ragni et al. *Journal of Acquired Immune Deficiency Syndromes and Human Retrovirology* 17: 42–45. When these assumptions hold, virulent strains of sexually transmitted pathogens should spread through a population most efficiently when rates of partner change are high, while benign strains should spread most efficiently when rates of partner change are low (see part c).

ted during a given episode of sexual intercourse. The cost of rapid growth is killing enough CD4 cells to make the host ill and less likely to engage in intercourse.

The logic behind the transmission rate hypothesis now becomes clear: If people change sexual partners rarely, then any single episode of intercourse is unlikely to involve a new partner. In this environment, transmission to a new host is unlikely during any single episode of intercourse. If infected people are monogamous, highly virulent forms of HIV are almost certain to kill both partners before

*Differences in the degree of
virulence among HIV strains
can be analyzed as competing
strategies, each with costs
and benefits. Which strategy
succeeds depends on
environmental conditions—in
this case, the degree of
sexual promiscuity.*

transmission to a new host can occur. If monogamy is widespread in the host population, then highly virulent forms of the virus should die out. Slowly reproducing forms of HIV should increase in frequency, instead. Why? These strains are present in fewer numbers in a host's bloodstream or semen at any time, but can be transmitted if monogamous individuals occasionally obtain new sexual partners after a divorce or the death of a spouse.

Conversely, if promiscuity is prevalent, then highly virulent strains will be transmitted to new hosts much more frequently than slowly replicating forms. Thus, they will increase in frequency in the total viral population.

Ewald's claim is that highly virulent strains came to dominate HIV populations because of changing sexual practices among heterosexual men and women in eastern and central Africa and homosexual men in the United States and Europe. Because of economic circumstances in the 1970s and 1980s, millions of African men migrated from their homes in rural areas to work in the large cities of Zaire, Uganda, and Kenya. A large commercial sex industry sprang up in response, with prostitutes in some cities having up to 1000 sexual partners per year (see Ewald 1994). Likewise, rates of partner change among homosexual men in the United States and Europe in the late 1970s and early 1980s were high—up to 10 per six months (Koblin et al. 1992). The transmission rate hypothesis contends that changes in transmission dynamics, caused by these behavioral changes in host populations, strongly favored the evolution of virulent strains of HIV.

Further, the hypothesis predicts that different populations of people should harbor HIV strains with different degrees of virulence, depending on how often host individuals change sexual partners. Two natural experiments now underway may help us to understand just how valid the hypothesis is.

- Since the start of the epidemic, rates of partner change among homosexual men in the United States and Europe have decreased dramatically (Adib et al. 1991), and rates of condom use have increased (Catania et al. 1992). In contrast, condom use and sexual practices in Africa seem to have changed little (Editorial staff 1995). The transmission rate hypothesis predicts that HIV will slowly become less deadly in North America and Europe, but continue to kill people, after relatively brief infections, in central and East Africa, and perhaps in Asia as well.
- HIV-2, a close relative to HIV-1, is similar in life cycle and gene composition, but is much more benign (De Cock et al. 1993). HIV-2 has recently moved from its historical center of incidence in West Africa, where sexual transmission rates due to partner changes are low, to India, where rates may be much higher (Ewald 1994). If research confirms that HIV-2 is being transmitted more rapidly in India than in West Africa because of more promiscuous sexual practices, the transmission rate hypothesis predicts that highly virulent Asian strains of HIV-2 should evolve.

Is the transmission rate hypothesis correct? Only time, and data, will tell.

1.4 Why Are Some People Resistant to Infection by HIV?

The principle of evolution by natural selection explains why strains of HIV have become drug resistant. It may also explain why the virus is deadly. Can the same principle clarify why some people who are repeatedly exposed to the virus do not become infected?

For researchers and physicians who are struggling to find strategies for controlling the AIDS epidemic, the existence of exposed, but uninfected, people is a ray of hope. If natural resistance to the virus exists, and if the molecular basis of this resistance can be identified, it might be possible to replicate the mechanism of resistance with new drug therapies.

Both of these "ifs" have now been confirmed. In the early 1990s, work from several laboratories demonstrated that some people remain uninfected even after repeated exposure to the virus, and that some people who are infected with the virus survive many years longer than expected (see Cao et al. 1995). A breakthrough in recognizing the molecular basis of resistance occurred when a team led by Edward Berger identified the coreceptor molecules that allow HIV to enter macrophages and T cells (see Feng et al. 1996; Alkhatib et al. 1996). Soon after, Rong Liu and co-workers (1996) and Michel Samson and associates (1998) suggested that resistant individuals might have unusual forms of the coreceptor molecules, and that these mutant proteins thwart HIV entry.

To test this hypothesis, Samson and colleagues sequenced the gene that codes for a particularly important coreceptor, called CCR5, from three HIV-infected individuals who are long-term survivors. One of the individuals had a mutant form of the gene, as predicted. Because this allele is distinguished by a 32-base-pair deletion in the normal sequence of DNA, Samson and co-workers named it the $\Delta 32$ allele (Δ is the Greek letter delta). Then, they showed that HIV cannot enter cells that have the $\Delta 32$ form of CCR5 on the surfaces. This experiment confirmed that the allele protects individuals from infection.

To follow up on this result, Samson and co-workers took DNA samples from a large number of individuals of northern European, Japanese, and African heritage, examined the gene for CCR5 in each individual, and calculated the frequency of the normal and $\Delta 32$ alleles in each population. A striking difference emerged: The mutant allele is present at a relatively high frequency of 9% in Caucasians, but is completely absent in individuals of Asian or African descent.

Why would one form of a gene be relatively common in one population, but absent in others? Samson and co-workers offered two possible explanations: Either the $\Delta 32$ allele was recently favored by natural selection in Caucasian populations, or it could have increased by chance due to a process called genetic drift. These competing hypotheses are now being tested. Stephen O'Brien, for example, favors an explanation based on natural selection. He proposes that selection led to an increase in the frequency of the $\Delta 32$ allele in Europeans during outbreaks of black plague in the 14th century. According to O'Brien, the $\Delta 32$ allele protects individuals against infection by the bacterium that causes black plague, as well as against infection by HIV. If so, then the experiments currently underway should show that T cells with the mutant form of the CCR5 protein resist infection by the plague bacterium, just as they resist infection by HIV.

The story of "resistance alleles" does not start and end with the $\Delta 32$ allele, however. After the study by Samson and associates was published, teams led by Luc Montagnier and Stephen O'Brien succeeded in finding two other mutant alleles associated with resistance to infection or slow progression to AIDS (Smith et al. 1997; Quillent et al. 1998; see also Carrington et al. 1999). Some of these alleles occur at equal frequencies in different ethnic groups; others vary in frequency from population to population.

These discoveries are inspiring work on two fronts: Molecular biologists are trying to design drugs that mimic the effect of the resistance alleles, while evolutionary

Alleles that confer resistance to HIV exist at different frequencies in different human populations. As a result...

biologists measure how common they are in various populations and analyze how their frequencies may change as the epidemic continues. From an evolutionary perspective, HIV is creating natural selection on human populations. Because humans vary in their ability to withstand infection, human populations should evolve in response to the epidemic. Specifically, as people with normal versions of the co-receptor genes die from AIDS, the frequency of the resistance alleles in the population should increase. If this predicted change in the genetic make-up of human populations occurs, it would become a prominent example of evolution by natural selection.

1.5 Could a Vaccine Provide Protection from the Diverse Strains of HIV?

The great success stories in controlling viral diseases—from polio to smallpox—have come through the development of vaccines. The difficulty of designing antiviral drugs, together with the rate at which HIV has evolved drug resistances, has made vaccine development an urgent priority for the AIDS research community. Is it possible to design a vaccine that would make people immune to HIV?

A recent study on the evolutionary history of the AIDS virus has reinforced a growing consensus about vaccine development. To understand the result and its implication for the future of the epidemic, we need to do two things: review how vaccines work, and grasp the basic logic behind efforts to reconstruct evolutionary history.

A Brief Look at How Vaccines Work

To respond to bacterial and viral infections, immune system cells called T cells have to identify a protein from the pathogen as foreign, or non-self. A fragment of foreign protein that is recognized as non-self and that triggers a response by T cells is called an **epitope**.

Vaccines consist of epitopes from killed or incomplete virions. Although no actual infection occurs after a vaccination, the immune system responds by activating cells that recognize the epitopes presented. If an authentic infection starts later, the immune system is "pre-primed" to respond quickly. In almost all cases, the invader is eliminated before the infection progresses to the point of causing disease.

In the case of HIV, most epitopes presented to the immune system are derived from the protein called gp120 that coats the virion's surface. To be effective, then, a vaccine would have to contain epitopes from the gp120 proteins found in many different strains of HIV. Just how diverse are these strains? To answer this question biologists have analyzed HIV gene sequences from all over the world, and used the data to reconstruct the virus' evolutionary history.

How Do Researchers Reconstruct Evolutionary History?

According to the theory of evolution by natural selection, examined in detail in Chapters 2 and 3, all organisms are related to one another by common ancestry. In the case of HIV, the theory predicts that the diversity of strains present today originated from a single ancestral population. What was the nature of this ancestor? How much has HIV diversified since then?

Just as the historical relationships among individuals are described by their genealogy, the historical relationships among populations or species are described by their **phylogeny**. A picture of these evolutionary relationships shows the family tree of a group of species or populations and is called a **cladogram** or **phylogenetic tree**. The methodology for reconstructing phylogenies is complex (we devote all of Chapter 13 to this topic), but the basic logic of the research program is simple: In general, closely related species should have more similar characteristics than distantly related forms. In the case of HIV, researchers infer the historical relationships among strains by comparing the nucleotide sequences of their genes. The working premise is that strains with extremely similar nucleotide sequences shared a common ancestor more recently than strains with extremely different nucleotide sequences.

A phylogenetic tree shows the historical relationships among a group of viruses or organisms.

To evaluate prospects for vaccine development in light of HIV's evolutionary history, we examine two phylogenies. The first tree shows the relationships between HIV and the viruses that infect immune system cells in other primates. The second tree is more narrowly focused, and shows relationships among strains of HIV.

The Origin of HIV

To reconstruct the history of HIV, Feng Gao and colleagues (1999) sequenced the gene that codes for reverse transcriptase in several simian immunodeficiency viruses (SIVs) and compared them to the sequences found in a variety of HIV strains. The SIVs are parasites that infect the immune systems of chimpanzees and monkeys. These viruses do not appear to cause serious disease in their hosts, however.

When Gao and associates used the sequence data to estimate which viruses are most closely related, the phylogeny shown in Figure 1.6a resulted. In this tree, the length of the horizontal lines indicates the percentage of bases that are different between viral strains. Short branches between species mean that their sequences are similar; longer branches mean that their sequences are more divergent. Because sequences diverge as a result of mutations that take place over a period of years, the length of the horizontal branches on this tree correlate roughly with time. (In contrast, the lengths of the vertical lines in the tree are arbitrary. They are drawn simply to make the tree more readable.)

To read the tree and understand what it implies about the history of HIV, start at the arrow on the lower left. The branching point, or **node**, at this arrow represents the common ancestor of all viruses included in the tree. Note that each of the distinct groups, or **lineages**, that branch off from the ancestral population lead to viruses that infect monkeys or chimpanzees. This observation suggests that HIV is descended from viruses that infected monkeys and chimpanzees. The branches drawn in blue diversified into viruses that infect a variety of nonhuman primates, while the branches drawn in red and in green produced viruses that parasitize both human and nonhuman primates.

Where did the human immunodeficiency viruses come from? Find the virus designated HIV-2 on the tree, and note that it is found next to a virus that infects a species of monkey called the sooty mangabey. HIV-2 is prevalent in West Africa and is much less virulent than the HIV that is causing the AIDS epidemic. Because sooty mangabeys are hunted for food and kept as pets in West Africa, and because their gene sequences are so closely related to HIV-2, researchers concur that the virus was probably transmitted from sooty mangabeys to humans in the recent past. Once it occupied humans, evolution by natural selection led to the strain known as HIV-2.

The two main types of HIV, HIV-2 and HIV-1, were transmitted to humans from different sources. HIV-2 originated in sooty mangabeys, while HIV-1 was originally transmitted to humans from chimpanzees.

In contrast, the red lineage at the top of the tree diversified into strains that infect humans and chimpanzees. These populations include HIV-1, the virus that is causing the AIDS epidemic. Because chimpanzees are hunted for food in Africa

a)

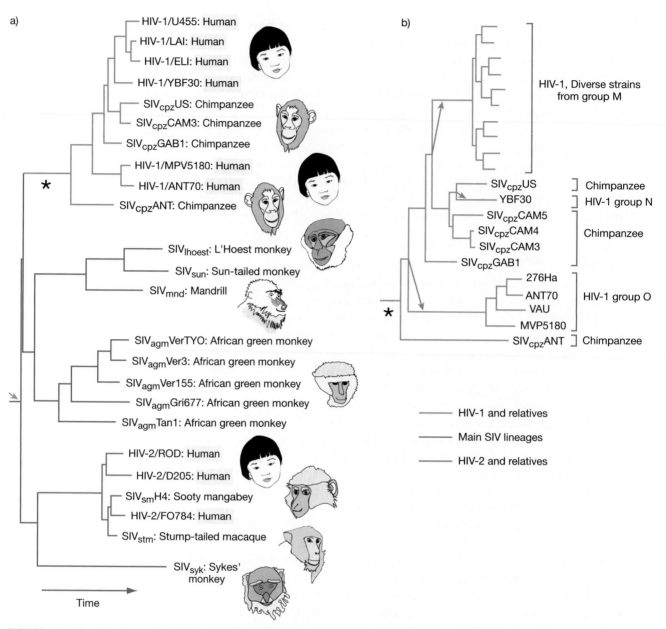

b)

FIGURE 1.6 **The "family tree" of HIV and related viruses** (a) This tree shows the evolutionary relationships among the two major forms of HIV, called HIV-1 and HIV-2, and the immunodeficiency viruses that afflict nonhuman primates. Note that the viruses that branch off near the arrow at the base of the tree parasitize monkeys. Based on this observation, researchers conclude that virus strains jumped from monkeys to humans. (b) This tree shows a more detailed analysis performed by Gao et al. (1999). (The asterisk marks the same branch point on both trees.) The arrows indicate the places on the tree where immunodeficiency viruses were transmitted from chimpanzees to humans. According to this tree, each major strain of HIV-1 originated in a different transmission event from a chimpanzee host. Modified from Figs. 1a and 3b in Gao et al. (1999). Copyright © 1999, Macmillan Magazines Ltd. Reprinted by permission from *Nature*.

and because their gene sequences are so closely related to HIV-1, Gao and colleagues propose that the SIV that infects chimpanzees (SIV_{cpz}) was transmitted from chimps to humans, where it evolved into HIV-1.

To look at this transmission event more closely, Gao and co-workers compared HIV-1 and SIV_{cpz} sequences from the gene that codes for the proteins found on the surface of the virions. The tree based on these data gives a more detailed view of the relationships among these viruses, and is reproduced in Figure 1.6b. Note that HIV strains cluster into three distinct groups, which HIV researchers call subgroups M, N, and O. Each HIV subgroup is closely related to a different strain of SIV$_{cpz}$. To Gao and associates, this is evidence that the virus "jumped" from chimps to humans on three different occasions. They claim that HIV-1 was transmitted to humans from chimps not once, but multiple times.

Each major subgroup of HIV-1 originated in an independent transmission event from chimpanzees to humans.

Using Evolutionary Trees to See the Forest

Biomedical researchers, public health officials, and physicians can draw several general conclusions from analyzing the phylogeny of HIV:

- Because SIV$_{cpz}$ is so closely related to HIV-1, chimpanzees are an important animal to study. How common is SIV$_{cpz}$ in the wild, and how is it transmitted? Why doesn't this virus make chimpanzees ill?

- Many different subgroups of HIV-1 exist as a result of independent transmission events from chimpanzees to humans. The resulting diversity of HIV strains poses a challenge to vaccine development. Further, because transmission of SIVs to humans has happened repeatedly in the past, it is likely to continue in the future.

- The branch lengths in Figure 1.6b suggest that sequence divergence is high, even within subgroups of HIV-1. To drive this point home, Tuofo Zhu and co-workers (1998) sequenced HIV-1 genes found in a blood sample taken from a Congolese man in 1959. This is the earliest HIV infection known to date. Their analysis shows that the 1959 sample is strikingly different from contemporary strains. In the researchers' own words (p. 596), "The diversification of HIV-1 in the past 40–50 years portends even greater viral heterogeneity in the coming decades." The rapid evolution of HIV, like the rapid genetic change commonly observed in flu and cold viruses, makes vaccine design difficult.

HIV evolves so rapidly that the search for an effective vaccine may be futile.

What, if Anything, Does Evolutionary Biology Have to Say About Ways to Stem the AIDS Epidemic?

Understanding evolutionary biology may help inform the fight against HIV, if only in a minor way.

First, many researchers in the field have concluded that the search for an AIDS vaccine may be futile (Korber et al. 1998; Letvin 1998; Baltimore and Heilman 1998). Because HIV's mutation rate is so high, the virus diversifies rapidly. As a result, different strains present a wide variety of epitopes to the human immune system, making vaccine design difficult if not impossible.

Second, resistant alleles exist and should increase in frequency in human populations as the epidemic continues. People who lack resistance alleles are at a higher risk of infection.

Third, if the transmission rate hypothesis is correct, the best defense against HIV is unquestionably education and changes in behavior. Increased condom use, sexual fidelity, abstinence, and use of clean needles would not only decrease rates of transmission directly, but also potentially help halt the epidemic indirectly by leading to the evolution of a less-virulent parasite. The transmission rate hypothesis argues that slowing transmission rates will have a multiplicative effect in stemming the epidemic—by lowering both the incidence *and* the severity of the disease.

Finally, HIV will undoubtedly continue to evolve resistance to antiviral drugs. Physicians now routinely prescribe "combination therapies"—meaning several drugs in combination—in an effort to slow the evolution of resistant strains. The intent is to buy time for the development of new drugs.

Summary

In this chapter we focused on adaptation and diversification in a virus and introduced topics that resonate throughout the text: mutation and variation, competition, natural selection, phylogeny reconstruction, lineage diversification, and applications of evolutionary theory to scientific and human problems. Our task now is to widen the fine focus of this introduction, and introduce the diversity of theory, experiments, and models that make up evolutionary biology.

This story begins, in Chapter 2, with the fact of evolution. Understanding that species change through time, and that the organisms alive today are descended from forms that lived in the past, is what motivated Darwin to seek an explanatory mechanism: a process that could create change through time. What, then, is the evidence for the fact of evolution?

Questions

1. AIDS is primarily a disease of young adults. How does a disease like this affect the size, age distribution, and growth rate of human populations over time?

2. In the early 1990s, researchers began to find AZT-resistant strains of HIV-1 in recently infected patients who had never received AZT. How can this be?

3. In this chapter, we discussed two different types of selection: selection of different virus strains within one host, and selection of those virus strains that are able to transmit themselves from host to host. Suppose an HIV counselor is talking to a patient who is worried that he or she might have a virulent strain of HIV. "Don't worry," the counselor says. "Even if you have a virulent strain now, as long as you are monogamous the virulent strain will die out." How has the counselor misinterpreted the transmission rate hypothesis?

4. An alternative to the transmission rate hypothesis, traditionally championed by biomedical researchers, is that disease-causing agents "naturally" evolve into more benign forms as the immune systems of their hosts evolve more efficient responses to them. What predictions does this "coevolution hypothesis" make about evolution of HIV virulence in the United States vs. Africa vs. South Asia? What data would help you decide whether the transmission rate hypothesis or coevolution hypothesis is correct? Do the hypotheses suggest different ways to spend the limited budget for HIV research and education?

5. Respond to the following quote, from the character named Mr. Spock in "Star Trek": "A truly successful parasite is commensal, living in amity with its host, or even giving it positive advantages, as, for instance, the protozoans who live in the digestive system of your termites and digest for them the wood they eat. A parasite that regularly and inevitably kills its host cannot survive long, in the evolutionary sense, unless it multiplies with tremendous rapidity . . . it is not pro-survival."

6. The text claims that human populations will evolve in response to the AIDS epidemic, because alleles that confer resistance to HIV infection should increase in frequency in the population over time. Do you agree with this prediction? How would you design a study capable of testing it?

7. Suppose that HIV was the ancestor of the SIVs, instead of the other way around. If immunodeficiency viruses were originally transmitted from humans to monkeys and chimpanzees, make a sketch of what Figure 1.6a would look like.

Exploring the Literature

8. Drug resistance has evolved in a wide variety of viruses, bacteria, and other parasites. The following papers introduce data on the evolution of drug resistance in the flu and hepatitis B viruses and in populations of the bacterium that causes tuberculosis:

Bishai, W. R., N. M. H. Graham, S. Harrington, C. Page, K. Moore-Rice, N. Hooper, and R. E. Chaisson. 1996. Rifampin-resistant tuberculosis in a patient receiving rifabutin prophylaxis. *New England Journal of Medicine* 334:1573–1576.

Carman, W., H. Thomas, and E. Domingo. 1993. Viral genetic variation: Hepatitis B virus as a clinical example. *Lancet* 341:349–353.

9. For tests of the transmission rate hypothesis in other organisms, see the following papers:

  Herre, E.A. 1993. Population structure and the evolution of virulence in nematode parasites of fig wasps. *Science* 259:1442–1445.

Lipsitch, M., S. Siller, and M.A. Nowak. 1996. The evolution of virulence in pathogens with vertical and horizontal transmission. *Evolution* 50:1729–1741.

Citations

Adib, S. M., J. G. Joseph, D. G. Ostrow, M. Tal, and S. A. Schwartz. 1991. Relapse in sexual behavior among homosexual men: A 2-year follow-up from the Chicago MACS/CCS. *AIDS* 5: 757-760.

Ala, P. J. et al. 1997. Molecular basis of HIV-1 protease drug resistance: Structural analysis of mutant proteases complexed with cyclic urea inhibitors. *Biochemistry* 36: 1573–1580.

Alkhatib, G., C. Combadiere, C.C. Broder, Y. Feng, P. E. Kennedy, P. M. Murphy, and E.A. Berger. 1996. CC CKR5: A RANTES, MIP-1α, MIP-1β receptor as a fusion cofactor for macrophage-tropic HIV-1. *Science* 272: 1955–1952.

Baltimore, D. and C. Heilman. 1998. HIV vaccines: prospects and challenges. *Scientific American* 279 (July): 98-103.

Cao, Y., L. Qin, L. Zhang, J. Safrit, and D. D. Ho. 1995. Virologic and immunologic characterization of long-term survivors of human immunodeficiency virus type 1 infection. *New England Journal of Medicine* 332: 201–208.

Carrington, M. et al. 1999. *HLA* and HIV-1: Heterozygote advantage and *B*35-CW*04* disadvantage. *Science* 283: 1748–1752.

Catania, J. A., T. J. Coates, R. Stall, H. Turner, J. Peterson, N. Hearst, M. M. Dolcini, E. Hudes, J. Gagnon, J. Wiley, and R. Groves. 1992. Prevalence of AIDS-related risk factors and condom use in the United States. *Science* 258: 1101–1106.

Culliton, B. 1990. Emerging viruses, emerging threat. *Science* 247:279–280.

De Cock, K. M., G. Adjorlolo, E. Ekpini, T. Sibailly, J. Kouadio, M. Maran, K. Brattegaard, K. M. Vetter, R. Doorly, and H. D. Gayle. 1993. Epidemiology and transmission of HIV-2. *Journal of the American Medical Association* 270: 208–2086.

Deacon, N. J., et al. 1995. Genomic structure of an attenuated quasi species of HIV-1 from a blood transfusion donor and recipients. *Science* 270: 988–991.

Deeks, S. G., M. Smith, M. Holodniy, and J. O. Kahn. 1997. HIV-1 protease inhibitors. *Journal of the American Medical Association* 277: 145–153.

Editorial staff. 1995. Death by denial. *The Lancet* 345: 1519-1520.

Ewald, P. W. 1994. *Evolution of Infectious Disease.* Oxford: Oxford University Press.

Feng, Y., C. C. Broder, P. E. Kennedy, and E. A. Berger. (1996). HIV-1 entry cofactor: Functional cDNA cloning of a seven-transmembrane, G protein-coupled reactor. *Science* 272: 872–877.

Gao, F., E. Bailes, D. L. Robertson, Y. Chen, C. M. Rodenburg, S. F. Michael, L. B. Cummins, L. O. Arthur, M. Peeters, G. M. Shaw, P. M. Sharp, and B. H. Hahn. 1999. Origin of HIV-1 in the chimpanzee *Pan troglodytes troglodytes. Nature* 397: 436–441.

Goldsmith, M. F. 1990. Science ponders whether HIV acts alone or has another microbe's aid. *Journal of the American Medical Association* 264: 665–666.

Groenink, M., R. A. M. Fouchier, S. Broersen, C. H. Baker, M. Koot, A. B. van't Wout, H. G. Huisman, F. Miedema, M. Tersmette, and H. Schuitemaker. 1993. Relation of phenotype evolution of HIV-1 to envelope V2 configuration. *Science* 260:1513–1515.

Hübner, A., M. Kruhoffer, F. Grosse, and G. Krauss. 1992. Fidelity of human immunodeficiency virus type 1 reverse transcriptase in copying natural RNA. *Journal of Molecular Biology* 223: 595–600.

Koblin, B. A., J. M. Morrison, P. E. Taylor, R. L. Stoneburner, and C. E. Stevens. 1992. Mortality trends in a cohort of homosexual men in New York City, 1978–1988. *American Journal of Epidemiology* 136: 646-656.

Korber, B., J. Theiler, and S. Wolinsky. 1998. Limitations of a molecular clock applied to considerations of the origin of HIV-1. *Science* 280: 1868-1870.

Larder, B. A., P. Kellam, and S. D. Kemp. 1993. Convergent combination therapy can select viable multidrug-resistant HIV-1 in vitro. *Nature* 365: 451–453.

Letvin, N. L. 1998. Progress in the development of an HIV-1 vaccine. *Science* 280: 1875–1880.

Liu, R. et al. 1996. Homozygous defect in HIV-1 coreceptor accounts for resistance of some multiply-exposed individuals to HIV-1 infection. *Cell* 86: 367–377.

Mann, J. M. and D. J. M. Tarantola. 1998. HIV 1998: The global picture. *Scientific American* 279 (July): 82–83.

Marlink, R., P. Kanki, I. Thior, K. Travers, G. Eisen, T. Siby, I. Traore, C-C. Hsieh, M. C. Dia, E-H. Gueye, J. Hellinger, A. Gueye-Ndiaye, J-L. Sankalé, I. Ndoye, S. Mboup, and M. Essex. 1994. Reduced rate of disease development after HIV-2 infection as compared to HIV-1. *Science* 265:1587–1590.

Mohri, H., M. K. Singh, W. T. W. Ching, and D. D. Ho. 1993. Quantitation of zidovudine-resistant human immunodeficiency virus type 1 in the blood of treated and untreated patients. *Proceedings of the National Academy of Sciences, USA* 90:25–29.

Piot, P. 1998. The science of AIDS: A tale of two worlds. *Science* 280: 1844–1845.

Quillent, C. et al. 1998. HIV-1-resistance phenotype conferred by combination of two separate inherited mutations of CCR5 gene. *The Lancet* 351: 14–18.

St. Clair, M. H., J. L. Martin, G. Tudor-Williams, M. C. Bach, C. L. Vavro, D. M. King, P. Kellam, S. D. Kemp, and B. A. Larder. 1991. Resistance to ddI and sensitivity to AZT induced by a mutation in HIV-1 reverse transcriptase. *Science* 253:1557–1559.

Samson, M. et al. 1998. Resistance to HIV-1 infection in caucasian individuals bearing mutant alleles of the CCR-5 chemokine receptor gene. *Nature* 382: 722–725.

Shirasaka, T., R. Yarchoan, M. C. O'Brien, R. N. Husson, B. D. Anderson, E. Kojima, T. Shimada, S. Broder, and H. Mitsuya. 1993. Changes in drug sensitivity of human immunodeficiency virus type 1 during therapy with azidothymidine, didioxycytidine, and dideoxyinosine: An in vitro comparative study. *Proceedings of the National Academy of Sciences, USA* 90: 562–566.

Smith, M. W. et al. 1997. Contrasting genetic influence of CCR2 and CCR5 variants on HIV-1 infection and disease progression. *Science* 277: 959–965.

Wain-Hobson, S. 1993. The fastest genome evolution ever described: HIV variation in situ. *Current Opinion in Genetics and Development* 3: 878–883.

Zhu, T., B. T. Korber, A. J. Nahmias, E. Hooper, P. M. Sharp, and D. D. Ho. 1998. An African HIV-1 sequence from 1959 and implications for the origin of the epidemic. *Nature* 391: 594–597.

CHAPTER 2

The Evidence for Evolution

These footprints were made by *Australopithecus afarensis*—a species that is related to modern humans. The fossils are from Laetoli, Tanzania and are 3.6 million years old. Note that two individuals are walking side by side. (John Reader/Science Photo Library/Photo Researchers, Inc.)

EVOLUTIONARY BIOLOGY ADDRESSES FUNDAMENTAL QUESTIONS ABOUT THE world. Where do living things come from? Why are there so many different kinds of organisms, and how have they come to be so proficient at tasks like finding food, acquiring mates, fighting disease, and avoiding predators?

Charles Darwin was an English naturalist who devoted his life to answering these questions. When he began to study biology seriously, as a college student in the early 1820s, the leading explanation in Europe for the origin of species was the Theory of Special Creation. This theory held that all organisms were created by God during the six days of creation as described in the Bible's Book of Genesis 1:1–2:4. The ideal types formed by this special process, including Adam and Eve, were the progenitors of all organisms living today. The theory stated that species are unchanged since their creation, or immutable, and that variation within each type is strictly limited. The creation event was also believed to be recent: In 1664, Archbishop James Ussher, of the Irish Protestant Church, used Old Testament genealogies to calculate that the Earth was precisely 5,668 years old. He wrote that "Heaven and Earth, Centre and substance were made in the same instant of time and clouds full of water and man were created by the Trinity on the 26th of October 4004 B.C. at 9:00 in the Morning."

By the time Darwin began working on the problem in the 1830s, however, dissatisfaction with the Theory of Special Creation had begun to grow. Research in the biological and geological sciences was advancing rapidly, and the data clashed with creationism's central tenets and predictions.

To understand the issues that Darwin addressed, it is important to recognize that scientific theories frequently have two components. The first component is a statement of fact—a claim about a pattern that exists in the natural world. The second identifies the process that is responsible for producing the pattern. The Theory of Special Creation, for example, makes three statements of fact: (1) Species were created independently of one another, (2) They do not change through time, and (3) They were created recently. According to the Theory of Special Creation, the process responsible for this pattern was a special, or supernatural, act of creation by God.

Scientific theories often have two components. The first is a statement about a pattern that exists in the natural world; the second is a process that explains the pattern.

The goal of this chapter is to review evidence that supports an alternative statement of fact that Darwin called "descent with modification," and which later came to be known as **evolution**. According to Darwin, species have changed through time and are related by descent from a common ancestor. Chapter 3 introduces the process, called **natural selection**, that Darwin proposed as the main agent responsible for this pattern.

The first three sections of this chapter explore data that challenge each claim of fact made by the Theory of Special Creation—that species are independent, immutable, and recent. The final section introduces datasets drawn from different fields of scientific inquiry that corroborate one another and support predictions made by the **theory of evolution by natural selection**.

2.1 Relatedness of Life Forms

Although Darwin's name is usually associated with the Theory of Evolution, he was not the first to recognize the theory's pattern component. The fact of evolution had been proposed by several workers in the late 1700s and early 1800s, including Comte de Buffon, Erasmus Darwin (Charles' grandfather), and the great French biologist, Jean-Baptiste Lamarck (Eiseley 1958; Desmond and Moore 1991). These early advocates challenged the hypothesis that each species was created independently and argued that species are related by common ancestry. Darwin compiled and synthesized much of the evidence for this conclusion in his book *On the Origin of Species by Means of Natural Selection,* published in 1859. What is the nature of these data?

The Theory of Special Creation contends that each species was created independently. In contrast, the Theory of Evolution contends that organisms are related by common ancestry.

Homology

As the fields of comparative anatomy and comparative embryology developed in the early 1800s, one of the most striking results to emerge was that fundamental similarities underlie the obvious physical differences among species. Early researchers called the phenomenon **homology**—literally, the study of likeness. For example, Richard Owen, Britain's leading anatomist, and Baron Georges Cuvier of Paris, the founder of comparative anatomy, described extensive homologies among vertebrate skeletons and organs. A few of these are illustrated in Figure 2.1a. Darwin (1859, p. 434) referred to their work when he wrote, "What could be more curious than that the hand of a man, formed for grasping, that of a mole

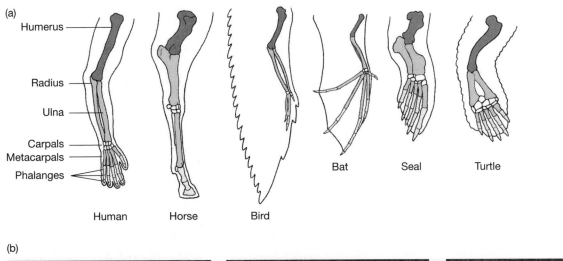

(a)

Humerus
Radius
Ulna
Carpals
Metacarpals
Phalanges

Human   Horse   Bird   Bat   Seal   Turtle

(b)

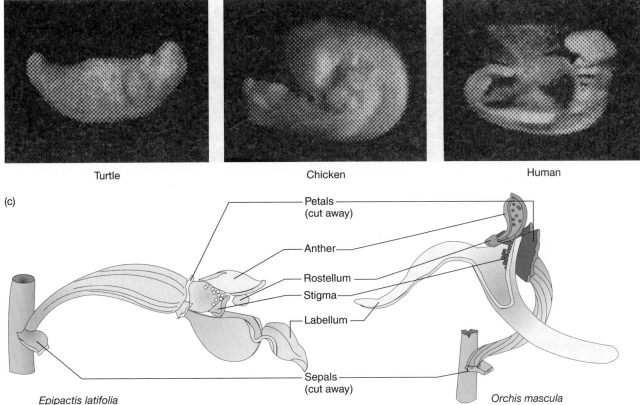

Turtle   Chicken   Human

(c)

Petals (cut away)
Anther
Rostellum
Stigma
Labellum
Sepals (cut away)

Epipactis latifolia   *Orchis mascula*

Figure 2.1 Structural homologies (a) These vertebrate forelimbs are used for different functions, but have the same sequence and arrangement of bones. In this illustration, homologous bones are colored or shaded in the same way, and are labeled on the human arm. (b) Embryos from different vertebrates are similar, especially early in development. The embryos pictured here are in the tailbud stage. (c) Orchid flowers are diverse in size and shape, but are comprised of elements that are similar in structure and orientation.

for digging, the leg of the horse, the paddle of the porpoise, and the wing of the bat, should all be constructed on the same pattern, and should include the same bones, in the same relative positions?" His point was that the underlying design of each structure is similar, even though their function and appearance are very different. Based on this observation, Darwin concluded that the structures were not created independently, but are related by common ancestry. His reasoning was that a conscious engineer would not design implements for grasping, digging, running, swimming, and flying using the same set of structural elements in the same arrangement. According to Darwin, homology supports the theory of evolution.

The data were not confined to vertebrates or to adult forms, however. The naturalist Louis Agassiz was among many who observed that the embryos of vertebrates ranging from fish to humans are strikingly similar, especially early in development (Figure 2.1b). Darwin (1862) himself analyzed the anatomy of orchid flowers and showed that, even though they are diverse in shape, they are actually constructed from the same set of component pieces. Like vertebrate limbs, each flower in Figure 2.1c has the same parts in the same relative positions.

Structural homologies are a product of shared developmental pathways.

What causes these similarities? Agassiz, Owen, Cuvier, and other early workers recognized that homologous structures in adults develop from the homologous groups of cells in embryos. As a consequence, homology was originally defined as similarity due to shared developmental pathways. But why should some organisms share developmental pathways? Darwin argued that descent from a common ancestor was the most logical explanation. He contended that the embryos in Figure 2.1c are similar because all vertebrates evolved from the same common ancestor, and because some developmental stages have remained similar as fish, amphibians, reptiles, and mammals diversified over time.

Similar developmental sequences are a product of homologous genetic programs.

Homologous genetic programs are a product of shared ancestry.

Today, biologists recognize that developmental and structural homologies are caused by genetic homologies. Embryonic and adult structures are similar among species because the genes that code for those structures are similar. For example, Figure 2.2 shows the amino acid sequences encoded by a gene that is involved in the development of the eye. When Rebecca Quiring and colleagues (1994) compared the sequence of this protein in fruit flies to its sequence in vertebrates, they found that over 90% of the amino acids are identical in certain segments. They hypothesize that genetic homologies like this are responsible for similarities in the light-gathering organs of animals.

Advances in molecular genetics have revealed other fundamental similarities among organisms. Prominent among these is the genetic code. With a few minor exceptions, all organisms studied to date use the same nucleotide triplets, or **codons,** to specify the same amino-acid-bearing transfer RNAs. Biologists interpret these similarities as evidence that all living organisms descended from an ancestor which used this same genetic code. Box 2.1 points out that the same evolutionary logic underlies the use of model organisms in biomedical research and drug testing.

Relationships Among Species

Darwin's recognition of relationship through shared descent extended to phenomena other than homology. His trip to the Galápagos Islands had a strong influence on his thinking about the relationships among species. While he was aboard the H.M.S. *Beagle* during a five-year mapping and exploratory mission, Darwin collected and cataloged the flora and fauna encountered during the voyage. He was especially impressed by the mockingbirds he found during his work in the Galápagos, because several islands had distinct populations. Although they were all similar in color, size, and shape—and thus clearly related to one another—each mockingbird population seemed distinct enough to be classified as a separate species. This was confirmed, later, by a taxonomist colleague of Darwin's back in England. Darwin and others followed up on this result with studies showing the same pattern in Galápagos tortoises and finches: The various islands hosted different, but closely related, species (see Desmond and Moore, 1991).

(a)

Second base

First base													Third base
		U			**C**			**A**			**G**		
U	UUU	Phenylalanine	F	UCU	Serine	S	UAU	Tyrosine	Y	UGU	Cysteine	C	U
	UUC	Phenylalanine	F	UCC	Serine	S	UAC	Tyrosine	Y	UGC	Cysteine	C	C
	UUA	Leucine	L	UCA	Serine	S	UAA	Stop		UGA	Stop		A
	UUG	Leucine	L	UCG	Serine	S	UAG	Stop		UGG	Tryptophan	W	G
C	CUU	Leucine	L	CCU	Proline	P	CAU	Histidine	H	CGU	Arginine	R	U
	CUC	Leucine	L	CCC	Proline	P	CAC	Histidine	H	CGC	Arginine	R	C
	CUA	Leucine	L	CCA	Proline	P	CAA	Glutamine	Q	CGA	Arginine	R	A
	CUG	Leucine	L	CCG	Proline	P	CAG	Glutamine	Q	CGG	Arginine	R	G
A	AUU	Isoleucine	I	ACU	Threonine	T	AAU	Asparagine	N	AGU	Serine	S	U
	AUC	Isoleucine	I	ACC	Threonine	T	AAC	Asparagine	N	AGC	Serine	S	C
	AUA	Isoleucine	I	ACA	Threonine	T	AAA	Lysine	K	AGA	Arginine	R	A
	AUG	Start (Methionine M)		ACG	Threonine	T	AAG	Lysine	K	AGG	Arginine	R	G
G	GUU	Valine	V	GCU	Alanine	A	GAU	Aspartic Acid	D	GGU	Glycine	G	U
	GUC	Valine	V	GCC	Alanine	A	GAC	Aspartic Acid	D	GGC	Glycine	G	C
	GUA	Valine	V	GCA	Alanine	A	GAA	Glutamic Acid	E	GGA	Glycine	G	A
	GUG	Valine	V	GCG	Alanine	A	GAG	Glutamic Acid	E	GGG	Glycine	G	G

Codon   Amino acid   Abbreviation

(b)

human	L Q R N R T S F T Q E Q I E A L E K E F E R T H Y P D V F A R E R L A A K I D L P E A R I Q V W F S N R R A K W R R E L
mouse	· ·
quail	· ·
zebrafish	· ·
fruit fly	· · · · · · · · N D · · D S · · · · · · · · · · · · · · · · · · · G · · G ·

Figure 2.2   Genetic homologies   (a) In almost every organism studied, the same nucleotide triplets, or codons, specify the same amino-acid bearing transfer RNAs. (b) This chart shows the amino acid sequences of a section called the homeodomain in a protein involved in the development of the eye (Quiring et al. 1994). Dots indicate the same amino acid as the one above.

BOX 2.1   Homology and model organisms

Homology may appear to be an abstract concept, but it is actually the guiding principle behind most biomedical research. Homology is the reason medical researchers can obtain valid results when testing the safety of new drugs in mice or studying the molecular basis of disease in rats. The results can be extrapolated to humans if the molecular or cellular basis of the phenomenon being studied is homologous.

Researchers choose a study organism—also called a "model" organism—based on the degree of homology required to study a particular process or disease. In psychiatry and the behavioral sciences, for example,

monkeys and apes are often the preferred experimental subject because aspects of their behavior and brain structures are homologous with those of humans. Because some of the genes involved in more basic processes, like the cell cycle, are homologous between even distant relatives, researchers can use baker's yeast (*Saccharomyces cerevisiae*) to study why certain malfunctioning genes cause cancer in humans. At an even more fundamental level, the genes involved in DNA repair are homologous between humans and the bacterium *Escherichia coli*. Primates, yeast, and bacteria share these characteristics with humans because they inherited them from a common ancestor.

Species that are extremely similar to one another tend to be clustered geographically. This suggests that they were not created independently, but are descended from a common ancestor that lived in the same area.

To explain this pattern, Darwin hypothesized that a small population of mockingbirds colonized the Galápagos from South America long ago. His thesis was that the population expanded in the new habitat and that subpopulations subsequently colonized different islands in the group. Once mockingbird populations had become physically isolated from one another in this way, they diverged enough to become distinct species. Like structural homologies, the existence of closely related forms in island groups was a logical outcome of descent with modification.

In contrast, both patterns were inconsistent with special creation, which predicted that organisms were created independently. Under special creation, no particular patterns are expected in the design or geographic relationships of organisms.

An Introduction to Tree Thinking

The data reviewed in this section have one fundamental message: Species are not independent, but are connected by descent from a common ancestor. This means that species have genealogical relationships analogous to the family trees of individual humans. In Chapter 1, we pointed out that the genealogy of a group of species is called its phylogeny, and that a graphical representation of these relationships is called a phylogenetic tree or cladogram.

Chapter 13 will introduce the techniques that evolutionary biologists use to reconstruct evolutionary relationships and estimate the shape of phylogenetic trees. Here, we ask more basic questions: How are phylogenies read and interpreted? and How do biologists use them to answer questions about evolution?

Reading a Tree

The only figure in Darwin's 490-page *Origin of Species* was a diagram that presented his view of how species change through time. The illustration was hypothetical, but it included the three major elements found in actual phylogenies: **tips**, **branches**, and **nodes**. Tips represent extinct species or species living today. Branches depict ancestral populations of these species through time. Nodes denote points where one species splits into two or more descendant populations.

Part of Darwin's drawing is reproduced in Figure 2.3. Note that the vertical axis on the diagram represents time and the horizontal axis indicates morphological divergence. To read the tree, start with the population marked A and begin reading up. The diagram indicates that species A immediately splits into six subpopulations. Four of these quickly go extinct, but the populations labeled a^1 and m^1 survive to time I. These populations continue to diverge over time and produce subpopulations. Although most of these new populations eventually go extinct, by time X, the ancestral species A has given rise to three new species, labeled a^{10}, f^{10}, and m^{10}.

Phylogenetic trees are a visual representation of the fact that species are related by descent from a common ancestor.

In creating this diagram, Darwin invented a technique for illustrating how species are related by descent with modification from a common ancestor. The phylogenies published in today's scientific literature are the direct descendants of this graphical device.

Figure 2.4a shows one way that phylogenies are drawn currently. The tips on this tree are labeled Taxon 1 through Taxon 6 to illustrate an important point: The tips, branches, and nodes on a phylogeny can represent populations of organisms at any taxonomic level—ranging from populations or species to phyla or kingdoms.

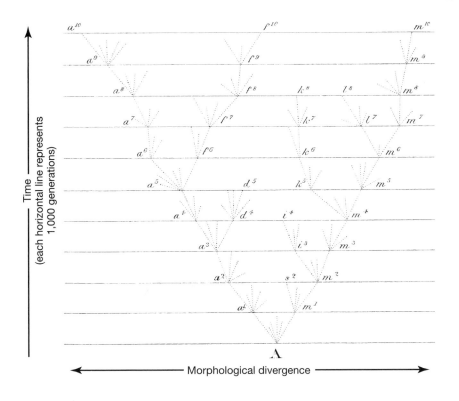

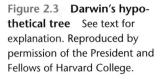

Figure 2.3 **Darwin's hypothetical tree** See text for explanation. Reproduced by permission of the President and Fellows of Harvard College.

(A **taxon** is any named group of organisms, such as a population, genus, or family; the plural is **taxa**.)

Another key to reading trees is to recognize that they can be oriented vertically, with the origin at the base and the tips at the top, or horizontally, with the origin at the left and the tips at the right. Branches can be represented by either diagonal or squared-off lines (Figure 2.4b). If the branch lengths on a particular tree are proportional to time or to the amount of genetic change that has occurred since taxa diverged, then a scale or labeled axis is provided. Otherwise, branch lengths are arbitrary and are drawn only to improve readability.

Finally, evolutionary trees may or may not be rooted. Because rooted trees identify where the lineage in question originated, they establish the order that divergence events occurred. Unrooted trees, in contrast, show the relationships among species but do not indicate which nodes and branches existed earlier or later in history (see Figure 2.4c).

Using Phylogenies

The effort that goes into estimating and reading phylogenies pays off when a tree allows researchers to answer questions about how species have changed through time. As an example of the insights that can come through "tree thinking," consider the evolution of the swim bladder in fish.

Swim bladders are organs that help fish float. Muscle and bone are heavier than water, so to avoid sinking constantly, many fish maintain a gas-filled chamber called a swim bladder. Structurally, these organs are homologous with the lungs found

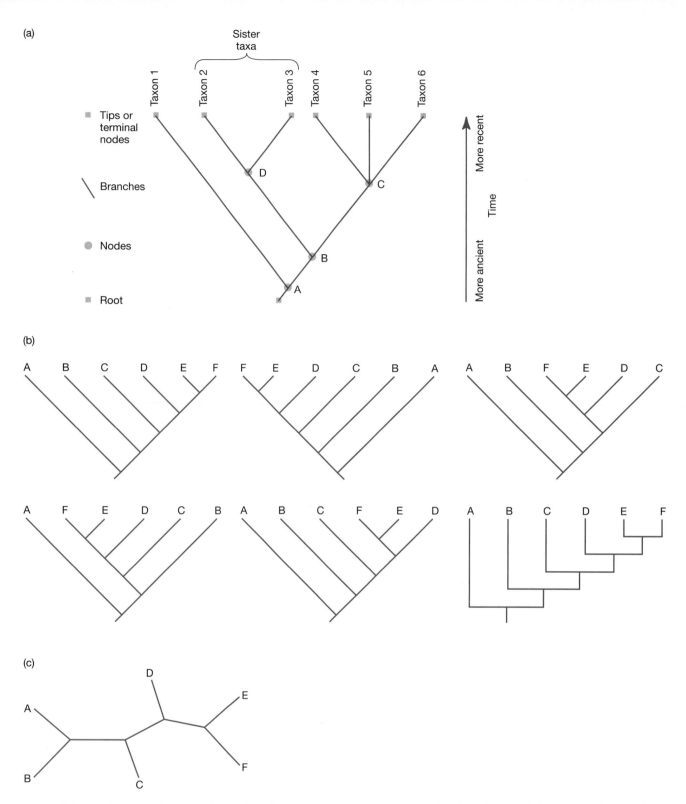

Figure 2.4 Reading a phylogenetic tree (a) To read this tree, start at the bottom and work up. The population at the node marked A is the common ancestor of taxa 1–6. It split into two groups. One of these groups evolved into Taxon 1; the other evolved into the population at node B, which is the common ancestor of taxa 2–5. What do the populations at nodes C and D represent, and what is their relationship to the population at node B? (b) These graphs show six ways to illustrate the same evolutionary relationships. The trees all happen to be oriented vertically, so that the basal taxon is at the bottom and derived groups are toward the top. These could also be tipped 90° and presented horizontally (from Mayden and Wiley 1992). (c) This unrooted tree shows the same relationships among taxa that are diagrammed in part (b). It does not have a root, however, so it does not indicate which branching events occurred earlier and which occurred later during evolution.

in fish and tetrapods. (The tetrapods, or "four footed," include amphibians, reptiles, and mammals). Developmentally, both lungs and swim bladders originate as out-pocketings of the gut.

Thanks to the efforts of earlier researchers, Darwin was aware of the homology between the lung and swim bladder. Because fish appear in the fossil record much earlier than tetrapods, and because swim bladders are much more common in fish than are lungs, he concluded that lungs must have been derived from swim bladders. Is this true?

A phylogeny of the tetrapods and the major groups of fish, based on an analysis of characteristics other than the lungs and swim bladder, is shown in Figure 2.5a. Note that lungs are indicated as present in the population at the base of the tree, which represents the common ancestor of all the groups shown. The biologist who did this study, Karl Liem, made this claim because lungs occur in one of the earliest groups of fish that have jaws, called the placoderms. Specimens of a placoderm common in rocks on Quebec's Gaspé Peninsula, called *Bothriolepis,* show impressions that clearly indicate a well-developed pair of lungs opening from the pharynx (Liem 1988; Colbert and Morales 1991). Lungs or swim bladders also occur in every descendant group on the phylogeny, save for the sharks and rays (elasmobranchs). Swim bladders are found only in two descendant groups, however: the superorders Chondrostei (an ancient group of ray-finned fish) and Teleostei (the group of ray-finned fish that includes most modern forms).

To decide how lungs and swim bladders had changed over the course of evolution, Liem relied on a logical principle called **parsimony**. Invoking parsimony in evolutionary biology means that, when drawing conclusions about what happened in the course of evolution, investigators favor simple explanations over complicated ones. That is, researchers prefer interpretations of data that minimize the amount of evolutionary change that has occurred. This is appropriate when events that occur infrequently are being studied. For example, Figure 2.5b illustrates contrasting hypotheses for how swim bladders evolved. Hypothesis 1 contends that swim bladders evolved early and then reverted to lungs four times (in the lineages leading to the Dipnoi et al. group, Cladistia, Ginglymodi, and Halecomorpha). Hypothesis 2 contends that swim bladders originated late and evolved independently twice (in the lineages leading to Chondrostei and Teleostei). Under parsimony, we accept Hypothesis 2 as much more likely because it requires just two changes instead of four.

The phylogenetic analysis, then, shows that lungs were present at the base of the tree, that lungs were lost completely in the lineage leading to the sharks and rays, and that lungs were transformed into swim bladders in two different lineages of fish. This last conclusion is bolstered by the observation that the swim bladders of chondrosteans and teleosts develop differently: In the former, they originate as outpocketings from the stomach, and in the latter, as outpocketings of the esophagus (Liem 1988).

The overall conclusion of the analysis is remarkable: Fish were good at breathing air before they were good at floating. What misled Darwin and many other workers was the assumption that ancestral forms must have had swim bladders because most living and many fossil fish do. It is not necessarily true that widespread traits are ancestral. Interpreting the evolution of the swim bladders in a phylogenetic context made the actual direction of change clear. Tree thinking is a powerful tool for understanding the history of life.

The Theory of Evolution is considered powerful because it suggests new ideas to test and because it leads to new insights. Analyzing the origin of the swim bladder in fish is an example of these insights. This analysis was based on the recognition that species are related by common descent.

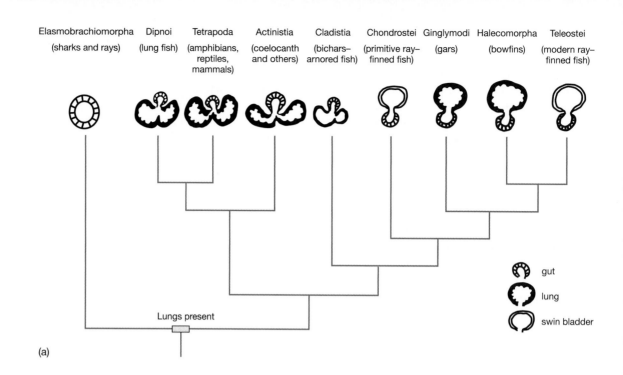

(a)

Hypothesis 1: Swim bladders evolved early

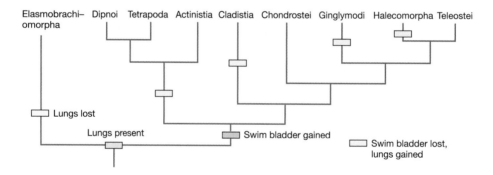

Hypothesis 2: Swim bladders evolved late

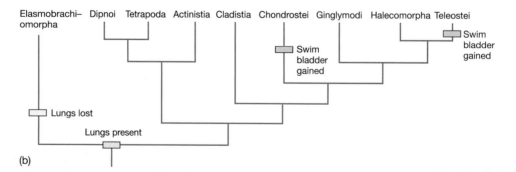

(b)

Figure 2.5   **The evolutionary history of the lung and swim bladder**   (a) This diagram shows an estimate for the relationships among the major groups of vertebrates living today. In the drawings above each group, the sacs in heavy black outline are lungs, the sacs with open outlines are swim bladders, and the smaller sacs with hatching represent the gastro-intestinal tract. In each drawing, the animal's dorsal side is above and the ventral side is below. (On an animal body, the dorsal side faces up and the ventral side faces down.) (b) These trees illustrate two contrasting hypotheses for the evolution of the lung and swim bladder. Note that the relationships among taxa are the same as in part (a). See text for explanation.

2.2 Change Through Time

One of the central tenets of the Theory of Special Creation was that species, once created, were immutable. This claim was challenged by several lines of evidence that supported the alternative hypothesis that organisms have changed through time. The data reviewed here come from living species as well as from extinct forms preserved in the fossil record.

Evidence from Living Species

By the time Darwin began his work on "the species question," comparative anatomists had described a series of curious traits called **vestigial structures**. A vestigial structure is a functionless or rudimentary homolog of a body part that has an important function in closely related species. Several examples are illustrated in Figure 2.6a. Some blind, cave-dwelling fish have eye sockets but no eyes; flightless birds and insects have reduced wings; some snakes have tiny hips and rear legs; humans have a reduced tailbone. Humans also have muscles that can make body hair stand on end when an individual is cold or excited. This phenomenon is homologous with the erectile fur of other mammals, which signals alarm or aggressive intent. In humans, the homologous response results in the vestigial trait called goose bumps.

Vestigial Traits Can Be Identified at Three Levels

Like the homologies described in Section 2.1, vestigial traits can be observed at the structural, developmental, and genetic levels. Figure 2.6b illustrates a vestigial trait that appears during the development of the "hands" and feet of birds. Adult chickens

(a)

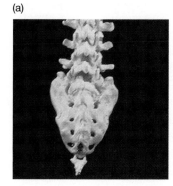

(b)

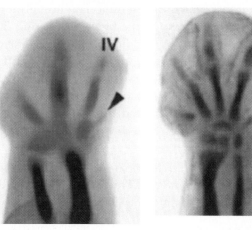

Figure 2.6 Vestigial traits Vestigial structures and developmental stages are common.
(a) Humans have a rudimentary tailbone, called the coccyx, at the base of their vertebral column. (Vincent Zuber/Custom Medical Stock Photo) Many blind cave-dwelling creatures, like this grotto salamander (*Typhlotriton spelaeus*), have functionless bulbs of tissue in place of eyes. (Nathan W. Cohen/Visuals Unlimited) (b) Adult chickens have three digits in their wings and four in their feet. But during development, an extra digit appears for a short time in the "hand" (left) and foot (right).
(A. C. Burke/Alan Feduccia 1997)

have three digits in their wings and four in their feet. But when chicken embryos are treated with a stain to mark the tissues that initiate bone development, an additional digit—marked with an arrow in the figure—appears and later disappears. Why? Darwin argued that the presence of vestigial traits like these is inexplicable under special creation. But under the Theory of Evolution, they are readily interpretable. In this case, the key observation is that most living and fossil tetrapods have five digits in both their forelimbs and hindlimbs. It is logical, then, to argue that the ancestors of birds also had five digits in each limb. The evolutionary hypothesis contends that the number of digits was reduced to three or four during the evolution of birds, and that a vestigial "lost digit" still appears for a short period during development.

A similar phenomenon is present at the molecular level, in the form of DNA sequences called **pseudogenes**. These "false genes" do not code for functional RNA or protein products. As an example, consider the genes that code for the oxygen-carrying protein called hemoglobin. Humans have a large family of loci that code for the polypeptide subunits of hemoglobin. Three of these loci are similar in structure and sequence to functioning genes, but do not produce a product. One of these, called the $\psi\alpha$ (psi-alpha) locus, resembles the normal α locus but contains a mutation that prevents normal transcription. The existence of a nonfunctional pseudogene like this is puzzling under the Theory of Special Creation, but readily understandable under the Theory of Evolution. The evolutionary explanation is that at some time in the past, a mutation created a stop codon in the middle of the normal sequence and effectively disabled the locus. The pseudogene has persisted as a molecular vestige of a normal trait. It provides evidence that organisms have changed through time.

Vestigial structures are a manifestation of change through time.

Direct Observation of Change Through Time

Change through time can also be observed directly. Over the past 60 years, biologists have documented evolutionary change in hundreds of different species. As an example, consider recent work on the soapberry bug, an insect native to the southern United States.

Soapberry bugs feed by piercing the fruits of their host plants with their beaks. As the illustrations in Figure 2.7 show, they reach in to penetrate the coats of the seeds inside the fruit, then suck up the contents of the seeds. The data in Figure 2.7 show that, in the first part of the 20th century, soapberry bugs that were collected in Florida and preserved in museum collections tended to have extremely long beaks. This is sensible, because at that time the most important host plant for the bugs was the large-fruited balloon vine shown in the figure. In the mid-20th century, however, most populations of soapberry bugs in Florida began feeding on the small fruits of a plant that had just been imported from Asia, called the flat-podded golden rain tree. As the figure shows, soapberry bugs collected after the host-plant switch tend to have much smaller beaks than older populations. The characteristics of soapberry bugs are not immutable. Instead, they have changed dramatically over time.

Evidence from the Fossil Record

A **fossil** is a trace of any organism that lived in the past. The total, worldwide collection of fossils, scattered among thousands of different institutions and individuals, is called the **fossil record**.

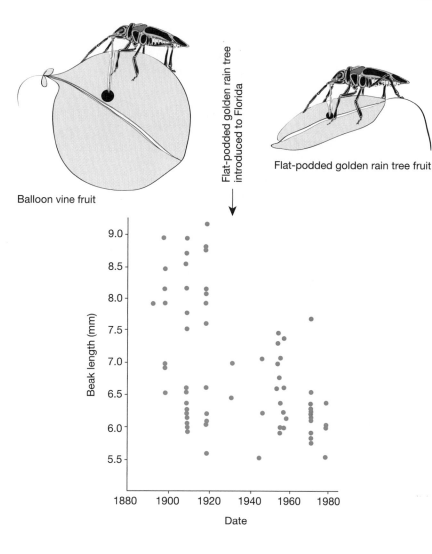

Balloon vine fruit

Flat-podded golden rain tree introduced to Florida

Flat-podded golden rain tree fruit

Figure 2.7   Evolutionary change in soapberry bugs   The drawings at the top of this figure show soapberry bugs feeding on the fruit of a balloon vine (left) and a flat-podded golden rain tree (right). The balloon vine is native to the state of Florida, while the flat-podded golden rain tree was introduced in the late 1920s from Asia. The scatterplot shows the beak lengths of female soapberry bugs from Florida in museum collections (each data point represents one individual). See text for explanation. From Figs. 1 and 6 in Carroll and Boyd (1992), Copyright © 1992 Evolution. Reprinted by permission.

The simple fact that fossils exist, and that the vast majority of fossil forms are unlike species that are living today, argues that life has changed through time. Four specific observations about the fossil record helped Darwin and other 19th-century scientists drive this point home.

The Fact of Extinction

In 1801, the comparative anatomist Baron Georges Cuvier published a list of 23 species that were no longer in existence. His paper was a direct challenge to the widely accepted hypothesis that unusual forms in the fossil record would eventually be found as living species, once European scientists had visited all parts of the globe. The list called this hypothesis into question, because it included mastodons and other enormous creatures excavated from the rocks of the Paris basin. These species were so large that it seemed highly unlikely that they had still escaped detection.

The fact of extinction stopped generating controversy after 1812, when Cuvier published a careful examination of fossils of the Irish elk—the huge ice-age deer shown in Figure 2.8. Fossils of this deer had been found throughout northern Europe and the British Isles. Cuvier's analysis proved that it was an extinct species, and not just a larger version of living species in the region.

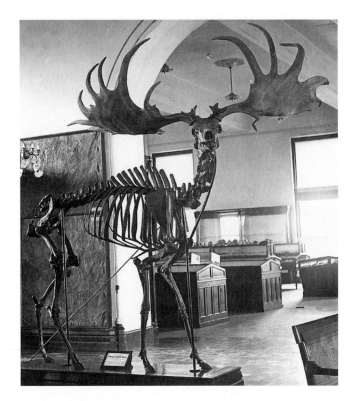

Figure 2.8 **The Irish Elk confirms the fact of extinction**
Cuvier confirmed that fossils of the huge ice-age deer called the Irish elk represented an extinct species (Neg. No. 22581. Photo AMNH. Courtesy Dept. of Library Services. American Museum of Natural History).

By the time Darwin wrote *On The Origin of Species,* extinct plants and animals were being found in rock layers that had been formed in many different times and places. Creationists contended that these species had perished in a series of floods akin to the biblical event at the time of Noah. In contrast, Darwin and other biologists interpreted extinct species as the relatives of living organisms. They pointed to the fact of extinction as more evidence that Earth's flora and fauna have changed through time.

The Law of Succession

An early 18th-century paleontologist named William Clift was the first to publish an observation later confirmed and expanded upon by Darwin (Darwin 1859; Eiseley 1958): Fossil and living organisms in the same geographic region are related to each other and are distinctly different from organisms found in other areas. Clift worked on the extinct marsupial fauna of Australia and noted its close relationship to forms alive today; Darwin analyzed the armadillos of Argentina and their relationship to the fossil glyptodonts he excavated there (Figure 2.9). This general correspondence between fossil and living forms came to be known as the **law of succession**. The law maintains that the fossil species found in a given area are succeeded by similar living species. The result was supported by analyses from a wide variety of locations and taxonomic groups, and provided strong documentation for change through time.

Transitional Forms

Darwin maintained that species have changed through time, and that fossils represent populations that are ancestral to species alive today. If this is true, then the fossil record should contain forms that are transitional between major groups of organisms. The prediction is that transitional species should have characteristics from ancestral populations as well as novel traits seen in descendant species. A

Figure 2.9   The "Law of Succession"   Early researchers observed close relationships between fossil and extant species from the same geographical area, and between fossil forms in adjacent rock strata, so routinely that the pattern became known as the law of succession. Darwin noted the similarities between the contemporary pygmy armadillo (*Zaedyus pichiy*) (left) and the fossil glyptodont (right) of Argentina. (Photo by Tom McHugh/Photo Researchers, Inc.).

good example is the most ancient bird in the fossil record, *Archaeopteryx*. The presence of feathers clearly identifies this species as a bird, but the skeleton is so dinosaur-like that a specimen was once mistakenly identified as the theropod dinosaur *Compsognathus*. *Archaeopteryx* represents a transition between ancestral dinosaur populations and their descendants, the birds.

Because few transitional forms had been discovered in his lifetime, Darwin took pains to explain that they should be rare in the fossil record in a section of his book titled "Difficulties on theory." In the intervening years, however, a large number of transitional forms have been found. Consider the fossil whales pictured in Figure 2.10. Because the earliest mammal fossils represent terrestrial species, biologists infer that the ancestors of whales also lived on land and had functioning limbs. (As the figure shows, some modern whales still have vestigial limbs.) Between these ancestral groups and modern whales, then, there should be intermediate forms that have functioning limbs, as well as features that identify them as marine-dwelling species. Two such transitional forms are shown in the figure. The first, called *Ambulocetus natans,* was discovered and described by J. G. M. Thewissen and co-workers (1994). This fossil is estimated to be about 50 million years old. The second is *Basilosaurus isis*—a fossil species analyzed by Philip Gingerich and colleagues (1990) that lived approximately 38 million years ago. As predicted, the fossil species are intermediate between limbed ancestors and limbless descendants. From their analysis of how the limb articulated with the body, Thewissen et al. suggest that *Ambulocetus* used its limbs in swimming much as modern otters do. Gingerich's group contends that the limbs of *Basilosaurus* were too reduced to function in swimming and may instead have served as grasping organs during copulation. Whatever the function of their limbs, the fossils mark a major evolutionary transition.

In the fossil record, transitional forms record both the gain and loss of prominent characteristics over time.

Environmental Change

The whale fossils illustrated in Figure 2.10 were found in the deserts of Egypt and the hills of Pakistan alongside clams, snails, and other marine species. In Darwin's time, fossils from marine organisms were being discovered in the Andes in South America, the Alps of Europe, and the Grand Canyon of America's arid southwest. These observations suggest that the Earth itself has changed over time. Darwin interpreted these data as evidence that environmental conditions have not

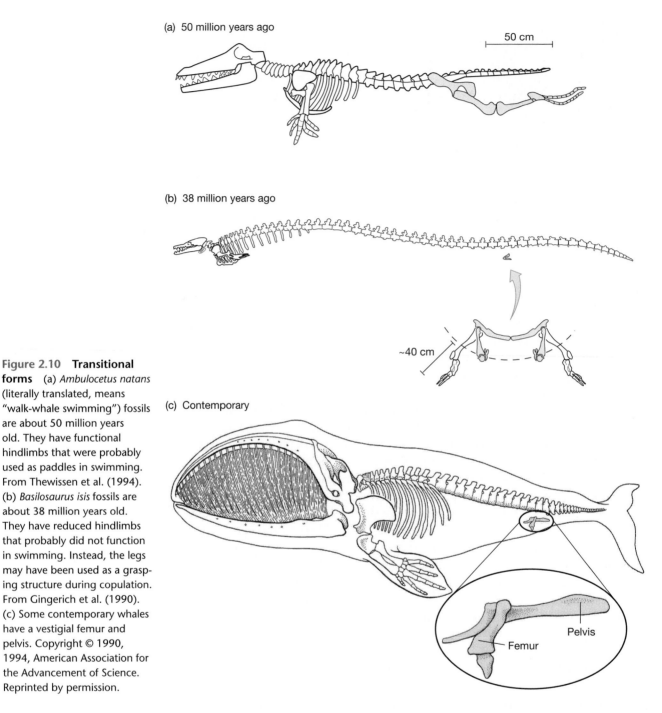

(a) 50 million years ago

50 cm

(b) 38 million years ago

~40 cm

(c) Contemporary

Pelvis

Femur

Figure 2.10 Transitional forms (a) *Ambulocetus natans* (literally translated, means "walk-whale swimming") fossils are about 50 million years old. They have functional hindlimbs that were probably used as paddles in swimming. From Thewissen et al. (1994). (b) *Basilosaurus isis* fossils are about 38 million years old. They have reduced hindlimbs that probably did not function in swimming. Instead, the legs may have been used as a grasping structure during copulation. From Gingerich et al. (1990). (c) Some contemporary whales have a vestigial femur and pelvis. Copyright © 1990, 1994, American Association for the Advancement of Science. Reprinted by permission.

been static. Instead, he suggested that habitats have changed in location and character throughout the course of history. He contended that landforms and habitats are continuously being modified, just as species are.

2.3 The Age of the Earth

By the time Darwin began working on the origin of species, data from the young science of geology had challenged a linchpin in the Theory of Special Creation: that the Earth was only about 6000 years old. Evidence was mounting that the Earth was ancient.

Much of this evidence was grounded in a principle called **uniformitarianism**, first articulated by James Hutton in the late 1700s. Uniformitarianism is the claim that geological processes taking place now operated similarly in the past. It was proposed in direct contrast to a hypothesis called **catastrophism**. This was the proposal that today's geological formations resulted from catastrophic events, like the biblical flood, which occurred in the past on a scale never observed today.

The assumption of uniformitarianism and rejection of catastrophism led Hutton, and later Charles Lyell, to infer that the Earth was unimaginably old in human terms. This conclusion was driven by data. These early geologists measured the rate of ongoing rock-forming processes like the deposition of mud, sand, and gravel at beaches and river deltas and the accumulation of marine shells (the precursors of limestone). Based on these observations, it was clear that vast stretches of time were required to produce the immense rock formations being mapped in the British Isles and Europe by these researchers.

The Geologic Time Scale

When Darwin began his work, Hutton and followers were already in the midst of a 50-year effort to put the major rock formations and fossil-bearing strata of Europe in a younger-to-older sequence. Their technique was called **relative dating** because its objective was to determine how old each rock formation was relative to other strata. Relative dating was an exercise in logic based on the following assumptions:

- Younger rocks are deposited on top of older rocks (this is called the principle of superposition).
- Lava and sedimentary rocks, like sandstones, limestones, and mudstones, were originally laid down in a horizontal position. As a result, any tipping or bending events in these types of rocks must have occurred after deposition (principle of original horizontality).
- Rocks that intrude into seams in other rocks, or as dikes, are younger than their host rocks (principle of cross-cutting relationships).
- Boulders, cobbles, or other fragments found in a body of rock are older than their host rock (principle of inclusions).
- Earlier fossil life forms are simpler than more recent forms, and more recent forms are most similar to existing forms (principle of faunal succession).

Using these rules, geologists established the chronology of relative dates known as the **geologic time scale** (Figure 2.11). They also created the concept of the **geologic column**, which is a geologic history of the Earth based on a composite, older-to-younger sequence of rock strata. (There is no one place on Earth where all rock strata that have formed through time are still present. Instead, there are always gaps where some strata have eroded completely away. But by combining data from different locations, geologists are able to assemble a complete record of geologic history.)

The principle of uniformitarianism, the geologic time scale, and the geologic column furnished impressive evidence for an ancient Earth. Geologists began working in time scales of tens of millions of years, instead of a few thousand years, long before Darwin published his ideas on change through time and descent with modification. These data were important to the Theory of Evolution. Special creation is an instantaneous process, but evolutionary change required long periods of time to produce the diversity of life seen today.

The young science of Geology confirmed that Earth had existed for vast stretches of time. Evolution is a time-dependent process, but special creation is not.

Eon	Era	Period		Epoch	Age Ma	Life Forms
Phanerozoic	Cenozoic	Quaternary		Holocene		
				Pleistocene		
		Tertiary	Neogene	Pliocene	1.8	Earliest *Homo*
				Miocene	5.2	First daisy–family plants
			Paleogene	Oligocene	23.8	First apes
				Eocene	33.5	First extensive grasslands / First whales
				Paleocene	55.6	First horses
					65	Extinction of dinosaurs
	Mesozoic	Cretaceous		Late		First placental mammals
				Early	98.9	First flowering plants
		Jurassic		Late	144	First birds
				Middle	160	
				Early	180	
		Triassic		Late	206	First mammals
				Middle	228	First dinosaurs
				Scythian		
	Paleozoic	Permian			251	First plants with water–conducting vessels
		Carboniferous	Pennsylvanian		290	First mammal-like reptiles
			Mississippian			First reptiles
		Devonian			353.7	First amphibians / First woody plants / First insects
		Silurian			408.5	First vascular plants / First fish with jaws
		Ordovician			439	First fish (no jaws) / First land plants
		Cambrian			495	
					543	First multicellular organisms
Proterozoic						First eukaryotes
					2500	
Archaean					3600	First bacteria / Origin of life?
Hadean						Oldest rocks
					4600	Formation of the Earth

Figure 2.11 **The geologic time scale** The sequence of eons, eras, periods, and epochs shown on the left part of this diagram was established through the techniques of relative dating. Each named interval of time is associated with a distinctive fossil flora and fauna. The absolute ages included here were added much later, when radiometric dating systems became available. The abbreviation Ma stands for millions of years ago.

Radiometric Dating

By the mid–19th century, Hutton, Lyell, and their followers had established, beyond a reasonable doubt, that the Earth was old. But how old? How much time has passed since life on Earth began?

Marie Curie's discovery of radioactivity, in the early 1900s, gave scientists a way to answer these questions. Using a technique called **radiometric dating**, physicists and geologists began to assign absolute ages to the relative dates established by the geologic time scale.

The technique for radiometric dating utilizes unstable isotopes of naturally occurring elements. These isotopes decay, meaning that they change into either different elements or different isotopes of the same element. Each isotope decays at a particular and constant rate, which is measured in a unit called a half-life. One half-life is the amount of time it takes for 50% of the parent isotope present to decay into its daughter isotope (Figure 2.12). The number of decay events observed in a rock sample over time depends only on how many radioactive atoms are present. Decay rates are not affected by temperature, moisture, or any other environmental factor. As a result, radioactive isotopes function as natural clocks. For more detail, see Box 2.2.

Because of their long half-lives, potassium–argon and uranium–lead systems are the isotopes of choice for determining the age of the Earth. Using these systems, what rocks can be tested to determine when the Earth first formed? Current models of Earth formation predict that the planet was molten for much of its early history, which makes answering this question difficult. If we assume that all of the components of our solar system were formed at the same time, however, two classes of candidate rocks become available to date the origin of the Earth: moon rocks and meteorites. Both uranium–lead and potassium–argon dating systems place the age of the moon rocks brought back by the Apollo astronauts at 4.53 billion years. Also, virtually every meteorite found on Earth that has been dated yields an age of 4.6 billion years. Therefore, scientists can infer that the planet is about 4.6 billion years old.

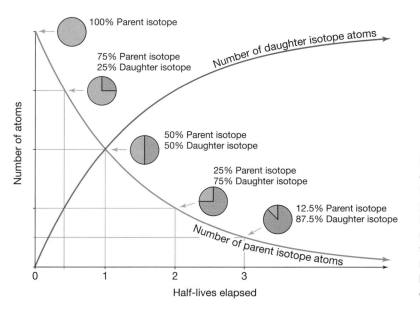

Figure 2.12   Radioactive decay   Many radioactive isotopes decay through a series of intermediates until a stable daughter isotope is produced. Researchers measure the ratio of parent isotope to daughter isotope in a rock sample, then use a graph like this to convert the measured ratio to the number of half-lives elapsed. Multiplying the number of half-lives that have passed by the number of years it takes for a half-life to elapse yields an estimate for the absolute age of the rock.

BOX 2.2 **A closer look at radiometric dating**

Radiometric dating allows geologists to assign absolute ages to rocks. Here is how the technique works: First, the half-life of a radioactive isotope is determined by putting a sample in an instrument that records the number of decay events over time. For long-lived isotopes, of course, researchers must extrapolate from data collected over a short time interval. Then the ratio of parent to daughter isotopes in a sample of rock is measured, often with an instrument called a mass spectrometer. Once the half-life of the parent isotope and the current ratio of parent to daughter isotopes is known, the number of years that have passed since the rock was formed can be calculated.

A critical assumption here is that the ratio of parent to daughter isotope when the rock was formed is known. This assumption can be tested. Potassium-argon dating, for example, is an important system for dating rocks of volcanic origin. We can predict that, initially, there will be zero of the daughter isotope, argon-40, present. That is because argon-40 is a gas that bubbles out of liquid rock and only begins to accumulate after solidification. Observations of recent lava flows confirm that this is true: Expressed as a ratio of percentages, the ratio of potassium-40 to argon-40 in newly minted basalts, lavas, and ashes is, as predicted, 100:0 (see Damon 1968, Faure 1986).

Of the many radioactive atoms present in the Earth's crust, the isotopes listed in Table 2.1 are the most useful. Not only are they common enough to be present in measurable quantities, but they are also stable in terms of not readily migrating into or out of rocks after their initial formation. If the molecules did move, it would throw off our estimate of the surrounding rock's age.

To choose an isotope suitable for dating rocks and fossils of a particular age, geochronologists and paleontologists look for a half-life short enough to allow a measurable amount of daughter isotope to accumulate, but long enough to ensure that a measurable quantity of parent isotope is still left. In many instances more than one isotope system can be used on the same rocks or fossils, providing an independent check on the date.

Table 2.1   Parent and daughter isotopes used in radiometric dating

Method	Parent isotope	Daughter isotope	Half-life of parent (years)	Effective dating range (years)	Materials commonly dated
Rubidium-strontium	Rb-87	Sr-87	47 billion	10 million–4.6 billion	Potassium-rich minerals such as biotite, potassium, muscovite, feldspar, and hornblende; volcanic and metamorphic rocks
Uranium-lead	U-238	Pb-206	4.5 billion	10 million–4.6 billion	Zircons, uraninite, and uranium ore such as pitchblende; igneous and metamorphic rock
Uranium-lead	U-235	Pb-207	713 million	10 million–4.6 billion	Same as above
Thorium-lead	Th-232	Pb-208	14.1 billion	10 million–4.6 billion	Zircons, uraninite
Potassium-argon	K-40	Ar-40	1.3 billion	100,000–4.6 billion	Potassium-rich minerals such as biotite, muscovite, and potassium feldspar; volcanic rocks
Carbon-14	C-14	N-14	5,730	100–100,000	Any carbon-bearing material, such as bones, wood, shells, charcoal, cloth, paper, and animal droppings

This leads to the question, How long has life on Earth been evolving? The oldest fossil organisms discovered to date are impressions of bacterial cells found in 3.5-billion-year-old rocks from Western Australia (Schopf 1993). The oldest chemical evidence of life is in rocks from Greenland dating more than 3.7 billion years old. This chemical evidence consists of carbon granules that have a distinctive ratio of the carbon isotopes ^{13}C and ^{12}C. Living cells preferentially sequester ^{12}C, and the carbon found in the Greenland rocks has much less ^{13}C and more ^{12}C than is expected from the actual abundance of the two isotopes in nature. As a result, investigators infer that the carbon granules represent the remains of living cells (Mojzsis et al. 1996; Rosing 1999). These data suggest that the Earth was lifeless for about 900 million years, and that evolution has been occurring for about 3.7 billion years.

In the 19th century, relative dating suggested that the Earth was much older than the 6000 years predicted by Bishop Ussher. In the 20th century, absolute dating confirmed that life has existed over 600,000 times longer than suggested by the Theory of Special Creation.

2.4 Correspondence Among Data Sets

The data reviewed in this chapter contradict the propositions that species were created independently, that species are immutable, and that the Earth is young. Ironically, much of the data that Darwin used to refute these hypotheses was gathered by ardent creationists like Cuvier, Agassiz, and Owen. These scientists contributed observations on structural and developmental homologies, vestigial traits, extinction, the law of succession, and the geologic time scale.

As persuasive as the evidence for evolution is, though, it is important to note that no single grand experiment swept aside belief in the Theory of Special Creation and opened the way for a new theory. Observations like structural and developmental homologies did not falsify special creation outright. Rather, evolution was simply much more powerful in explaining the data.

Direct observation of evolution has now provided overwhelming evidence for change through time. But there is another type of data that also supports the pattern component of the theory of evolution. This is a correspondence among independent sources of data on Earth history and the history of life. In the example summarized here, data from analyses of Earth's magnetic field, radiometric dating, and the fossil record tell a cohesive story about change through time.

Geologic Change and Plate Movement

Geologists interpret Earth history through the Theory of Plate Tectonics. This set of hypotheses has served as the central organizing principle of the geological sciences since the late 1960s. The pattern component of this theory contends that the Earth's crust is broken into plates and that the positions of these plates—and hence the locations of continents and oceans—have changed through time. The process responsible for this pattern is the unequal distribution of heat in the Earth's interior, which drives plate movement by causing certain rocks to expand (see Chernicoff 1995 to learn more about how this theory was developed and tested).

Geologists can use data on the history of Earth's magnetic field to track where the continents have moved over time. This is possible because magnetic minerals lose their magnetism when they are heated sufficiently. When demagnetized minerals reach the Earth's surface during volcanic events, they cool and become remagnetized in a direction dictated by Earth's magnetic field. In this way, magnetic fields are frozen in volcanic rocks. As the plate containing these rocks moves, the direction of the "frozen" magnetic field begins to differ from Earth's actual magnetic field. Because the frozen magnetic field does not change, it faithfully records where a particular plate was at a particular moment in Earth's history. By collecting volcanic rocks that were created at different times on the same plate, and then figuring out where the plate had to be to produce each of the ancient magnetic fields, geologists can reconstruct the history of plate movement. Figure 2.13 shows the types of maps which result. The maps reproduced here show the positions of the continents over the past 90 million years.

The History of Marsupial Mammals

How do data on plate movements support the pattern component of evolution? When the geologic data are combined with radiometric dates and data from the fossil record, a coherent picture of change through time emerges. For example, consider the history of the mammal group called the marsupials. Marsupials are distinguished by their mode of reproduction: After a short pregnancy, mothers give birth to tiny young, who then climb into a pouch where they continue their development. Marsupials are the dominant type of mammal found in Australia; a few species are still found in South America and one species, the opossum, lives in North America.

Data from plate tectonics, radiometric dating, and the fossil record can be combined to reconstruct how marsupials moved from one continent to another over time.

The fossil record shows that marsupials once lived throughout the world, however. Note, from Figure 2.13, that marsupials first appeared in North America about 80 million years ago. About 10 million years later they began to be found in the fossil record of northern South America. In rocks dated to about 56 million years ago, marsupial fossils are present in Europe and North Africa as well as in Antarctica. Somewhat later, marsupial fossils showed up in Australia. By about 30 million years ago, marsupials disappeared from the fossil record of Asia, North Africa, Europe, and Antarctica. By that time, Australia had broken its connection with Antarctica and had became an island continent.

The key point here is that the three independent sources of data describe movements of marsupial species that are logical in terms of their timing, direction, and the availability of connections among continents. The data sets combine to provide strong, corroborative evidence that Earth and its life forms have changed through time.

How does evolutionary change occur? That is the question we take up in Chapter 3.

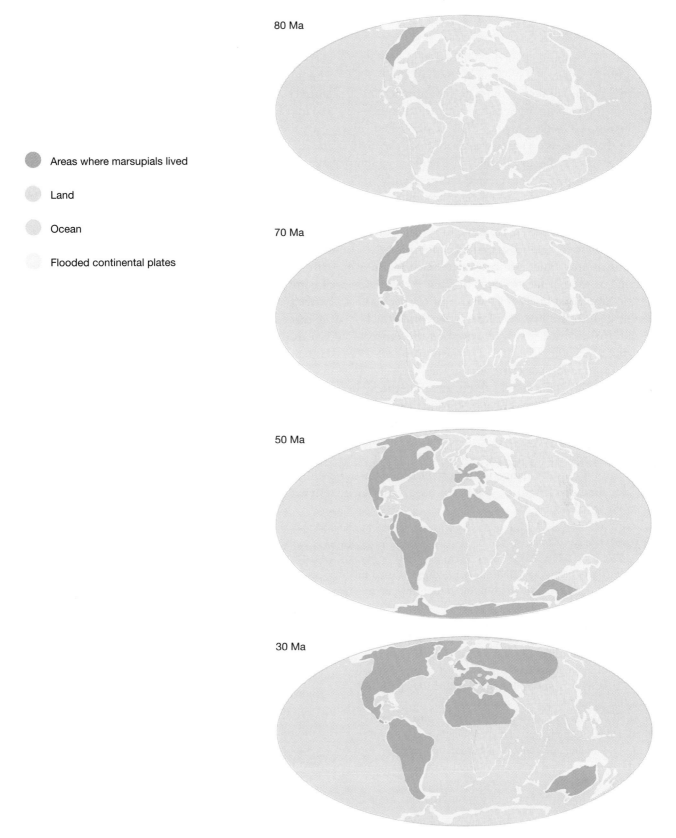

80 Ma

Areas where marsupials lived

Land

Ocean

Flooded continental plates

70 Ma

50 Ma

30 Ma

Figure 2.13   **The evolutionary history of marsupial mammals**   See text for explanation.

Summary

The pattern component of the Theory of Evolution contends that species have changed through time and are related to one another through descent from a common ancestor. Darwin argued forcefully for this theory in his book *On the Origin of Species,* published in 1859. At that time, one leading explanation for the history of life was the Theory of Special Creation, which maintained that species were created independently and recently, and that they do not change through time.

Several lines of evidence argue that species were not created independently. For example, extensive structural, developmental, and genetic homologies exist among organisms. These similarities are most logically explained as the product of descent from a common ancestor. Similarly, closely related groups of species in the same geographic area, like the mockingbirds, finches, and tortoises that Darwin observed on the Galápagos islands, are readily interpreted as the descendants of populations that colonized the area in the past.

Data sets from living and fossil species refute the hypothesis that species do not change through time. The presence of rudimentary structures, transitory developmental stages, and vestigial DNA sequences in contemporary organisms is readily understood as a result of change through time. Change in important characteristics, like the beak length of soapberry bugs, has also been observed directly in hundreds of different species. The hypothesis of change through time is further supported by the extensive extinctions, "law of succession," and transitional forms documented in the fossil record.

By the mid–1800s, the principle of uniformitarianism and the completion of the geologic time scale had persuaded most scientists that the Earth is much older than the few thousand years posited by the Theory of Special Creation. This result was verified in the early 1900s by radiometric dating. The best data available suggest that the Earth formed about 4.6 billion years ago. The first fossil evidence for life is 3.7 billion years old.

In many cases, independent data sets on the history of Earth and the history of life combine to provide a cohesive picture of change through time. The history of the marsupial mammals is one example. By combining data from the fossil record of marsupials, radiometric dating, and maps of continental movements through time, biologists have reconstructed how the group originated in North America and then spread—east to Europe and Asia and south to South America, Antarctica, and finally Australia.

The Theory of Evolution is successful because it provides a logical explanation for a wide variety of observations, and because it makes predictions that can be tested and verified.

Questions

1. Analogy and homology are important concepts used in comparing species. Traits are homologous if they are derived, evolutionarily and developmentally, from the same source structure. Traits are analogous if they have similar functions but are derived, evolutionarily and developmentally, from different source structures. A classic example of analogous structures is insect wings and bat wings. Which of the following pairs of structures are analogous, and which are homologous?
   a. The dorsal fins of a porpoise and a salmon.
   b. The jointed leg of a ladybird beetle and a robin.
   c. A siamang ape's tail and a human's coccyx.
   d. The bright red bracts (modified leaves) of a poinsettia and the green leaves of a rose.
   e. The bright red bracts of a poinsettia and the red petals of a rose.

2. In the early 20th century, radiometric dating allowed geologists to assign absolute ages to most fossil-bearing strata. The absolute dates turned out to be entirely consistent with the relative dating done in the early 19th century. What does this result say about the assumptions behind relative dating listed on page 37?

3. Review the evidence for evolution analyzed in Sections 2.1–2.3. List the sources of evidence that were available to Darwin, and which appeared later. Indicate which evidence you consider strongest and weakest. Explain why.

4. Draw a simple phylogenetic tree showing what the relationships among five living species might be. Then, draw a genealogy of your family or a friend's family, starting with the oldest and continuing to the youngest generation. Label the parts of each diagram. How are phylogenetic trees and pedigrees similar? How are they different?

5. Review the alternative hypotheses illustrated in Figure 2.5. Is it truly more parsimonius to infer that swim bladders evolved from lungs, and not the other way around? Explain.

Suppose that we didn't know anything about the biology of gars, bichirs, and bowfins. Would you arrive at a different conclusion about whether swim bladders or lungs had evolved first? Why?

6. Make a flow chart that illustrates the relationships between genetic, developmental, and structural homologies. If two organisms share a homologous structure, do you think they must also share a corresponding homologous gene(s)? If they share a homologous gene(s), do you think they must also share a corresponding homologous structure or protein?

7. Based on the assumption that extinctions were caused by catastrophic, worldwide floods of the type described in the Bible, what predictions does the Theory of Special Creation make about the nature of the fossil record? What predictions does the Theory of Evolution make about the nature of the fossil record?

8. The text maintains that correspondence between independent data sets is particularly strong evidence for evolution. Do you agree with this claim? In what sense are the fossil record, radiometric dates, and maps of continental positions independent?

Exploring the Literature

9. Phylogenies that are estimated from morphological data, phylogenies that are estimated from molecular data, radiometric dating, and the fossil record represent another suite of independent data sets that combine to corroborate an evolutionary view of the history of life. Research on the evolution of tetrapods presents a good example of combining data sets. The following citations will help you get started on this literature:

Ahlberg, P. E. and Z. Johanson. 1998. Osteolepiforms and the ancestry of tetrapods. *Nature* 395: 792–794.

Hedges, S. B. and L. L. Poling. 1999. A molecular phylogeny of reptiles. *Science* 283: 998–1001.

Shubin, N. 1998. Evolutionary cut and paste. *Nature* 394: 12–13.

Zardoya, R. and A. Meyer. 1996. Evolutionary relationships of the coelacanth, lungfishes, and tetrapods based on the 28S ribosomal RNA gene. *Proceedings of the National Academy of Sciences, USA* 93: 5449–5454.

10. Based largely on the strength of the evidence that Darwin compiled, there has been little, if any, scientific debate about the fact of evolution since the 1870s. Biologists continue to debate the evidence for evolution with nonscientists, however. To participate in the discussion, explore http://www.talkorigins.org.

Citations

Burke, A. C. and A. Feduccia. 1997. Developmental patterns and the identification of homologies in the avian hand. *Science* 278: 666–668.

Carroll, S. P. and C. Boyd. 1992. Host race radiation in the soapberry bug: natural history with the history. *Evolution* 46: 1052–1069.

Chernicoff, S. and R. Venkatakrishnan 1995. *Geology: An Introduction to Physical Geology.* New York: Worth Publishers.

Colbert, E. H. and M. Morales. 1991. *Evolution of the Vertebrates.* New York: Wiley-Liss.

Damon, P. E. 1968. Potassium-argon dating of igneous and metamorphic rocks with applications to the basin ranges of Arizona and Sonora. In Hamilton, E. I., and R. M. Farquhar, eds. *Radiometric Dating for Geologists* London: Interscience Publishers.

Darwin, C. 1859. *On the Origin of Species by Means of Natural Selection.* London: John Murray.

Darwin, C. 1862. *The Various Contrivances by Which Orchids are Fertilized by Insects.* London: John Murray.

Desmond, A. and J. Moore. 1991. *Darwin.* New York: W.W. Norton.

Eiseley, L. 1958. *Darwin's Century* Garden City, NY: Anchor Books.

Faure, G. 1986. *Principles of Isotope Geology.* New York: John Wiley & Sons.

Gingerich, P. D., B. H. Smith, and E. L. Simons. 1990. Hind limbs of Eocene *Basilosaurus:* evidence of feet in whales. *Science* 249: 154–156.

Liem, K. F. 1988. Form and function of lungs: the evolution of air breathing mechanisms. *American Zoologist* 28: 739–759.

Mayden, R. L. and E. O. Wiley. 1992. The fundamentals of phylogenetic systematics. Pp. 114–185 in R. L. Mayden, ed. *Systematics, Historical Ecology, and North American Freshwater Fishes.* Stanford, CA, Stanford Univ. Press.

Mojzsis, S. J., G. Arrhenius, K. D. McKeegan, T. M. Harrison, A. P. Nutman, and C. R. L. Friend. 1996. Evidence for life on Earth before 3,800 million years ago. *Nature* 384: 55–59.

Quiring, R., U. Walldorf, U. Kloter, and W. J. Gehring. 1994. Homology of the *eyeless* gene of *Drosophila* to the *small eye* gene in mice and *aniridia* in humans. *Science* 265: 785–789.

Richardson, M. K., J. Hanken, M. L. Gooneratne, C. Pieau, A. Raynaud, L. Selwood, and G. M. Wright. 1997. There is no highly conserved embryonic stage in the vertebrates: implications for current theories of evolution and development. *Anatomy and Embryology* 196: 91–106.

Rosing, M. T. 1999. ^{13}C-depleted carbon microparticles in >3700-Ma seafloor sedimentary rocks from West Greenland. *Science* 283: 674–676.

Scotese, C. R. 1998. Paleomap project. http://www.scotese.com.

Schopf, J. W. 1993. Microfossils of the early archean apex chert: New evidence of the antiquity of life. *Science* 260: 640–646.

Szalay, F. S. 1994. *Evolutionary History of the Marsupials and an Analysis of Osteological Characters.* New York: Cambridge University Press.

Thewissen, J. G. M., S. T. Hussain, and M. Arif. 1994. Fossil evidence for the origin of aquatic locomotion in archaeocete whales. *Science* 263: 210–212.

CHAPTER 3

Darwinian Natural Selection

The blue leg bands on this male medium ground finch mark it for study. Researchers have observed natural selection in action by studying this species on the Galápagos islands. (Peter R. Grant, Princeton University)

IN THE INTRODUCTION TO *ON THE ORIGIN OF SPECIES,* DARWIN (1859, p. 3) wrote that "a naturalist, reflecting on the mutual affinities of organic beings, on their embryological relations, their geographical distribution, geological succession, and other such facts, might come to the conclusion that each species had not been independently created, but had descended ... from other species. Nevertheless, such a conclusion, even if well founded, would be unsatisfactory, until it could be shown *how* the innumerable species inhabiting this world have been modified ..." (emphasis added).

With this statement, Darwin pinpointed the relationship between the pattern and process components of a scientific theory. The early evolutionists had discovered an important phenomenon. A growing body of facts indicated that both fossilized and living organisms had descended from a common ancestor. The evidence that Darwin amassed to support this hypothesis was indirect, but persuasive enough that scientific controversy over the pattern component of the theory of evolution had virtually ended by the mid-1870s. Thanks to Darwin and his intellectual forebears, evolution became a well-established fact.

But what process could produce the pattern called evolution? Understanding the mechanism that produces a pattern in nature is the heart and soul of a scientific explanation. Chapter 2 focused on the evidence for the pattern called descent with modification; this chapter introduces a process, called natural selection, that produces the pattern.

3.1 Natural Selection: Darwin's Four Postulates

Natural selection is the logical outcome of four postulates, which Darwin laid out in his introduction to *On the Origin of Species by Means of Natural Selection*. He considered the rest of the book "one long argument" in their support (Darwin 1859, p. 459). The postulates are as follows:

1. Individuals within species are variable.
2. Some of these variations are passed on to offspring.
3. In every generation, more offspring are produced than can survive.
4. The survival and reproduction of individuals are not random: The individuals who survive and go on to reproduce, or who reproduce the most, are those with the most favorable variations. They are naturally selected.

Natural selection is a process that produces descent with modification, or evolution.

As a result of this process, the characteristics of populations change from one generation to the next. The logic is clear: If there is variation among the individuals in a population that can be passed on to offspring, and if there is differential success among those individuals in surviving and/or reproducing, then some traits will be passed on more frequently than others. As a result, the characteristics of the population will change slightly with each succeeding generation. This is Darwinian evolution: gradual change in populations over time. Changes in populations result from natural selection on individuals.

To drive this point home, recall the HIV virions described in Chapter 1. Individual virions within the same host varied in their ability to synthesize DNA in the presence of AZT, because of differences in the amino acid sequences of the reverse transcriptase active site. Virions with forms of reverse transcriptase that were less likely to bind AZT reproduced more than virions with forms that bound AZT readily. In the next generation, then, a higher percentage of virions had the modified form of reverse transcriptase than in the generation before. This is evolution by natural selection.

Darwin referred to the individuals who win this competition (it is rarely an actual head-to-head contest), and whose offspring make up a greater percentage of the population in the next generation, as more fit. In doing so he gave the everyday English words "fit" and "fitness" a new meaning. **Darwinian fitness** is the ability of an individual to survive and reproduce in its environment.

An adaptation is a characteristic that increases the fitness of an individual compared to individuals without the trait.

An important aspect of fitness is its relative nature. Fitness refers to how well an individual survives and how many offspring it produces compared to other individuals of its species. Biologists use the word **adaptation** to refer to a trait or characteristic of an organism, like a modified form of reverse transcriptase, that increases its fitness relative to individuals without the trait.

The same theory had, incidentally, been developed independently by a colleague of Darwin's named Alfred Russel Wallace. Though trained in England, Wallace had been making his living in Malaysia by selling natural history specimens to private collectors. While recuperating from a bout with malaria in 1858, he wrote a manuscript explaining natural selection and sent it to Darwin. Darwin, who had written his first draft on the subject in 1842 but never published it, immediately realized that he and Wallace had formulated the same theory independently. Brief papers by Darwin and by Wallace were read together before the Linnean Society of London, and Darwin then rushed *On the Origin of Species* into publication (17 years after he had written the first draft). Today, Darwin's name is more prominently associated with the Theory of Evolution by Natural Selection for two reasons: He had clearly thought of it first, and his book provided a full exposition of the idea, along with massive documentation.

One of the most attractive aspects of the Darwin-Wallace theory is that each of the four postulates—and their logical consequence—can be verified independently. That is, the theory is testable. There are neither hidden assumptions nor anything that has to be accepted uncritically. In the next section, we examine each of the four assertions by reviewing an ongoing study of finches in the Galápagos Islands off the coast of Ecuador. Can the Theory of Evolution by Natural Selection be tested rigorously, by direct observation?

The Theory of Evolution by Natural Selection is testable.

3.2 The Evolution of Beak Shape in Galápagos Finches

Peter Grant and Rosemary Grant and their colleagues have been studying several species of Galápagos finches continuously and on various islands in the Galápagos archipelago since 1973 (see Boag and Grant 1981; Grant 1981a, 1991, 1999; Grant and Grant 1989). The 14 finch species found on the islands are similar in size and coloration. They range from four to six inches in length and from brown to black in color. Two traits do show remarkable variation among species, however: the size and shape of their beaks.

The beak is the primary tool used by birds in feeding, and the enormous range of beak morphologies among the Galápagos finches reflects the diversity of foods they eat (Figure 3.1). The warbler finch (*Certhidea olivacea*) feeds on insects, spiders, and nectar; woodpecker and mangrove finches (*C. pallida* and *C. heliobates*) use twigs or cactus spines as tools to pry insect larvae or termites from dead wood; several ground finches in the genus *Geospiza* pluck ticks from iguanas and tortoises in addition to eating seeds; the vegetarian finch (*Platyspiza crassirostris*) eats leaves and fruit.

Because it is the most important trait used in acquiring food, the size and shape of a bird's beak has important consequences for its fitness.

To test the Theory of Evolution by Natural Selection, we focus on data Grant and Grant and colleagues have gathered on the medium ground finch, *Geospiza fortis,* on Isla Daphne Major.

Daphne Major's size and location make it a superb natural laboratory. The island is tiny. It is just under 40 hectares (about 80 football fields) in extent, with a maximum elevation of 120 meters. Like all of the islands in the Galápagos, it is the top of a volcano. The climate is seasonal even though the location is equatorial (Figure 3.2). A warmer, wetter season from January through May alternates with a cooler, drier season from June through December. The vegetation consists of dry forest and scrub, with several species of cactus present.

The *Geospiza fortis* on Daphne Major make an ideal study population because few finches migrate onto or off of the island, and the population is small enough to be studied exhaustively. In an average year, there are about 1200 individual finches on the island. By 1977, Grant and Grant's team had captured and marked over half of them; since 1980, virtually 100% of the population has been marked.

The medium ground finch is primarily a seed eater. The birds crack seeds by grasping them at the base of the bill and then applying force. Grant and Grant and their colleagues have shown that both within and across finch species, beak size is

(a)

(b)

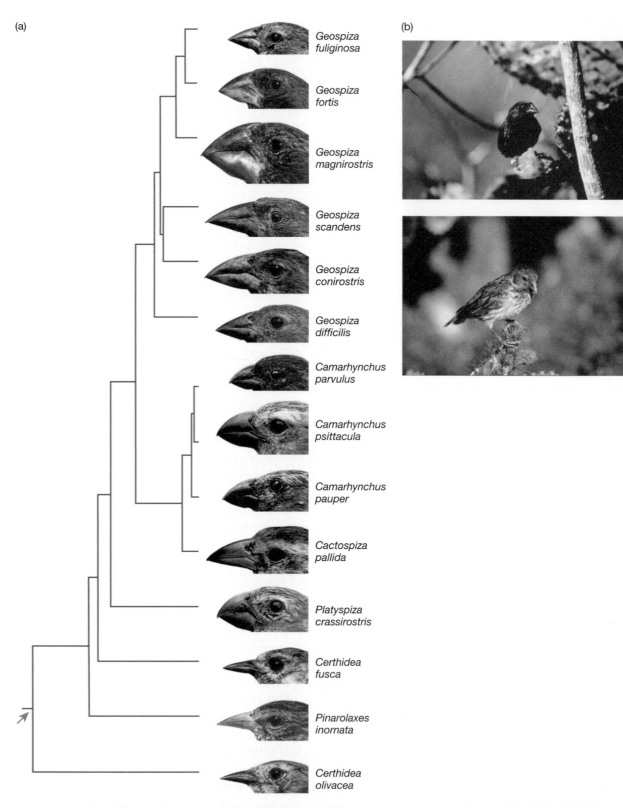

Figure 3.1 The Galápagos finches, and the medium ground finch *Geospiza fortis* (a) This phylogeny was estimated from similarities and differences in DNA sequences by Kenneth Petren and colleagues (1999), and shows the evolutionary relationships among 14 species of Darwin's finches. The photos show the extensive variation in beak size and shape among species. From Petren et al. (1999). (b) A male (top) and female (bottom) medium ground finch. (Photos: Peter R. Grant, Princeton University)

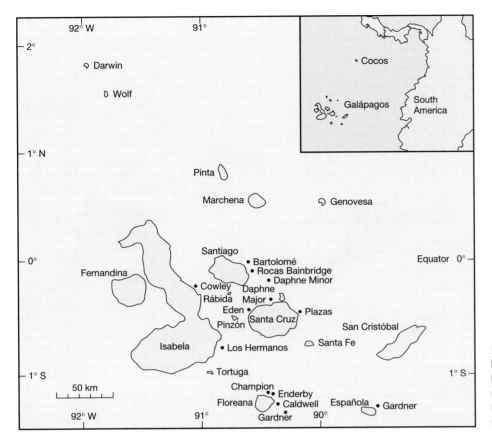

Figure 3.2   The Galápagos islands   Fourteen species of finch inhabit the Galapágos, and one finch resides on Cocos. On this map, Isla Daphne Major is a tiny speck between Santa Cruz and Santiago.

correlated with the size of seeds harvested. In general, birds with bigger beaks eat larger seeds, and birds with smaller beaks eat smaller seeds. This is because birds with different beak sizes are able to handle different sizes of seeds more efficiently (Bowman 1961; Grant et al. 1976; Abbott et al. 1977; Grant 1981b).

Testing Postulate 1: Are Populations Variable?

The researchers mark every finch they catch by placing colored aluminum bands around each of its legs. This allows them to identify individual birds in the field. The scientists also weigh each finch and measure its wing length, tail length, beak width, beak depth, and beak length. All of the traits they have investigated are variable. For example, when Grant and Grant plotted measurements of beak depth in the Isla Daphne Major population of *G. fortis,* the data indicated that beak depth varies (Figure 3.3). All of the finch characteristics they have measured clearly conform to Darwin's first postulate. As we will see in Chapter 4, variation among the individuals within populations is virtually universal.

Testing Postulate 2: Is Some of the Variation Among Individuals Heritable?

Within a population, individual finches could vary in beak depth because the environments they have experienced are different or because their genotypes are different, or both. There are several ways that environmental variation could cause the variation in beak depth recorded in Figure 3.3. Variation in the amount of food that individual birds happened to have received as chicks can lead to variation in beak

Some Geospiza fortis have beaks that are only half as deep as other individuals.

Figure 3.3 **Beak depth in medium ground finches** This histogram shows the distribution of beak depth in medium ground finches on Daphne Major in 1976, at the start of the Grant study. A few birds have shallow beaks, less than 8 mm deep. Most birds have medium beaks, 8 to 11 mm deep. A few birds have deep beaks, more than 11 mm deep. (*N* stands for sample size; the blue arrow along the *x* axis indicates the mean, or average.)

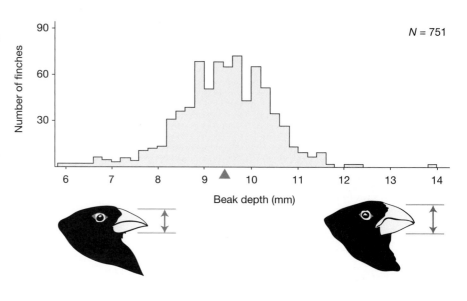

When variation in the characteristics of organisms is plotted, it is common to observe a bell-shaped curve like the one formed by this histogram. The extent, cause, and nature of variation in traits is a focus of Chapter 4 and Chapter 7.

BOX 3.1 Issues that complicate how heritabilities are estimated

Heritabilities are estimated by measuring the similarity of traits among closely related individuals. But relatives share their environment as well as their genes, and any correlation that is due to their shared environment inflates the estimate of heritability. For example, it is well known that birds tend to grow larger when they have abundant food as young. But the most food-rich breeding territories are often won and defended by the largest adults in the population. Young from these territories will tend to become the largest adults in the next generation. As a result, a researcher might measure a strong relationship between parent and offspring beak and body size and claim a high heritability for these traits when in reality there is none. In this case, the real relationship is between the environments that parents and their young each experienced as chicks.

In many species, this problem can be circumvented by performing what are called cross-fostering, common garden, or reciprocal-transplant experiments. In birds, these experiments involve taking eggs out of their original nest and placing them in the nests of randomly assigned foster parents. Measurements in the young, taken when they are fully grown, are then compared with the data from their biological parents. This ex-

perimental treatment removes any bias in the analysis created by the fact that parents and offspring share environments. Unfortunately, even cross-fostering experiments cannot remove environmental effects that are due to differences in the nutrient stores of eggs or seeds. We simply have to assume that these maternal effects are too small to seriously bias the result.

Cross-fostering experiments in a wide variety of bird species have confirmed large heritabilities in most or all of the morphological characters analyzed. It was not possible for Boag and Grant (1978) to perform this critical experiment on the Galápagos finches, however. Because the Galápagos are a national park, experiments that manipulate individuals beyond catching and marking are forbidden. To get around this limitation, Peter Grant (1991) compared individuals that were raised in food-poor versus food-rich territories to see if they bred in a similar environment once they reached adulthood. The data showed no correlation between the quality of the territory a bird was raised in and the quality of the territory where it, in turn, raised its young. This result bolsters the claim that the large heritabilities estimated in finch traits, like beak dimensions, are real, and not an artifact of parents and offspring sharing a similar environment.

depth among adults. Injuries or abrasion against hard seeds or rocks can also affect beak size and shape.

To determine whether at least part of the variability among finch beaks is genetically based, and thus capable of being passed from parents to offspring, a colleague of Peter Grant and Rosemary Grant's named Peter Boag estimated a quantity known as **heritability**.

Heritability is the proportion of the variation observed in a population that is due to variation in the effects of genes. Because it is a proportion, heritability varies between 0 and 1. We will develop the theory behind how heritability is estimated much more fully in Chapter 7. For now, we simply point out that it is usually estimated by measuring the similarity between pairs of relatives. This is a valid approach, because similarities between relatives are caused, at least in part, by the alleles they share (for more detail, see Box 3.1). Data are usually collected from siblings, or from parents and offspring.

Boag compared the beak depth of *G. fortis* young after they had attained adult size to the average bill depth of their mother and father, and found a strong correspondence between relatives. As the data plotted in Figure 3.4 show, parents with deep beaks tend to have offspring with deep beaks, and parents with shallow beaks tend to have chicks with shallow beaks. This is evidence that a large proportion of the observed variation in beak depth is genetic, and can be transmitted to offspring (Boag and Grant 1978; Boag 1983). The result is consistent with hundreds of similar studies. In most traits of most organisms, a significant amount of the variation that exists within populations is due to variation in the genetic makeup of individuals.

In finches, the beak depth of parents and offspring are similar. This observation suggests that some alleles tend to produce shallow beaks, while other alleles tend to produce deeper beaks.

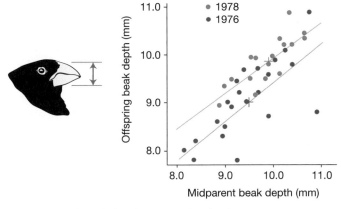

Figure 3.4   Heritability of beak depth in *Geospiza fortis*   This graph shows the relationship between the beak depth of parents and their offspring. Midparent value is the average of the maternal and paternal measurements. Using this measurement is important because male *G. fortis* are bigger than females.

The lines in the graph are the result of a statistical procedure called regression analysis. In the type of regression used here, the line is placed in a way that minimizes the sum of the squared vertical distances between each point in the plot and the line. This is called the best-fit line. If the slope of the line is flat, or 0, then there is no relationship between the two variables plotted—meaning that all variation among individuals is caused by differences in the environments they have experienced. If the slope is 1, then all of the variation among individuals is caused by variation in their genotypes.

The red line and circles are from 1978 data, and the blue line and circles are from 1976 data. The results from the two years are consistent. Both show a strong relationship between the beak depth of parents and their offspring. We can infer that the association is a product of their shared genes. From Boag (1983).

Testing Postulate 3: Is There an Excess of Offspring, So That Only Some Individuals Live to Reproduce?

A drought on Daphne Major produced a dramatic selection event.

Because they routinely censused the ground finch population over several years, the researchers were able to observe a dramatic event. In 1977, there was a severe drought at the study site. Instead of the normal 130 mm of rainfall during the wet season, only 24 mm fell. Over the course of 20 months, 84% of the Darwin's medium ground finch population on Daphne Major disappeared (Figure 3.5a). The team inferred that most died of starvation: There was a strong correspondence between population size and seed availability (Figure 3.5b); 38 emaciated birds were actually found dead, and none of the missing birds reappeared the following year. It is clear that only a fraction of the population survived to reproduce. This sort of mortality is not unusual. For example, Rosemary Grant has shown that 89% of *Geospiza conirostris* individuals die before they breed (Grant 1985). Trevor Price and co-workers (1984) determined that an additional 19% and 25% of the *G. fortis* on Daphne Major died during subsequent drought events in 1980 and 1982, respectively.

In fact, in every natural population studied, more offspring are produced each generation than survive to breed. If a population is not increasing in size, then

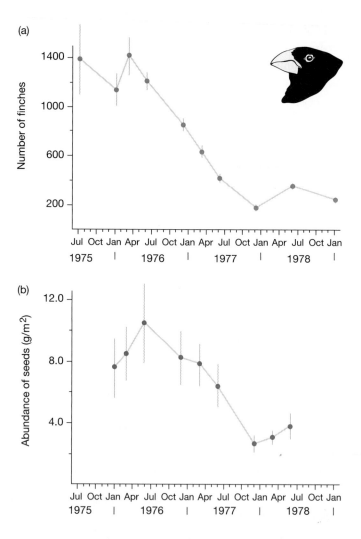

Figure 3.5 Decline of ground finch population and available seeds, during the 1977 drought (a) This graph shows the number of ground finches found on Daphne Major before, during, and after the drought. The vertical lines through each data point represent a quantity called the standard error, which indicates the amount of variation in census estimates. The lines in this graph are simply drawn from point to point to make the trend easier to see.

(b) This graph shows the abundance of seeds on Daphne Major before, during, and after the drought. Reprinted with permission from Boag and Grant (1981). Copyright © 1981, American Association for the Advancement of Science.

each parent will, in the course of its lifetime, leave an average of one offspring that survives to breed. But the reproductive capacity (or biotic potential) of organisms is astonishing (Table 3.1).

Similarly, data show that, in most populations, some individuals are more successful at mating and producing offspring than others. Variation in reproductive success represents an opportunity for selection, as does variation in survival.

Testing Postulate 4: Are Survival and Reproduction Nonrandom?

Darwin's fourth claim was that the individuals who survive and go on to reproduce, or who reproduce the most, are those with certain, favorable variations. Did a nonrandom, or selected, subset of the ground finch population survive the 1977 drought? By measuring the same traits they had measured in 1976 on a large and random sample of surviving birds early in 1978, the Grant team found that a distinct subset of the population had survived best: those with the deepest beaks (Figure 3.6). Because the average survivor had a deeper beak than the average nonsurvivor, the average bill size in the population changed.

In what way were deep beaks favorable? Can we link ecological cause with evolutionary effect? The answer is yes: Not only the number, but also the types of seeds available during the 1977 drought changed dramatically (Figure 3.7). Specifically, the large, hard fruits of an annual plant called *Tribulus cistoides* became a key food item. These seeds are largely ignored in normal years, but during the drought the supply of small, soft seeds quickly became exhausted. Only large birds with deep, narrow beaks can crack and eat *Tribulus* fruits successfully. In addition, large birds defend food sources more successfully during conflicts. Because large size and deep beaks are positively correlated, the two traits responded to selection together.

The 1977–1978 selection event, as dramatic as it was, was not an isolated occurrence. In 1980 and 1982 there were similar droughts, and selection again favored individuals with large body size and deep beaks (Price et al. 1984). Then, in 1983, an

Natural selection occurs because (1) only a fraction of offspring survive long enough to breed, and (2) of the individuals that do breed, some are much more successful than others.

During the drought, finches with larger, deeper beaks had an advantage.

| Table 3.1 | **Reproductive potential** |

This table gives the number of offspring that a single individual (or pair of individuals, for sexual species) can produce under optimal conditions, assuming that all progeny survive to breed, over various time intervals. Darwin picked the elephant for his calculations because it was the slowest breeder then known among animals.

Organism	Reproductive potential	Citation
Aphis fabae (an aphid)	524 billion in one year	Gould 1977
Elephant	19 million in 750 years	Darwin 1859
Housefly	191×10^{18} in 5 months	Keeton 1972
Mycophila speyeri (a fly that feeds on mushrooms)	20,000/square foot in 35 days	Gould 1977
Staphylococcus aureus (a bacterium)	Cells would cover the Earth 7 feet deep in 48 hours	Audesirk and Audesirk 1993
Starfish	10^{79} in 16 years*	Dodson 1960

*10^{79} is the estimated number of electrons in the visible universe.

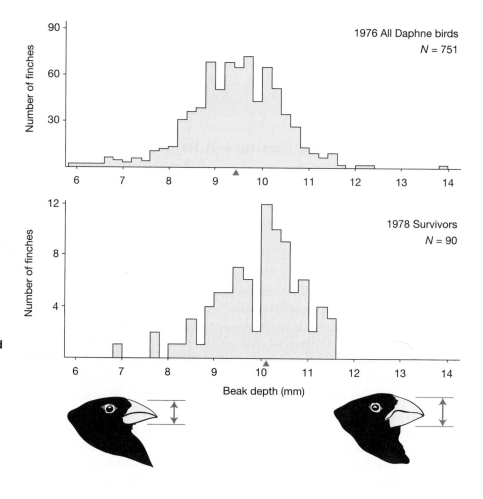

Figure 3.6 Beak depth before and after natural selection These histograms show the distribution of beak depth in medium ground finches on Daphne Major, before and after the drought of 1977 (Grant 1986). Copyright © 1986, Princeton University Press. Reprinted by permission of Princeton University Press.

influx of warm surface water off the South American coast, called an El Niño event, created a wet season with 1359 mm of rain on Daphne Major. This dramatic environmental change (almost 57 times as much rain as in 1977) led to a superabundance of small, soft seeds and, subsequently, to strong selection for smaller body size (Gibbs and Grant 1987). After wet years, small birds with shallow beaks survive better and reproduce more because they harvest small seeds much more efficiently than large birds with deep beaks. Larger birds were favored in drought conditions, but smaller birds were favored in wet years. Natural selection—as we pointed out in our analysis of HIV evolution in Chapter 1—is dynamic.

Did Evolution Occur?

Selection occurs within generations; evolution occurs between generations.

The changes observed in finch beaks are examples of natural selection in action. But we said earlier that evolution was a *response* to selection—a change in the characteristics of a population from one generation to the next. Selection produces a distinct change in trait distributions within a generation; evolution is a change in trait distributions between generations. Did evolution occur in the Galápagos finches? The answer is yes: As the data in Table 3.2 show, the offspring of birds surviving the 1977 drought were significantly larger, on average, than the population that existed before the drought.

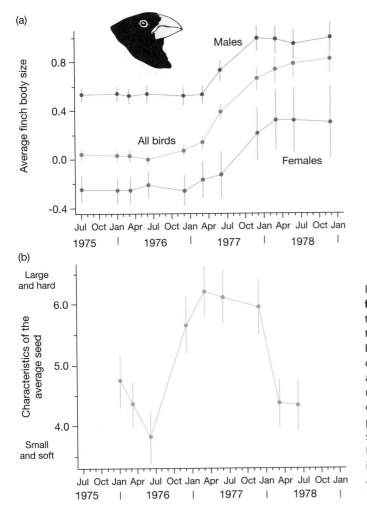

(a)

(b)

Figure 3.7   Changes in the overall body size of ground finches and in seed characteristics   (a) These plots show the overall body size of finches caught before, during, and after the drought on Daphne Major. (The axis has no units because body size was calculated as a composite measure called a principal component score.) The "All birds" line represents the average of males and females. As in Figure 3.5, the data points represent population means, the vertical lines represent standard errors, and the horizontal lines are drawn between data points simply to make the trends easier to see. (b) This graph shows changes in the hardness of seeds available on Daphne Major before, during, and after the drought. The hardness index plotted on the *y*-axis is a special measure created by Boag and Grant (1981). Reprinted with permission. Copyright © 1981, American Association for the Advancement of Science.

3.3 The Nature of Natural Selection

Although the Theory of Evolution by Natural Selection can be stated concisely, tested rigorously in natural populations, and validated, it can be difficult to understand thoroughly. One reason is that it is essentially a statistical process: a change in the trait distributions of populations. Statistical thinking does not come naturally to most people, and there are a number of widely shared ideas about natural selection that are incorrect. In this section our goal is to cover some key points about how selection does and does not operate.

Natural Selection Acts on Individuals, but Its Consequences Occur in Populations

When HIV strains were selected by exposure to AZT, or finch populations were selected by changes in seed availability, none of the selected individuals (virions or finches) changed in any way. They simply lived through the selection event while others died, or reproduced more than competing virions or birds. What changed after the selection process were the characteristics of the populations of virions and finches, not the affected individuals themselves. Specifically, a higher frequency of

Table 3.2 Evolutionary response to selection

These data summarize changes in the population means of body and beak traits in *Geospiza fortis*, before and after the 1976–1977 drought. "SE" is an abbreviation for standard error, which quantifies the amount of variation around the average, or mean, value. The delta (Δ) column shows the differences between generations.

Trait	Before Selection 1976		Next Generation 1978		
	Mean	SE	Mean	SE	Δ
Weight (g)	16.06	0.06	17.13	0.13	+1.07
Wing length (mm)	67.88	0.10	68.87	0.20	+0.99
Tarsus length (mm)	19.08	0.03	19.29	0.07	+0.21
Bill length (mm)	10.63	0.03	10.95	0.06	+0.32
Bill depth (mm)	9.21	0.03	9.70	0.06	+0.49
Bill width (mm)	8.58	0.02	8.83	0.05	+0.25
Sample size	634		135		

Source: Grant and Grant 1995

Natural selection does not change the characteristics of individuals. It changes the characteristics of populations.

HIV virions in the population were able to replicate in the presence of AZT, and a higher proportion of finches had deep beaks.

To state this point another way, the effort of cracking *Tribulus* seeds did not make finch beaks become deeper and their bodies larger, and the effort of transcribing RNA in the presence of AZT did not change the amino acid composition of the reverse transcriptase active site. Instead, the average beak depth and body size in the finch population increased because more smaller finches died than larger ones, and the average active site sequence in reverse transcriptase changed because certain mutants did a better job of making new virions.

Natural Selection Acts on Phenotypes, but Evolution Consists of Changes in Allele Frequencies

Finches with large bodies and deep beaks would have been favored during the drought even if all of the variation in the population had been environmental in origin (that is, if heritabilities had been zero). But no evolution would have occurred. The frequencies of the phenotypes observed before and after selection would have changed, but in the next generation the phenotype distribution might have gone back to what it was before selection occurred.

Because evolution is the response to selection, it occurs only when the selected traits have a genetic basis. The variation in finch phenotypes that selection acted on had a genetic basis. As a result, the phenotype distribution changed in the next generation.

Natural Selection Is Backward Looking, Not Forward Looking

Each generation is a product of selection by the environmental conditions that prevailed in the generation before. The offspring of the HIV virions and finches that underwent natural selection are better adapted to environments dominated by

AZT and drought conditions, respectively, than their parents' generation was. If the environment changed again during the lifetime of these offspring, however, they would not necessarily be adapted to the new conditions.

There is a common misconception that organisms can be adapted to future conditions, or that selection can look ahead in the sense of anticipating environmental changes during future generations. This is impossible. Evolution is always a generation behind any changes in the environment.

Natural selection adapts populations to conditions that prevailed in the past, not conditions that might occur in the future.

Natural Selection Can Produce New Traits, Even Though It Acts on Existing Traits

Natural selection can select only from the variations that already exist in a population. Selection cannot, for example, instantly create a new and optimal beak for cracking *Tribulus* fruits. It only selects from the range of beaks already present in the population.

Over time, however, natural selection *can* produce new traits. This seems paradoxical, given that the process acts only on existing traits. The evolution of new traits is possible because, in each generation, mutations produce new variants, and thus a new suite of existing traits, for selection to act upon. To understand why this is important, consider the results of an artificial selection experiment conducted at the University of Illinois (Leng 1962). A research team started with 163 ears of corn, tested the oil content in the kernels of each, and found that the amount of oil ranged from 4–6%. They selected the 24 ears with the highest oil content to be the parents of the next generation, raised the offspring, tested their kernels for oil content, and again selected the individuals with the highest oil content to be the parents of the next generation. By continuing this selection regime for 60 years, the researchers succeeded in producing corn plants whose kernels had an oil content of about 16% (Figure 3.8). There is no overlap between the distribution of oil content in the ancestral and descendant populations. Evolution produced a new trait value.

Natural selection can also lead to novel characteristics. This is possible because selection is able to "repurpose" existing behaviors, structures, or genes for new functions. The giant panda's thumb is a good example (Gould 1980). Pandas use this

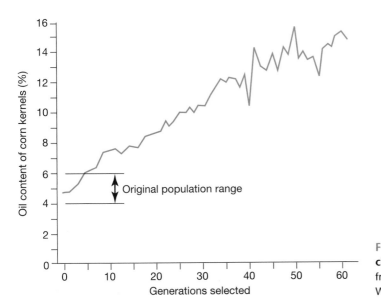

Figure 3.8 Continued selection can result in dramatic changes in traits See text for explanation. Modified from Leng (1962). Reprinted by permission of Blackwell Wissenschafts-Verlag GmbH.

structure like a sixth finger when eating their favorite food, bamboo. As Figure 3.9 shows, they pass the stalks through the slot between the thumb and their other five digits to strip off the leaves and expose the shoots, which they eat. But this sixth digit is not a true thumb at all. Anatomically, the bone that forms the "thumb" is a highly modified radial sesamoid, which in closely related species is a component of the wrist. Knowing how natural selection works in contemporary populations, we surmise that when panda populations first began exploiting bamboo, there was variation among individuals in the length of the radial sesamoid bone. As a result of strong and continued selection over many generations, the average length of the bone increased in the populations until it reached its present proportions.

A trait that is used in a novel way and is eventually elaborated by selection into a completely new structure, like the radial sesamoid of the ancestral panda, is known as a **preadaptation**. An important point about preadaptations is that they represent a happenstance. A preadaptation improves an individual's fitness by accident—not because natural selection is conscious or forward looking.

Natural Selection Is Not "Perfect"

The previous paragraphs stressed that natural selection continually improves adaptation. Although this is true, it is equally important to realize that evolution does not result in "perfect" traits.

To drive this point home, consider that when Boag and Grant analyzed their data on the survival of finches after the 1977 drought, they noticed that individuals with relatively narrow beaks also survived better. This makes sense, because finches push down and to either side when cracking *Tribulus* fruits, and narrower beaks concentrate these twisting forces more effectively. But beak width is positively correlated with beak depth and large body size. As a result, birds with deep beaks, which are good for applying downward force, tend also to have wider beaks, which are less effective at applying the twisting force. Presumably, this correlation

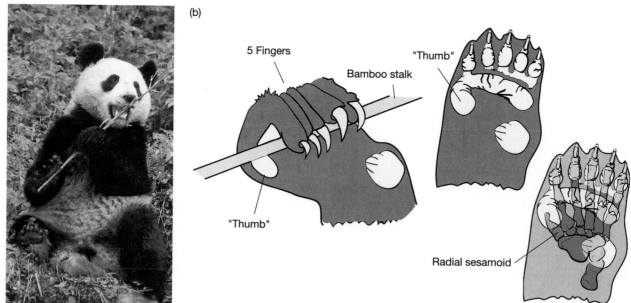

Figure 3.9 The panda's thumb (a) Giant pandas strip the leaves from bamboo by passing the stalk through their hands. (Bill Kamin/Visuals Unlimited) (b) This drawing shows how the panda's "thumb" forms a slot for bamboo stalks to pass through. After Endo et al. 1999.

exists because the same genes affect beak depth and width and overall body size, making all three traits smaller, or larger, together. The upshot is that selection for larger size and deeper beaks resulted in wider beaks, even when narrower beaks should have been favored.

This is an important point. Because of the genetic correlations among characters, natural selection did not optimize all of the traits involved. Natural selection results in adaptation, not perfection.

Natural Selection Is Nonrandom, but It Is Not Progressive

Evolution by natural selection is sometimes characterized as a random or chance process, but nothing could be further from the truth. Evolution by natural selection is nonrandom because it increases adaptation to the environment.

As the HIV, finch, and panda examples demonstrate, however, the process of nonrandom selection is completely free of any entity's conscious intent. Darwin actually came to regret coining the phrase "naturally selected," because people thought the word selection implied a conscious act or choice by some entity. Nothing of the sort happens.

Also, although evolution has tended to increase the complexity, degree of organization, and specialization of organisms over time, it is not progressive in the sense of leading toward some predetermined goal. Evolution makes organisms "better" only in the sense of increasing their adaptation to their environment. There is no inexorable trend toward more advanced forms of life. For example, contemporary tapeworms have no digestive system, and have actually evolved to be simpler than their ancestors. Snakes evolved from ancestors that had limbs. The earliest birds in the fossil record had teeth.

Unfortunately, a progressivist view of evolution dies hard. Even Darwin had to remind himself to "never use the words higher or lower" when discussing evolutionary relationships. It is true that some organisms are the descendants of ancient lineages and some are the descendants of more recent lineages, but all organisms in the fossil record and those living today were adapted to their environments. They are all able to survive and reproduce. None is "higher" or "lower" than any other.

There is no such thing as a higher or lower plant or animal.

Fitness Is Not Circular

The Theory of Evolution by Natural Selection is often criticized—by nonbiologists—as tautological, or circular in its reasoning. That is, after reviewing Darwin's four postulates, one could claim, "Of course individuals with favorable variations are the ones that survive and reproduce, because the theory defines favorable as the ability to survive and reproduce."

The key to resolving the issue is to realize that the word "favorable," although a convenient shorthand, is misleading. The only requirement for natural selection is for certain variants to do better than others, as opposed to random ones. As long as a nonrandom subset of the population survives better and leaves more offspring, natural selection will result. In the examples we have been analyzing, research not only determined that nonrandom groups survived a selection event, but also uncovered why those groups were favored.

It should also make sense by now that Darwinian fitness is not an abstract quantity. Fitness can be measured in nature. This is done by counting the offspring that individuals produce during their lifetime, or by observing the ability of individuals to survive a selection event, and comparing each individual's performance to that of other individuals in the population. These are independent, measurable, and objective criteria for assessing fitness.

Natural Selection Acts on Individuals, Not Groups

One of the most pervasive misconceptions about natural selection, especially selection on animal behavior, is that individual organisms will perform actions for the good of the species. Self-sacrificing, or altruistic, acts do occur in nature. Prairie dogs give alarm calls when predators approach, which draws attention to themselves. Lion mothers sometimes nurse cubs that are not their own. But selection cannot favor traits unless they increase the bearer's fitness relative to competing individuals. If an allele existed that produced a truly altruistic behavior—that is, a behavior that reduced the bearer's fitness and increased the fitness of others—it would be strongly selected against. As we will see in Chapter 10, every altruistic behavior that has been studied in detail has been found to increase the altruist's fitness either because the beneficiaries of the behavior are close genetic relatives (as in prairie dogs) or because the beneficiaries reciprocate (as in nursing lions).

Individuals do not do things for the good of the species. They behave in a way that maximizes their individual fitness.

The idea that animals will do things for the good of the species is so ingrained, however, that we will make the same point a second way. Consider lions again. Lions live in social groups called prides. Coalitions of males fight to take over prides. If a new group of males defeats the existing pride males in combat, the newcomers quickly kill all of the pride's nursing cubs. These cubs are unrelated to them. Killing the cubs increases the new males' fitness, because pride females become fertile again sooner and will conceive offspring by the new males (Packer and Pusey 1983, 1984). Infanticide is widespread in animals. Clearly, behavior like this does not exist for the good of the species. Rather, infanticide exists because, under certain conditions, it enhances the fitness of the individuals who perform the behavior relative to individuals who do not.

3.4 The Evolution of Darwinism

Because evolution by natural selection is a general organizing feature of living systems, Darwin's theory ranks as one of the great ideas in intellectual history. Its impact on biology is analogous to that of Newton's laws on physics, Copernicus's Sun-Centered Theory of the Universe on astronomy, and the Theory of Plate Tectonics on geology. In the words of evolutionary geneticist Theodosius Dobzhansky (1973), "Nothing in biology makes sense except in the light of evolution."

For all its power, though, the Theory of Evolution by Natural Selection was not universally accepted by biologists until some 70 years after it was initially proposed. There were three serious problems with the theory, as originally formulated by Darwin, that had to be resolved.

1. Because Darwin knew nothing about mutation, he had no idea how variability was generated in populations. As a result, he could not answer critics who maintained that the amount of variability in populations was strictly limited, and that natural selection would grind to a halt when variability ran out. It was not until the early 1900s, when geneticists such as Thomas Hunt Morgan began experimenting with fruit flies, that biologists began to appreciate the continuous and universal nature of mutation. Morgan and colleagues showed that mutations occur in every generation and in every trait.

2. Because Darwin knew nothing about genetics, he had no idea how variations are passed on to offspring. It was not until Mendel's experiments with peas were rediscovered and verified, 35 years after their original publication, that biologists understood how parental traits are passed on to offspring. Mendel's laws of segregation and independent assortment confirmed the mechanism behind postulate 2, which states that some of the variation observed in populations is heritable.

Until then, many biologists proposed that genes acted like pigments in paint. Advocates of this hypothesis, called **blending inheritance**, argued that favorable mutations would simply merge into existing traits and be lost. In 1867, a Scottish engineer named Fleeming Jenkin published a mathematical treatment of blending inheritance, along with a famous thought experiment concerning the offspring of light-skinned and dark-skinned people. For example, if a dark-skinned sailor became stranded on an equatorial island inhabited by light-skinned people, Jenkins' model predicted that no matter how advantageous dark skin might be (in reducing skin cancer, for example), the population would never become dark-skinned because traits like skin color blended. If the dark-skinned sailor had children by a light-skinned woman, their children would be brown-skinned. If they, in turn, had children with light-skinned people, their children would be light-brown-skinned, and so on. Conversely, if a light-skinned sailor became stranded on a northern island inhabited by dark-skinned people, blending inheritance argued that, no matter how advantageous light skin might be (in facilitating the synthesis of vitamin D from UV light, for example), the population would never become light. Under blending inheritance new variants are swamped, and new mutations diluted, until they cease to have a measurable effect. For natural selection to work, favorable new variations have to be passed on to offspring intact, and remain discrete.

We understand now, of course, that phenotypes blend in some traits, like skin color, but genotypes never do. Jenkins's hypothetical population would, in fact, become increasingly darker or lighter skinned if selection were strong and mutation continually added darker- or lighter-skinned variants to the population via changes in the genes involved in the production of melanin (Figure 3.10).

Darwin himself struggled with the problem of inheritance, and eventually adopted an entirely incorrect view based on the work of Jean-Baptiste Lamarck. Lamarck was a great French biologist of the early 19th century who proposed that species evolve through the inheritance of changes wrought in individuals. Lamarck's idea was a breakthrough for two reasons: It recognized that species have changed through time, and proposed a mechanism for producing change. His theory was wrong, however, because offspring do not inherit phenotypic changes acquired by their parents. If people build up muscles lifting weights, their offspring are not more powerful; if giraffes stretch their necks reaching for leaves in treetops, it has no consequence for the neck length of their offspring.

3. Lord Kelvin, the foremost physicist of the 19th century, published an important series of papers in the early 1860s calculating the age of the Earth at a maximum of 15–20 million years. Kelvin's analyses were based on measurements of the Sun's heat and the current temperature of the Earth. Because fire was the only known source of heat at the time, Kelvin assumed that the Sun was combusting like an enormous lump of coal. This had to mean that the Sun was gradually burning down, releasing progressively less heat with each passing millennium. Likewise, both geologists and physicists had come to believe that

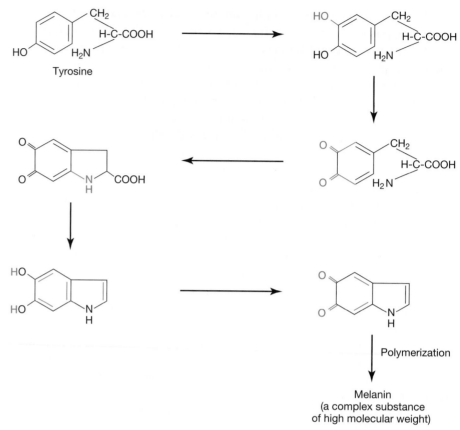

Figure 3.10 Why blending inheritance does not occur This diagram shows the biochemical pathway for melanin, the dark pigment in human skin. The pathway involves a series of steps starting with phenylalanine or tyrosine, which are common dietary amino acids. The enzymes responsible for catalyzing each reaction shown are coded for by several different genes. Because different alleles of these genes have different levels of activity, and because the pathway to melanin involves numerous steps, skin color is a product of the activities of many genes and alleles. In this way, the actions of genes blend to form a phenotype. Because each allele is distinct, however, each enzyme's effect on the biochemical pathway is passed on to offspring intact. This is why there is no such thing as blending inheritance.

the surface of the Earth was gradually cooling. This was based on the assumption that the Earth was changing from a molten state to a solid one by radiating heat to the atmosphere. This assumption seemed to be supported by measurements of progressively higher temperatures deeper down in mineshafts. These data allowed Kelvin to calculate the rate of radiant cooling.

The bottom line from Kelvin's calculations was that the transition from a hot to cold Sun and hot to cold Earth created a narrow window of time when life on Earth was possible. The window was clearly too narrow to allow the gradual changes of Darwinism to accumulate, and strongly supported a role for instantaneous and special creation in explaining adaptation and diversity.

The discovery of radioactive isotopes early in the 20th century changed all that. Kelvin's calculations were unassailable, but his assumptions were dead wrong. Geologists and physicists confirmed that the Earth's heat is a by-product of radioactive decay and not radiant cooling, and that the Sun's energy is from nuclear fusion, not combustion.

The Modern Synthesis

The Modern Synthesis resolved decades of controversy over the validity of evolution by natural selection.

Understanding variability, inheritance, and time was so difficult that the first 70 years of evolutionary biology were characterized by turmoil (see Provine 1971; Mayr 1980, 1991). But between 1932 and 1953 a series of landmark books was published that successfully integrated genetics with Darwin's four postulates and led to a reformulation of the Theory of Evolution. This restatement, known as the Modern Synthesis or the Evolutionary Synthesis, was a consensus grounded in two propositions:

- Gradual evolution results from small genetic changes that are acted upon by natural selection.
- The origin of species and higher taxa, or macroevolution, can be explained in terms of natural selection acting on individuals, or microevolution.

With the synthesis, Darwin's original four postulates—and their outcome—could be restated along the following lines:

1. As a result of mutation creating new alleles, and segregation and independent assortment shuffling alleles into new combinations, individuals within populations are variable for nearly all traits.
2. Individuals pass their alleles on to their offspring intact.
3. In most generations, more offspring are produced than can survive.
4. The individuals that survive and go on to reproduce, or who reproduce the most, are those with the alleles and allelic combinations that best adapt them to their environment.

The outcome is that alleles associated with higher fitness increase in frequency from one generation to the next.

This View of Life

Darwin ended the introduction to the first edition of *On the Origin of Species* with a statement that still represents the consensus view of evolutionary biologists (Darwin 1859, p. 6): "Natural Selection has been the main but not exclusive means of modification." We now think of modification in terms of changes in the frequencies of the alleles responsible for traits like beak depth and AZT resistance. We are more keenly aware of other processes that cause evolutionary change in addition to natural selection. (Chapters 4–6 explore these processes in detail.) But the Darwinian view of life, as a competition between individuals with varying abilities to survive and reproduce, has proven correct in almost every detail.

As Darwin wrote in his concluding sentence (1859, p. 490): "There is grandeur in this view of life, with its several powers, having been originally breathed into a few forms or into one; and that, whilst this planet has gone cycling on according to the fixed law of gravity, from so simple a beginning endless forms most beautiful and most wonderful have been, and are being, evolved."

3.5 The Debate over "Scientific Creationism"

Scientific controversy over the fact of evolution ended in the late 1800s, when the evidence reviewed in Chapter 2 simply overwhelmed the critics. Whether natural selection was the primary process responsible for both adaptation and diversity was still being challenged until the 1930s, when the works of the Modern Synthesis provided a mechanistic basis for Darwin's four postulates and unified micro- and macroevolution. Evolution by natural selection is now considered the great unifying idea in biology. Although scientific discourse about the validity of evolution by natural selection ended well over a half-century ago, a political and philosophical controversy in the United States and Europe still continues (Holden 1995; Kaiser 1995). What is this debate, and why is it occurring?

Creationists want the Theory of Special Creation to be taught in public schools, even though it was dismissed as a viable alternative to the Theory of Evolution by Natural Selection over a century age.

History of the Controversy

The Scopes Monkey Trial of 1925 is perhaps the most celebrated event in a religious debate that has raged since Darwin first published (see Gould 1983, essay 20). John Scopes was a biology teacher who assigned reading about Darwinism in violation of the State of Tennessee's Butler Act, which prohibited the teaching of evolution in public schools. William Jennings Bryan, a famous politician and a fundamentalist orator, was the lawyer for the prosecution; Clarence Darrow, the most renowned defense attorney of his generation, represented Scopes. Although Scopes was convicted and fined $100, the trial was widely perceived as a great triumph for evolution because Bryan had suggested, on the witness stand, that the six days of creation described in Genesis 1:1–2:4 may each have lasted far longer than 24 hours. This was considered a grave inconsistency and a blow to the integrity of the creationist viewpoint. But far from ending the debate over teaching evolution in U.S. schools, the Scopes trial was merely a way station.

The Butler Act, in fact, stayed on the books until 1967; it was not until 1968, in *Epperson v. Arkansas,* that the U.S. Supreme Court struck down laws that prohibit the teaching of evolution. The court's ruling was made on the basis of the U.S. Constitution's separation of church and state. In response, fundamentalist religious groups in the United States reformulated their arguments as "creation science" and demanded equal time for what they insisted was an alternative theory for the origin of species. By the late 1970s, 26 state legislatures were debating equal-time legislation (Scott 1994). Arkansas and Louisiana passed such laws only to have them struck down in state courts. The Louisiana law was then appealed all the way to the U.S. Supreme Court, which decided in 1987 (*Edwards v. Aquillard*) that because creationism is essentially a religious idea, teaching it in the public schools was a violation of the first amendment. Two justices, however, formally wrote that it would still be acceptable for teachers to present alternative theories to evolution (Scott 1994).

One response from opponents of evolution has been to drop the words creation and creator from their literature and call either for equal time for teaching that no evolution has occurred, or for teaching a proposal called Intelligent Design Theory, which infers the presence of a creator from the perfection of adaptation in contemporary organisms (Scott 1994; Schmidt 1996). The complexity and perfection of organisms is actually a time-honored objection to evolution by natural selection. Darwin was aware of it; in his *Origin* he devoted a section of the chapter titled "Difficulties on Theory" to "Organs of extreme perfection." How can natural selection, by sorting random changes in the genome, produce elaborate and integrated traits like the vertebrate eye?

Perfection and Complexity in Nature

The English cleric William Paley, writing in 1802, promoted the Theory of Special Creation with a now-classic argument. If a person found a watch and discovered that it was an especially complex and accurate instrument, they would naturally infer that it had been made by a highly skilled watchmaker. Paley then drew a parallel between the watch and the perfection of the vertebrate eye, and asked his readers to infer the existence of a purposeful and perfect Creator. He contended that organisms are so well-engineered that they have to be the work of a conscious designer. This logic, still used by creationists today, is called the Argument from Design (Dawkins 1986).

Because we perceive perfection and complexity in the natural world, evolution by natural selection seems to defy credulity. There are actually two concerns here. The first is how random changes can lead to order. Mutations are chance events, so the generation of variation in a population is random. But the selection of those variants, or mutants, is nonrandom: It is directed in the sense of increasing fitness. And adaptations—structures or behaviors that increase fitness—are what we perceive as highly ordered, complex, or even perfect in the natural world. But there is nothing conscious or intelligent about the process. The biologist Richard Dawkins captured this point by referring to natural selection as a "blind watchmaker."

A second, and closely related, concern is, How can complex, highly integrated structures, like the vertebrate eye, evolve through the Darwinian process of gradual accumulation of small changes? Each evolutionary step would have to increase the fitness of individuals in the population. A creationist named Michael Behe (1996), for example, claims that biological systems such as biochemical pathways are "irreducibly complex," and that it is not possible for them to result from natural selection. Darwinism, in contrast, predicts that complex structures have evolved through a series of intermediate stages, or graded forms. Is this true? For example, when we consider a structure like the eye, do we find a diversity of forms, some of which are more complex than others?

The answer to these questions is yes. In some unicellular species there are actually subcellular organelles with functions analogous to the eye. The eyespots of a group of protozoans called euglenoids, for example, contain light-absorbing molecules that are shaded on one side by a patch of pigment. When these molecules absorb light, they undergo structural changes. Because light can reach them from one side only, a change in the light-absorbing molecule contains useful information about where light is coming from. Some dinoflagellates even have a subcellular, lenslike organelle that can concentrate light on a pigment cup. It is unlikely that these single-celled protists can form an image, however, because they are not capable of neural processing. Rather, their eye probably functions in transmitting information about the cell's depth in the water column, helping the cell orient itself and swim toward light.

More complex eyes have a basic unit called the photoreceptor. This is a cell that contains a pigment capable of absorbing light. The simplest type of multicellular eye, consisting of a few photoreceptor cells in a cup or cuplike arrangement, is shown in Figures 3.11a and 3.11b. This type of eye is found in a wide diversity of taxa, including flatworms, polychaetes (segmented worms in the phylum Annelida), some crustaceans (the shrimps, crabs, and allies), and some vertebrates. These organs are used in orientation and day-length monitoring (Willson 1984; Brusca and Brusca 1990). Slightly more complex eyes, like those illustrated in Figure 3.11c, have optic cups with a narrow aperture acting as a lens, and may be capable of forming images in at least some species. These are found in a few nemerteans (ribbon worms) and annelids (segmented worms), copepod crustaceans, and abalone and nautiloids (members of the phylum Mollusca). The most complex eyes (Figure 3.11d) fall into two functional categories based on whether the photoreceptor cells are arrayed on a retina that is concave, like the eyes of vertebrates and octopuses, or convex, like the compound eyes of insects and other arthropods (Goldsmith 1990). These eyes have lenses, and in most cases are capable of forming images.

It is important to recognize, however, that the simpler eyes we have just reviewed do not represent intermediate forms on the way to more advanced structures. The

The Argument from Design contends that adaptations must result from the actions of a conscious entity.

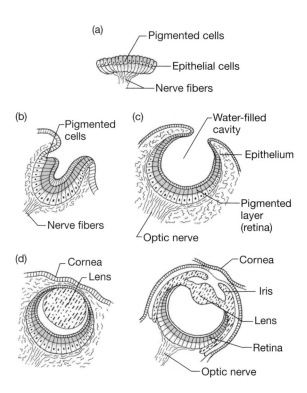

Figure 3.11   Variation in mollusc eyes   (a) A pigment spot; (b) a simple pigment cup; (c) the simple optic cup found in abalone; (d) the complex lensed eyes of a marine snail called *Littorina* and the octopus. Pigmented cells are shown in color.

eyespots, pigment cups, and optic cups found in living organisms are contemporary adaptations to the problem of sensing light. They are not ancestral forms. It is, however, sensible to argue that the types of eyes discussed here form an evolutionary pathway (Gould 1983, essay 1). That is, it is conceivable that eyes like these formed intermediate stages in the evolution of the complex eyes found in vertebrates, octopuses, and insects. This is exactly what Darwin argued in his section on "organs of extreme perfection." (To learn more about the evolution of the eye, see Salvini-Plawen and Mayr 1977; Nilsson and Pelger 1994; Quiring et al. 1994; Dawkins 1994.)

Other Objections

We have found four additional arguments that creationists use regularly, and have added responses from an evolutionary perspective (see Gish 1978; Kitcher 1982; Futuyma 1983; Gould 1983 essays 19, 20, 21; Dawkins 1986; Swinney 1994):

1. Evolution by natural selection is unscientific because it is not falsifiable and because it makes no testable predictions. Each of Darwin's four postulates is independently testable, so the theory meets the classical criterion that ideas must be falsifiable to be considered scientific. Also, the claim that evolutionary biologists do not make predictions is not true. Paleontologists routinely (and correctly) predict which strata will bear fossils of certain types (a spectacular example was that fossil marsupial mammals would be found in Antarctica); Peter Grant and Rosemary Grant have used statistical techniques based on evolutionary theory to correctly predict the amount and direction of change in finch characteristics during selection events in the late 1980s and early 1990s (Grant and Grant 1993, 1995). Scientific creationism, on the other hand, amounts to an oxymoron; in the words of one of its leading advocates, Dr. Duane Gish (1978, p. 42): "We cannot discover by scientific investigations anything about the creative processes used by God."

2. Because the Earth was created as little as 6000–8000 years ago, there has not been enough time for Darwinian evolution to produce the adaptation and diversity observed in living organisms. Creation scientists present short-Earth theories and argue that most geological landforms and strata resulted from the flood during the time of Noah. (For example, see Gish 1978 and Swinney 1994.) Most simply disbelieve the assumptions behind radiometric dating and deny the validity of the data. The assumption of uniformitarianism in the evolution of life and landforms is also rejected by creation scientists. Again, we quote Gish (1978, p. 42): "We do not know how God created, what processes He used, *for God used processes which are not now operating anywhere in the natural universe*" (emphasis original).

The assumptions of radiometric dating have been tested, however, and demonstrated to be correct. Radiometric dating has demonstrated that rock strata differ in age, and that the Earth is about 4.6 billion years old.

3. Because organisms progress from simpler to more complex forms, evolution violates the Second Law of Thermodynamics. Although the Second Law has been stated in a variety of ways since its formulation in the late 19th century, the most general version is: "Natural processes tend to move toward a state of greater disorder" (Giancoli 1995). The Second Law is focused on the concept of entropy. This is a quantity that measures the state of disorder in a system. The Second Law, restated in terms of entropy, is "The entropy of an isolated system never decreases. It can only stay the same or increase" (Giancoli 1995).

The key to understanding the Second Law's relevance to evolution is the word "isolated." The Second Law is true only for closed systems. Organisms, however, live in an open system: the Earth, where photosynthetic life-forms capture the radiant energy of the sun and convert it to chemical energy that they and other organisms can use. Because energy is constantly being added to living systems, the Second Law does not apply to their evolution.

4. No one has ever seen a new species formed, so evolution is unproven. And because evolutionists say that speciation is too slow to be directly observed, evolution is unprovable and thus based on faith. Although speciation is a slow process, it is ongoing and can be studied. In Chapter 12 we explore one of the best-studied examples: the divergence of apple maggot flies into separate host races. The two forms of these flies lay their eggs in different fruits, which serve as a food source for their maggots. As a result of different patterns of natural selection on traits like food preferences and timing of breeding, marked genetic differences between the two populations are beginning to emerge. Research on these organisms is documenting key events early in the process of one species splitting into two.

Because there are tens of millions of insect species, and because most insects are specialist plant eaters, what is happening in apple maggot flies is of general interest: They may now be undergoing an event that has occurred many times in the course of evolution. Chapter 12 introduces other experimental and observational studies of "speciation in action" as well.

What Motivates the Controversy?

For decades, evolution by natural selection has been considered one of the best documented and most successful theories in the biological sciences. Many scientists see no conflict between evolution and religious faith (Easterbrook 1997; Scott

1998), and many Christians agree. In 1996, for example, Pope John Paul II acknowledged that Darwinian evolution was a firmly established scientific result, and stated that accepting Darwinism was compatible with traditional Christian understandings of God.

If the fact of evolution and the validity of natural selection are utterly uncontroversial, and if belief in evolution is compatible with belief in God, then why does the creationist debate continue?

During a discussion about whether material on evolution should be included in high school textbooks, a member of the Alabama State School Board named David Byers said, "It's foolish and naive to believe that what children are taught about who they are, how they got here, doesn't have anything to do with what they conclude about why they are here and what their obligations are, if, in fact, they have any obligations, and how they should live" (National Public Radio 1995). This statement suggests that, for some creationists, the controversy is not about the validity of the scientific evidence or its compatibility with religion. Instead, the concern is about what evolution means for human morality and behavior.

Creationists and evolutionists share the desire that children should grow up to become morally responsible adults. Creationists fight evolution because they believe it is morally dangerous. Evolutionary biologists, on the other hand, believe that children should learn what science says about how living things came to be, and let them sort out the moral implications, if any, on their own.

Summary

Before Darwin began to work on the origin of species, many scientists had become convinced that species change through time. The unique contribution made by Darwin and Wallace was to realize that the process of natural selection provided a mechanism for this pattern, which Darwin termed descent with modification.

Evolution by natural selection is the logical outcome of four facts: (1) Individuals vary in most or all traits; (2) some of this variation is genetically based and can be passed on to offspring; (3) more offspring are born than can survive to breed, and of those that do breed, some are more successful than others; and (4) the individuals that reproduce the most are a nonrandom, or more fit, subset of the general population. This selection process causes changes in the genetic makeup of populations over time, or evolution.

Questions

1. In everyday English, the word adaptation means "an adjustment to environmental conditions." How is the evolutionary definition of adaptation different from the everyday English sense?
2. Think about how the finch bill data demonstrate Darwin's postulates.
   - What would Figure 3.3 have looked like if bill depth was not variable?
   - What would the data in Table 3.2 look like if bill depth was variable, but the variation was not heritable?
   - In Figure 3.4, why is the line drawn from 1978 data, after the drought, higher on the y-axis than the line drawn from 1976 data, before the drought?
3. According to the text, it is legitimate to claim that most finches died from starvation during the 1977 drought because "there was a strong correspondence between population size and seed availability." Do you accept this hypothesis? If so, why don't the data in Figure 3.5 show a perfect correspondence between when seed availability started declining and when population size started declining?

4. Suppose that you are starting a long-term study of a population of annual, flowering plants isolated on a small island. Reading some recent papers has convinced you that global warming is real and will lead to significant, long-term changes in the amount of rain the island receives. Outline the observations and experiments you would need to do in order to document whether natural selection occurs in your study population over the course of your research. What traits would you measure, and why?

5. At the end of an article on how mutations in variable number tandem repeat (VNTR) sequences of DNA are associated with disease, Krontiris (1995, p. 1683) writes: "the VNTR mutational process may actually be positively selected; by culling those of us in middle age and beyond, evolution brings our species into fighting trim." This researcher proposes that natural selection on humans favors individuals who die relatively early in life. His logic is that the trait of dying from VNTR mutations is beneficial and should spread, because the population as a whole becomes younger and healthier as a result. Can this hypothesis be true, given that selection acts on individuals? Explain.

6. Many working scientists are relatively uninterested in the history of their fields. Did the historical development of Darwinism, reviewed in Section 3.4, help you understand the theory better? Why or why not? Do you think it is important for practicing scientists to spend time studying history?

7. Recently, the Alabama School Board, after reviewing high school biology texts, voted to require that this disclaimer be posted on the inside front cover of the approved book (National Public Radio 1995):

> This textbook discussed evolution, a controversial theory some scientists present as a scientific explanation for the origin of living things, such as plants, animals, and humans. No one was present when life first appeared on Earth; therefore, any statement about life's origins should be considered as theory, not fact.

Do you accept the last sentence in this statement? Does the insert's point of view pertain to other scientific theories, such as the Cell Theory, the Atomic Theory, the Theory of Plate Tectonics, and the Germ Theory of Disease?

8. A 1991 Gallup poll of U.S. adults found that 47% of the respondents believed that God created man within the last 10,000 years (Root-Bernstein, 1995). Given the evidence for evolution by natural selection, comment on why so few people in the United States accept it.

Exploring the Literature

9. During the past 50 years, hundreds of viruses, bacteria, fungi, and insects have evolved resistance to drugs, herbicides, fungicides, or pesticides. These are outstanding examples of evolution in action. In several of these cases, we know the molecular mechanisms of the evolutionary changes involved. To explore this topic further, look up the following papers. Think about how the evidence from these studies compares with the evidence for evolution in Darwin's finches and HIV.

Anthony, R. G., T. R. Waldin, J. A. Ray, S. W. J. Bright, and P. J. Hussey. 1998. Herbicide resistance caused by spontaneous mutation of the cytoskeletal protein tubulin. *Nature* 393: 260–263.

Cohen, M. L. 1992. Epidemiology of drug resistance: Implications for a post-antimicrobial era. *Science* 257: 1050–1055.

Davies, J. 1994. Inactivation of antibiotics and the dissemination of resistance genes. *Science* 264: 375–382.

Van Rie, J., W. H. McGaughey, D. E. Johnson, B. D. Barnett, and H. Van Melleart. 1990. Mechanism of insect resistance to the microbial insecticide *Bacillus thuringiensis*. *Science* 247: 72–74.

10. It seems unlikely that selection of traits "for the good of the group" can occur. However, some evolutionary biologists contend that under certain conditions, group selection of altruistic behaviors may in fact be possible. Look up the following papers to learn more about this topic:

Avilés, L., and P. Tufino. 1998. Colony size and individual fitness in the social spider *Anelosimus eximius*. *American Naturalist* 152: 403–418.

Morell, V. 1996. Genes vs. Teams: Weighing group tactics in evolution. *Science* 273: 739–740. (News perspective.)

Wilson, D. F., and E. Sober. 1994. Reintroducing group selection to the human behavioral sciences. *Behavioral Brain Sciences* 17: 585–609.

Wilson, D. S., and L. A. Dugatkin. 1997. Group selection and assortative interactions. *American Naturalist* 149: 336–351.

Citations

Abbott, I., L. K. Abbott, and P. R. Grant. 1977. Comparative ecology of Galápagos ground finches (*Geospiza* Gould): Evaluation of the importance of floristic diversity and interspecific competition. *Ecological Monographs* 47: 151–184.

Audesirk, G., and T. Audesirk. 1993. *Biology: Life on Earth.* New York: Macmillan.

Behe, M. 1996. *Darwin's Black Box: The Biochemical Challenge to Evolution.* New York: Free Press/Simon and Schuster.

Boag, P. T. 1983. The heritability of external morphology in Darwin's ground finches (*Geospiza*) on Isla Daphne Major, Galápagos. *Evolution* 37: 877–894.

Boag, P. T., and P. R. Grant. 1978. Heritability of external morphology in Darwin's finches. *Nature* 274: 793–794.

Boag, P. T., and P. R. Grant. 1981. Intense natural selection in a population of Darwin's finches (Geospizinae) in the Galápagos. *Science* 214: 82–85.

Bowman, R. I. 1961. Morphological differentiation and adaptation in the Galápagos finches. *University of California Publications in Zoology* 58: 1–302.

Brusca, R. C., and G. J. Brusca. 1990. *Invertebrates.* Sunderland, MA: Sinauer.

Darwin, C. 1859. *On the Origin of Species by Means of Natural Selection.* London: John Murray.

Dawkins, R. 1986. *The Blind Watchmaker.* Essex: Longman Scientific.

Dawkins, R. 1994. The eye in a twinkling. *Nature* 368: 690–691.

Dobzhansky, T. 1973. Nothing in biology makes sense except in the light of evolution. *American Biology Teacher* 35: 125–129.

Dodson, E. O. 1960. *Evolution: Process and Product.* New York: Reinhold Publishing.

Easterbrook, G. 1997. Science and God: A warming trend? *Science* 277: 890–893.

Endo, H., D. Yamagiwa, Y. Hayashi, H. Koie, Y. Yamaya, and J. Kimura. 1999. Role of the giant panda's 'pseudo-thumb.' *Nature* 397: 309–310.

Futuyma, D. J. 1983. *Science on Trial: The Case for Evolution.* New York: Pantheon.

Giancoli, D. C. 1995. *Physics: Principles with Applications.* Englewood Cliffs, NJ: Prentice Hall.

Gibbs, H. L., and P. R. Grant. 1987. Oscillating selection on Darwin's finches. *Nature* 327: 511–513.

Gish, D. T. 1978. *Evolution: The Fossils Say No!* San Diego: Creation-Life Publishers.

Goldsmith, T. H. 1990. Optimization, constraint, and history in the evolution of eyes. *Quarterly Review of Biology* 65: 281–322.

Gould, S. J. 1977. *Ever Since Darwin: Reflections in Natural History.* New York: W. W. Norton.

Gould, S. J. 1980. *The Panda's Thumb.* New York: W. W. Norton.

Gould, S. J. 1983. *Hen's Teeth and Horse's Toes.* New York: W. W. Norton.

Grant, B. R. 1985. Selection on bill characters in a population of Darwin's finches: *Geospiza conirostris* on Isla Genovesa, Galápagos. *Evolution* 39: 523–532.

Grant, B. R., and P. R. Grant. 1989. *Evolutionary Dynamics of a Natural Population.* Chicago: University of Chicago Press.

Grant, B. R., and P. R. Grant. 1993. Evolution of Darwin's finches caused by a rare climatic event. *Proceedings of the Royal Society of London,* Series B 251: 111–117.

Grant, P. R. 1981a. Speciation and adaptive radiation on Darwin's finches. *American Scientist* 69: 653–663.

Grant, P. R. 1981b. The feeding of Darwin's finches on *Tribulus cistoides* (L.) seeds. *Animal Behavior* 29: 785–793.

Grant, P. R. 1991. Natural selection and Darwin's finches. *Scientific American* October: 82–87.

Grant, P. R. 1999. *Ecology and Evolution of Darwin's Finches,* 2nd ed. Princeton: Princeton University Press.

Grant, P. R., and B. R. Grant. 1995. Predicting microevolutionary responses to directional selection on heritable variation. *Evolution* 49: 241–251.

Grant, P. R., B. R. Grant, J. N. M. Smith, I. J. Abbott, and L. K. Abbott. 1976. Darwin's finches: Population variation and natural selection. *Proceedings of the National Academy of Sciences, USA* 73: 257–261.

Holden, C. 1995. Alabama schools disclaim evolution. *Science* 270: 1305.

Kaiser, J. 1995. Dutch debate tests on evolution. *Science* 269: 911.

Keeton, W. T. 1972. *Biological Science.* New York: W. W. Norton.

Kitcher, P. 1982. *Abusing Science: The Case Against Creationism.* Cambridge, MA: MIT Press.

Krontiris, T. G. 1995. Minisatellites and human disease. *Science* 269: 1682–1683.

Leng, E. R. 1962. Results of long-term selection for chemical composition in maize and their significance in evaluating breeding systems. *Zeitschrift für Pflanzenzüchtung* 47: 67–91.

Mayr, E. 1980. Prologue. In Mayr, E., and W. B. Provine, eds., *The Evolutionary Synthesis.* Cambridge, MA: Harvard University Press.

Mayr, E. 1991. *One Long Argument: Charles Darwin and the Genesis of Modern Evolutionary Thought.* Cambridge, MA: Harvard University Press.

National Public Radio. 1995. Evolution disclaimer to be placed in Alabama textbooks. Morning Edition, Transcript #1747, Segment #13.

Nilsson, D.-E., and S. Pelger. 1994. A pessimistic estimate of the time required for an eye to evolve. *Proceedings of the Royal Academy of London,* Series B 256: 53–58.

Packer, C., and A. E. Pusey. 1983. Adaptations of female lions to infanticide by incoming males. *American Naturalist* 121: 716–728.

Packer, C., and A. E. Pusey. 1984. Infanticide in carnivores. Pp. 31–42 in G. Hausfater and S. B. Hrdy, eds. *Infanticide.* New York: Aldine Publishing Company.

Petren, K., B. R. Grant, and P. R. Grant. 1999. A phylogeny of Darwin's finches based on microsatellite DNA length variation. *Proceedings of the Royal Society of London,* Series B 266: 321–329.

Price, T. D., P. R. Grant, H. L. Gibbs, and P. T. Boag. 1984. Recurrent patterns of natural selection in a population of Darwin's finches. *Nature* 309: 787–789.

Provine, W. B. 1971. *The Origins of Theoretical Population Genetics.* Chicago: University of Chicago Press.

Quiring, R., U. Walldorf, U. Kloter, and W. J. Gehring. 1994. Homology of the *eyeless* gene of *Drosophila* to the *small eye* gene in mice and *aniridia* in humans. *Science* 265: 785–789.

Root-Bernstein, R. S. 1995. "Darwin's Rib." *Discover* (September) 38–41.

Salvini-Plawen, L. v., and E. Mayr. 1977. On the evolution of photoreceptors and eyes. *Evolutionary Biology* 10: 207–263.

Schmidt, K. 1996. Creationists evolve new strategy. *Science* 273: 420–422.

Scott, E. C. 1994. The struggle for the schools. *Natural History* 7: 10–13.

Scott, E. C. 1998. Two kinds of materialism. *Free Inquiry* 18: 20.

Swinney, S. 1994. "Evolution: Fact or Fiction." Kansas City, MO: 1994 Staley Lecture Series, KLJC Audio Services.

Willson, M. F. 1984. *Vertebrate Natural History.* Philadelphia: Saunders.

PART II

Although they vary in color, most of the seastars in this photograph belong to the same species. The chapters in this unit explore how evolutionary processes act on genetic variation to produce changes in the characteristics of populations through time. (Carr Clifton/Minden Pictures)

MECHANISMS OF EVOLUTIONARY CHANGE

Throughout Part I we focused on natural selection as a mechanism of evolutionary change. By the end of Chapter 3, we had defined evolution as a change in allele frequencies within populations and probed how natural selection functions as an agent of evolutionary change.

Natural selection is not the only process that alters allele frequencies, however. There are three additional evolutionary forces: mutation, migration, and genetic drift.

Because none of the four forces of evolution can act unless genetic variation exists, we open Part II with a chapter devoted to mutation—the process that continually introduces new alleles, and occasionally new genes, into populations. Using a combination of algebraic models and experimental tests, Chapters 5 and 6 probe how selection, mutation, migration, and drift act on this variation to produce evolutionary change. Chapter 6 also investigates how inbreeding and other forms of nonrandom mating affect the fate of alleles in populations. Chapter 7 concludes the unit by focusing on interactions that occur among loci as the four forces act, and by exploring how biologists study evolutionary change in traits that are molded by large numbers of genes.

The overall goal of this unit is to offer a broad perspective on evolutionary processes. The four chapters should provide a solid understanding of how and why evolution occurs, and lay the groundwork for an in-depth look at what evolution has produced: adaptation and change through time. These are the subjects of Part III and Part IV.

CHAPTER 4

Mutation and Genetic Variation

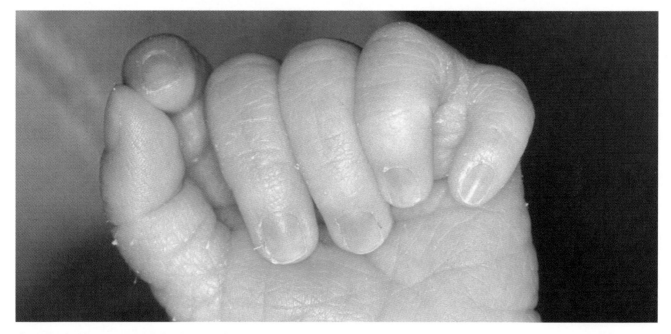

This individual has a mutation that leads to the development of six fingers on each hand. (Science Photo Library/Photo Researchers, Inc.)

Mutations are the raw material of evolution. Without mutation there are no new genes, no new alleles, and eventually, no evolution. Mutation is the ultimate source of the heritable variation acted upon by natural selection and other evolutionary processes.

This chapter has two goals: to investigate the mechanisms responsible for generating new alleles and new genes, and to explore how biologists quantify the amount of genetic variation that exists in natural populations. We begin by reviewing how single-base mutations and other types of small-scale changes occur in DNA sequences. These processes produce new alleles. Later in the chapter we consider larger-scale changes that can produce new genes, change the organization of individual chromosomes, or alter the number of chromosome sets in a species. The chapter closes by considering how researchers analyze genetic variation within species.

The chapter does not attempt an encyclopedic review of all mutations that can affect gene sequences and organization, however. The roster of mutation types, especially at the level of chromosomes, is simply too large. Instead, we focus on the subset of mutations that have the greatest evolutionary impact (Table 4.1).

Table 4.1 **Types of mutation with significant evolutionary impact**

This table summarizes the types of mutation reviewed in this chapter.

Name	Description	Cause	Significance
Point mutation	Base-pair substitutions in DNA sequences	Chance errors during DNA synthesis or during repair of damaged DNA	Creates new alleles
Chromosome inversion	Flipping of a chromosome segment, so that the order of genes along the chromosome is altered	Breaks in DNA caused by radiation	Alleles inside the inversion are "locked together" into a unit
Gene duplication	Duplication of a short stretch of DNA, creating an additional copy of a gene	Unequal crossing over during meiosis (see Fig. 4.3)	The "extra" gene is free to mutate and perhaps gain new function
Polyploidy	Addition of a complete set of chromosomes	Errors in meiosis or (in plants) mitosis	Can create new species

4.1 Where New Alleles Come From

The instructions for making and running an organism are encoded in its hereditary material—the molecule called deoxyribonucleic acid, or DNA. As Figure 4.1a shows, DNA is made up of smaller molecules called deoxyribonucleotides. The four deoxyribonucleotides found in DNA are similar in structure: Each contains the 5-carbon sugar called deoxyribose, a phosphate group, and a distinctive nitrogen-containing base. The four bases in DNA belong to two discrete chemical groups: Cytosine and thymine are pyrimidines, while adenine and guanine are purines. The four deoxyribonucleotides are routinely abbreviated to C, T, A, and G.

Figure 4.1b illustrates how these molecules are linked into long strands by phosphodiester bonds that form between the 5' carbon of one deoxyribonucleotide and the 3' carbon of another. A single strand of DNA, then, consists of a sequence of bases attached to a sugar-phosphate "backbone." In cells, however, DNA normally consists of two such strands. These are wound around one another in the double helix diagrammed in Figure 4.1c. This structure is stabilized by hydrogen bonds that form between the bases on either strand. Due to the geometry of the bases and the amount of space available inside the helix, hydrogen bonds only form when adenine and thymine (A—T) or guanine and cytosine (G—C) bases line up on opposite strands. These purine–pyrimidine combinations are called complementary base pairs. As Figure 4.1d shows, three hydrogen bonds form between G and C, but only two are made between A and T.

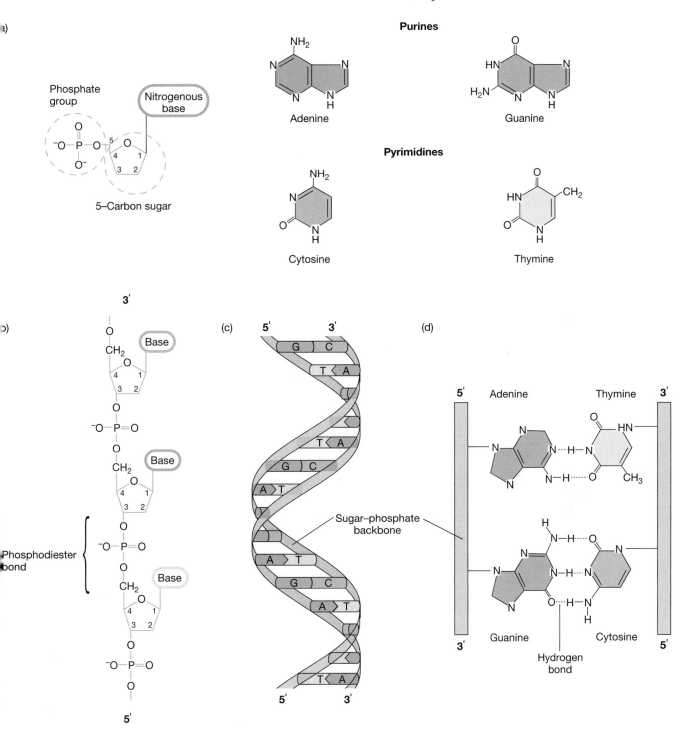

Figure 4.1 **The structure of the genetic material** (a) The diagram on the left shows the generalized form of a nucleotide. Note that each carbon atom in the sugar is numbered, and that the hydrogen and oxygen atoms bonded to these carbons are not shown. The diagrams on the right show the structure of the four nitrogen-containing bases. (b) Nucleotides are linked into long chains by phosphodiester bonds between the 5′ carbon on one nucleotide and the 3′ carbon on another. (c) When complementary bases on opposite DNA strands form hydrogen bonds, the molecule twists into a double helix like the one shown here. (d) Adenine and thymine form two hydrogen bonds; cytosine and guanine form three.

The Nature of Mutation

Once James Watson and Francis Crick (1953) had deduced the double-helical structure of DNA shown in Figure 4.1c, they immediately realized that complementary base pairing provided a mechanism for copying the hereditary material. As Figure 4.2 illustrates, one strand serves as the template for making a copy of the other strand. In 1960, Arthur Kornberg succeeded in isolating the first of several proteins, called DNA polymerases, that are responsible for copying the DNA in cells.

In the late 1950s and early 1960s, a series of experiments succeeded in clarifying how the sequence of bases in DNA encodes information and how the genetic information is transformed into the proteins that make and run cells. The central result was that DNA is transcribed into messenger RNA (mRNA), which is then translated into protein (Figure 4.3a). Researchers established that the genetic code is read in triplets called codons. They also deciphered which amino acid is specified by each of the 64 different codons (Figure 4.3b). Because only 20 amino acids need to be specified by the 64 codons, the genetic code is highly redundant—meaning that the same amino acid can be specified by more than one codon.

These results inspired an explicitly molecular view of the gene and mutation. Genes became defined as stretches of DNA that code for a distinct RNA or protein product. Alleles became defined as versions of the same gene that differ in their base sequence. Mutations were understood as changes in the base sequence of DNA.

A mutation is a change in the base sequence of DNA.

To drive this point home, consider the first mutation ever characterized on a molecular level: the change in human hemoglobin that results in the sometimes-fatal disease sickle-cell anemia. Hemoglobin is the oxygen-carrying protein found in red blood cells. In 1949, Linus Pauling's lab reported that people suffering from sickle-cell anemia had a form of hemoglobin different from that of people without the disease. In 1958, Vernon Ingram showed that the difference between normal and sickle-cell hemoglobin was due to a single amino acid change at position number 6 in the protein chain, which is 146 amino acids long. Instead of having glutamic acid at this position, the sickling allele has valine. Further work established that the amino acid replacement is caused by a single base substitution in the hemoglobin gene. The mutant allele has an adenine instead of a thymine at nu-

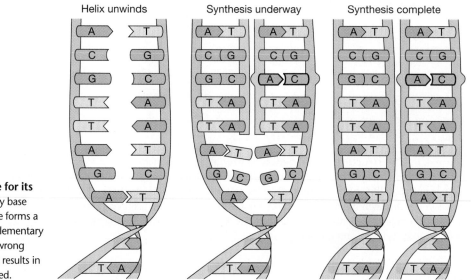

Figure 4.2 DNA forms a template for its synthesis Because of complementary base pairing, each strand in a DNA molecule forms a template for the synthesis of the complementary strand. If DNA polymerase inserts the wrong base, as in the strand at the far right, it results in a mismatched pair that must be repaired.

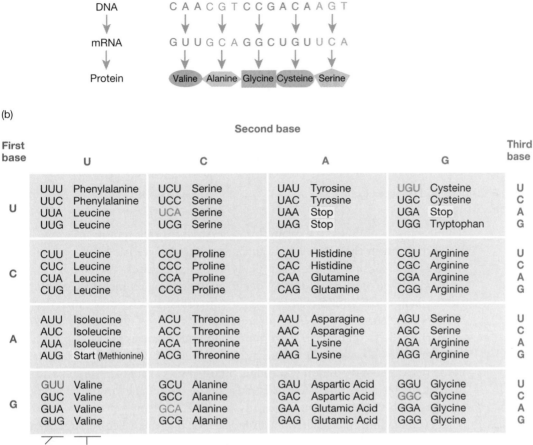

Figure 4.3 In organisms, information flows from DNA to RNA to proteins (a) In cells, the sequence of bases in DNA is transcribed to a sequence of bases in a strand of messenger RNA (mRNA), which is then translated into a sequence of amino acids in a protein. Note that RNA contains a nitrogenous base called uracil instead of thymine. An adenine in DNA specifies a uracil in RNA. (b) This is the genetic code. Each of the 64 mRNA codons shown here specifies an amino acid or the start or end of a transcription unit. Note that in many instances, changing the third base in a codon does not change the message.

cleotide 2 in the codon for amino acid 6. A change like this is called a **point mutation**, because it alters a single point in the base sequence of a gene.

Point Mutations

Point mutations are single-base substitutions in DNA caused by one of two processes: random errors in DNA synthesis, or random errors in the repair of sites damaged by chemical mutagens or high-energy radiation. Both types of changes result from reactions catalyzed by DNA polymerase.

If DNA polymerase mistakenly substitutes a purine (A or G) for another purine, or a pyrimidine (T or C) for another pyrimidine during normal synthesis or the synthesis that occurs during a repair, the point mutation is called a **transition** (Figure 4.4). If a purine is substituted for a pyrimidine or a pyrimidine for a purine, the mutation is called a **transversion**. Of the two kinds of point mutation, transitions are far more common. They routinely outnumber transversions by at least 2:1.

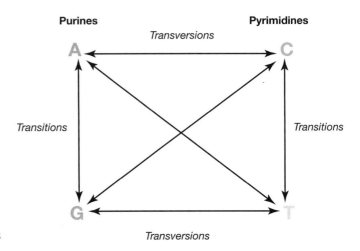

Figure 4.4   Transitions and transversions

The leading hypothesis to explain this observation is that transitions cause much less disruption in the DNA helix during synthesis, so are less likely to be recognized as an error and therefore less likely to be immediately corrected.

If either type of base substitution occurs in the coding region of a gene, the mutation changes the codon read by the protein, called RNA polymerase, that synthesizes RNA from the DNA template. For example, the substitution of an A for a T in the gene for hemoglobin is a transversion that changes the message in codon number 6. Look back at the genetic code in Figure 4.3b, and note that changes in the first or second position of a codon almost always change the amino acid specified by the resulting mRNA. But because of the redundancy in the genetic code, changes in the third position frequently produce no change at all.

Point mutations are single-base substitutions in DNA. They are classified in two ways: as transitions or transversions and as replacement or silent substitutions.

Point mutations that result in an amino acid change are called **replacement** (or **nonsynonymous**) **substitutions**; those that result in no change are called **silent site** (or **synonymous**) **substitutions**. Both types of point mutations create new alleles. Now the question becomes, How do these new alleles affect the fitness of organisms?

The Fitness Effects of Mutations

Silent site substitutions in DNA do not affect the phenotype of the organism because they do not alter gene products. As a result, silent site mutations are not subject to natural selection based on protein or RNA function. Alleles that have no effect on fitness are said to be **neutral**.

But what about replacement substitutions, which *do* produce a change in protein structure? Because they change the organism's phenotype, replacement substitutions are exposed to natural selection. For example, the sickle-cell mutation produces a dramatic change in phenotype. The mutant hemoglobin molecules tend to crystallize, forming long fibers. Hemoglobin is found inside red blood cells, and when the molecule crystallizes it distorts the normally disc-shaped cells into the sickled cells pictured in Figure 4.5. The sickled cells tend to get stuck in capillaries. This blocks the flow of blood, deprives tissues of oxygen, and causes severe pain. Sickled red cells are also more fragile than normal cells and more rapidly destroyed. This continuous loss of red blood cells causes anemia.

The sickling and anemia are most severe in people homozygous for the mutant allele, because all of their hemoglobin molecules are prone to crystallization. Heterozygous people get some sickling, particularly when the concentration of dissolved

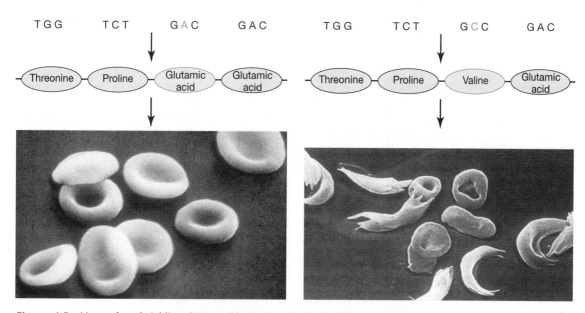

Figure 4.5 Normal and sickling forms of human red blood cells The diagram on the left shows a short section near the start of the DNA and protein sequence of normal hemoglobin; the photo on the left shows normal red blood cells. The diagram on the right shows the same section of the DNA and protein sequence of a mutant hemoglobin; the photo on the right shows the sickled red blood cells that result from this mutation. (Photos from Photo Researchers, Inc.)

oxygen in their blood gets low. It would appear, then, that the allele has a strong adverse affect on the fitness of individuals. Natural selection should quickly eliminate the mutant alleles from the population.

The situation is not this simple, however. People who have one copy of the normal allele and one copy of the mutant allele enjoy an unexpected benefit: They are resistant to malaria. The resistance apparently results because in heterozygotes, red blood cells containing malaria parasites are much more prone to sickling than unparasitized cells. This means that red cells containing parasites are selectively destroyed. In environments where malaria is common, resistance to malaria is so valuable that heterozygotes have higher fitness than either kind of homozygote. The mutant allele is thus beneficial in environments where malaria is common, but deleterious in environments where malaria is rare. The genetic basis of sickle-cell anemia provides an example of heterozygote superiority. We will discuss heterozygote superiority in more detail in Chapter 5.

To summarize, the sickle-cell mutation created a new allele that is beneficial in some environments and deleterious in others. This is unusual, because the overwhelming majority of replacement substitutions create new alleles that have little to no effect on fitness (Keightley and Caballero 1997; García-Dorado 1997). It is not surprising that many of the random changes in the amino acid sequences of proteins do not improve their ability to function, because most proteins have been under selection for millions of years. We would not expect a random change to improve a protein's function any more than we would expect a random change in a computer's circuitry to improve processing performance. It is important, though, to recognize that both replacement and silent substitutions produce a wide range of effects on fitness—from highly deleterious to neutral to beneficial—and that the fitness effects of alleles depend on their environment. We will return to these points in Chapters 5 and 18.

In coding sequences, the fitness effects of replacement substitutions range from highly deleterious to beneficial. Most mutations have very small effects on fitness.

Mutation Rates

How often are new alleles formed? Many of our best data on mutation rates are for a class of changes known as **loss-of-function mutations**. In these types of mutations, the lack of a normal protein product leads to an easily recognizable phenotype. For example, researchers may survey a large human population and count the incidence of an autosomal-dominant syndrome like Achondroplasia (dwarfism) or an X-linked recessive disorder like Hemophilia A (in which blood clotting is impaired). The idea is to pick a trait that is easy to detect and whose transmission genetics allow researchers to identify new mutations. For example, an individual with Achondroplasia, neither of whose parents has the condition, must represent a new mutation. From data like these, a researcher can report mutation rates in units of per gene per generation.

The problem with this method is that loss-of-function mutations result from any process that inactivates a gene. As Figure 4.6 shows, genes can be knocked out by base-pair substitutions that produce a chain-terminating codon or a dysfunctional amino acid sequence. They can also be knocked out by the insertion of a mobile genetic element, chromosome rearrangements, or a disruption in the reading sequence of codons caused by the addition or deletion of one or two base pairs (these are called frameshift mutations). Further, many interesting mutations (most replacement substitutions, for example) are not detectable when researchers assess offspring phenotypes because their effects are less subtle than a loss of function. Because of these difficulties, most existing data seriously underestimate the actual rate at which mutations occur.

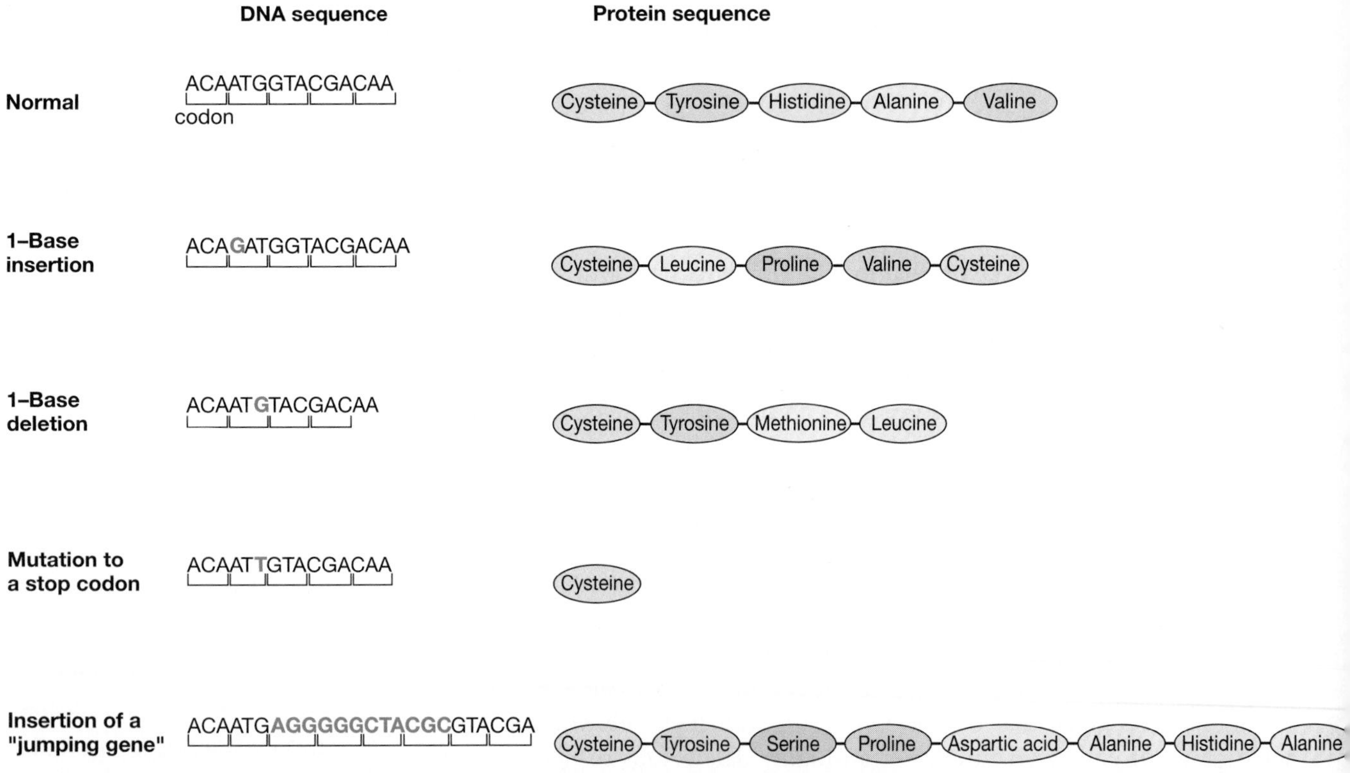

Figure 4.6   **Loss-of-function mutations**   "Knock-out" mutants result from any event that inactivates a gene. Note that the insertion or deletion of a single base changes all subsequent codons in a gene, creating a dysfunctional protein.

Even with these limitations, we can still say some interesting things about mutation rates. For example, consider the data in Table 4.2 on rates and frequencies of loss-of-function mutations and other changes with major effects. The mutation rates reported are very low on a per-gene basis. But there are so many loci in organisms (at least 60,000 in humans, for example) that perhaps 10% of all gametes carry a phenotypically detectable mutation. This is a large percentage, considering that the reported rates undoubtedly underestimate the number of

Table 4.2 Variation in mutation rates among genes and species

(a) Rates of mutation to recessive phenotypes among genes in corn. L. J. Stadler (1942) bred a large number of corn plants and scored offspring for a series of recessive conditions.

Gene	Number of gametes tested	Number of mutations found	Average number of mutations per million gametes	Mutation rate (frequency per gamete)
$R \rightarrow r$	554,786	273	492.0	4.9×10^{-4}
$I \rightarrow i$	265,391	28	106.0	1.1×10^{-4}
$Pr \rightarrow pr$	647,102	7	11.0	1.1×10^{-5}
$Su \rightarrow su$	1,678,736	4	2.4	2.4×10^{-6}
$Y \rightarrow y$	1,745,280	4	2.2	2.2×10^{-6}
$Sh \rightarrow sh$	2,469,285	3	1.2	1.2×10^{-6}
$Wx \rightarrow wx$	1,503,744	0	0.0	0.0

(b) These data, summarizing mutation rates for a variety of genes and species, are taken from R. Sager and F. J. Ryan, *Heredity*. New York: John Wiley, 1961.

Organism	Mutation	Value	Units
Bacteriophage T2 (bacterial virus)	Lysis inhibition $r \rightarrow r^+$	1×10^{-8}	*Rate:* mutant genes per gene replication
	Host range $h^+ \rightarrow h$	3×10^{-9}	
Escherichia coli (bacterium)	Lactose fermentation $lac^- \rightarrow lac^+$	2×10^{-7}	*Rate:* mutant cells per cell division
	Histidine requirement $his^- \rightarrow his^+$	4×10^{-8}	
	$his^+ \rightarrow his^-$	2×10^{-6}	
Chlamydomonas reinhardtii (alga)	Streptomycin sensitivity $str^s \rightarrow str^r$	1×10^{-6}	
Neurospora crassa (fungus)	Inositol requirement $inos^- \rightarrow inos^+$	8×10^{-8}	*Frequency:* per asexual spore
	Adenine requirement $ad^- \rightarrow ad^+$	4×10^{-8}	
Drosophila melanogaster (fruit fly)	Eye color $W \rightarrow w$	4×10^{-5}	*Frequency:* per gamete
Mouse	Dilution $D \rightarrow d$	3×10^{-5}	
Human			
to autosomal dominants	Huntington's chorea	1×10^{-6}	
	Nail-patella syndrome	2×10^{-6}	
	Epiloia (predisposition to a type of brain tumor)	4–8×10^{-6}	
	Multiple polyposis of large intestine	1–3×10^{-5}	
	Achondroplasia (dwarfism)	4–12×10^{-5}	
	Neurofibromatosis (predisposition to tumors of nervous system)	3–25×10^{-5}	
to X-linked recessives	Hemophilia A	2–4×10^{-5}	
	Duchenne's muscular dystrophy	4–10×10^{-5}	
in bone-marrow tissue-culture cells	Normal $\rightarrow$ azaguanine resistance	7×10^{-4}	*Rate:* mutant cells per cell division

replacement substitutions and thus new alleles. One message of these data sets is that it is conceivable, or even probable, that the majority of all offspring carry at least one new allele somewhere in their genome.

A second message in these data concerns variation in mutation rates. Rates of phenotypically detectable mutations vary by 500-fold among genes within species (Table 4.2a), and by as much as five orders of magnitude, or 100,000-fold, among species (Table 4.2b). An obvious question is, why?

Why Are Mutation Rates Variable?

The rate at which new alleles are produced varies at three levels: among individuals within populations, among genes within individuals, and among species. According to the research done to date, it appears that different mechanisms may be responsible for the variation observed at each level.

Variation Among Individuals

Mutation rate varies among individuals for two reasons: Alleles of DNA polymerase can vary in their error rate, and the alleles involved in repairing damaged DNA can vary in their efficiency.

Frances Gillin and Nancy Nossal (1976a, b) were responsible for documenting that DNA polymerases vary in their accuracy. They did this by investigating single-base substitutions in the DNA polymerase of bacteriophage T4 (a virus that parasitizes bacteria). Some of the mutations Gillin and Nossal isolated decreased the rate at which polymerase made errors during DNA replication, and reduced the overall mutation rate. Other mutations in polymerase increased the error rate, and heightened the overall mutation rate. In a key finding, they also determined that the more error-prone polymerase mutants were significantly faster than the more accurate form of the enzyme. This finding implies that there is a trade-off between the speed and accuracy of DNA replication.

Mutation rates vary among individuals because of variation in the base sequences of DNA polymerase and DNA repair loci.

Point mutation rates also depend on how efficiently mistakes are corrected. Repair of mismatched bases on complementary strands can take place after synthesis or after DNA has been damaged by chemical mutagens or radiation. Research on mismatch repair has been intense because mutations in the genes responsible for repair have been implicated in aging and in the development of certain cancers. There are several different mismatch repair systems, and in mammals at least 30 different proteins are involved (Mellon et al. 1996). At least some of the repair systems are highly conserved; mismatch repair genes in humans were first identified through their homology with genes in yeast and the bacterium *Escherichia coli* (Friedberg et al. 1995). In *E. coli* and *Salmonella enteritidis,* mutations in these loci produce strains with mutation rates 100 to 1000 times higher than normal (LeClerc et al. 1996). The general message of these studies is that the efficiency of DNA mismatch repair, like the error rate of DNA polymerase, is a trait with heritable variation.

Variation Among Species

The data in Table 4.2b suggest that mutation rates vary among species. Fruit flies, mice, and humans, for example, appear to have lower mutation rates than viruses and bacteria. John Drake (1991) has published similar data implying that mutation rates vary widely among viruses, bacteria, and yeast. One problem with these studies, however, is that they do not directly compare mutation rates for homologous genes. As a result, it is not clear whether the differences observed are due to rate variation among genes or among species.

Edward Klekowski and Paul Godfrey (1989) solved this problem by studying the rate of mutation to albinism in the mangrove *Rhizophora mangle,* and comparing it to the rate of mutation to albinism in well-studied domesticated species like barley and buckwheat. Albinism in plants occurs because of loss-of-function mutations in the genes responsible for the synthesis of chlorophyll. Klekowski and Godfrey chose the mangrove as their study organism because it is a long-lived tree with an unusual trait: The seedlings germinate on the parent. To estimate a mutation rate, then, the researchers could count the number of albino offspring germinating on normal parents. Their data showed that the rate was 25 times greater in mangroves than the rate previously calculated for mutations to albinism in barley and buckwheat.

Why the difference? Klekowski and Godfrey's explanation was inspired by the realization that, in plants, germline cells differentiate late in development. (In animals, germline and somatic cells separate early in development. As a result, mutations that occur in the somatic tissues of animals are not passed on to offspring.) In large, long-lived plants like mangroves, somatic cells in stems and shoots accumulate mutations through many somatic cell divisions before differentiating as germline tissue and undergoing meiosis. As a result, long-lived plants should have higher mutation rates per generation than short-lived plants. Klekowski and Godfrey's data are consistent with this prediction, because barley and buckwheat are annuals and small in size. The result suggests an interesting generalization: Generation time may be a key factor influencing variation in mutation rates among species.

Mutation rates may vary among species because of differences in the number of cell divisions that take place prior to gamete formation.

Variation Among Genes

Compared to variation among individuals and species, we know much less about why mutation rates vary among genes. Even so, two strong generalizations have emerged from studies on DNA repair systems: Coding regions are repaired much more efficiently than noncoding regions (Bohr et al. 1985), and several of the repair systems work on transcriptionally active genes only. As a result, accuracy appears to be greatest where mutations could be most damaging.

Mutation rates vary among loci because the most transcriptionally active genes are repaired most efficiently.

4.2 Where New Genes Come From

The creation of new alleles is fairly straightforward, but where do new genes come from? As with novel alleles, several kinds of mutations can create new genes. We will review only a subset. **Gene duplications** are probably the most important source of new genes. Duplications result from a phenomenon known as unequal cross-over, diagrammed in Figure 4.7. **Unequal cross-over** is a chance mistake caused by the proteins involved in managing recombination (crossing over) during meiosis.

As Figure 4.7 shows, one of the products of unequal cross-over is a redundant stretch of DNA. The genome now has an extra copy of the sequences located in the duplicated segment. Because the parent copy still produces a normal product, the redundant sequences are free to accumulate mutations without consequences to the phenotype. The new sequence might even change function over time, thereby becoming a new locus. This is an important point. Because it creates additional DNA,

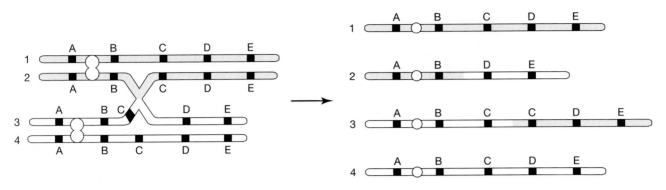

Figure 4.7 Unequal cross-over and the origin of gene duplications The letters and bars on each chromosome in the diagram indicate the location of loci; the open circles indicate the location of the centromere. The chromosomes on the left have synapsed, but cross-over has occurred at nonhomologous points. As a result, one of the cross-over products has a duplication of gene C and one a deletion of gene C. Unequal cross-over events like this are thought to be the most common mechanism for producing gene duplications.

gene duplication is the first mechanism we have encountered that results in entirely new possibilities for gene function. The globin gene family provides a superb example of how duplicated genes diverge in function.

Gene Duplication Events in the Globin Gene Family

In humans, the globin gene family consists of two major clusters of loci that code for the protein subunits of hemoglobin. The groups are the α–like cluster on chromosome 16 and the β–like cluster on chromosome 11 (α and β are the Greek letters alpha and beta). A completed hemoglobin molecule is made up of an iron-binding heme group surrounded by four protein subunits—two coded by loci from the α–like cluster and two coded by loci from the β–like cluster.

The data plotted in Figure 4.8 show that each locus in the α- and β-like families is expressed at a different time in development. In first-trimester human em-

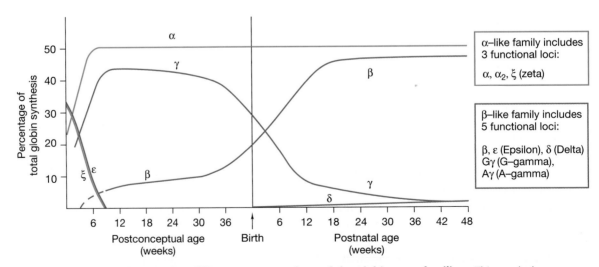

Figure 4.8 Timing of expression differs among members of the globin gene families This graph shows changes in the expression of globin genes from the α and β families in humans, during pregnancy and after birth. In embryos, hemoglobin is made up of ζ-globin from the α-like gene cluster and ε-globin from the β-like gene cluster. In the fetus, hemoglobin is made up of α-globin from the α-like gene cluster and δ-globin from the β-like gene cluster. In adults, hemoglobin is made up of α-globin from the α-like gene cluster and β-globin from the β-like gene cluster. Each of these hemoglobins has important functional differences.

bryos, for example, hemoglobin is made up of two ζ (zeta) chains and two ε (epsilon) chains, while in adults it is made up of two α chains and two β chains (recall that the sickle-cell mutation occurs in one of the β chains). Different combinations of globin polypeptides result in hemoglobin molecules with important functional differences. For example, fetal hemoglobin has a higher affinity for oxygen molecules than adult hemoglobin. This enhances oxygen transfer from the mother to the embryo.

The globin-family loci are thought to be a product of gene duplication events. The hypothesis is supported by the high structural similarity of transcription units among loci, including the remarkable correspondence in the length and position of their exons and introns diagrammed in Figure 4.9. The logic behind this claim is that it is extremely unlikely that such high structural resemblance could occur in loci that do not share a recent common ancestor. The duplication hypothesis is also supported by the observation of high sequence similarity among globin loci, as well as similarity in function.

The general model, then, is that an ancestral sequence was duplicated several times during the course of vertebrate evolution. In several of these new loci, mutations changed the function of the protein product in a way that was favored by natural selection, leading to the formation of the gene family. Because the α- and β-like clusters also contain nonfunctional loci called **pseudogenes**, which are not transcribed, biologists infer that some duplicated loci were rendered functionless by mutation.

Duplicated loci can (1) retain their original function and provide an additional copy of the parent locus, (2) gain a new function through mutation and selection, or (3) become functionless pseudogenes.

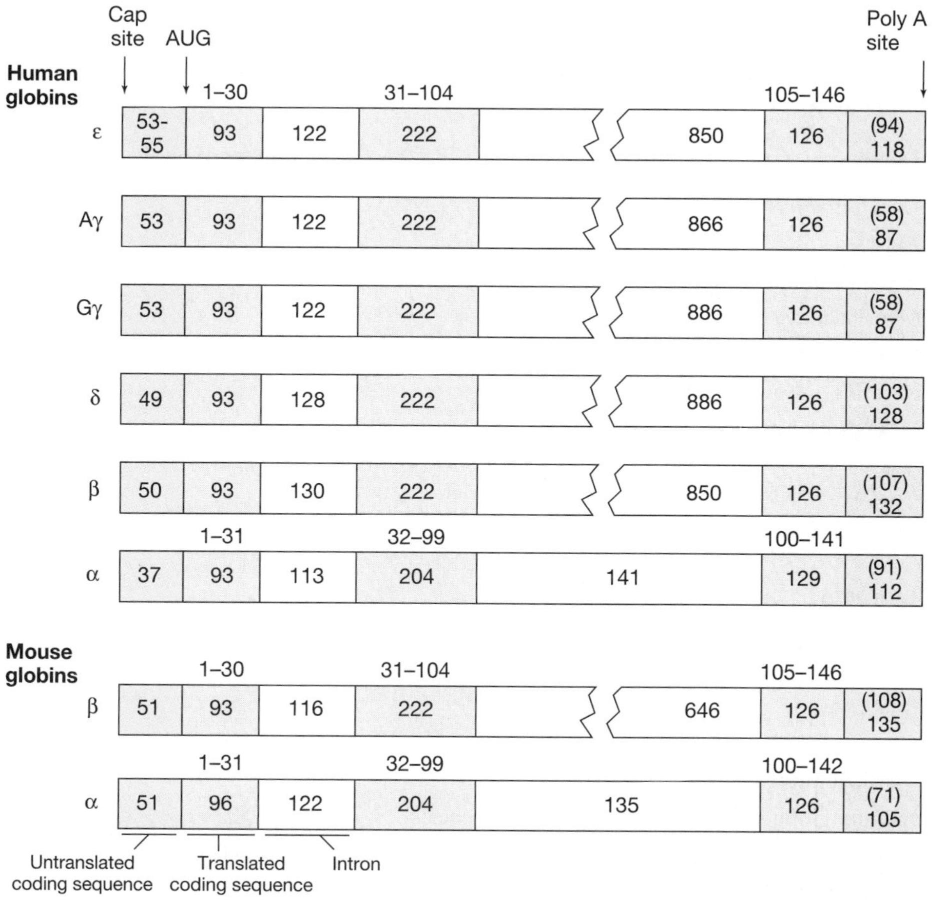

Figure 4.9 **Transcription units in the globin gene family** In these diagrams, the yellow-orange boxes represent coding sequences that are untranslated, the greenish boxes stand for coding sequences that are translated, and the white segments signify introns. The numbers inside the boxes denote the number of nucleotides present in the primary transcript, while the numbers above the boxes give the amino acid positions in the resulting polypeptide. AUG is the start codon. The lengths and positions of introns and exons in loci throughout the α- and β-like clusters are virtually identical.

The gene families listed in Table 4.3 share several important features with the globin loci: They contain structurally homologous genes with similar functions, clustered together on the same chromosome and accompanied by an occasional pseudogene. It is important to note, however, that not all gene duplications result in loci with different functions. In some important cases, like rRNA genes, multiple copies of the same gene have an identical or nearly identical base sequence, and produce a product with the same function.

Other Mechanisms for Creating New Genes

In addition to duplication, there are other mechanisms that can create new genes or radically new functions for duplicated genes. One example—called overprinting—results from point mutations that produce new start codons and new reading frames for translation. Paul Keese and Adrian Gibbs (1992) investigated the phenomenon in the tymoviruses, which cause mosaic disease in a variety of plants. The tiny genome of one tymovirus, turnip yellow mosaic virus, consists of three genes. Two of these overlap, meaning that they are translated from different reading frames in the same stretch of nucleotides. Keese and Gibbs estimated the phylogeny of five tymoviruses and nine close relatives based on amino acid sequences in their replicase protein (Figure 4.10), and found that the tymoviruses form their

Table 4.3   Some gene families

In this table, the "Number of duplicate genes" column refers to the number of loci in various gene families. These loci are presumed to be the result of duplication events. They have high sequence homology, code for products with closely related functions, and are often clustered close to one another on the same chromosome.

Family	Number of duplicate genes
Loci found in many organisms	
Actins	5–30
Tubulins (*a* and *b*)	5–15
Myosin, heavy chain	5–10
Histones	100–1000
Keratins	> 20
Heat-shock proteins	3
Insects	
Eggshell proteins (silk moth and fruit fly)	> 50
Vertebrates	
Globins (many species)	
α-like	1–5
β-like	≥ 50
Ovalbumin (chicken)	3
Vitellogenin (frog, chicken)	5
Immunoglobulins, variable regions (many species)	> 500
Transplantation antigens (mouse and human)	50–100

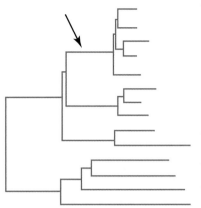

Turnip yellow mosaic virus
Kennedya yellow mosaic tymovirus
Eggplant mosaic tymovirus
Ononis yellow mosaic tymovirus
Erysimum latent tymovirus
Potato X potexvirus
White clover mosaic potexvirus
Narcissus mosaic potexvirus
Apple chlorotic leafspot closterovirus
Potato M carlavirus
Alfalfa mosaic alfamovirus
Sindbis
Tobacco mosaic tobamovirus
Beet necrotic yellow vein furovirus

Figure 4.10   Estimated phylogeny of the tymoviruses and their relatives   Of the species represented in this figure, only the tymoviruses have overprinted genes. This implies that the mutation which led to an overprinted gene occurred somewhere along the branch indicated by the arrow. From Keese and Gibbs (1992).

own branch on the phylogeny. The tymoviruses are also the only species in the phylogeny with an overlapping gene overprinted on the replicase gene.

Based on these data, Keese and Gibbs propose that the replicase gene is ancestral, and that the overlapping gene was created in the common ancestor of the tymoviruses. Their hypothesis is that a mutation created a new start codon, in a different reading frame, in the stretch of nucleotides encoding the replicase protein. They note that evolution at the new locus is likely to be tightly constrained, because any mutation that improves the function of the overlapping protein would probably be deleterious to the function of the replicase protein. Based on their survey of the literature on viral genomes, they also propose that overprinting has been a common mechanism for creating new genes during viral evolution.

Charles Langley and colleagues have investigated the origin of new genes by yet another mechanism. The ancestral gene they studied, found in fruit flies from the genus *Drosophila,* codes for the enzyme alcohol dehydrogenase (*Adh*). This locus is located on chromosome 2. Langley et al. (1982) discovered a similar locus on chromosome 3 in two (and only two) species of flies, *D. teissieri* and *D. yakuba.* Jeffs and Ashburner (1991) sequenced this chromosome 3 locus and found that it lacks the introns found in the *Adh* gene on chromosome 2. Jeffs and Ashburner propose that the new locus on chromosome 3 was created when a messenger RNA from the *Adh* gene was reverse transcribed, and the resulting complementary DNA (cDNA) was inserted into chromosome 3. As we will see in Chapter 18, this mechanism of gene duplication is not unusual. Reverse transcriptase is common in the nuclei of eukaryotic cells.

The question now becomes, Does this new locus have some function, or is it merely a pseudogene? Long and Langley (1993) sequenced the DNA in the chromosome 3 locus from a number of individuals in both *D. teissieri* and *D. yakuba,* with the goal of analyzing alleles of the new locus that had arisen by point mutation. They discovered that most of the alleles differed from each other only in silent site substitutions. This implies that natural selection is acting to conserve the amino acid sequence of a protein encoded by the new locus. In contrast, the common pattern in pseudogenes is that replacement substitutions are as common as silent site substitutions. As a result, Long and Langley concluded that the new locus is a functional gene. They named it *jingwei,* after the protagonist in a Chinese reincarnation myth.

Long and Langley were able to isolate mRNA transcribed from the *jingwei* gene. Upon sequencing the mRNA, they found that the gene contains additional exons

In many genomes, reverse transcription of mRNAs and insertion of the resulting DNA at a new location is an important source of new genes.

not found in its *Adh* ancestor. These additional exons were apparently annexed from a flanking region of chromosome 3 after the *Adh* reverse transcript was inserted. *Jingwei* is thus a hybrid locus, stitched together from pieces of genes on two different chromosomes. Although this mechanism of gene duplication sounds exotic, it illustrates a very general point: Genomes are dynamic. The amount, location, and makeup of the genetic material change through time.

4.3 Chromosome Alterations

A wide variety of changes can occur in the gross morphology of chromosomes. Some of these mutations affect only gene order and organization; others produce duplications or deletions that affect the total amount of genetic material. They can also involve the entire DNA molecule or just segments. Here we focus on two types of chromosome alterations that are particularly important in evolution.

Inversions

Chromosome inversions involve much larger stretches of DNA than the mutation types reviewed in Sections 4.1 and 4.2. They also produce very different consequences. Inversions result from a multistep process that starts when ionizing radiation causes two double-strand breaks in a chromosome. After breakage, a chromosome segment can detach, flip, and reanneal in its original location. As Figure 4.11 shows, gene order along the chromosome is now inverted.

What is the evolutionary impact? Inversions affect a phenomenon known as genetic **linkage**. Linkage is the tendency for alleles of different genes to assort together at meiosis. For obvious reasons, genes on the same chromosome tend to be more tightly linked (that is, more likely to be inherited together) than genes on nonhomologous chromosomes. Similarly, the closer together genes are on the same chromosome, the tighter the linkage. Crossing over at meiosis, on the other hand, breaks up allele combinations and reduces linkage (see Chapter 7).

Because inverted sequences cannot align properly with their normal homolog during synapsis, a crossing-over event that takes place within an inversion results in the duplication or loss of chromosome regions and the production of dysfunctional gametes. When inversions are heterozygous, successful crossing-over

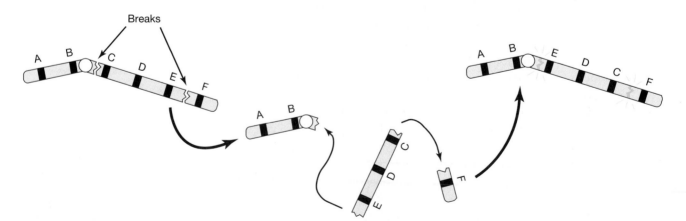

Figure 4.11   Chromosome inversion   Inversions result when a chromosome segment breaks in two places, flips, and reanneals. Note that after the event, the order of the genes labeled C, D, and E is inverted.

events are extremely rare. The result is that alleles inside the inversion are locked so tightly together that they are inherited as a single "supergene."

Inversions are common in *Drosophila*—the most carefully studied of all insects. Are they important in evolution? Consider a series of inversions found in populations of *Drosophila subobscura*. This fruit fly is native to Western Europe, North Africa, and the Middle East, and has six chromosomes. Five of these chromosomes are polymorphic for at least one inversion (Prevosti et al. 1988), meaning that chromosomes with and without the inversions exist. Biologists have known since the 1960s that the frequencies of these inversions vary regularly with latitude and climate. This type of regular change in the frequency of an allele or an inversion over a geographic area is called a **cline**. Several authors have argued that different inversions must contain specific combinations of alleles that function well together in cold, wet weather or hot, dry conditions. But is the cline really the result of natural selection on the supergenes? Or could it be an historical accident, caused by differences in the founding populations long ago?

A natural experiment has settled the issue. In 1978 *D. subobscura* showed up in the New World for the first time, initially in Puerto Montt, Chile, and then four years later in Port Angeles, Washington, USA. Several lines of evidence argue that the North American population is derived from the South American one. For example, of the 80 inversions present in Old World populations, precisely the same subset of 19 is found in both Chile and Washington State. Also, *Drosophila* are frugivores, Chile is a major fruit exporter, and Port Angeles is a seaport. Within a few years of their arrival on each continent, the *D. subobscura* populations had expanded extensively along each coast and developed the same clines in inversion frequencies found in the Old World (Figure 4.12). The clines are even correlated with the same general changes in climate type: from wet marine environments, to Mediterranean climates, to desert and dry steppe habitats (Prevosti et al. 1988; Ayala et al. 1989). This is strong evidence that the clines result from natural selection, and are not due to historical accident.

Which genes are locked in the inversions, and how do they affect adaptation to changes in climate? In the lab, *D. subobscura* lines that are bred for small body size tend to become homozygous for the inversions found in the dryer, hotter part of the range (Prevosti 1967). Recent research has confirmed that pronounced and parallel clines in body size exist in fly populations from North America and Europe (Huey et al. 2000). These results hint that alleles inside the inversions affect body size, with natural selection favoring large flies in cold, wet climates and small flies in hot, dry areas. Research into this natural experiment is continuing. In the meantime, the fly study illustrates a key point about inversions: They are an important class of mutations because they affect selection on groups of alleles. We will return to the topic of selection on multiple loci in Chapter 7.

> *Inversions change gene order and lessen the frequency of crossing over...*

> *...as a result, the alleles found inside inversions tend to be inherited as a unit.*

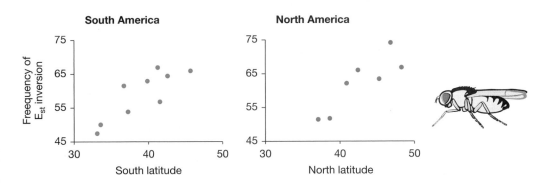

South America

North America

Figure 4.12 **Inversion frequencies form clines in *Drosophila subobscura*** These graphs plot the frequency of an inversion called E_{st} in South American and North American populations of *Drosophila subobscura*. From data in Prevosti et al. 1988.

Polyploidy

The final type of mutation that we will examine occurs at the largest scale possible: entire sets of chromosomes. Instead of being haploid (n) or diploid ($2n$), **polyploid** organisms can be tetraploid ($4n$) up to octoploid ($8n$) or higher.

Polyploidy is common in plants and rare in animals. Nearly half of all angiosperm (flowering plant) species are polyploid, as are the vast majority of the ferns. But in animals, polyploidy is rare. It occurs in taxa like earthworms and some flatworms where individuals contain both male and female gonads and can self-fertilize (these species are called self-compatible hermaphrodites). It is also present in groups that are capable of producing offspring without fertilization, through a process called parthenogenesis. In some species of beetles, sow bugs, moths, shrimp, goldfish, and salamanders, a type of parthenogenesis occurs that can lead to chromosomal doubling.

In plants, polyploidy can result from several different events. Perhaps the most frequent is errors at meiosis that result in diploid gametes (Ramsey and Schemske 1998). When plants produce diploid gametes, one of two things happens. If individuals that produce diploid gametes contain both male and female reproductive structures and self-fertilize, a tetraploid ($4n$) offspring can result (see Figure 4.13a). If this offspring self-fertilizes when it matures, or if it mates with a tetraploid sibling that produces diploid gametes, a population of tetraploids can become established.

Alternatively, individuals that produce diploid gametes can mate with normal individuals that produce haploid gametes. As Figure 4.13b shows, this type of mating results in a triploid offspring. Triploid individuals have poor fertility, however. Because their homologous chromosomes are present in an odd number, they cannot synapse correctly during meiosis. As a result, the majority of the gametes produced by triploids end up with the wrong number of chromosomes (see the uppermost histogram in Figure 4.13b). But the bottom histogram of Figure 4.13b shows that if triploid individuals self-fertilize, most of the offspring that result are tetraploid. These data show that the few offspring that escape from the "triploid block" can go on to establish viable populations of tetraploids.

Polyploidy is important because it can result in new species being formed. To understand why, imagine the outcome of matings between individuals in a tetraploid population, established by one of the mechanisms described above, and the parental diploid population. If individuals from the two populations mate, their triploid offspring are semisterile. But if tetraploid individuals continue to self-fertilize or mate among themselves, fully fertile tetraploid offspring will result. In this way, natural selection should favor polyploids that are reproductively isolated from their parent population. If the diploid and tetraploid populations became genetically isolated, they would be considered a separate species.

Populations of polyploid individuals are often isolated, genetically, from their parental species.

It is also important to recognize that doubled chromosome sets, like the duplicated individual genes analyzed earlier in the chapter, are free to gain new function as a result of mutation and natural selection. Polyploidy is a key source of genetic variation because it produces hundreds or thousands of duplicated loci.

What is the mutation rate to polyploidy in plants? Justin Ramsey and Douglas Schemske (1998) answered this question by calculating how frequently the two major mechanisms of tetraploid formation occur. To estimate how frequently diploid gametes combine to form tetraploid offspring (the route to polyploidy summarized in Figure 4.13a), they surveyed published studies on the rate of diploid

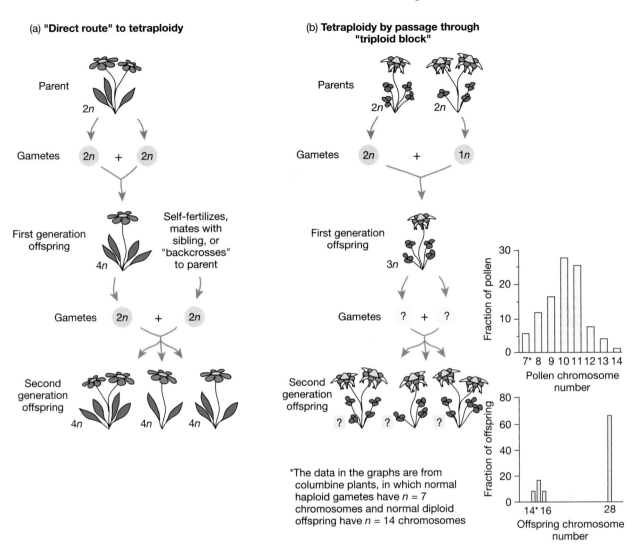

(a) "Direct route" to tetraploidy

(b) Tetraploidy by passage through "triploid block"

*The data in the graphs are from columbine plants, in which normal haploid gametes have *n* = 7 chromosomes and normal diploid offspring have *n* = 14 chromosomes

Figure 4.13 Mechanisms for producing tetraploid individuals in plants See text for explanation.

gamete formation in flowering plants. Because the data in the literature suggested that diploid gametes are produced at an average frequency of 0.00465 in most flowering plant species, the frequency of tetraploids produced by this route should be $0.00465 \times 0.00465 = 2.16 \times 10^{-5}$. In each generation, then, this mechanism of polyploid formation occurs in about 2 out of every 10,000 offspring produced by a typical flowering plant species. Ramsey and Schemske pursued the same strategy to estimate the frequency at which triploid offspring arise and go on to produce tetraploid descendants. (This is the route to polyploidy summarized in Figure 4.13b.) Using published data sets on the frequency of each step in the sequence of events leading to polyploid formation by this mechanism, they calculated that tetraploids are produced at a frequency of at least 1.16×10^{-5} each generation.

The overall message of the Ramsey and Schemske study is striking: In flowering plants, polyploid formation occurs about as frequently as point mutations in individual loci. Along with replacement substitutions, gene duplications, and chromosome inversions, polyploidy is an important source of genetic variation in natural populations.

In plants, polyploidy occurs often enough to be considered an important type of mutation.

4.4 Measuring Genetic Variation in Natural Populations

In the previous three sections of this chapter, we discussed the processes that generate new alleles, genes, and chromosomes. These processes create the genetic variation that is the raw material for evolution. In this last section, we turn to the methods biologists use to measure the amount of genetic variation present in natural populations.

Traditionally, biologists thought that allelic variation in populations would be limited...

We will focus on measurements of allelic variation at individual loci. Before looking at data on genetic variation in populations, however, it is worth considering how much variation we might expect to find. The classical view was that populations should contain little genetic variation. The reasoning behind this view was that among the alleles possible at any given locus, one should be better than all the others. Natural selection should preserve the one allele most conducive to survival and reproduction, and eliminate the rest. The one best allele was called the wild type; any other alleles present were considered mutants.

...but research has revealed that it is actually extensive.

As we will see, modern methods for assessing genetic variation have revealed that the classical view was mistaken. Starting with pioneering work by Harris (1966) and Lewontin and Hubby (1966), evolutionary biologists have looked at the proteins encoded by alleles, and at the DNA of the alleles themselves. The deeper biologists have looked, the more genetic variation they have found. Today, evolutionary biologists recognize that natural populations harbor substantial genetic variation.

Determining Genotypes

To measure the diversity of alleles at a particular locus, we must determine the genotypes of individuals. For some loci, it is possible to infer the genotypes of individuals from their phenotypes. For example, intestinal schistosomiasis is a human disease caused by infection with the parasitic flatworm *Schistosoma mansoni*. Several lines of evidence indicate that susceptibility to infection by *S. mansoni* is strongly influenced by a single locus on chromosome 5, called SM1. (For a review, see Online Mendelian Inheritance in Man 1999.) Laurent Abel and colleagues (1991) analyzed the pedigrees of 20 Brazilian families, and found that their data were consistent with a model in which SM1 has two codominant alleles. Individuals homozygous for one of the alleles are susceptible, whereas individuals homozygous for the other allele are resistant. Heterozygotes have intermediate resistance. In areas where everyone is exposed to water contaminated with *S. mansoni,* it is possible, with reasonable accuracy, to infer a person's genotype from the intensity of his or her infection. Abel and colleagues estimated that in the Brazilian population they studied, about 60% of the people were homozygous resistant, about 5% were homozygous susceptible, and about 35% were heterozygous.

In contrast to SM1, for most loci it is difficult or impossible to infer the genotypes of individuals simply by noting their phenotypes. As a result, most biologists studying allelic diversity look directly at the proteins encoded by alleles, or at the DNA of the alleles themselves. There are a variety of methods for doing so, most of which rely on gel electrophoresis. Gel electrophoresis uses a slab of gelatin-like material and an electric field to separate molecules by size, mass, and electric charge. (See Box 4.1.)

BOX 4.1   Gel electrophoresis

Gel electrophoresis is a widely used technique for assessing the amount of genetic variation in populations. At heart, electrophorsesis is simply a method of sorting molecules. A basic electrophoresis setup appears in Figure 4.14a. The gel itself is a porous slab of gelatin-like material made of any of a number of ingredients, including starch, agarose, or polyacrylamide. At one end of the gel, or sometimes in the center, is

(a)

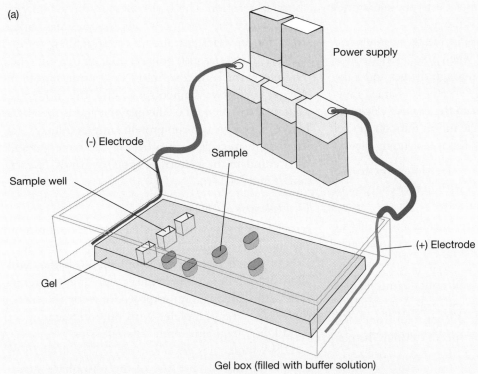

Power supply

(-) Electrode

Sample

Sample well

Gel

(+) Electrode

Gel box (filled with buffer solution)

(b)

(c)

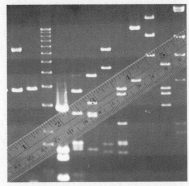

Figure 4.14   *Gel electrophoresis* (a) The diagram shows the minimal equipment needed for electrophoresis. In the setup shown, the power supply is simply five 9-volt batteries snapped together to make a single 45-volt battery. Researchers typically use commerical power supplies with variable voltage, timers, etc. (b) This photograph shows the result of electrophoresis of proteins. The gel in the photo has been stained for the PGM enzyme. If an individual has two forms of the enzyme, two bands appear in the same lane Because we assume that the alternative forms are the products of different alleles (Feder et al. 1989a), we can infer that an individual with two bands is a heterozygote. This gel indicates that three different alleles exist, and that 8 of the 20 flies in this sample are heterozygotes at the PGM locus. (Jeff Feder, University of Notre Dame) (c) This photograph shows the result of electrophoresis of DNA fragments. The DNA fragments in this gel have been stained with ethidium bromide, which makes them visible in ultraviolet light. Smaller fragments are closer to the bottom of the gel. (National Institutes of Health/Custom Medical Stock Photo, Inc.)

BOX 4.1 Continued

a series of slots, or wells. A researcher fills each well with a solution containing a mixture of molecules to be sorted. The researcher bathes the gel in a buffer solution, which keeps the gel moist and conducts electricity. A power supply connected to the electrodes creates an electric field across the gel.

Most biological molecules, including proteins and DNA, are electrically charged when in solution. Because they are charged, they move through the gel under the influence of the electric field. Negatively charged molecules, for example, move toward the positive electrode. The speed at which a molecule moves through the gel is determined by a number of factors, including these:

1. The ratio of the molecule's electric charge to the molecule's mass. Molecules with higher charge and lower mass move more quickly.

2. The molecule's physical size. Smaller molecules fit more easily through the pores in the gel, and therefore move more quickly.

If the mixture placed in a well contains molecules with different charge-to-mass ratios, and/or different sizes, the different molecules will separate from each other as they move at different speeds through the gel.

Electrophoresis of proteins

Suppose there are two alleles at a locus encoding an enzyme. Recall that enzymes are proteins, and that proteins are chains of amino acids. Imagine that our two alleles encode versions of the enzyme that have different amino acids at a few positions. These distinct versions of the enzyme, encoded by alleles at the same locus, are called allozymes. If one allozyme has a negatively charged amino acid where the other has a neutral or positively charged one, then the two versions of the protein will have different net electric charges. They will move at different speeds through an electrophoresis gel.

To determine an individual's genotype at our enzyme locus, we can extract a sample of the individual's proteins, and separate them on an electrophoresis gel. We then place the gel in a bath containing a substrate for a chemical reaction catalyzed by the enzyme in question, and a stain that binds to a product of the chemical reaction. This bath will stain the gel only in places where the enzyme we are studying has ended up. If the gel yields just one stained spot, then our individual contains just one version of the enzyme. Therefore, the individual must be a homozygote. If the gel yields two stained spots, then the individual contains two versions of the enzyme and is thereby a heterozygote. We can run protein samples from several individuals next to each other on the same gel and then compare the pattern of spots, or bands, in each lane (Figure 4.14b).

Electrophoresis of DNA

Suppose there are two alleles at a locus. By definition, the alleles have different DNA sequences. There are a variety of electrophoresis-based methods that will allow us to distinguish individuals with different genotypes. All of them rely on procedures for preparing the DNA wherein alleles with different sequences yield DNA fragments with different sizes.

DNA molecules are negatively charged when in solution, primarily because of the phosphate group on each nucleotide. All DNA molecules have approximately the same charge-to-mass ratio, regardless of their length. However, smaller DNA molecules move faster through an electrophoresis gel. If we run a mixture of DNA fragments on a gel, the fragments will sort themselves by size. If we render the fragments visible, perhaps by staining them or making them fluorescent, then we will see a band on the gel corresponding to each fragment size (Figure 4.14c). We can run DNA prepared from several different individuals next to each other on the gel. Individuals with different genotypes give different patterns of bands. See the main text for an example.

Our example of how researchers use electrophoresis to determine genotypes concerns a gene in humans called *CC-CKR-5*. This gene, located on chromosome 3, encodes a protein called C-C chemokine receptor-5, often abbreviated CCR5. CCR5 is a cell surface protein found on white blood cells. As its name suggests, CCR5's function is to bind chemokines, which are molecules released as signals by other immune system cells. When a white blood cell is stimulated by chemokines binding to its receptors, the cell moves into inflamed tissues to help fight an infection. What makes CCR5 particularly interesting is that it is also exploited as a coreceptor by most sexually transmitted strains of HIV-1.

As we mentioned in Chapter 1, HIV-1 virions use a protein of their own, called Env, to infiltrate host cells. Env appears to work by first binding to a cell surface protein called CD4, and then binding to CCR5. When Env binds to CCR5, it initiates the fusion of the viral envelope with the host cell's membrane. This fusion delivers the viral core into the host cell's cytoplasm.

In 1996, Rong Liu and colleagues discovered allelic variation at the *CCR5* locus that influences susceptibility to infection by sexually transmitted strains of HIV-1. Liu and colleagues were studying two individuals who remained uninfected despite multiple unprotected sexual encounters with partners known to be HIV-positive. The researchers discovered that these individuals were homozygous for a 32-base pair deletion in a coding region of the gene for CCR5. As a result of the deletion, the encoded protein is severely shortened, and nonfunctional. Lacking CCR5 on their surface, the cells of deletion homozygotes offer no handles to HIV-1 virions that must bind to CCR5 to initiate an infection.

We will call the functional allele *CCR5+*, or just *+*, and the allele with the 32-base pair deletion *CCR5-Δ32*, or just *Δ32*. Individuals with genotype *+/+* are susceptible to infection with HIV-1, individuals with genotype *+/Δ32* are susceptible, but may progress to AIDS more slowly, and individuals with genotype *Δ32/Δ32* are resistant to most sexually transmitted strains of the virus.

Upon learning of the *CCR5-Δ32* allele, AIDS researchers immediately wanted to know how common it is. Michel Samson and colleagues (1996), who discovered the allele independently, developed a genotype test that works as follows. Researchers first extract DNA from a sample of the subject's cells. Then the researchers use a polymerase chain reaction (PCR) to make many copies of a region of the gene, several hundred base pairs long, that contains the site of the 32-base pair deletion. (PCR duplicates a targeted sequence many times over by employing a test-tube DNA replication system in combination with specifically tailored primer sequences that direct the DNA polymerase to copy just the locus of interest.) Finally, the researchers cut the duplicated DNA sequences with a restriction enzyme, and run the resulting fragments on an electrophoresis gel.

The results appear in Figure 4.15. Both alleles yield two DNA fragments. The fragments from a *CCR5-+* allele are 332- and 403-base pairs long. The fragments from a *CCR5-Δ32* allele are 332- and 371-base pairs long. Homozygotes have just two bands in their lane on the gel, whereas heterozygotes have three bands.

Several laboratories have completed surveys of *CCR5* genotypes in various indigenous populations from around the world. Data excerpted from a survey by Jeremy Martinson and colleagues (1997) appear in Table 4.4.

Calculating Allele Frequencies

We have noted that a pressing question concerning the *CCR5-Δ32* allele is, How common is it? To answer this question precisely, we need to use the data on genotypes in Table 4.4 to calculate the frequency of the *Δ32* allele in the various populations tested. The frequency of an allele is its fractional representation among all the alleles present in a population.

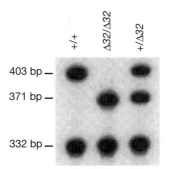

Figure 4.15 Determining CCR5 genotypes by electrophoresis of DNA. Each lane of this gel contains DNA fragments prepared from the *CCR5* alleles of a single individual. The locations of the dark spots, or bands, on the gel indicate the sizes of the fragments. Each genotype yields a unique pattern of bands. From Samson et al. (1996). Reprinted with permission from Nature. © 1996, Macmillan Magazines Ltd.

Table 4.4   Diversity of *CCR5* genotypes in various populations

Population	Number of people tested	Number with each genotype			Allele frequency (%)	
		+/+	+/Δ32	Δ32/Δ32	CCR5-+	CCR5-Δ32
Europe						
Ashkenazi	43	26	16	1	79.1	20.9
Iceland	102	75	24	3	85.3	14.7
Britain	283	223	57	3		
Spain: Basque	29	24	5	0		
Italy	91	81	10	0		
Ireland	44	40	4	0		
Greece	63	60	3	0		
Middle East						
Caucasus: Daghestan	110	96	14	0		
Saudi Arabia	241	231	10	0		
Yemen	34	34	0	0		
Asia						
Russia: Udmurtia	46	38	7	1		
Pakistan	34	32	2	0		
Punjab	34	33	1	0		
Bengal	25	25	0	0		
Hong Kong	50	50	0	0		
Filipino	26	26	0	0		
Mongolia	59	59	0	0		
Thailand	101	100	1	0		
Borneo	151	151	0	0		
Africa						
Nigeria	111	110	1	0		
Central African Repub.	52	52	0	0		
Kenya	80	80	0	0		
Ivory Coast	87	87	0	0		
Zambia	96	96	0	0		
Kalahari San	36	36	0	0		
Oceania						
New Guinea Coast	96	96	0	0		
French Polynesia	94	94	0	0		
Aboriginal Australian	98	96	2	0		
Guam	59	58	1	0		
Fiji	17	17	0	0		
Americas						
Nuu-Chah-Nulth	38	37	1	0		
Mexican (Huicholes)	52	52	0	0		
Brazilian Amerindian	98	98	0	0		
Jamaica	119	119	0	0		

As an example, we will calculate the frequency of the *Δ32* allele in the Ashkenazi population in Europe, from the data in the first row of Table 4.4. The simplest way to calculate allele frequencies is to count allele copies. Martinson and colleagues tested 43 individuals. Each individual carries 2 allele copies, so the researchers tested a total of 86 allele copies. Of these 86 allele copies, 18 were copies of the *Δ32* allele: one from each of the 16 heterozygotes, and 2 from the single homozygote. Thus the frequency of the *Δ32* allele in the Ashkenazi sample is

To estimate the amount of genetic variation in a population, researchers compute the frequencies of each allele present.

$$\frac{18}{86} = 0.209$$

or 20.9%. We can check our work by calculating the frequency of the + allele. It is

$$\frac{(52 + 16)}{86} = 0.791$$

or 79.1%. If our calculations are correct, the frequencies of the two alleles should sum to one, which they do.

An alternative method of figuring the allele frequencies in the Ashkenazi population is to calculate them from the genotype frequencies. Martinson and colleagues tested 43 individuals, so the genotype frequencies are as follows:

+/+	+/Δ32	Δ32/Δ32
$\frac{26}{43} = 0.605$	$\frac{16}{43} = 0.372$	$\frac{1}{43} = 0.023$

The frequency of the *Δ32* allele is the frequency of *Δ32/Δ32* plus half the frequency of +/*Δ32*:

$$0.023 + \frac{1}{2}(0.372) = 0.209$$

This is the same value we got by the first method.

We have filled in the allele frequencies for the first two rows in Table 4.4. We leave it to readers to calculate the allele frequencies for the rest of the populations listed in the table, and to note the locations of the various populations on a world map. Readers who do so will discover an intriguing pattern. The *CCR5-Δ32* allele is common in populations of Northern European extraction, with frequencies as high as 21%. As we move away from northern Europe, both to the east and to the south, the frequency of the *Δ32* allele declines. Outside of Europe, the Middle East, and western Asia, the *Δ32* allele is virtually absent. We will return to this pattern in Chapters 5 and 7.

Documenting allele frequencies in a variety of populations can reveal interesting patterns.

How Much Genetic Diversity Exists in a Typical Population?

Since the mid-1960s, evolutionary biologists have used gel electrophoresis of enzymes to assess genetic diversity in populations representing hundreds of species of plants and animals. Data from two studies appear in Figure 4.16.

J. G. Oakeshott and colleagues (1982) studied allelic diversity at the alcohol dehydrogenase locus in fruit flies. Alcohol dehydrogenase, or Adh, breaks down ethanol, the poisonous active ingredient in wine, beer, and, most important to fruit flies, rotting fruit. There are two electrophoretically distinguishable allozymes of Adh: Adh^F and Adh^S. The F and S are short for fast and slow, the speeds at which the allozymes move through an electrophoresis gel. Oakeshott and colleagues

If more than one allele exists at a particular locus, a population is said to be polymorphic at that locus.

determined the frequencies of the two Adh allozymes in 34 Australian fly populations (Figure 4.16a). This survey showed that almost all fly populations are **polymorphic** at the Adh locus. That is, nearly all populations harbor both alleles. The map in Figure 4.16a also reveals a pattern that is repeated in Europe and North America: Adh^S is generally at higher frequency at low latitudes (that is, closer to the equator), whereas Adh^F is at higher frequency at high latitudes. The significance of this pattern is unclear, although it may relate to the fact that Adh^S is more stable at higher temperatures.

Dennis Powers and colleagues (1991, 1998) studied allelic diversity at the lactate dehydrogenase-B locus in populations of the mummichog (*Fundulus heteroclitus*), a

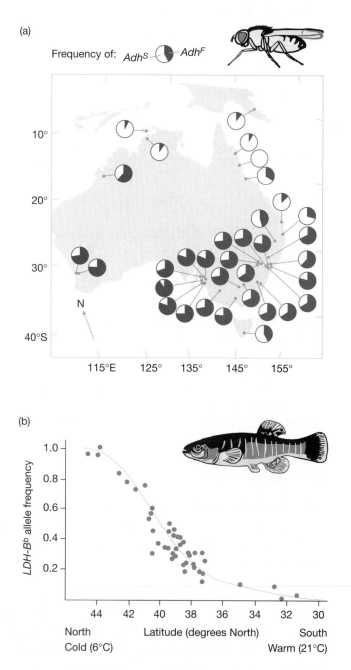

Figure 4.16 **Gel electrophoresis of enzymes reveals allelic diversity in natural populations** (a) The pie diagrams on this map show the frequencies of two alleles of alcohol dehydrogenase in Australian populations of fruit flies: Adh^F (black) and Adh^S (white). From Oakeshott et al. (1982), Copyright © 1991, Evolution. Reprinted by permission. (b) This scatterplot shows the frequency of the Ldh-B^b allele as a function of latitude in populations of the mummichog fish (*Fundulus heteroclitus*) along the east coast of the United States. Ldh-B^b is one of two electrophoretically distinguishable alleles of lactate dehydrogenase-B; the other is Ldh-B^a. Reprinted with permission from Powers et al. (1991). © 1991, by Annual Reviews, www.AnnualReviews.org.

5- to 10-cm-long fish that lives in inlets, bays, and estuaries along the Atlantic coast of North America. Lactate dehydrogenase-B, or Ldh-B, is an enzyme that converts lactate to pyruvate; it is important in both glucose production and aerobic metabolism. There are two electrophoretically distinguishable allozymes of Ldh-B: $Ldh\text{-}B^a$, and $Ldh\text{-}B^b$. Powers and colleagues determined the frequencies of the two Ldh-B allozymes in mummichog populations from Maine to Georgia. This survey shows that most populations are polymorphic. Furthermore, as the scatterplot in Figure 4.16b shows, there is a strong geographic pattern: $Ldh\text{-}B^b$ is at higher frequency in northern populations, while $Ldh\text{-}B^a$ is at higher frequency in southern populations. This pattern makes sense, because $Ldh\text{-}B^b$ has higher catalytic efficiency at low temperatures, whereas $Ldh\text{-}B^a$ has higher catalytic efficiency at high temperatures.

To draw general conclusions from studies like those reviewed in the preceding two paragraphs, we need to summarize data on allelic diversity across loci within populations. There are two commonly used summary statistics: the mean heterozygosity, and the percentage of polymorphic loci. The mean heterozygosity can be interpreted in two equivalent ways: as the average frequency of heterozygotes across loci, or as the fraction of loci that are heterozygous in the genotype of the average individual. The percentage of polymorphic loci is the fraction of loci in a population that have at least two alleles.

Electrophoretic studies of enzymes have demonstrated that most natural populations harbor substantial genetic variation. Figure 4.17 summarizes data on mean heterozygosities from invertebrates, vertebrates, and plants. As a broad generalization, in a typical natural population, between 33 and 50% of the enzyme loci are polymorphic, and the average individual is heterozygous at 4 to 15% of its loci (Mitton 1997).

Methods that directly examine the DNA of alleles are even more powerful at revealing genetic diversity. This is because not every change in the DNA sequence at a locus produces an electrophoretically distinguishable protein. Among the most intensively studied loci to date is that of the gene associated with cystic fibrosis in humans. This locus, on chromosome 7, encodes a protein called the cystic fibrosis transmembrane conductance regulator (CFTR). CFTR is a cell surface protein expressed in the mucus membrane lining of the intestines and lungs. Gerald Pier and colleagues (1997) demonstrated that one of CFTR's key functions is to enable the cells of the lung lining to ingest and destroy *Pseudomonas aeruginosa* bacteria. Individuals homozygous for loss-of-function mutations in the CFTR gene have cystic fibrosis. They suffer chronic infections with *P. aeruginosa,* eventually leading to severe lung damage. Molecular geneticists have examined the DNA sequence in the CFTR alleles of more than 15,000 cystic fibrosis patients, for a total of more than 30,000 copies of disease alleles. They have discovered over 500 different loss-of-function mutations at this one locus (Figure 4.18). We will return to the CFTR gene in Chapter 5.

Analysis of protein variation suggest that in a typical population, between a third and a half of all coding loci are polymorphic...

...while early surveys of DNA sequence variation suggest that polymorphism may be even more extensive.

Why Are Populations Genetically Diverse?

As we noted at the beginning of this section, the classical view of genetic diversity, under which little diversity was expected in most populations, was clearly wrong. So how can we explain the substantial diversity present in most populations? Two

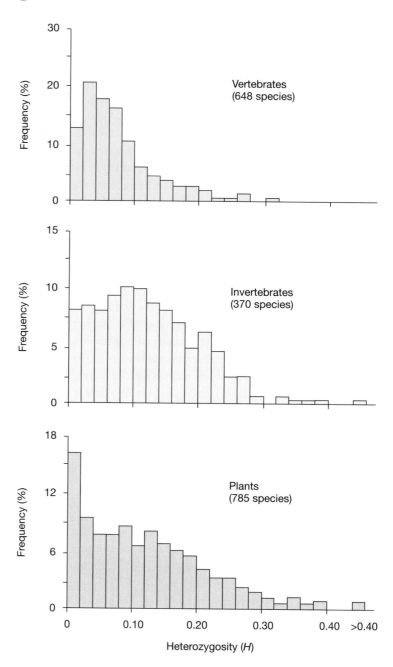

Figure 4.17   Electrophoresis of enzymes reveals that most populations harbor considerable genetic diversity   These histograms show the distribution of enzyme heterozygosities among species of animals and plants. For example, about 7% of all plant species have a heterozygosity between 0.08 and 0.10. Heterozygosity can be interpreted in two ways: as the mean percentage of individuals heterozygous per locus, or as the mean percentage of loci heterozygous per individual. From Fig. 2.2, p. 19, of Avise (1994). © 1994, Chapman and Hall. Reprinted by permission of Kluwer Academic Publishers.

modern views have replaced the classic theory. According to the balance, or selectionist theory, genetic diversity is maintained by natural selection—in favor of rare individuals, in favor of heterozygotes, or in favor of different alleles at different times and places. According to the neutral theory, most of the alleles at most polymorphic loci are functionally and selectively equivalent. In effect, genetic diversity is maintained because it is not eliminated by selection. We will consider the selectionist and neutral theories in more detail in Chapters 5, 7, and 18.

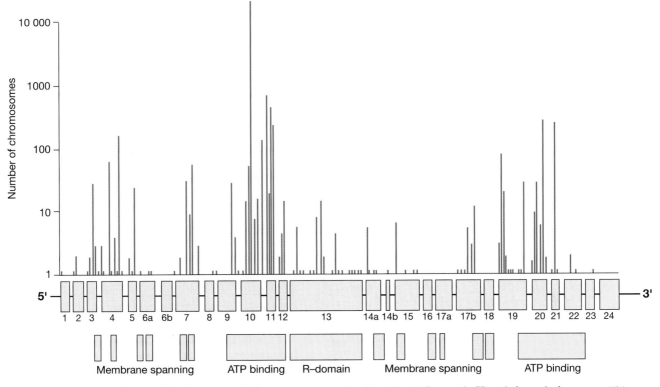

Figure 4.18   Sequencing studies have revealed enormous genetic diversity at the cystic fibrosis locus in humans   This graph shows the abundance and location of the loss-of-function mutations discovered in an examination of over 30,000 disease-causing alleles at the cystic fibrosis locus. The histogram shows the number of copies of each mutation found. The locus map below it, in which the boxes represent exons, shows the location of each mutation within the CFTR gene. The boxes at the bottom of the graph give the functions of the coding regions of the gene. From Fig. 2, p. 395, in Tsui (1992). Copyright © 1992, Elsevier Science. Reprinted with permission of Elsevier Science.

Summary

Mutations range from single base-pair substitutions to the duplication of entire chromosome sets, and vary in impact from no amino acid sequence change to single amino acid changes to gene creation and genome duplication.

Point mutations result from errors made by DNA polymerase during DNA synthesis or errors made by DNA repair enzymes after sequences have been damaged by chemical mutagens or radiation. Point mutations in first and second positions of codons frequently result in replacement substitutions that lead to changes in the amino acid sequence of proteins. Point mutations in third positions of codons usually result in silent substitutions that do not lead to changes in the amino acid sequence of proteins. Point mutations create new alleles.

Mutation rates vary both among genes within genomes and among species. Both DNA polymerase and the many loci involved in mismatch repair exhib-

it heritable variation. As a result, mutation rate is a trait that can respond to natural selection and other evolutionary processes.

The most common sources of new genes are duplication events that result from errors during crossing over. A duplicated gene can diverge from its parent sequence and become a new locus with a different function, or a nonfunctional pseudogene. New genes can also arise when mRNAs are reverse-transcribed and inserted into the genome in a new location.

Chromosome alterations form a large class of mutations. Chromosome inversions have interesting evolutionary implications because they reduce the frequency of recombination between loci within the inversion. As a result, alleles within inversions tend to be inherited as a group instead of independently. Polyploidy is a condition characterized by duplication of an entire chromosome set. Polyploidy is common in plants, and is important because polyploid individuals are genetically isolated from the population that gave rise to them.

Evolutionary biologists typically measure allelic diversity in populations by using gel electrophoresis to look directly at the proteins encoded by alleles or at the DNA of the alleles themselves. Such studies have revealed that most natural populations contain substantial genetic diversity.

Now that we know something about where alleles and genes come from, and about the genetic diversity present in most populations, we are ready to shift our focus and ask a different question: What determines the fate of these new alleles and genes in a population? This is the subject of Chapter 5 and Chapter 6.

Questions

1. The evolutionary biologist Graham Bell (1997) has said, "Most mutations are not very deleterious. The assault on adaptedness is not carried out primarily by a storm of mutations that kill or maim, but rather by a steady drizzle of mutations with slight or inappreciable effects on health and vigor" (Bell 1997). This statement is supported by classical experiments in fruit flies and other organisms. Yet most known human mutations cause severe disease (for example, see Table 4.2b). Is it possible that mutations in humans are qualitatively different from mutations in other organisms? Or has our traditional perception of mutations as highly deleterious been colored by intensive studies on a small subset of human mutations? For a gene like the β-globin gene, how would you go about determining how often mutations with small effects occur?

2. Schlager & Dickie (1971) set out to determine the rate of coat-color mutations in mice. They spent six years studying five coat-color genes in over seven million mice, examining thousands of brother–sister matings from 28 inbred strains. For each gene they studied two mutation rates: (1) the rate at which a normal gene mutated to a form resulting in loss of function, and (2) the rate at which a mutant gene would then mutate back to the normal form. For example, in 67,395 tested crosses, the "albino" gene mutated from normal (colored) to nonfunctional (albino) just three times, for a mutation rate of 44.5×10^{-6}. Interestingly, in all the coat-color genes, the back-mutation rates (e.g., from albino to colored) were typically around 2.5×10^{-6}, which was always lower than the rates for loss of function. Think about the different mutations that can all cause loss of function of a single gene, versus the mutations that can cause regain of function. Why were the mutation rates for loss of function always higher than the back-mutation rates for regain of function?

3. Voelker, Schaffer, & Mukai (1980) studied 1000 lines of flies for 220 generations to estimate the mutation rate of a certain protein. In 3,111,598 crosses, they found 16 flies with new replacement substitution mutations in that protein (detected by a slight change in its electrical charge), a mutation rate of 5.1×10^{-6}. However, only 4 of these mutations actually altered protein function; the remaining 12 did not detectably affect protein function. If Voelker et al. had simply been measuring loss of function, would their estimate of mutation rate have been lower or higher? If they had been able to measure silent site substitutions as well, would their estimate of mutation rate have been lower or higher? Do you think that by measuring changes in protein charge, they were able to measure all of the replacement substitutions that occurred? What do you think is the most informative measure of mutation rate?

4. In this chapter, we introduced the consequences of mutations in two sorts of traits: changes in phenotypic traits such as hemoglobin structure, and changes in mutation rate itself. To clarify the distinction between these two kinds of traits, examine the following list. Which of these proteins can affect mutation rate itself? Which do not affect mutation rate, but instead affect some other trait of the organism?

Protein	Example of mutation in the protein
• β-globin	• Increased tendency to sickle
• Mismatch repair proteins	• Increased repair of damaged DNA
• Melanin (coat-color protein)	• Red coat instead of black coat
• Growth hormone	• Dwarfism or gigantism
• DNA polymerase	• Increased speed and decreased accuracy during DNA replication

5. The discovery of "overprinting" shows that it is possible for one stretch of DNA to code for two entirely different, and functional, proteins. Let's examine this unlikely phenomenon further. Suppose we discover a tiny gene, 12 base pairs long, that codes for a tiny polypeptide just 3 amino acids long:
   - DNA   ACUGCUGUCUAA
   - Amino acids   thr-ala-val-stop

Now suppose this organism would benefit greatly if it had another little polypeptide composed of leu-leu-ser. Would this be possible if the organism started transcribing the gene from the second base pair? Look back at the genetic code in Fig. 4.3b to figure this out. Suppose further that the organism would benefit even more if instead of leu-leu-ser, it had pro-leu-ser. What mutations would be necessary to do this, and would they destroy the amino acid sequence of the original protein shown above? In general, what sort of mutations can occur that will not alter the original protein, but will allow changes in amino acids to occur in the other, "overprinted" protein?

6. The amino acid sequences encoded by the red and green visual pigment genes found in humans are 96% identical (Nathans et al. 1986). These two loci are found close together on the X chromosome, while the locus for the blue pigment is located on chromosome 7. Among primates, only Old World monkeys, the great apes, and humans have a third pigment gene—New World monkeys have only one X-linked pigment gene. Comment on the following three hypotheses:

   • One of the two visual pigment loci on the X chromosome originated in a gene duplication event.

   • The gene duplication event occurred after New World and Old World monkeys had diverged from a common ancestor, which had two visual pigment genes.

   • Human males with a mutated form of the red or green pigment gene experience the same color vision of our male primate ancestors.

7. Chromosome number can evolve by smaller-scale changes than duplication of entire chromosome sets. For example, domestic horses have 64 chromosomes per diploid set while Przewalski's horse, an Asian subspecies, has 66. Przewalski's horse is thought to have evolved from an ancestor with $2n = 64$ chromosomes. The question is, Where did its extra chromosome pair originate? It seems unlikely that an entirely new chromosome pair was created *de novo* in Przewalski's horse. To generate a hypothesis explaining the origin of the new chromosome in Przewalski's horse, examine the adjacent figure.

The drawing shows how certain chromosomes synapse in the hybrid offspring of a domestic horse–Przewalski's horse mating (Short et al. 1974). The remaining chromosomes show a normal 1:1 pairing.

Do you think this sort of gradual change in chromosome number involves a change in the actual number of genes present, or just rearrangement of the same number of genes?

8. If you have not already done so, complete Table 4.4 by calculating the frequencies of the *CCR5-+* and *CCR5-Δ32* alleles in each population. Can you suggest any hypotheses to explain the global distribution of the CCR-*Δ32* allele? What additional questions are raised by the data presented in the table? List as many as you can. Then, pick one and describe a research project that might answer it.

9. We have seen that in plants, generation time can affect the number of mutations seen in the offspring—individual plants that had long lifespans had fruits with more mutations than individual plants with short lifespans. This is apparently because in plants, numerous cell divisions of somatic cells precede the production of gamete cells.

   a. Do you think that the same might be true in animals—could generation time affect accumulation of mutations in the gametes of *individual* animals? Could generation time affect accumulation of mutations *per year* in a *population* of animals? Why or why not?

   b. In mammals, sperm cells are produced by parent cells (spermatogonia) that undergo constant cell division throughout life, whereas egg cells are produced only during fetal development. Do you think the average number of mutations per gamete might differ in males vs. females? Why or why not? How could you test this theory? Look up

Shimmin, L. C., B. H.-J. Chang, W.-H. Li. 1993. Male-driven evolution of DNA sequences. *Nature* 362:745–747.

Exploring the Literature

10. The directed-mutation hypothesis has been the most controversial topic in recent research on mutation. This hypothesis, inspired by experimental work on the bacterium *Escherichia coli,* maintains that organisms can generate specific types of mutations in response to particular environmental challenges. For example, if the environment grew hotter over time, the hypothesis maintains that organisms would respond by specifically generating mutations in genes involved in coping with hot temperatures.

This implies that mutations do not occur randomly, but are directed by the environment. Look up the following papers to learn more about the controversy:

Cairns, J., J. Overbaugh, and S. Miller. 1988. The origin of mutants. *Nature* 335: 142–145.

Foster, P. L., and J. M. Trimarchi. 1994. Adaptive reversion of a frameshift mutation in *Escherichia coli* by simple base deletions in homopolymeric runs. *Science* 265: 407–409.

Galitski, T., and J. R. Roth. 1995. Evidence that F plasmid transfer replication underlies apparent adaptive mutation. *Science* 268: 421–423.

Radicella, J. P., P. U. Park, and M. S. Fox. 1995. Adaptive mutation in *Escherichia coli:* A role for conjugation. *Science* 268: 418–420.

Rosenberg, S. M., S. Longerich, P. Gee, and R. S. Harris. 1994. Adaptive mutations by deletions in small mononucleotide repeats. *Science* 265: 405–407.

Sniegowski, P. D., and R. E. Lenski. 1995. Mutation and adaptation: The directed mutation controversy in evolutionary perspective. *Annual Review of Ecology and Systematics* 26: 553–578.

11. Some evolutionary geneticists have suggested that the genetic code has been shaped by natural selection to minimize the deleterious consequences of mutations. For an entry into the literature on this issue, see

Knight, R. D., S. J. Freeland, and L. F. Landweber. 1999. Selection, history and chemistry: The three faces of the genetic code. *Trends in Biochemical Sciences* 24: 241–247.

Freeland S. J., and L. D. Hurst. 1998. Load minimization of the genetic code: History does not explain the pattern. *Proceedings of the Royal Society London,* Series B 265: 2111–2119.

Freeland S. J., and L. D. Hurst. 1998. The genetic code is one in a million. *Journal of Molecular Evolution* 47: 238–248.

Citations

Abel, L., F. Demenais, et al. 1991. Evidence for the segregation of a major gene in human susceptibility/resistance to infection by *Schistosoma mansoni. American Journal of Human Genetics* 48: 959–970.

Avise, John C. 1994. *Molecular Markers, Natural History and Evolution.* New York: Chapman & Hall.

Ayala, F. J., L. Serra, and A. Prevosti. 1989. A grand experiment in evolution: The *Drosophila subobscura* colonization of the Americas. *Genome* 31: 246–255.

Bell, G. 1997. *Selection: The Mechanism of Evolution.* New York: Chapman & Hall.

Bohr, V. A., C. A. Smith, D. S. Okumoto, and P. C. Hanawalt. 1985. DNA repair in an active gene: Removal of pyrimidine dimers from the DHRF gene of CHO cells is much more efficient than in the genome overall. *Cell* 40: 359–369.

Drake, J. W. 1991. A constant of rate of spontaneous mutation in DNA-based microbes. *Proceedings of the National Academy of Sciences, USA* 88: 7160–7164.

Feder, J. L., C. A. Chilcote, and G. L. Bush. 1989. Inheritance and linkage relationships of allozymes in the apple maggot fly. *Journal of Heredity* 80: 277–283.

Friedberg, E. C., G. C. Walker, and W. Siede. 1995. *DNA Repair and Mutagenesis.* Washington D.C.: ASM Press.

García-Dorado, A. 1997. The rate and effects distribution of viability mutation in *Drosophila:* minimum distance estimation. *Evolution* 51: 1130–1139.

Gillin, F. D., and N. G. Nossal. 1976a. Control of mutation frequency by bacteriophage T4 DNA polymerase I. The ts CB120 antimutator DNA polymerase is defective in strand displacement. *Journal of Biological Chemistry* 251: 5219–5224.

Gillin, F. D., and N. G. Nossal. 1976b. Control of mutation frequency by bacteriophage T4 DNA polymerase II. Accuracy of nucleotide selection by L8 mutator, CB120 antimutator, and wild type phage T4 DNA polymerases. *Journal of Biological Chemistry* 251: 5225–5232.

Harris, H. 1966. Enzyme polymorphisms in man. *Proceedings of the Royal Society London,* Series B 164: 298–310.

Huey, R. B., G. W. Gilchrist, M. L. Carlson, D. Berrigan, and L. Serra. 2000. Rapid evolution of a geographic cline in size in an introduced fly. *Science* 287: 308–309

Ingram, V. M. 1958. How do genes act? *Scientific American* 198: 68–76.

Jeffs, P., and M. Ashburner. 1991. Processed pseudogenes in *Drosophila. Proceedings of the Royal Society of London,* Series B 244: 151–159.

Keese, P. K., and A. Gibbs. 1992. Origins of genes: "Big bang" or continuous creation? *Proceedings of the National Academy of Sciences, USA* 89: 9489–9493.

Keightley, P. D., and A. Caballero. 1997. Genomic mutation rates for lifetime reproductive output and lifespan in *Caenorhabditis elegans. Proceedings of the National Academy of Sciences, USA* 94: 3823–3827.

Klekowski, E. J., Jr., and P. J. Godfrey. 1989. Aging and mutation in plants. *Nature* 340: 389–391.

Langley, C. H., E. Montgomery, and W. F. Quattlebaum. 1982. Restriction map variation in the *Adh* region of *Drosophila. Proceedings of the National Academy of Sciences, USA* 78: 5631–5635.

LeClerc, J. E., B. Li, W. L. Payne, and T. A. Cebula. 1996. High mutation frequencies among *Escherichia coli* and *Salmonella* pathogens. *Science* 274: 1209–1211.

Lewontin, R. C., and J. L. Hubby. 1966. A molecular approach to the study of genetic heterozygosity in natural populations. II. Amount of variation and degree of heterozygosity in natrual populations of *Drosophila pseudoobscura. Genetics* 54: 595–609.

Liu, R., W. A. Paxton, et al. 1996. Homozygous defect in HIV-1 coreceptor accounts for resistance in some multiply-exposed individuals to HIV-1 infection. *Cell* 86: 367–377.

Long, M., and C. H. Langley. 1993. Natural selection and the origin of *jingwei,* a chimeric processed functional gene in *Drosophila. Science* 260: 91–95.

Martinson, J. J., N. H. Chapman, et al. 1997. Global distribution of the CCR5 gene 32-base-pair deletion. *Nature Genetics* 16: 100–103.

Mellon, I., D. K. Rajpal, M. Koi, C. R. Boland, and G. N. Champe. 1996. Transcription-coupled repair deficiency and mutations in the human mismatch repair genes. *Science* 272: 557–560.

Mitton, J. B. 1997. *Selection in Natural Populations.* Oxford: Oxford University Press.

Nathans, J., D. Thomas, and D. S. Hogness. 1986. Molecular genetics of human color vision: The genes encoding blue, green, and red pigments. *Science* 232: 193–202.

Oakeshott, J. G., J. B. Gibson, et al. 1982. Alcohol dehydrogenase and glycerol-3-phosphate dehydrogenase clines in *Drosophila melanogaster* on different continents. *Evolution* 36: 86-96.

Online Mendelian Inheritance in Man, OMIM(TM). 1999. Johns Hopkins University, Baltimore, MD. MIM Number: 181460. World Wide Web URL: http://www.ncbi.nlm.nih.gov/omim/

Pauling, L., H. A. Itano, S. J. Singer, and I. C. Wells. 1949. Sickle-cell anemia, a molecular disease. *Science* 110: 543–548.

Pier, G. B., M. Grout , and T. S. Zaidi. 1997. Cystic fibrosis transmembrane conductance regulator is an epithelial cell receptor for clearance of *Pseudomonas aeruginosa* from the lung. *Proceedings of the National Academy of Science, U.S.A.* 94: 12088–12093.

Powers, D. A., T. Lauerman, et al. 1991. Genetic mechanisms for adapting to a changing environment. *Annual Review of Genetics* 25: 629–659.

Powers, D. A., P. M. Schulte, D. Crawford, and L. DiMichele. 1998. Evolutionary adaptations of gene structure and expression in natural populations in relation to a changing environment: A multidisciplinary approach to address the million-year saga of a small fish. *The Journal of Experimental Zoology* 282: 71–94.

Prevosti, A. 1967. Inversion heterozygosity and selection for wing length in *Drosophila subobscura*. *Genetical Research Cambridge* 10: 81–93.

Prevosti, A., G. Ribo, L. Serra, M. Aguade, J. Balaña, M. Monclus, and F. Mestres. 1988. Colonization of America by *Drosophila subobscura:* Experiment in natural populations that supports the adaptive role of chromosomal-inversion polymorphism. *Proceedings of the National Academy of Sciences, USA* 85: 5597–5600.

Ramsey, J., and D. W. Schemske. 1998. Pathways, mechanisms, and rates of polyploid formation in flowering plants. *Annual Review of Ecology and Systematics* 29: 467–501.

Samson, M., F. Libert, et al. 1996. Resistance to HIV-1 infection in caucasian individuals bearing mutant alleles of the CCR5 chemokine receptor gene. *Nature* 382: 722–725.

Schlager, G., and M.M. Dickie. 1971. Natural mutation rates in the house mouse. Estimates for 5 specific loci and dominant mutations. *Mutation Research* 11: 89–96.

Short, R. V., A.C. Chandley, R. C. Jones, and W. R. Allen. 1974. Meiosis in interspecific equine hybrids. II. The Przewalski horse/domestic horse hybrid. *Cytogenetics and Cell Genetics* 13: 465–478.

Stadler, L. J. 1942. Some observations on gene variability and spontaneous mutation. Spragg Memorial Lectures. East Lansing: Michigan State College.

Tsui, L.-C. 1992. The spectrum of cystic fibrosis mutations. *Trends in Genetics* 8: 392–398.

Voelker, R. A., H. E. Schaffer, and T. Mukai. 1980. Spontaneous allozyme mutations in *Drosophila melanogaster:* Rate of occurrence and nature of the mutants. *Genetics* 94: 961–968.

Watson, J. D., and F. H. C. Crick 1953. A structure for deoxyribose nucleic acid. *Nature* 171: 737–738.

CHAPTER 5

Mendelian Genetics in Populations I: Selection and Mutation as Mechanisms of Evolution

A population of mice, including newborns, juveniles, and adults. (C. C. Lockwood/Animals Animals/Earth Scenes)

Most people are susceptible to HIV. Their best hope of avoiding infection is to avoid contact with the virus. There are, however, a few individuals who remain uninfected despite repeated exposure. In 1996, AIDS researchers discovered that at least some of the variation in susceptibility to HIV has a genetic basis (see Chapters 1 and 4). The gene responsible encodes a cell surface protein called CCR5. CCR5 is exploited by most sexually transmitted strains of HIV-1 as a means of infiltrating white blood cells. There is a mutant allele of the CCR5 gene, called *CCR5-Δ32*, with a 32-base-pair deletion that destroys the encoded protein's ability to function. Individuals who inherit two copies of this allele have no CCR5 on the surface of their white blood cells, and are therefore highly resistant to HIV-1. The fact that individuals homozygous for *CCR5-Δ32* are much less likely to contract AIDS raises a question: Will the AIDS epidemic cause an increase in the frequency of the *Δ32* allele in human populations?

Consider, too, this question. In the 1927 case of *Buck v. Bell,* the United States Supreme Court upheld by a vote of eight to one the state of Virginia's sterilization statute. Drafted on the advice of eugenicists, the law was intended to improve the genetic quality of future generations by allowing the forced sterilization of individuals afflicted with hereditary forms of insanity, feeblemindedness, and other mental defects. The court's decision in *Buck v. Bell* reinvigorated a compulsory sterilization movement dating from 1907 (Kevles 1995). By 1940, thirty states had enacted sterilization laws, and by 1960 over 60,000 people had been sterilized without their consent (Reilly 1991; Lane 1992). In hindsight, the evidence that these individuals suffered from hereditary diseases was weak. If the genetic assumptions had been correct, would sterilization have been an effective means of reducing the incidence of undesirable traits?

And, finally, consider this question. Cystic fibrosis is among the most common serious genetic diseases among people of European ancestry, affecting approximately 1 newborn in 2500. Cystic fibrosis is inherited as an autosomal recessive trait (see Chapter 4). Affected individuals suffer chronic infections with the bacterium *Pseudomonas aeruginosa* and ultimately sustain severe lung damage (Pier et al., 1997). At present, most individuals with cystic fibrosis live into their thirties or forties (Elias et al. 1992), but until recently few survived to reproductive age. In spite of the fact that cystic fibrosis was lethal for most of human history, in some populations as many as 4% of individuals are carriers. How can alleles that cause a lethal genetic disease remain this common?

The questions posed in the preceding paragraphs all concern forces that influence the frequencies of alleles in populations. Such questions can be addressed using the tools of **population genetics**. Population genetics, the subject of Chapters 5 and 6, integrates Darwin's theory of evolution by natural selection with Mendelian genetics (for a history, see Provine 1971). The crucial insight of population genetics is that changes in the relative abundance of traits in a population can be tied to changes in the relative abundance of the alleles that influence them. From a population-genetic perspective, evolution is a change across generations in the frequencies of alleles. Population genetics provides the theoretical foundation for much of our modern understanding of evolution.

5.1 Mendelian Genetics in Populations: The Hardy–Weinberg Equilibrium Principle

Population genetics begins with a model of what happens to allele and genotype frequencies in an idealized population. Once we know how Mendelian genes behave in the idealized population, we will be able to explore how they behave in real populations.

Before we can hope to predict whether the AIDS epidemic will cause an increase in the frequency of the *CCR5-Δ32* allele in human populations, we need to understand how the allele will behave without the AIDS epidemic. In other words, we need to develop a null model for the behavior of genes in populations. This null model should specify, under the simplest possible assumptions, what will happen across generations to the frequencies of alleles and genotypes. The model should apply not just to humans, but to any population of organisms that are both diploid and sexual.

A population is a group of interbreeding individuals and their offspring (Figure 5.1). The crucial events in the life cycle of a population are these: Adults produce gametes, the gametes combine to make zygotes, and the zygotes grow up to become the next generation of adults. We want to track the fate, across generations, of

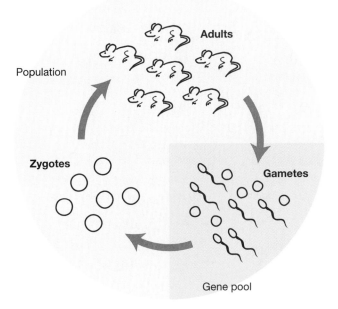

Figure 5.1   The life cycle of an imaginary population of mice, highlighting the stages that will be important in our development of population genetics

Mendelian genes in such a population. We want to know whether a particular allele or genotype will become more common or less common over time, and why.

Imagine that the mice in Figure 5.1 have in their genome a particular locus, the A locus, with two alleles: *A* and *a*. We can begin tracking these alleles at any point in the life cycle. We will then follow them through one complete turn of the cycle, from one generation to the next, to see if their frequencies change.

A Numerical Example

Our task of following alleles will be simpler if we start with the gametes produced by the adults when they mate. We will assume that the adults choose their mates at random. A useful mental trick is to picture the process of random mating happening like this: We take all the eggs and sperm produced by all the adults in the population and mix them together in a barrel. This barrel is known as the gene pool. Each sperm then swims about at random until it collides with an egg to make a zygote. Something rather like this actually happens in sea urchins and other marine organisms that simply release their gametes onto the tide. For other organisms, like mice and humans, it is obviously a simplification.

The adults in our mouse population are diploid, so each carries two alleles for the A locus. But the adults made their eggs and sperm by meiosis. Following Mendel's law of segregation, each gamete received just one allele for the A locus. Imagine that 60% of the eggs and sperm received allele *A,* and 40% received allele *a*. In other words, the frequency of allele *A* in the gene pool is 0.6, and the frequency of allele *a* is 0.4 (Figure 5.2).

What happens when eggs meet sperm? For example, what fraction of the zygotes they produce will have genotype *AA*? Figure 5.2 shows the four possible combinations of egg and sperm, the zygotes they produce, and a calculation specifying the probability of each. For example, if we pick an egg to watch at random, there is a 60% chance that it will have genotype *A*. When a sperm comes along

Starting with the eggs and sperm that constitute the gene pool, our model tracks alleles through zygotes and adults and into the next generation's gene pool.

Figure 5.2 Random mating in the gene pool of our model mouse population produces zygotes with predictable genotype frequencies

A gene pool with allele frequencies of 0.6 and 0.4...

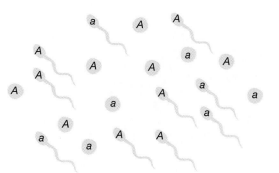

...yields zygotes with genotype frequencies of 0.36, 0.48, and 0.16

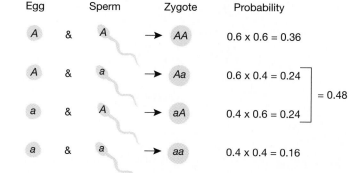

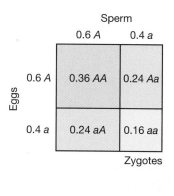

to fertilize the egg, there is a 60% chance that the sperm will have genotype A. The probability that we will witness the production of an AA zygote is therefore

$$0.6 \times 0.6 = 36 \qquad \text{(see Box 5.1)}$$

If we watched the formation of all the zygotes, 36% of them would have genotype AA. The calculations in Figure 5.2 show that random mating in the gene pool produces zygotes in the following proportions:

$$\begin{array}{ccc} \textbf{AA} & \textbf{Aa} & \textbf{aa} \\ 0.36 & 0.48 & 0.16 \end{array}$$

(The Aa category includes heterozygotes produced by combining either an A egg with an a sperm or an a egg with an A sperm.) Notice that

$$0.36 + 0.48 + 0.16 = 1$$

This confirms that we have accounted for all of the zygotes. Figure 5.3 shows a geometrical representation of the same calculations.

We now let the zygotes grow to adulthood, and we let the adults produce gametes to make the next generation's gene pool. Will the frequencies of alleles A and a in the new gene pool be different from what they were before?

We can calculate the frequency of allele A in the new gene pool as follows. Because adults of genotype AA constitute 36% of the population, they will make 36% of the gametes. All of these gametes carry allele A. Likewise, adults of genotype Aa constitute 48% of the population, and will make 48% of the gametes. Half of these gametes carry allele A. So the total fraction of the gametes in the gene pool that carry allele A is

Figure 5.3 A geometrical representation of the genotype frequencies produced by random mating The fractions along the left and top edges of the box represent the frequencies of A and a eggs and sperm in the gene pool. The fractions inside the box represent the genotype frequencies among zygotes formed by random encounters between gametes in the gene pool.

BOX 5.1 Combining probabilities

The combined probability that two independent events will occur together is equal to the product of their individual probabilities. For example, the probability that a tossed penny will come up heads is $\frac{1}{2}$. The probability that a tossed nickel will come up heads is also $\frac{1}{2}$. If we toss both coins together, the outcome for the penny is independent of the outcome for the nickel. Thus the probability of getting heads on the penny and heads on the nickel is

$$\frac{1}{2} \times \frac{1}{2} = \frac{1}{4}$$

The combined probability that either of two mutually exclusive events will occur is the sum of their individual probabilities. When rolling a die we can get a one or a two (among other possibilities), but we cannot get both at once. Thus, the probability of getting either a one or a two is

$$\frac{1}{6} + \frac{1}{6} = \frac{1}{3}$$

$$0.36 + \left(\frac{1}{2}\right) 0.48 = 0.6$$

Figure 5.4 shows this calculation graphically. The figure also shows a calculation establishing that the fraction of the gametes in the gene pool that carry allele a is 0.4. Notice that

$$0.6 + 0.4 = 1$$

This confirms that we have accounted for all of the gametes. Figure 5.5 shows a geometrical representation of the same calculations.

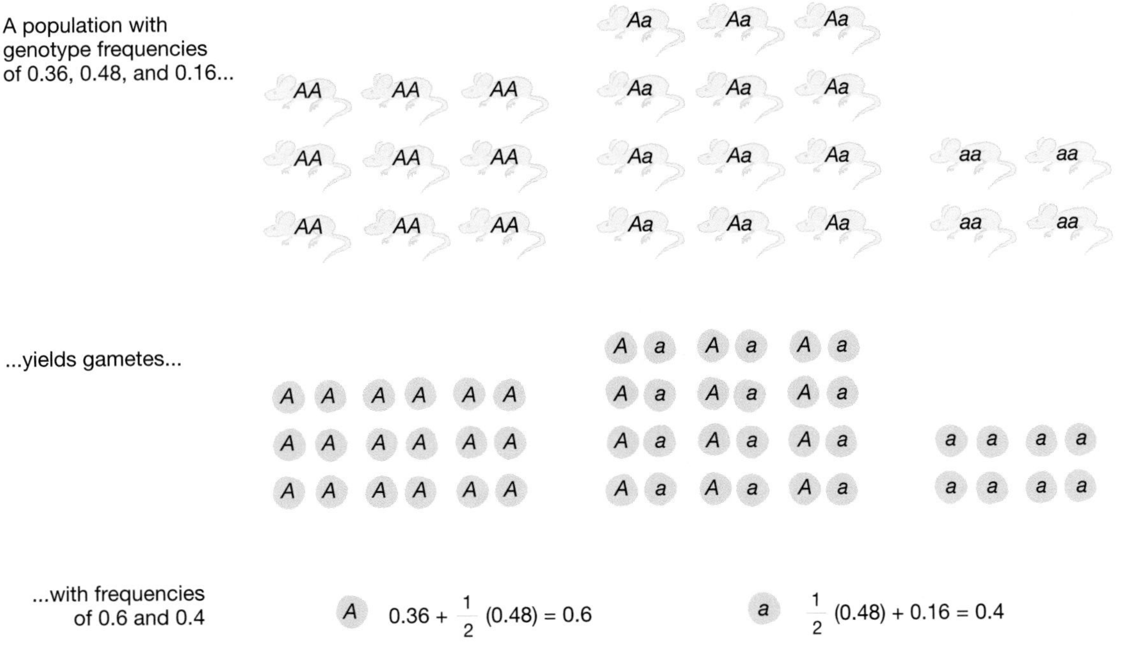

A population with genotype frequencies of 0.36, 0.48, and 0.16...

...yields gametes...

...with frequencies of 0.6 and 0.4

A $0.36 + \dfrac{1}{2}(0.48) = 0.6$

a $\dfrac{1}{2}(0.48) + 0.16 = 0.4$

Figure 5.4 When the adults in our model mouse population make gametes, they produce a gene pool in which the allele frequencies are identical to the ones we started with a generation ago

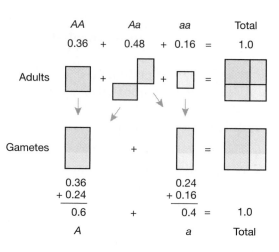

	AA		Aa		aa		Total
	0.36	+	0.48	+	0.16	=	1.0

Adults

Gametes

0.36		0.24
+ 0.24		+ 0.16
0.6	+	0.4 = 1.0
A		a Total

Figure 5.5 A geometrical representation of the allele frequencies produced when adults in our model population make gametes The area of each box represents the frequency of an adult or gamete genotype. Note that half the gametes produced by *Aa* adults carry allele *A,* and half carry allele *a.*

Numerical examples show that in our idealized population, allele frequencies remain constant from one generation to the next.

We have come full circle, and arrived right where we began. We started with allele frequencies of 60% for *A* and 40% for *a* in our population's gene pool. We followed the alleles through zygotes and adults and into the next generation's gene pool. The allele frequencies in the new gene pool are still 60% and 40%. The allele frequencies for *A* and *a* in our population are thus in equilibrium: They do not change from one generation to the next. The population does not evolve.

The first biologist to work a numerical example, tracing the frequencies of Mendelian alleles from one generation to the next in an ideal population, was G. Udny Yule in 1902. He started with a gene pool in which the frequencies of two alleles were 0.5 and 0.5, and showed that in the next generation's gene pool the allele frequencies were still 0.5 and 0.5. (Readers should reproduce his calculations as an exercise.)

Like us, Yule concluded that the allele frequencies in his imaginary population were in equilibrium. Yule's conclusion was both ground breaking and correct, but he took it a bit too literally. He had worked only one example, and he believed that allele frequencies of 0.5 and 0.5 represented the only possible equilibrium state for a two–allele system. For example, Yule believed that if a single *A* allele appeared as a mutation in a population whose gene pool otherwise consisted only of *a*'s, then the *A* allele would automatically increase in frequency until it constituted one half of the gene pool. Yule argued this claim during the discussion that followed a talk given in 1908 by R. C. Punnett (of Punnett-square fame). Punnett thought that Yule was wrong, but he did not know how to prove it.

We have already demonstrated, of course, that Punnett was correct in rejecting Yule's claim. Our calculations showed that a population with allele frequencies of 0.6 and 0.4 is in equilibrium too. What Punnett wanted, however, is a general proof. This proof should show that any allele frequencies, so long as they sum to 1, will remain unchanged from one generation to the next.

Punnett took the problem to his mathematician friend G. H. Hardy, who produced the proof in short order (Hardy 1908). Hardy simply repeated the calculations that Yule had performed, using variables in place of the specific allele frequencies that Yule had assumed. Hardy's calculation of the general case indeed showed that any allele frequencies will be in equilibrium.

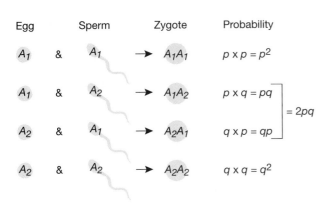

Figure 5.6   **The general case for random mating in the gene pool of our model mouse population**

The General Case

For our version of Hardy's general case, we will again work with an imaginary population. We are concerned with a single locus with two alleles: A_1 and A_2. We use capital letters with subscripts because we want our calculation to cover cases in which the alleles are codominant as well as cases in which they are dominant and recessive. The three possible diploid genotypes are A_1A_1, A_1A_2, and A_2A_2.

As in our numercial example, we will start with the gene pool, and follow the alleles through one complete turn of the life cycle. The gene pool will contain some frequency of A_1 gametes and some frequency of A_2 gametes. We will call the frequency of A_1 in the gene pool p and the frequency of A_2 in the gene pool q. There are only two alleles in the population, so

$$p + q = 1$$

The first step is to let the gametes in the gene pool combine to make zygotes. Figure 5.6 shows the four possible combinations of egg and sperm, the zygotes they produce, and a calculation specifying the probability of each. For example, if we pick an egg to watch at random, the chance is p that it will have genotype A_1. When a sperm comes along to fertilize the egg, the chance is p that the sperm will have genotype A_1. The probability that we will witness the production of an A_1A_1 zygote is therefore

$$p \times p = p^2$$

If we watched the formation of all the zygotes, p^2 of them would have genotype A_1A_1. The calculations in Figure 5.6 show that random mating in our gene pool produces zygotes in the following proportions:

A_1A_1	A_1A_2	A_2A_2
p^2	$2pq$	q^2

Figure 5.7 shows a geometrical representation of the same calculations. The figure also shows that

$$p^2 + 2pq + q^2 = 1$$

This confirms that we have accounted for all the zygotes. The same result can be demonstrated algebraically by substituting $(1 - p)$ for q in the expression $p^2 + 2pq + q^2$, then simplifying.

The challenge now is to prove algebraically that there was nothing special about our numerical examples. Any allele frequencies will remain constant from generation to generation.

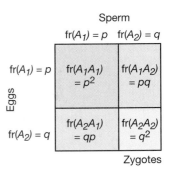

Figure 5.7   **A geometrical representation of the general case for the genotype frequencies produced by random mating**   The variables along the left and top edges of the box represent the frequencies of A and a eggs and sperm in the gene pool. The expressions inside the box represent the genotype frequencies among zygotes formed by random encounters between gametes in the gene pool.

We have gone from the allele frequencies in the gene pool to the genotype frequencies among the zygotes. We now let the zygotes grow up to become adults, and let the adults produce gametes to make the next generation's gene pool.

We can calculate the frequency of allele A_1 in the new gene pool as follows. Because adults of genotype A_1A_1 constitute a proportion p^2 of the population, they will make p^2 of the gametes. All of these gametes carry allele A_1. Likewise, adults of genotype A_1A_2 constitute a proportion $2pq$ of the population, and will make $2pq$ of the gametes. Half of these gametes carry allele A_1. So the total fraction of the gametes in the gene pool that carry allele A_1 is

$$p^2 + \left(\frac{1}{2}\right)2pq = p^2 + pq$$

We can simplify the expression on the right by substituting $(1 - p)$ for q. This gives

$$p^2 + pq = p^2 + p(1 - p)$$
$$= p^2 + p - p^2$$
$$= p$$

Figure 5.8 shows this calculation graphically. The figure also shows a calculation establishing that the fraction of the gametes in the gene pool that carry allele A_2 is q. We assumed at the outset that p and q sum to 1, so we know that we have accounted for all of the gametes.

Our model has shown that our idealized population does not evolve. This conclusion is known as the Hardy–Weinberg equilibrium principle.

Once again we have come full circle, and arrived back where we started. We started with allele frequencies of p and q in our population's gene pool. We followed the alleles through zygotes and adults and into the next generation's gene pool. The allele frequencies in the new gene pool are still p and q. The allele frequencies p and q can be stable at any values at all between 0 and 1, as long as they sum to 1. In other words, *any* allele frequencies will be in equilibrium, not just $p = q = 0.5$ as Yule thought.

This is a profound result. At the beginning of the chapter we defined evolution as change in allele frequencies in populations. The calculations we just performed show, given simple assumptions, that in populations following the rules of Mendelian genetics, allele frequencies do not change.

We have presented this result as the work of Hardy (1908). It was derived independently by Wilhelm Weinberg (1908) and has become known as the

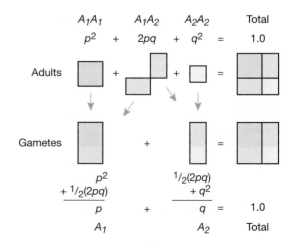

Figure 5.8   A geometrical representation of the general case for the allele frequencies produced when adults in our model population make gametes   The area of each box represents the frequency of an adult or gamete genotype.

Hardy–Weinberg equilibrium principle. (Some evolutionary biologists refer to it as the Hardy–Weinberg–Castle equilibrium principle, because William Castle [1903] worked a numerical example and stated the general equilibrium principle nonmathematically five years before Hardy and Weinberg explicitly proved the general case [see Provine 1971].) The Hardy–Weinberg equilibrium principle yields two fundamental conclusions:

- **Conclusion 1:** The allele frequencies in a population will not change, generation after generation.
- **Conclusion 2:** If the allele frequencies in a population are given by p and q, the genotype frequencies will be given by p^2, $2pq$, and q^2.

We get an analogous result if we generalize the analysis from the two-allele case to the usual case of a population containing many alleles at a locus (see Box 5.2).

What Use Is the Hardy–Weinberg Equilibrium Principle?

It may seem puzzling that in a book about evolution we have devoted so much space to a proof that apparently shows that evolution does not happen. What makes the Hardy–Weinberg equilibrium principle useful is that it rests on a specific set of simple assumptions. When one or more of these assumptions is violated, the Hardy–Weinberg conclusions no longer hold.

We left most of the assumptions unstated when we developed our null model for how Mendelian alleles behave in populations. We can now state them explicitly. The crucial assumptions are:

The Hardy–Weinberg equilibrium principle becomes useful when we list the assumptions we made about our idealized population. By providing a set of explicit conditions under which evolution does not happen, the Hardy–Weinberg analysis identifies the forces that can cause evolution in real populations.

1. **There is no selection.** All members of our model population survived at equal rates and contributed equal numbers of gametes to the gene pool. When this assumption is violated—when individuals with some genotypes survive and reproduce at higher rates than others—the frequencies of alleles may change from one generation to the next.
2. **There is no mutation.** In the model population, no copies of existing alleles were converted by mutation into other existing alleles, and no new alleles were

---

BOX 5.2   The Hardy–Weinberg equilibrium principle with more than two alleles.

Imagine a single locus with several alleles. We can call the alleles A_i, A_j, A_k, and so on, and we can represent the frequencies of the alleles in the gene pool with the variables p_i, p_j, p_k, and so on. The formation of a zygote with genotype A_iA_i requires the union of an A_i egg with an A_i sperm. Thus the frequency of any homozygous genotype A_iA_i is p_i^2. The formation of a zygote with genotype A_iA_j requires either the union of an A_i egg with an A_j sperm, or an A_j egg with an A_i sperm. Thus, the frequency of any heterozygous genotype A_iA_j is $2p_ip_j$.

For example, if there are three alleles with frequencies p_1, p_2, and p_3, such that

$$p_1 + p_2 + p_3 = 1$$

then the genotype frequencies are given by

$$(p_1 + p_2 + p_3)^2 = p_1^2 + p_2^2 + p_3^2 + 2p_1p_2 + 2p_1p_3 + 2p_2p_3$$

and the allele frequencies do not change from generation to generation.

created. When this assumption is violated, and, for example, some alleles have higher mutation rates than others, allele frequencies may change from one generation to the next.

3. **There is no migration.** No individuals moved into or out of the model population. When this assumption is violated, and individuals with some alleles move into or out of the population at higher rates than individuals with other alleles, allele frequencies may change from one generation to the next.

4. **There were no chance events** that caused individuals with some genotypes to pass more of their alleles to the next generation than others. We avoided this type of random event by assuming that the eggs and sperm in the gene pool collided with each other at their actual frequencies of p and q, with no deviations caused by chance. Another way to state this assumption is that the model population was infinitely large. When this assumption is violated, and by chance some individuals contribute more alleles to the next generation than others, allele frequencies may change from one generation to the next. This kind of allele frequency change is called **genetic drift**.

5. **Individuals choose their mates at random.** We explicitly set up the gene pool to let gametes find each other at random. In contrast to assumptions 1 through 4, when this assumption is violated—when, for example, individuals prefer to mate with other individuals of the same genotype—allele frequencies do not change from one generation to the next. Genotype frequencies may change, however. Such shifts in genotype frequency, in combination with a violation of one of the other four assumptions, can influence the evolution of populations.

By furnishing a list of specific ideal conditions under which populations will not evolve, the Hardy–Weinberg equilibrium principle identifies the set of forces that can cause evolution in the real world. This is the sense in which the Hardy–Weinberg equilibrium principle serves as a null model. Biologists can measure allele and genotype frequencies in nature, and determine whether the Hardy–Weinberg conclusions hold. A population in which conclusions 1 and 2 hold is said to be in **Hardy–Weinberg equilibrium**. If a population is not in Hardy–Weinberg equilibrium—if the allele frequencies change from generation to generation or if the genotype frequencies cannot, in fact, be predicted by multiplying the allele frequencies—then one or more of the Hardy–Weinberg model's assumptions is being violated. Such a discovery does not, by itself, tell us which assumptions are being violated, but it does tell us that further research may be rewarded with interesting discoveries.

In the following sections of Chapter 5, we consider how violations of assumptions 1 and 2 affect the two Hardy–Weinberg conclusions, and we explore empirical research on selection and mutation as forces of evolution. In Chapter 6, we consider violations of assumptions 3, 4, and 5.

Changes in the Frequency of the *CCR5-Δ32* Allele

We began this chapter by asking whether we can expect the frequency of the *CCR5-Δ32* allele to change in human populations. Now that we have developed a null model for how Mendelian alleles behave in populations, we can give a partial answer. As long as individuals of all CCR5 genotypes survive and reproduce at equal rates, and as long as no mutations convert some CCR5 alleles into others, and as long as no one moves from one population to another, and as long as populations are infinitely large, and as long as people choose their mates at random, then no, the frequency of the *CCR5-Δ32* allele will not change.

This answer is, of course, thoroughly unsatisfing. It is unsatisfing because none of the assumptions will be true in any real population. We asked the question in the first place precisely because we expect *Δ32/Δ32* individuals to survive the AIDS epidemic at higher rates than individuals with either of the other two genotypes. In the next two sections, we will see that our null model, the Hardy–Weinberg equilibrium principle, provides a framework that will allow us to assess with precision the importance of differences in survival.

5.2 Selection

When we worked with a model population to derive the Hardy–Weinberg equilibrium principle, first on our list of assumptions was that all individuals survive at equal rates and contribute equal numbers of gametes to the gene pool. Systematic violations of this assumption are examples of **selection**. Selection happens when individuals with particular phenotypes survive to reproductive age at higher rates than individuals with other phenotypes, or when individuals with particular phenotypes produce more offspring during reproduction than individuals with other phenotypes. The bottom line in either kind of selection is differential reproductive success: Some individuals have more offspring than others. Selection can lead to evolution when the phenotypes that exhibit differences in reproductive success are heritable—that is, when certain phenotypes are associated with certain genotypes.

Population geneticists often assume that phenotypes are determined strictly by genotypes. They might, for example, think of pea plants as being either tall or short, such that individuals with the genotypes *TT* and *Tt* are tall and individuals with the genotype *tt* are short. Such a view is at least roughly accurate for some traits, including the examples we use in this chapter.

When phenotypes fall into discrete classes that appear to be determined strictly by genotypes, we can think of selection as if it acts directly on the genotypes. We can then assign a particular level of lifetime reproductive success to each genotype. In reality, most phenotypic traits are not, in fact, strictly determined by genotype. Pea plants with the genotype *TT,* for example, vary in height. This variation is due to genetic differences at other loci and to differences in the environments in which the pea plants grew. We will consider such complications in Chapter 7. For the present, however, we adopt the simple view.

When we think of selection as if it acts directly on genotypes, its defining feature is that some genotypes contribute more alleles to future generations than others. In other words, there are differences among genotypes in fitness.

Our task in this section is to incorporate selection into the Hardy–Weinberg analysis. We begin by asking whether selection can change the frequencies of alleles in the gene pool from one generation to the next. In other words, can violation of the no-selection assumption lead to a violation of conclusion 1 of the Hardy–Weinberg equilibrium principle?

First on the list of assumptions about our idealized population was that individuals survive at equal rates and have equal reproductive success. We now explore what happens to allele frequencies when this assumption is violated.

Adding Selection to the Hardy–Weinberg Analysis: Changes in Allele Frequencies

We start with a numerical example that shows that selection can indeed change the frequencies of alleles. Imagine that in our population of mice there is a locus, the B locus, that affects the probability of survival. Assume, as we did for the A locus in Figure 5.2, that the frequency of allele B_1 in the gene pool is 0.6 and the frequency of allele B_2 is 0.4 (Figure 5.9). After random mating, we get genotype frequencies for B_1B_1, B_1B_2, and B_2B_2, of 0.36, 0.48, and 0.16. The rest of our calculations will be simpler if we give the population of zygotes a finite size, so imagine that there are 1,000 zygotes:

B_1B_1	B_1B_2	B_2B_2
360	480	160

These zygotes are represented by a bar graph in the figure. We will follow the individuals that develop from these zygotes as they grow to adulthood. Those that survive will breed to produce the next generation's gene pool.

We incorporate selection by stipulating that the genotypes differ in their rates of survival. All of the B_1B_1 individuals survive, 75% of the B_1B_2 individuals survive, and 50% of the B_2B_2 individuals survive. As shown in Figure 5.9, there are now 800 adults in the population:

B_1B_1	B_1B_2	B_2B_2
360	360	80

The frequencies of the three genotypes among the adults are as follows:

B_1B_1	B_1B_2	B_2B_2
360/800	360/800	80/800
= 0.45	= 0.45	= 0.1

When these adults produce gametes, the frequency of allele B_1 in the new gene pool is equal to the frequency of B_1B_1 among the surviving adults, plus half the

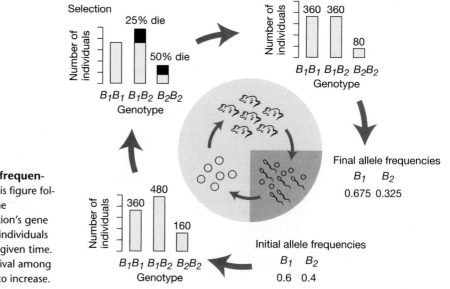

Figure 5.9 Selection can cause allele frequencies to change across generations This figure follows our model mouse population from one generation's gene pool to the next generation's gene pool. The bar graphs show the number of individuals of each genotype in the population at any given time. Selection, in the form of differences in survival among juveniles, causes the frequency of allele B_1 to increase.

frequency of B_1B_2. This calculation, and that for the frequency of allele B_2, are as follows:

$$
\begin{array}{cc}
\boldsymbol{B_1} & \boldsymbol{B_2} \\
0.45 + \left(\dfrac{1}{2}\right)0.45 & \left(\dfrac{1}{2}\right)0.45 + 0.1 \\
= 0.675 & = 0.325
\end{array}
$$

The frequency of allele B_1 has risen by an increment of 7.5 percentage points. The frequency of allele B_2 has dropped by the same amount.

Violation of the no-selection assumption has resulted in violation of conclusion 1 of the Hardy–Weinberg analysis. The population has evolved in response to selection.

We used strong selection to make a point in our numerical example. Rarely in nature are differences in survival rates large enough to cause such dramatic change in allele frequencies in a single generation. If selection continues for many generations, however, even small changes in allele frequency in each generation can add up to substantial changes over the long run. Figure 5.10 illustrates the cumulative change in allele frequencies that can be wrought by selection. The figure is based on a model population similar to the one we used in the preceding numerical example, except that the initial allele frequencies are 0.01 for B_1 and 0.99 for B_2. The yellow line shows the change in allele frequencies when the survival rates are 100% for B_1B_1, 90% for B_1B_2, and 80% for B_2B_2. The frequency of allele B_1 rises from 0.01 to 0.99 in less than 100 generations. Under weaker selection schemes, the frequency of B_1 rises more slowly, but still inexorably. (See Box 5.3 for a general algebraic treatment incorporating selection into the Hardy–Weinberg analysis.)

A numerical example shows that when individuals with some genotypes survive at higher rates than individuals with other genotypes, allele frequencies can change from one generation to the next. In other words, our model shows that natural selection causes evolution.

Empirical Research on Allele Frequency Change by Selection

Douglas Cavener and Michael Clegg (1981) documented a cumulative change in allele frequencies over many generations in a laboratory-based natural selection experiment on the fruit fly (*Drosophila melanogaster*). Fruit flies, like most other animals, make an enzyme that breaks down ethanol, the poisonous active ingredient in beer, wine, and rotting fruit. This enzyme is called alcohol dehydrogenase, or ADH. Cavener and Clegg worked with populations of flies that had two alleles at the ADH locus: Adh^F and Adh^S. (The *F* and *S* refer to whether the protein encoded by the allele moves quickly or slowly through an electrophoresis gel [see Box 4.1 in Chapter 4]).

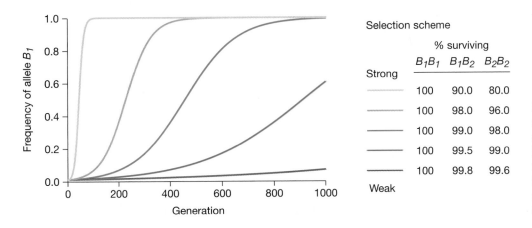

Figure 5.10 Persistent selection can produce substantial changes in allele frequencies over time Each curve shows the change in allele frequency over time under a particular selection intensity.

BOX 5.3   A general treatment of selection

Here we develop equations that predict allele frequencies in the next generation, given allele frequencies in this generation and the fitnesses of the different genotypes. We start with a population that has a gene pool in which allele A_1 is at frequency p and allele A_2 is at frequency q. We allow gametes to pair at random to make zygotes of genotypes A_1A_1, A_1A_2, and A_2A_2 at frequencies p^2, $2pq$, and q^2, respectively.

We incorporate selection by imagining that A_1A_1 zygotes survive to adulthood at rate w_{11}, A_1A_2 zygotes survive at rate w_{12}, and A_2A_2 zygotes survive at rate w_{22}. All individuals that survive produce the same number of offspring. Therefore, a genotype's survival rate is proportional to the genotype's lifetime reproductive success, or fitness. We thus refer to the survival rates as fitnesses. The average fitness for the whole population, $\overline{w}$, is given by the expression:

$$\overline{w} = p^2 w_{11} + 2pq w_{12} + q^2 w_{22}$$

[To see this, note that we can calculate the average of the numbers 1, 2, 2, and 3 as $\frac{(1 + 2 + 2 + 3)}{4}$ or as $(\frac{1}{4} \times 1) + (\frac{1}{2} \times 2) + (\frac{1}{4} \times 3)$. Our expression for the average fitness is of the second form: We multiply the fitness of each genotype by its frequency in the population and then sum the results.]

We now calculate the genotype frequencies among the surviving adults (right before their gametes go into the gene pool). The new frequencies of the genotypes are

A_1A_1	A_1A_2	A_2A_2
$\dfrac{p^2 w_{11}}{\overline{w}}$	$\dfrac{2pq w_{12}}{\overline{w}}$	$\dfrac{q^2 w_{22}}{\overline{w}}$

(We have to divide by the average fitness in each case to ensure that the new frequencies still sum to 1.)

Finally, we let the adults breed, and calculate the allele frequencies in the gene pool:

• For the A_1 allele: A_1A_1 individuals contribute $\frac{p^2 w_{11}}{\overline{w}}$ of the gametes, all of them A_1; A_1A_2 individuals contribute $\frac{2pq w_{12}}{\overline{w}}$ of the gametes, half of them A_1. So the new frequency of A_1 is

$$\frac{p^2 w_{11} + pq w_{12}}{\overline{w}}$$

• For the A_2 allele: A_1A_2 individuals contribute $\frac{2pq w_{12}}{\overline{w}}$ of the gametes, half of them A_2; A_2A_2 individuals contribute $\frac{q^2 w_{22}}{\overline{w}}$ of the gametes, all of them A_2. So the new frequency of A_2 is

$$\frac{pq w_{12} + q^2 w_{22}}{\overline{w}}$$

Readers should confirm that the new frequencies of A_1 and A_2 sum to 1.

It is instructive to calculate the change in the frequency of allele A_1 from one generation to the next. This value, Δp, is the new frequency of allele A_1 minus the old frequency of A_1:

$$\Delta p = \frac{p^2 w_{11} + pq w_{12}}{\overline{w}} - p$$

$$= \frac{p^2 w_{11} + pq w_{12}}{\overline{w}} - \frac{p\overline{w}}{\overline{w}}$$

$$= \frac{p^2 w_{11} + pq w_{12} - p\overline{w}}{\overline{w}}$$

$$= \frac{p}{\overline{w}}(p w_{11} + q w_{12} - \overline{w})$$

The final expression is a useful one, because it shows that the change in frequency of allele A_1 is proportional to $(p w_{11} + q w_{12} - \overline{w})$. The expression $(p w_{11} + q w_{12} - \overline{w})$ is equal to the average fitness of allele A_1 when paired at random with other alleles $(p w_{11} + q w_{12})$ minus the average fitness of the population $(\overline{w})$. In other words, if the average A_1-carrying individual has higher-than-average fitness, then allele A_1 will increase in frequency.

The change in the frequency of allele A_2 from one generation to the next is

$$\Delta q = \frac{pq w_{12} + q^2 w_{22}}{\overline{w}} - q$$

$$= \frac{q}{\overline{w}}(p w_{12} + q w_{22} - \overline{w})$$

The scientists maintained two experimental populations of flies on food spiked with ethanol, and two control populations of flies on normal, nonspiked food. The researchers picked the breeders for each generation at random. This is why we are calling the project a natural selection experiment: Cavener and Clegg set up different environments for their different populations, but the researchers did not themselves directly manipulate the survival or reproductive success of individual flies.

Every several generations, Cavener and Clegg took a random sample of flies from each population, determined their ADH genotypes, and calculated the allele frequencies. The results appear in Figure 5.11. The control populations showed no large or consistent long-term change in the frequency of the Adh^S allele. The experimental populations, in contrast, showed a rapid and largely consistent decline in the frequency of Adh^S (and, of course, a corresponding increase in the frequency of Adh^F). Hardy–Weinberg conclusion 1 appears to hold in the control populations, but is clearly not in force in the experimental populations.

Empirical research on fruit flies is consistent with our conclusion that natural selection can cause allele frequencies to change.

Can we identify for certain which of the assumptions of the Hardy–Weinberg analysis is being violated? The only difference between the two kinds of populations is that the experimentals have ethanol in their food. This suggests that it is the assumption of no selection that is being violated in the experimental populations. Flies with the Adh^F allele appear to have higher lifetime reproductive success (higher fitness) than flies with the Adh^S allele when ethanol is present in the food. Cavener and Clegg note that this outcome is consistent with the fact that alcohol dehydrogenase extracted from Adh^F homozygotes breaks down ethanol at twice the rate of alcohol dehydrogenase extracted from Adh^S homozygotes. Whether flies with the Adh^F allele have higher fitness because they have higher rates of survival or because they produce more offspring is unclear.

Adding Selection to the Hardy–Weinberg Analysis: The Calculation of Genotype Frequencies

The calculations and example we have just discussed show that selection can cause allele frequencies to change across generations. Selection invalidates conclusion 1 of the Hardy–Weinberg analysis. We now consider how selection affects conclusion 2 of the Hardy–Weinberg analysis. In a population under selection, can we still calculate the genotype frequencies by multiplying the allele frequencies?

We often cannot. As before, we use a population with two alleles at a locus affecting survival: B_1 and B_2. We assume that the initial frequency of each allele in

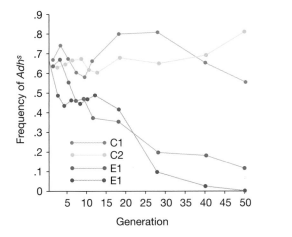

Figure 5.11 **Frequencies of the Adh^S allele in four populations of fruit flies over 50 generations** The red and yellow dots and lines represent control populations living on normal food; the blue and green dots and lines represent experimental populations living on food spiked with ethanol. From Cavener and Clegg (1981). Copyright © Evolution. Reprinted by permission.

the gene pool is 0.5 (Figure 5.12). After random mating, we get genotype frequencies for B_1B_1, B_1B_2, and B_2B_2, of 0.25, 0.5, and 0.25. The rest of our calculations will be simpler if we give the population of zygotes a finite size, so imagine that there are 1000 zygotes:

B_1B_1	B_1B_2	B_2B_2
250	500	250

These zygotes are represented by a bar graph in the figure. We will follow the individuals that develop from these zygotes as they grow to adulthood. Those that survive will breed to produce the next generation's gene pool.

We incorporate selection by stipulating that the genotypes differ in their rates of survival. Fifty percent of the B_1B_1 individuals survive, all of the B_1B_2 individuals survive, and 50% of the B_2B_2 individuals survive. As shown in Figure 5.12, there are now 750 adults in the population:

B_1B_1	B_1B_2	B_2B_2
125	500	125

The frequencies of the three genotypes among the adults are as follows:

B_1B_1	B_1B_2	B_2B_2
125/750	500/750	125/750
= 0.167	= 0.667	= 0.167

When these adults produce gametes, the frequencies of the two alleles in the new gene pool are

$$
\begin{array}{cc}
B_1 & B_2 \\
0.167 + \left(\dfrac{1}{2}\right)0.667 & \left(\dfrac{1}{2}\right)0.667 + 0.167 \\
= 0.5 & = 0.5
\end{array}
$$

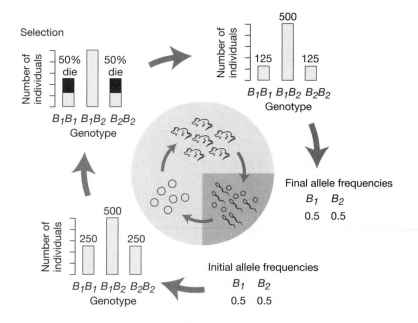

Figure 5.12 Selection can change genotype frequencies so that they cannot be calculated by multiplying the allele frequencies When half of the homozygotes in this population die, the allele frequencies do not change. But among the survivors, there are more heterozygotes then predicted under Hardy–Weinberg equilibrium.

In spite of strong selection against homozygotes, the frequencies of the alleles have not changed; the population has not evolved.

Note, however, that violation of the no-selection assumption *has* resulted in violation of conclusion 2 of the Hardy–Weinberg analysis. We can no longer calculate the genotype frequencies among the adult survivors by multiplying the frequencies of the alleles. For example:

$$\text{Frequency of } B_1B_1 \qquad (\text{Frequency of } B_1)^2$$
$$0.167 \qquad \neq \qquad (0.5)^2 = 0.25$$

We used strong selection in our numerical example to make a point. In fact, selection is rarely strong enough to produce, in a single generation, such a large violation of Hardy–Weinberg conclusion 2. Even if it does, a single bout of random mating will immediately put the genotypes back into Hardy–Weinberg equilibrium. Nonetheless, researchers sometimes find violations of Hardy–Weinberg conclusion 2 that seem to be the result of selection.

Natural selection can also drive genotype frequencies away from the values predicted under the Hardy–Weinberg equilibrium principle.

Empirical Research on Selection and Genotype Frequencies

Our example comes from research on a serious human genetic disease known as Jaeken syndrome (also known as carbohydrate-deficient-glycoprotein syndrome type 1). Patients with Jaeken sydrome are severely developmentally disabled, with skeletal deformities and an abnormal distribution of subcutaneous fat (Matthijs et al. 1998). They have inadequate liver function, and 20% die before the age of five.

Jaeken syndrome is inherited as an autosomal recessive condition. Most cases are caused by loss-of-function mutations in a gene on chromosome 16. The gene, called *PMM2,* encodes an enzyme called phosphomannomutase (PMM). Biochemically, reduced PMM activity results in an insufficient ability to attach carbohydrates to proteins to make glycoproteins.

There are at least 24 different loss-of-function mutations in the *PMM2* gene that can cause Jaeken sydrome. Most are missense mutations; one is a single-base-pair deletion. Gert Matthijs and colleagues (1998), in research led by Jaak Jaeken, investigated whether some of these loss-of-function mutations are more serious than others. The researchers hypothesized that some mutations preserve less of PMM's catalytic activity than others, and are therefore especially severe.

The researchers performed a test of their hypothesis that was based on the Hardy–Weinberg equilibrium principle. The logic of the test runs as follows. Individuals with Jaeken syndrome are all homozygotes, in the sense that they all carry two mutant alleles. However, many carry two different loss-of-function mutations. We can think of affected individuals as being homozygous or heterozygous with respect to the specific disease alleles they carry. Within the affected individuals in a population, the different disease alleles should be in Hardy–Weinberg equilibrium with each other (see Box 5.4.). If a particular disease allele causes an especially severe loss of function, then individuals homozygous for that allele might have so little PMM activity that they would be unable to survive. Because selection would remove such homozygotes from the affected population, there will be fewer of them than we would predict by multiplying the frequencies of the different disease alleles.

The researchers determined the identity of the disease alleles present in 54 caucasian patients with Jaeken syndrome. The most common mutation they found was

---

BOX 5.4   Hardy–Weinberg equilibrium among different mutant alleles that cause a recessive genetic disease

Consider an autosomal recessive genetic disease. At the first level of analysis, there are two alleles, which we will call D and d. Individuals with genotype DD or Dd are phenotypically normal; dd individuals have the disease.

Let p represent the frequency of allele D, and let z represent the frequency of allele d, such that $p + z = 1$. In a population obeying the Hardy–Weinberg assumptions, the genotype frequencies are

DD	Dd	dd
p^2	$2pz$	z^2

Now, imagine that there are, in fact, two different disease alleles, d_1 and d_2. In other words, dd individuals can have genotype d_1d_1, d_1d_2, or d_2d_2. Let q represent the frequency of d_1, and let r represent the

frequency of allele d_2. Because d_1 and d_2 are the two possible versions of d, we know that

$$z = q + r$$

Thus, the frequency of affected individuals can be expressed as

$$\text{Frequency of } dd = z^2$$
$$= (q + r)^2$$
$$= q^2 + 2qr + r^2$$

The terms of this expression are the frequencies that we would predict from the Hardy–Weinberg equilibrium principle simply by multiplying the frequencies of d_1 and d_2. In other words, within the population of affected individuals, the different disease alleles are in Hardy–Weinberg equilibrium with each other.

an amino acid substitution called *R141H*. If we divide the disease alleles into those containing *R141H* and those containing other mutations, the allele frequencies in the population of 54 affected individuals are

Other	*R141H*
0.6	0.4

From these allele frequencies, we can calculate that if the population is in Hardy–Weinberg equilibrium, then the genotype frequencies will be

Other/Other	*Other/R141H*	*R141H/R141H*
0.36	0.48	0.16

The discovery that genotype frequencies in a population are not in Hardy–Weinberg equilibrium may be a clue that natural selection is at work.

In fact, the actual genotype frequencies among the affected individuals are

Other/Other	*Other/R141H*	*R141H/R141H*
$\frac{11}{54} = 0.2$	$\frac{43}{54} = 0.8$	0

The allele and genotype frequencies among the affected individuals are in violation of conclusion 2 of the Hardy–Weinberg analysis. There is a striking deficit of homozygotes, most notably *R141H* homozygotes. The deficit of homozygotes is statistically significant (see Box 5.5.).

Jaeken and his team believe the most plausible interpretation for the absence of *R141H* homozygotes is that *R141H* is an especially severe loss-of-function mutation. Under this hypothesis, zygotes with genotype *R141H/R141H* are produced at the frequency predicted under the Hardy–Weinberg equilibrium principle, but the individuals that develop from these zygotes die before or shortly after birth. The deficit of *Other/Other* homozygotes may be due to additional severe loss-of-function mutations among the other disease alleles.

BOX 5.5 Statistical analysis of allele and genotype frequencies using the χ^2 (chi-square) test

Here we use the data from Matthijs and colleagues (1998) to illustrate one method for determining whether genotype frequencies deviate significantly from what they would be under Hardy–Weinberg equilibrium. The researchers surveyed a population of 54 individuals affected with Jaeken syndrome. The numbers of individuals with each genotype were as follows:

Other/Other	*Other/R141H*	*R141H/R141H*
11	43	0

From these numbers, we will calculate the allele frequencies of *Other* and *R141H,* and determine whether the genotype frequencies observed are those we would expect according to the Hardy–Weinberg equilibrium principle. There are five steps:

1. Calculate the allele frequencies. The sample of 54 individuals is also a sample of 108 alleles. All 22 of the alleles carried by the 11 *Other/Other* individuals are *Other,* as are 43 of the alleles carried by the 43 *Other/R141H* individuals. Thus, the frequency of *Other* alleles is

$$\frac{(22 + 43)}{108} = 0.6$$

and the frequency of allele *R141H* is

$$\frac{(43 + 0)}{108} = 0.4$$

2. Calculate the genotype frequencies expected under the Hardy–Weinberg principle, given the allele frequencies calculated in Step 1. According to the Hardy–Weinberg equilibrium principle, if the frequencies of two alleles are p and q, then the frequencies of the genotypes are p^2, $2pq$, and q^2. Thus, the expected frequencies of genotypes in a population of individuals with Jaeken syndrome are

Other/Other	*Other/R141H*	*R141H/R141H*
$(0.6)^2$	$2(0.6)(0.4)$	$(0.4)^2$
= 0.36	= 0.48	= 0.16

3. Calculate the expected number of individuals of each genotype under Hardy–Weinberg equilibrium. This is simply the expected frequency of each genotype multiplied by the number of individuals in the sample, 54. The expected values are

Other/Other	*Other/R141H*	*R141H/R141H*
(0.36)(54)	(0.48)(54)	(0.16)(54)
= 19.44	= 25.92	= 8.64

The numbers of individuals expected are different from the number of individuals actually observed (11, 43, and 0). The actual sample contains more heterozygotes and fewer homozygotes than expected. Is it plausible that this large a difference between expectation and reality could arise by chance? Or is the difference statistically significant? Our null hypothesis is that the difference is simply due to chance.

4. Calculate a test statistic. We will use a test statistic that was devised in 1900 by Karl Pearson. It is called the chi-square (χ^2). The chi-square is defined as

$$\chi^2 = \sum \frac{(\text{observed} - \text{expected})^2}{\text{expected}}$$

where the symbol Σ indicates a sum taken across all the classes considered. In our data there are three classes: the three genotypes. For our data set:

$$\chi^2 = \frac{(11 - 19.44)^2}{19.44} + \frac{(43 - 25.92)^2}{25.92}$$
$$+ \frac{(0 - 8.64)^2}{8.64} = 23.56$$

5. Determine whether the value of the test statistic is signficant. The chi-square is defined in such a way that the chi-square gets larger as the difference between the observed and expected values gets larger. How likely is it that we could get a chi-square as large as 23.56 by chance? Most statistical textbooks have a table that provides the answer. In Zar (1996) this table is called "Critical values of the chi-square distribution."

To use this table, we need to calculate a number called the degrees of freedom for the test statistic. The number of degrees of freedom for the chi-square is equal to the number of classes minus the number of independent values we calculated from the data for use in determining the expected values. For our chi-square there are three classes: the three genotypes. We calculated two values from the data for use in determining the expected values:

BOX 5.5   Continued

the total number of individuals, and the frequency of allele *R141H*. (We also calculated the frequency of *Other* alleles, but it is not independent of the frequency of allele *R141H*, because the frequency of *Other* is one minus the frequency of *R141H*).Thus the number of degrees of freedom is 1. (Another formula for calculating the degrees of freedom in chi-square tests for Hardy–Weinberg equilibrium is

$$df = k - 1 - m$$

where *k* is the number of classes and *m* is the number of independent allele frequencies estimated from the data).

According to the table in the statistics book, the critical value of chi-square for one degree of free- dom and *P* = 0.05 is 3.841. This means that there is a 5% chance under the null hypothesis of get- ting $\chi^2 \geq 3.841$. The probability under the null hy- pothesis of getting $\chi^2 \geq 23.56$ is therefore (considerably) less than 5%. We reject the null hy- pothesis, and assert that our value of chi-square is statistically significant at $P < 0.05$. (In fact, in this case $P < 0.00001$.)

The chi-square test tells us that the alleles of the *PMM2* locus in the population of individuals with Jaeken syndrome are not in Hardy–Weinberg equi- librium. This indicates that one or more of the as- sumptions of the Hardy–Weinberg analysis has been violated. By itself, however, it does not tell us which assumptions are being violated, or how.

This study of Jaeken syndrome is a case in which a population-genetic analy- sis led to a potentially important medical discovery. The analysis suggests that some PMM activity is necessary for life. Furthermore, it indicates that parents who are both carriers of *R141H* can expect a different distribution of phenotypes among their children than parents who are carriers of two different disease alleles.

Changes in the Frequency of the *CCR5-Δ32* Allele Revisited

Our exploration of natural selection has given us tools we can use to predict the future of human populations.

We are now in a position to give a more satisfying answer to the medical ques- tion we raised at the beginning of the chapter: Will the AIDS epidemic cause the frequency of the *CCR5-Δ32* allele to increase in human populations? The AIDS epidemic could, in principle, cause the frequency of the allele to increase rapidly, but at present it appears that it will probably not do so in any real population. This conclusion is based on the three model populations depicted in Figure 5.13 (See Box 5.6 for the algebra.) Each model is based on different assumptions about the initial frequency of the *CCR5-Δ32* allele and the prevalence of HIV infec- tion. Each graph shows the predicted change in the frequency of the *Δ32* allele over 40 generations, or approximately 1000 years of evolution.

The model population depicted in Figure 5.13a provides a scenario in which the frequency of the *Δ32* allele could increase rapidly. In this scenario, the initial frequency of the *CCR5-Δ32* allele is 20%. One quarter of the individuals with genotype +/+ or +/*Δ32* contract AIDS and die without reproducing, whereas all of the *Δ32/Δ32* individuals survive. The 20% initial frequency of *Δ32* is ap- proximately equal to the highest frequency reported for any population, a sample of Ashkenazi Jews studied by Martinson et al. (1997). The mortality rates approx- imate the situation in Botswana, Namibia, Swaziland, and Zimbabwe, where up to 25% of individuals between the ages of 15 and 49 are infected with HIV (UN- AIDS 1998). In this model population, the frequency of the *Δ32* allele increases by as much as a few percentage points each generation. By the end of 40 generations, the allele is at a frequency of virtually 100%. Thus, in a human population that combined the highest reported frequency of the *Δ32* allele with the highest reported rates of infection, the AIDS epidemic could cause the frequency of the allele to increase rapidly.

At present, however, no known population combines a high frequency of the *Δ32* allele with a high rate of HIV infection. In northern Europe, many populations have *Δ32* frequencies between 0.1 and 0.2 (Martinson et al. 1997; Stephens et al. 1998), but HIV infection rates are well under 1% (UNAIDS 1998). A model population reflecting these conditions is depicted in Figure 5.13b. The initial frequency of the *Δ32* allele is 0.2, and 0.5% of the *+/+* and *+/Δ32* individuals contract AIDS and die without reproducing. The frequency of the *Δ32* allele hardly changes at all. Selection is too weak to cause appreciable evolution in such a short time.

In parts of sub-Saharan Africa, as many as a quarter of all individuals of reproductive age are infected with HIV. However, the *Δ32* allele is virtually absent (Martinson et al. 1997). A model population reflecting this situation is depicted in Figure 5.13c. The initial frequency of the *Δ32* allele is 0.01, and 25% of the *+/+* and *+/Δ32* individuals contract AIDS and die without reproducing. Again, the frequency of the *Δ32* allele hardly changes at all. When the *Δ32* allele is at low frequency, most copies are in heterozygotes. Because heterozygotes are susceptible to infection, these copies are hidden from selection.

The analysis we have just described is based on a number of simplifying assumptions. For example, we have assumed that all HIV-infected individuals die without reproducing, and that the death rate is the same in heterozygotes as in *+/+* homozygotes. In reality, although heterozygotes are susceptible to HIV infection, they appear to progress more slowly to AIDS (Dean et al. 1996). As a result, the fitness of heterozygotes may actually be higher than that of *+/+* homozygotes. We challenge our readers to explore the evolution of human populations under a variety of selection schemes, to see how strongly our simplifying assumptions affect the predicted course of evolution.

5.3 Patterns of Selection

At the start of the chapter we asked whether a compulsory sterilization law could achieve its eugenic goal of reducing the incidence of a genetic disease. Before attempting to provide an answer, it will be helpful to consider how the details of Mendelian genetics at a particular locus might affect the course of evolution. For example, does the rate of evolution in response to selection against a particular allele depend on whether the allele is dominant or recessive? Does the rate of evolution depend on whether the allele is common or rare?

Selection on Recessive and Dominant Alleles

Data collected by Peter Dawson (1970) illustrate how recessiveness and dominance affect the course of evolution. Dawson had been studying a laboratory colony of flour beetles (*Tribolium castaneum*), and had identified a gene we will call the *l* locus. This locus has two alleles: *+* and *l*. Individuals with genotype *+/+* or *+/l* are phenotypically normal, whereas individuals with genotype *l/l* do not survive. In other words, *l* is a recessive lethal allele.

Dawson collected heterozygotes from his flour beetle colony and used them to establish two new experimental populations. Because all the founders were heterozygotes,

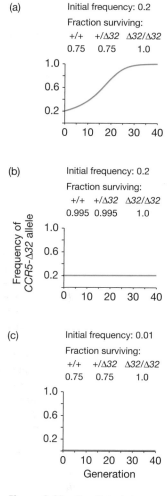

Figure 5.13 Predicted change in allele frequencies at the CCR5 locus due to the AIDS epidemic under three different scenarios (a) When the initial frequency of the *CCR5-Δ32* allele is high and a large fraction of the population becomes infected with HIV, the allele frequencies can change rapidly. However, no real population combines these characteristics. (b) In European populations allele frequencies are high, but only a small fraction of individuals become infected. (c) In parts of Africa there are high infection rates, but allele frequencies are low.

BOX 5.6 Predicting the frequency of the *CCR5-Δ32* allele in future generations

Let q_g be the frequency of the *CCR5-Δ32* allele in the present generation. Based on Box 5.3, we can write an equation predicting the frequency of the allele in the next generation, given estimates of the survival rates (fitnesses) of individuals with each genotype. The equation is

$$q_{g+1} = \frac{(1 - q_g)q_g w_{+\Delta} + q_g^2 w_{\Delta\Delta}}{(1 - q_g)^2 w_{++} + 2(1 - q_g)q_g w_{+\Delta} + q_g^2 w_{\Delta\Delta}}$$

where q_{g+1} is the frequency of the *Δ32* allele in the next generation, w_{++} is the fitness of individuals

homozygous for the normal allele, $w_{+\Delta}$ is the fitness of heterozygotes, and $w_{\Delta\Delta}$ is the fitness of individuals homozygous for the *CCR5-Δ32* allele.

After choosing a starting value for the frequency of the *Δ32* allele, we plug it and the estimated fitnesses into the equation to generate the frequency of the *Δ32* allele after one generation. We then plug this resulting value into the equation to get the frequency of the allele after two generations, and so on.

Empirical research on flour beetles shows that predictions made with population genetic models are accurate, at least under laboratory conditions.

the initial frequency of the two alleles was 0.5 in both populations. Because *l/l* individuals have zero fitness, Dawson expected his populations to evolve toward ever lower frequencies of the *l* allele and ever higher frequencies of the *+* allele. Dawson used the equations derived in Box 5.3 and the method described in Box 5.6 to make a quantitative prediction of the course of evolution. He then let his two populations evolve for a dozen generations, each generation measuring the frequencies of the two alleles.

The results appear in Figure 5.14. Dawson's data match his theoretical predictions closely. At first, the frequency of the recessive lethal allele declined rapidly (Figure 5.14a). By the second generation it had dropped from 0.5 to about 0.25. As evolu-

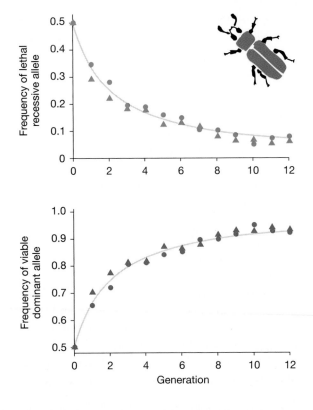

Figure 5.14 Evolution in laboratory populations of flour beetles
(a) The decline in frequency of a lethal recessive allele (red symbols) matches the theoretical prediction (gray curve) almost exactly. As the allele becomes rare, the rate of evolution slows dramatically. (b) This graph plots the increase in frequency of the corresponding dominant allele. Redrawn from Dawson (1970).

tion progressed, however, further reductions in the frequency of the lethal allele came more slowly. Between generations 10 and 12 the frequency did not drop at all. Dawson's experiment demonstrates that the theory of population genetics, in spite of its simplifying assumptions, allows us to accurately predict the course of evolution—at least under controlled conditions in the lab.

The experiment also shows that dominance and allele frequency interact to determine the rate of evolution. When a recessive allele is common (and a dominant allele is rare), evolution by natural selection is rapid. In contrast, when a recessive allele is rare, and a dominant allele is common, evolution by natural selection is slow.

The Hardy–Weinberg equilibrium principle explains why. First imagine a recessive allele that is common: Its frequency is, say, 0.95. The dominant allele thus has a frequency of 0.05. By multiplying the allele frequencies, we can calculate the genotype frequencies:

AA	**Aa**	**aa**
$(0.05)^2$	$2(0.05)(0.95)$	$(0.95)^2$
$= 0.0025$	$= 0.095$	$= 0.9025$

Roughly 10% of the individuals in the population have the dominant phenotype, while 90% have the recessive phenotype. Both phenotypes are reasonably well represented, and if they differ in fitness then the allele frequencies in the next generation may be substantially different. Now imagine a recessive allele that is rare: Its frequency is 0.05. The dominant allele thus has a frequency of 0.95. The genotype frequencies are

AA	**Aa**	**aa**
$(0.95)^2$	$2(0.95)(0.05)$	$(0.05)^2$
$= 0.9025$	$= 0.095$	$= 0.0025$

Approximately 100% of the population has the dominant phenotype, while approximately 0% has the recessive phenotype. Even if the phenotypes differ greatly in fitness, there are so few of the minority phenotype that there will be little change in allele frequencies in the next generation. In a random mating population, most copies of a rare recessive allele are phenotypically hidden inside heterozygous individuals. For an algebraic treatment of selection on recessive and dominant alleles see Box 5.7.

Selection on Heterozygotes and Homozygotes

When one allele is recessive and the other is dominant, the fitness of heterozygotes is equal to that of one kind of homozygote. Other scenarios are possible, of course. Often the fitness of heterozygotes is between that of the two homozygotes. This may change the rate of evolution, but it does not alter the ultimate outcome. Eventually, one allele becomes fixed in the population and the other is lost. Figure 5.10 shows several examples.

It is also possible for the fitness of heterozygotes to be superior or inferior to that of either homozygote. Heterozygote superiority and inferiority produce dramatically different outcomes.

Our first example comes from research on laboratory populations of fruit flies (*Drosophila melanogaster*) by Terumi Mukai and Allan Burdick (1959). Like Dawson, Mukai and Burdick studied evolution at a single locus with two alleles. Homozygotes for one allele were viable, whereas homozygotes for the other allele were

Natural selection is most potent as a force of evolution when it is acting on common recessive alleles (and rare dominant alleles).

BOX 5.7   An algebraic treatment of selection on recessive and dominant alleles

Here we develop equations that illuminate the differences between selection on recessive versus dominant alleles. Imagine a single locus with two alleles. Let p be the frequency of the dominant allele A, and let q be the frequency of the recessive allele a.

Selection on the recessive allele

Let the fitnesses of the genotypes be given by:

w_{AA}	w_{Aa}	w_{aa}
1	1	$1 + s$

where s, called the **selection coefficient**, gives the strength of selection on homozygous recessives relative to the other genotypes. Positive values of s represent selection in favor of the recessive allele; negative values of s represent selection against the recessive allele.

Based on Box 5.3, the following equation gives the frequency of allele a in the next generation, q', given the frequency of a in this generation and the fitnesses of the three genotypes:

$$q' = \frac{pqw_{Aa} + q^2w_{aa}}{\overline{w}} = \frac{pqw_{Aa} + q^2w_{aa}}{p^2w_{AA} + 2pqw_{Aa} + q^2w_{aa}}$$

Substituting the fitness values from the table above, and $(1 - q)$ for p, and then simplifying gives

$$q' = \frac{q(1 + sq)}{1 + sq^2}$$

If a is a lethal recessive, then s is equal to -1. Substituting this value into the preceding equation gives

$$q' = \frac{q(1 - q)}{1 - q^2} = \frac{q(1 - q)}{(1 - q)(1 + q)} = \frac{q}{(1 + q)}$$

A little experimentation shows that once a recessive lethal allele becomes rare, further declines in frequency are slow. For example, if the frequency of allele a in this generation is 0.01, then in the next generation its frequency will be approximately 0.0099.

Selection on the dominant allele

Let the fitnesses of the genotypes be given by:

w_{AA}	w_{Aa}	w_{aa}
$1 + s$	$1 + s$	1

where s, the selection coefficient, gives the strength of selection on genotypes containing the dominant

allele relative to homozygous recessives. Positive values of s represent selection in favor of the dominant allele; negative values of s represent selection against the dominant allele.

Based on Box 5.3, we can write an equation that predicts the frequency of allele A in the next generation, p', given the frequency of A in this generation and the fitnesses of the three genotypes:

$$p' = \frac{p^2w_{AA} + pqw_{Aa}}{\overline{w}} = \frac{p^2w_{AA} + pqw_{Aa}}{p^2w_{AA} + 2pqw_{Aa} + q^2w_{aa}}$$

Substituting the fitnesses from the table, and $(1 - p)$ for q, and then simplifying gives

$$p' = \frac{p(1 + s)}{1 + 2sp - sp^2}$$

If A is a lethal dominant, s is equal to -1. Substituting this value into the foregoing equation shows that a lethal dominant is eliminated from a population in a single generation.

Selection on recessive alleles versus selection on dominant alleles

Selection on recessive alleles and selection on dominant alleles are opposite sides of the same coin. Selection against a recessive allele is selection in favor of the dominant allele and vice versa.

Figure 5.15a (left) shows 100 generations of evolution in a model population under selection against a recessive allele and in favor of the dominant allele. At first, the frequencies of the alleles change rapidly. As the recessive allele becomes rare, however, the rate of evolution slows dramatically. When the recessive allele is rare, most copies in the population are in heterozygous individuals, where they are effectively hidden from selection.

The figure also shows (right) the mean fitness of the population (see Box 5.3) as a function of the frequency of the dominant allele. As the dominant allele goes from rare to common, the mean fitness of the population rises. Mean fitness is maximized when the favored allele reaches a frequency of 100%. Graphs of mean fitness as a function of allele frequency are often referred to as adaptive landscapes.

BOX 5.7   **Continued**

Figure 5.15b (left) shows 100 generations of evolution in a model population under selection in favor a recessive allele and against the dominant allele. At first, the frequencies of the alleles change slowly. The recessive allele is rare, most copies present are in heterozygotes, and selection cannot see it. However, as the recessive allele becomes common enough that a substantial fraction of homozygotes appear, the rate of evolution increases dramatically. Once the pace of evolution accelerates, the favorable recessive allele quickly achieves a frequency of 100%. That is, the recessive allele becomes fixed in the population.

The figure also shows (right) the mean fitness of the population (see Box 5.3) as a function of the frequency of the recessive allele. As the recessive allele goes from rare to common, the mean fitness of the population rises. Mean fitness is maximized when the favored allele reaches a frequency of 100%.

(a) Selection against a recessive allele and for a dominant allele

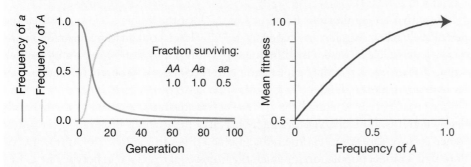

(b) Selection for a recessive allele and against a dominant allele

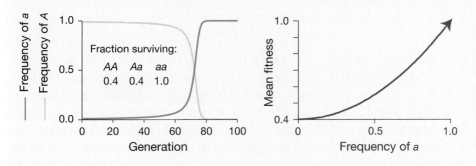

Figure 5.15   Evolution in model populations under selection on recessive and dominant alleles Graphs on the left show changes in allele frequencies over time. Graphs on the right show adaptive landscapes: changes in population mean fitness as a function of allele frequencies.

not. The researchers used heterozygotes as founders to establish two experimental populations with initial allele frequencies of 0.5. They allowed the populations to evolve for 15 generations, each generation measuring the frequency of the viable allele.

Mukai and Burdick's results appear in Figure 5.16, represented by the red symbols. As expected, the frequency of the viable allele increased rapidly over the first few generations. However, the rate of evolution slowed long before the viable allele approached a frequency of 1.0. Instead, the viable allele seemed to reach an equilibrium, or unchanging state, at a frequency of about 0.79.

To investigate further, Mukai and Burdick established two more experimental populations, this time with the initial frequency of the viable allele at 0.975. Evolution in these populations is represented by the blue symbols in Figure 5.16. Instead

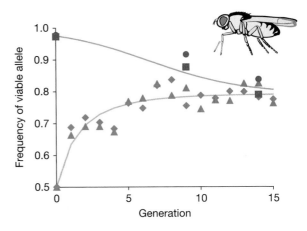

Figure 5.16   Evolution in four laboratory populations of fruit flies   In homozygous state, one allele is viable and the other is lethal. Nonetheless, all four populations evolved toward an equilibrium in which both alleles are maintained. The likely explanation is that heterozygotes enjoy superior fitness to either homozygote. Drawn from data presented in Mukai and Burdick (1958).

Research on fruit flies shows that natural selection can act to maintain two alleles at a stable equilibrium. One way this can happen is when heterozygotes have superior fitness.

of rising toward 1.0, the frequency of the viable allele dropped, again reaching an equilibrium near 0.79.

Note that an equilibrium frequency of 0.79 for the viable allele means that the lethal allele has an equilbrium frequency of 0.21. How could natural selection maintain a lethal allele at such a high frequency in this population? Mukai and Burdick argue that the most plausible explanation is **heterozygote superiority**, also known as **overdominance**. Under this hypothesis, heterozygotes have higher fitness than either homozygote. At equilibrium, the selective advantage enjoyed by the lethal allele when it is in heterozygotes exactly balances the obvious disadvantage it suffers when it is in homozygotes. The red and blue curves in Figure 5.16 represent evolution in a model population in which the fitnesses of the three genotypes are as follows (V represents the viable allele; L represents the lethal allele):

VV	VL	LL
0.735	1.0	0

The theoretical curves match the data closely. By keeping a population at an equibrium in which both alleles are present, heterozygote superiority can maintain genetic diversity. For an algebraic treatment of heterozygote superiority, see Box 5.8.

Our second example, from work by G. G. Foster and colleagues (1972), demonstrates how populations evolve when heterozygotes have lower fitness than either homozygote. Foster and colleagues used fruit flies with compound chromosomes. Compound chromosomes are homologous chromosomes that have swapped entire arms, so that one homolog has two copies of one arm, and the other homolog has two copies of the other arm (Figure 5.17a,b). During meiosis, compound chromosomes may or may not segregate. As a result, four kinds of gametes are produced in equal numbers: gametes with both homologous chromosomes, gametes with just one member of the pair, gametes with the other member of the pair, and gametes with neither member of the pair (Figure 5.17c). When two flies with compound chromosomes mate with each other, one quarter of their zygotes have every chromosome arm in the correct dose and are viable. The other three quarters have too many or too few of copies of one or both chromosome arms, and are inviable (Figure 5.17d). When a fly with compound chromosomes mates with a fly with normal chromosomes, none of the zygotes they make are viable (Figure 5.17e).

It is also possible for heterozygotes to have inferior fitness.

Foster and colleagues established laboratory populations in which some of the founders had compound second chromosomes [*C(2)*] and others had normal second chromosomes [*N(2)*]. For purposes of analysis, we can treat each chromo-

(a) A normal pair of homologous chromosomes (each has one blue arm and one green arm).

(b) A pair of compound chromosomes (one has two blue arms, the other has two green arms).

(c) Gametes made by an individual with compound chromosomes may contain both chromosomes, one, or neither.

(d) When individuals with compound chromosomes mate, one quarter of their zygotes are viable.

(e) When an individual with compound chromosomes mates with an individual with normal chromosomes, none of their zygotes are viable.

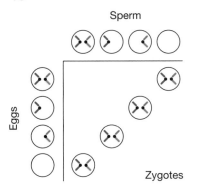

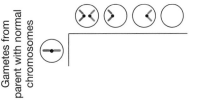

(f) Evolution in 13 populations of *Drosophila melanogaster* containing a mixture of compound second chromosomes [*C(2)*] and normal second chromosomes [*N(2)*]. The initial frequency of *C(2)* ranged from 0.71 to 0.96.

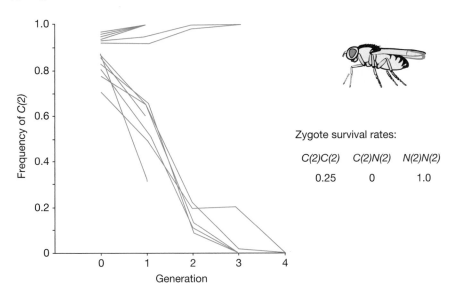

Zygote survival rates:

C(2)C(2)	C(2)N(2)	N(2)N(2)
0.25	0	1.0

Figure 5.17 **An experiment designed to show how populations evolve when heterozgyotes have lower fitness than either homozygote** The experimental design makes clever use of compound chromosomes. Redrawn with permission from Foster et al. (1972). Copyright © 1972, American Association for the Advancement of Science.

some as though it were a single allele. Thus, the founders consisted of *C(2)C(2)* homozygotes and *N(2)N(2)* homozygotes. Based on the zygote viablities we just described, the fitnesses of the genotypes in the mixed population are

C(2)C(2)	**C(2)N(2)**	**N(2)N(2)**
0.25	0	1.0

In other words, the genotypes exhibit strong **underdominance**.

BOX 5.8   Stable equilibria with heterozygote superiority and unstable equilibria with heterozygote inferiority

Here we develop algebraic and graphical methods for analyzing evolution at loci with overdominance and underdominance. Imagine a population in which allele A_1 is at frequency p and allele A_2 is at frequency q. In Box 5.3, we developed an equation describing the change in p from one generation to the next under selection:

$$\Delta p = \frac{p}{\overline{w}}(pw_{11} + qw_{12} - \overline{w})$$

$$= \frac{p}{\overline{w}}(pw_{11} + qw_{12} - p^2w_{11} - 2pqw_{12} - q^2w_{22})$$

Substituting $(1 - q)$ for p in the first and third terms in the expression in parentheses gives

$$\Delta p = \frac{p}{\overline{w}}((1 - q)w_{11} + qw_{12}$$
$$- (1 - q)^2w_{11} - 2pqw_{12} - q^2w_{22})$$

which, after simplifying and factoring out q, becomes

$$\Delta p = \frac{pq}{\overline{w}}(w_{12} + w_{11} - qw_{11} - 2pw_{12} - qw_{22})$$

Now, by definition, the frequency of allele A_1 is at equilibrium when $\Delta p = 0$. The equation above shows that $\Delta p = 0$ when $p = 0$ or $q = 0$. These two equilibria are unsurprising. They occur when one allele or the other is absent from the population. The equation also gives a third condition for equilibrium, which is

$$w_{12} + w_{11} - qw_{11} - 2pw_{12} - qw_{22} = 0$$

Substituting $(1 - p)$ for q and solving for p gives

$$\hat{p} = \frac{w_{22} - w_{12}}{w_{11} - 2w_{12} + w_{22}}$$

where $\hat{p}$ is the frequency of allele A_1 at equilibrium. Finally, let the genotype fitnesses be as follows:

A_1A_1	A_1A_2	A_2A_2
$1 + s$	1	$1 + t$

Positive values of the selection coefficients s and t represent underdominance; negative values represent overdominance. Substituting the fitnesses into the previous equation and simplifying gives

$$\hat{p} = \frac{t}{s + t}$$

For example, when $s = -0.4$ and $t = -0.6$, heterozygotes have superior fitness, and the equilbrium frequency for allele A_1 is 0.6. When $s = 0.4$ and $t = 0.6$, heterozygotes have inferior fitness, and the equilibrium frequency for allele A_1 is also 0.6.

Another useful method for analyzing equilibria is to plot Δp as a function of p. Figure 5.18a shows such a plot for the two numerical examples we just calculated. Both curves show that $\Delta p = 0$ when $p = 0$, $p = 1$, or $p = 0.6$.

The curves in Figure 5.18a also allow us to determine whether an equilibrium is stable or unstable. Look at the red curve; it describes a locus with heterozygote superiority. Notice that when p is greater than 0.6, Δp is negative. This means that when the frequency of allele A_1 exceeds its equilibrium value, the population will move back towards equilibrium in the next generation. Likewise, when p is less than 0.6, Δp is positive. When the frequency of allele A_1 is below its equilibrium value, the population will move back towards equilibrium in the next generation. The "internal" equilibrium for a locus with heterozyote superiority is stable.

Figure 5.18b shows an adaptive landscape for a locus with heterozygote superiority. The graph plots population mean fitness as a function of the frequency of allele A_1. Mean fitness is low when A_1 is absent, and relatively low when A_1 is fixed. As the allele frequency moves from either direction towards its stable equilibrium, the population mean fitness rises to a maximum.

Now, look at the blue curve in Figure 5.18a. It describes a locus with heterozygote inferiority. If p rises even slightly above 0.6, p will continue to rise toward 1.0 in subsequent generations; if p falls even slightly below 0.6, p will continue to fall toward 0 in subsequent generations. The internal equilibrium for a locus with heterozygote inferiority is unstable.

Figure 5.18c shows an adaptive landscape for a locus with heterozygote inferiority. Population mean fitness is lowest when the frequency of allele A_1 is at its unstable internal equilibrium. As the allele frequency moves away from this equilibrium in either direction mean fitness rises.

BOX 5.8 Continued

A comparison of the adaptive landscape in Figure 5.18c with those in Figure 5.18b and Figure 5.15 offers a valuable insight. As a population evolves in response to selection, the mean fitness of the individuals in the population tends to rise. Selection does not, however, always maximize mean fitness in a global sense. Depending on the intial allele frequencies, the population depicted in Figure 5.18c may evolve toward either fixation or loss of A_1. If the allele becomes fixed, the population will be at a stable equilibrium, but the population's mean fitness will be substantially lower than it would be if the allele were lost.

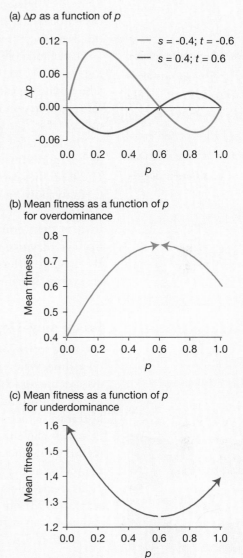

(a) Δp as a function of p

— $s = -0.4; t = -0.6$
— $s = 0.4; t = 0.6$

(b) Mean fitness as a function of p for overdominance

(c) Mean fitness as a function of p for underdominance

Figure 5.18 A graphical analysis of stable and unstable equilibria at loci with overdominance and underdominance a) A plot of Δp as a function of p. (b) and (c) Adaptive landscapes.

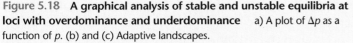

The algebraic analysis described in Box 5.8 predicts that a mixed population will be in genetic equilibrium, with both alleles present, when the frequency of *C(2)* is exactly 0.8. This equilibrium is unstable, however. If the frequency of *C(2)* ever gets above 0.8, then it should quickly rise to 1.0. Likewise, if the frequency of *C(2)* ever dips below 0.8, it should quickly fall to zero.

Intuitively, the reason for this prediction is as follows. Heterozygotes are inviable, so the adults in the population are all homozygotes. Imagine first that *C(2)C(2)* individuals are common and *N(2)N(2)* individuals are rare. If the flies mate at random, then almost all matings will involve *C(2)C(2)* flies mating with each other, or *C(2)C(2)* flies mating with *N(2)N(2)* flies. Only very rarely will *N(2)N(2)* flies mate with their own kind. Consequently, most *N(2)N(2)* flies will have zero reproductive success, and the frequency of *C(2)* will climb to 1.0. Now imagine that there are enough *N(2)N(2)* flies present that appreciable numbers of them *do* mate with each other. These matings will produce four times as many offspring as matings between *C(2)C(2)* flies. Consequently, the frequency of *N(2)* will climb to 1.0 and the frequency of *C(2)* will fall to zero.

When heterozygotes have inferior fitness, one allele tends to go to fixation while the other allele is lost. However, different populations may lose different alleles.

Foster and colleagues set up 13 mixed populations, with *C(2)* frequencies ranging from 0.71 to 0.96, then monitored their evolution for up to four generations. The results appear in Figure 5.17f. Qualitatively, the outcome matches the theoretical prediction nicely. In populations with higher initial *C(2)* frequencies, *C(2)* quickly rose to fixation, while in populations with lower initial *C(2)* frequencies, *C(2)* was quickly lost. The exact location of the unstable equilibrium turned out to be approximately 0.9 instead of 0.8. Foster and colleagues note that their *C(2)C(2)* flies carried recessive genetic markers that the biologists had bred into them to allow for easy identification. They suggest that these markers reduced the relative fitness of the *C(2)C(2)* flies below the value of 0.25 inferred solely on the basis of their compound chromosomes. Foster et al.'s experiment demonstrates that heterozygote inferiority leads to a loss of genetic diversity within populations. By driving different alleles to fixation in different populations, however, heterozygote inferiority may help maintain genetic diversity among populations.

Frequency-Dependent Selection

We have so far looked at examples in which the pattern of selection is constant over time. When selection consistently favors a particular allele, the population evolves inexorably toward fixation of that allele and loss of the others. When selection consistently favors heterozygotes, the population evolves to a stable equilibrium at which both alleles are present. The last scenario we will examine is one in which allele frequencies in a population remain near an equilibrium, but the reason is that the direction of selection fluctuates. First, selection favors one allele, then it favors the other. This scenario is called **frequency-dependent selection**.

Our example of frequency-dependent selection comes from Michio Hori's research (1993) on the scale-eating fish, *Perissodus microlepis,* in Africa's Lake Tanganyika. As its name suggests, the scale-eating fish makes its living by biting scales off of other fish. The scale eater attacks from behind, grabs scales off the victim's flank, then darts away. Stranger still, within the species *P. microlepis,* there are right-handed (dextral) fish, whose mouths are twisted to the right, and left-handed (sinistral) fish, whose mouths are twisted to the left (Figure 5.19). Hori showed that to a first approximation, handedness is determined by a single locus with two alleles. Right-handedness is dominant over left-handedness.

Figure 5.19 Scale-eating fish (Perissodus microlepis) from Lake Tanganyika (top) A right-handed (dextral) fish, shown from both sides. The mouth of this fish twists to the fish's right. (bottom) A left-handed (sinistral) fish, shown from both sides. The mouth of this fish twists to the fish's left. These two individuals belong to the same species and are members of the same population. (Dr. Michio Hori, Kyoto University, Kyoto, Japan)

Hori observed attacks on prey fish used as lures, and in addition examined scales recovered from the stomachs of scale-eating fish. These observations show that right-handed fish always attack their victim's left flank, and left-handed fish always attack their victim's right flank. (Readers can visualize the reason for this by first twisting their lips to the right or to the left, then imagining trying to bite scales off the flank of a fish.) The prey species are wary and alert, and scale eaters are successful in only about 20% of their attacks.

Hori reasoned that if right-handed scale eaters were more abundant than left-handers, then prey species would be more vigilant for attacks from the left. This would give left-handed scale eaters, who attack from the right, an advantage in their efforts to catch their prey unawares. Left-handed scale eaters would get more food than right-handers, have more offspring, and pass on more of their left-handed genes. This would increase the frequency of left-handed scale eaters in the population.

After left-handed scale eaters had become more abundant than right-handers, the prey fish would start to be more vigilant for attacks from the right. This would give right-handed scale eaters, who attack from the left, an advantage. The right-handers would get more food and have more offspring, and the frequency of right-handers in the population would increase.

The result is that left- and right-handed fish should be just about equally abundant in the population at any given time. The dominant right-handed allele should be held at a frequency of just under 0.3, and the recessive left-handed allele at a frequency of just over 0.7; under the Hardy–Weinberg equilibrium principle, these allele frequencies give phenotype frequencies of 0.5 and 0.5. This phenomenon, in which the rare phenotype is favored by natural selection, is an example of frequency-dependent selection.

Selection can also maintain two alleles in a population if each allele is advantageous when it is rare.

Hori's fish story is an elegant hypothesis, but is it true? To find out, Hori used a gill net to catch a sample of scale-eating fish in Lake Tanganyika every year or two for 11 years. The frequencies of the two phenotypes indeed appear to oscillate around 0.5 for each (Figure 5.20). In some years, slightly more than half of the fish were left-handed. Invariably, a year or two later, the pendulum had swung back, and slightly more than half of the fish were right-handed.

Is it possible that the frequencies of the two forms were just drifting at random rather than being held near 0.5 by selection? In three different years, Hori examined adult fish that were known to be breeding because they were caught in the act of brooding their young by scuba-diving scientists. The frequencies of handedness in these samples oscillated too, but the most abundant handedness among the breeders was always the opposite of the most abundant handedness in the population as a whole. (The breeders are indicated by the squares in Figure 5.20.) In other words, at any given time, the rare form seems to be doing more breeding. This is consistent with Hori's hypothesis of frequency-dependent selection.

Furthermore, Hori has evidence to support his hypothesis about the mechanism of selection. He examined bite wounds on a prey species in 1980, when left-handed

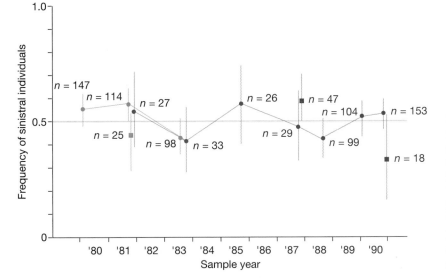

Figure 5.20 Frequency of left-handed (sinistral) fish over time   Circles represent scale-eating fish captured with a gill net; red versus blue circles represent fish caught at different locations. Squares (in 1981, 1987, and 1990) represent actively breeding adults, selectively captured by scuba-diving scientists. The numbers (*n*) report the sample sizes. Reprinted with permission from Hori (1993). Copyright © 1993, American Association for the Advancement of Science.

scale eaters were more common, and in 1983, when right-handed scale eaters were more common. In both years, the prey had more bite marks inflicted by whichever of the two forms was rarer at the time. This is consistent with the idea that the prey are indeed more vigilant for attacks from whichever form of scale eater is more common. It appears that Hori's hypothesis is correct and that over the long term, the allele, genotype, and phenotype frequencies in scale-eating fish are being held constant by frequency-dependent selection. Frequency-dependent selection, like heterozygote superiority, maintains genetic diversity in populations.

Compulsory Sterilization

We can use population genetic models to evaluate whether eugenic sterilization could have accomplished the aims of its proponents, had their assumptions about the heritability of traits been correct. The answer depends on the frequency of the alleles in question, and on the criteria for success.

Having discussed a variety of patterns of selection, we can now consider the evolutionary consequences of a eugenic compulsory sterilization program. The proponents of eugenic sterilization sought to reduce the fitness of particular genotypes to zero, and thereby to reduce the frequency of alleles responsible for undesirable phenotypes.

The phenotype that caught the eugenicists' attention perhaps more than any other was feeblemindedness. The Royal College of Physicians in England defined a feebleminded individual as "One who is capable of earning his living under favorable circumstances, but is incapable from mental defect existing from birth or from an early age (a) of competing on equal terms with his normal fellows or (b) of managing himself and his affairs with ordinary prudence" (see Goddard 1914). Evidence presented in 1914 by Henry H. Goddard, who was the director of research at the Training School for Feebleminded Girls and Boys in Vineland, New Jersey, convinced many eugenicists that strength of mind behaved like a simple Mendelian trait (see Paul and Spencer 1995). Normalmindedness was believed to be dominant, and feeblemindedness recessive.

A recessive genetic disease is not a promising target for a program that would eliminate it by sterilizing affected individuals. As Figure 5.14 and Figure 5.15 show, rare recessive alleles decline in frequency slowly, even under strong selection. On the other hand, eugenicists did not believe that feeblemindedness was especially rare (Paul and Spencer 1995). Indeed, they believed that feeblemindedness was alarmingly common and increasing in frequency. Edward M. East (1917) estimated the frequency of feeblemindedness at three per thousand. Henry H. Goddard reported a frequency of 2% among New York school children. Tests of American soldiers during World War I suggested a frequency of nearly 50% among white draftees.

We will assume a frequency for feeblemindedness of 1%, and reproduce a calculation reported by R. C. Punnett (1917) and revisited by R. A. Fisher (1924). Let f be the purported allele for feeblemindedness, with frequency q. If 1% of the population has genotype ff, then, by the Hardy–Weinberg equilibrium principle, the initial frequency of f is

$$q = \sqrt{0.01} = 0.1$$

If all affected individuals are sterilized, then the fitness of genotype ff is zero (or, equivalently, the selection coefficient for genotype ff is -1). Using the equation developed in Box 5.7, we can calculate the value of q in successive generations, and from q we can calculate the frequency of genotype ff.

The result appears in Figure 5.21. Over 10 generations, about 250 years, the frequency of affected individuals declines from 0.01 to 0.0025.

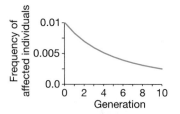

Figure 5.21 **Predicted change in the frequency of homozygotes for a putative allele for feeblemindedness under a eugenic sterilization program that prevents homozygous recessive individuals from reproducing**

Whether a geneticist saw this calculation as encouraging or discouraging depended on whether he or she saw the glass as partially empty or partially full. Some looked at the numbers, saw that it would take a very long time to completely eliminate feeblemindedness, and argued that compulsory sterilization was such a hopelessly slow solution that it was not worth the effort. Others, such as Fisher, dismissed this argument as "anti-eugenic propaganda." Fisher noted that after just one generation, the frequency of affected individuals would drop from 100 per ten thousand to 82.6 per ten thousand. "In a single generation," he wrote, "the load of public expenditure and personal misery caused by feeblemindedness...would be reduced by over 17 percent." Fisher also noted that most copies of the allele for feeblemindedness are present in heterozygous carriers rather than affected individuals. Along with East, Punnett, and others, Fisher called for research into methods for identifying carriers.

It is not entirely fair to use modern standards to criticize Goddard's research on the genetics of feeblemindedness. Mendelian genetics was in its infancy. Nonetheless, looking back after nearly a century, Goddard's evidence is deeply flawed.

First, the individuals whose case studies he reports are a highly diverse group. Some have Down syndrome; some have other forms of mental retardation. At least one is deaf and appears to be the victim of a woefully inadequate education. Some appear to have been deposited at Goddard's training school by widowed fathers who felt that children from a prior marriage were a liability in finding a new wife. Some may just have behaved differently than the directors of the school thought they should. Concluding the first case report in his book, Goddard writes of a 16-year-old who has been at the school for seven years:

> "Gertrude is a good example of that type of girl who, loose in the world, makes so much trouble. Her beauty and attractiveness and relatively high [intelligence] would enable her to pass almost anywhere as a normal child and yet she is entirely incapable of controlling herself and would be led astray most easily. It is fortunate for society that she is cared for as she is."

Second, Goddard's methods for collecting data were prone to distortion. He sent caseworkers to collect pedigrees from the families of the students at the training school. The caseworkers relied on hearsay and subjective judgements to assess the strength of mind of family members—many of whom were long since deceased.

Third, Goddard's method of analysis stacked the cards in favor of his conclusion. He first separated his 327 cases into a variety of categories: definitely hereditary cases; probably hereditary cases; cases caused by accidents; and cases with no assignable cause. He apparently placed cases in his "definitely hereditary" group only when they had siblings, recent ancestors, or other close kin also classified as feebleminded. When he later analyzed the data to determine whether feeblemindedness was a Mendelian trait, Goddard analyzed only the data from his "definitely hereditary" group. Given how he had filtered the data ahead of time, it is not too surprising that he concluded that feeblemindedness is Mendelian.

Although feeblemindedness is not among them, many genetic diseases are now known to be inherited as simple Mendelian traits. Yet eugenic sterilization has few advocates. One reason is that most serious genetic diseases are recessive and very rare; sterilization of affected individuals would have little impact on the frequency at which new affected individuals are born. A second reason is that mainstream attitudes about reproductive rights have changed to favor individual autonomy over societal mandates (Paul and Spencer 1995). A third reason is that, as we will discuss in the next section, there is a growing list of disease alleles that are suspected or known to be maintained in populations by heterozygote superiority. It would be futile and possibly ill-advised to try to reduce the frequency of such alleles by preventing reproduction by affected individuals.

5.4 Mutation

Second on the list of assumptions for the Hardy–Weinberg equilibrium principle was that there are no mutations. We now explore what happens to allele frequencies when this assumption is violated.

The last of the three questions posed at the outset of the chapter was how a highly deleterious allele, like the one that causes cystic fibrosis, could remain at relatively high frequency in a population. Our consideration of heterozygote superiority in the previous section hinted at one possible answer. Another potential answer is that new disease alleles are constantly introduced into populations by mutation. Before we can evaluate the merits of these two hypotheses for explaining the persistence of any particular disease allele, we need to discuss mutation in more detail.

In Chapter 4, we presented mutation as the source of all new alleles and genes. In its capacity as the ultimate source of all genetic variation, mutation provides the raw material for evolution. Here, we consider the importance of mutation as a force of evolution. How effective is mutation at changing allele frequencies over time? How strongly does mutation affect the conclusions of the Hardy–Weinberg analysis?

Adding Mutation to the Hardy–Weinberg Analysis: Mutation as an Evolutionary Force

Mutation by itself is generally not a potent force of evolution. To see why, return to our model population of mice. Imagine a locus with two alleles, A and a, with initial frequencies of 0.9 and 0.1. A is the wild-type allele, and a is a recessive loss-of-function mutation. Furthermore, imagine that copies of A are converted by mutation into new copies of a the rate of 1 copy per 10,000 per generation. This is a very high mutation rate, but it is within the range of mutation rates known (see Table 4.2 in Chapter 4). Back mutations that restore function are much less common than loss-of-function mutations, so we will ignore mutations that convert copies of a into new copies of A. Finally, imagine that all mutations happen while the gametes are in the gene pool.

Figure 5.22 follows the genotype and allele frequencies from one generation's adults through the gene pool to the next generation's zygotes. The adult genotypes are in the Hardy–Weinberg proportions:

AA	*Aa*	*aa*
0.81	0.18	0.01

When the adults produce gametes, the allele frequencies in the gene pool are as we specified:

A	*a*
0.9	0.1

Now, one of every ten thousand copies of allele A is converted into a new copy of allele a. The new frequency of A is given by the old frequency minus the fraction lost to mutation; the new frequency of a is given by the old frequency plus the fraction gained by mutation. That is,

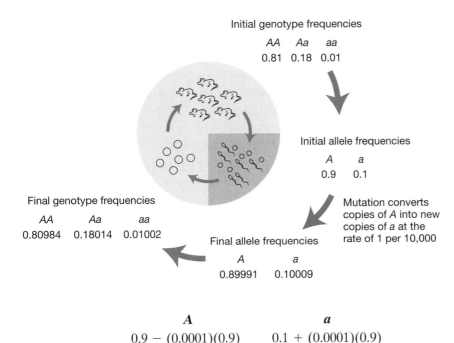

Figure 5.22   **Mutation is a weak force of evolution**   In a single generation in our model population, mutation produces virtually no change in allele and genotype frequencies.

$$\begin{array}{cc} \mathbf{A} & \mathbf{a} \\ 0.9 - (0.0001)(0.9) & 0.1 + (0.0001)(0.9) \\ = 0.89991 & = 0.10009 \end{array}$$

Finally, when the gametes combine at random to make zygotes, the zygote genotypes are in the new Hardy–Weinberg proportions:

$$\begin{array}{ccc} \mathbf{AA} & \mathbf{Aa} & \mathbf{aa} \\ 0.80984 & 0.18014 & 0.01002 \end{array}$$

Note that the new allele and genotype frequencies are almost identical to the old allele and genotype frequencies. As a force of evolution, mutation has had virtually no effect.

But virtually no effect is not the same as exactly no effect. Could mutation of *A* into *a,* occurring at the rate of 1 copy per 10,000 every generation for many generations, eventually result in an appreciable change in allele frequencies? The graph in Figure 5.23 provides the answer (see Box 5.9 for a mathematical treatment). After one thousand generations, the frequency of allele *A* in our model population will be about 0.81. Mutation can cause substantial change in allele frequencies, but it does so slowly.

As mutation rates go, the value we used in our model, 1 per 10,000 per generation, is very high. For most genes, mutation is an even less efficient mechanism of allele frequency change.

Hardy–Weinberg analysis shows that mutation is a week force of evolution.

Mutation and Selection

Although mutation alone usually cannot cause appreciable changes in allele frequencies, this does not mean that mutation is unimportant in evolution. In combination with selection, mutation becomes a potent evolutionary force. This point is demonstrated by an experiment conducted in Richard Lenski's lab (Lenski and Travisano 1994; Elena et al. 1996). Lenski and co-workers studied the evolution of a strain of *Escherichia coli* that is incapable of recombination (here, recombination means conjugation and exchange of DNA among cells). For *E. coli* populations of this strain, mutation is the only source of genetic variation. The researchers

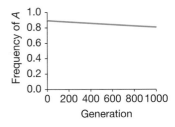

Figure 5.23   **Over very long periods of time, mutation can eventually produce appreciable changes in allele frequency**

BOX 5.9 A mathematical treatment of mutation as an evolutionary force

Imagine a single locus with two alleles: a wild-type allele, A, and a recessive loss-of-function mutation, a. Let μ be the rate of mutation from A to a. Assume that the rate of back mutation from a to A is negligible. If the frequency of A in this generation is p, then its frequency in the next generation is given by

$$p' = p - \mu p$$

If the frequency of a in this generation is q, then its frequency in the next generation is given by

$$q' = q + \mu p$$

The change in p from one generation to the next is

$$\Delta p = p' - p$$

which simplifies to

$$\Delta p = -\mu p$$

After n generations, the frequency of A is approximately

$$p_n = p_0 e^{-\mu n}$$

where p_n is the frequency of A in generation n, p_0 is the frequency of A in generation 0, and e is the base of the natural logarithms.

Readers familiar with calculus can derive the last equation as follows. First, assume that a single generation is an infinitesimal amount of time, so that we can rewrite the equation $\Delta p = -\mu p$ as

$$\frac{dp}{dg} = -\mu p$$

Now divide both sides by p, and multiply both sides by dg to get

$$\left(\frac{1}{p}\right) dp = -\mu \, dg$$

Finally, integrate the left side from frequency p_0 to p_n and the right side from generation 0 to n, then solve for p_n.

started 12 replicate populations with single cells placed in a glucose-limited, minimal salts medium—a demanding environment for these bacteria. After allowing each culture to grow to about 5×10^8 cells, Lenski and colleagues removed an aliquot (containing approximately 5 million cells) and transferred it to fresh medium. The researchers performed these transfers daily for 1500 days, or approximately 10,000 generations.

At intervals throughout the experiment, the researchers froze samples of the transferred cells for later analysis. Because *E. coli* are preserved but not killed by freezing, Lenski and colleagues could take ancestors out of the freezer and grow them up in a culture flask with an equivalent number of cells from descendant populations. These experiments allowed the team to directly measure the relative fitness of ancestral and descendant populations, as the growth rate of each under competition. In addition to monitoring changes in fitness over time in this way, the Lenski team measured cell size.

During the course of the study, both fitness and cell size increased dramatically in response to natural selection. The key point for our purposes is that these increases occurred in jumps (Figure 5.24). The step-like pattern resulted from a simple process: the occurrence of beneficial mutations that swept rapidly through the population. Each new mutation enabled the bacteria that carried it to divide at a faster rate. The frequency of the mutants quickly increased as they out-reproduced the other members of the population. Eventually, each new mutation became fixed in the population. The time from the appearance of each mutation to its fixation in the population was so short that we cannot see it in the figure. Most of the beneficial mutations caused larger cell size. Thus the plot of cell size over

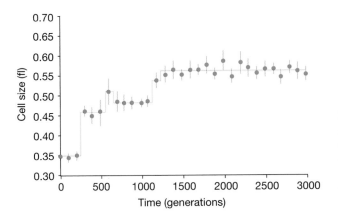

Figure 5.24 Changes over time in cell size of experimental *E. coli* populations. Each point on the plot represents the average cell size in the 12 replicate populations. The vertical lines are error bars; 95% of the observations fall within the range indicated by the bars. Reprinted with permission from Elena et al. (1996). Copyright © 1996, American Association for the Advancement of Science.

time also shows abrupt jumps. Between the appearance of one beneficial mutation and the next, all of the 12 replicate populations stood still. Why larger cells were beneficial in the nutrient-poor laboratory environment is a focus of ongoing research (Mongold and Lenski, 1996; Lenski et al. 1998).

The experiment by Lenski and colleagues reinforces one of the messages of Chapter 4. Without mutation, evolution would eventually grind to a halt. Mutation is the ultimate source of genetic variation.

Research with bacteria illustrates that while mutation itself is only a weak force of evolution, it nonetheless supplies the raw material on which natural selection acts.

Mutation-Selection Balance

Unlike the minority of mutations that led to increased cell size and higher fitness in Lenski et al.'s *E. coli* populations, most mutations are deleterious. Selection acts to eliminate such mutations from populations. Deleterious alleles persist, however, because they are continually created anew. When the rate at which copies of a deleterious allele are being eliminated by selection is exactly equal to the rate at which new copies are being created by mutation, the frequency of the allele is at equilibrium. This situation is called **mutation-selection balance**.

What is the frequency of the deleterious allele at equilibrium? If the allele is recessive, its equilibrium frequency, $\hat{q}$, is given by

$$\hat{q} = \sqrt{\frac{\mu}{s}}$$

where μ is the mutation rate, and s, the selection coefficient, is a number between 0 and 1 expressing the strength of selection against the allele (see Box 5.10 for a derivation). This equation captures with economy what intuition tells us about mutation-selection balance. If the selection coefficient is small (the allele is only mildly deleterious) and the mutation rate is high, then the equilibrium frequency of the allele will be relatively high. If the selection coefficient is large (the allele is highly deleterious) and the mutation rate is low, then the equilibrium frequency of the allele will be low.

Research by Brunhilde Wirth and colleagues (1997) on patients with spinal muscular atrophy provides an example. Spinal muscular atrophy is a neurodegenerative disease characterized by weakness and wasting of the muscles that control voluntary movement. It is caused by deletions in a locus on chromosome 5 called the telomeric survival motor neuron gene (*telSMN*). In some cases, the disease may be exacerbated by additional mutations in a nearby gene. Spinal muscular atrophy

At the same time selection removes deleterious alleles from a population, mutation constantly supplies new copies. In some cases, this balance between mutation and selection may explain the persistence of deleterious alleles in populations.

BOX 5.10 Allele frequencies under mutation-selection balance

Here we derive equations for predicting the equilibrium frequencies of deleterious alleles under mutation-selection balance. Imagine a single locus with two alleles, A_1 and A_2, with frequencies p and q. A_1 is the wild type; A_2 is deleterious. Let μ be the rate at which copies of A_1 are converted into copies of A_2 by mutation. Assume that the rate of back mutation is negligible.

Selection will continuously remove copies of A_2 from the population, while mutation will continuously create new copies. We want to calculate the frequency of A_2 at which these processes cancel each other. Following Felsenstein (1997), we will perform our calculation in a roundabout way. We will develop an equation in terms of p that describes mutation-selection balance for allele A_1. Then, we will solve the equation for q to get the equilibrium frequency of A_2. This approach may seem perverse, but it greatly simplifies the algebra.

Mutation-selection balance for a deleterious recessive allele

Imagine that A_2 is a deleterious recessive allele, such that the genotype fitnesses are given by

w_{11}	w_{12}	w_{22}
1	1	$1-s$

where the selection coefficient s gives the strength of selection against A_2.

First, we will write an equation for p^*, the frequency of allele A_1 after selection has operated, but before mutations occur. From Box 5.3, this equation is

$$p^* = \frac{p^2 w_{11} + pq w_{12}}{p^2 w_{11} + 2pq w_{12} + q^2 w_{22}}$$

Substituting the fitnesses from the table above, and $(1-p)$ for q, then simplifying gives

$$p^* = \frac{p}{1 - s(1-p)^2}$$

Next we will write an expression for p', the frequency of allele A_1 after mutations occur. These mutations convert a fraction μ of the copies of A_1 into copies of A_2, leaving behind a fraction $(1-\mu)$. Thus

$$p' = (1-\mu)p^* = \frac{(1-\mu)p}{1 - s(1-p)^2}$$

Finally, when mutation and selection are in balance, p' is equal to p, the frequency of allele A_1 that we started with:

$$\frac{(1-\mu)p}{1 - s(1-p)^2} = p$$

This simplifies to

$$(1-p)^2 = \frac{\mu}{s}$$

Substituting q for $(1-p)$ and solving for q yields an equation for $\hat{q}$, the equilibrium frequency of allele A_2 under mutation-selection balance:

$$\hat{q} = \sqrt{\frac{\mu}{s}}$$

If A_2 is a lethal recessive, then $s = 1$, and the equilibrium frequency of A_2 is equal to the square root of the mutation rate.

Mutation-selection balance for a lethal dominant allele

Imagine that A_2 is a lethal dominant allele, such that the genotype fitnesses are given by

w_{11}	w_{12}	w_{22}
1	0	0

Now the expression for p^* simplifies to

$$p^* = 1$$

which makes sense because, by definition, selection removes all copies of the lethal dominant A_2 from the population. Now the expression for p' is

$$p' = 1 - \mu$$

and the equilibrium condition is

$$1 - \mu = p$$

Substituting $(1-q)$ for p and simplifying gives

$$\hat{q} = \mu$$

In other words, the equilibrium frequency of A_2 is equal to the mutation rate.

is, after cystic fibrosis, the second most common lethal autosomal recessive disease in Caucasians (McKusick et al. 1999).

Collectively, the loss-of-function alleles of *telSMN* have a frequency of about 0.01 in the Caucasian population. Wirth and colleagues estimate that the selection coefficient is about 0.9. With such strong selection against them, we would expect that disease-causing alleles would slowly but inexorably disappear from the population. How, then, do they persist at a frequency of 1 in 100?

One possibility is that the disease alleles are being kept in the population by a balance between mutation and selection. If we substitute the allele frequency and selection coefficient for $\hat{q}$ and s in the equation on page 145, and then solve for μ, we find that this scenario requires a mutation rate of about 0.9×10^{-4} mutations per *telSMN* allele per generation. Wirth et al. analyzed the chromosomes of 340 individuals with spinal muscular atrophy, and the chromosomes of their parents and other family members. They found that 7 of the 340 affected individuals carried a new mutation not present in either parent. These numbers allowed the scientists to estimate directly the mutation rate at the *telSMN* locus (see Box 5.11). Their estimate is 1.1×10^{-4}. This directly measured mutation rate is in good agreement with the rate predicted under the hypothesis of a mutation-selection balance. Wirth et al. conclude that mutation-selection balance provides a sufficient explanation for the persistence of spinal muscular atrophy alleles.

Are the Alleles That Cause Cystic Fibrosis Maintained by a Balance Between Mutation and Selection?

Cystic fibrosis is caused by recessive loss-of-function mutations in a locus on chromosome 7 that encodes a protein called the cystic fibrosis transmembrane conductance regulator (CFTR). CFTR is a cell surface protein expressed in the mucus membrane lining the intestines and lungs. Gerald Pier and colleagues (1997) demonstrated that one of CFTR's key functions is to enable cells of the lung lining to ingest and destroy *Pseudomonas aeruginosa* bacteria. These bacteria cause chronic lung infections in individuals with cystic fibrosis, eventually leading to severe lung damage (Figure 5.25). Selection against the alleles that cause cystic fibrosis appears to be strong. Until recently, few affected individuals survived to reproductive age; those that do survive are often infertile. And yet the alleles that cause

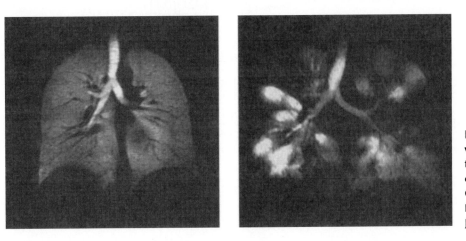

Figure 5.25   A normal lung (left) versus a lung ravaged by the bacterial infections that accompany cystic fibrosis (right)   (Photos by G. Allan Johnson, Duke University Medical Center; *Scientific American* June 1999, page 34).

BOX 5.11 Estimating mutation rates for recessive alleles

Here, we present the method used by Brunhilde Wirth and colleagues (1997) to estimate mutation rates for recessive alleles. The key information required is the fraction of affected individuals that carry a brand-new mutant allele. With modern molecular techniques, this fraction can be obtained by direct examination of the chromosomes of affected individuals and their relatives.

Let q be the frequency of recessive loss-of-function allele a. Ignoring the extremely rare individuals with two new mutant copies, there are two ways to be born with genotype aa:

1. An individual can be the offspring of two carriers. The probablity of this outcome for a given birth is the product of: (a) the probability that an offspring of two carriers that will be affected; (b) the probability that the mother is a carrier; and (c) the probability that the father is a carrier. This probability is given by

$$\left[\frac{1}{4}\right] \times [2q(1-q)] \times [2q(1-q)]$$

2. An individual can be the offspring of one carrier and one homozygous dominant parent *and* can receive allele a from the affected parent and a new mutant copy of a from the unaffected parent. The probability of this outcome for a given birth is the product of (a) the probability that an offspring of one carrier will receive that carrier's mutant allele; (b) the probability that the mother is a carrier; (c) the probability that the father is a homozygous dominant; and (d) the mutation rate *plus* the same probability for the scenario in which the father is the carrier and the mother is the homozygous dominant:

$$\left\{\left[\frac{1}{2}\right] \times [2q(1-q)] \times [(1-q)^2] \times [\mu]\right\}$$

$$+ \left\{\left[\frac{1}{2}\right] \times [2q(1-q)] \times [(1-q)^2] \times [\mu]\right\}$$

$$= [2q(1-q)] \times [(1-q)^2] \times [\mu]$$

With these probabilities, we can write an expression for r, the fraction of affected individuals that carry one new mutant allele. This is the second probability divided by the sum of the second probability and the first. Simplified just a bit, we have

$$r = \frac{2q(1-q)(1-q)^2\mu}{2q(1-q)(1-q)^2\mu + q(1-q)q(1-q)}$$

Simplifying further yields

$$r = \frac{2(1-q)\mu}{2(1-q)\mu + q}$$

Finally, assume that q is small, so that $(1-q)$ is approximately equal to one. This assumption gives

$$r = \frac{2\mu}{2\mu + q}$$

which can be solved for μ:

$$\mu = \frac{rq}{2 - 2r}$$

The mutation rate for spinal muscular atrophy

In Caucasian populations, spinal muscular atrophy affects about 1 infant in 10,000, implying that the frequency of the mutant allele is

$$q = \sqrt{0.0001} = 0.01$$

Wirth and colleagues examined the chromosomes of 340 affected patients and their family members. The researchers discovered that 7 of their patients had a new mutant allele not present in either parent. Thus,

$$r = \frac{7}{340} = 0.021$$

Substituting these values for q and r into the equation for μ gives the estimate

$$\mu = \frac{(0.021)(0.01)}{2 - 2(0.021)} = 0.00011$$

The mutation rate for cystic fibrosis

In Caucasian populations, cystic fibrosis affects about 1 infant in 2500. Wirth and colleagues cite data from other authors to establish that only 2 of about 30,000 cystic fibrosis patients studied proved to have a new mutant allele not present in either parent. These figures give an estimated mutation rate of

$$\mu = 6.7 \times 10^{-7}$$

cystic fibrosis have a collective frequency of approximately 0.02 among people of European ancestry.

Could cystic fibrosis alleles be maintained at a frequency of 0.02 by mutation-selection balance? If we assume a selection coefficient of 1 and use the equation derived in Box 5.10, the mutation rate creating new disease alleles would have to be 4×10^{-4}. The actual mutation rate for cystic fibrosis alleles appears to be considerably lower than that: approximately 6.7×10^{-7} (see Box 5.11). We can conclude that a steady supply of new mutations cannot, by itself, explain the maintenance of cystic fibrosis alleles at a frequency of 0.02.

Our discussion of heterozygote superiority suggests an alternative explanation (Figure 5.16 and Box 5.8). Perhaps the the fitness cost suffered by cystic fibrosis alleles when they are in homozygotes is balanced by a fitness advantage they enjoy when they are in heterozygotes.

Gerald Pier and colleagues (1998) hypothesized that cystic fibrosis heterozygotes might be resistant to typhoid fever and therefore have superior fitness. Typhoid fever is caused by *Salmonella typhi* bacteria. The bacteria initiate an infection by crossing the layer of epithelial cells that line the gut. Pier and colleagues suggested that *S. typhi* bacteria infiltrate the gut by exploiting the CFTR protein as a point of entry. If so, then heterozygotes, which have fewer copies of CFTR on the surface of their cells, should be less vulnerable to infiltration.

Pier and colleagues tested their hypothesis by constructing mouse cells with three different CFTR genotypes: homozygous wild-type cells; heterozygotes with one functional CFTR allele and one allele containing the most common human cystic fibrosis mutation, a single base-pair deletion called *ΔF508;* and homozygous *ΔF508* cells. The researchers exposed these cells to *Salmonella typhi,* then measured the number of bacteria that got inside cells of each genotype. The results were dramatic (Figure 5.26). As the researchers predicted, homozygous *ΔF508* cells were almost totally resistant to infiltration by *S. typhi,* while homozygous wild-type cells were highly vulnerable. Heterozygous cells were partially resistant; they accumulated 86% fewer bacteria than did the wild-type cells. These results are consistent with the hypothesis that cystic fibrosis disease alleles are maintained in human populations because heterozygotes have superior fitness during typhoid fever epidemics. Pier et al.'s research serves as another example in which an evolutionary analysis proved valuable in biomedical research.

In other cases, the frequency of a deleterious allele may be too high to explain by mutation-selection balance. This may be a clue that heterozygotes have superior fitness.

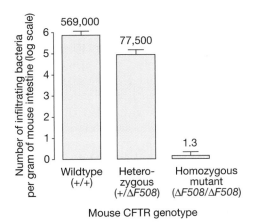

Figure 5.26   Cultured mouse cells heterozygous for cystic fibrosis show substantial resistance to infiltration by the bacteria that cause typhoid fever   Cells homozygous for the most common human disease allele are almost totally resistant. From Pier et al. (1998). Reprinted by permission from Nature 393: 79–82. © 1998, Macmillan Magazines Ltd.

Summary

Population genetics represents a synthesis of Mendelian genetics and Darwinian evolution, and is concerned with the mechanisms that cause allele frequencies to change from one generation to the next. The Hardy–Weinberg equilibrium principle is a null model that provides the conceptual framework for population genetics. It shows that under simple assumptions—no selection, no mutation, no migration, no genetic drift, and random mating—allele frequencies do not change. Furthermore, genotype frequencies can be calculated from allele frequencies.

When any one of the first four assumptions is violated, allele frequencies may change across generations. Selection, mutation, migration, and genetic drift are thus the four mechanisms of evolution. Nonrandom mating does not cause allele frequencies to change, and is thus not a mechanism of evolution. It can, however, alter genotype frequencies and thereby affect the course of evolution.

Population geneticists can measure allele and genotype frequencies in real populations. Thus, biologists can test whether allele frequencies are stable across generations and whether the genotype frequencies conform to Hardy–Weinberg expectations. If either of the conclusions of the Hardy–Weinberg analysis is violated, it means that one or more of the assumptions does not hold. The nature of the deviation from Hardy–Weinberg expectations does not, by itself, identify the faulty assumption. We can, however, often infer which mechanisms of evolution are at work based on other characteristics of the populations under study.

Selection occurs when individuals with different genotypes differ in their success at getting copies of their genes into future generations. It is a powerful force of evolution. Selection can cause allele frequencies to change from one generation to the next, and can take genotype frequencies away from Hardy–Weinberg equilibrium. Some patterns of selection tend to drive some alleles to fixation and others to loss; other patterns of selection serve to maintain allelic diversity in populations.

Alone, mutation is a weak evolutionary force. Mutation does, however, provide the genetic variation that is the raw material for evolution. In some cases a steady supply of new mutant alleles can balance selection against those same alleles, and thereby serve to hold allele frequencies at equilibrium.

Questions

1. Black color in horses is governed primarily by a recessive allele at the A locus. *AA* and *Aa* horses are nonblack colors such as bay, while *aa* horses are black all over. (Other loci can override the effect of the A locus, but we will ignore that complication.) A few years ago, a reader of the Usenet newsgroup "rec.equestrian" asked why there are relatively few black horses of the Arabian breed. One response was "Black is a rare color because it is recessive. More Arabians are bay or grey because those colors are dominant." What is wrong with this explanation? (Assume that the *A* and *a* alleles are in Hardy–Weinberg equilibrium, which was probably true at the time of this discussion.) Generally, what does the Hardy–Weinberg model show us about the impact that an allele's dominance or recessiveness has on its frequency? Should an allele become more common (or less common) simply because it is dominant (or recessive)?

2. In humans, the COL1A1 locus codes for a certain collagen protein found in bone. The normal allele at this locus is denoted with *S*. A recessive allele *s* is associated with reduced bone mineral density and increased risk of fractures in both *Ss* and *ss* women. A recent study of 1778 women showed that 1194 were *SS*, 526 were *Ss*, and 58 were *ss* (Uitterlinden et al. 1998).

   Are these two alleles in Hardy–Weinberg equilibrium in this population? How do you know? What information would you need to determine whether the alleles will be in Hardy–Weinberg equilibrium in the next generation?

3. We used Figure 5.11 as an example of how the frequency of an allele (*Adh*^{S} in fruit flies) does not change in unselected (control) populations, but does change in response to selection. However, look again at the unselected control lines in Figure 5.11. The frequency of the *Adh*^{S} allele in the two control populations did change a little, moving up and down over time. Which assumption of the Hardy–Weinberg model is most probably being violated? If this experiment were repeated, what change in experimental design would reduce this deviation from Hardy–Weinberg equilibrium?

4. Most animal populations have a 50:50 ratio of males to females. This does not have to be so; it is theoretically

possible for parents to produce predominantly male offspring, or predominantly female offspring. Imagine a monogamous population with a male-biased sex ratio, say, 70% males and 30% females. Which sex will have an easier time finding a mate? As a result, which sex will probably have higher average fitness? Which parents will have higher fitness—those that produce mostly males, or those that produce mostly females? Now imagine the same population with a female-biased sex ratio, and answer the same questions. What sort of selection is probably maintaining the 50:50 sex ratio seen in most populations?

5. Discuss how each of the following recent developments may affect the frequency of alleles that cause cystic fibrosis (CF).

   a. Many women with CF now survive long enough to have children. (CF causes problems with reproductive ducts, but many CF women can bear children nonetheless. CF men are usually sterile.)

   b. Typhoid fever in developed nations has declined to very low levels since 1900.

   c. In some populations, couples planning to have children are now routinely screened for the most common CF alleles.

   d. Drug-resistant typhoid fever has recently appeared in several developing nations.

6. Consider what makes a new mutant allele dominant or recessive. To guide your thinking, imagine an enzyme that changes substance A to substance B. If B is a nutrient that is needed only in minimal amounts, will a loss-of-function mutation be dominant or recessive? If A is a toxin that must be entirely broken down, will a loss-of-function mutation be dominant or recessive? How about a new mutant allele that results in a form of the protein that can catalyze an entirely new reaction (say, from A to new substance C?) Can you think of other examples of protein function that will affect whether a new allele is dominant or recessive?

7. There are two common alleles for the human muscle enzyme ACE (angiotensin-converting enzyme)—a shorter "*D*" allele, and a longer "*I*" allele that has an insertion of 287 base pairs. The ACE coded by the *I* allele has lower activity, but it also associated with superior muscular performance after physical training. One study (Williams et al. 2000) of 35 *II* and 23 *DD* men found that though they didn't differ in muscular efficiency before training, after 11 weeks of aerobic training the *II* homozygotes had 8% greater muscular efficiency. The *I* allele is also associated with greater endurance and greater muscular growth after strength training. Speculate on why the *D* allele still remains at relatively high frequency in the human population. How could you test your ideas?

8. In our discussion of Jaeken syndrome in Section 5.2, we asserted that parents who are both carriers of the *R141H* allele can expect a different distribution of phenotypes among their children than parents who are carriers of two different disease alleles. Explain the logic behind this assertion. What genotypes and phenotypes, and in what ratios, should the following pairs of parents expect among their liveborn children? What would you tell these parents if you were a genetic counselor?

   Mother's genotype: +/*R141H*; Father's genotype: +/*R141H*

   Mother's genotype: +/*R141H*; Father's genotype: +/*Other*

   Mother's genotype: +/*Other*; Father's genotype: +/*Other*

Exploring the Literature

9. In the example of the scale-eating fish, we saw that frequency-dependent selection tends to maintain an even ratio of left-handed fish to right-handed fish. See the following references for some interesting cases of possible frequency-dependent selection in other species. How plausible do you find each scenario?

   Raymond, M., D. Pontier, A. B. Dufour, and A. P. Møller. 1996. Frequency-dependent maintenance of left-handedness in humans. *Proceedings of the Royal Society of London,* Series B 263: 1627–1633.

   Sinervo, B., and C. M. Lively. 1996. The rock-paper-scissors game and the evolution of alternative male strategies. *Nature* 380: 240–243.

   Smithson, A., and M. R. MacNair. 1996. Frequency-dependent selection by pollinators: Mechanisms and consequences with regard to behaviour of bumblebees *Bombus terrestris* (L.) (Hymenoptera: Apidae). *Journal of Evolutionary Biology* 9: 571–588.

10. The version of the adaptive landscape presented in Box 5.7 and Box 5.8, in which the landscape is a plot of mean fitness as a function of allele frequency, is actually somewhat different from the original version of the concept that Sewall Wright presented in 1932. Furthermore, there is even a third common interpretation of the adaptive landscape idea. For a discussion of the differences among the three versions, see Chapter 9 in

   Provine, W. B. 1986. *Sewall Wright and Evolutionary Biology.* Chicago: University of Chicago Press.

For Sewall Wright's response to Provine's history, see
Wright, S. 1988. Surfaces of selective value revisited. *The
American Naturalist* 131: 115–123.

Wright's original 1932 paper is reprinted in Chapter 11 of
Wright, S. 1986. *Evolution: Selected Papers,* William B.
Provine, ed. Chicago: University of Chicago Press.

11. If your library has the earliest volumes of the Journal of
Heredity, read

Bell, Alexander Graham. 1914. How to improve the race.
Journal of Heredity 5: 1–7.

Keep in mind that population genetics was in its infancy; Mendelism had yet to be integrated with natural se-

lection. What was accurate and inaccurate in Bell's understanding of the mechanisms of evolution? Would the
policy Bell advocated actually have accomplished his
aims? Why or why not? If so, would it have done so for
the reasons Bell thought it would?

12. For an example in which strong natural selection caused
rapid change in allele frequencies in a wild population, see
Johannesson, K., B. Johannesson, and U. Lundgren. 1995.
Strong natural selection causes microscale allozyme
variation in a marine snail. *Proceedings of the National
Academy of Sciences, USA* 92: 2602–2606.

Citations

*Much of the population genetics material in this chapter is modeled after presenta-
tions in the following:*

Crow, J. F. 1983. *Genetics Notes.* Minneapolis, MN: Burgess Publishing.
Felsenstein, J. 1997. *Theoretical Evolutionary Genetics.* Seattle, WA: ASUW
Publishing, University of Washington.
Griffiths, A. J. F., J. H. Miller, D. T. Suzuki, R. C. Lewontin, and W. M. Gel-
bert. 1993. *An Introduction to Genetic Analysis.* New York: W.H. Freeman.
Templeton, A. R. 1982. Adaptation and the integration of evolutionary
forces. In R. Milkman, ed., *Perspectives on Evolution.* Sunderland, MA:
Sinauer, 15–31.

Here is the list of all other citations in this chapter:

Castle, W. E. 1903. The laws of heredity of Galton and Mendel, and some
laws governing race improvement by selection. *Proceedings of the American Academy of Arts and Sciences* 39: 223–242.
Cavener, D. R., and M. T. Clegg. 1981. Multigenic response to ethanol in
Drosophila melanogaster. Evolution 35: 1–10.
Dawson, P. S. 1970. Linkage and the elimination of deleterious mutant genes
from experimental populations. *Genetica* 41: 147–169.
Dean, M., M. Carrington, et al. 1996. Genetic restriction of HIV-1 infec-
tion and progression to AIDS by a deletion allele of the CKR5 struc-
tural gene. *Science* 273: 1856–1862.
East, E. M. 1917. Hidden feeblemindedness. *Journal of Heredity* 8: 215–217.
Elena, S. F., V. S. Cooper, and R. E. Lenski. 1996. Punctuated evolution
caused by selection of rare beneficial mutations. *Science* 272:
1802–1804.
Elias, S., M. M. Kaback, et al. 1992. Statement of The American Society of
Human Genetics on cystic fibrosis carrier screening. *American Journal of
Human Genetics* 51: 1443–1444.
Fisher, R. A. 1924. The elimination of mental defect. *The Eugenics Review* 16:
114–116.
Foster, G. G., M. J. Whitten, T. Prout, and R. Gill. 1972. Chromosome re-
arrangements for the control of insect pests. *Science* 176: 875–880.
Goddard, H. H. 1914. *Feeblemindedness: Its Causes and Consequences.* New York:
The Macmillan Company.
Hardy, G. H. 1908. Mendelian proportions in a mixed population. *Science* 28:
49–50.
Hori, M. 1993. Frequency-dependent natural selection in the handedness of
scale-eating cichlid fish. *Science* 260: 216–219.
Kevles, D. J. 1995. *In the Name of Eugenics: Genetics and the Uses of Human
Heredity.* Cambridge, MA: Harvard University Press.

Lane, H. 1992. *The Mask of Benevolence: Disabling the Deaf Community.* New
York: Vintage Books.
Lenski, R. E., and M. Travisano. 1994. Dynamics of adaptation and diversification: A 10,000-generation experiment with bacterial populations.
Proceedings of the National Academy of Sciences, USA 91: 6808–6814.
Lenski, R. E., J. A. Mongold, et al. 1998. Evolution of competitive fitness in
experimental populations of *E. coli:* What makes one genotype a better
competitor than another? *Antonie Van Leeuwenhoek International Journal
of General and Molecular Microbiology* 73: 35–47.
Martinson, J. J., N. H. Chapman, et al. 1997. Global distribution of the CCR5
gene 32-base-pair deletion. *Nature Genetics* 16: 100–103.
Matthijs, G., E. Schollen, E. Van Schaftingen, J.-J. Cassiman, and J. Jaeken.
1998. Lack of homozygotes for the most frequent disease allele in car-
bohydrate-defcient-glycoprotein syndrome type 1A. *American Journal of
Human Genetics* 62: 542–550.
McKusick, Victor A., et al. 1999. Spinal muscular atrophy I. Record #253300
in Online Mendelian Inheritance in Man. Center for Medical Genetics, Johns Hopkins University (Baltimore, MD) and National Center for
Biotechnology Information, National Library of Medicine (Bethesda,
MD). World Wide Web URL: http://www.ncbi.nlm.nih.gov/omim/
Mongold, J. A., and R. E. Lenski. 1996. Experimental rejection of a non-
adaptive explanation for increased cell size in *Escherichia coli. Journal of Bacteriology* 178: 5333–5334.
Mukai, T., and A. B. Burdick. 1959. Single gene heterosis associated with a
second chromosome recessive lethal in *Drosophila melanogaster. Genetics*
44: 211–232.
Paul, D. B., and H. G. Spencer. 1995. The hidden science of eugenics. *Na-
ture* 374: 302–304.
Pier, G. B., M. Grout, and T. S. Zaidi. 1997. Cystic fibrosis transmembrane
conductance regulator is an epithelial cell receptor for clearance of
Pseudomonas aeruginosa from the lung. *Proceedings of the National Academy
of Sciences, USA* 94: 12088–12093.
Pier, G. B., M. Grout, et al. 1998. *Salmonella typhi* uses CFTR to enter in-
testinal epithelial cells. *Nature* 393: 79–82.
Provine, W. B. 1971. *The Origins of Theoretical Population Genetics.* Chicago:
The University of Chicago Press.
Punnett, R. C. 1917. Eliminating feeblemindedness. *Journal of Heredity* 8: 464–465.
Reilly, P. 1991. *The Surgical Solution: A History of Involuntary Sterilization in
the United States.* Baltimore: Johns Hopkins University Press.
Stephens, J. C., D. E. Reich, et al. 1998. Dating the origin of the CCR5-Δ32
AIDS-resistance allele by the coalescence of haplotypes. *American Jour-
nal of Human Genetics* 62: 1507–1515.

Uitterlinden A. G., H. Burger, et al. 1998. Relation of alleles of the collagen type IA1 gene to bone density and the risk of osteoporotic fractures in postmenopausal women. *New England Journal of Medicine* 338: 1016–21.

UNAIDS. 1998. AIDS epidemic update: December 1998. (Geneva, Switzerland). World Wide Web URL: http://www.unaids.org

United States Supreme Court. 1927. *Buck v. Bell,* 274 U.S. 200.

Weinberg, W. 1908. Ueber den nachweis der vererbung beim menschen. *Jahreshefte des Vereins für Vaterländische Naturkunde in Württemburg* 64: 368–382. English translation in Boyer, S. H. 1963. Papers on Human Genetics. Englewood Cliffs, NJ: Prentice Hall.

Williams, A. G., M. P. Rayson, et al. 2000. The ACE gene and muscle performance. *Nature* 403: 614.

Wirth, B., T. Schmidt, et al. 1997. De novo rearrangements found in 2% of index patients with spinal muscular atrophy: Mutational mechanisms, parental origin, mutation rate, and implications for genetic counseling. *American Journal of Human Genetics* 61: 1102–1111.

Yule, G. U. 1902. Mendel's laws and their probable relations to intra-racial heredity. *New Phytologist* 1: 193–207; 222–238.

Zar, J. H. 1996. *Biostatistical Analysis,* 3rd ed. Upper Saddle, NJ: Prentice Hall.

CHAPTER 6

Mendelian Genetics in Populations II: Migration, Genetic Drift, and Nonrandom Mating

Male greater prairie chickens advertising for mates. (Richard Day/Daybreak Imagery)

THE GREATER PRAIRIE CHICKEN, *TYMPANUCHUS CUPIDO PINNATUS*, IS A two-pound bird with a ten-pound mating display (Figure 6.1). Each spring, the males congregate in communal breeding areas, called leks, where they stake out small territories and advertise for mates. They spread their tail feathers, stomp their feet, and inflate the bright orange air sacs on their throats. As the birds draw air into the sacs it makes a booming noise that is audible for miles—like the sound created when a person blows air across the mouth of a large empty bottle, but much louder (Thomas 1998). Females visit the lek, inspect the displaying males, and choose a mate.

Two hundred years ago, the state of Illinois was almost entirely covered with prairie (Figure 6.2a) and was home to millions of greater prairie chickens. In 1837, however, the steel plow was introduced (Thomas 1998). With the first blade that could break through the dense roots of prairie plants, the steel plow allowed the conversion of prairie into farmland. As the Illinois prairie shrank, the range of the Illinois greater prairie chicken contracted with it (Figure 6.2b-d). And as the bird's

Figure 6.1   A greater prairie chicken
This male has inflated his air sacs and fanned his feathers as part of his courtship display. (Richard Day/Daybreak Imagery)

Efforts to save remnant populations of greater prairie chickens by restoring prairie habitat appeared at first to be succeeding. Soon, however, population sizes resumed their decline. Why?

range contracted, its numbers crashed: to 25,000 in 1933; 2000 in 1962; 500 in 1972; 76 in 1990 (Westemeier et al. 1991; Bouzat et al. 1998). By 1994, there were fewer than 50 greater prairie chickens left in Illinois (Westemeier et al. 1998). These remaining birds belonged to two remnant populations—one in Marion County and the other in Jasper County.

Efforts to save the Illinois greater prairie chicken began with a ban on hunting in 1933 (Thomas 1998). In 1962 and 1967, respectively, the habitats occupied by the Jasper County and Marion County populations were established as sanctuaries, and as sites for the restoration and management of grasslands (Westemeier et al. 1998). Figure 6.3 tracks the number of males displaying on leks in Jasper County from 1963 to 1997. From the mid-1960s through the early 1970s, the number of birds increased steadily. The conservation measures appeared to be working. In the mid-1970s, however, the population began to crash again. The population hit its all-time low of five or six males in 1994, despite the fact that there was now more managed grassland available to the birds than there had been in 1963.

Why did the Jasper County prairie chicken population continue to decline from the mid-1970s to the mid-1990s, even though the amount of habitat available was increasing? And what did wildlife managers do to finally reverse the decline?

The answers to these questions involve three phenomena introduced in Chapter 5, but not discussed there: migration, genetic drift, and nonrandom mating. We identified these processes as potentially important factors in the evolution of populations when we developed the Hardy–Weinberg equilibrium principle. When an ideal population has no selection, no mutation, no migration, and an infinitely large size, and when individuals choose their mates at random, then (1) the al-

Figure 6.2   Habitat destruction and the shrinking range of Illinois greater prairie chickens   Map (a) shows the extent of prairie in Illinois before the introduction of the steel plow. Maps (b), (c), and (d) show the distribution of the greater prairie chickens in 1940, 1962, and 1994. Source: From Fig. 1 in Westemeier et al. (1998). Derived from R. C. Anderson, *Transactions of the Illinois State Academy of Science* 63, 214 (1970). Reprinted by permission of the Illinois State Academy of Science.

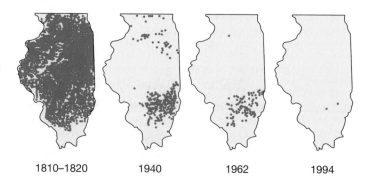

1810–1820   1940   1962   1994

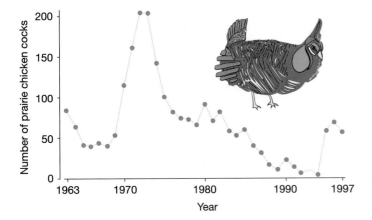

Figure 6.3   A greater prairie chicken population in danger of extinction   This graph plots the number of male prairie chickens displaying each year on booming grounds in Jasper County, Illinois from 1963 to 1997. Redrawn with permission from Westemeier et al. (1998). Copyright © 1998, American Association for the Advancement of Science.

lele frequencies do not change from one generation to the next, and (2) the genotype frequencies can be calculated by multiplying the allele frequencies. In Chapter 5 we looked at what happens when we relax the assumptions of no selection and no mutation. In this chapter we will explore what happens when we relax the assumptions of no migration, infinite population size, and random mating. We will then return to the case of the Illinois greater prairie chicken, and address the questions it poses.

6.1   Migration

Migration, in an evolutionary sense, is the movement of alleles between populations. This use of the term migration is distinct from its more familiar meaning, which refers to the seasonal movement of individuals. To evolutionary biologists, migration means gene flow: The transfer of alleles from the gene pool of one population to the gene pool of another population. Migration can be caused by anything that moves alleles far enough to go from one population to another. Mechanisms of gene flow range from the occasional long-distance dispersal of juvenile animals to the transport of pollen, seeds, or spores by wind, water, or animals. The actual amount of migration among populations in different species varies enormously, depending on how mobile individuals or propagules are at various stages of the life cycle.

Adding Migration to the Hardy–Weinberg Analysis: Migration as an Evolutionary Force

To investigate the effects of migration on the two conclusions of the Hardy–Weinberg analysis, we consider a simple model of migration, called the one-island model. Imagine two populations: one on a continent, and the other on a small island offshore (Figure 6.4). Because the island population is so small relative to the continental population, any migration from the island to the continent will be inconsequential for allele and genotype frequencies on the continent. So migration, and the accompanying gene flow, is effectively one way, from the continent to the island. As usual, consider a single locus with two alleles, A_1 and A_2. Can migration from the continent to the island take the allele and genotype frequencies on the island away from Hardy–Weinberg equilibrium?

To see that the answer is yes, imagine that before migration, the frequency of A_1 on the island is 1.0 (that is, A_1 is fixed in the island population—see Figure 6.5).

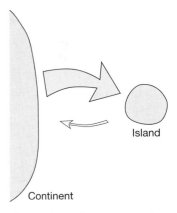

Island

Continent

Figure 6.4   The one-island model of migration   The arrows in the diagram show the relative amount of gene flow between the island and continental populations. Alleles arriving on the island from the continent represent a relatively large fraction of the island gene pool, whereas alleles arriving on the continent from the island represent a relatively small fraction of the continental gene pool.

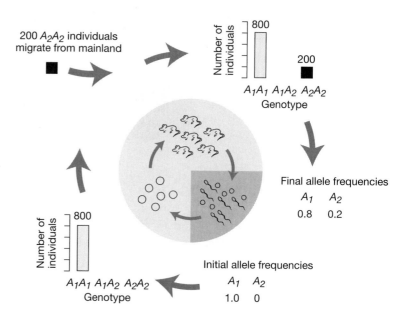

200 A_2A_2 individuals
migrate from mainland

Final allele frequencies
A_1 A_2
0.8   0.2

Initial allele frequencies
A_1 A_2
1.0   0

Figure 6.5   Migration can alter allele and geno-type frequencies   This diagram follows an imaginary island population of mice from one generation's gene pool (initial allele frequencies) to the next generation's gene pool (final allele frequencies). The bar graphs show the number of individuals of each genotype in the population at any given time. Migration, in the form of individuals arriving from a continental population fixed for allele A_2, increases the frequency of allele A_2 in the island population.

When gametes in a gene pool in which A_1 is fixed combine at random to make zygotes, the genotype frequencies among the zygotes are 1.0 for A_1A_1, 0 for A_1A_2, and 0 for A_2A_2. Our calculations will be simpler if we give our population a fixed size, so imagine that there are 800 zygotes, which we will let grow up to become adults.

Now suppose that the continental population is fixed for allele A_2, and that before the individuals on the island reach maturity, 200 individuals migrate from the

---

Box 6.1   An algebraic treatment of migration as an evolutionary force

Let p_I be the frequency of allele A_1 in an island population, and p_C be the frequency of A_1 in the mainland population. Imagine that every generation a group of individuals moves from the mainland to the island, where they constitute a fraction m of the island population. We want to know how the frequency of allele A_1 on the island changes as a result of migration, and whether there is an equilibrium frequency for A_1 at which there will be no further change even if migration continues.

We first write an expression for p_I', the frequency of A_1 on the island in the next generation. A fraction $(1 - m)$ of the individuals in the next generation were already on the island. Among these individuals, the frequency of A_1 is p_I. A fraction m of the individuals in the next generation came from the mainland. Among them, the frequency of A_1 is p_C. Thus the new frequency of A_1 in the island population is a weighted average of the frequency among the residents and the frequency among the immigrants:

$$p_I' = (1 - m)(p_I) + (m)(p_C)$$

We can now write an expression for Δp_I, the change in the allele frequency on the island from one generation to the next:

$$\Delta p_I = p_I' - p_I$$

Substituting our earlier expression for p_I' and simplifying gives

$$\Delta p_I = (1 - m)(p_I) + (m)(p_C) - p_I = m(p_C - p_I)$$

Finally, we can determine the equilibrium frequency of allele A_1 on the island. The equilibrium condition is no change in p_I. That is,

$$\Delta p_I = 0$$

If we set our expression for Δp_I equal to zero, we have

$$m(p_C - p_I) = 0$$

This expression shows that the frequency of A_1 will remain constant on the island if there is no migration ($m = 0$), or if the frequency of A_1 on the island is already identical to its frequency on the mainland ($p_I = p_C$). In other words, without any opposing force, migration will eventually equalize the frequencies of the island and mainland populations.

continent to the island. After migration, 80% of the island population is from the island, and 20% is from the continent. The new genotype frequencies are 0.8 for A_1A_1, 0 for A_1A_2, and 0.2 for A_2A_2. When individuals on the island reproduce, their gene pool will have allele frequencies of 0.8 for A_1 and 0.2 for A_2.

Migration has changed the allele frequencies in the island population, violating Hardy–Weinberg conclusion 1. Before migration, the island frequency of A_1 was 1.0; after migration, the frequency of A_1 is 0.8. The island population has evolved as a result of migration. (For an algebraic treatment of migration as a mechanism of allele frequency change, see Box 6.1).

Migration has also produced genotype frequencies among the adults on the island that are not consistent with Hardy–Weinberg conclusion 2. Under the Hardy–Weinberg equilibrium principle, a population with allele frequencies of 0.8 and 0.2 should have genotype frequencies of 0.64, 0.32, and 0.04. Compared to these expected values, the post-migration island population has an excess of homozygotes and a deficit of heterozygotes. A single bout of random mating will, of course, put the population back into Hardy–Weinberg equilibrium for genotype frequencies.

Migration is a potent force of evolution. In practice, migration is most important in preventing populations from diverging.

Empirical Research on Migration as a Mechanism of Evolution

The water snakes of Lake Erie (Figure 6.6) provide an empirical example of migration from a mainland population to an island population. These snakes (*Nerodia sipedon*) live on the mainland surrounding Lake Erie and on the islands in the lake.

(a)

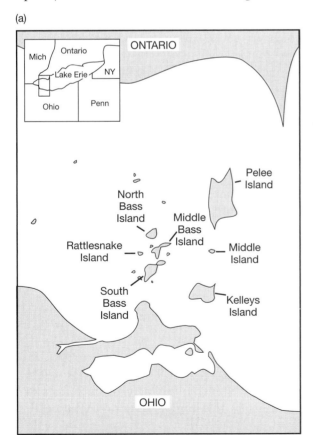

(b)

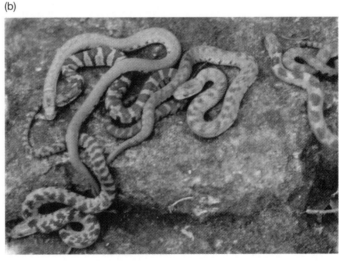

Figure 6.6   Water snakes and where they live   The map in (a) shows the island and mainland areas in and around Lake Erie where Richard King and colleagues studied migration as a force of evolution in water snakes. From King and Lawson (1995). Copyright © 1995 *Evolution*. Reprinted by permission. The photo in (b) shows unbanded, banded, and intermediate forms of the Lake Erie water snake (*Nerodia sipedon*). (Richard B. King)

Figure 6.7   Variation in color pattern within and between populations   These histograms show frequency of different color patterns in various populations. Category A snakes are unbanded; category B and C snakes are intermediate; category D snakes are strongly banded. Snakes on the mainland tend to be banded; snakes on the islands tend to be unbanded or intermediate. From Camin and Ehrlich (1958). Copyright © 1958 Evolution. Reprinted by permission.

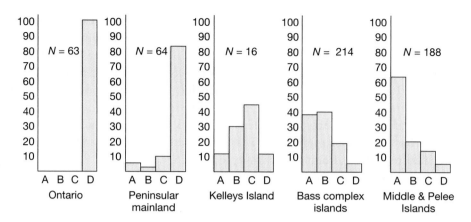

Individuals vary in appearance, ranging from strongly banded to unbanded. To a rough approximation, color pattern is determined by a single locus with two alleles, with the banded allele dominant over the unbanded allele (King 1993a).

On the mainland virtually all the water snakes are banded, whereas on the islands many snakes are unbanded (Figure 6.7). The difference in the composition of mainland versus island populations appears to be the result of natural selection caused by predators. On the islands, the snakes bask on limestone rocks at the water's edge. Following up on earlier work by Camin and Ehrlich (1958), Richard B. King (1993b) showed that among very young snakes unbanded individuals are more cryptic on island rocks than are banded individuals. The youngest and smallest snakes are presumably most vulnerable to predators. King (1993b) used mark-recapture studies, among other methods, to show that on the islands unbanded snakes indeed have higher rates of survival than banded snakes.

Migration of individuals from the mainland to islands appears to be preventing the divergence of island versus mainland populations of Lake Erie water snakes.

If selection favors unbanded snakes on the islands, then we would expect that the island populations would consist entirely of unbanded snakes. Why is this not the case? The answer, at least in part, is that every generation several banded snakes move from the mainland to the islands. The migrants bring with them alleles for banded coloration. When the migrant snakes interbreed with the island snakes, they contribute copies of the banded allele to the island gene pool. In this example, migration is acting as an evolutionary force in opposition to natural selection, preventing the island population from becoming fixed for the unbanded allele. (For an algebraic treatment of the influence of opposing forces of selection and migration on the Lake Erie water snakes, see Box 6.2.)

Migration as a Homogenizing Evolutionary Force Across Populations

Migration of water snakes from the mainland to the islands makes the island population more similar to the mainland population than it otherwise would be. This is the general effect of migration: It tends to homogenize allele frequencies across populations. In the water snakes, this homogenization is opposed by selection.

How far would the homogenization go if selection did not oppose it? The algebraic model developed in Box 6.1 shows that gene flow from a continent to an island will eventually drive the allele frequency on the island to a value exactly equal

Box 6.2   Selection and migration in Lake Erie water snakes

As described in the main text, the genetics of color pattern in Lake Erie water snakes can be roughly approximated by a single locus, with a dominant allele for the banded pattern and a recessive allele for the unbanded pattern (King 1993a). Selection by predators on the islands favors unbanded snakes. If the fitness of unbanded individuals is defined as 1, then the relative fitness of banded snakes is between 0.78 and 0.90 (King and Lawson 1995). So why has selection not eliminated banded snakes from the islands? Here we calculate the effect that migration has when it introduces new banded alleles into the island population every generation.

King and Lawson (1995) lumped all the island snakes into a single population, because snakes appear to move among islands much more often than they move from the mainland to the islands. King and Lawson used genetic techniques to estimate that 12.8 snakes move from the mainland to the island every generation. The scientists estimated that the total island snake population is between 523 and 4064 individuals, with a best estimate of 1262. This means that migrants represent a fraction of 0.003 to 0.024 of the population each generation, with a best estimate of 0.01.

With King and Lawson's estimates of selection and migration, we can calculate the equilibrium allele frequencies in the island population, at which the effects of selection and migration exactly balance each other. Let A_1 represent the dominant allele for the banded pattern, and A_2 the recessive allele for the unbanded pattern. Let p represent the frequency of A_1, and q the frequency of A_2. Following Box 5.3, we create individuals by random mating, then let selection act. After selection (but before migration), the new frequency of allele A_2 is

$$q^* = \frac{pqw_{12} + q^2w_{22}}{\overline{w}}$$

where w_{12} is the fitness of A_1A_2 heterozygotes, w_{22} is the fitness of A_2A_2 homozygotes, and $\overline{w}$ is the mean fitness of all the individuals in the population, given by $(p^2w_{11} + 2pqw_{12} + q^2w_{22})$.

For our first calculation, we will use $w_{11} = w_{12} = 0.84$, and $w_{22} = 1$. A relative fitness of 0.84 for banded snakes is the midpoint of the range within which King and Lawson (1995) estimated the true value to fall. This gives

$$q^* = \frac{pq(0.84) + q^2}{[p^2(0.84) + 2pq(0.84) + q^2]}$$

Substituting $(1 - q)$ for p gives

$$q^* = \frac{(1 - q)q(0.84) + q^2}{[(1 - q)^2(0.84) + 2(1 - q)q(0.84) + q^2]}$$

$$= \frac{0.84q + 0.16q^2}{0.84 + 0.16q^2}$$

Now we allow migration, with the new migrants representing, in this first calculation, a fraction 0.01 of the island's population (King and Lawson's best estimate). None of the new migrants carry allele A_2, so the new frequency of A_2 is

$$q' = (0.99)\frac{0.84q + 0.16q^2}{0.84 + 0.16q^2}$$

The change in q from one generation to the next is

$$\Delta q = q' - q = (0.99)\frac{0.84q + 0.16q^2}{0.84 + 0.16q^2} - q$$

Plots of Δq as a function of q appear in Figure 6.8. Look first at the green curve (b). This curve is for the function we just calculated. It shows that if q is greater than 0.05 and less than 0.94 in this generation, then q will be larger in the next generation (Δq is positive). If q is less than 0.05 or greater than 0.94 in this generation, then q will be smaller in the next generation (Δq is negative). The points where the curve crosses the horizontal axis, where $\Delta q = 0$, are the equilibrium points. The upper equilibrium point is stable: if q is less than 0.94, then q will rise in the next generation; if q is greater than 0.94, then it will fall in the next generation. Thus a middle-of-the-road prediction, given King and Lawson's estimates of selection and gene flow, is that the equilibrium frequency of the unbanded allele in the island population will be 0.94.

Curve (a) is a high-end estimate; it uses fitnesses of 0.78 for A_1A_1, 0.78 for A_1A_2, and 1 for A_2A_2, and a migration rate of 0.003 (0.3% of every generation's population are migrants). It predicts an equilibrium at $q = 0.99$. Curve (c) is a low-end estimate; it uses fitnesses of 0.90 for A_1A_1, 0.90 for A_1A_2, and 1 for A_2A_2, and a migration rate of 0.24 (2.4% of every generation's population are migrants). It predicts an equilibrium at $q = 0.64$.

Box 6.2   Continued

King and Lawson's best estimate of the actual frequency of A_2 is 0.73. This value is toward the low end of our range of predictions. Our calculation is a relatively simple one, and leaves out many factors, including recent changes in the population sizes of both the water snakes and their predators, as well as recent changes in the frequencies of banded versus unbanded snakes. For a detailed treatment of this example, see King and Lawson (1995).

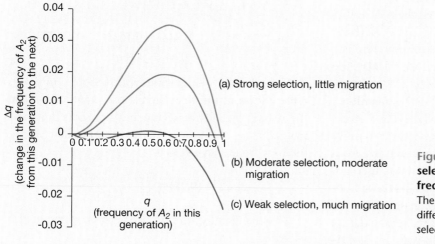

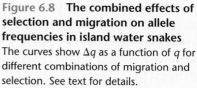

Figure 6.8   The combined effects of selection and migration on allele frequencies in island water snakes The curves show Δq as a function of q for different combinations of migration and selection. See text for details.

to what it is on the continent. In other words, if allowed to proceed unopposed by any other force of evolution, migration will eventually homogenize allele frequencies across populations completely.

Barbara Giles and Jérôme Goudet (1997) documented the homogenizing effect of gene flow on populations of red bladder campion, *Silene dioica*. Red bladder campion is an insect-pollinated perennial wildflower (Figure 6.9a). The populations that Giles and Goudet studied occupy islands in the Skeppsvik Archipelago, Sweden. These islands are mounds of material deposited by glaciers during the last ice age and left underwater when the ice melted. The area on which the islands sit is rising at a rate of 0.9 centimeters per year. As a result of this geological uplift, new islands are constantly rising out of the water. The Skeppsvik Archipelago thus contains dozens of islands of different ages.

Red bladder campion seeds are transported by wind and water, and the plant is among the first to colonize new islands. Campion populations grow to sizes of several thousand individuals. There is gene flow among islands as a result of both seed dispersal and the transport of pollen by insects. After a few hundred years, campion populations are invaded by other species of plants and by a pollinator-borne disease. Establishment of new seedlings ceases, and populations dwindle as individuals die.

Giles and Goudet predicted that young populations, having been founded by the chance transport of just a few seeds, would vary in their allele frequencies at a variety of loci. (We will consider why in more detail in Section 6.2.) Populations of intermediate age should be more homogeneous in their allele frequencies as a result of migration—that is, as a result of gene flow among populations via seed

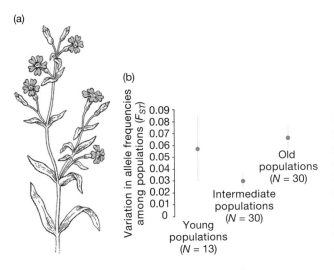

Figure 6.9   Variation in allele frequencies among populations of red bladder campion, *Silene dioica*   (a) Red bladder campion, a perennial wildflower. (b) Giles and Goudet's (1997) measurements of variation in allele frequencies among populations. The red dots represent values of F_{ST} (see text); the vertical gray lines represent standard errors (larger standard errors represent less certain estimates of F_{ST}). There is less variation in allele frequencies among intermediate age populations than among young populations ($P = 0.05$). There is more variation in allele frequencies among old populations than among intermediate populations ($P = 0.04$). After Giles and Goudet (1997).

dispersal and pollen transport. Finally, the oldest populations, structured mainly by the fortuitous survival of a few remaining individuals, should again become more variable in their allele frequencies.

The researchers tested their predictions by collecting leaves from many individual red bladder campions on 52 islands of different ages. By analyzing proteins in the leaves, Giles and Goudet determined each individual's genotype at six enzyme loci. They divided their populations by age into three groups: young, intermediate, and old. For each of these groups, they calculated a test statistic called F_{ST}. A value for F_{ST} refers to a group of populations and reflects the variation in allele frequencies among the populations in the group. The value of F_{ST} can be anywhere from 0 to 1. Larger values represent more variation in allele frequency among populations.

The results confirm Giles and Goudet's predictions (Figure 6.9b). There is less variation in allele frequencies among populations of intermediate age than among young and old populations. The low diversity among intermediate populations probably reflects the homogenizing influence of gene flow. The higher diversity of young and old populations probably represents genetic drift, the subject of the next section.

In summary, migration is the movement of alleles from population to population. Within a single population, migration can cause allele frequencies to change from one generation to the next. For small populations receiving immigrants from large source populations, migration can be a potent mechanism of evolution. Across groups of populations, gene flow tends to homogenize allele frequencies. Thus migration tends to prevent the evolutionary divergence of populations.

6.2 Genetic Drift

In Chapter 2, we refuted the misconception that evolution by natural selection is a random process. To be sure, Darwin's mechanism of evolution depends on the generation of random variation by mutation. The variation generated by mutation is random in the sense that when mutation substitutes one amino acid for another in a protein, it does so without regard to whether the change will improve the protein's ability to function. But natural selection itself is anything but random. It is precisely the nonrandomness of selection in sorting among mutations that leads to adaptation.

We are now in a position to revisit the role of chance in evolution. Arguably, the most important insight from population genetics is that natural selection is not the only mechanism of evolution. Among the nonselective mechanisms of evolution, there is one that is absolutely random. That mechanism is genetic drift. Genetic drift does not lead to adaptation, but it does lead to changes in allele frequencies. In the Hardy–Weinberg model, genetic drift results from violation of the assumption of infinite population size.

A Model of Genetic Drift

To see how genetic drift works, imagine an ideal population similar to the ones we have worked with before, but finite—in fact, small—in size. As usual, we will focus on a single locus with two alleles, A_1 and A_2. Imagine that in the present generation's gene pool, allele A_1 is at frequency 0.6, and allele A_2 is at frequency 0.4 (Figure 6.10). We will let the gametes in this gene pool combine at random to make exactly ten zygotes. These ten zygotes will constitute the entire population for the next generation.

We can simulate the production of ten zygotes from our gene pool with a physical model. A bag containing 100 beans represents the gene pool. Sixty of the beans are black, representing allele A_1; forty of the beans are white, representing allele A_2. We make each zygote by shaking the bag, closing our eyes, and drawing out beans. First we draw a bean to represent the egg, note its genotype, and return it to the bag. Then we draw a bean to represent the sperm, note its genotype and return it to the bag. We are drawing beans from a bag as we write. The genotypes of the ten zygotes are

$$A_2A_1 \quad A_1A_1 \quad A_1A_1 \quad A_1A_1 \quad A_2A_2$$

$$A_1A_1 \quad A_2A_2 \quad A_1A_2 \quad A_1A_1 \quad A_1A_1$$

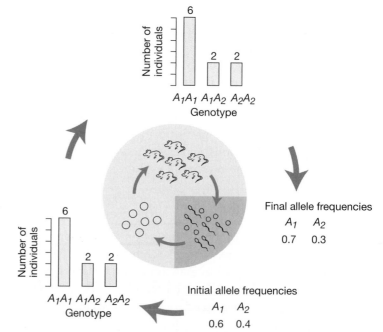

Figure 6.10 Chance events can alter allele and genotype frequencies This diagram follows an imaginary population of ten mice from one generation's gene pool (initial allele frequencies) to the next generation's gene pool (final allele frequencies). The bar graphs show the number of individuals of each genotype in the population at any given time. Genetic drift, in the form of sampling error in drawing gametes from the initial gene pool to make zygotes, increases the frequency of allele A_1. Note that many other outcomes are also possible.

Counting the genotypes, we have A_1A_1 at a frequency of 0.6, A_1A_2 at a frequency of 0.2, and A_2A_2 at a frequency of 0.2 (Figure 6.10). Counting the alleles, we see that when these zygotes grow up and reproduce, the frequency of allele A_1 in the new gene pool will be 0.7, and the frequency of allele A_2 will be 0.3 (Figure 6.10).

We have completed one turn of the life cycle of our model population. Nothing much seems to have happened, but note that both conclusions of the Hardy–Weinberg equilibrium principle have been violated. The allele frequencies have changed from one generation to the next, and we cannot calculate the genotype frequencies by multiplying the allele frequencies. The reason our population has failed to conform to the Hardy–Weinberg principle is simply that the population is small.

In a small population, chance events produce outcomes that differ from theoretical expectations. The chance events in our simulated population were the draws of beans from the bag to make zygotes. We picked black beans and white beans not in their exact predicted ratio of 0.6 and 0.4, but in a ratio that just happened to be a bit richer in black beans and a bit poorer in white beans. This kind of random discrepancy between theoretical expectations and actual results is called sampling error. Sampling error in the production of zygotes from a gene pool is **genetic drift**. Because it is nothing more than cumulative effect of random events, genetic drift cannot produce adaptation. But it can, as we have seen, cause allele frequencies to change. Blind luck is, by itself, a mechanism of evolution.

Sometimes it is difficult to see the difference between genetic drift and natural selection. In our model small population, copies of allele A_1 were more successful at getting into the next generation than copies of allele A_2. Differential reproductive success is selection, is it not? In this case, it is not. If it had been selection, the differential success of alleles in our model population would have been explicable in terms of the phenotypes the alleles confer on the individuals that carry them. Individuals with one or two copies of A_1 might have been better at surviving, finding food, or attracting mates. In fact, however, individuals carrying copies of allele A_1 were none of these things. They were just lucky; their alleles happened to get drawn from the gene pool more often. Selection is differential reproductive success that happens for a reason. Genetic drift is differential reproductive success that just happens.

Another way to see that genetic drift is different from selection is to recognize that the genotype and allele frequencies among our ten zygotes could easily have been different from what they turned out to be. To prove it, we can repeat the exercise drawing beans from our bag to make ten zygotes. This time, the genotypes of the zygotes are

$$A_1A_1 \quad A_1A_1 \quad A_1A_1 \quad A_2A_1 \quad A_1A_2$$
$$A_2A_2 \quad A_1A_2 \quad A_1A_1 \quad A_2A_1 \quad A_2A_2$$

Among this set of zygotes the genotype frequencies are 0.4 for A_1A_1, 0.4 for A_1A_2, and 0.2 for A_2A_2. The allele frequencies are 0.6 for A_1 and 0.4 for A_2.

Repeating the exercise a third time produces these zygotes:

$$A_1A_1 \quad A_1A_1 \quad A_1A_1 \quad A_1A_2 \quad A_1A_1$$
$$A_1A_2 \quad A_2A_1 \quad A_2A_2 \quad A_2A_2 \quad A_2A_2$$

Now the genotype frequencies are 0.4 for A_1A_1, 0.3 for A_1A_2, and 0.3 for A_2A_2, and the allele frequencies are 0.55 for A_1 and 0.45 for A_2.

In populations of finite size, chance events—in the form of sampling error in drawing gametes from the gene pool—can cause evolution.

Selection is differential reproductive success that happens for a reason; genetic drift is differential reproductive success that just happens.

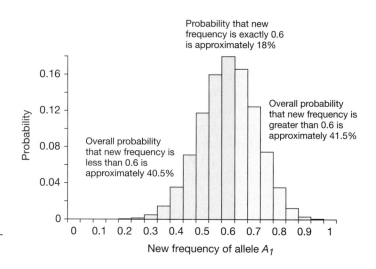

Figure 6.11   The range of possible outcomes in our model population of ten mice   This graph shows the possible outcomes, and the probability of each, when we make 10 zygotes by drawing alleles from a gene pool in which alleles A_1 and A_2 have frequencies of 0.6 and 0.4. The single most probable outcome is that the allele frequencies will remain unchanged. However, the chance of this happening is only about 18%.

Here is a summary of the results from our model population:

	Frequency of A_1
In the gene pool	0.6
In the first set of 10 zygotes	0.7
In the second set of 10 zygotes	0.6
In the third set of 10 zygotes	0.55

The three sets of zygotes have shown us that if we start with a gene pool in which allele A_1 is at a frequency of 0.6 and make a population of just ten zygotes, the frequency of A_1 may rise, stay the same, or fall. In fact, the new frequency of A_1 among a set of ten zygotes drawn from our gene pool could turn out to be anywhere from 0 to 1.0, although outcomes at the extremes of this range are not likely. The graph in Figure 6.11 shows the theoretical probability of each possible outcome. Overall, there is about an 18% chance that the frequency of allele A_1 will stay at 0.6, about a 40.5% chance that it will drop to a lower value, and about a 41.5% chance that it will rise to a higher value. Readers should not just take our word for it; they should set up their own bag of beans and make a few batches of zygotes. Again, the point is that genetic drift is evolution that simply happens by chance.

Genetic Drift and Population Size

Genetic drift is fundamentally the result of finite population size. If we draw beans from our bag to make a population of more than ten zygotes, the allele frequencies among the zygotes will get closer to the values predicted by the Hardy–Weinberg equilibrium principle. Drawing beans out of a bag quickly becomes tedious, so we used a computer to simulate drawing gametes to make not just 10, but 250 zygotes (Figure 6.12a). As the computer drew each gamete, it gave a running report of the frequency of A_1 among the zygotes it had made so far. At first this running allele frequency fluctuated wildly. For example, the first zygote turned out to have genotype A_2A_2, so the running frequency of allele A_1 started at zero. The next several zygotes were mostly A_1A_1 and A_1A_2, which sent the running frequency of allele A_1 skyrocketing to 0.75. As the cumulative number of zygotes made increased, the frequency of allele A_1 in the new generation bounced around

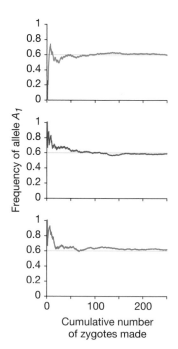

Figure 6.12   A simulation of drawing alleles from a gene pool, run three times   At first the new frequency of allele A_1 fluctuates considerably, in a unique trajectory for each run. As the number of zygotes made increases, however, the new frequency of A_1 settles toward the expected value of 0.6.

less and less, gradually settling toward the expected value of 0.6. The deviations from expectation along the way to a large number of zygotes were random, as illustrated by the graphs in Figure 6.12b and (c). These graphs show two more sets of draws to make 250 zygotes. In each, the allele frequency in the new generation fluctuates wildly at first, but in a unique pattern. As in the first graph, however, the allele frequency in the new generation always eventually settles toward the theoretically predicted value of 0.6.

Our simulations demonstrate that sampling error diminishes as sample size increases. If we kept drawing gametes forever, to make an infinitely large population of zygotes, the frequency of allele A_1 among the zygotes would be exactly 0.6. Genetic drift is a powerful evolutionary mechanism in small populations, but its power declines in larger populations. We will return to this point in later sections.

Genetic drift is most important in small populations.

Empirical Research on Sampling Error as a Mechanism of Evolution: The Founder Effect

If we want to observe genetic drift in nature, the best place to look is in small populations. Populations are often small when they have just been founded by a group of individuals that have moved, or been moved, to a new location. The allele frequencies in the new population are likely, simply by chance, to be different than they are in the source population. This is called the **founder effect**.

The founder effect is a direct result of sampling error. For example, if 25 different alleles are present at a single locus in a continental population of insects, but just 10 individuals are on a log that rafts to a remote island, the probability is zero that the new island population will contain all of the alleles present on the continent. If, by chance, any of the founding individuals are homozygotes, allele frequencies in the new population will have shifted even more dramatically. In any founder event, some degree of random genetic differentiation is almost certain between old and new populations. In other words, the founding of a new population by a small group of individuals typically represents not only the establishment of a new population but also the instantaneous evolution of differences between the new population and the old population.

Peter Grant and Rosemary Grant (1995) watched the establishment of a new population of large ground finches (*Geospiza magnirostris*) in the Galápagos Islands. Grant and Grant had been working on the island of Daphne Major since 1973. Each year, large ground finches visited the island, with anywhere from 10 to 50 juveniles arriving after the breeding season and staying through the dry months. For the first several years that Grant and Grant were there, the visiting birds all left the island before the next breeding season began. The visiting finches were all members of some other island's population. Then in the fall of 1982, three males and two females, apparently enticed by early rains, stayed on Daphne Major to breed. These five birds formed pairs (one female bred with two different males), built eight nests over the course of the breeding season, and fledged 17 young during early 1983.

The five 1982–1983 breeders were the founders of a new population. Since 1983, large ground finches have bred on Daphne Major every year, with the exception of three drought years. By 1993, the Daphne Major breeding population included 23 males and 23 females. Through at least 1992, the majority of the Daphne Major breeders had been hatched on the island. They were natives of the new population.

Is the newly founded finch population genetically different from the source population? Although Grant and Grant do not have direct data on allele frequencies at specific loci, they do have extensive morphological measurements of the 238 ground finches that visited Daphne Major prior to the founding event, and of the five original Daphne Major breeders and 22 of their offspring. Grant and Grant assumed that the 238 nonbreeding visitors were representative of the source population, and compared them to the members of the newly founded population. In at least two morphological traits, bill width and bill shape in relation to body size, the members of the new population were significantly larger than the source population. Research on the new population of large ground finches, and on other Darwin's finches, suggests that these traits are heritable. Thus it appears that the founding event created a new population that is measurably different from the source population. Evolution occurred, not through selection but by random sampling error. This was genetic drift in the form of the founder effect.

When a new population is founded by a small number of individuals, it is likely that chance alone will cause the allele frequencies in the new population to be different from those in the source population. This is the founder effect.

Founder effects are often seen in genetically isolated human populations. For example, the Amish population of eastern Pennsylvania is descended from a group of about 200 European settlers who came to the United States in the 18th century. One of the founders—either the husband or wife (or both) in a couple named King—was a carrier for Ellis–van Creveld syndrome. Ellis–van Creveld syndrome is a rare form of dwarfism caused by a recessive allele on chromosome 4 (Bodmer and McKie 1995). The frequency of this allele is about 0.001 in most populations, but it is about 0.07 in the present-day Amish (Postlethwait and Hopson 1992). The high frequency of the allele in the Amish population is probably not due to any selective advantage conferred by the allele in either heterozygotes or homozygotes. Instead, the high frequency of the allele is simply due to chance. The allele was at a high frequency in the small population of founders and has continued to drift upward in subsequent generations. In the next section we will consider in detail the cumulative effect of genetic drift over the course of many generations.

Random Fixation of Alleles and Loss of Heterozygosity

We have seen that genetic drift can produce substantial change in allele frequencies in a single generation. Drift is even more powerful as a mechanism of evolution when its effects are compounded over many generations. We can investigate the cumulative effects of genetic drift with the same physical model we used before: black beans and white beans in a bag. We again set up a bag with 60 black beans and 40 white beans, representing a gene pool in which alleles A_1 and A_2 are at frequencies of 0.6 and 0.4. We will call the parents who produced this gene pool generation zero. As we did before, we now draw beans from the bag to simulate the production of ten zygotes by random mating. This time, the allele frequencies among the newly formed zygotes turn out to be 0.5 for A_1 and 0.5 for A_2. We will call these zygotes generation one.

To continue the simulation for another generation, we need to set up a new bag, with 50 black beans and 50 white beans, to represent generation one's gene pool. Drawing beans from this gene pool, we get the zygotes for generation two. Generation two's allele frequencies happen to be 0.4 for A_1 and 0.6 for A_2.

We now set up a bag with 40 black beans and 60 white beans and draw zygotes to make generation three. Generation three's allele frequencies are 0.45 for A_1 and 0.55 for A_2.

Now, we need a bag with 45 black beans and 55 white beans, and so on. The advantage of using a computer to simulate drawing beans from bags is rapidly becoming apparent.

Graphs (a), (b), and (c) in Figure 6.13 show the results of 100 successive generations of genetic drift in simulated populations of different sizes. Each graph

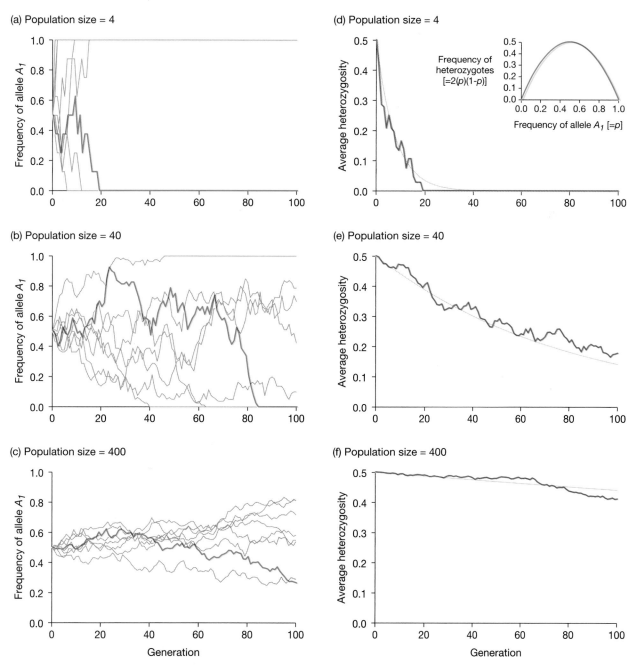

Figure 6.13   Simulations of genetic drift in populations of different sizes   Plots (a), (b), and (c) show the frequency of allele A_1 over 100 generations. Eight populations are tracked in each plot, with one population highlighted in red. Plots (d), (e), and (f) show the average frequency of heterozygotes over 100 generations in the same sets of simulated populations. The gray curves represent the rate of decline predicted by theory. The inset in plot (d) shows the frequency of heterozygotes in a population, calculated as $2(p)(1 - p)$, as a function of p, the frequency of allele A_1. Collectively, the graphs in this figure show that genetic drift leads to random fixation of alleles and loss of heterozygosity, and that drift is a more powerful force of evolution in small populations.

tracks allele frequencies in eight populations. Every population starts with allele frequencies of 0.5 for A_1 and 0.5 for A_2. The populations tracked in graph (a) have just four individuals each, the populations tracked in graph (b) have 40 individuals each, and the populations tracked in graph (c) have 400 individuals each. Three patterns are evident:

Under genetic drift, every population follows a unique evolutionary path. Genetic drift is rapid in small populations and slow in large populations.

1. Because the fluctuations in allele frequency from one generation to the next are caused by random sampling error, every population follows a unique evolutionary path.
2. Genetic drift has a more rapid and dramatic effect on allele frequencies in small populations than in large populations.
3. Given sufficient time, genetic drift can produce substantial changes in allele frequencies, even in populations that are fairly large.

Note that if genetic drift is the only evolutionary force at work in a population—if there is no selection, no mutation, and no migration—then sampling error causes allele frequencies to wander between 0 and 1. This wandering is particularly apparent in the population whose evolution is highlighted in the graph in Figure 6.13b. During the first 25 generations allele A_1's frequency rose from 0.5 to over 0.9. Between generations 25 and 40 it dropped back to 0.5. Between generations 40 and 80 the frequency bounced between 0.5 and 0.8. Then the frequency of A_1 dropped precipitously, so that by generation 85 it hit zero and A_1 disappeared from the population altogether. The wandering of allele frequencies produces two important and related effects: (1) Eventually alleles drift to fixation or loss, and (2) the frequency of heterozygotes declines.

Random Fixation of Alleles

As any allele drifts between frequencies of 0 and 1.0, sooner or later the allele will meet an inevitable fate: Its frequency will hit one boundary or the other. If the allele's frequency hits 0, then the allele is lost forever (assuming that mutation or migration do not reintroduce it). If the allele's frequency hits 1, then the allele is said to be fixed, also forever. Among the eight populations tracked in Figure 6.13a, allele A_1 drifted to fixation in five and to loss in three. Among the eight populations tracked in Figure 6.13b, A_1 drifted to fixation in one and to loss in three. It is just a matter of time before A_1 will become fixed or lost in the other populations as well. As some alleles drift to fixation and others drift to loss, the allelic diversity present in a population declines.

If genetic drift is the only evolutionary force at work, eventually one allele will drift to a frequency of 1 (that is, to fixation) and all other alleles will be lost.

Now imagine a finite population in which there are several alleles present at a particular locus: A_1, A_2, A_3, A_4, and so on. If genetic drift is the only evolutionary mechanism at work, then eventually one of the alleles will drift to fixation. At the same moment one allele becomes fixed, the last of the other alleles will be lost.

We would like to be able to predict which alleles will meet which fate. We cannot do so with certainly, but we can give odds. Sewall Wright (1931) proved that the probability that any given allele in a population will be the one that drifts to fixation is equal to that allele's initial frequency (see Box 6.3). If, for example, we start with a finite population in which A_1 is at a frequency of 0.73, and A_2 is at a frequency of 0.27, there is a 73% chance that the allele that drifts to fixation will be A_1.

Box 6.3   The probability that a given allele will be the one that drifts to fixation

Sewall Wright (1931) developed a detailed theory of genetic drift. Among many other results, he showed that the probability that a given allele will be the one that drifts to fixation is equal to that allele's initial frequency. Wright's model of genetic drift is beyond the scope of this book, but we can provide an intuitive explanation of fixation probabilities.

Imagine a population of N individuals. This population contains a total of $2N$ alleles. Imagine that every one of these alleles is unique. Assume that genetic drift is the only mechanism of evolution at work.

At some point in the future, one of the $2N$ alleles will have drifted to fixation and all the others will have been lost. Each allele must have an equal chance of being the one that drifts to fixation; that is what we meant by our assumption that genetic drift is the only mechanism of evolution at work. So we have $2N$ alleles, each with an equal probability of becoming fixed. Each allele's chance of becoming fixed must therefore be $\frac{1}{2N}$.

Now imagine that instead of each allele being unique, there are x copies of allele A_1, y copies of allele A_2, and z copies of allele A_3. Each copy of allele A_1 has a $\frac{1}{2N}$ chance of being the one that drifts to fixation. Therefore, the overall probability that a copy of allele A_1 will be the allele that drifts to fixation is

$$x \times \frac{1}{2N} = \frac{x}{2N}$$

Likewise, the probability that a copy of allele A_2 will be the allele that drifts to fixation is $\frac{y}{2N}$, and the probability that a copy of allele A_3 will be the allele that drifts to fixation is $\frac{z}{2N}$.

Notice that, $\frac{x}{2N}$, $\frac{y}{2N}$, and $\frac{z}{2N}$ are also the initial frequencies of A_1, A_2, and A_3 in the population. We have shown that the probability that a given allele will be the one that drifts to fixation is equal to that allele's initial frequency.

Loss of Heterozygosity

As allele frequencies in a finite population drift toward fixation or loss, the frequency of heterozygotes in the population decreases. Graphs (d), (e), and (f) in Figure 6.13 show the decline in the frequency of heterozygotes in our simulated populations.

To see why the frequency of heterozygotes declines, look first at the inset in graph (d). The inset plots the frequency of heterozygotes, calculated as $2(p)(1-p)$, in a random mating population as a function of p, the frequency of allele A_1. The frequency of heterozygotes has its highest value, 0.5, when A_1 is at frequency 0.5. As the frequency of A_1 drops toward 0 or rises toward 1, the frequency of heterozygotes falls. And, of course, if the frequency of A_1 reaches 0 or 1, the frequency of heterozygotes falls to 0.

Now look at graphs (a), (b), and (c). In any given generation, the frequency of A_1 may move toward or away from 0.5 in any particular population (so long as A_1 has not already been fixed or lost). Thus the frequency of heterozygotes in any particular population may rise or fall. However, the overall trend across all populations is for allele frequencies to drift away from intermediate values and toward 0 or 1. So the average frequency of heterozygotes, across populations, should tend to fall.

Now look at graphs (d), (e), and (f). The heavy blue line in each graph tracks the frequency of heterozygotes averaged across the eight populations in question. The frequency of heterozygotes does indeed tend to fall, rapidly in small populations and slowly in large populations. Eventually one allele or the other will become fixed in every population, and the average frequency of heterozygotes will fall to 0.

The frequency of heterozygotes in a population is sometimes called the population's **heterozygosity**. We would like to be able to predict just how fast the heterozygosity of finite populations can be expected to decline. Sewell Wright (1931) showed that, averaged across many populations, the frequency of heterozygotes obeys the relationship

As alleles drift to fixation or loss, the frequency of heterozygotes in the population declines.

$$H_{g+1} = H_g \left[1 - \frac{1}{2N} \right]$$

where H_{g+1} is the heterozygosity in the next generation, H_g is the heterozygosity in this generation, and N is the number of individuals in the population. The value of $\left[1 - \frac{1}{2N} \right]$ is always between $\frac{1}{2}$ and 1, so the expected frequency of heterozygotes in the next generation is always less than the frequency of heterozygotes in this generation. The gray curves in Figure 6.13d, (e), and (f) show the declines in heterozygosity predicted by Wright's equation.

To appreciate just one of the implications of the inevitable loss of heterozygosity in finite populations, imagine that you are responsible for managing a captive population of an endangered species. Suppose there are just 50 breeding adults in zoos around the world. Even if you could arrange the shipment of adults or semen to accomplish random mating, you would still see a loss in heterozygosity of 1% every generation due to genetic drift.

An Experimental Study on Random Fixation and Loss of Heterozygosity

Our discussion of random fixation and loss of heterozygosity has so far been based on simulated populations and mathematical equations. Peter Buri (1956) studied these phenomena empirically, in small laboratory populations of the fruit fly, *Drosophila melanogaster*. Adopting an approach that had been used by Kerr and Wright (1954), Buri established 107 populations of flies, each with eight females and eight males. All the founding flies were heterozygotes for alleles of an eye-color gene called *brown*. All the flies had the same genotype: bw^{75}/bw. Thus, in all 107 populations, the initial frequency of the bw^{75} allele was 0.5. Buri maintained these populations for 19 generations. For every population in every generation, Buri kept the population size at 16 by picking eight females and eight males at random to be the breeders for the next generation.

What results would we predict? If neither allele bw^{75} nor allele bw confers a selective advantage, then we expect the frequency of allele bw^{75} to wander at random by genetic drift in every population. Nineteen generations should be enough, in populations of 16 individuals, for many populations to become fixed for one allele or the other. Because allele bw^{75} has an initial frequency of 0.5, we expect this allele to be lost about as often as it becomes fixed. As the bw^{75} allele is drifting toward fixation or loss in each population, we expect the average heterozygosity across all populations to decline. The rate of decline in heterozygosity should follow Wright's equation, given in the previous section.

Buri's results confirm these predictions. Each small graph in Figure 6.14 is a histogram summarizing the allele frequencies in all 107 populations in a particular generation. The horizontal axis represents the frequency of the bw^{75} allele, and the vertical axis represents the number of populations showing each frequency. The frequency of bw^{75} was 0.5 in all populations in generation zero, which is not shown

Generation

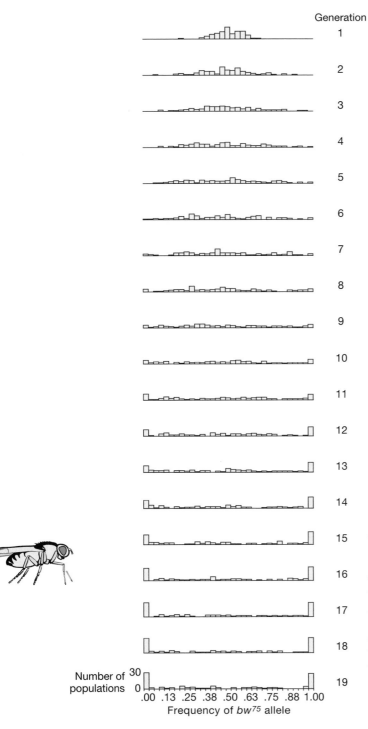

Number of populations

Frequency of bw^{75} allele

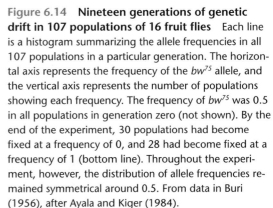

Figure 6.14   Nineteen generations of genetic drift in 107 populations of 16 fruit flies   Each line is a histogram summarizing the allele frequencies in all 107 populations in a particular generation. The horizontal axis represents the frequency of the bw^{75} allele, and the vertical axis represents the number of populations showing each frequency. The frequency of bw^{75} was 0.5 in all populations in generation zero (not shown). By the end of the experiment, 30 populations had become fixed at a frequency of 0, and 28 had become fixed at a frequency of 1 (bottom line). Throughout the experiment, however, the distribution of allele frequencies remained symmetrical around 0.5. From data in Buri (1956), after Ayala and Kiger (1984).

in the figure. After one generation of genetic drift, most populations still had an allele frequency near 0.5, although one population had an allele frequency as low as 0.22 and another had an allele frequency as high as 0.69. As the frequency of the bw^{75} allele rose in some populations and fell in others, the distribution of allele frequencies rapidly spread out. In generation four, the frequency of bw^{75} hit 1 in a population for the first time. In generation six, the frequency of bw^{75} hit 0 in a population for the first time. As the allele frequency reached 0 or 1 in more and

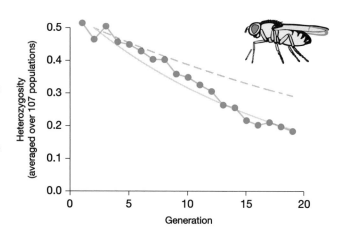

Figure 6.15 The frequency of heterozygotes declined with time in Buri's experimental populations The red dots show the frequency of heterozygotes in each generation, averaged across all 107 populations. The dashed gray curve shows the theoretical prediction for a population of 16 flies. The solid gray curve shows the theoretical prediction for a population of 9 flies. The graph demonstrates that (1) heterozygosity decreases across generations in small populations, and (2) although all the populations had an actual population of 16 flies, their effective population size was roughly 9. Replotted from data in Buri (1956), after Hartl (1981).

Empirical studies confirm that under genetic drift alleles become fixed or lost and the frequency of heterozygotes declines. Indeed, these processes often happen faster than predicted.

more populations, the distribution of frequencies became U-shaped. By the end of the experiment, bw^{75} had been lost in 30 populations and had become fixed in 28. The 30:28 ratio of losses to fixations is very close to the 1:1 ratio we would predict under genetic drift. During Buri's experiment there was dramatic evolution in nearly all 107 of the fruit fly populations, but natural selection had nothing to do with it.

The genetic properties of the *brown* locus were such that Buri could identify all three genotypes from their phenotypes. Thus Buri was able to directly assess the frequency of heterozygotes in each population. The frequency of heterozygotes in generation zero was 1, so the heterozygosity in generation one was 0.5. Every generation thereafter, Buri noted the frequency of heterozygotes in each population, then took the average heterozygosity across all 107 populations. Figure 6.15 tracks these values for average heterozygosity over the 19 generations of the experiment. Look first at the red dots, which represent the actual data. Consistent with our theoretical prediction, the average frequency of heterozygotes steadily declined.

The fit between theory and results is not perfect, however. The dashed gray curve in the figure shows the predicted decline in heterozygosity, using Wright's equation and a population size of 16. The actual decline in heterozygosity was more rapid than expected. The solid gray curve shows the predicted decline for a population size of 9; it fits the data well. Buri's populations lost heterozygosity as though they contained only 9 individuals instead of 16. In other words, the **effective population size** in Buri's experiment was 9 (see Box 6.4). Among the explanations are that some of the flies in each population may have died due to accidents before reproducing, or some males may have been rejected as mates by the females.

Buri's experiment with fruit flies shows that the theory of genetic drift allows us to make accurate qualitative predictions, and reasonably accurate quantitative predictions, about the behavior of alleles in finite populations—at least in the lab. In the next section we will consider evidence on random fixation of alleles and loss of heterozygosity in natural populations.

Random Fixation and Loss of Heterozygosity in Natural Populations

Alan Templeton and colleagues (1990) tested predictions about the random fixation of alleles by documenting the results of a natural experiment in Missouri's Ozark Mountains. Although now covered in oak-hickory forest, the Ozarks were

Box 6.4 Effective population size

The effective population size is the size of an ideal theoretical population that would lose heterozygosity at the same rate as an actual population of interest. The effective population size is virtually always smaller than the actual population size. In Buri's experiment, two possible reasons for the difference in effective versus actual population size are that (1) some of the flies in each bottle died (by accident) before reproducing and (2) fruit flies exhibit sexual selection by both male–male combat and female choice (see Chapter 9)—either of which could have prevented some males from reproducing.

The effective population size is particularly sensitive to differences in the number of reproductively active females versus males. When there are different numbers of each sex in a population, the effective population size N_e can be estimated as

$$N_e = \frac{4N_m N_f}{(N_m + N_f)}$$

where N_m is the number of males and N_f is the number of females.

To see how strongly an imbalanced sex ratio can reduce the effective population size, use the formula to show that: when there are 5 males and 5 females, $N_e = 10$; when there is 1 male and 9 females, $N_e = 3.6$; and when there is 1 male and 1000 females, $N_e = 4$. Consider the logistical problems involved in maintaining a captive breeding program for a species in which the males are extremely aggressive and will not tolerate each other's presence.

part of a desert during an extended period of hot, dry climate that lasted from 8000 to 4000 years ago. The desert that engulfed the Ozarks was contiguous with the desert of the American Southwest. Many southwestern desert species expanded their ranges eastward into the Ozarks. Among them was the collared lizard (*Crotaphytus collaris*). When the warm period ended, the collared lizard's range retracted westward again and the oak-hickory forest reinvaded the Ozarks. Within this forest, however, on exposed rocky outcrops, are small remnants of desert habitat called glades. Living in some of these glades are relict populations of collared lizards. Most populations are sufficiently isolated from each other that there is little or no gene flow among them. The relict populations are tiny; most harbor no more than a few dozen individuals.

Because of the small size of the glade populations, Templeton and colleagues predicted that the collard lizards of the Ozarks would bear a strong imprint of genetic drift. Within each population, most loci should be fixed for a single allele, and genetic variation should be very low. Which allele became fixed in any particular population should be a matter of chance, however, so there should be considerable genetic diversity among populations.

Templeton and colleagues assayed several glade populations for genetic variation. The researchers screened lizards for their genotypes at a variety of enzyme loci, for their ribosomal DNA genotypes, and for their mitochondrial DNA genotypes. The researchers identified among the lizards seven distinct multilocus genotypes. Confirming the predicted consequences of isolation and small population size, most glade populations are fixed for a single multilocus genotype, with different genotypes fixed in different populations (Figure 6.16).

Andrew Young and colleagues (1996) tested predictions about the effect of population size on heterozygosity with a comparative study of plants. The researchers compiled data from the literature and plotted two measures of overall genetic diversity against population size in three flowering herbs and a tree. The first measure of genetic diversity was genetic polymorphism, the fraction of loci within

Figure 6.16 Genetic variation in Ozark glade populations of the collared lizard
(a) The pie diagram gives a key to the seven distinct multilocus genotypes that Templeton et al. (1990) found in Ozark collared lizards. Each multilocus genotype is characterized by a malate dehydrogenase (MDH) genotype (the two alleles are "slow" (S) and "fast" (F), a mitochondrial DNA haplotype (designated A–D), and a ribosomal DNA genotype (designated I–III). (b) This is a map of southern Missouri, showing the locations and genetic compositions of nine glade populations. The shading of each pie diagram represents the frequency in a single population of each multilocus genotype present. (c) This is an expanded map of a small piece of the map in (b). It gives the locations and genetic compositions of five more glade populations. From Templeton et al. (1990). Copyright © 1990 Alan R. Templeton. Reprinted by permission of the author.

Crotaphytus collaris

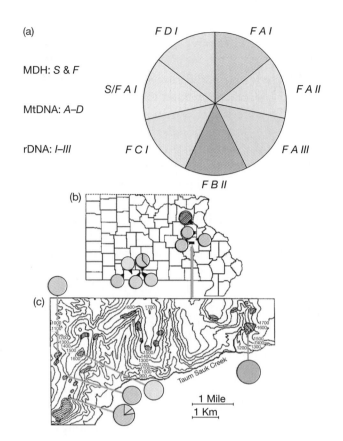

the genome that have at least two alleles with frequencies higher than 0.01. The second was allelic richness, the average number of alleles per locus. Both of these measures are related to heterozygosity. If a given locus has more than one allele, and if a substantial number of individuals in the population are heterozygotes, then the locus contributes to high values of polymorphism and allelic richness. If, on the other hand, the locus is fixed for a single allele and no individual in the population is a heterozygote, then the locus lowers the population's polymorphism and allelic richness. Because genetic drift is more pronounced in small populations than in large ones, and because genetic drift results in the loss of heterozygosity, Young and colleagues predicted that small populations would have lower levels of polymorphism and allelic richness. Young et al.'s plots appear in Figure 6.17. Consistent with their prediction, in almost every case smaller populations did indeed harbor less genetic diversity.

The studies by Templeton et al. and Young et al. show that in at least some natural populations genetic drift leads, as predicted, to random fixation and reduced heterozygosity. The loss of genetic diversity in small populations is of particular concern to conservation biologists, for two reasons. First, genetic diversity is the raw material for adaptive evolution. Imagine a species reduced to a few remnant populations by habitat destruction or some other environmental change. Genetic drift may rob the remnant populations of their potential to evolve in response to a changing environment at precisely the moment the environment is changing most drastically. Second, a loss of heterozygosity also entails an increase in homozygosity. Increased homozygosity often leads to reduced fitness in experimental populations (see, for example, Polans and Allard 1989; Barrett and Charlesworth 1991). Presumably this in-

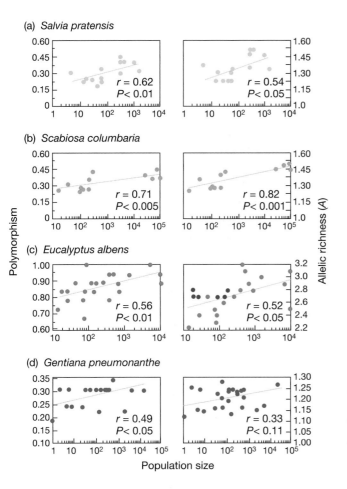

Figure 6.17   Population size and genetic diversity Each data point on these scatterplots represents a population of flowering plants. Polymorphism, plotted on the vertical axis of the graphs at left, is the proportion of allozyme loci at which the frequency of the most common allele in the population is less than 0.99. In other words, polymorphism is the fraction of alleles that are substantially polymorphic. Allelic richness, plotted on the vertical axis of the graphs at right, is the average number of alleles per locus. The statistic *r* is a measure of association, called the Pearson correlation coefficient, which varies from 0 (no association between variables) to 1 (perfect correlation). *P* specifies the probability that the correlation coefficient is significantly different from zero. *Salvia pratensis, Scabiosa columbaria,* and *Gentiana pneumonanthe* are all flowering herbs; *Eucalyptus albens* is a tree. Reprinted from Young et al. (1996). Copyright © 1996 Elsevier Science. Reprinted with permission of Elsevier Science.

volves the same mechanism as inbreeding depression: It exposes deleterious alleles to selection. We will consider inbreeding depression in Section 6.3.

The Rate of Evolution by Genetic Drift

The theory and experiments we have discussed in this section establish that sampling error can be an important mechanism of evolution. The final aspect of drift that we shall consider here is the rate of evolution when genetic drift is the only force at work.

First, we need to define what we mean by the rate of evolution at a single locus. We will take the rate of evolution to be the rate at which new alleles created by mutation are substituted for other alleles already present. Figure 6.18 illustrates the process of **substitution** and distinguishes substitution from mutation. The figure follows a gene pool of 10 alleles for 20 generations. Initally, all of the alleles are identical (white dots). In the fourth generation, a new allele appears (light orange dot), created by a mutation in one of the original alleles. Over several generations, this allele drifts to high frequency. In generation fifteen, a second new allele appears (blue dot), created by a mutation in a descendant of the first light orange allele. In generation nineteen, the last of the white alleles is lost. At this point, we can say that the light orange allele has been subsituted for the white allele. Thus, by evolutionary substitution, we mean the fixation of a new mutation, with or without additional mutational change.

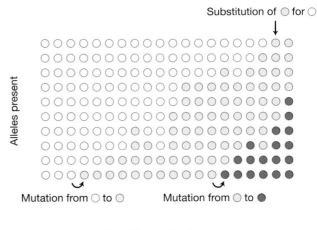

Figure 6.18 Mutation is the creation of a new allele; substitution is the fixation of the new allele, with or without additional mutational change This graph shows the 10 alleles present in each of 20 successive generations, in a hypothetical population of five individuals. During the time covered in the graph, the light orange allele was substituted for the white allele. The blue allele may ultimately be substituted for the light orange allele, or it may be lost.

When mutation, genetic drift, and selection interact, three processes occur: (1) Deleterious alleles appear and are eliminated by selection; (2) neutral mutations appear and are fixed or lost by chance; and (3) advantageous alleles appear and are swept to fixation by selection. The relative importance of (2) and (3) in determining the overall substitution rate is a matter of debate.

When genetic drift is the only mechanism of evolution at work, the rate of substitution is simply equal to the mutation rate (see Box 6.5). This is true regardless of the population size, because two effects associated with population size cancel each other out: More mutations occur in a larger population, but in a large population each new mutation has a smaller chance of drifting to fixation. Under genetic drift, large populations generate and maintain more genetic variation than small populations, but populations of all sizes accumulate substitutions at the same rate.

Of course, other mechanisms of evolution than drift are often at work. We can allow some natural selection into our model and still get a similar result. Imagine that some mutations are deleterious, while others are selectively neutral. The deleterious mutations will be eliminated by natural selection and will never become fixed. The rate of substitution will then be equal to the rate at which neutral mutations occur.

Evolutionary biologists are divided on the relevance of this calculation to real populations. All agree that one kind of mutation and one kind of selection have been left out (see Box 6.5). Some mutations are selectively advantageous, and are swept to fixation by natural selection more surely and much faster than drift would ever carry them. Evolutionists are of two minds, however, over how often this happens.

Proponents of the **neutral theory**, long championed by Motoo Kimura (1983), hold that advantageous mutations are exceedingly rare, and that most alleles of most genes are selectively neutral. Neutralists predict that for most genes in most populations, the rate of evolution will, indeed, be equal to the neutral mutation rate.

Proponents of the **selectionist theory**, most strongly championed by John Gillespie (1991), hold that advantageous mutations are common enough that they cannot be ignored. Selectionists predict that for many genes in most populations, the rate of substitution will reflect the action of natural selection on advantageous mutations.

The dispute between the neutralists and the selectionists is unresolved, and we will defer the presentation of empirical evidence until Chapter 18. The neutral theory is central, however, to many aspects of modern evolutionary biology. It provides a null model for the detection of selection at the level of DNA and protein sequences (see Chapters 18 and 19). And it provides the theoretical basis for the molecular clocks used to infer the ages of common ancestors from sequence data (see Chapter 13).

Box 6.5   The rate of evolutionary substitution under genetic drift

Here we show a calculation establishing that when genetic drift is the only mechanism of evolution at work, the rate of evolutionary substitution is equal to the mutation rate (Kimura 1968).

Imagine a diploid population of size N. Within this population are $2N$ alleles of the locus of interest, where by "alleles" we mean copies of the gene, regardless of whether they are identical or not. Let v be the rate of selectively neutral mutations per allele per generation, and assume that each mutation creates an allele that has not previously existed in the population. Then every generation, there will be

$$2Nv$$

new alleles created by mutation. Because by assumption all new alleles are selectively neutral, genetic drift is the only force at work. Each new allele has the same chance of drifting to fixation as any other allele in the population. That chance, equal to the frequency of the new allele, is

$$\frac{1}{2N}$$

Therefore, each generation, the number of new alleles that are created by mutation and are destined to drift to fixation, is

$$2Nv \times \frac{1}{2N} = v$$

The same argument applies to every generation. Therefore, the rate of evolution at the locus of interest is v substitutions per generation.

It will be useful for discussions in other chapters to explore in more detail what we mean by v, the rate of neutral mutations. Imagine that the locus of interest is a gene encoding a protein that is L amino acids long. Let u be the rate of mutations per codon per generation. The overall rate of mutation at our locus is given by

$$\mu = uL(d + a + f) = uLd + uLa + uLf$$

where d is the fraction of codon changes that are deleterious, a is the fraction that are selectively advantageous, f is the fraction that are selectively neutral, and $d + a + f = 1$. Note that the rightmost term, uLf, is equal to our earlier v.

In showing that the rate of substitution is equal to v, we assumed that d and a are both equal to zero. In any real population, of course, many mutations are deleterious and d is not zero. This does not change our calculation of the substitution rate. Deleterious alleles are eliminated by natural selection, and do not contribute to the rate of evolutionary substitution.

Proponents of the neutral theory hold that a is approximately equal to zero, and that f is much larger than a. Therefore they predict that evolutionary subsitution will be dominated by neutral mutations and drift, and will occur at the rate $v = uLf$, as we have calculated.

Proponents of the selectionist theory hold that a is too large to ignore, and that the rate of evolutionary substitution will be significantly influenced by the action of natural selection in favor of advantageous alleles.

In summary, genetic drift is a nonadaptive mechanism of evolution. Simply as a result of chance sampling error, allele frequencies can change from one generation to the next. Genetic drift is most powerful in small populations and when compounded over many generations. Ultimately, genetic drift leads to the fixation of some alleles and the loss of others, and to an overall decline in genetic diversity.

6.3 Nonrandom Mating

The final assumption of the Hardy–Weinberg analysis is that individuals in the population mate at random. In this section we relax that assumption and allow individuals to mate nonrandomly. Nonrandom mating does not, by itself, cause evolution. Nonrandom mating can nonetheless have profound indirect effects on evolution.

The most common type of nonrandom mating, and the kind we will focus on here, is inbreeding. Inbreeding is mating among genetic relatives. The effect of inbreeding on the genetics of a population is to increase the frequency of homozygotes compared to what is expected under Hardy–Weinberg assumptions.

Inbreeding decreases the frequency of heterozygotes and increases the frequency of homozygotes compared to expectations under Hardy–Weinberg assumptions.

To show how this happens, we will consider the most extreme example of inbreeding: self-fertilization, or selfing. Imagine a population in Hardy–Weinberg equilibrium with alleles A_1 and A_2 at initial frequencies of 0.5 each. The frequency of A_1A_1 individuals is 0.25, the frequency of A_1A_2 individuals is 0.5, and the frequency of A_2A_2 individuals is 0.25 (Figure 6.19a). Imagine that there are 1000 individuals in the population: 250 A_1A_1, 500 A_1A_2, and 250 A_2A_2. If all the individuals in the population reproduce by selfing, homozygous parents will produce all homozygous offspring, while heterozygous parents will produce half homozygous and half heterozygous offspring. Among 1000 offspring in our population, there will be 375 A_1A_1, 250 A_1A_2, and 375 A_2A_2. If the population continues to self for two more generations, then, among every 1000 individuals in the final generation, there will be 468.75 homozygotes of each type and 62.5 heterozygotes (Figure 6.19b). The frequency of heterozygotes has been halved every generation, and the frequency of homozgyotes has increased.

Conclusion 2 of the Hardy–Weinberg analysis is violated when individuals self: We cannot predict the genotype frequencies by multiplying the allele frequencies. Note that in generation three, in Figure 6.19b, the allele frequencies are still 0.5 for A_1 and 0.5 for A_2. Yet the frequency of heterozygotes is far less than $2(0.5)(0.5)$. Compared to Hardy–Weinberg expectations, there is a deficit of heterozygotes and an excess of homozygotes. The general case under selfing is shown algebraically in Table 6.1.

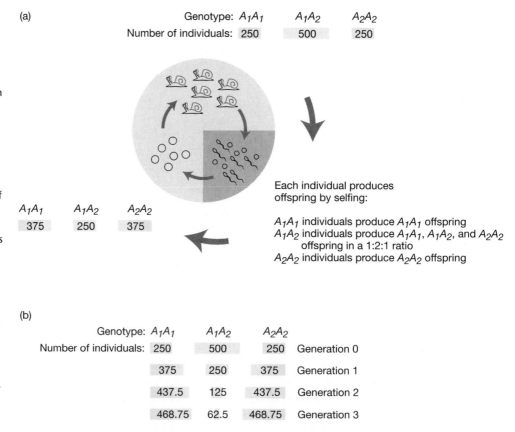

Figure 6.19 Inbreeding alters genotype frequencies (a) This figure follows the genotype frequencies in an imaginary population of 1000 snails from one generation's adults (top) to the next generation's zygotes (bottom left). The frequencies of both allele A_1 and A_2 are 0.5. The colored bar charts show the number of individuals with each genotype. Every individual reproduces by selfing. Homozygotes produce homozygous offspring and heterozygotes produce both heterozygous and homozygous offspring, so the frequency of homozygotes goes up, and the frequency of heterozygotes goes down. (b) These bar charts show what will happen to the genotype frequencies if this population continues to self for two more generations.

Table 6.1	**Changes in genotype frequencies with successive generations of selfing**		

The frequency of allele A_1 is p and the frequency of allele A_2 is q. Note that allele frequencies do not change from generation to generation—only the genotype frequencies. After Crow (1983).

Generation	A_1A_1	Frequency of A_1A_2	A_2A_2
0	p^2	$2pq$	q^2
1	$p^2 + (pq/2)$	pq	$q^2 + (pq/2)$
2	$p^2 + (3\,pq/4)$	$pq/2$	$q^2 + (3\,pq/4)$
3	$p^2 + (7\,pq/8)$	$pq/4$	$q^2 + (7\,pq/8)$
4	$p^2 + (15\,pq/16)$	$pq/8$	$q^2 + (15\,pq/16)$

What about Hardy–Weinberg conclusion 1? Do the allele frequencies change from generation to generation under inbreeding? They did not in our numerical example. We can check the general case by calculating the frequency of allele A_1 in the gene pool produced by the population shown in the last row of Table 6.1. The frequency of allele A_1 in the gene pool is equal to the frequency of A_1A_1 adults in the population $\left(= p^2 + \frac{15pq}{16} \right)$ plus half the frequency of A_1A_2 $\left(= \frac{1}{2}[\frac{pq}{8}] \right)$. That gives

$$p^2 + \frac{15pq}{16} + \frac{1}{2}\left[\frac{pq}{8}\right] = p^2 + \frac{15pq}{16} + \frac{pq}{16} = p^2 + pq$$

Now substitute $(1 - p)$ for q to give $p^2 + p(1 - p) = p$. This is the same frequency for allele A_1 that we started out with at the top of Table 6.1. Although inbreeding does cause genotype frequencies to change from generation to generation, it does not cause allele frequencies to change. Inbreeding by itself, therefore, is not a mechanism of evolution. As we will see, however, inbreeding can have important evolutionary consequences.

Empirical Research on Inbreeding: The Malaria Parasite

Because inbreeding can produce a large excess of homozygotes, Hardy–Weinberg analysis can be used to detect inbreeding in nature. As an example, we consider research on a malaria parasite in New Guinea (Paul et al. 1995). The life cycle of this protozoan, *Plasmodium falciparum,* alternates between stages that live in mosquitoes and stages that live in humans (Figure 6.20). The only diploid part of the parasite's life cycle occurs in the mosquito; there, a stage called the oocyst resides in the midgut wall. The other stages, which infect the human liver and red blood cells and include the cells transmitted to mosquitoes, are haploid.

The biology of the *P. falciparum* malaria parasite suggests that inbreeding may be common in this species. Here is the logic: Years ago, W. D. Hamilton (1967) observed that unusual sex ratios are common when a single female parasite or parasitoid colonizes a new host. In the fig wasps Hamilton was studying, for example, females colonizing figs alone tended to produce many more female young than male young. Hamilton went on to develop the mathematics showing why this strategy evolved, and predicted that it would occur in any parasite where single foundresses are common (Hamilton 1967, 1979; see also Read et al. 1992). The

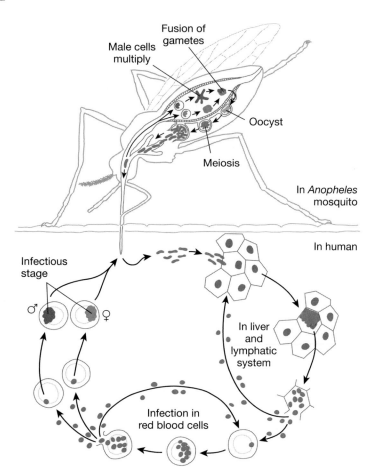

Figure 6.20 Life cycle of the malaria parasite *Plasmodium falciparum* The oocyst (orange) is diploid; all other stages of the parasite's life cycle (blue, red, and green) are haploid.

essence of the argument is this: In species in which brother–sister mating is the rule, females will have more grand-offspring if they produce only enough males to ensure that all their daughters will be fertilized.

What does this have to do with the malaria parasite? R. E. L. Paul and colleagues (1995), working in Karen Day's lab, observed a female-biased sex ratio (more females than males) in the infectious stages of *P. falciparum* . The researchers hypothesized that the phenomenon Hamilton discovered was at work in the parasite population they studied. Although malaria is common in New Guinea, the rate of transmission is relatively low. This means that every human infected tends to have only one to a few *P. falciparum* genotypes present. In addition, it appears that each cycle of infection in a human's red blood cells, which produces the type of *Plasmodium* cells picked up by mosquitoes, may be restricted to cells of a single genotype. The result is that each mosquito tends to become infected with one, or at most two, genotypes of *P. falciparum* cells. In effect, all the cells that the mosquito picks up are offspring of a single female. This infectious stage is where Paul et al. found the female-biased sex ratio. The few male cells that occur multiply inside the mosquito and then fuse with the female cells—often their own siblings—to form diploid oocysts. If this hypothesis is correct, then the oocysts should be strongly inbred, and should show a large excess of homozygotes.

Paul and colleagues estimated allele and genotype frequencies among oocysts for three *P. falciparum* protein-encoding genes: *MSP-1, MSP-2,* and *GLURP.* The

Table 6.2 Allele frequencies for three polymorphic genes in the malaria parasite *Plasmodium falciparum*

	MSP-2		MSP-1		GLURP	
Allele	Observed Frequency	Allele	Observed Frequency	Allele	Observed Frequency	
A_1	0.02	A_1	0.02	A_1	0.07	
A_2	0.06	A_2	0.26	A_2	0.42	
A_3	0.18	A_3	0.19	A_3	0.28	
A_4	0.27	A_4	0.32	A_4	0.08	
A_5	0.12	A_5	0.12	A_5	0.15	
A_6	0.08	A_6	0.09			
A_7	0.05					
A_8	0.07					
A_9	0.09					
A_{10}	0.06					

Source: Calculated from Figure 1 in Paul et al. (1995).

scientists first dissected female mosquitoes and isolated the malaria oocysts encased in their stomach linings. The researchers then extracted DNA from the oocysts and directly analyzed it for allelic variation. All three protein-encoding loci were polymorphic, meaning that each had more than one allele. The data provide an estimate for the frequency of different alleles in the population (Table 6.2), and a count of the number of homozygotes and heterozygotes in the sample (Table 6.3).

Do the allele and genotype frequencies the researchers measured conform to those expected under Hardy–Weinberg conditions? The answer, resoundingly, is no (Table 6.3). Consistent with Paul and colleagues' prediction, there is an enormous excess of homozygotes in the parasite population and a corresponding deficit of heterozygotes.

Strictly speaking, the excess of homozygotes shows only that one or more of the Hardy–Weinberg assumptions is being violated in the malaria population. However, only nonrandom mating can easily produce homozygote excesses as large as the one Paul et al. found. In combination with the researchers' observations on the parasite's reproductive biology, the data in Tables 6.2 and 6.3 make a persuasive case that *P. falciparum* in New Guinea are indeed inbreeding.

Table 6.3 The observed number of homozygotes and heterozygotes at three *P. falciparum* loci

In each case, the observed number of individuals with a particular kind of genotype is compared to the number expected under Hardy–Weinberg conditions of random mating and no mutation, selection, migration, or genetic drift. In all three cases, the differences are statistically significant ($P < 0.01$; see Box 5.5).

	(a) MSP-2		(b) MSP-1		(c) GLURP	
	Observed	Expected	Observed	Expected	Observed	Expected
Homozygotes	55	10	38	9	40	12
Heterozygotes	9	54	1	30	1	29

Box 6.6 Genotype frequencies in an inbred population

Here we add inbreeding to the Hardy–Weinberg analysis. Imagine a population with two alleles at a single locus: A_1 and A_2, with frequencies p and q. We can calculate the genotype frequencies in the next generation by letting gametes find each other in the gene pool, as we would for a random mating population. The twist added by inbreeding is that the gene pool is not thoroughly mixed. Once we have picked an egg to watch, for example, we can think of the sperm in the gene pool as consisting of two fractions: a fraction $(1 - F)$ carrying alleles that are not identical by descent to the one in the egg; and the fraction F carrying alleles that are identical by descent to the one in the egg (because they were produced by relatives of the female that produced the egg). The calculations of genotype frequencies are as follows:

- A_1A_1 **homozygotes:** There are two ways we might witness the creation of an A_1A_1 homozygote. The first way is that we pick an egg that is A_1 (an event with probability p) and watch it get fertilized by a sperm that is A_1 by chance, rather than by common ancestry. The frequency of unrelated A_1 sperm in the gene pool is $p(1 - F)$, so the probability of getting a homozygote by chance is

$$p \times p(1 - F) = p^2(1 - F)$$

The second way to get a homozygote is to pick an egg that is A_1 (an event with probability p) and watch it get fertilized by a sperm that is A_1 because of common ancestry (an event with probability F). The probability of getting a homozygote this way is pF. The probability of getting an A_1A_1 homozygote by either the first way or the second way is the sum of their individual probabilities:

$$p^2(1 - F) + pF$$

- A_1A_2 **heterozygotes:** There are two ways to get an A_1A_2 heterozygote. The first way is to pick an egg that is A_1 (an event with probability p) and watch it get fertilized by an unrelated sperm that is A_2. The frequency of A_2 unrelated sperm is $q(1 - F)$, so the probability of getting a heterozygote this first way is $pq(1 - F)$. The second way is to pick an egg that is A_2 (probability: q) and watch it get fertilized by an unrelated sperm that is A_1 [probability: $p(1 - F)$]. The probability of getting a heterozygote the second way is $qp(1 - F)$. The probability of getting a heterozygote by either the first way or the second way is the sum of their individual probabilities:

$$pq(1 - F) + qp(1 - F) = 2pq(1 - F)$$

- A_1A_2 **homozygotes:** We can get an A_2A_2 homozygote either by picking an A_2 egg (probability: q) and watching it get fertilized by an unrelated A_2 sperm [probability: $q(1 - F)$], or by picking an A_2 egg (probability: q) and watching it get fertilized by a sperm that is A_2 because of common ancestry (probability: F). The overall probability of getting an A_2A_2 homozygote is

$$q^2(1 - F) + qF$$

Readers may wish to verify that the genotype frequencies sum to 1.

General Analysis of Inbreeding

So far our treatment of inbreeding has been limited to self-fertilization and sibling mating. But inbreeding can also occur as matings among more distant relatives, such as cousins. Inbreeding that is less extreme than selfing produces the same effect as selfing—it increases the proportion of homozygotes—but at a slower rate. For a general mathematical treatment of inbreeding, population geneticists use a conceptual tool called the **coefficient of inbreeding**. This quantity is symbolized by F, and is defined as the probability that the two alleles in an individual are identical by descent (meaning that both alleles came from the same ancestor allele in some previous generation). Box 6.6 shows that in an inbred population that otherwise obeys Hardy–Weinberg assumptions, the genotype frequencies are

$$\begin{array}{ccc} \mathbf{A_1A_1} & \mathbf{A_1A_2} & \mathbf{A_2A_2} \\ p^2(1-F)+pF & 2pq(1-F) & q^2(1-F)+qF \end{array}$$

Readers can verify these expressions by substituting the values $F=0$, which gives the original Hardy–Weinberg genotype ratios, and $F=0.5$, which represents selfing and gives the ratios shown for generation 1 in Table 6.1.

The same logic applies when many alleles are present in the gene pool. Then, the frequency of any homozygote A_iA_i is given by

$$p_i^2(1-F)+p_iF$$

and the frequency of any heterozygote A_iA_j is given by

$$2p_ip_j(1-F)$$

where p_i is the frequency of allele A_i and p_j is the frequency of allele A_j.

The last expression states that the fraction of individuals in a population that are heterozygotes (that is, the population's heterozygosity) is proportional to $(1-F)$. If we compare the heterozygosity of an inbred population, H_F, with that of a random mating population, H_0, then the relationship will be

$$H_F=H_0(1-F)$$

Anytime F is greater than 0, the frequency of heterozygotes is lower in an inbred population than it is in a random mating population.

Computing F

To measure the degree of inbreeding in actual populations, we need a way to calculate F. Doing this directly requires a pedigree—a diagram showing the geneological relationships of individuals. Figure 6.21 shows a pedigree leading to a female who is the daughter of half-siblings. There are two ways this female could receive alleles that are identical by descent. One is that she could receive two copies of her

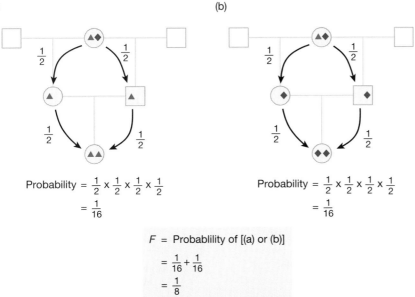

(a)

Probability $= \frac{1}{2} \times \frac{1}{2} \times \frac{1}{2} \times \frac{1}{2}$

$\qquad\quad = \frac{1}{16}$

(b)

Probability $= \frac{1}{2} \times \frac{1}{2} \times \frac{1}{2} \times \frac{1}{2}$

$\qquad\quad = \frac{1}{16}$

F = Probablility of [(a) or (b)]

$\quad = \frac{1}{16} + \frac{1}{16}$

$\quad = \frac{1}{8}$

Figure 6.21   Calculating F from a pedigree   In parts (a) and (b), squares represent males; circles represent females; arrows represent the movement of alleles from parents to offspring via gametes. The green triangles and blue diamonds represent alleles at a particular locus.

grandmother's "green triangle" allele (Figure 6.21a). This will happen if the grandmother passes the triangle allele to her daughter and to her son, and the daughter passes it to the granddaughter, and the son passes it to the granddaughter. The total probability of this scenario is $\frac{1}{16}$. The second way is that she could receive two copies of her grandmother's "blue diamond" allele (Figure 6.21b). The total probability of this scenario is $\frac{1}{16}$. The probability that the daughter of half-siblings will have two alleles identical by descent by either the first scenario or the second scenario is $\frac{1}{16} + \frac{1}{16} = \frac{1}{8}$. Thus, F for an offspring of half-siblings is $\frac{1}{8}$.

Inbreeding Depression

Inbreeding may lead to reduced mean fitness if it generates offspring homozygous for deleterious alleles.

Although inbreeding does not directly change allele frequencies, it can still affect the evolution of a population. Among the most important consequences of inbreeding for evolution is inbreeding depression.

Inbreeding depression usually results from the exposure of deleterious recessive alleles to selection. To see how this works, consider the extreme case illustrated by loss–of–function mutations. These alleles are often recessive, because a single wild-type allele can still generate enough functional protein, in most instances, to produce a normal phenotype. Even though they may have no fitness consequences at all in heterozygotes, loss-of-function mutations can be lethal in homozygotes. By increasing the proportion of individuals in a population that are homozygotes, inbreeding increases the frequency with which deleterious recessives affect phenotypes. Inbreeding depression refers to the effect these alleles have on the average fitness of offspring in the population.

Studies on humans have shown that inbreeding does, in fact, expose deleterious recessive alleles, and data from numerous studies consistently show that children of first cousins have higher mortality rates than children of unrelated parents (Figure 6.22). Strong inbreeding depression has also been frequently observed in captive populations of animals (for example, Hill 1974; Ralls et al. 1979).

Perhaps the most powerful studies of inbreeding depression in natural populations concern flowering plants, in which the inbreeding can be studied experimentally. In many angiosperms, selfed and outcrossed offspring can be produced

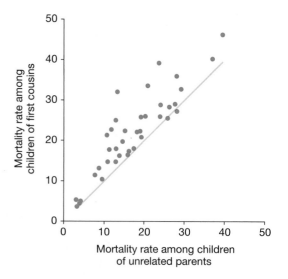

Figure 6.22 Inbreeding depression in humans Each dot on this graph represents childhood mortality rates for a human population. The horizontal axis represents mortality rates for children of unrelated parents; the vertical axis represents mortality rates for children of first cousins. The gray line shows where the points would fall if mortality rates for the two kinds of children were equal. Although childhood mortality rates vary widely among populations, the mortality rate for children of cousins is almost always higher than the rate for children of unrelated parents—usually by about four percentage points. Plotted from data in Bittles and Neel (1994).

from the same parent through hand pollination. In experiments like these, inbreeding depression can be defined as

$$\delta = 1 - \frac{w_s}{w_o}$$

where w_s and w_o are the fitnesses of selfed and outcrossed progeny, respectively. This definition makes levels of inbreeding depression comparable across species. Three patterns are starting to emerge from experimental studies.

First, inbreeding effects are often easiest to detect when plants undergo some sort of environmental stress. For example, when Michele Dudash (1990) compared the growth and reproduction of selfed and outcrossed rose pinks (*Sabatia angularis*), the plants showed some inbreeding depression when grown in the greenhouse or garden, but their performance diverged more strongly when they were planted in the field. Lorne Wolfe (1993) got a similar result with a waterleaf (*Hydrophyllum appendiculatum*): Selfed and outcrossed individuals had equal fitness when grown alone, but differed significantly when grown under competition. And in the common annual called jewelweed (*Impatiens capensis*), McCall et al. (1994) observed the strongest inbreeding effects on survival when an unplanned insect outbreak occurred during the course of their experiment.

Second, inbreeding effects are much more likely to show up later in the life cycle (not, for example, during the germination or seedling stage). This pattern is striking (Figure 6.23). Why does it exist? Wolfe (1993) suggests that maternal effects—specifically, the seed mother's influence on offspring phenotype through provisioning of seeds—can mask the influence of deleterious recessives until later in the life cycle.

Third, inbreeding depression varies among family lineages. Michele Dudash and colleagues (1997) compared the growth and reproductive performance of inbred versus outcrossed individuals from each of several families in two annual populations of the herb *Mimulus guttatus*. Some families showed inbreeding depression; others showed no discernable effect of type of mating; still others showed improved performance under inbreeding.

Inbreeding depression has been documented in natural populations of animals as well. Long-term studies in two separate populations of a bird called the great tit (*Parus major*) have shown that inbreeding depression can have strong effects on reproductive success. When Paul Greenwood and co-workers (1978) defined inbred matings as those between first cousins or more closely related individuals, they

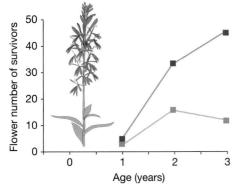

Figure 6.23 Inbreeding depression in flowering plants increases as individuals age This graph compares the number of flowers produced (a measure of fitness) as a function of time for outcrossed (blue boxes) versus selfed (red boxes) individuals in *Lobelia cardinalis,* a perennial in the bluebell family. The disparity in performance increases with time, indicating that inbreeding depression becomes more pronounced with age. From Johnston (1992). Copyright © 1992 Evolution. Reprinted by permission.

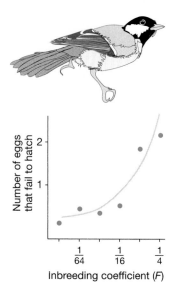

Figure 6.24 Inbreeding increases egg failure in great tits From van Noordwijk and Scharloo (1981). Copyright © 1992 Evolution. Reprinted by permission.

found that the survival of inbred nestlings was much lower than that of outbred individuals. Similarly, A. J. van Noordwijk and W. Scharloo (1981) showed that in an island population of tits, there is a strong relationship between the level of inbreeding in a pair and the number of eggs in a clutch that fail to hatch (Figure 6.24). More recently, Keller et al. (1994) found that outbred individuals in a population of song sparrows in British Columbia, Canada, were much more likely than inbred individuals to survive a severe winter.

Given the theory and data we have reviewed on inbreeding depression, it is not surprising that animals and plants have evolved mechanisms to avoid it. Mechanisms of inbreeding avoidance include mate choice, genetically controlled self-incompatibility, and dispersal. But under some circumstances, inbreeding may be unavoidable. In small populations, for example, the number of potential mates for any particular individual is limited. If a population is small and remains small for many generations, and if the population receives no migrants from other populations, then eventually all the individuals in the population will be related to each other even if mating is random. Thus, small populations eventually become inbred, and the individuals in them may suffer inbreeding depression. This can be a problem for rare and endangered species, and it creates a challenge for the managers of captive breeding programs, as we will see in Section 6.4.

In summary, nonrandom mating does not, by itself, alter allele frequencies. It is not, therefore, a mechanism of evolution. Nonrandom mating does, however, alter the frequencies of genotypes. It can thereby change the distribution of phenotypes in a population and alter the pattern of natural selection and the evolution of the population. For example, inbreeding increases the frequency of homozygotes and decreases the frequency of heterozygotes. This can expose deleterious recessive alleles to selection, leading to inbreeding depression.

6.4 Conservation Genetics of the Illinois Greater Prairie Chicken

We opened this chapter with the case of the Illinois greater prairie chicken (Figure 6.1), a once abundant bird that, in the mid-1990s, appeared to be destined for extinction. Like a great many other vulnerable and endangered species, the prairie chicken's worst enemy is habitat destruction (Figure 6.2). Before the introduction of the steel plow, prairie covered more than 60% of Illinois; today less than one hundredth of a percent of that prairie remains (Westemeier et al. 1998). Yet habitat destruction is not the prairie chicken's only problem. Beginning in the early 1960s, conservationists established prairie chicken reserves and worked to restore and maintain prairie habitats. From the late 1960s to the early 1970s, their efforts appeared to be working, as the prairie chicken's numbers began to rebound. But the apparent success was short-lived: By the mid-1970s, the prairie chicken population fell once again into a steady decline (Figure 6.3). Something else was now threatening the survival of the Illinois greater prairie chicken, but what? Our discussion of migration, genetic drift, and nonrandom mating have given us the tools to understand the likely answer.

Ronald Westemeier and colleagues (1998) developed a hypothesis that runs as follows: Destruction of the prairie did two things to the prairie chicken population. First, it directly reduced the size of the birds' population. Second, it fragmented the population that remained. By 1980, the few prairie chickens that survived in Illinois were trapped on small islands of prairie in a sea of farmland. Each island had its own small population of birds. These small populations were geographically isolated from each other and from populations in other states.

Small populations with little or no gene flow are precisely the setting in which genetic drift is most powerful. And genetic drift results in random fixation and declining heterozygosity. If some of the alleles that become fixed are deleterious recessives, then the average fitness of individuals will be reduced. A reduction in fitness due to genetic drift is reminiscent of inbreeding depression. In fact, it *is* inbreeding depression. Reduced heterozygosity due to drift and increased homozygosity due to inbreeding are two sides of the same coin. In a small population all individuals are related, and there is no choice but to mate with kin.

Michael Lynch and Wilfried Gabriel (1990) have proposed that an accumulation of deleterious recessives (a phenomenon known as genetic load) can lead to the extinction of small populations. They noted that when exposure of deleterious mutations produces a reduction in population size, the effectiveness of drift is increased. The speed and proportion of deleterious mutations going to fixation subsequently increases, which further decreases population size. Lynch and Gabriel termed this synergistic interaction between mutation, population size, and drift a "mutational meltdown."

Westemeier and colleagues suggested that the remnant populations of Illinois greater prairie chickens were trapped in just such a scenario. As the populations lost their genetic diversity, the birds began to suffer inbreeding depression. This inbreeding depression reduced individual reproductive success, and caused the remnant populations to continue their decline even as the amount of suitable habitat increased. The continued decline in population size led to even more drift, which led to worse inbreeding depression, and so on. The birds had fallen into an "extinction vortex" (see Soulé and Mills 1998).

To test their hypothesis, the researchers first used data from a long-term study of the Jasper County population to look for evidence of inbreeding depression. The researchers plotted the hatching success of prairie chicken eggs, a measure of individual fitness, as a function of time (Figure 6.25). Throughout the 1960s, over 90% of greater prairie chicken eggs in Jasper County hatched. This rate is comparable to what it was in the 1930s, and to what is today in larger prairie chicken populations in other states. By 1970, however, a steady decline in hatching

Early efforts to conserve remnant populations of greater prairie chickens apparently failed because the birds were suffering from inbreeding depression.

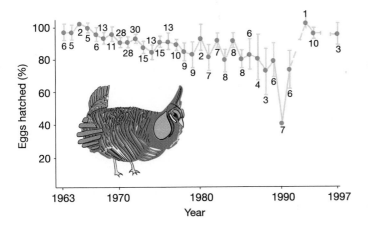

Figure 6.25 Declining hatching success in a greater prairie chicken population This graph plots, for greater prairie chickens in Jasper County, Illinois, the fraction of eggs hatching each year from 1963 to 1997. The small number below each data point indicates the number of nests followed; the wiskers indicate ±1 standard error (a statistical measure of uncertainty in the estimated hatching rate). The decline in hatching success from the mid-1960s to the early 1990s appears to reflect inbreeding depression. Redrawn with permission from Westemeier et al (1998). Copyright © 1998, American Association for the Advancement of Science.

success had begun. By the late 1980s, the hatching rate dipped below 80%. The all-time low came in 1990, with fewer than 40% of the eggs hatching. The decline in hatching success is statistically significant, and represents a substantial reduction in individual fitness. In other words, it looks like inbreeding depression.

If the decline in hatching success in Jasper County greater prairie chickens was, in fact, due to inbreeding depression caused by genetic drift, then it should be accompanied by a genetic signature. The Jasper County population should show less genetic diversity than larger populations in other states, and less genetic diversity now than it had in the past. Juan Bouzat and colleagues (1998) analyzed the DNA of a number of greater prairie chickens from Illinois, Kansas, Minnesota, and Nebraska, and determined each bird's genotype at six selectively neutral loci (these were noncoding regions with variable numbers of short tandem repeats). As predicted, the Illinois birds had an average of just 3.67 alleles per locus, signficantly fewer than the 5.33 to 5.83 alleles per locus shown by the other populations (Table 6.4). The researchers were even able to extract DNA from 10 museum specimens that had been collected in Jasper County in the 1930s, plus 5 from the 1960s. As shown in the last column in Table 6.4, Bouzat and colleagues used the data from the museum specimens to estimate that the Jasper County population once had an average of at least 5.12 alleles per locus. Consistent with the extinction vortex hypothesis, the greater prairie chickens of Jasper County are genetically depauperate, compared to both their own ancestral population and other present-day populations.

Inbreeding depression in remnant populations of greater prairie chickens was caused by a loss of allelic diversity under genetic drift.

The final test of the extinction vortex hypothesis was to use it to develop a practical conservation strategy. If the problem for the Jasper County prairie chicken population is reduced genetic diversity, then the solution is gene flow. Migrants from other populations should carry with them the alleles that have been lost in

Table 6.4   Number of alleles per locus found in each of the current populations of Illinois, Kansas, Minnesota, and Nebraska and estimated for the Illinois prebottleneck population

Locus	Illinois	Kansas	Minnesota	Nebraska	Illinois prebottleneck*
ADL42	3	4	4	4	3
ADL23	4	5	4	5	5
ADL44	4	7	8	8	4
ADL146	3	5	4	4	4
ADL162	2	5	4	4	6
ADL230	6	9	8	10	9
Mean	3.67	5.83	5.33	5.83	5.12
SE	0.56	0.75	0.84	1.05	0.87
Sample size	32	37	38	20	15

Note: SE indicates standard error of mean number of alleles per locus. The Illinois population in column 1 shows signficantly less allelic diversity than the rest of the populations ($P < 0.05$).

*Number of alleles in the Illinois prebottleneck population include both extant alleles that are shared with the other populations and alleles detected in the museum collection.

Source: From Bouzat et al. (1998).

Jasper County. Reintroduction of these lost alleles should reverse the effects of drift and eliminate inbreeding depression. Natural migration of greater prairie chickens into Jasper County ceased long ago. But in 1992, conservation biologists began trapping greater prairie chickens in Minnesota, Kansas, and Nebraska, and moving them to Jasper County. The plan seems to be working. Westemeier and colleagues (1998) report that in 1993, 1994, and 1997, hatching rates in Jasper County were over 90%—higher than they had been in 25 years (Figure 6.25). And the Jasper County population is growing (Figure 6.3).

All of the data we have presented on Illinois greater prairie chickens come from observational studies, so it is always possible that some uncontrolled and unknown environmental variable is responsible for the variation in hatching success. On present evidence, however, Westhesier et al.'s extinction vortex hypothesis—involving migration, genetic drift, and nonrandom mating—appears to be the best explanation.

Migration, in the form of birds transported by biologists, appears to be restoring genetic diversity to remnant populations and alleviating inbreeding depression.

Summary

Among the important implications of the Hardy–Weinberg equilibrium principle is that natural selection is not the only mechanism of evolution. In this chapter, we examined violations of three assumptions of the Hardy–Weinberg analysis first introduced in Chapter 5, and considered their effects on allele and genotype frequencies.

Migration, in its evolutionary meaning, is the movement of alleles from one population to another. When allele frequencies are different in the source population than in the recipient population, migration causes the recipient population to evolve. As a mechanism of evolution, migration tends to homogenize allele frequencies across populations. In doing so, it may tend to eliminate adaptive differences between populations that have been produced by natural selection.

Genetic drift is evolution that occurs as a result of sampling error in the production of a finite number of zygotes from a gene pool. Just by chance, allele frequencies change from one generation to the next. Genetic drift is more dramatic in smaller populations than in large ones. Over many generations, drift results in an inexorable loss of genetic diversity. If some of the alleles that become fixed are deleterious recessives, genetic drift can result in a reduction in the fitness of individuals in the population.

Nonrandom mating does not directly change allele frequencies and is thus not strictly speaking a mechanism of evolution. However, nonrandom mating does influence genotype frequencies. For example, inbred populations have more homozygotes and fewer heterozygotes than otherwise comparable populations in which mating is random. An increase in homozygosity often exposes deleterious recessive alleles, and results in a reduction in fitness known as inbreeding depression.

As illustrated by the case of the Illinois greater prairie chicken, the phenomena discussed in this chapter find practical application in conservation efforts. Drift can rob small remnant populations of genetic diversity, resulting in inbreeding depression and greater risk of extinction. Migration can sometimes restore lost genetic diversity, improving a population's chances for long-term survival.

Questions

1. Conservation managers often try to purchase corridors of undeveloped habitat so that larger preserves are linked into networks. Why? What genetic goals do you think the conservation managers are aiming for?

2. The graph in Figure 6.26 shows F_{ST}, a measure of genetic differentiation between populations, as a function of geographic distance. The data are from human populations in Europe. Genetic differentiation has been calculated based on loci on the autosomes (inherited from both parents), the mitochondrial chromosome (inherited only from the mother), and the Y chromosome (inherited only from the father). Note that the patterns are different for

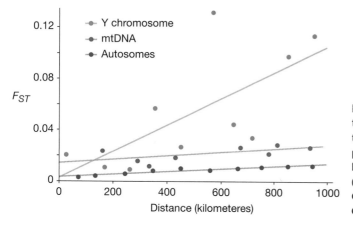

Figure 6.26   **Genetic distance between human populations as a function of geographic distance**   Genetic distance (F_{ST}) is a measure of genetic differentiation among populations. Here it has been calculated based on autosomal loci (blue), mithochondrial loci (green), and Y-chromosome loci (red). From Seielstad et al. (1998). Copyright © 1998 Nature Genetics. Reprinted by permission of the Nature Publishing Group, New York, NY, and M. T. Seielstad, Ph.D.

the three different kinds of loci. Keep in mind that migration tends to homogenize allele frequencies across populations. Develop a hypothesis to explain why allele frequencies are more homogenized across populations for autosomal and mitochondrial loci than for Y-chromosome loci. Then go to the library and look up the following paper, to see if your hypothesis is similar to the one favored by the biologists who prepared the graph:

Seielstad, M. T., E. Minch, and L. L. Cavalli-Sforza. 1998. Genetic evidence for [....] in humans. *Nature Genetics* 20:278–280. [Part of title deleted to encourage readers to develop their own hypotheses.]

3. Loss of heterozygosity may be especially detrimental at MHC loci, because allelic variability at these loci increases disease resistance. Surveys of microsatellite loci show that the gray wolves on Isle Royale, Michigan are highly inbred (Wayne et al. 1991). This wolf population crashed during an outbreak of canine parvovirus during the 1980s. How might these disparate facts be linked? How could you test your ideas?

4. If you were a manager charged with conserving the collared lizards of the Ozarks, one of your tasks might be to reintroduce the lizards into glades in which they have gone extinct. When reintroducing lizards to a glade, you will have a choice between using only individuals from a single extant glade populations, or from several extant glade populations. What would be the evolutionary consequences of each choice, for both the donor and recipient populations? Which strategy will you follow, and why?

5. Recall Ellis–van Creveld syndrome, a genetic disease that is unusually common among the Amish. The Amish population in Pennsylvania now numbers well over 10,000.

   a. What evolutionary forces are presently acting on the allele for Ellis–van Creveld syndrome in this population? Why do you think the allele has remained at relatively high frequency? How could you test your ideas?

   b. Suppose hundreds of non-Amish people began marrying into the Amish population each year, and raised their children within the Amish community. What would happen to the frequency of Ellis–van Creveld syndrome, and why?

6. Bodmer and McKie (1995) review several cases, in addition to Ellis–van Creveld syndrome in the Amish, in which genetic diseases occur at unusually high frequency in populations that are, or once were, relatively isolated. An enzyme deficiency called hereditary tyrosinemia, for example, occurs at an unusually high rate in the Chicoutimi region north of Quebec City in Canada. A condition called porphyria is unusually common in South Africans of Dutch descent. Why are genetic diseases so common in isolated populations? What else do these populations all have in common?

7. Remote oceanic islands are famous for their endemic species—unique forms that occur nowhere else (see Quammen 1996 for a gripping and highly readable account). Consider the roles of migration and genetic drift in the establishment of new species on remote islands.

   a. How do plant and animal species become established on remote islands? Do you think island endemics are more likely to evolve in some groups of plants and animals than others?

   b. Consider a new population that has just arrived at a remote island. Is the population likely to be large or small? Will founder effects, genetic drift, and additional waves of migration from the mainland play a relatively large or a small role in the evolution of the new island population (compared to a similar population on an island closer to the mainland)? Do your answers help explain why unusual endemic species are more common on remote islands than on islands close to the mainland?

8. As we have seen, inbreeding can reduce offspring fitness by exposing deleterious recessive alleles. However, some

animal breeders practice generations of careful inbreeding within a family, or "line breeding," and surprisingly many of the line-bred animals, from champion dogs to prize cows, have normal health and fertility. How can it be possible to continue inbreeding for many generations without experiencing inbreeding depression due to recessive alleles? (*Hint:* Responsible animal breeders do not breed animals known to carry deleterious traits.) Generally, if a small population continues to inbreed for many generations, what will happen to the frequency of the deleterious recessive alleles over time?

9. In the mid-1980s, conservation biologists reluctantly recommended that zoos should not try to preserve captive populations of all the endangered species of large cats. For example, some biologists recommended ceasing efforts to breed the extremely rare Asian lion, the beautiful species seen in Chinese artwork. In place of the Asian lion, the biologists recommended increasing the captive populations of other endangered cats, such as the Siberian tiger and Amur leopard. By reducing the number of species kept in captivity, the biologists hoped to increase the captive population size of each species to several hundred, preferably at least 500. Why did the conservation biologists think that this was so important as to be worth the risk of losing the Asian lion forever?

10. In this chapter we saw that in many cases, gene frequencies in small populations change at different rates than in large populations. As a review, state whether the following processes will typically have greater, smaller, or similar effects on evolution in small vs. large populations:

    Selection

    Migration

    Genetic drift

    Inbreeding

    New mutations per individual

    New mutations per generation in the whole population

    Substitution of a new mutation for an old allele

    Fixation of a new mutation

Exploring the Literature

11. For a paper that explores migration as a homogenizer of allele frequencies among human populations, see

    Parra, E. J., Marcini, A., et al. 1998. Estimating African-American admixture proportions by use of population-specific alleles. *American Journal of Human Genetics* 63: 1839–1851.

12. For another example like the research on collared lizards by Templeton and colleagues (1990) in which biologists took advantage of a natural experiment to make test predictions about the effect of genetic drift on genetic diversity, see

    Eldridge, M. D. B., King, J. M., et al. 1999. Unprecedented low levels of genetic variation and inbreeding depression in an island population of the black-footed rock-wallaby. *Conservation Biology* 13: 531–541.

13. We mentioned in Section 6.3 that inbreeding depression is a concern for biologists trying to conserve endangered organisms with small population sizes. Geneticists have recently discovered that inbreeding depression varies among environments and among families. For papers that explore the implications of this discovery for conservation efforts, see

    Pray, L. A., J. M. Schwartz, C. J. Goodnight, and L. Stevens. 1994. Environmental dependency of inbreeding depression: Implications for conservation biology. *Conservation Biology* 8: 562–568.

    Pray, L. A., and C. J. Goodnight. 1995. Genetic variation in inbreeding depression in the red flour beetle *Tribolium castaneum*. *Evolution* 49: 176–188.

14. Determining the minimum population size necessary to make the extinction of a species unlikely over the long term is an active area of research in conservation genetics. The following papers explore this question:

    Lande, R. 1995. Mutation and conservation. *Conservation Biology* 9: 782–791.

    Lynch, M. 1996. A quantitative genetic perspective on conservation issues. In J. C. Avise and J. Hamrick, eds. *Conservation Genetics: Case Histories from Nature.* New York: Chapman and Hall, 471–501.

15. Cheetahs have long been used as a classic example of a species whose low genetic diversity put it at increased risk of extinction. Other researchers have debated the validity of this view. For a start on the literature, see

    Menotti-Raymond, M., and S. J. O'Brien. 1993. Dating the genetic bottleneck of the African cheetah. *Proceedings of the National Academy of Sciences, USA* 90: 3172–3176.

    Merola, M. 1994. A reassessment of homozygosity and the case for inbreeding depression in the cheetah, *Acinonyx jubatus:* Implications for conservation. *Conservation Biology* 8: 961–971.

16. For other attempts to determine whether low genetic diversity threatens the survival of populations, see

Ledberg, P. L. 1993. Strategies for population reintroduction: Effects of genetic variability on population growth and size. *Conservation Biology* 7: 194–199.

Jimenez, J. A., K. A. Hughes, G. Alaks, L. Graham, and R. C. Lacy. 1994. An experimental study of inbreeding depression in a natural habitat. *Science* 266: 271–273.

Sanjayan, M. A., K. Crooks, G. Zegers, and D. Foran. 1996. Genetic variation and the immune response in natural populations of pocket gophers. *Conservation Biology* 10: 1519–1527.

Literature Cited

Much of the population genetics material in this chapter is modeled after presentations in the following:

Crow, J. F. 1983. *Genetics Notes.* Minneapolis, MN: Burgess Publishing.

Felsenstein, J. 1997. *Theoretical Evolutionary Genetics.* Seattle, WA: ASUW Publishing, University of Washington.

Griffiths, A. J. F., J. H. Miller, D. T. Suzuki, R. C. Lewontin, and W. M. Gelbert. 1993. *An Introduction to Genetic Analysis.* New York: W. H. Freeman.

Maynard Smith, J. 1998. *Evolutionary Genetics.* Oxford University Press, Oxford.

Roughgarden, J. 1979. *Theory of Population Genetics and Evolutionary Ecology: An Introduction.* MacMillan Publishing, New York.

Templeton, A. R. 1982. Adaptation and the integration of evolutionary forces. In R. Milkman, ed., *Perspectives on Evolution.* Sunderland, MA: Sinauer, 15–31.

Here is the list of all other citations in this chapter:

Ayala, F. J., and J. A. Kiger, Jr. 1984. *Modern Genetics.* Menlo Park, CA: Benjamin/Cummings.

Barrett, S. C. H., and D. Charlesworth. 1991. Effects of a change in the level of inbreeding on the genetic load. *Nature* 352: 522–524.

Bittles, A. H., and J. V. Neel. 1994. The costs of human inbreeding and their implications for variations at the DNA level. *Nature Genetics* 8: 117–121.

Bodmer, W., and R. McKie. 1995. *The Book of Man.* New York: Scribner.

Bouzat, J. L., H. A. Lewin, and K. N. Paige. 1998. The ghost of genetic diversity past: Historical DNA analysis of the greater prairie chicken. *American Naturalist* 152: 1–6.

Buri, P. 1956. Gene frequency in small populations of mutant *Drosophila. Evolution* 10: 367–402.

Camin, J. H., and P. R. Ehrlich. 1958. Natural selection in water snakes (*Natrix sipedon* L.) on islands in Lake Erie. *Evolution* 12: 504–511.

Dudash, M. R. 1990. Relative fitness of selfed and outcrossed progeny in a self-compatible, protandrous species, *Sabatia angularis* L. (Gentianaceae): A comparison in three environments. *Evolution* 44: 1129–1139.

Dudash, M. R., D. E. Carr, and C. B. Fenster. 1997. Five generations of enforced selfing and outcrossing in *Mimulus guttatus*: Inbreeding depression variation at the population and family level. *Evolution* 51: 54–65.

Giles, B. E., and J. Goudet. 1997. Genetic differentiation in *Silene dioica* metapopulations: Estimation of spatiotemporal effects in a successional plant species. *American Naturalist* 149: 507–526.

Gillespie, J. H. 1991. *The Causes of Molecular Evolution.* New York: Oxford University Press.

Grant, P. R., and B. R. Grant. 1995. The founding of a new population of Darwin's finches. *Evolution* 49: 229–240.

Greenwood, P. J., P. H. Harvey, and C. M. Perrins. 1978. Inbreeding and dispersal in the great tit. *Nature* 271: 52–54.

Hamilton, W. D. 1967. Extraordinary sex ratios. *Science* 156: 477–488.

Hamilton, W. D. 1979. Wingless and fighting males in fig wasps and other insects. In M. S. Blum and N. A. Blum, eds., *Sexual Selection and Reproductive Competition in Insects.* New York: Academic Press, 167–220.

Hartl, D. L. 1981. *A Primer of Population Genetics.* Sunderland, MA: Sinauer.

Hill, J. L. 1974. *Peromyscus*: Effect of early pairing on reproduction. *Science* 186: 1042–1044.

Johnston, M. 1992. Effects of cross and self-fertilization on progeny fitness in *Lobelia cardinalis* and *L. siphilitica. Evolution* 46: 688–702.

Keller, L., P. Arcese, J. N. M. Smith, W. M. Hochachka, and S. C. Stearns. 1994. Selection against inbred song sparrows during a natural population bottleneck. *Nature* 372: 356–357.

Kerr, W. E., and S. Wright. 1954. Experimental studies of the distribution of gene frequencies in very small populations of *Drosophila melanogaster.* I. Forked. *Evolution* 8: 172–177.

Kimura, M. 1968. Evolutionary rate at the molecular level. *Nature* 217: 624–626.

Kimura, M. 1983. *The Neutral Theory of Molecular Evolution.* New York: Cambridge University Press.

King, R. B. 1987. Color pattern polymorphism in the Lake Erie water snake, *Nerodia sipedon insularum. Evolution* 41: 241–255.

King, R. B. 1993a. Color pattern variation in Lake Erie water snakes: Inheritance. *Canadian Journal of Zoology* 71: 1985–1990.

King, R. B. 1993b. Color-pattern variation in Lake Erie water snakes: Prediction and measurement of natural selection. *Evolution* 47: 1819–1833.

King, R. B., and R. Lawson. 1995. Color-pattern variation in Lake Erie water snakes: The role of gene flow. *Evolution* 49: 885–896.

King, R. B., and R. Lawson. 1997. Microevolution in island water snakes. *BioScience* 47: 279-286.

Lynch, M., and W. Gabriel. 1990. Mutation load and the survival of small populations. *Evolution* 44: 1725–1737.

McCall, C., D. M. Waller, and T. Mitchell-Olds. 1994. Effects of serial inbreeding on fitness components in *Impatiens capensis. Evolution* 48: 818–827.

Paul, R. E. L., M. J. Packer, M. Walmsley, M. Lagog, L. C. Ranford-Cartwright, R. Paru, and K. P. Day. 1995. Mating patterns in malaria parasite populations of Papua, New Guinea. *Science* 269: 1709–1711.

Polans, N. O., and R. W. Allard. 1989. An experimental evaluation of the recovery potential of ryegrass populations from genetic stress resulting from restriction of population size. *Evolution* 43: 1320–1324.

Postlethwait, J. H., and J. L. Hopson. 1992. *The Nature of Life,* 2nd ed. New York: McGraw-Hill.

Quammen, D. 1996. *The Song of the Dodo.* New York: Touchstone.

Ralls, K., K. Brugger, and J. Ballou. 1979. Inbreeding and juvenile mortality in small populations of ungulates. *Science* 206: 1101–1103.

Read, A. F., A. Narara, S. Nee, A. E. Keymer, and K. P. Day. 1992. Gametocyte sex ratios as indirect measures of outcrossing rates in malaria. *Parasitology* 104: 387–395.

Seielstad, M. T., E. Minch, and L. L. Cavalli-Sforza. 1998. Genetic evidence for a higher female migration rate in humans. *Nature Genetics* 20: 278–280.

Soulé, M. E., and L. S. Mills. 1998. No need to isolate genetics. *Science* 282: 1658–1659.

Templeton, A. R., K. Shaw, E. Routman, and S. K. Davis. 1990. The genetic consequences of habitat fragmentation. *Annals of the Missouri Botanical Garden* 77: 13–27.

Thomas, J. 1998. A bird's race toward extinction is halted. *The New York Times* 29 December: D3.

van Noordwijk, A. J., and W. Scharloo. 1981. Inbreeding in an island population of the great tit. *Evolution* 35: 674–688.

Wayne, R. K., N. Lehman, D. Girman, P. J. P. Gogan, D. A. Gilbert, K. Hansen, R. O. Peterson, U. S. Seal, A. Eisenhawer, L. D. Mech, and R. J. Krumenaker. 1991. Conservation genetics of the endangered Isle Royale gray wolf. *Conservation Biology* 5: 41–51.

Wright, S. 1931. Evolution in Mendelian populations. *Genetics* 16: 97–159.

Westemeier, R. L., S. A. Simpson, and D. A. Cooper. 1991. Successful exchange of prairie-chicken eggs between nests in two remnant populations. *Wilson Bulletin* 103:717–720.

Westemeier, R. L., J. D. Brawn, et al. 1998. Tracking the long-term decline and recovery of an isolated population. *Science* 282: 1695-1698.

Wolfe, L. M. 1993. Inbreeding depression in *Hydrophyllum appendiculatum*: Role of maternal effects, crowding, and parental mating history. *Evolution* 47: 374–386.

Young, A., T. Boyle, and T. Brown. 1996. The population genetic consequences of habitat fragmentation for plants. *Trends in Ecology and Evolution* 11: 413–418.

CHAPTER 7

Evolution at Multiple Loci: Linkage, Sex, and Quantitative Genetics

A living histogram. These students and faculty at the University of Connecticut have sorted themselves into columns by height. (Peter Morenus, University of Connecticut)

IN CHAPTERS 5 AND 6 WE INTRODUCED BASIC POPULATION GENETICS, BUILT ON the Hardy–Weinberg equilibrium principle. The models we discussed are elegant and powerful. Figure 5.14 (page 130), for example, documents a case in which a researcher used population genetics to accurately predict the course of evolution 12 generations into the future. In human terms, that is equivalent to accurately predicting what will happen 300 years from now. What a theory!

As with many theories, however, basic population genetics buys its elegance at the price of simplification. The models we have used until now track allele frequencies at just one locus at a time. We have only been able to consider the evolution of traits that are (or appear to be) controlled by a single gene. The genomes of real organisms, of course, contain hundreds or thousands of loci. And many traits are determined by the combined influence of numerous genes.

In Chapter 7 we will take our models of the mechanics of evolution closer to real organisms by considering two or more loci simultaneously. Our first step in that direction will be an extension of the Hardy–Weinberg analysis that follows two loci at a time. The two-locus model will tell us when we can use the single-locus models developed in Chapters 5 and 6 to make predictions, and when we must take into account the confounding influence of selection at other loci.

Our discussion of the two-locus version of Hardy–Weinberg analysis, which uses terms like "linkage disequilibrium," may at first seem hopelessly abstract. But effort invested in understanding it will produce two surprising payoffs. First, the two-locus model provides tools we can use to reconstruct the history of genes and populations. We will use these tools to address an unresolved question from the discussion in Chapters 4 and 5 of *CCR5-Δ32*, the allele that protects against HIV: Where did the *Δ32* allele come from, and why does it occur only in Europe? Second, the two-locus model provides insight into the adaptive significance of one of the most striking and puzzling characteristics of organisms: sexual reproduction.

The two-locus model is more realistic than the one-locus model, but it is still inadequate for analyzing the evolution of traits determined by the combined effects of alleles at many loci. When studying such traits, we often do not know the identities of the loci involved. The last section of this chapter introduces quantitative genetics, the branch of evolutionary biology that provides tools for analyzing multilocus traits. Again there will be a surprising practical payoff. Our discussion of quantitative genetics will allow us to debunk erroneous claims about differences in IQ scores among ethnic groups.

7.1 Evolution at Two Loci: Linkage Equilibrium and Linkage Disequilibrium

In this section we will expand the one-locus version of Hardy–Weinberg analysis to consider two loci simultaneously. In principle, we could focus on any pair of loci in an organism's genome. Our discussion will be easier to understand, however, if we focus on a pair of loci located on the same chromosome. That is, we will consider two loci that are physically linked (Figure 7.1). We will imagine that locus A has two alleles, *A* and *a,* and that locus B has two alleles, *B* and *b.*

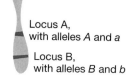

Locus A, with alleles *A* and *a*

Locus B, with alleles *B* and *b*

Figure 7.1 A pair of linked loci

In the single-locus version of Hardy–Weinberg analysis, we were concerned primarily with tracking allele frequencies. In the two-locus version, we will be concerned with tracking both allele frequencies and chromosome frequencies. Note that the assumptions we made in the previous paragraph allow four different chromosome genotypes: *AB, Ab, aB,* and *ab.* The multilocus genotype of a chromosome or gamete is sometimes referred to as its **haplotype** (a term that comes from the contraction of 'haploid genotype').

Our main goal will be to determine whether selection at the A locus will interfere with our ability to use the models of Chapters 5 and 6 to make predictions about evolution at the B locus. The answer will be, "Sometimes—depending on whether the loci are in linkage equilibrium or linkage disequilibrium." We will define linkage equilibrium and disequilibrium shortly.

When we use population-genetic models to analyze evolution at a particular locus, do we need to worry about the effects of selection at other loci? Only if the locus of interest and the other loci are in linkage disequilibrium.

A Numerical Example

A numerical example will illustrate key concepts and help us define terms. Figure 7.2 shows two hypothetical populations, each with a gene pool containing 25 chromosomes. In studying the genetic structure of these populations, the first thing we might do is calculate allele frequencies. In the top population, for example, 15 of the 25 chromosomes carry allele *A* at locus A. Thus the frequency of allele *A* is 15/25 = 0.6.

(a) A population in linkage equilibrium

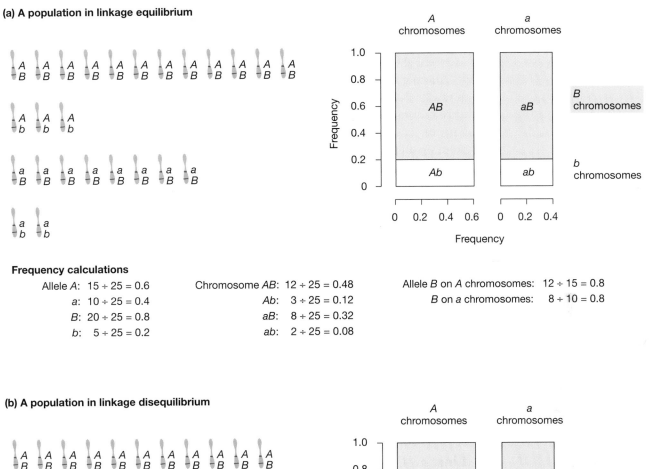

Frequency calculations

Allele *A*: 15 ÷ 25 = 0.6	Chromosome *AB*: 12 ÷ 25 = 0.48	Allele *B* on *A* chromosomes: 12 ÷ 15 = 0.8
a: 10 ÷ 25 = 0.4	*Ab*: 3 ÷ 25 = 0.12	*B* on *a* chromosomes: 8 ÷ 10 = 0.8
B: 20 ÷ 25 = 0.8	*aB*: 8 ÷ 25 = 0.32	
b: 5 ÷ 25 = 0.2	*ab*: 2 ÷ 25 = 0.08	

(b) A population in linkage disequilibrium

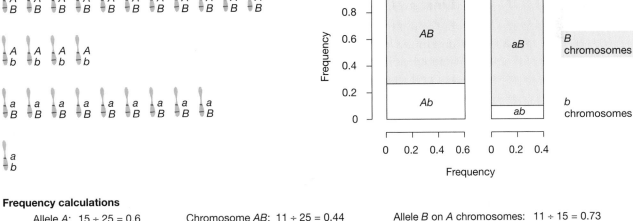

Frequency calculations

Allele *A*: 15 ÷ 25 = 0.6	Chromosome *AB*: 11 ÷ 25 = 0.44	Allele *B* on *A* chromosomes: 11 ÷ 15 = 0.73
a: 10 ÷ 25 = 0.4	*Ab*: 4 ÷ 25 = 0.16	*B* on *a* chromosomes: 9 ÷ 10 = 0.9
B: 20 ÷ 25 = 0.8	*aB*: 9 ÷ 25 = 0.36	
b: 5 ÷ 25 = 0.2	*ab*: 1 ÷ 25 = 0.04	

Figure 7.2   Two populations: one in linkage equilibrium; one in linkage disequilibrium

The same is true for the bottom population. In fact, the allele frequencies at both loci are identical in the two populations. If we were studying locus A only, or locus B only, we would conclude that the two populations are identical.

But the populations are not identical. This we discover when we calculate the chromosome frequencies. In the top population, for example, 12 of the 25 chromosomes carry haplotype AB, giving it a frequency of 0.48. In the bottom population, on the other hand, the frequency of AB chromosomes is 11 of 25, or 0.44. This is the first lesson of two-locus Hardy–Weinberg analysis: A pair of populations can have identical allele frequencies but different chromosome frequencies.

Another way to see the difference between the two populations in Figure 7.2 is to calculate the frequency of allele B on chromosomes carrying allele A versus chromosomes carrying allele a. In the top population there are 15 chromosomes carrying allele A, 12 of which carry allele B. The frequency of B on A chromosomes is thus $12/15 = 0.8$. In the same population there are 10 chromosomes carrying allele a, 8 of which carry allele B. The frequency of B on a chromosomes is thus $8/10 = 0.8$. In the top population, then, the frequency of allele B is the same on chromosomes carrying A as it is on chromosomes carrying a. The same is not true for the bottom population. There, the frequency of B is 0.73 on A chromosomes, but 0.9 on a chromosomes.

To understand linkage disequilibrium, it is helpful to recognize that when we consider two linked loci at once, populations can have identical allele frequencies, but different chromosome (that is, haplotype) frequencies.

The bar graphs in Figure 7.2 provide a visual representation of the difference between the two populations. The widths of the two bars in each graph represent the frequencies of A-bearing chromosomes versus a-bearing chromosomes. Note that the combined widths of the two bars must equal 1, so if one bar gets wider, the other must get narrower. The shaded versus unshaded shaded portion of each bar represents the frequency of allele B versus allele b on the chromosomes in question. The bar graphs let us see at a glance what we discovered by calculation in the previous paragraph. In the top population the frequency of B is the same on A chromosomes as on a chromosomes—the same fraction is shaded in both bars. In the bottom population the frequency of B is lower on A chromosomes than on a chromosomes.

Linkage Disequilibrium Defined

In the top population in Figure 7.2, locus A and locus B are in linkage equilibrium. In the bottom population, the loci are in linkage *dis*equilibrium. Two loci in a population are in **linkage equilibrium** when the genotype of a chromosome at one locus is independent of its genotype at the other locus. This means that knowing the genotype of the chromosome at one locus is of no use at all in predicting the genotype at the other. Two loci are in **linkage disequilibrium** when there is a nonrandom association between a chromosome's genotype at one locus and its genotype at the other locus. If we know the genotype of a chromosome at one locus, it provides a clue about the genotype at the other.

These definitions are rather abstract. More concretely, the following conditions are true for a pair of loci if, and only if, they are in linkage equilibrium:

When genotypes at one locus are independent of genotypes at another locus, the two loci are in linkage equilibrium. Otherwise, the loci are in linkage disequilibrium.

1. The frequency of B on chromosomes carrying allele A is equal to the frequency of B on chromosomes carrying allele a.

2. The frequency of any chromosome haplotype can be calculated by multiplying the frequencies of the constituent alleles. For example, the frequency of AB chromosomes can be calculated by multiplying the frequency of allele A and the frequency of allele B.

3. The quantity D, known as the coefficient of linkage disequilibrium, is equal to zero. D is calculated as

$$g_{AB}g_{ab} - g_{Ab}g_{aB}$$

where g_{AB}, g_{ab}, g_{Ab}, and g_{aB} are the frequencies of *AB, ab, Ab,* and *aB* chromosomes (see Box 7.1).

We have already established, by calculation and with bar graphs, that the first condition is true for the top population in Figure 7.2, but false for the bottom population. The reader should verify that the second and third conditions are likewise true for the top population but false for the bottom one.

The Two-Locus Version of Hardy–Weinberg Analysis

We can perform a two-locus version of Hardy–Weinberg analysis that is analogous to the single-locus version we performed in Chapter 5. We assume Hardy–Weinberg conditions of no selection, no mutation, no migration, infinite population size and random mating, and we follow chromosome frequencies through one complete turn of our population's life cycle, from gametes in the gene pool to zygotes to adults and back to gametes in the gene pool. This calculation is given in Box 7.2. It provides our first piece of evidence that linkage equilibrium is important in evolution. If the two loci in our ideal population are in linkage equilibrium, then under Hardy–Weinberg conditions chromosome frequencies will not change from one generation to the next. If, instead, the loci are in linkage disequilibrium, then the chromosome frequencies will change.

Under Hardy–Weinberg assumptions, chromosome frequencies remain unchanged from one generation the next, but only if the loci in question are in linkage equilibrium. If the loci are in linkage disequilibrium, the chromosome frequencies move closer to linkage equilibrium each generation.

What Creates Linkage Disequilibrium in a Population?

Three mechanisms can create linkage disequilibrium in a random-mating population: selection on multilocus genotypes, genetic drift, and population admixture. We will consider each of these mechanisms in turn.

To see how selection on multilocus genotypes can create linkage disequilibrium, start with the population whose gene pool is shown in Figure 7.2a. Locus A and locus B are in linkage equilibrium. Imagine that the gametes in this gene pool combine at random to make zygotes. The 10 kinds of zygotes produced, and their

---

Box 7.1 The coefficient of linkage disequilibrium

The coefficient of linkage disequilibrium, symbolized by *D,* is defined as

$$g_{AB}g_{ab} - g_{Ab}g_{aB}$$

where g_{AB}, g_{ab}, g_{Ab}, and g_{aB} are the frequencies of *AB, ab, Ab,* and *aB* chromosomes.

To see why *D* is called the coefficient of linkage disequilibrium, recall that when two loci are in linkage equilibrium, the allele frequencies at one locus are independent of allele frequencies at the other locus. Let *p* and *q* be the frequencies of *A* and *a,* and let *s* and *t* be the frequencies of *B* and *b.* If a population is in linkage equilibrium, then $g_{AB} = ps$, $g_{Ab} = pt$, $g_{aB} = qs$, and $g_{ab} = qt$. And furthermore,

$$D = psqt - ptqs = 0$$

If, on the other hand, the population is in linkage disequilibrium, then $g_{AB} \neq ps$, $g_{Ab} \neq pt$, $g_{aB} \neq qs$, and $g_{ab} \neq qt$. And $D \neq 0$.

The maximum value that *D* can assume is 0.25, when *AB* and *ab* are the only chromosomes present and each has a frequency of 0.5. The minimum value that *D* can assume is -0.25, when *Ab* and *aB* are the only chromosomes present and each is at a frequency of 0.5. Thus calculating *D* is a useful way to quantify the degree of linkage disequilibrium in a population.

Box 7.2 Hardy–Weinberg analysis for two loci

Here we develop the two-locus version of the Hardy–Weinberg equilibrium principle. We will show that when an ideal population is in linkage equilibrium, the chromosome frequencies do not change from one generation to the next.

In the single-locus version of Hardy–Weinberg analysis, introduced in Chapter 5, we followed allele frequencies around a complete turn of the life cycle of a population, from one generation's gene pool into zygotes, from zygotes into adults, and from adults into the next generation's gene pool. We will use a similar strategy here, except that we will track not allele frequencies but chromosome frequencies. The chromosomes in our organisms contain two loci: the A locus, with alleles *A* and *a*; and the B locus, with alleles *B* and *b*. (We do not intend these symbols to necessarily imply a dominant/recessive relationship between alleles. We use them only because they make the equations easier to read than alternative notations.) There are four kinds of chromosomes: *AB*, *Ab*, *aB*, and *ab*.

Imagine an ideal population in whose gene pool chromosomes *AB*, *Ab*, *aB*, and *ab* are present at frequencies g_{AB}, g_{Ab}, g_{aB}, and g_{ab}, respectively. If the gametes in the gene pool combine at random to make zygotes, among the possible zygote genotypes is *AB/AB*. Its frequency will equal the probability that a randomly chosen egg contains an *AB* chromosome multiplied by the probability that a randomly chosen sperm contains an *AB* chromosome, or $g_{AB} \times g_{AB}$. Another possible zygote genotype is *AB/Ab*. Its frequency will be $2 \times g_{AB} \times g_{Ab}$. This expression contains a 2 because there are two ways to make an *AB/Ab* zygote: An *AB* egg can be fertilized by an *Ab* sperm, or an *Ab* egg can be fertilized by an *AB* sperm. Overall, there are 10 possible zygotes. The zygotes and their frequencies are

AB/AB	**Ab/Ab**	**aB/aB**	**ab/ab**	**AB/Ab**
$g_{AB}g_{AB}$	$g_{Ab}g_{Ab}$	$g_{aB}g_{aB}$	$g_{ab}g_{ab}$	$2g_{AB}g_{Ab}$

AB/aB	**AB/ab**	**Ab/aB**	**Ab/ab**	**aB/ab**
$2g_{AB}g_{aB}$	$2g_{AB}g_{ab}$	$2g_{Ab}g_{aB}$	$2g_{Ab}g_{ab}$	$2g_{aB}g_{ab}$

If we allow these zygotes to grow up without selection, then the genotype frequencies among the adults will be the same as they are among the zygotes.

We have followed the chromosome frequencies from gene pool to zygotes to adults. We can now calculate the chromosome frequencies in the next generation's gene pool. Consider chromosome *AB*. Gametes containing *AB* chromosomes can be produced by 5 of the 10 adult genotypes. The adults that can make *AB* gametes, together with the allotment of *AB* gametes they contribute to the new gene pool, are

Adult	Allotment of *AB* gametes contributed	Notes
AB/AB	$g_{AB}g_{AB}$	
AB/Ab	$\left(\frac{1}{2}\right)(2g_{AB}g_{Ab})$	
AB/aB	$\left(\frac{1}{2}\right)(2g_{AB}g_{aB})$	
AB/ab	$(1-r)\left(\frac{1}{2}\right)(2g_{AB}g_{ab})$	where r = recombination rate
Ab/aB	$(r)\left(\frac{1}{2}\right)(2g_{Ab}g_{aB})$	where r = recombination rate

The first row in this table is straightforward: *AB/AB* adults constitute a fraction $g_{AB}g_{AB}$ of the population, and thus contribute $g_{AB}g_{AB}$ of the gametes in the gene pool, all of them *AB*. The second row is also straightforward: *AB/Ab* adults constitute a fraction $g_{AB}g_{Ab}$ of the population, and thus contribute $g_{AB}g_{Ab}$ of the gametes in the gene pool, half of them *AB*. It is the last two rows of the table that require explanation.

Adults of genotype *AB/ab* will produce gametes containing *AB* chromosomes only when meiosis occurs *without* crossing-over between the A locus and the B locus. When no crossing-over occurs, half of the gametes produced by *AB/ab* adults carry *AB* chromosomes. If *r* is the rate of crossing-over, or recombination, between the A locus and the B locus, then the allotment of *AB* gametes contributed to the gene pool by *AB/ab* individuals is $(1-r)\left(\frac{1}{2}\right)(2g_{AB}g_{ab})$.

Adults of genotype *Ab/aB* produce gametes containing *AB* chromosomes only when meiosis occurs *with* crossing-over between the A locus and the B locus. When crossing-over occurs, half of the gametes produced by *Ab/aB* adults carry *AB* chromosomes. If *r* is the rate of crossing-over, then the allotment of *AB* gametes contributed to the gene pool by *Ab/aB* individuals is $(r)\left(\frac{1}{2}\right)(2g_{Ab}g_{aB})$.

We can now write an expression for g_{AB}', the frequency of *AB* chromosomes in the new gene pool:

Box 7.2 **Continued**

$$g_{AB}' = g_{AB}g_{AB} + (\tfrac{1}{2})(2g_{AB}g_{Ab}) + (\tfrac{1}{2})(2g_{AB}g_{aB}) +$$
$$(1 - r)(\tfrac{1}{2})(2g_{AB}g_{ab}) + (r)(\tfrac{1}{2})(2g_{Ab}g_{aB})$$
$$= g_{AB}g_{AB} + g_{AB}g_{Ab} + g_{AB}g_{aB} +$$
$$g_{AB}g_{ab} - rg_{AB}g_{ab} + rg_{Ab}g_{aB}$$
$$= g_{AB}(g_{AB} + g_{Ab} + g_{aB} + g_{ab}) - r(g_{AB}g_{ab} - g_{Ab}g_{aB})$$

We can simplify this expression further by noting that $(g_{AB} + g_{Ab} + g_{aB} + g_{ab}) = 1$, and that $g_{AB}g_{ab} - g_{Ab}g_{aB}$ is D, defined in the text and Box 7.1. This gives us

$$g_{AB}' = g_{AB} - rD$$

We leave it to the reader to derive the expressions for the other three chromosome frequencies, which are

$$g_{Ab}' = g_{Ab} + rD \quad g_{aB}' = g_{aB} + rD \quad g_{ab}' = g_{ab} - rD$$

The expressions for g_{AB}', g_{Ab}', g_{aB}', and g_{ab}' show that when a population is in linkage equilibrium—when $D = 0$—the chromosome frequencies do not change from one generation to the next. When, on the other hand, the population is in linkage disequilibrium—when $D \neq 0$—the chromosome frequencies do change from one generation to the next. The first population geneticist to report this result was H. S. Jennings (1917).

We should note that allele frequencies at a pair of loci can be in linkage disequilibrium even when the loci are on different chromosomes. For loci on different chromosomes, it is appropriate to speak of gamete frequencies rather than chromosome frequencies. The Hardy–Weinberg analysis for such a situation is identical to the one we have developed here, except that r is always equal to exactly $(\tfrac{1}{2})$.

expected frequencies, appear in the grid in Figure 7.3a. Because 32% of the eggs are aB, for example, and 32% of the sperm are aB, we predict that the frequency of aB/aB zygotes will be $0.32 \times 0.32 = 0.1024$. Now let the zygotes develop into adults and assign phenotypes as follows: Individuals with genotype ab/ab have a size of 10. For other genotypes, every copy of A or B adds 1 unit to the individual's size. For example, aB/aB individuals have a size of 12, and AB/Ab individuals have

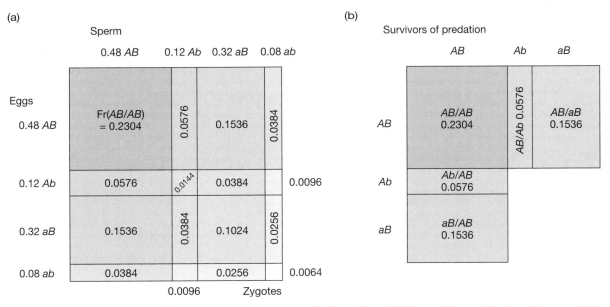

Figure 7.3 Selection on multilocus genotypes can create linkage disequilibrium Diagram (a) shows the expected frequencies of zygotes produced by random mating in the linkage-equilibrium population from Figure 7.2a. Diagram (b) shows the genotypes that survive after predators kill all individuals with fewer than three capital-letter alleles in their genotype. The population of survivors is in linkage disequilibrium.

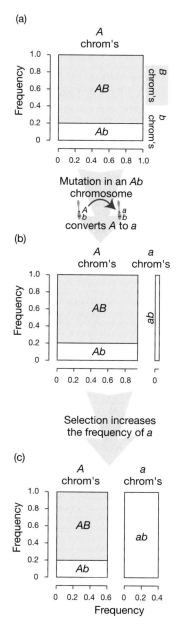

Figure 7.4 Genetic drift can create linkage disequilibrium
(a) Chromosome frequencies in a finite population in which only one allele, *A*, is present at locus A; (b) chromosome frequencies after a mutation creates a single copy of allele *a*; (c) chromosome frequencies after selection in favor of *a* increases the frequency of *ab* chromsomes. The population in (c) is in linkage disequilibrium. Drift is the crucial mechanism that created the disequilibrium, because this scenario could only happen in a finite population.

a size of 13. Finally, imagine that predators catch and eat every individual whose size is less than 13. The survivors, which represent 65.28% of the original population, appear in the grid in Figure 7.3b.

In the population of survivors, locus A and locus B are in linkage disequilibrium. Perhaps the easiest way to see this is to calculate the frequency of allele *a* and allele *b*. One way to calculate the frequency of *a* is this: Individuals carrying *a* alleles constitute a fraction (0.1536 + 0.1536)/0.6528 ≈ 0.47 of the survivors. Thus, they carry 47% of the locus A alleles. Half of these alleles are *a*. Therefore, the frequency of *a* among the survivors is 0.5 × 0.47 ≈ 0.24. The frequency of *b* is approximately 0.09. If our two loci were in linkage equilibrium, then, by criterion 2 on our list, the frequency of *ab* chromosomes among the survivors would be 0.24 × 0.09 ≈ 0.02. In fact, the frequency of *ab* chromosomes is 0. Our two loci are in linkage disequilibrium. As an exercise, the reader should demonstrate that the loci are in linkage disequilibrium by criteria 1 and 3 as well.

To see how genetic drift can create linkage disequilibrium, look at the scenario diagrammed in Figure 7.4. This scenario starts with a gene pool in which the only chromosomes present are *AB* and *Ab* (Figure 7.4a). In other words, copies of allele *a* do not exist in this population. Locus A and locus B are in linkage equilibrium. Now imagine that in a single *Ab* chromosome, a mutation converts allele *A* into allele *a*. This creates a single *ab* chromosome (Figure 7.4b). It also puts the population in linkage disequilibrium, because there is now a possible chromosome haplotype—*aB*—that is missing. Finally, imagine that selection favors allele *a* over allele *A,* so that *a* increases in frequency and *A* decreases (Figure 7.4c). This increases the degree of linkage disequilibrium between locus A and locus B.

The reader may wonder why we are ascribing the linkage disequilibrium created in this scenario to genetic drift, when the key events seem to be mutation and selection. The reason is that the scenario, as we described it, could only happen in a finite population. In an infinite population, the mutation converting allele *A* into allele *a* would happen not once but many times each generation, on both *AB* and *Ab* chromosomes. At no point would *aB* chromosomes be missing. Selection favoring *a* over *A* would simultaneously increase the frequency of both *ab* and *aB* chromosomes. Locus A and locus B would never be in linkage disequilibrium. Because our scenario can only create linkage disequilibrium in a finite population, the crucial evolutionary mechanism at work is genetic drift. It was sampling error that caused the mutation creating allele *a* to happen only once, and in an *Ab* chromosome.

Finally, to see how population admixture can create linkage disequilibrium, imagine two gene pools (Figure 7.5). In one, there are 60 *AB* chromosomes, 20 *Ab* chromosomes, 15 *aB* chromosomes, and 5 *ab* chromosomes. In the other, there are 10 *AB*, 40 *Ab*, 10 *aB*, and 40 *ab*. Locus A and locus B are in linkage equilibrium in both gene pools, as the top two bar graphs in Figure 7.5 show. Now combine the two gene pools. This produces a new gene pool in which there are 70 *AB* chromosomes, 60 *Ab*, 25 *aB*, and 45 *ab*. In this new gene pool, locus A and locus B are in linkage disequilibrium.

Selection on multilocus genotypes, genetic drift, and population admixture can all create linkage disequilibrium because they can all produce populations in which some chromosome haplotypes are underrepresented, and others overrepresented, compared to what their frequencies would be under linkage equilibrium. In our multilocus selection scheme, for example, selection acted more strongly against *ab* than any other haplotype, because no individual containing an *ab* chromosome survived. In our drift scenario, a chance event led to the creation of an *ab* chromo-some but no *aB* chromosome. In our population admixture example, a simple combination of populations with different allele and chromosome frequencies created a new population with an excess of *AB* and *ab* chromosomes.

What Eliminates Linkage Disequilibrium from a Population?

At the same time that selection, drift, and admixture may be creating linkage dis-equilibrium in a population, sexual reproduction inexorably reduces it. By sexu-al reproduction, we mean meiosis with crossing-over and outbreeding. The union of gametes from unrelated parents brings together chromosomes with different haplotypes. When the zygotes grow to adulthood and themselves reproduce, cross-ing-over during meiosis breaks up old combinations of alleles and creates new ones. The creation of new combinations of alleles during sexual reproduction is called **genetic recombination**. Because genetic recombination tends to ran-domize genotypes at one locus with respect to genotypes at another, it tends to reduce the frequency of overrepresented chromosome haplotypes and to increase the frequency of underrepresented haplotypes. In other words, genetic recombi-nation reduces linkage disequilibrium.

The action of sexual reproduction in reducing linkage disequilibrium is demon-strated algebraically in Box 7.3. The analysis in the box shows that under Hardy–Weinberg assumptions the rate of decline in linkage disequilibrium between a pair of loci is proportional to the rate of recombination between them. Predictions of the rate of decline for several different rates of recombination appear in Figure 7.6.

Michael Clegg and colleagues (1980) documented the decay of linkage dise-quilibrium in experimental populations of fruit flies. Every population they stud-ied harbored two alleles at each of two loci on chromosome 3. One locus encodes the enzyme esterase-c; we will call it locus A, and its alleles *A* and *a*. The other locus encodes the enzyme esterase-6; we will call it locus B, and its alleles *B* and *b*.

Clegg and colleagues set up fruit fly populations with only *AB* and *ab* chro-mosomes, each at a frequency of 0.5. The researchers also set up populations with only *Ab* and *aB* chromosomes, again each at a frequency of 0.5. Thus every pop-ulation was initially in complete linkage disequilibrium, either with $D = 0.25$, or with $D = -0.25$.

The researchers maintained their fly populations for 48 to 50 generations, at sizes of approximately 1000 individuals, and let the flies mate as they pleased. Every

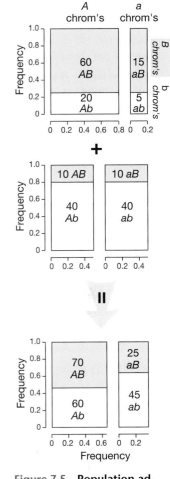

Figure 7.5 Population ad-mixture can create linkage disequilibrium The top two bar charts represent chromosome frequencies in two distinct popu-lations, each in linkage equilibri-um. Mixed together, these two populations yield the population shown in the bottom bar chart; it is in linkage disequilibrium.

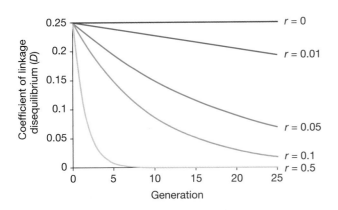

Figure 7.6 With sexual reproduction and random mating, linkage disequilibrium falls over time This graph shows the level of linkage disequilibrium between two loci over 25 genera-tions in a random-mating population. The population starts with linkage disequilibrium at its maximum possible value, 0.25. Each curve shows the decline in linkage disequilibrium, according to the equation $D' = D(1 - r)$, for a different value of *r*. With $r = 0.5$, which corresponds to the free recombination of loci on dif-ferent chromosomes, the population reaches linkage equilibrium in less than 10 generations. With $r = 0.01$, which corresponds to closely linked loci, linkage disequilibrium persists for many gener-ations. After Hedrick (1983).

Box 7.3   Sexual reproduction reduces linkage disequilibrium

Here we show that the level of linkage disequilibrium inexorably declines in a random-mating sexual population. We do so by starting with the definition of D, given in the text and Box 7.1, and deriving an expression for D', the coefficient of linkage disequilibrium in the next generation.

By the definition of D,

$$D' = g_{AB}'g_{ab}' - g_{Ab}'g_{aB}'$$

Substituting the expressions for g_{AB}', g_{ab}', g_{Ab}', and g_{aB}' that were derived in Box 7.2 gives

$$D' = [(g_{AB} - rD)(g_{ab} - rD)] -$$
$$[(g_{Ab} + rD)(g_{aB} + rD)]$$
$$= [g_{AB}g_{ab} - g_{AB}rD - g_{ab}rD + (rD)^2] -$$
$$[g_{Ab}g_{aB} + g_{Ab}rD + g_{aB}rD + (rD)^2]$$
$$= g_{AB}g_{ab} - g_{AB}rD - g_{ab}rD + (rD)^2 -$$
$$g_{Ab}g_{aB} - g_{Ab}rD - g_{aB}rD - (rD)^2$$

Canceling and rearranging terms gives

$$D' = g_{AB}g_{ab} - g_{Ab}g_{aB} - g_{AB}rD -$$
$$g_{ab}rD - g_{Ab}rD - g_{aB}rD$$
$$= (g_{AB}g_{ab} - g_{Ab}g_{aB}) - rD(g_{AB} + g_{ab} + g_{Ab} + g_{aB})$$

Finally, the expression $(g_{AB}g_{ab} - g_{Ab}g_{aB})$ is equal to D, and the expression $(g_{AB} + g_{ab} + g_{Ab} + g_{aB})$ is equal to 1, so we have

$$D' = D - rD = D(1 - r)$$

Recall that r is the rate of recombination during meiosis, which is always between 0 and $\frac{1}{2}$. This means that $(1 - r)$ is always between $\frac{1}{2}$ and 1. Thus, unless there is no recombination at all between a pair of loci, the linkage disequilibrium between them will move closer to 0 every generation. The higher the rate of recombination between the loci, the faster the population reaches linkage equilibrium.

In random mating populations, three forces create linkage disequilibrium: selection on multilocus genotypes, genetic drift, and population admixture. One force reduces linkage disequilibrium: genetic recombination resulting from meiosis and outbreeding (that is, sex).

generation or two, the researchers sampled each population to determine the frequencies of the four chromosome haplotypes and calculate the level of linkage disequilibrium between the two loci. For reasons beyond the scope of our discussion, the researchers measured linkage disequilibrium not with D, but with a related statistic called the correlation of allelic state. There is no one-to-one relationship between values of D and the correlation of allelic state, but as a general rule we can say that as linkage disequilibrium in a population declines, and as D moves from 0.25 or -0.25 toward 0, the correlation of allelic state declines as well, moving toward 0 from 1.0 or -1.0. Clegg and colleagues predicted that this is just what they would see in their freely-mating fruit fly populations.

The results appear in Figure 7.7. The smooth gray curves show the predicted pattern of decline; the jagged colored lines show the data. As predicted, crossing-over during meiosis created the missing chromosome haplotypes, and the linkage disequilibrium between the loci declined. Indeed, linkage disequilibrium declined somewhat faster than predicted. Clegg and colleagues believe that the more-rapid-than-expected decline was the result of heterozygote superiority at the enzyme loci they were studying. Heterozygote superiority would increase the frequency of individuals heterozygous for both loci, and thus provide more opportunities for crossing-over to break down nonrandom associations between alleles at one locus and alleles at the other.

Why Does Linkage Disequilibrium Matter?

We have defined linkage disequilibrium as a nonrandom association between genotypes at different loci. We have identified multilocus selection, genetic drift, and population admixture as evolutionary mechanisms that can create it. We have

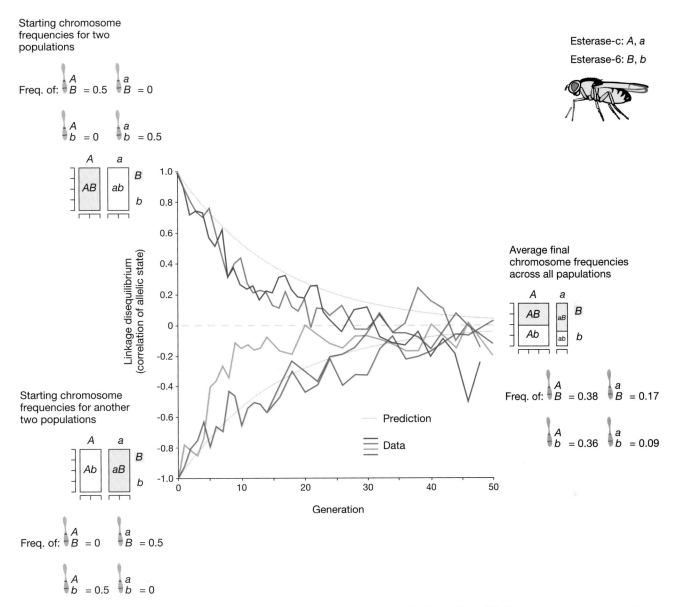

Figure 7.7 An empirical demonstration that sexual reproduction reduces linkage disequilibrium Each of several populations of fruit flies began in complete linkage disequilibrium (upper and lower bar graphs on the left). Over 50 generations, all populations approached linkage equilibrium (bar graph on the right). Redrawn from Clegg, et al. (1980), with frequencies in the bar graph on the right inferred from data therein.

seen that sexual reproduction reduces it, restoring populations to a state of linkage equilibrium. And we have demonstrated that in an ideal Hardy–Weinberg population that is in linkage equilibrium, chromosome frequencies do not change from one generation to the next. We have not, however, addressed what we said at the outset was to be our primary question: Can selection at one locus interfere with our ability to use single-locus models to predict the course of evolution at other loci? We are ready address this question now.

The bad news is that if locus A and locus B are in linkage disequilibrium, then selection at locus A changes the frequencies of the alleles at locus B. This means that a single-locus population genetic model looking only at locus B will make inaccurate predictions about evolution.

Figure 7.8a illustrates how selection on locus A can change allele frequencies at locus B. Before selection, allele *B* is at high frequency. Most copies of *B* are on *aB* chromosomes. Selection in favor of allele *A* decreases the frequency of *aB* chromosomes. As *aB* chromosomes disappear, they take copies of *B* with them. Because the frequency of *B* is much lower among *A*-bearing chromosomes than among *a*-bearing chromosomes, many of the lost copies of *B* are replaced by copies of *b*. The end result is that the frequency of *B* declines.

Note that in the scenario we just described, selection was acting only at locus A, not at locus B. Genotypes at locus B had no effect on fitness. Instead, the frequency of allele *B* dropped simply because it got dragged along for the ride. But if we were monitoring locus B only, and watching the frequency of allele *B* decline over time, we might erroneously conclude that the target of selection was locus B itself. This is the most depressing lesson of the two-locus version of Hardy–Weinberg analysis: Single-locus studies can be derailed by linkage disequilibrium.

The good news is that if locus A and locus B are in linkage equilibrium, selection on locus A has no effect whatsoever on allele frequencies at locus B. Look at Figure 7.8b. Selection in favor of allele *A* again eliminates many *aB* chromosomes. But because the frequency of *B* is the same among *A*-bearing chromosomes as among *a*-bearing chromosomes, every copy of allele *B* that is lost is replaced by another copy of *B*. If selection at locus A has no effect on allele frequencies at locus B, then it will not interfere with our use of single-locus models to analyze locus B's evolution.

Still better news is that in random-mating populations, sex is so good at eliminating linkage disequilibrium that most pairs of loci are in linkage equilibrium most of the time. Work by Gavin Huttley and colleagues (1999) illustrates this claim. Huttley and colleagues surveyed the human genome for linkage disequilibrium among short tandem repeat loci. A short tandem repeat locus is a spot on a chromosome where a short nucleotide sequence is repeated several times; such loci typically have several alleles. Huttley and colleagues conducted over 200,000 pairwise tests of linkage disequilibrium involving over 5000 loci distributed across all 22 autosomes.

When a pair of loci are in linkage disequilibrium, selection at one locus can change allele frequencies at the other locus. This means that single-locus models may make inaccurate predictions.

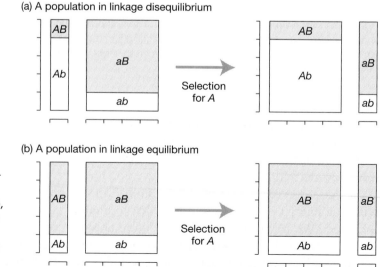

(a) A population in linkage disequilibrium

(b) A population in linkage equilibrium

Figure 7.8   Linkage equilibrium, selection at a single locus, and allele frequencies at a linked locus In a population in linkage disequilibrium, as shown in (a), selection in favor of allele *A* at locus A causes a decline in the frequency of allele *B* at locus B. In a population in linkage equilibrium, as shown in (b), selection in favor of allele *A* has no effect on the frequency of allele *B*.

Huttley and colleagues did find several places in the human genome where neighboring loci exhibit substantial linkage disequilibrium. One region of high linkage disequilibrium, already known from prior studies, is an area on chromosome 6 containing the human leukocyte antigen (HLA) loci. The HLA loci encode proteins that immune system cells use in recognizing foreign invaders. The HLA loci are under strong selection, and the linkage disequilibrium among them is probably a result of selection on multilocus genotypes.

Overall, however, pairs of loci exhibiting linkage disequilibrium were in the minority. The pairs most likely to show linkage disequilibrium were those that are closely linked physically—that is, situated near enough to each other on the same chromosome that crossing-over between them is rare. Huttley and colleagues focussed on pairs of loci close enough that crossing-over occurs between them in 4% or fewer of meiotic cell divisions. Of these pairs, just 4% exhibited linkage disequilibrium.

In a similar study, Naohiko Miyashita and colleagues (1999) surveyed the genome of the plant *Arabidopsis thaliana,* a small member of the mustard family. One might expect that the *Arabidopsis* genome would harbor considerable linkage disequilibrium, even among loci that are far apart or on different chromosomes. This is because *Arabidopsis* self-fertilizes, drastically reducing the opportunity for genetic recombination. Miyashita and colleagues conducted nearly 70,000 pairwise tests of linkage disequilibrium among almost 8000 loci. Among these pairwise tests, about 12% revealed linkage disequilibrium. The researchers conclude that *Arabidopsis* must occasionally outbreed. Even a relatively small amount of recombination goes a long way toward reducing linkage disequilibrium.

We can summarize the take-home message in our exploration of two-locus Hardy–Weinberg analysis as follows. Population geneticists need to be aware that any particular locus of interest may be in linkage disequilibrium with other loci, especially other loci located nearby. If the locus of interest is, in fact, in linkage disequilibrium with another, then single-locus population genetic models may yield inaccurate predictions. Nonetheless, in freely mating populations, most pairs of loci can be expected to be in linkage equilibrium. In general, we can expect that single-locus models will work well most of the time.

When a pair of loci are in linkage equilibrium, selection at one locus has no effect on allele frequences at the other, and we can use single-locus models with confidence. Fortunately, sex is so good at reducing linkage disequilibrium that most pairs of loci are in linkage equilibrium most of the time.

A Practical Reason for Measuring Linkage Disequilibrium

In the introduction to this chapter, we promised rewards awaiting readers who mastered the abstractions of the previous sections. One such reward is this: Measurements of linkage disequilibrium provide clues that are useful in reconstructing the history of genes and populations.

Recall from Chapters 4 and 5 our discussion of the *CCR5-Δ32* allele. This allele is a loss-of-function mutation at the CCR5 locus. It protects homozygotes against sexually transmitted strains of HIV-1. Among the unresolved questions from the earlier discussion are these: Where did the *Δ32* allele come from? and Why is it common only in Europe?

J. Claiborne Stephens and colleagues (1998) addressed these questions by measuring linkage disequilibrium between the CCR5 locus and two loci located nearby on the same chromosome. The nearby loci are short tandem repeat sites called GAAT and AFMB. GAAT has three alleles and AFMB has four. GAAT and AFMB are noncoding, and their alleles appear to have no effect on fitness. Stephens and colleagues determined the haplotypes of 192 chromosomes from a sample of Europeans. As Figure 7.9a shows, GAAT and AFMB are close to being in linkage equilibrium with each other. The frequencies of the various alleles at the AFMB locus are nearly the same among chromosomes carrying any one of the alleles at the GAAT locus as they are among chromosomes carrying the other GAAT alleles. As Figures 7.9b and 7.9c show, however, the CCR5 locus is in strong linkage disequilibrium with both the GAAT locus and the AFMB locus. Almost all chromosomes carrying the *Δ32* allele at CCR5 also carry allele *197* at GAAT (Figure 7.9b) and allele *215* at AFMB (Figure 7.9c).

How did the disequilibrium between CCR5 and its neighbors arise? As we showed in our discussion above on what creates linkage disequilibrium, there are

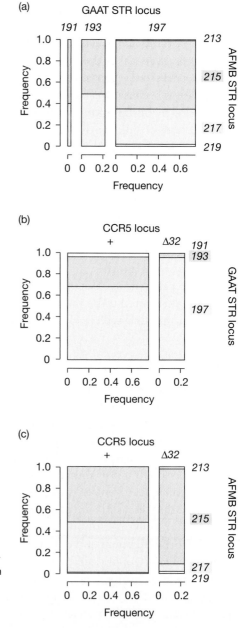

Figure 7.9   Where did the *CCR5-Δ32* allele come from?   All charts show haplotype frequencies on chromosome 3 in a European population. The bar graph in (a) shows that two neutral loci near the CCR5 locus are close to linkage equilibrium. The bar graphs in (b) and (c) show that each of these neutral loci is in strong linkage disequilibrium with the CCR5 locus. Almost all chromosomes carrying the *CCR5-Δ32* allele also carry allele *197* at the GAAT locus and allele *215* at the AFMB locus. These data are consistent with the creation of the *CCR5-Δ32* allele in a unique mutation event that occurred between 275 and 1875 years ago. Plotted from data in Stephens et al. (1998).

three possibilities: selection on multilocus genotypes, genetic drift, and population admixture. Selection on multilocus genotypes is an unlikely candidate, because, as we just noted, GAAT and AFMB are noncoding loci and their alleles appear to be selectively neutral. Population admixture is also an unlikely candidate, because it would require a source population in which the frequency of the *Δ32* allele is much higher than it is in Europe, and no such population exists. That leaves genetic drift.

Stephens and colleagues believe that the linkage disequilibrium between CCR5 and its neighbors arose by a scenario similar to the one shown in Figure 7.4. At some time in the past, the European population had only one allele for CCR5, the *CCR5-+* allele. Then, in a chromosome with the CCR5-GAAT-AFMB haplotype *+ - 197 - 215,* a mutation occured that created the *Δ32* allele. Therefore, the ancestor of all *Δ32* alleles was on a chromosome with the haplotype *Δ32 - 197 - 215.* Finally, the newly created *Δ32* allele was favored by natural selection. It rose to high frequency, carrying alleles *197* and *215* at the neighboring loci with it.

The association between the *Δ32* allele at CCR5, the *197* allele at GAAT, and the *215* allele at AFMB is no longer perfect. Since the *Δ32* allele first appeared, recombination and/or additional mutations have put *Δ32* into other haplotypes, such as *Δ32 - 197 - 217.* In other words, the linkage disequilibrium between the CCR5 locus and its neighbors is breaking down. Stephens and colleagues used estimates of the rates of crossing-over and mutation to calculate how fast the linkage disequilibrium would be expected to break down. Then they used this calculation to estimate how long it has been since the *Δ32* allele first appeared. The researchers concluded that *Δ32* first appeared between 275 and 1875 years ago, with a best estimate of about 700 years ago (see Box 7.4).

The answer to our question about the origin of the *Δ32* allele appears to be this: The allele was created by a unique mutation that occurred in Europe within the last several hundred years. And the allele does not occur outside Europe either because the mutation creating it has never occurred in a non-European population, or because when the mutation *has* occurred outside Europe, it has not been favored by selection.

This answer, of course, raises new questions. For example, selection favoring the *Δ32* allele in Europe must have been strong. We know this because only strong selection could have carried the allele from a frequency of virtually zero to a frequency of 10 to 20 percent in just 700 years. What was the selective agent responsible? The most obvious suspects would be epidemic diseases. One intriguing possibility is bubonic plague, the disease responsible for the Black Death that swept Europe during the 14th century, killing between one quarter and one half of the population. Bubonic plague is caused by the bacterium *Yersinia pestis.* Perhaps the *Δ32* allele also protects against *Yersinia pestis.* As of this writing, the bubonic plague hypothesis is being tested by researchers in the laboratory of Stanley Falkow at Stanford University. Definitive results are not yet in. Recently Alshad Lalani and colleagues (1999) reported that myxoma, a rabbit virus in the poxvirus family, can, like HIV-1, use CCR5 to gain entry into host cells. Lalani and colleagues speculate that the disease responsible for the high frequency of the *Δ32* allele in Europe might have been smallpox.

If it turns out that the selective agent favoring the *Δ32* allele in Europe was bubonic plague or smallpox, then we might expect that loss-of-function mutations in the CCR5 gene would have been favored in other parts of the world that have

Because the list of forces that create linkage disequilibrium is short, the presence of linkage disequilibrium in a population provides clues to the population's past.

Box 7.4   **Estimating the age of the *CCR5-Δ32* mutation**

Here we outline the calculation J. Claiborne Stephens and colleagues (1998) used to make a rough estimate of the age of the *CCR5-Δ32* mutation. First, look at Figure 7.10. It shows the arrangement of the three crucial loci on the short arm of chromosome 3: CCR5 is first, then GAAT, then AFMB. We will refer to haplotypes for chromosome 3 by listing their genotypes at these loci with dashes in between. For example, haplotype *Δ32 - 197 - 215* has allele *Δ32* at the CCR5 locus, allele *197* at the GAAT locus, and allele *215* at the AFMB locus.

Note that haplotype *Δ32 - 197 - 215* accounts for 84.8% of all chromosomes carrying the *Δ32* allele at the CCR5 locus. Furthermore, haplotype *+ - 197 - 215* is, at a frequency of 36%, the most common haplotype among chromosomes carrying the *+* allele at the CCR5 locus. Based on these numbers, Stephens et al. concluded that the mutation creating the first copy of the *Δ32* allele probably occurred on a *+ - 197 - 215* chromosome, converting it to a *Δ32 - 197 - 215* chromosome. Since the original mutation occurred, 84.8% of *Δ32 - 197 - 215* chromosomes have remained unchanged, while 15.2% have been converted to other haplotypes by either (1) crossing-over between CCR5 and GAAT or between GAAT and AFMB, or (2) mutation at GAAT or AFMB.

Stephens and colleagues developed an equation for P_g, the probability that any given *Δ32 - 197 - 215* chromosome has remain unchanged across the *g* generations that have passed since the first *Δ32 - 197 - 215* chromosome was created. Their reasoning was as follows: The chance that a *Δ32 - 197 - 215* chromosome will remain unchanged across a single generation is equal to

$$1 - c - \mu$$

where *c* is the probability that crossing-over will convert the chromosome to a different haplotype, and μ is the probability that mutation at either the GAAT locus or the AFMB locus will convert the chromosome to a different haplotype. Thus,

$$P_g = (1 - c - \mu)^g$$

All we have to do to estimate *g* is develop independent estimates of P_g, *c*, and μ, substitute them into this equation, and solve for *g*.

We already have an estimate for P_g. It is 0.848, the fraction of *Δ32* chromosomes that still have the *Δ32 - 197 - 215* haplotype.

Stephens and colleagues estimated the value of *c* by assuming that *Δ32 - 197 - 215* chromosomes have virtually always been paired with *CCR5-+* chromosomes, and that among *CCR5-+* chromosomes the frequencies of various haplotypes have remained constant over time. The rate of crossing-over between CCR5 and GAAT is 0.0021. And because 64% of *CCR5-+* chromosomes have a haplotype other than *197 - 215* at GAAT and AFMB, 64% of such crossovers will disrupt the *Δ32 - 197 - 215* haplotype. The rate of crossing-over between GAAT and AFMB is 0.0072. And because 48% of *CCR5-+* chromosomes have a genotype other than *215* at AFMB, 48% of such cross-overs will disrupt the *Δ32 - 197 - 215* haplotype. Thus

$$c = (0.64 \times 0.0021) + (0.48 \times 0.0072) \approx 0.005$$

Based on other researchers' estimates of mutation rates at loci similar to GAAT and AFMB, Stephens et al. estimated that

$$\mu = 0.001$$

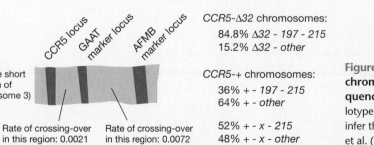

CCR5-Δ32 chromosomes:
84.8% Δ32 - 197 - 215
15.2% Δ32 - other

CCR5-+ chromosomes:
36% + - 197 - 215
64% + - other

52% + - x - 215
48% + - x - other

(On the short arm of chromosome 3)

Rate of crossing-over in this region: 0.0021   Rate of crossing-over in this region: 0.0072

Figure 7.10   Three loci on the short arm of chromosome 3, and selected haplotype frequencies in a European population   These haplotype frequencies provide the information needed to infer the age of the *CCR5-Δ32* allele. After Stephens et al. (1998).

Box 7.4 Continued

Substituting these values for P_g, c, and μ into the equation for P_g gives

$$0.848 = (0.994)^g$$

Taking the logarithm of both sides gives

$$\log[0.848] = \log[(0.994)^g] = g \log[0.994]$$

Finally, solving for g gives

$$g \approx 27.5$$

If we take 25 years to be the length of a human generation, then the mutation creating the original $\Delta 32$ allele occurred about $27.5 \times 25 = 688$ years ago. Based on other calculations, Stephens and colleagues placed a 95% confidence interval around this estimate that ranges from 275 to 1875 years ago.

---

also seen epidemics of these diseases. In fact, researchers have found several other mutations in CCR5 (Carrington et al. 1999). Among them is a loss-of-function mutation with a frequency of 3 to 4% in Chinese and Japanese populations (Ansari-Lari et al. 1997). The origin and selective significance of this allele have not been established.

We have shown that an understanding of linkage disequilibrium yields powerful tools for reconstructing the history of alleles. Another reward we promised readers was that understanding linkage disequilibrium would help them understand the adaptive significance of sexual reproduction. The mystery of sex is the subject of the next section.

7.2 The Adaptive Significance of Sex

Sexual reproduction is complicated, costly, and dangerous. Searching for a mate takes time and energy, and may increase the searcher's risk of being killed by a predator. Once found, a potential mate may demand additional exertion or investment before agreeing to cooperate. Sex itself may expose the parties to sexually transmitted diseases. And after all that, the mating may prove to be infertile. Why not avoid all the trouble and risk, and simply reproduce asexually instead?

This question sounds odd to human ears, because we have no choice: We inherited from our ancestors the inability to reproduce by any other means than sex. But many organisms do have a choice, at least in a physiological sense: They are capable of both sexual and asexual reproduction, and regularly switch between the two. Most aphid species, for example, have spring and summer populations composed entirely of asexual females. These females feed on plant juices and, without the participation of males, produce live-born young genetically identical to their mothers (Figure 7.11). This mode of reproduction, in which offspring develop from unfertilized eggs, is called **parthenogenesis**. In the fall aphids change modes, producing males and sexual females. These mate, and the females lay overwintering eggs from which a new generation of parthenogenetic females hatch the following spring.

Many other organisms are capable of both sexual and asexual reproduction. Examples include *Volvox* and hydra (Figure 7.12), and the many species of plants that can reproduce both by sending out runners and by developing flowers that exchange pollen with other individuals.

Many species are capable of both sexual and asexual reproduction.

Figure 7.11 **Asexual reproduction in an aphid** The large aphid is giving birth to a daughter, produced by parthenogenesis, that is genetically identical to its mother. (Peter J. Bryant/University of California at Irvine/Biological Photo Service)

Which Reproductive Mode Is Better: Sexual or Asexual?

The existence of two different modes of reproduction in the same population raises the question of whether one mode will replace the other over time. John Maynard Smith (1978) approached this question by developing a null model. The null model explores, under the simplest possible assumptions, the evolutionary fate of a population in which some females reproduce sexually and others reproduce asexually. Maynard Smith made just two assumptions:

1. A female's reproductive mode does not affect the number of offspring she can make.

2. A female's reproductive mode does not affect the probability that her offspring will survive.

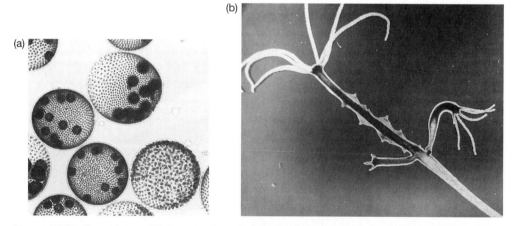

Figure 7.12 **Organisms with two modes of reproduction** (a) *Volvox aureus,* a freshwater alga. Each large sphere is a single adult individual. Before maturity, any individual has the potential to develop as a sexual male, a sexual female, or an asexual. The (entirely visible) individual at lower right is a male. The randomly oriented disks are packets of sperm. The large individual above and to his left is a female. Each of the dark fuzzy spheres inside her is an encysted zygote. The individual directly to the left of the male is an asexual. Each of the dark spheres inside it is an offspring, developing by mitosis into a clone of the parent. (Jon C. Herron). (b) A hydra. This individual is reproducing both sexually and asexually. The crown of tentacles at the upper left surround the hydra's mouth. Along the body below the mouth are rows of testes. Below the testes are two asexual buds. (Photo by P. S. Tice from Buchsbaum and Pearse, *Animals Without Backbones,* 3rd ed., University of Chicago Press, 1987.)

Maynard Smith also noted that all the offspring of a parthenogenetic female are themselves female, whereas the offspring of a sexual female are a mixture, typically with equal numbers of daughters and sons.

In a population conforming to Maynard Smith's assumptions, asexual females produce twice as many grandchildren as sexual females (Figure 7.13). This means that asexual females will constitute a larger and larger fraction of the population every generation. Ultimately, asexual females should completely take over. In principle, all that would be required is for a mutation to produce a single asexual female in an otherwise exclusively sexual population. From the moment the mutation occurred, the population would be destined to be overwhelmed by asexuals.

And yet such asexual takeovers do not seem to have happened very often. The vast majority of multicellular species are sexual, and there are many species, like aphids, *Volvox,* and hydra, in which sexual and asexual reproduction stably coexist. Maynard Smith's model demonstrates, as he intended it to, that these facts represent a paradox for evolutionary theory.

Obviously, sex must confer benefits that allow it to persist in spite of the strong reproductive advantage offered by parthenogenesis. But what are these benefits? The mathematical logic of Maynard Smith's model is correct, so the benefits of sex must lie in the violation of one or both of the assumptions. This is the model's greatest value. By making a short list of explicit assumptions, Maynard Smith focussed the inquiry on just a few essential facts of biology.

The first assumption, that the number of offspring a female can make does not depend on whether she is sexual or asexual, is violated in species in which fathers provide resources or other forms of parental care essential for producing young. With no male to provide help, asexual females are likely to produce fewer offspring. Species in which female reproductive success is limited by male parental care certainly exist. Examples include humans, many birds, and pipefish (see Chapter 9). However, species with male parental care are in the minority. In most species—most mammals and most insects, for example—males contribute only genes. A general advantage to sex is thus more likely to be found in the violation of the second

The persistence of sex is a paradox, because a simple model shows that asexual females should rapidly take over any population.

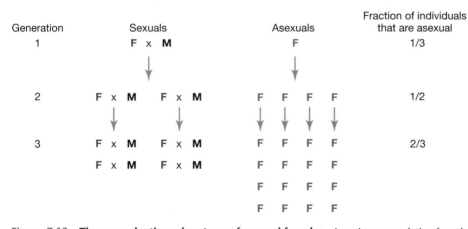

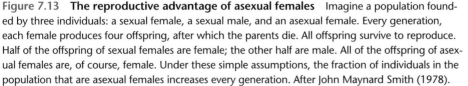

Figure 7.13   The reproductive advantage of asexual females   Imagine a population founded by three individuals: a sexual female, a sexual male, and an asexual female. Every generation, each female produces four offspring, after which the parents die. All offspring survive to reproduce. Half of the offspring of sexual females are female; the other half are male. All of the offspring of asexual females are, of course, female. Under these simple assumptions, the fraction of individuals in the population that are asexual females increases every generation. After John Maynard Smith (1978).

assumption, that the probability that a female's offspring will survive does not depend on whether she produces them asexually or sexually.

R.L. Dunbrack and colleagues (1995) tested this second assumption experimentally. They showed that it is wrong, at least under the conditions of their experiments. Dunbrack and colleagues studied laboratory populations in the flour beetle *Tribolium castaneum*. In each of a series of trials, the researchers established a mixed population founded with equal numbers of red beetles and black beetles. The beetles of one color were designated the "sexual" strain, and the beetles of the other color were designated the "asexual" strain. For example, in half of the trials the red beetles were the sexual strain, and the black beetles the asexual strain. We will stay with these designations for the rest of our description of the protocol.

Flour beetles are not actually capable of asexual reproduction, so the researchers had to manage the black population so that it was numerically and evolutionarily equivalent to a population in which individuals really do reproduce asexually. Every generation, the researchers counted the adults of the black strain and threw them out. They then replaced each one of the discarded black adults with three new black adults from a reservoir population of pure-bred black beetles unexposed to competition with reds. This procedure effectively gave the black strain a threefold reproductive advantage over the red strain, but prevented them from adapting to the new environment. Thus, the black strain was analogous to an asexual subpopulation in which every generation is genetically identical to the generation before, but in which individuals enjoy an even greater reproductive edge than the twofold advantage that an actual asexual strain would have in nature. Because every generation of the black population was (except for drift) genetically identical to the generation before, the black strain could not evolve in response to selection imposed by competition with red.

The red adults were allowed to breed among themselves and remain in the experimental culture. Thus they constituted a sexual population that could evolve in response to competitive interactions with the black (asexual) strain.

Dunbrack and colleagues added an environmental challenge for their beetle populations by spiking the flour in which the beetles lived with the insecticide malathion. This imposed selection on the flour beetle populations that favored the evolution of insecticide resistance. Finally, the researchers used a clever procedure, the details of which need not concern us here, to prevent the red and black beetles from mating with each other. The researchers maintained the experiment for 30 generations, which took two years.

Maynard Smith's null model predicts that, in each trial, the asexual strain should occupy an ever-increasing fraction of the population, until eventually the sexual strain is eliminated altogether. The model's first assumption is built into the experiment; in fact, asexual individuals in the experiment produced *more* offspring each generation than sexual individuals. The only way asexuals could fail to take over is if the model's second assumption is incorrect.

Dunbrack and colleagues performed eight replicates of their experiment. Four were as we described, with red as the sexual (evolving) strain and black as the asexual (nonevolving) strain. Each of these replicates used a different concentration of malathion. The other four replicates used black as the sexual (evolving) strain and red as the asexual (nonevolving) strain. Again, each replicate used a different concentration of malathion. The researchers also performed a control for each of the eight replicates. In the controls, neither the red nor the black beetles were allowed to evolve, but one color or the other had a threefold reproductive advantage.

The results appear in Figure 7.14. In the control cultures [Figure 7.14b and (d)], the outcome was always consistent with Maynard Smith's null model: The strain with the threefold reproductive advantage quickly eliminated the other strain. In the experimental cultures, however [Figure 7.14a and (c)], the outcome was always contrary to the prediction of the null model. Initially, the asexual strain appeared to be on its way to taking over, but within about 20 generations, depending on the concentration of malathion, the evolving sexual strain recovered. Eventually, the evolving sexual strain completely eliminated the nonevolving asexual strain, in spite of the asexual strain's threefold reproductive advantage.

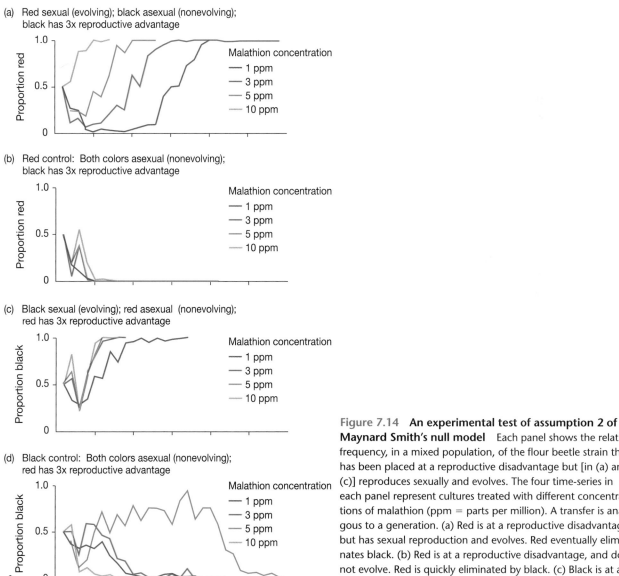

Figure 7.14 An experimental test of assumption 2 of Maynard Smith's null model Each panel shows the relative frequency, in a mixed population, of the flour beetle strain that has been placed at a reproductive disadvantage but [in (a) and (c)] reproduces sexually and evolves. The four time-series in each panel represent cultures treated with different concentrations of malathion (ppm = parts per million). A transfer is analogous to a generation. (a) Red is at a reproductive disadvantage, but has sexual reproduction and evolves. Red eventually eliminates black. (b) Red is at a reproductive disadvantage, and does not evolve. Red is quickly eliminated by black. (c) Black is at a reproductive disadvantage, but has sexual reproduction and evolves. Black quickly eliminates red. (d) Black is at a reproductive disadvantage, and does not evolve. Black is eventually eliminated by red. Redrawn from Dunbrack et al. (1995). Copyright © 1995, The Royal Society.

At least under some conditions, descendants produced by sexual reproduction achieve higher fitness than descendants produced by asexual reproduction.

We can conclude that assumption 2 of the null model is incorrect. Over time spans of just a few generations, descendants produced by sexual reproduction achieve higher fitness than descendants produced by asexual reproduction.

The next question is, Why? The only inherent difference between offspring that a female makes sexually versus asexually is that asexual offspring are genetically identical to their mother and to each other, whereas sexual offspring are genetically different from their mother and from each other. Most theories about the benefits of sex are concerned with reasons why females that produce genetically diverse offspring will see more of them survive and reproduce than will females that produce genetic copies of themselves.

We should note at this point that there is a tremendous diversity of theories about the advantages of sex. We have space here to focus only on population genetic models and tests, and a small number of them at that. For a more comprehensive overview of the field, see Michod and Levin (1988).

Sex in Populations Means Genetic Recombination

When population geneticists talk about sex, what they usually mean, and what we mean here, is reproduction involving (1) meiosis with crossing-over and (2) matings between unrelated individuals, such as occur during random mating. The consequence of these processes acting in concert is genetic recombination. If we follow a particular allele through several generations of a pedigree, every generation the allele will be part of a different multilocus genotype. For example, a particular allele for blue eyes may be part of a genotype that includes genes for blond hair in one generation, and part of a genotype that includes genes for brown hair in the next generation.

In a population-genetic analysis, sex has exactly one effect: It reduces linkage disequilibrium.

In a population-genetic analysis, genetic recombination, by shuffling multilocus genotypes, reduces linkage disequilibrium. This was a central conclusion of Section 7.1.

In fact, the reduction of linkage disequilibrium is the *only* consequence of sex at the level of population genetics (Felsenstein 1988). In a population that is already in linkage equilibrium, sex has no effect. If sex has no effect, it can confer no benefits. Therefore, any population-genetic model for the evolutionary benefits of sex must, at minimum, include two things. First, the model must include a mechanism that eliminates particular multilocus genotypes or produces an excess of others, thereby creating linkage disequilibrium. Second, the model must include a reason why genes that tend to reduce linkage disequilibrium—by promoting sex—are favored.

Based on this analysis, Joe Felsenstein neatly divides nearly all population-genetic models for the benefit of sex into two general theories. These general theories are distinguished by the evolutionary force they posit for the creation of linkage disequilibrium. Some models posit genetic drift as the factor that creates linkage disequilibrium; other models posit selection on multilocus genotypes.

Genetic Drift, in Combination with Mutation, Can Make Sex Beneficial

According to the drift theory of sex, mutation and drift create problems that sex can solve. Imagine, for example, that an asexual female sustains a deleterious genetic mutation in her germ cells. She will pass the mutation to all of her offspring, which in turn will pass it to all of their offspring. The female's lineage is forever hobbled by the deleterious mutation. The only hope of escape is if one of her descendants is lucky enough to experience either a back-mutation or an additional mutation that compensates for the first. If the female were sexual, however, she could produce mutation-free offspring immediately, simply by mating with a mutation-free male.

The role of drift in this scenario becomes clear when we scale the model up to the level of populations. The most famous drift model is called Muller's ratchet; it argues that asexual populations are doomed to accumulate deleterious mutations. H. J. Muller (1964) imagined a finite asexual population in which individuals occasionally sustain deleterious mutations. Because the mutations envisioned by Muller are deleterious, they will be selected against. The frequency of each mutant allele in the population will reflect the mutation rate, the strength of selection, and genetic drift (see Chapters 5 and 6).

At any given time, Muller's population may include individuals that carry no mutations, individuals that carry one mutation, individuals that carry two mutations, and so on. Because the population is asexual, we can think of these groups as distinct subpopulations, and plot the relative number of individuals in each subpopulation in a histogram (Figure 7.15). The number of individuals in each group may be quite small, depending on the size of the entire population and on the balance between mutation and selection (see Chapter 5). The group with zero mutations is the one whose members, on average, enjoy the highest fitness; but if this group is small, then in any given generation chance events may conspire to prevent the reproduction of all individuals in the group. If this happens just once, then the zero-mutation subpopulation is lost, and the members of the one-mutation group are now the highest-fitness individuals. The only way the zero-mutation group will reappear is if a member of the one-mutation group sustains the back-mutation that converts into a zero-mutation individual.

With the demise of the zero-mutation group, the members of the one-mutation subpopulation enjoy the highest mean fitness. But this group may also be quite small and may be lost by chance in any given generation. Again, the loss of the group by drift is much easier than its recreation by back-mutation. As the ratchet clicks away, and highest-fitness group after highest-fitness group is lost from the population, the average fitness of the population declines over time. The burden imposed by the accumulating mutations is known as the **genetic load**. Eventually, the genetic load carried by the asexual population becomes so high that the population becomes extinct.

Sex breaks the ratchet. If the no-mutation group is lost by chance in any given generation, it can be quickly reconstituted by outcrossing and recombination. If two individuals mate, each carrying a single copy of a deleterious mutation, one quarter of their offspring will be mutation-free. In Muller's view, the genes responsible for sex are maintained in populations because they help to create zero-mutation genotypes. As these zero-mutation genotypes increase in frequency, the genes for sex increase in frequency with them—in effect going along for the ride.

In Muller's scenario, linkage disequilibrium is created by drift. Particular multilocus genotypes are at lower-than-linkage-equilibrium frequencies because chance events have eliminated them. These missing multilocus genotypes are the zero-mutation genotype, then the one-mutation genotype, and so on. Sex reduces linkage disequilibrium by recreating the missing genotypes.

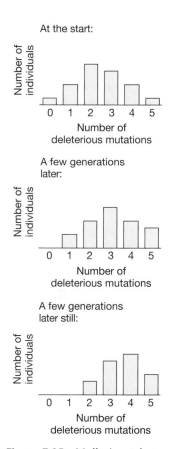

Figure 7.15 Muller's ratchet: Asexual populations accumulate deleterious mutations Each histogram shows a snapshot of a finite asexual population. In any given generation, the class with the fewest deleterious mutations may be lost by drift. Because forward-mutation to deleterious alleles is more likely than back-mutation to wild-type alleles, the distribution slides inexorably to the right. After Maynard Smith (1988).

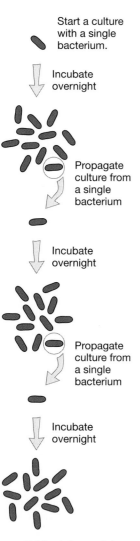

Start a culture with a single bacterium.

Incubate overnight

Propagate culture from a single bacterium

Incubate overnight

Propagate culture from a single bacterium

Incubate overnight

Figure 7.16 A bacterial population subjected to periodic bottlenecks The bacterial population is propagated from a single individual (the bottleneck), allowed to grow and divide to produce a large colony, then propagated again from a single individual. The bottlenecks provide an opportunity for genetic drift to operate.

Haigh (1978, reviewed in Maynard Smith 1988) developed and explored an explicit mathematical model of Muller's ratchet. Not surprisingly, the most critical parameter in the model is population size. In populations of 10 or fewer individuals, drift is a potent mechanism of evolution and the ratchet turns rapidly. In populations of more than 1000, drift is a weak mechanism of evolution and the ratchet does not turn at all. Also important are the mutation rate and the impact of deleterious mutations. The ratchet operates fastest with mildly deleterious mutations. This is because severely deleterious mutations are eliminated by selection before drift can carry them to fixation.

Dan Andersson and Diarmid Hughes (1996) tested Muller's ratchet experimentally with populations of the bacterium *Salmonella typhimurium*. From a standard wild-type strain, Andersson and Hughes set up 444 experimental cultures, each established from a single individual (Figure 7.16). After letting them grow overnight, the researchers propagated the cultures, again from a single individual. After another night of growth, they propagated the cultures again, and so on. The bacteria reproduced asexually by binary fission. The periodic bottlenecks, during which the population size of each culture was reduced to one, exposed the cultures to genetic drift. The researchers maintained the experiment for two months, giving drift a total of about 1700 generations to operate. Based on Muller's ratchet, Andersson and Hughes predicted that the bacterial cultures would accumulate deleterious mutations.

Andersson and Hughes checked this prediction by comparing the fitness of each experimental culture to the wild-type strain that was their common ancestor. The researchers assessed fitness by measuring population growth rate. Among their 444 cultures, Andersson and Hughes found 5, or 1%, with significantly reduced fitness. The generation time of these five cultures ranged from 25.0 to 47.5 minutes, compared to 23.2 minutes for the wild-type ancestor. None of the 444 cultures had higher fitness than the wild-type. These results are consistent with Muller's ratchet.

J. David Lambert and Nancy Moran (1998) took advantage of a natural experiment to assess whether Muller's ratchet operates in nature. Lambert and Moran studied nine species of bacteria that live inside the cells of insects. These bacteria are obligate endosymbionts, which means that they live *only* inside the cells of insects. The bacteria are transmitted from mother to offspring by travelling in the cytoplasm of the egg, just as mitochondria do. Note that the endosymbiotic bacteria in Lambert and Moran's study are propagated in a manner closely analogous to the protocol Andersson and Hughes used in their lab study. The only difference is that, in Lambert and Moran's study, the bacteria had been propagated that way for many millions of years.

The nine species of bacteria that Lambert and Moran studied represent at least four separate inventions of the endosymbiotic lifestyle, and each group of independently derived endosymbionts has close relatives that are free-living. Lambert and Moran checked whether, compared to their free-living relatives, the endosymbionts have accumulated deleterious mutations. The researchers focused on small-subunit ribosomal RNA (rRNA) genes. From the sequence of each species' rRNA gene, Lambert and Moran calculated the thermal stability of the encoded rRNA. Stability is a good thing in rRNAs, and deleterious mutations in an rRNA gene would reduce stability. In every case, Lambert and Moran found that the endosymbiotic bacteria have rRNAs that are 15 to 25% less stable than their free-living relatives. Again, this result is consistent with Muller's ratchet.

Thus Muller's ratchet works, both in principle and in practice. By counteracting the ratchet, sex could confer benefits. There is a problem, however, with models that explain sex by positing drift as the source of linkage disequilibrium. The benefits conferred by sex in these models accumulate only over the long term. If an asexual female appeared in a sexual population, it would take many generations for Muller's ratchet to catch up with her descendants, and lower their fitness enough to drive them to extinction. In the meantime, the asexual female's descendants would enjoy the twofold reproductive benefit identified by Maynard Smith. Yet the rarity of asexual species suggests, and the experiment by Dunbrack and colleagues demonstrates, that the advantage of sex accrues over just a few generations. This reasoning has prompted a search for short-term benefits of sex.

Sex may be advantageous because it recreates favorable multilocus genotypes that have been lost due to drift. The genes for sex then ride to high frequency in the high-fitness genotypes they help to create.

Selection Imposed by a Changing Environment Can Make Sex Beneficial

To see the logic of the changing-environment theories for sex, first imagine an asexual female and a sexual female living in a constant environment. If these females themselves survived to reproduce, and their offspring live in the same environment, then the offspring of the asexual female will probably survive to reproduce also. After all, they will receive exactly their mother's already-proven genotype. The genetically diverse offspring of the sexual female, however, may or may not survive to reproduce, depending on the nature of the genetic differences between their mother and themselves. By this reasoning, in a constant environment asexual reproduction is a safer bet.

In a variable environment, however, all bets are off. If the environment is changing in such a way that the asexual female herself might not survive in the new conditions, then her offspring will have poor prospects too. If the environment changes for a sexual female, however, there is always a chance that some of her diverse offspring will have genotypes that allow them to thrive in the new conditions. Some changing-environment theories focus on changes in the physical environment, whereas others focus on changes in the biological environment. Note that all changing-environment theories assume trade-offs, such that genotypes that do relatively well in some environments necessarily do relatively poorly in others.

A. H. Sturtevant and K. Mather (1938, reviewed in Felsenstein 1988) were the first to consider an explicit population-genetic model of varying selection. They imagined a population in which selection favors some multilocus genotypes in some generations (say, genotypes *AABB* and *aabb* in a two-locus model) and other genotypes (*AAbb* and *aaBB*) a few generations later. The population would alternate between a selection regime that generates linkage disequilibrium with positive values of D and a selection regime that generates linkage disequilibrium with negative values of D (see Box 7.1). Under these conditions, sex could be favored for its ability to recreate genotypes that were recently eliminated by selection but are now favored. As in Muller's ratchet, the genes for sex get a ride to high frequency in the high-fitness multilocus genotypes they help to create.

The variable pattern of selection required by the changing-environment theories can be imposed either by physical factors in the environment or by biological interactions. Currently, the most popular changing-environment theory of sex, sometimes called the "Red Queen hypothesis," involves evolutionary arms races between parasites and their hosts (for reviews see Seger and Hamilton 1988;

Sex may be advantageous because it recreates now-favorable multilocus genotypes that were recently eliminated by selection. The genes for sex then ride to high frequency in the high-fitness genotypes they help to create.

Lively 1996). Parasites and their hosts are locked in a perpetual struggle, with the host evolving to defend itself and the parasite evolving to evade the host's defenses. It is easy to imagine that a population of parasites would select in favor of some multilocus host genotypes in some generations, and in favor of other multilocus host genotypes in other generations. Figure 7.17 presents a scenario for such an ongoing evolutionary interaction.

Curtis Lively (1992) investigated whether parasites, in fact, select in favor of sex in their hosts. Lively studied the freshwater snail *Potamopyrgus antipodarum*. This snail, which lives in lakes and streams throughout New Zealand, is the host of over a dozen species of parasitic trematode worms. The trematodes typically castrate their host by eating its gonads. In an evolutionary sense, castration is equivalent to death: It prevents reproduction. Trematodes thus exert on snail populations strong selection for resistance to infection. Most populations of the snail contain two kinds of females: obligately sexual females that produce a mixture of male and female offspring, and obligately parthenogenetic females whose daughters are clones of their mother. (Note that both kinds of female must have an ovary to reproduce; the difference is that the eggs of the parthenogenetic females do not have to be fertilized.) The proportion of sexuals versus asexuals varies from population to population. So does the frequency of trematode infection. If an evolutionary arms race between the snails and the trematodes selects in favor of sex in the snails (see Figure 7.17), then sexual snails should be more common in the populations with higher trematode infection rates.

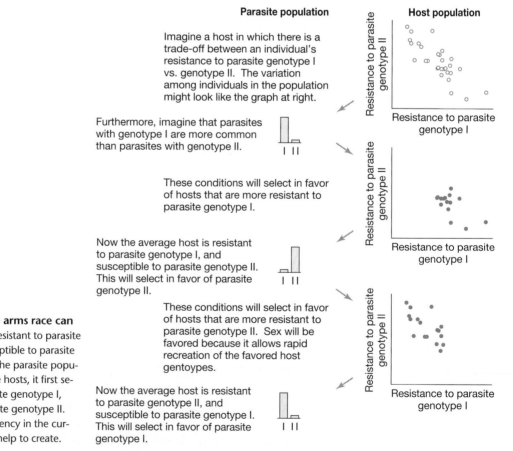

Figure 7.17   A host parasite arms race can make sex beneficial   Hosts resistant to parasite genotype I are necessarily susceptible to parasite genotype II, and vice versa. As the parasite population evolves in response to the hosts, it first selects for hosts resistant to parasite genotype I, then for hosts resistant to parasite genotype II. Genes for sex ride to high frequency in the currently-more-fit genotypes they help to create.

Lively took samples of snails from 66 lakes, and determined the sex of each snail and whether it was infected with parasites. Lively used the frequency of males in each population as an index of the frequency of sexual females, on the logic that males are produced only by sexual females. Lively found that a higher proportion of the females are indeed sexual in more heavily parasitized populations (Figure 7.18). This result is consistent with the varying-selection theory of sex.

Lively noted that because his study was observational, alternative explanations for the association he found should be considered. For example, if

1. Trematode infection rates are higher in more dense populations of snails, because high host density facilitates parasite transmission, and
2. The frequency of parthenogenetic females is higher in less dense populations of snails, because the real benefit of parthenogenesis is that it allows females to reproduce even when mates are hard to find,

then these two effects in combination would produce a positive association between the frequency of sexuals and the frequency of trematode infection. Lively rejected this alternative explanation by showing that although there is a positive correlation between infection rate and snail density (effect 1 is true), there is also a positive correlation between the frequency of parthenogenetic females and snail density (effect 2 is false). After considering this and other alternative explanations, Lively concluded that the simplest explanation for the pattern he found is that the trematodes indeed select in favor of sexual reproduction by the snails.

In summary, in the context of population genetics the effect of sex is to reduce linkage disequilibrium. A population-genetic model for the adaptive value of sex must therefore have two components: a mechanism for the creation of linkage disequilibrium, and a reason why selection favors traits that tend to reduce linkage disequilibrium. There are two classes of models for sex. In the first class, genetic drift creates linkage disequilibrium. Sex is then favored because it helps to recreate high-fitness genotypes lost by drift. In the second class, natural selection creates linkage disequilibrium. Then the pattern of selection changes, and sex is favored because it helps to recreate the now-favorable genotypes recently selected against. The various scenarios favoring sex are mutually compatible with each other. It is likely that varying selection imposed by changes in both the biological and physical environments

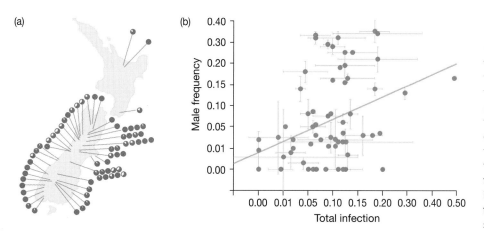

(a)

(b)

Figure 7.18 The frequency of sexual individuals in populations of a host snail is positively correlated with the frequency of its trematode parasites (a) The map shows the locations of the 66 lakes Lively sampled. In each population's pie diagram, the white slice represents the frequency of males. (b) The scatterplot shows the frequency of males in each population versus the proportion of snails infected with trematodes. The graph includes a best-fit line. Males are more frequent in populations in which more snails are infected. From Lively (1992).

combines with Muller's ratchet to create an advantage for sex that is greater than the advantage any one factor can produce alone (Howard and Lively 1998).

7.3 Selection on Quantitative Traits

The conceptual tools we developed in Chapters 5 and 6, and the tools we have developed so far in Chapter 7, allow us to analyze the evolution of traits controlled by one or two loci. Most traits in most organisms, however, are determined by the combined effects of many different loci, and are also influenced by the environment. Examples include human height, lizard sprint speed, and plant flower size. Such traits show continuous variation among individuals, and are called **quantitative traits**. [This is in contrast to qualitative traits like cystic fibrosis (Chapters 4 and 5), in which a person either has the affected phenotype or does not.] We need tools that allow us to analyze and understand the genetics and evolution of quantitative traits, even when we do not know the identities of the many specific genes involved. These tools are the province of quantitative genetics, and are the subject of this last section of the chapter.

Quantitative genetics allows us to analyze evolution by natural selection in traits controlled by many loci.

Recall the basic tenets of Darwin's theory of evolution by natural selection: If there is heritable variation among the individuals in a population, and if there are differences in survival and/or reproductive success among the variants, then the population will evolve. Quantitative genetics includes tools for measuring heritable variation, tools for measuring differences in survival and/or reproductive success, and tools for predicting the evolutionary response to selection.

Measuring Heritable Variation

Imagine a population of organisms in which there is continuous variation among individuals in some trait. For example, imagine a population of humans in which there is continuous variation among individuals in height. Continuously variable traits are typically normally distributed, so that a histogram of a trait has the familiar bell-curve shape. For example, in a typical human population a few people are very short, many people are more or less average in height, and a few people are very tall (Figure 7.19). We want to know: Is height heritable?

It is worth thinking carefully about exactly what this question means. Questions about heritability are often expressed in terms of nature versus nurture. But such questions are meaningful only if they concern comparisons among individuals. It makes no sense to focus only on the student on the far left of Figure 7.19a and ask, without reference to the other individuals, whether this student is 4 feet 10 inches tall because of his genes (nature) or because of his environment (nurture). He had to have both genes and an environment to be alive and of any height at all. He did not get 3 feet of his height from his genes, and 1 foot 10 inches from his environment, so that $3' + 1'10'' = 4'10''$. Instead, he got all of his 4 feet 10 inches from the activity of his genes operating within his environment. Within this single student we cannot, even in principle, disentangle the influence of nature and nurture.

The only kind of question it makes sense to ask is a comparative one: Is the shortest student shorter than the tallest student because they have different genes, or because they grew up in different environments, or both? This is a question we can answer. In principle, for example, we could take an identical twin of the short student and raise him in the environment experienced by the tall student. If this

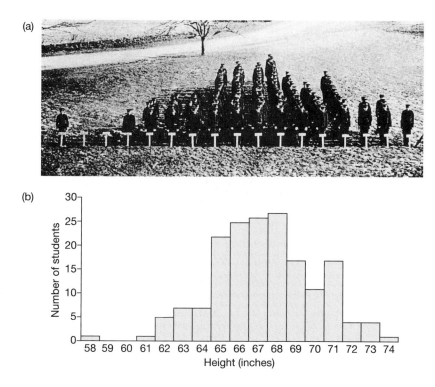

(a)

(b)

Figure 7.19 Normally distributed variation in a trait (a) A photograph, published in the Journal of Heredity in 1914 by Albert Blakeslee, of a group of students at Connecticut Agricultural College sorted by height. The arrangement of the students forms a living histogram. (Compare Blakeslee's photo to the one on page 197, taken at the same school in 1996.) Reproduced by permission of Oxford University Press (b) A graphical histogram representing the distribution of heights among the students shown in (a).

twin still grew up to be 4 feet 10 inches tall, then we would know that the difference between the shortest and tallest students is due entirely to differences in their genes. If the twin grew up to be 6 feet 2 inches tall, then we would know that the difference between the shortest and tallest students is due entirely to differences in their environments. In fact, the twin would probably grow up to be somewhere between 4'10" and 6'2". This would indicate that the difference between the two students is partly due to differences in their genes and partly due to differences in their environments. Considering the whole population, rather than just two individuals, we can ask: What fraction of the variation in height among the students is due to variation in their genes, and what fraction is due to variation in their environments?

The fraction of the total variation in a trait that is due to variation in genes is called the **heritability** of the trait. The total variation in a trait is referred to as the **phenotypic variation**, and is symbolized by V_P. Variation among individuals that is due to variation in their genes is called **genetic variation**, and is symbolized by V_G. Variation among individuals due to variation in their environments is called **environmental variation**, and is symbolized by V_E. Thus, we have

$$\text{heritability} = \frac{V_G}{V_P} = \frac{V_G}{V_G + V_E}$$

More precisely, this fraction is known as the broad-sense heritability, or degree of genetic determination. We will define the narrow-sense heritability shortly. Heritability is always a number between 0 and 1.

Before wading any deeper into symbolic abstractions, we note the simple truth that if the variation among individuals is due to variation in their genes, then offspring will resemble their parents. It is easy, in principle, to check whether they do. We first make a scatterplot with offspring's trait values represented on the y-axis, and their parents' trait values on the x-axis (Figure 7.20). We have two parents for

The first step in a quantitative genetic analysis is to determine the extent to which the trait in question is heritable. That is, we must partition the total phenotypic variation (V_P) into a component due to genetic variation (V_G) and a component due to environmental variation (V_E).

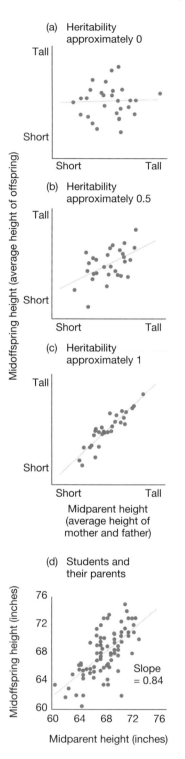

(a) Heritability approximately 0

(b) Heritability approximately 0.5

(c) Heritability approximately 1

Midparent height (average height of mother and father)

(d) Students and their parents

Slope = 0.84

Midoffspring height (average height of offspring)

Midoffspring height (inches)

Midparent height (inches)

The heritability, h², is a measure of the (additive) genetic variation in a trait.

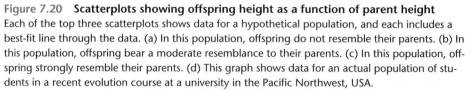

Figure 7.20   Scatterplots showing offspring height as a function of parent height Each of the top three scatterplots shows data for a hypothetical population, and each includes a best-fit line through the data. (a) In this population, offspring do not resemble their parents. (b) In this population, offspring bear a moderate resemblance to their parents. (c) In this population, offspring strongly resemble their parents. (d) This graph shows data for an actual population of students in a recent evolution course at a university in the Pacific Northwest, USA.

every offspring, so we use the midparent value, which is the average of the two parents. If we have more than one offspring for each family, we use a midoffspring value as well. We then draw the best-fit line through the data points. If offspring do not resemble their parents, then the slope of the best-fit line through the data will be near 0 (Figure 7.20a); this is evidence that the variation among individuals in the population is due to variation in their environments, not variation in their genes. If offspring strongly resemble their parents, the slope of the best-fit line will be near 1 (Figure 7.20c); this is evidence that variation among individuals in the population is due to variation in their genes, not variation in their environments. Most traits in most populations fall somewhere in the middle, with offspring showing a moderate resemblance to their parents (Figure 7.20b); this is evidence that the variation among individuals is partly due to variation in their environments and partly due to variation in their genes. Figure 7.20d shows data for an actual population of students. (For another example, look back at the scatterplot in Figure 3.4, which analyzes the heritability of beak depth in Darwin's finches.)

The examples in Figure 7.20 illustrate that the slope of the best-fit line for a plot of midoffspring versus midparents is a number between 0 and 1 that reflects the degree to which variation in a population is due to variation in genes. In other words, we can take the slope of the best-fit line as an estimate of the heritability. If we determine the best-fit line using the method of least-squares linear regression, then the slope represents a version of the heritability symbolized by h^2 and known as the narrow-sense heritability. Least-squares linear regression is the standard method taught in introductory statistics texts and used by statistical software packages to determine best-fit lines. (For readers familiar with statistics, it may prevent some confusion if we note here that h^2 is *not* the fraction of the variation among the offspring that is explained by variation in the parents. That quantity would be r^2. Instead, h^2 is an estimate of the fraction of the variation among the *parents* that is due to variation in their genes.)

To explain the difference between narrow-sense heritability and broad-sense heritability, we need to distinguish between two components of genetic variation: additive genetic variation versus dominance genetic variation. Additive genetic variation (V_A) is variation among individuals due to the additive effects of genes, whereas dominance genetic variation (V_D) is variation among individuals due to gene interactions such as dominance (see Box 7.5). The total genetic variation is the sum of the additive and dominance genetic variation: $V_G = V_A + V_D$. The broad-sense heritability, defined earlier, is V_G/V_P. The narrow-sense heritability, h^2, is defined as follows:

$$h^2 = \frac{V_A}{V_P} = \frac{V_A}{V_A + V_D + V_E}$$

When evolutionary biologists mention heritablity without noting whether they are using the term in the broad or narrow sense, they almost always mean the nar-

Box 7.5 Additive genetic variation versus dominance genetic variation

Here we use a numerical example to distinguish between additive genetic variation and dominance genetic variation. To simplify the discussion, we will analyze genetic variation at a single locus with two alleles as though were analyzing a quantitative trait. We will assume that there is no environmental variation in the trait in question: An individual's phenotype is determined solely and exactly by its genotype. The alleles at the locus are A_1 and A_2; each has a frequency of 0.5, and the population is in Hardy–Weinberg equilibrium. We will consider two situations: (1) the alleles are codominant; (2) allele A_2 is dominant over allele A_1.

Situation (1): Alleles A_1 and A_2 are codominant.

A_1A_1 individuals have a phenotype of 1. In A_1A_2 and A_2A_2 individuals, each copy of allele A_2 adds 0.5 to the phenotype. At the left in Figure 7.21a is a histogram showing the distribution of phenotypes in the population. At the center and right are scatterplots that allow us to analyze the genetic variation in the population. The x-axis represents the genotype, calculated as the number of copies of allele A_2. The y-axis represents the phenotype. The horizontal gray line shows the mean phenotype for the population ($= 1.5$). The plot at center shows that the total genetic variation V_G is a function of the deviations of the datapoints from the population mean (green arrows). We can quantify V_G by calculating the sum of the squared deviations. The plot

at right shows the best-fit line through the datapoints (red). The additive genetic variation V_A is defined as that fraction of the total genetic variation that is explained by the best-fit line (blue arrows). In this case, the best-fit line explains all of the genetic variation, so $V_G = V_A$. There is no dominance genetic variation.

Situation (2) Allele A_2 is dominant over allele A_1.

This time, A_1A_1 individuals again have a phenotype of 1. The effects of substituting copies of A_2 for copies of A_1 are not strictly additive, however: The first copy of A_2 (which makes the genotype A_1A_2) changes the phenotype from 1 to 2. The second copy of A_2 (which makes the genotype A_2A_2) does not alter the phenotype any further. At left in Figure 7.21b is a histogram showing the distribution of phenotypes in the population. At center and right are scatterplots that allow us to analyze the genetic variation in the population. The plot at center shows that the total genetic variation V_G is a function of the deviations of the datapoints (green arrows) from the population mean (gray line; $= 1.75$). The plot at right shows the best-fit line through the datapoints (red). The additive genetic variation V_A is that fraction of the total genetic variation that is explained by the best-fit line (blue arrows). The dominance genetic variation V_D is that fraction of the total genetic variation left unexplained by the best-fit line (yellow arrows). In this case, the best-fit line explains only part of the genetic variation, so $V_G = V_A + V_D$.

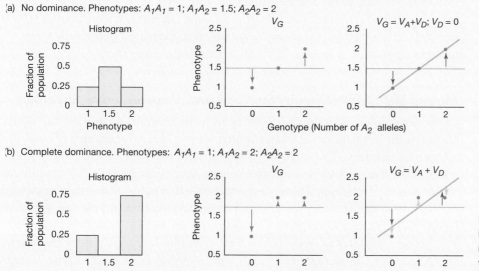

(a) No dominance. Phenotypes: $A_1A_1 = 1$; $A_1A_2 = 1.5$; $A_2A_2 = 2$

(b) Complete dominance. Phenotypes: $A_1A_1 = 1$; $A_1A_2 = 2$; $A_2A_2 = 2$

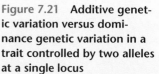

Figure 7.21 Additive genetic variation versus dominance genetic variation in a trait controlled by two alleles at a single locus

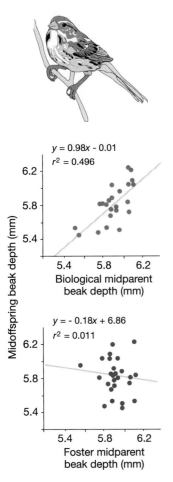

$y = 0.98x - 0.01$
$r^2 = 0.496$

Midoffspring beak depth (mm)
6.2
5.8
5.4

5.4   5.8   6.2
Biological midparent
beak depth (mm)

$y = -0.18x + 6.86$
$r^2 = 0.011$

6.2
5.8
5.4

5.4   5.8   6.2
Foster midparent
beak depth (mm)

Figure 7.22   A field experiment on the heritability of beak size in song sparrows The top scatterplot shows the relationship between midoffspring beak size (mean family beak depth) and the biological parents' beak size (biological midparent beak depth). The bottom scatterplot shows the relationship between midoffspring beak size (mean family beak depth) and the foster parents' beak size (foster midparent beak depth). Chicks resemble their biological parents strongly, and their foster parents not at all. From Smith and Dhondt (1980). Copyright © 1980, Evolution. Reprinted by permission of Evolution.

row-sense heritability. We will use the narrow-sense heritability in the rest of this discussion. It is the narrow-sense heritability, h^2, that allows us to predict how a population will respond to selection.

When estimating the heritability of a trait in a population, it is important to keep in mind that offspring can resemble their parents for reasons other than the genes the offspring inherit (see Box 3.1). Environments run in families too. Among humans, for example, some families exercise more than others, and different families eat different diets. Our estimate of heritability will be accurate only if we can make sure that there is no correlation between the environments experienced by parents and those experienced by their offspring. We obviously cannot do so in a study of humans. In an animal study, however, we could collect all the offspring at birth, then foster them at random among the parents. In a plant study, we could place seeds at random locations in a field.

James Smith and André Dhondt (1980), for example, studied song sparrows (*Melospiza melodia*) to determine the heritability of beak size. They collected young from natural nests, sometimes as eggs and sometimes as hatchlings, and moved them to the nests of randomly chosen foster parents. When the chicks grew up, Smith and Dhondt calculated midoffspring values for the chicks, and midparent values for both the biological and foster parents. Graphs of offspring beak depth versus parental beak depth appear in Figure 7.22. The chicks resembled their biological parents strongly, and their foster parents not at all. These results show that virtually all the variation in beak depth in this population is due to variation in genes. Smith and Dhondt estimated that the heritability of beak depth is 0.98.

There are a variety of other methods for estimating heritability besides calculating the slope of the best-fit line for offspring versus parents. For example, studies of twins can be used. The logic of twin studies works as follows. Monozygotic (identical) twins share their environment and all of their genes, whereas dizygotic (fraternal) twins share their environment and half of their genes. If heritability is high, and variation among individuals is due mostly to variation in genes, then monozygotic twins will be more similar to each other than are dizygotic twins. If heritability is low, and variation among individuals is due mostly to variation in environments, than monozygotic twins will be as different from each other as dizygotic twins. For a detailed introduction to methods of measuring heritability, see Falconer (1989). Data on the heritability of traits are frequently misinterpreted, particularly when the species under study is humans. We will return to this issue later.

Measuring Differences in Survival and Reproductive Success

In the preceding paragraphs we developed techniques for measuring the heritable variation in quantitative traits, the first tenet of Darwin's theory of evolution by natural selection. The next tenet of Darwin's theory is that there are differences in survival and/or reproductive success among individuals. We now discuss techniques for measuring differences in success—that is, for measuring the strength of selection. Once we can measure both heritable variation and the strength of selection, we will be able to predict evolutionary change in response to selection.

The kind of differences in success envisioned by Darwin's theory are systematic differences. On average, individuals with some values of a trait survive at higher rates, or produce more offspring, than individuals with other values of a trait. To measure the strength of selection, we first note who survives or reproduces and who fails to do so. Then we quantify the difference between the winners and the losers in the trait of interest.

In selective breeding experiments, the strength of selection is easy to calculate. Consider, for example, an experiment conducted by R. J. Di Masso and colleagues (1991). These researchers set out to breed mice with longer tails. They wanted to know how the developmental program that constructs a mouse's tail would change under selection. Would a mouse embryo make a longer tail by elongating the individual vertebrae, or by adding extra ones? Every generation, the researchers measured the tails of all the mice in their population. Then they picked the mice with the longest tails, and let them breed among themselves to produce the next generation.

To see how to quantify the strength of selection, suppose the researchers picked as breeders the one third of the mice whose tails are the longest. The simplest measure of the strength of selection is the difference between the mean tail length of the breeders and the mean tail length of the entire population (Figure 7.23a). This measure of selection is called the selection differential and is symbolized by *S*.

There is a second measure of the strength of selection that is useful because of its broad applicability. This measure is called the selection gradient (Lande and Arnold 1983). The selection gradient is calculated as follows:

1. Assign absolute fitnesses to the mice in the population. We will think of fitness as survival to reproductive age. In our population, $\frac{1}{3}$ of the mice survived long enough to reproduce. (This does not necessarily mean that the short-tailed mice were actually killed, just that they were removed from the breeding population; as far as our breeding population is concerned, the short-tails did not breed so they did not survive long enough to reproduce.) The short-tailed $\frac{2}{3}$ of the mice have a fitness of 0, and the long-tailed $\frac{1}{3}$ have a fitness of 1.

The second step in a quantitative genetic anlaysis is to measure the strength of selection on the trait in question. One measure is the selection differential, S, equal to the difference between the mean of the selected individuals and the mean of the entire population.

A second (and related) measure of the strength of selection is the selection gradient.

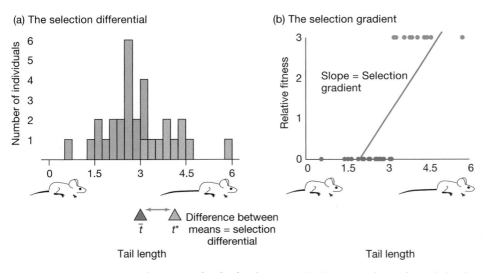

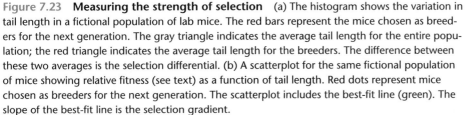

Figure 7.23 Measuring the strength of selection (a) The histogram shows the variation in tail length in a fictional population of lab mice. The red bars represent the mice chosen as breeders for the next generation. The gray triangle indicates the average tail length for the entire population; the red triangle indicates the average tail length for the breeders. The difference between these two averages is the selection differential. (b) A scatterplot for the same fictional population of mice showing relative fitness (see text) as a function of tail length. Red dots represent mice chosen as breeders for the next generation. The scatterplot includes the best-fit line (green). The slope of the best-fit line is the selection gradient.

2. Convert the absolute fitnesses to relative fitnesses. The mean fitness of the population is 0.33 (if, for example, there are 30 mice in the population, the mean is $[\{20 \times 0\} + \{10 \times 1\}]/30 = 0.33$). We calculate each mouse's relative fitness by dividing its absolute fitness (0 or 1) by the mean fitness (0.33). The short-tailed mice have a relative fitness of 0; the long-tailed mice have a relative fitness of 3.

3. Make a scatterplot of relative fitness as a function of tail length, and calculate the slope of the best-fit line (Figure 7.23b). The slope of the best-fit line is the selection gradient.

Box 7.6.　**The selection gradient and the selection differential**

The selection differential is an intuitively straightforward measure of the strength of selection: It is the difference between the mean of a trait among the survivors and the mean of the trait among the entire population. The selection gradient, while more abstract, has several advantages. Among them is that the selection gradient can be calculated for a wider variety of fitness measures. The fact that the selection gradient is closely related to the selection differential lends the former some of the intuitive appeal of the latter.

Here we show that in our example on tail length in mice (Figure 7.23), the selection gradient for tail length t is equal to the selection differential for tail length divided by the variance of tail length. Imagine that we have 30 mice in our population. First, note that the selection differential is

$$S = t^* - \bar{t}$$

where t^* is the mean tail length of the 10 mice we kept as breeders, and $\bar{t}$ is the mean tail length of the entire population of 30 mice.

The selection gradient is the slope of the best-fit line for relative fitness w as a function of tail length. The slope of the best-fit line in linear regression is given by the covariance of y and x divided by the variance of x:

$$\text{slope} = \frac{\text{cov}(y,x)}{\text{var}(x)}$$

The covariance of y and x is defined as

$$\text{cov}(y,x) = \frac{1}{n}\sum_{i=1}^{n}(y_i - \bar{y})(x_i - \bar{x})$$

and the variance of x is defined as

$$\text{var}(x) = \frac{1}{n}\sum_{i=1}^{n}(x_i - \bar{x})^2$$

where n is the number observations, $\bar{y}$ is the mean value of y, and $\bar{x}$ is the mean value of x. The selection gradient for t is therefore:

$$\text{selection gradient} = \frac{\text{cov}(w, t)}{\text{var}(t)}$$

Thus, what we need to show is that $\text{cov}(w, t) = t^* - \bar{t}$

Because (by definition) the mean relative fitness is 1, we can write

$$\text{cov}(w,t) = \frac{1}{30}\sum_{i=1}^{30}(w_i - 1)(t_i - \bar{t})$$

$$= \frac{1}{30}\sum_{i=1}^{30}(w_i t_i) - \frac{1}{30}\sum_{i=1}^{30}(w_i \bar{t}) - \frac{1}{30}\sum_{i=1}^{30}(t_i) + \frac{1}{30}\sum_{i=1}^{30}(\bar{t})$$

$$= \frac{1}{30}\sum_{i=1}^{30}(w_i t_i) - \bar{t} - \bar{t} + \bar{t}$$

$$= \frac{1}{30}\sum_{i=1}^{30}(w_i t_i) - \bar{t}$$

$$= t^* - \bar{t}$$

The last step may not be transparent. To see that

$$\frac{1}{30}\sum_{i=1}^{30}(w_i t_i) = t^*$$

note that for the first 20 mice $w_i = 0$, and for the last 10 mice $w_i = 3$. This means that

$$\frac{1}{30}\sum_{i=1}^{30}(w_i t_i) = \frac{1}{30}\sum_{i=21}^{30}(3t_i)$$

$$= \frac{3}{30}\sum_{i=21}^{30}(t_i)$$

$$= \frac{1}{10}\sum_{i=21}^{30}(t_i) = t^*$$

It may not appear at first glance that the selection gradient and the selection differential have much to do with each other. In fact, they are closely related, and each can be converted into the other. If we are analyzing selection on a single trait like tail length, then the selection gradient is equal to the selection differential divided by the variance in tail length (see Box 7.6). One advantage of the selection gradient is that we can calculate it for any measure of fitness, not just survival. We might, for example, measure fitness in a natural population of mice as the number of offspring weaned. If we first calculate each mouse's relative fitness (by dividing its number of offspring by the mean number of offspring), then plot relative fitness as a function of tail length and calculate the slope of the best-fit line, then that slope is the selection gradient. (For an introduction to other advantages of the selection gradient, see Box 7.7.)

In their mice, Di Masso et al. selected for long tails in 18 successive generations. The mice in the 18th generation had tails more than 10% longer than mice in a control population. The long-tailed mice had 28 vertebrae in their tails, compared to 26 or 27 vertebrae for the controls. The developmental program had been altered to make more vertebrae, not to elongate the individual vertebrae.

Box 7.7   Selection on multiple traits and correlated characters

In the main text we analyze selection on just one quantitative trait at a time. Selection in nature, however, often acts on several traits at once. Here we provide a brief introduction to how the techniques of quantitative genetics can be extended to analyze selection on multiple traits. For mathematical details, see Lande and Arnold (1983), Phillips and Arnold (1989), and Brodie, Moore, and Janzen (1995).

In Chapter 3 we discussed natural selection on beak size in Darwin's finches. During the drought of 1976–1977 on Daphne Major Island, medium ground finches with deeper beaks survived at higher rates. Beak depth is heritable, so the population evolved. Peter Grant and Rosemary Grant have reanalyzed the data from this selection episode, looking at selection on several traits simultaneously. We will discuss their analysis of two traits: beak depth and beak width. We will present a qualitative overview only; for the numbers see Grant and Grant (1995).

The medium ground finches of Daphne Major vary both in beak depth and beak width. These two traits are strongly correlated. Deep beaks tend to be wide, and shallow beaks tend to be narrow. The reasons why this might be true are beyond the scope of this discussion. We will just take it as given that it is difficult or impossible to build a finch with a beak that is deep and narrow, or a beak that is shallow and wide.

During the drought of 1976–1977, when food was scarce and many finches starved, selection acted on both beak depth and beak width. If we were looking at just one of these characteristics, we could measure the strength of selection as the slope of the best-fit line relating fitness to beak size. This is the selection gradient introduced in the main text. To look at both characteristics at once, we can measure the strength of selection as the slope of the best-fit *plane* relating fitness to both beak depth *and* beak width. This slope is the two-dimensional selection gradient.

Look at the three-dimensional graph in Figure 7.24a. Beak depth is represented on one horizontal axis; beak width is represented on the other horizontal axis. Fitness is represented on the vertical axis. The surface given by the blue grid is the best-fit plane. The fitness at each corner of the best-fit plane, or selection surface, is marked with a blue triangle. Selection favored birds with beaks that were both deep *and* narrow. The bird with the highest chance of surviving the drought would be a bird with a very deep, very narrow beak—located at the right rear corner of the selection surface.

Recall that beak depth and width are correlated, however. It is impossible to build the perfect bird. The correlation between depth and width is represented by the double-headed black arrow on the floor of the 3-D

Box 7.7 Continued

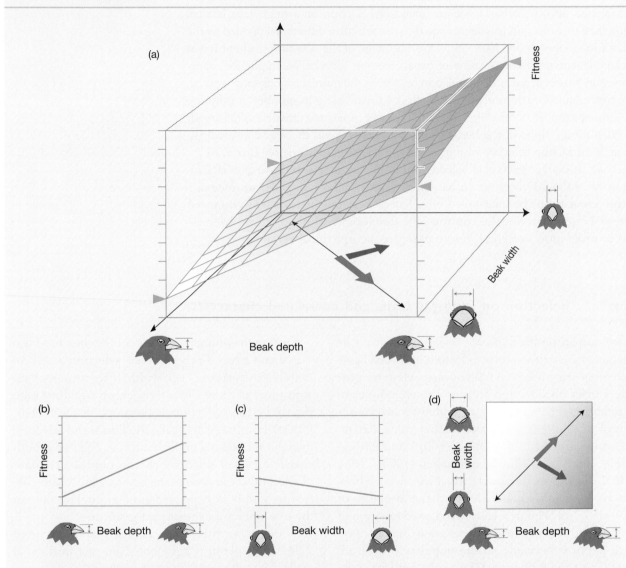

Figure 7.24 A multidimensional analysis of selection on beak size in medium ground finches (a) The gridded plane shows the relationship between fitness and both beak depth and beak width. Birds with deep and narrow beaks had highest fitness. (b), (c), and (d) show the same scenario in two-dimensional graphs.

graph. During the drought, selection pushed most strongly in the direction indicated by the wide dark blue arrow. This would have taken the population average from the center of the graph directly up the steepest path along the selection surface toward the best possible beak shape. But because of the correlation between beak depth and beak width, the population could not move in the direction selection pushed; it could only move along the double-headed black arrow. Selection favored deep beaks more strong-

ly than it favored narrow beaks. As a result, the population average moved toward a beak that was deeper and *wider* than it was before the drought. This change is represented by the wide green arrow.

Three-dimensional graphs can be difficult to interpret, so we have included Figures 7.24b, (c), and (d), which illustrate the same analysis with two-dimensional graphs. Figure 7.24b shows the selection gradient on beak depth, with beak width held constant. Selection favored birds with deeper beaks. Figure 7.24c shows the

Box 7.7 Continued

selection gradient on beak width, with beak depth held constant. Selection favored narrower beaks. Figure 7.24d shows the correlation between beak width and beak depth (double-headed black arrow), with fitness represented by the intensity of blue color across the graph. Selection pushed the population average toward a bird with a deep, narrow beak (wide dark blue arrow), but because of the correlation between depth and width, the population could not go there. Selection favored increased depth more strongly than decreased width, so the population average moved toward a deeper, wider beak (wide green arrow).

Grant and Grant's analysis of selection on finch beaks illustrates the advantages of looking at several traits at once, of using selection gradients to measure the strength of selection, and of recognizing that traits may be correlated with each other. Imagine that we were to look only at beak width, and calculate the selection differential. The average survivor had a wider beak than did the average bird alive before the drought. The selection differential, the difference between the population mean before and after selection, would suggest that selection favored wider beaks. But the multidimensional analysis reveals that this was not the case. Beak width was selected against, but was dragged along for the ride anyway as a result of stronger selection on beak depth.

Grant and Grant assumed for their analysis that the relationship between beak depth, beak width, and fitness was linear, as shown by the planar selection surface in Figure 7.24a. However, the relationship between a pair of traits and fitness is not always linear. Work by Edmund Brodie (1992) provides an example. Brodie monitored the survival of several hundred individually marked juvenile garter snakes. He estimated the effect on fitness of two traits that help the snakes evade predators: color pattern (striped versus unstriped or spotted) and escape behavior (straight-line escape versus many reversals of direction). Brodie's analysis produced the selection surface shown in Figure 7.25. The snakes with the highest rates of survival were those with stripes that fled in a straight line, and those without stripes that performed many reversals of direction. Other combinations of traits were selected against.

Given a selection surface, we can follow the evolution of a population by tracking the position of the average individual. In general, a population is expected to evolve so as to move up the steepest slope from its present location. As Grant and Grant's study of finch beaks demonstrated, however, correlations among traits may prevent a population from following this predicted route. Selection surfaces like those shown in Figures 7.24 and 7.25 are sometimes referred to as adaptive landscapes, but this term has a complex history and several different meanings (see Chapter 9 in Provine 1986; Chapter 11 of Wright 1986; Wright 1988).

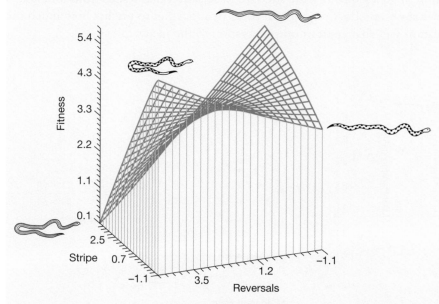

Figure 7.25 A multidimensional analysis of selection on antipredator defenses in garter snakes The gridded surface shows the relationship between fitness and both color pattern and evasive behavior. From Fig. 1 from Brodie (1992). Copyright © 1992, Evolution. Reprinted by permission of Evolution.

Predicting the Evolutionary Response to Selection

Once we know the heritability and the selection differential, we can predict the evolutionary response to selection. Here is the equation for doing so:

$$R = h^2 S$$

where R is the predicted response to selection, h^2 is the heritability, and S is the selection differential.

The logic of this equation is shown graphically in Figure 7.26. This figure shows a scatterplot of midoffspring values as a function of midparent values, just like the scatterplots in Figure 7.20. The scatterplot in Figure 7.26 represents tail lengths in a population of 30 families of mice. The plot includes a best–fit line, whose slope estimates the heritability h^2.

Look first at the x-axis. $\overline{P}$ is the average midparent value for the entire population. P^* is the average of the 10 largest midparent values. The difference between P^* and $\overline{P}$ is the selection differential (S) that we would have applied to this population had we picked as our breeders only the 10 pairs of parents with the largest midparent values.

Now look at the y-axis. $\overline{O}$ is the average midoffspring value for the entire population. O^* is the average midoffspring value for the 10 pairs of parents with the largest midparent values. The difference between O^* and $\overline{O}$ is the evolutionary response (R) we would have gotten as a result of selecting as breeders the 10 pairs of parents with the largest midparent values.

The slope of a line can be calculated as the rise over the run. If we compare the population averages with selection versus without selection, we have a rise of $(O^* - \overline{O})$ over a run of $(P^* - \overline{P})$ so

$$h^2 = \frac{(O^* - \overline{O})}{(P^* - \overline{P})} = \frac{R}{S}$$

In other words, $R = h^2 S$.

We now have a set of tools for studying the evolution of multilocus traits under natural selection. We can estimate how much of the variation in a trait is due to variation in genes, quantify the strength of selection that results from differences in survival or reproduction, and put these two together to predict how much the population will change from one generation to the next.

Once we know h^2 and S, we can use them to predict the response to selection, R.

Figure 7.26 The response to selection is equal to the heritability multiplied by the selection differential The midoffspring and midparent values are indicated both as dots on the scatterplot and as diamonds on the y- and x-axes. The red symbols represent the 10 families with the largest midparent values. $\overline{P}$ is the average midparent value for the entire population; P^* is the average midparent value of the families with the largest midparent values. $\overline{O}$ is the average midoffspring value for the entire population; O^* is the average midoffspring value for the families with the largest midparent values. After Falconer (1989).

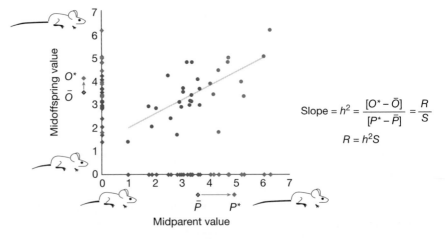

Slope $= h^2 = \dfrac{[O^* - \overline{O}]}{[P^* - \overline{P}]} = \dfrac{R}{S}$

$R = h^2 S$

(a)

(b)

Figure 7.27 **An alpine skypilot (*Polemonium viscosum*) (a) and a bumblebee (*Bombus* sp.) (b)** [(a) Richard Parker/Photo Researchers, Inc.; (b) Stephen Dalton/Photo Researchers, Inc.]

Alpine Skypilots and Bumblebees

As an example of the kinds of questions evolutionary biologists ask and answer with quantitative genetics, we review Candace Galen's (1996) research on flower size in the alpine skypilot (*Polemonium viscosum*). The alpine skypilot is a perennial Rocky Mountain wildflower (Figure 7.27a). Galen studied populations of skypilots on Pennsylvania Mountain in Colorado, including populations growing at the timberline and populations growing in the higher-elevation tundra. At the timberline, skypilots are pollinated by a diversity of insects, including flies, small solitary bees, and some bumblebees. In the tundra, skypilots are pollinated almost exclusively by bumblebees (Figure 7.27b). The flowers of tundra skypilots are, on average, 12% larger in diameter than those of timberline skypilots. Previously, Galen (1989) had documented that larger flowers attract more visits from bumblebees, and that skypilots that attract more bumblebees produce more seeds.

Galen wanted to know whether the selection on flower size imposed by bumblebees is responsible for the larger flowers of tundra skypilots. If it is, she also wanted to know how long it would take for selection by bumblebees to increase the average flower size in a skypilot population by 12%—the difference between tundra and timberline flowers.

Galen worked with a small-flowered timberline population. Her first step was to estimate the heritability of flower size. She measured the flower diameters of 144 skypilots and collected their seeds. She germinated the seeds in her laboratory, and planted the 617 resulting seedlings at random locations in the same habitat their parents had lived in. Seven years later, 58 of the seedlings had matured, so that Galen could measure their flowers. This allowed Galen to plot offspring flower diameter (corolla flare) as a function of maternal, or seed-parent, flower diameter (Figure 7.28). The slope of the best-fit line is approximately 0.5. For reasons beyond the scope of this discussion, the slope of the best-fit line for offspring versus a single parent (as opposed to the midparent) is an estimate of $\frac{1}{2} h^2$ (see Falconer 1989). Thus, the heritability of flower size in the timberline skypilot population is approximately $2 \times 0.5 = 1$. Note, however, that the graph in Figure 7.28 shows considerable scatter. Galen's statistical analysis indicated that she could safely conclude only that the heritability of flower size is between 0.2 and 1. In other words, at least 20% of the phenotypic variation in skypilot flower size is due to additive genetic variation.

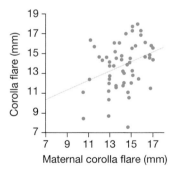

Figure 7.28 **Estimating the heritability of flower size (corolla flare) in alpine skypilots** This scatterplot shows offspring corolla flare as a function of maternal plant corolla flare for 58 skypilots. The slope of the best-fit line is 0.5. From Galen (1996).

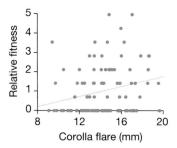

Figure 7.29 Estimating the selection gradient in alpine skypilots pollinated by bumblebees This scatterplot shows relative fitness (number of surviving 6-year-old offspring divided by average number of surviving 6-year-old offspring) as a function of maternal flower size (corolla flare). The slope of the best-fit line is 0.13. Prepared with data provided by Candace Galen.

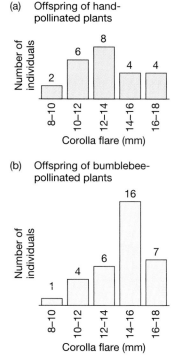

Figure 7.30 Measuring the evolutionary response to selection in alpine skypilots These histograms show the distribution of flower size (corolla flare) in the offspring of hand-pollinated skypilots (a; average = 13.1 mm) and bumblebee-pollinated skypilots (b; average = 14.4 mm). Redrawn from Galen (1996).

Galen's next step was to estimate the strength of the selection differential imposed by bumblebee pollinators. Recall that bumblebees prefer to visit larger flowers, and that more bumblebee pollinators means more seeds. Galen built a large screen-enclosed cage at her study site, transplanted 98 soon-to-flower skypilots into it, and added bumblebees. The cage kept the bumblebees in and all other pollinators out. When the caged skypilots flowered, Galen measured their flowers. Later she collected their seeds, germinated them in the lab, and planted the resulting seedlings at random locations back out in the parental habitat. Six years later, Galen counted the number of surviving offspring that had been produced by each of the original caged plants. Using the number of surviving 6-year-old offspring as her measure of fitness, Galen plotted relative fitness as a function of flower size, and calculated the slope of the best-fit line (Figure 7.29). The slope of this line, 0.13, is the selection gradient resulting from bumblebee pollination. Multiplying the selection gradient by the variance in flower size, 5.66, gives the selection differential: $S = 0.74$ millimeter. The average flower size was 14.2 mm. Thus the selection differential can also be expressed as $\frac{0.74}{14.2} = 0.05$, or 5%. Roughly speaking, this means that when skypilots attempt to reproduce by enticing bumblebees to visit them, the plants that win have flowers that are 5% larger than those of the average plant in the population.

Galen performed two control experiments to confirm that bumblebees select in favor of larger flowers. In one control, she pollinated skypilots by hand (without regard to flower size); in the other, she allowed skypilots to be pollinated by all other natural pollinators except bumblebees. In neither control was there any relationship between flower size and fitness; only bumblebees select for larger flowers.

Galen's data allowed her to predict how the population of timberline skypilots should respond to selection by bumblebees. The scenario she imagined was that a population of timberline skypilots that had been pollinated by a variety of insects moves (by seed dispersal) to the tundra, where the plants are now pollinated exclusively by bumblebees. Using the low-end estimate that $h^2 = 0.2$, and the estimate that $S = 0.05$, Galen predicted that the response to selection would be $R = h^2 S = 0.2 \times 0.05 = 0.01$. Using the high-end estimate that $h^2 = 1$, and the estimate that $S = 0.05$, Galen predicted that the response to selection would be $R = h^2 S = 1 \times 0.05 = 0.05$. In other words, a single generation of selection by bumblebees should produce an increase of 1 to 5% in the average flower size of a population of timberline skypilots moved to the tundra.

Galen's prediction was, therefore, that flower size would evolve rapidly under selection by bumblebees. Is this prediction correct? Recall the experiment described earlier, in which Galen reared offspring of timberline skypilots that had been pollinated by hand and offspring of timberline skypilots that been pollinated exclusively by bumblebees. Galen calculated the mean flower size of each group and found that the offspring of bumblebee-pollinated skypilots had flowers that were, on average, 9% larger than those of hand-pollinated skypilots (Figure 7.30). Her prediction was correct: Skypilots show a strong and rapid response to selection. In fact, the response is even larger than Galen predicted.

Galen concluded that the 12% difference in flower size between timberline and tundra skypilots can be plausibly explained by the fact that timberline skypilots are pollinated by a diversity of insects, whereas tundra skypilots are pollinated almost exclusively by bumblebees. Timberline skypilots can set seed even if bumblebees avoid them, but tundra skypilots cannot. Furthermore, it would take only a few generations of bumblebee-only pollination for a population of timberline skypilots to evolve flowers that that are as large as those of tundra skypilots.

The Bell-Curve Fallacy, and Other Misinterpretations of Heritability

We promised, in the introduction to this chapter, that our discussion of quantitative genetics would allow us to debunk erroneous claims about differences in IQ scores among ethnic groups. We are now ready to make good on that promise.

A key point is that the formula for heritability includes both genetic variation, V_G, and environmental variation, V_E. Any estimate of heritability is, therefore, specific to a particular population living in a particular environment. As a result, heritability tells us nothing about the causes of differences between populations that live in different environments.

We can illustrate this point with a thought experiment. Pea aphids reproduce asexually during spring and early summer. Imagine that we collect an aphid from each of a dozen different fields and bring them into the lab. Because the aphids came from different fields, they are genetically different from each other; our lab population contains genetic variation. But because each aphid in the population reproduces asexually (by mitosis), her offspring are genetically identical to their mother. In other words, the descendants of any one of our original dozen aphids constitute a clone. By taking one member from each of our dozen clones, we can create duplicates of our original lab population.

Imagine that we set up two such duplicate populations. We rear one population in a carefully controlled environment on pea plants, and the aphids grow to large sizes. Because all the aphids in this population shared the same environment, any differences in size among them are entirely due to genetic variation. When grown on peas, the heritability of size in our lab population is 1.

Now we rear the other duplicate aphid population on clover, and the aphids in this population do not get so big (they are pea aphids, after all). Nonetheless, all the aphids in this population shared the same environment, so any differences among them in size will be due genetic variation. When grown on clover, the heritability of size in our lab population is 1.

We have high heritability in both populations, and a difference in mean size between populations. Does this mean that the population reared on peas is genetically superior to the clover population with respect to size? Of course not; we intentionally set up the populations to be identical in genetic composition. *The fact that heritability is high in each population tells us nothing about the cause of differences between the populations, because the populations were reared in different environments.* (For an actual experiment similar to the one we have described, see Via 1991.)

The mistaken notion that heritability can tell us something about the causes of differences between populations has been particularly persistent in studies of human intelligence. In 1994, Charles Murray and Richard J. Herrnstein sold thousands of copies of their book, *The Bell Curve.* Murray and Herrnstein claimed that the difference in average IQ scores between African Americans and European Americans is due to genetic differences between these groups (our analysis of their argument is based on an extract of their book published in *The New Republic* [Murray and Herrnstein 1994]).

Studies of heritability are often misinterpreted as implying that differences between populations are due to differences in genes.

Murray and Herrnstein take note of the point we just made with aphids. They state, "Most scholars accept that I.Q. in the human species as a whole is substantially heritable, somewhere between 40 percent and 80 percent, meaning that much of the observed variation in I.Q. is genetic. And yet this information tells us nothing for sure about the origin of the differences between groups." Having said that, however, Murray and Herrnstein proceed to develop the following erroneous argument.

Based on various sources, Murray and Herrnstein assume that the average IQ of African Americans is 85, that the average IQ of European Americans is 100, and that the variance (a statistical measure of the variation) in each group is 225. Bell curves representing these assumptions appear in Figure 7.31a. Murray and Herrnstein further assume that the heritability of IQ in each group is 0.6. There are reasons to dispute each of Murray and Herrnstein's assumptions, but we will grant them here—for the sake of argument only. (There are also reasons to dispute whether IQ tests measure anything meaningful at all, but we leave that argument to others.)

Next, Murray and Herrnstein imagine what the bell curves for IQ would look like if all the *genetic* variation among individuals within each population were re-

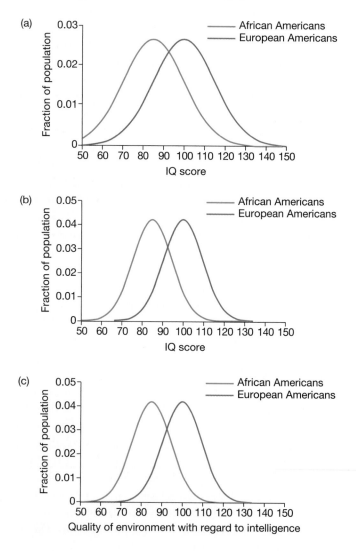

Figure 7.31 **Illustrations of Murray and Herrnstein's erroneous argument concerning IQ and ethnicity** See text for explanation.

moved. In other words, they imagine that all African Americans have been made genetically identical to the average African American, and all European Americans have been made genetically identical to the average European American. On the assumption that 60% of the variation within each group is due to genetic variation, this leaves 40% of the original variation within each group. Bell curves representing this thought experiment appear in Figure 7.31b.

Now Murray and Herrnstein consider the proposition that the difference between the average IQ of the African Americans and the average IQ of the European Americans in Figure 7.31b is solely due to differences in environment. Under this proposition, Murray and Herrnstein say, we could replace the label "IQ score" with the label "Quality of environment with regard to intelligence," as shown in Figure 7.31c. Murray and Herrnstein find it implausible that the difference in the quality of the environment experienced by African Americans versus European Americans is as great as that shown Figure 7.31c. They conclude that at least part of the difference between the mean IQ score of African Americans versus European Americans must be due to genetic differences between the groups.

There are at least two serious flaws in Murray and Herrnstein's argument. First, by replacing the label on the horizontal axis in Figure 7.31b with the label in Figure 7.31c, Murray and Herrnstein are implicitly assuming that there is a linear relationship between environment and IQ. This assumption is almost certainly wrong.

Second, Murray and Herrnstein's argument from incredulity amounts to rhetorical technique, not science. A scientific approach to Murray and Herrnstein's hypothesis would be to conduct a common garden experiment like we did with aphids: Rear European Americans and African Americans together in an environment typically experienced by European Americans, and then compare their IQ scores. This design, and the reciprocal experiment, in which everyone is reared in an environment typically experienced by African Americans, is shown in Figure 7.32.

We obviously cannot do this experiment with humans. But experiments like it have been done with plants and animals. For example, Clausen, Keck, and Hiesey (1948) conducted a series of common garden experiments with the plant *Achillea*

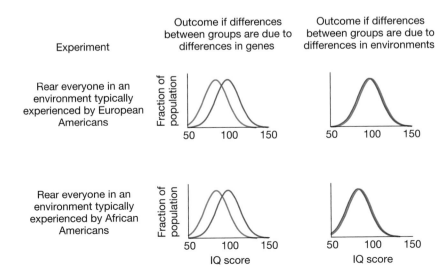

Figure 7.32 An experiment that would test Murray and Herrnstein's claim The left column describes two experimental treatments. The middle and right columns show the predicted outcomes under the hypothesis that differences between groups are due to differences in genes versus the predicted outcomes under the hypothesis that differences between groups are due to differences in environments.

The only way to determine the cause of differences between populations is to rear individuals from each of the populations in identical environments.

(Figure 7.33). Plants in this genus collected from low-altitude populations make more stems than plants collected from high-altitude populations (Figure 7.33a). Is the difference between the low- versus high-altitude plants due to differences in their genes or differences in their environments? When plants from low altitude and plants from high altitude are grown together at low altitude, the low-altitude plants make more stems (Figure 7.33b). This result is consistent with the hypothesis that plants from low altitude are genetically programmed to make more stems. When plants from low altitude and plants from high altitude are grown together at high altitude, however, the high-altitude plants make more stems (Figure 7.33c). This result was wholly unanticipated in the experimental design. It reveals genetic differences between low- and high-altitude plants in the way each responds to the environment. It also reveals that each population of plants is superior in its own environment of origin. This unanticipated outcome demonstrates that hypothetical claims about the causes of differences between populations are no substitute for experimental results.

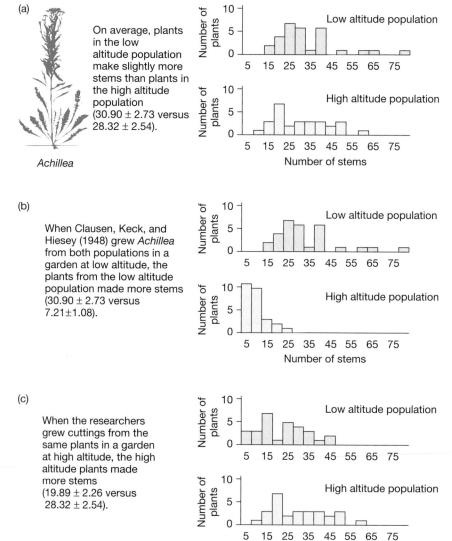

(a) *Achillea*

On average, plants in the low altitude population make slightly more stems than plants in the high altitude population (30.90 ± 2.73 versus 28.32 ± 2.54).

(b) When Clausen, Keck, and Hiesey (1948) grew *Achillea* from both populations in a garden at low altitude, the plants from the low altitude population made more stems (30.90 ± 2.73 versus 7.21±1.08).

(c) When the researchers grew cuttings from the same plants in a garden at high altitude, the high altitude plants made more stems (19.89 ± 2.26 versus 28.32 ± 2.54).

Figure 7.33 Data from experiments by Clausen, Keck, and Hiesey (1948) (a) A comparison between *Achillea* populations from low altitude (San Gregorio, California) versus high altitude (Mather, California). (b) Plants from low and high altitude grown in a common garden at low altitude (Stanford, California). (c) Plants from low and high altitude grown in a common garden at high altitude (Mather, California).

What would happen if we did this kind of experiment with African Americans and European Americans? No one has the slightest idea. It is misleading to say that high heritabilities for IQ within groups tell us "nothing for sure about the origin of the differences between groups" (Murray and Herrnstein 1994). In fact, high heritabilities within groups tell us *nothing at all* about the origin of differences between groups.

Finally, it is worth noting that heritability also tells us nothing about the role of genes in determining traits that are shared by all members of a population. There is no variation among humans in number of noses. The heritability of nose number is undefined, because $V_A/V_P = 0/0$. This obviously does not mean that our genes are not important in determining how many noses we have.

So what good *does* it do us to measure the heritability of a trait? Precisely and only this: It allows us to predict whether selection on the trait will cause a population to evolve.

Modes of Selection and the Maintenance of Genetic Variation

In our discussions of selection on quantitative traits, we have assumed that the relationship between phenotype and fitness is simple. In our mice, short tails were better than long tails; in skypilots, bigger flowers were better than smaller flowers. Before leaving the topic of selection on quantitative traits, we note that the relationship between phenotype and fitness may be complex. A variety of patterns, or modes of selection, are possible.

Figure 7.34 shows three distinct modes of selection acting on a hypothetical population. Each column represents a different mode. The histogram in the top row shows the distribution of values for a phenotypic trait before selection. The graph in the middle row shows the relationship between phenotype and fitness, plotted as the probability of surviving as a function of phenotype. The histogram in the bottom row shows the distribution of phenotypes among the survivors. The triangle and bar below each histogram show the mean and variation in the population. (The bar representing variation encompasses ± 2 standard deviations around the mean, or approximately 95% of the individuals in the population.)

In **directional selection**, fitness consistently increases (or decreases) with the value of a trait (Figure 7.34, first column). Directional selection on a continuous trait changes the average value of the trait in the population. In the hypothetical population shown in Figure 7.34, the mean phenotype before selection was 6.9, whereas the mean phenotype after selection was 7.4. Directional selection also reduces the variation in a population, although often not dramatically. In our hypothetical population, the standard deviation before selection was 1.92, whereas the standard deviation after selection was 1.89.

In **stabilizing selection**, individuals with intermediate values of a trait have highest fitness (Figure 7.34, middle column). Stabilizing selection on a continuous trait does not alter the average value of the trait in the population. Stabilizing selection does, however, trim off the tails of the trait's distribution, thereby reducing the variation. In our hypothetical population, the standard deviation before selection was 1.92, whereas the standard deviation after selection was 1.04.

In **disruptive selection**, individuals with extreme values of a trait have the highest fitness (Figure 7.34, last column). Disruptive selection on a continuous trait does not alter the average value of the trait in the population. Disruptive se-

Selection on a population may take any of a variety of forms. Directional selection and stabilizing selection tend to reduce the amount of variation in a population; disruptive selection tends to increase the amount of variation.

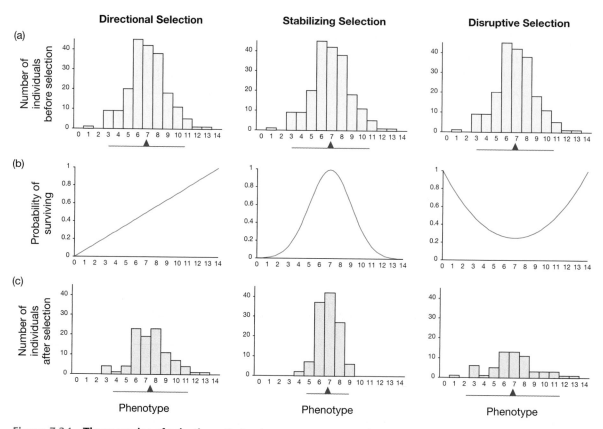

Figure 7.34 **Three modes of selection** Each column represents a mode of selection. The graphs in row (a) are histograms showing the distribution of a phenotypic trait in a hypothetical population before selection. The graphs in row (b) show different patterns of selection; they plot the probability of survival (a measure of fitness) as a function of phenotype. The graphs in row (c) are histograms showing the distribution of the phenotypic trait in the survivors. The blue triangle under each histogram shows the mean of the population. The blue bar under each histogram shows the variation (± 2 standard deviations from the mean). After Cavalli-Sforza and Bodmer (1971).

lection does, however, trim off the top of the trait's distribution, thereby increasing the variance. In our hypothetical population, the standard deviation before selection was 1.92, whereas the standard deviation after selection was 2.33.

All three modes of selection eliminate individuals with low fitness and preserve individuals with high fitness. As a result, all three modes of selection increase the mean fitness of the population.

We have already seen examples of directional selection. In alpine skypilots pollinated by bumblebees, for instance, larger flowers have higher fitness. And in medium ground finches, the drought of 1976–77 on Daphne Major selected for birds with larger beaks (see Chapter 3).

Research by Arthur Weis and Warren Abrahamson (1986) provides an elegant example of stabilizing selection. Weis and Abrahamson studied a fly called *Eurosta solidaginis*. The female in this species injects an egg into a bud of the tall goldenrod, *Solidago altissima*. After hatching, the fly larva digs into the stem and induces the plant to form a protective gall. As it develops inside its gall, the larva may fall victim to two kinds of predators. First, a female parasitoid wasp may inject *her* egg into the gall, where the wasp larva will eat the fly larva. Second, a bird may spot the gall and break it open, again to eat the larval fly. Weis and Abrahamson established that genetic variation among the flies is partly responsible for the variation

Figure 7.35 Stabilizing selection on a gall-making fly (a) Parasitoid wasps kill fly larvae inside small galls at higher rates than they kill larvae inside large galls. (b) Birds kill fly larvae inside large galls at higher rates than they kill larvae inside small galls. (c) The distribution of gall sizes before (light orange + blue portion of bars) and after (light orange portion of bars) selection by parasitoids and birds. Overall, fly larvae inside medium-sized galls survived at the highest rates. From Fig. 3 in Weis and Abrahamson (1986). Copyright © 1986, American Naturalist. Reprinted by permission of The University of Chicago Press.

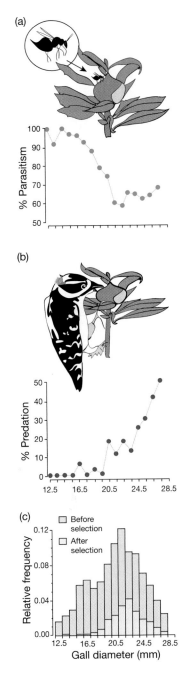

in the size of the galls they induce. The researchers also collected several hundred galls and determined, by dissecting them, the fate of the larva inside each one.

Weis and Abrahamson discovered that parasitoid wasps impose on the gall-making flies strong directional selection favoring larger galls (Figure 7.35a). Nearly all larvae in galls under about 16 mm in diameter were killed by wasps, whereas larvae in larger galls had at least a fighting chance to survive. However, the researchers also found that birds impose on the gall-makers strong directional selection favoring *smaller* galls (Figure 7.35b). Together, selection by wasps and selection by birds add up to stablizing selection on gall size. Figure 7.35c shows the distribution of sizes among the galls before and after selection.

Research by Thomas Bates Smith (1993) provides an example of disruptive selection. Bates Smith studied an African finch called the black-bellied seedcracker. Birds in this species exhibit two distinct beak sizes: large and small. The birds in the two groups specialize on different kinds of seeds. Bates Smith followed the fate of over 200 juvenile birds. The graphs in Figure 7.36 show the distribution of beak sizes among all juveniles, and among juveniles that survived to adulthood. The graphs reveal disruptive selection: The survivors were the birds with bills that were either relatively large or relatively small. Birds with beaks of intermediate size did not survive. (Note that an element of stabilizing selection appears to be at work here too: Except in the case of birds with extremely long bills, the birds with the most extreme phenotypes did not survive.)

Evolutionary biologists generally assume that directional selection and stabilizing selection are common, whereas disruptive selection is rare. If the preponderance of directional and stabilizing selection is real, however, it creates a puzzle. Recall from Figure 7.34 that both directional and stabilizing selection reduce the phenotypic variation present in a population. If the trait in question is heritable, then these modes of selection will reduce the genetic variation in the population as well. Eventually, the genetic variation in any trait related to fitness should be eliminated altogether, and the population should reach an equilibrium at which the mean value of the trait, the variation in the trait, and the mean fitness of the population will all cease to change. The puzzle is that populations typically exhibit significant genetic variation, even in traits closely related to fitness. How is this genetic variation maintained?

Here are three possible solutions to the puzzle of how genetic variation for fitness is maintained:

1. Most populations are not in evolutionary equilibrium with respect to directional and/or stabilizing selection. In any population there is a steady, if slow, supply of new favorable mutations creating genetic variation for fitness–related traits. While favorable mutations are rising in frequency, but have not yet become fixed, the population will exhibit genetic variation for fitness. This can be called the "Fisher's Fundamental Theorem Hypothesis." It was Ronald Fisher who first showed mathematically that the rate at which the mean fitness of a population

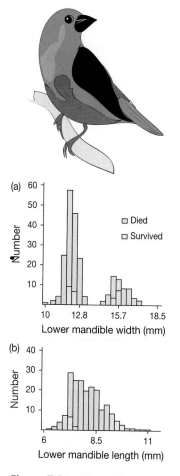

Figure 7.36 Disruptive selection on bill size in the black-bellied seedcracker (*Pyrenestes o. ostrinus*) Each graph shows the distribution of lower bill widths (a) or lengths (b) in a population of black-bellied seed-crackers, an African finch. The blue portion of each bar represents juveniles that did not survive to adulthood; the light orange portion represents juveniles that did survive. The survivors were those individuals with bills that were either relatively large or relatively small. From Bates Smith (1993).

increases is proportional to the additive genetic variation for fitness, a result he called the Fundamental Theorem of Natural Selection.

2. In most populations there is a balance between deleterious mutations and selection. In any population, there is a steady supply of new deleterious mutations. In Chapter 5 we showed that unless the mutation rate is high or selection is weak, selection will keep any given deleterious allele at low frequency. But quantitative traits are determined by the combined influence of many loci of small effect. Selection on the alleles at any single locus affecting a quantitative trait may be very weak, allowing substantial genetic variation to persist at the equilibrium between mutation and selection.

3. Disruptive selection, or patterns of selection with similar effects, may be more common than is generally recognized. Other patterns of selection that can maintain genetic variation in populations include frequency-dependent selection in which rare phenotypes (and genotypes) have higher fitness than common phenotypes, and selection imposed by a fluctuating environment.

All of these hypotheses are controversial, and have been the subject of considerable theoretical and empirical research (see, for example, Barton and Turelli 1989). A detailed discussion is beyond the scope of this text. We can, however, provide a brief review of an intriguing experiment by Santiago Elena and Richard Lenski.

Elena and Lenski (1997) studied six populations of the bacterium *Escherichia coli*. These populations were established from a common ancestral culture, so they were closely related to each other. Each population was founded by a single bacterium, so within any given culture all genetic variation had arisen as a result of new mutations. The six populations had been evolving in a constant lab environment for 10,000 generations. During this time, the mean fitness of each population, assessed via competition experiments, had increased by over 50% relative to their common ancestor. However, much of the increase in fitness had occurred in the first few thousand generations. After 10,000 generations, the populations appeared to have arrived at an evolutionary equilibrium. Elena and Lenski assessed the genetic variation in fitness among the various strains present in each of their six populations and found it to be significant. On average, two strains selected from the same population differed in fitness by about 4%.

Elena and Lenski tested the Fisher's Fundamental Theorem hypothesis by using the standing genetic variation in fitness within each population to predict how much additional improvement in fitness should occur in a further 500 generations of evolution. Depending on the assumptions used, the researchers predicted an additional increase in fitness of between 4 and 50%. In fact, between generations 10,000 and 10,500, none of the six populations showed any significant increase in mean fitness. Elena and Lenski concluded that the genetic variation in fitness within their *E. coli* populations is not the result of a continuous supply of new favorable mutations in the process of rising to fixation.

Elena and Lenski tested the mutation-selection balance hypothesis by noting that two of their six populations had evolved extraordinarily high mutation rates—on the order of 100 times higher than the other four populations and the common ancestor. If the genetic variation in fitness within each population is maintained by mutation-selection balance, then the two populations with high mutation rates should show by far the highest standing genetic variation in fitness. One of the two high-mutation populations did exhibit much higher genetic variation for fitness than the four low-mutation populations. But the other high-mutation population did not. Elena and Lenski concluded that the genetic variation in fitness within their *E. coli* populations is probably not the result of a balance between deleterious mutations and selection.

Finally, Elena and Lenski tested the hypothesis that frequency-dependent selection is maintaining the genetic variation for fitness in their populations. The researchers used competition experiments to determine whether the various strains present in each population enjoyed a fitness advantage when rare. They found that a typical *E. coli* strain, when rare, did indeed have a fitness edge of about 2% relative to its source population. Furthermore, across the six bacterial populations, the intensity of frequency-dependent selection was significantly correlated with the amount of standing genetic variation for fitness. Elena and Lenksi noted that the three hypotheses they tested are mutually compatible. Nonetheless, the researchers concluded that the best explanation for the variation in fitness in their populations is frequency-dependent selection. Whether this conclusion applies to other populations and other organisms remains to be seen.

Summary

The single-locus models of Chapters 5 and 6 are powerful, but potentially oversimplified. Extension of the Hardy–Weinberg analysis to two loci reveals complications. When genotypes at one locus are nonrandomly associated with genotypes at the other, the loci are in linkage disequilibrium. Even under Hardy–Weinberg assumptions, chromosome frequencies change across generations. Furthermore, selection on one locus can alter allele frequencies at the other, and single-locus models may make inaccurate predictions. When genotypes at one locus are independent of genotypes at the other locus, however, the loci are in linkage equilibrium. Chromosome frequencies do not change across generations. Selection on one locus has no effect on allele or genotype frequencies at the other, and we can use single-locus models to make predictions about evolution.

In a random-mating population, linkage disequilibrium can be created by selection on multilocus genotypes, genetic drift, and population admixture. All three mechanisms create an excess of some chromosome haplotypes and a deficit of others. Linkage disequilibrium is reduced by sexual reproduction. Sex brings together chromosomes with different haplotypes, and crossing-over during meiosis allows the chromosomes to exchange genes. This genetic recombination tends to break up overrepresented haplotypes and create underrepresented haplotypes.

The fact that sexual reproduction reduces linkage disequilibrium provides a key to understanding why sexual reproduction persists in populations. Simple theoretical arguments suggest that asexual reproduction should sweep to fixation in any population in which it appears. Empirical observations and experiments indicate, however, that sex confers substantial benefits. These benefits can be found in the population-genetic consequences of sex. When drift or selection has reduced the frequency of particular multilocus genotypes below their expected levels under linkage equilibrium, sexual reproduction can be favored because it recreates the missing genotypes.

Most traits involve many more than two loci, and we often do not know the identity of these loci. Quantitative genetics gives us tools for analyzing the evolution of such traits. Heritability can be estimated by examining similarities among relatives. The strength of selection can be measured by analyzing the relationship between phenotypes and fitness. When we know both the heritability of a trait and the strength of selection on the trait, we can predict how the population will evolve in response to selection.

Selection on quantitative traits can follow a variety of patterns, including directional selection, stabilizing selection, and disruptive selection. Directional selection and stabilizing selection reduce genetic variation in populations. Nonetheless, genetic variation persists in most populations, even for traits closely related to fitness. Variation may persist because most populations are not in equilibrium, because there is a balance between mutation and selection, or because disruptive selection (and related patterns, like frequency-dependent selection) are more common than generally recognized.

Questions

1. In horses, the basic color of the coat is governed by the E locus. *EE* and *Ee* horses can make black pigment, while *ee* horses are a reddish chestnut. A different locus, the R locus, can cause "roan," a scattering of white hairs throughout the basic coat color. However, the roan allele has a serious drawback: *RR* embryos always die during fetal development. *Rr* embryos survive and are roan, while *rr* horses survive and are not roan. The E locus and the R locus are tightly linked.

Suppose that several centuries ago, a Spanish galleon with a load of conquistadors' horses was shipwrecked by a large grassy island. Just by chance, the horses that survived the shipwreck and swam to shore were 20 chestnut roans (*eeRr*) and 20 nonroan homozygous blacks (*EErr*). On the island, they interbred with each other and established a wild population. The island environment exerts no direct selection on the E locus.

a. What was *D*, the coefficient of linkage disequilibrium, in the initial population of 20 horses? Was the initial population in linkage equilibrium or not? If not, what chromosomal genotypes were underrepresented?

b. Do you expect the frequency of the chestnut allele, *e*, to increase or decrease in the first crop of foals? Would your answer be different if the founding population had been just 10 horses (5 of each color)? Explain your reasoning.

c. If you could travel to this island today, can you predict what *D* would be now? Do you have predictions about whether more horses will be roan vs. nonroan, or chestnut vs. black? If not, explain what further information would you need.

2. Imagine a population of pea plants that is in linkage equilibrium for two linked loci, flower color (*P* = purple, *p* = red) and pollen shape (*L* = long, *l* = round).

a. What sort of selection event would create linkage disequilibrium? For example, will selection at just one locus (e.g., all red-flowered plants die) create linkage disequilibrium? How about selection at two loci (e.g., red-flowered plants die, and long-pollen plants die)? How about selection on a certain combination of genotypes at two loci (e.g., only plants that are both red-flowered and have long pollen grains die)?

b. Now imagine a population that is already in linkage disequilibrium for these two loci. Will selection for purple flowers affect evolution of pollen shape? How is your answer different from your answer to part a, and why?

3. Figure 7.3 shows how selection on multilocus genotypes can create linkage disequilibrium. From the post-selec-

tion population in Figure 7.3b, develop a bar graph like the ones in Figure 7.2. Does this bar graph confirm that the post-selection population is in linkage disequilibrium?

4. Box 7.1 shows that when $g_{AB} = ps, g_{Ab} = pt, g_{aB} = qs$, and $g_{ab} = qt$, then $D = 0$. Show that when $D = 0, ps = g_{AB}$. (*Hint: p*, the frequency of allele *A*, is equal to $g_{AB} + g_{Ab}$. Likewise, *s*, the frequency of allele *B*, is equal to $g_{AB} + g_{aB}$. Multiply these quantities and simplify the expression. Knowing that $D = g_{AB}g_{ab} - g_{Ab}g_{aB} = 0$ will allow you to make a key substitution.)

5. a. In the beetle evolution experiment (Figure 7.14), Dunbrack et al. did not actually use sexual beetles and asexual beetles. Instead, they used two colors of sexual beetles, but forced one color of beetles to grow in population size as if it were asexual. They also did the experiment again with the other color as "asexual." Why was it important that the researchers ran the experiment both ways? Compare the graphs of the two different sets of experiments (red = asexual, and black = asexual). Did the two strains of beetles perform differently?

b. Actual asexual beetles would reproduce twice as fast as sexual beetles, but in the experiment the authors made the "asexual" beetles reproduce three times as fast. Why do you think they did this?

c. The researchers' simulated asexual population was not allowed to evolve at all in response to selection imposed by competition. Is this different from what would have happened in an actual population of asexual beetles? Do you think Dunbrack et al.'s experiment is a valid test of asexual reproduction versus sexual reproduction? Briefly describe the next experiment that you think Dunbrack et al. should do to follow up on this topic.

6. In 1992, Spolsky, Phillips, and Uzzell reported genetic evidence that asexually reproducing lineages of a salamander species have persisted for about 5 million years, an unusually long time. Is this surprising? Why or why not? Speculate about what sort of environment these asexual salamanders live in, and whether their population sizes are typically small (say, under 100) or large (say, over 1000).

7. *Volvox* (Figure 7.12a) are abundant and active in lakes during the spring and summer. During winter they are inactive, existing in a resting state in encysted zygotes called zygospores. During most of the spring and summer, *Volvox* reprodue asexually; but at times they switch and reproduce sexually instead. When would you predict that *Volvox* would be sexual: spring, early summer, or late summer? Explain your reasoning.

8. Suppose you're telling your roommate that you learned in biology class that within any given human population, height is highly heritable. Your roommate, who is studying nutrition, says "That doesn't make any sense, because just a few centuries ago, most people were shorter than they are now, and it's clearly because of diet. If most variation in human height is due to genes, how could diet makes such a big difference?" Your roommate is obviously correct that poor diet can dramatically affect height. How do you explain this apparent paradox to your roommate?

9. Now consider heritability in more general terms. Suppose heritability is extremely high for a certain trait in a certain population. There are two important questions:
   a. First, can the trait be strongly affected by the environment despite the high heritability value? To answer this question, suppose that all the individuals within a certain population have been exposed all their lives to the same level of a critical environmental factor. Will the heritability value reflect the fact that the environment is very important?
   b. Second, can the heritability value itself change if the environment changes? To answer this question, imagine that the critical environmental factor changes, such that different individuals are now exposed to different levels of this environmental factor. What happens to variation in the trait in the whole population? What happens to the heritability value?

10. In our discussion of Weis and Abrahamson's work on goldenrod galls (data plotted in Figure 7.35), we mentioned that the researchers established that there is heritable variation among flies in the size of the galls they induce. How do you think Weis and Abrahamson did this? Describe the necessary experiment in as much detail as possible.

Exploring the Literature

11. If the endosymbiotic bacteria studied by Lambert and Moran (1998) suffer reduced fitness as a result of Muller's ratchet, then mitochondria should too. Do they? See

Lynch, M. 1996. Mutation accumulation in transfer RNAs: Molecular evidence for Muller's ratchet in mitochondrial genomes. *Molecular Biology and Evolution* 13: 209–220.

Lynch, M. 1997. Mutation accumulation in nuclear, organelle, and prokaryotic transfer RNA genes. *Molecular Biology and Evolution* 14: 914–925.

12. Many human pathogens, including bacteria and eukaryotes, are capable of both asexual reproduction and genetic recombination (that is, sex in the population-genetic sense). The frequency of recombination in a pathogen population can have medical implications. (Think, for example, about how fast resistance to multiple antibiotics will evolve in a population of bacteria that does have recombination versus a population that does not.) How can we tell whether a given pathogen population is engaging in genetic recombination or is predominantly clonal? Genetic recombination is such a powerful force in reducing linkage disequilibrium that the amount of disequilibrium in a pathogen population provides a clue. See

Maynard Smith, J., N. H. Smith, M. O'Rourke, and B. G. Spratt. 1993. How clonal are bacteria? *Proceedings of the National Academy of Sciences, USA* 90: 4384–4388.

Burt, A., D. A. Carter, G. L. Koenig, T. J. White, and J. W. Taylor. 1996. Molecular markers reveal cryptic sex in the human pathogen *Coccidioides immitis*. *Proceedings of the National Academy of Sciences, USA* 93: 770–773.

Go, M. F., V. Kapur, D. Y. Graham, and J. M. Musser. 1996. Population genetic analysis of *Helicobacter pylori* by multilocus enzyme electrophoresis: Extensive allelic diversity and recombinational population structure. *Journal of Bacteriology* 178: 3934–3938.

Gräser, Y. et al. 1996. Molecular markers reveal that population structure of the human pathogen *Candida albicans* exhibits both clonality and recombination. *Proceedings of the National Academy of Sciences, USA* 93: 12473–12477.

Jiménez, M., J. Alvar, and M. Tibayrenc. 1997. *Leishmania infantum* is clonal in AIDS patients too: Epidemiological implications. *AIDS* 11: 569–573.

13. How far and how fast can directional selection on a quantitative trait shift the distribution of the trait in a population? For one answer, see

Weber, K. E. 1996. Large genetic change at small fitness cost in large populations of *Drosophila melanogaster* selected for wind tunnel flight: Rethinking fitness surfaces. *Genetics* 144: 205–213.

Literature Cited

Please note that much of the population and quantitative genetics in this chapter is modeled after presentations in the following:

Cavalli-Sforza, L. L., and W. F. Bodmer. 1971. *The Genetics of Human Populations.* San Francisco: W. H. Freeman and Company.

Falconer, D. S. 1989. *Introduction to Quantitative Genetics.* New York: John Wiley & Sons.

Felsenstein, J. 1997. *Theoretical Evolutionary Genetics.* Seattle, WA: ASUW Publishing, University of Washington.

Felsenstein, J. 1988. Sex and the evolution of recombination. In R. E. Michod and B. R. Levin, eds. *The Evolution of Sex.* Sunderland, MA: Sinauer, 74–86.

Hartl, D. L. 1981. *A Primer of Population Genetics.* Sunderland, MA: Sinauer.

Maynard Smith, J. 1998. *Evolutionary Genetics,* 2nd edition. Oxford: Oxford University Press.

Here is the listing of all citations in this chapter:

Andersson, D. I., and D. Hughes. 1996. Muller's ratchet decreases fitness of a DNA-based microbe. *Proceedings of the National Academy of Sciences, USA* 93: 906–907.

Ansari-Lari, M. A., X.-M. Liu, et al. 1997. The extent of genetic variation in the CCR5 gene. *Nature Genetics* 16: 221–222.

Barton, N. H., and M. Turelli. 1989. Evolutionary quantitative genetics: How little do we know? *Annual Review of Genetics* 23: 337–370.

Bates Smith, T. 1993. Disruptive selection and the genetic basis of bill size polymorphism in the African finch *Pyrenestes. Nature* 363: 618–620.

Blakeslee, A. F. 1914. Corn and men. *Journal of Heredity* 5: 511–518.

Brodie, E. D., III. 1992. Correlational selection for color pattern and antipredator behavior in the garter snake *Thamnophis ordinoides. Evolution* 46: 1284–1298.

Brodie, E. D., III, A. J. Moore, and F. J. Janzen. 1995. Visualizing and quantifying natural selection. *Trends in Ecology and Evolution* 10: 313–318.

Carrington, M., M. Dean, et al. 1999. Genetics of HIV-1 infection: Chemokine receptor CCR5 polymorphism and its consequences. *Human Molecular Genetics* 8: 1939–1945.

Cavalli-Sforza, L. L., and W. F. Bodmer. 1971. *The Genetics of Human Populations.* San Francisco: W. H. Freeman and Company.

Clausen, J., D. D. Keck, and W. M. Hiesey. 1948. *Experimental Studies on the Nature of Species. III. Environmental Responses of Climatic Races of Achillea.* Washington, DC: Carnegie Institution of Washington Publication No. 581, 45–86.

Clegg, M. T., J. F. Kidwell, and C. R. Horch. 1980. Dynamics of correlated genetic systems. V. Rates of decay of linkage disequilibria in experimental populations of *Drosophila melanogaster. Genetics* 94: 217–234.

Di Masso, R. J., G. C. Celoria, and M. T. Font. 1991. Morphometric traits and femoral histomorphometry in mice selected for body conformation. *Bone and Mineral* 15: 209–218.

Dunbrack, R. L., C. Coffin, and R. Howe. 1995. The cost of males and the paradox of sex: An experimental investigation of the short-term competitive advantages of evolution in sexual populations. *Proceedings of the Royal Society of London,* Series B 262: 45–49.

Elena, S. F., and R. E. Lenski. 1997. Long-term experimental evolution in *Escherichia coli.* VII. Mechanisms maintaining genetic variability within populations. *Evolution* 51: 1058–1067.

Falconer, D. S. 1989. *Introduction to Quantitative Genetics.* New York: John Wiley & Sons.

Felsenstein, J. 1988. Sex and the evolution of recombination. In R. E. Michod and B. R. Levin, eds. *The Evolution of Sex.* Sunderland, MA: Sinauer, 74–86.

Galen, C. 1989. Measuring pollinator-mediated selection on morphometric floral traits: Bumblebees and the alpine sky pilot, *Polemonium viscosum. Evolution* 43: 882–890.

Galen, C. 1996. Rates of floral evolution: Adaptation to bumblebee pollination in an alpine wildflower, *Polemonium viscosum. Evolution* 50: 120–125.

Grant, P. R., and B. R. Grant. 1995. Predicting microevolutionary responses to directional selection on heritable variation. *Evolution* 49: 241–251.

Haigh, J. 1978. The accumulation of deleterious mutations in a population: Muller's ratchet. *Theoretical Population Biology* 14: 251–267.

Hedrick, P. W. 1983. *Genetics of Populations.* Boston: Science Books International.

Howard, R. S., and C. M. Lively. 1998. The maintenance of sex by parasitism and mutation accumulation under epistatic fitness functions. *Evolution* 52: 604–610.

Huttley, G. A., M. W. Smith, et al. 1999. A scan for linkage disequilibrium across the human genome. *Genetics* 152: 1711–1722.

Jennings, H. S. 1917. The numerical results of diverse systems of breeding, with respect to two pairs of characters, linked or independent, with special relation to the effects of linkage. *Genetics* 2: 97–154.

Lalani, A. S., J. Masters, et al. 1999. Use of chemokine receptors by poxviruses. *Science* 286: 1968–1971.

Lambert, J. D., and N. A. Moran. 1998. Deleterious mutations destabilize ribosomal RNA in endosymbiotic bacteria. *Proceedings of the National Academy of Sciences, USA* 95: 4458–4462.

Lande, R., and S. J. Arnold. 1983. The measurement of selection on correlated characters. *Evolution* 37: 1210–1226.

Lively, C. M. 1992. Parthenogenesis in a freshwater snail: Reproductive assurance versus parasitic release. *Evolution* 46: 907–913.

Lively, C. M. 1996. Host-parasite coevolution and sex. *BioScience* 46: 107–114.

Maynard Smith, J. 1978. *The Evolution of Sex.* Cambridge: Cambridge University Press.

Maynard Smith, J. 1988. The evolution of recombination. In R. E. Michod and B. R. Levin, eds. *The Evolution of Sex.* Sunderland, MA: Sinauer, 106–125.

Michod, R. E., and B. R. Levin, eds. 1988. *The Evolution of Sex.* Sunderland, MA: Sinauer.

Miyashita, N. T., A. Kawabe, and H. Innan. 1999. DNA variation in the wild plant *Arabidopsis thaliana* revealed by amplified fragment length polymorphism analysis. *Genetics* 152: 1723–1731.

Muller, H. J. 1964. The relation of recombination to mutational advance. *Mutation Research* 1: 2–9.

Murray, C., and R. J. Herrnstein. 1994. Race, genes and I.Q.—An apologia. *The New Republic* October: 27–37.

Phillips, P. C., and S. J. Arnold. 1989. Visualizing multivariate selection. *Evolution* 43: 1209–1222.

Provine, W. B. 1986. *Sewall Wright and Evolutionary Biology.* University of Chicago Press, Chicago.

Seger, J., and W. D. Hamilton. 1988. Parasites and sex. In R. E. Michod and B. R. Levin, eds. *The Evolution of Sex.* Sunderland, MA: Sinauer, 176–193.

Smith, J. M. N., and A. A. Dhondt. 1980. Experimental confirmation of heritable morphological variation in a natural population of song sparrows. *Evolution* 34: 1155–1160.

Spolsky, C. M., C. A. Phillips, and T. Uzzell. 1992. Antiquity of clonal salamander lineages revealed by mitochondrial DNA. *Nature* 356: 706–710.

Stephens, J. C., D. E. Reich, et al. 1998. Dating the origin of the CCR5-Δ32 AIDS-resistance allele by the coalescence of haplotypes. *American Journal of Human Genetics* 62: 1507–1515.

Sturtevant, A. H., and K. Mather. 1938. The interrelations of inversions, heterosis, and recombination. *American Naturalist* 72: 447–452.

Via, S. 1991. The genetic structure of host plant adaptation in a spatial patchwork: Demographic variability among reciprocally transplanted pea aphid clones. *Evolution* 45: 827–852.

Weis, A. E., and W. G. Abrahamson. 1986. Evolution of host-plant manipulation by gall makers: Ecological and genetic factors in the *Solidago-Eurosta* system. *American Naturalist* 127: 681–695.

Wright, S. 1986. *Evolution: Selected papers.* William B. Provine, editor. University of Chicago Press, Chicago.

Wright, S. 1988. Surfaces of selective value revisited. *American Naturalist* 131: 115–123.

PART III

A female giraffe, and her calf, feeding on leaves. (Walt Anderson/Visuals Unlimited)

ADAPTATION

THE POPULATION GENETIC MODELS AND EXPERIMENTS THAT WE INTRODUCED IN PART II delivered a fundamental message: Each of the four evolutionary forces has different consequences. Mutation alters DNA sequences randomly. As the source of all genetic variation, mutation supplies the raw material that makes evolution possible. Migration homogenizes allele frequencies among populations. Drift produces random changes in allele frequencies, especially in small populations. Natural selection, in contrast, is the only evolutionary process that results in adaptation. An adaptation is a trait that allows an individual to leave more offspring than individuals without the trait.

Our goal in Part III is to explore adaptation in depth. We begin, in Chapter 8, by surveying techniques that evolutionary biologists use to study adaptation. How can a researcher rigorously test the hypothesis that a particular trait is adaptive? This discussion is followed, in Chapter 9, by an exploration of how selection acts on morphological traits and behaviors that allow individuals to attract mates. This type of selection is called sexual selection. It can produce striking differences in the behavior and appearance of males and females of the same species. Chapter 10 asks how interactions among individuals living in social groups affect the fitness of the individuals. It explores the biological basis of altruism and introduces an important conceptual advance called kin selection. The unit concludes by probing questions about how long individuals live, how often they reproduce, and how much they invest in each of their offspring. As a whole, the chapters in Part III explore the consequences of natural selection in all of its forms.

CHAPTER 8

Studying Adaptation: Evolutionary Analysis of Form and Function

The flowers of *Fuchsia excorticata* remain green while producing nectar and receiving pollen, then change to deep purple or red. (Lynda Delph, Indiana University)

Giraffes demand an explanation. A tiny head on a massive neck, with the whole arrangement perched on stilts; to watch their gymnastic contortions when they bend to take a drink is to wonder if they weren't put together by a committee (Figure 8.1).

Everyone knows, of course, why giraffes have long necks. The ancestors of giraffes had to compete with other species of leaf-eating browsers. The giraffes with the longest necks got more food, and consequently had more offspring. Today their descendants reach heights of five meters or more and are the only grazers in Africa that can reach the tops of tall acacia trees.

The explanation of organismal design is among the triumphs of the theory of evolution by natural selection. Individuals in previous generations varied in their design, and the ones with the best designs passed on their genes in greater numbers. A trait, or integrated suite of traits, that increases the fitness of its possessor is called an **adaptation** and is said to be **adaptive**.

Demonstrating that the traits of organisms are indeed adaptations has been one of the major activities of evolutionary biology since the time of Darwin (Mayr

251

Figure 8.1   The giraffe's long neck makes getting a drink look difficult   (Joe McDonald/Animals Animals/Earth Scenes)

1983). This research effort is sometimes called the adaptationist program. Roughly speaking, in order to demonstrate that a trait is an adaptation, we need first to determine what a trait is for and then show that individuals possessing the trait contribute more genes to future generations than individuals lacking it.

The adaptive signficance of some traits is obvious: Eyes are manifestly devices for detecting objects at a distance by gathering and analyzing light; in many animal species, individuals with good eyesight will be better able to find food and avoid predators than individuals with poor eyesight. Other traits offer more subtle advantages, and understanding their adaptive significance requires effort.

Obvious explanations, in particular, can be dangerously seductive. As we shall see in the first section of this chapter, our adaptive hypothesis for the giraffe's neck has a serious flaw: The giraffes themselves do not seem to believe it. The moral is that no explanation for the adaptive value of a trait should be accepted simply because it is plausible and charming (Gould and Lewontin 1979). All hypotheses must be tested. Hypotheses can be tested by using them to make predictions, then checking to see if the predictions are correct.

This chapter explores a variety of methods evolutionary biologists use to test hypotheses about adaptations, including experiments, observational studies, and the comparative method. In the last sections of the chapter we also explore complexities of biological form and function that continue to make the adaptationist program a challenging and active area of research.

8.1   All Hypotheses Must Be Tested: The Giraffe's Neck Reconsidered

A plausible hypothesis about the adaptive value of a trait is the beginning of a careful study, not the end.

Robert Simmons and Lue Scheepers (1996) question the conventional wisdom on how the giraffe got its neck. Simmons and Scheepers watched wild giraffes eat, and reviewed observations published by other biologists. Simmons and Scheepers

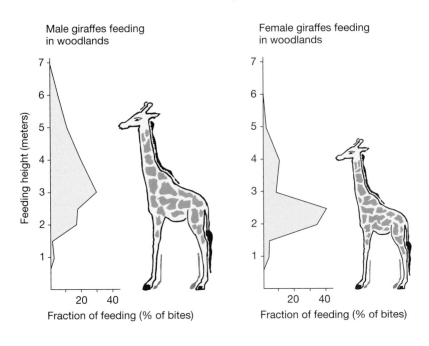

Figure 8.2 **Do giraffes take advantage of their long necks when foraging?** These graphs show the proportion of time males (left) and females (right) spend feeding at different heights in Kenyan woodlands. The wider the graph at any height, the more time giraffes spend foraging there. The cartoons show the heights of a typical male and female. Redrawn from Figures 5 and 6 in Young and Isbell (1991). Reprinted by permission of Blackwell Wissenschafts-Verlag, Berlin.

reasoned that if the foraging-competition hypothesis is correct, then during the dry season, when food is scarce, giraffes should spend most of their time foraging above the reach of their competitors. In fact, giraffes spend much of their dry-season foraging time browsing on low bushes, not tall trees. Even when giraffes forage above the reach of their competitors, they seldom forage at a level anywhere near their maximum height. In Kenya, for example, Truman Young and Lynne Isbell (1991) found that giraffes do most of their foraging in woodlands, and that they prefer to feed at shoulder height (Figure 8.2).

Simmons and Scheepers offer an alternative scenario for the evolution of the giraffe's neck. They note that bull giraffes sometimes fight viciously, employing their necks and heads as clubs, and occasionally even kill each other (Figure 8.3). Simmons and Scheepers suggest that the giraffe's neck evolved as a weapon, used by the males in combat over opportunities to mate. Consistent with this hypothesis, male giraffes have necks that are 30 to 40 centimeters longer and 1.7 times heavier than the necks of females of the same age. In addition, male giraffes have skulls that are more stoutly armored and 3.5 times heavier.

Figure 8.3 **Male giraffes use their heads and necks as clubs when they fight over opportunities to mate** (Patti Murray/Animals Animals/Earth Scenes)

Also consistent with the neck-as-a-weapon hypothesis are data on giraffe behavior collected by David Pratt and Virginia Anderson. Pratt and Anderson (1985) made extensive observations of social interactions among giraffes. The first time Pratt and Anderson saw each of the adult males in their study population, they placed him in one of three categories. Class C males were young adults; males in classes A and B were more mature. Class A males were often larger than class B males, but more importantly, class A males had thicker necks, more massive horns, and more heavily armored skulls. Pratt and Anderson witnessed 127 interactions in which one bull displaced another from a social group. This happened when, "by walking toward him with head held high, or sometimes by merely staring hard at him, the dominant individual oblige[d] the inferior to move off." Class A bulls were dominant over classes B and C, and Bs were dominant over Cs (Table 8.1a).

Pratt and Anderson also watched interactions between males and females. Bull giraffes monitor females to determine when they are in heat. To check on a female's reproductive condition, a male nuzzles her on the rump. If she is cooperative, the female urinates, and the male tastes her urine. Pratt and Anderson witnessed numerous attempts by males of each class to urine-test females. Females were much more cooperative with A and B males than with C males (Table 8.1b). Because urine testing sometimes leads to courtship and mating, these data suggest that females prefer to mate with older, larger, and larger-necked males.

Simmons and Scheepers argue that among the ancestors of today's giraffes, the long-necked males had higher reproductive success not because they got more to eat, but because they intimidated their rivals and attracted more mates. Why, then, do females have long necks too? Perhaps it is simply because males pass genes for long necks to their daughters as well as to their sons.

Table 8.1   Neck size and social interactions in giraffes

Males in class C are young adults; males in classes A and B are more mature. Class A males are often larger than class B males, but more importantly, class A males have stouter necks, more massive horns, and more heavily armored skulls.

(a) Neck size and male social interactions. These numbers represent observations of one male displacing another from a social group.

A displaces B, A displaces C, or B displaces C	A displaces A, B displaces B, or C displaces C	B displaces A, C displaces A, or C displaces B
82	**39**	**6**

(b) Neck size and female choice. These numbers represent observations of a male attempting to determine whether a female is in heat by tasting her urine. Urine testing of a female by a male requires the female's cooperation.

	Successful	Unsuccessful	% Successful
A bulls	**34**	**22**	**60.7**
B bulls	**76**	**61**	**55.5**
C bulls	**45**	**89**	**33.6**

Source: From Pratt and Anderson 1985. Copyright © 1985, Journal of Natural History. Reprinted by permission of Taylor and Francis Ltd. http://www.tandf.co.uk

The giraffe example demonstrates that we cannot uncritically accept a hypothesis about the adaptive significance of a trait simply because it is plausible. We must subject all hypotheses to rigorous tests.

Here are some other caveats to keep in mind when studying adaptations:

- Differences among populations or species are not always adaptive. Giraffes from different populations have different spot patterns on their coats. It is possible that each coat pattern is adaptive in the region where it occurs. It is also possible, however, that the regional differences in coat pattern are not adaptive at all. Mutations causing variant patterns could have become fixed in different giraffe populations by genetic drift, perhaps via the founder effect (see Chapter 6). At the molecular level, much of the variation among individuals, populations, and species may be selectively neutral (see Chapters 6 and 18).

- Not every trait of an organism, or every use of a trait by an organism, is an adaptation. Giraffes do sometimes forage at their full height, above the reach of their competitors. This does not necessarily mean that long necks evolved because they provide novel feeding opportunities. Long necks may have evolved because of their value as weapons, and are only incidentally exploited for the advantages they provide in feeding.
- Not every adaptation is perfect. Long necks may help male giraffes obtain mates, but as we have already observed, they also make it difficult for giraffes to get a drink.

In the next three sections, we review three methods evolutionary biologists use to test hypotheses about the adaptive significance of traits. The first of these sections concerns experiments, the second looks at observational studies, and the third explores the comparative method.

8.2 Experiments

Experiments are among the most powerful tools in science. A well-designed experiment allows us to isolate and test the effect that a single, well-defined factor has on the phenomenon in question. We have already reviewed a variety of experiments in earlier chapters. Figures 5.11 and 7.14, for example, illustrate the results of experiments on laboratory insect populations. Here our focus is on the process of planning and interpreting experiments. We have chosen our example because it illustrates several aspects of good experimental design.

What is the Function of the Wing Markings and Wing-Waving Display of the Tephritid Fly *Zonosemata*?

The tephritid fly *Zonosemata vittigera* has distinctive dark bands on its wings. When disturbed, the fly holds its wings perpendicular to its body and waves them up and down (Figure 8.4). Entomologists had noticed that this display seems to mimic the leg-waving, territorial threat display of jumping spiders (species in the family Salticidae). These entomologists suggested that, because jumping spiders are fast and have a nasty bite, a fly mimicking a jumping spider might be avoided by a wide variety of other predators. Erick Greene and colleagues (1987) had a different

Figure 8.4 A sheep in wolf's clothing? This photograph shows the tephritid fly *Zonosemata vittigera* (right) facing one of its predators, the jumping spider *Phidippus apacheanus* (left) (Erick Greene, University of Montana).

idea. Because jumping spiders are *Zonosemata*'s major predators, Greene et al. proposed that the fly uses its wing markings and wing-waving display to intimidate the jumping spiders themselves. The fly, in other words, is a sheep in wolf's clothing. Mimicry of a predator's behavior by its own prey had never before been recorded.

Both mimicry explanations are plausible hypotheses about the adaptive value of the fly's wing-waving display, but unless we test them they are just good stories. Can these hypotheses be tested rigorously? Greene and his co-workers (1987) set out to do so with an experiment.

The first step in any evolutionary analysis is to phrase the question as precisely as possible. In this case, Do the wing markings and the wing waving of *Zonosemata vittigera* mimic the threat displays that jumping spiders use on each other, and thereby allow the flies to escape predation? Stating a question precisely makes it easier to design an experiment that will provide a clear answer.

Greene et al.'s next step was to list alternative explanations for the behavior. Good experiments test as many competing hypotheses as possible. Note that each of the following is a viable and biologically realistic competing explanation, not an implausible straw man proposed simply to give the impression of rigor.

Good experimental designs test the predictions made by several alternative hypotheses.

H_1: The flies do not mimic jumping spiders. This is a distinct possibility, because other fly species have dark wing bands and wing-flicking displays that do not deter predators. In many species, the flies use their markings and displays during courtship.

H_2: The flies mimic jumping spiders, but the flies behave like spiders to deter other, nonspider predators. Other fly predators that might be intimidated by a jumping spider, or a jumping spider mimic, include other kinds of spiders, assassin bugs, preying mantises, and lizards.

H_3: The flies mimic jumping spiders, and this mimicry functions specifically to deter predation by the jumping spiders themselves.

To test these alternatives, Greene and colleagues needed flies with some parts, but not all, of the *Zonosemata* display. The researchers found that they could cut the wings off a *Zonosemata* fly and glue them back on with Elmer's glue. And they could cut the wings of a *Zonosemata* fly and replace them with the wings of a housefly (*Musca domestica*), which are clear and unmarked. Remarkably, the surgically altered *Zonosmata* continued to wave their wings in the normal way and could even fly.

By performing various surgeries, Greene and colleagues created a total of five experimental groups of flies (Figure 8.5). The five treatments distinguish among the three hypotheses, because each hypothesis makes a different suite of predictions about what will happen in encounters between predators and flies. The treatments also allow the researchers to determine whether both the wing markings and the wing-waving display are important in mimicry. This is a powerful experimental design.

To make the test, Greene et al. needed to measure the responses of jumping spiders and other predators to the five types of experimental flies. When confronted with a test fly, would the spiders retreat, stalk and attack, or kill? To answer this question, the researchers starved 20 jumping spiders from 11 different species for two days. Then they presented one of each of the five experimental fly types to each spider, in random order. The researchers made these presentations in a test arena, and recorded each jumping spider's most aggressive response during a five-minute interval. There was a clear difference: Jumping spiders tended to retreat from flies that gave the wing-waving display with marked wings, but attacked flies that lacked either wing markings, wing waving, or both (Figure 8.6).

When the researchers tested treatments A, C, and E against other predators (nonsalticid spiders, assassin bugs, mantises, and whiptail lizards), all of the test flies

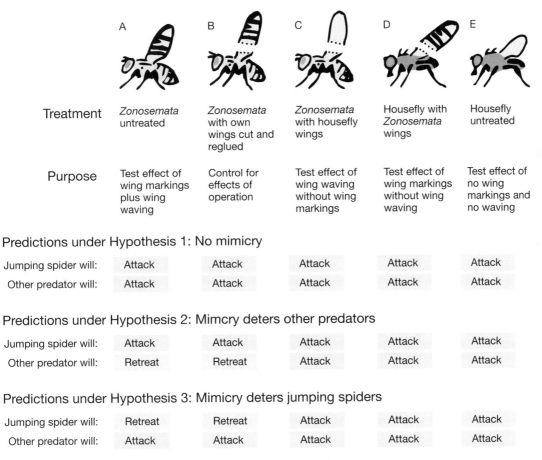

	A	B	C	D	E
Treatment	*Zonosemata* untreated	*Zonosemata* with own wings cut and reglued	*Zonosemata* with housefly wings	Housefly with *Zonosemata* wings	Housefly untreated
Purpose	Test effect of wing markings plus wing waving	Control for effects of operation	Test effect of wing waving without wing markings	Test effect of wing markings without wing waving	Test effect of no wing markings and no waving

Predictions under Hypothesis 1: No mimicry

Jumping spider will:	Attack	Attack	Attack	Attack	Attack
Other predator will:	Attack	Attack	Attack	Attack	Attack

Predictions under Hypothesis 2: Mimcry deters other predators

Jumping spider will:	Attack	Attack	Attack	Attack	Attack
Other predator will:	Retreat	Retreat	Attack	Attack	Attack

Predictions under Hypothesis 3: Mimicry deters jumping spiders

Jumping spider will:	Retreat	Retreat	Attack	Attack	Attack
Other predator will:	Attack	Attack	Attack	Attack	Attack

Figure 8.5 **Surgical treatments used in experiments testing the function of *Zonosemata's* wing-waving display** This table shows the predicted outcomes when different predators encounter flies with different treatments. Note that each hypothesis makes a unique suite of predictions. (The predictions listed for hypotheses 2 and 3 assume that both *Zonosemata's* wing markings and wing waving are necessary for effective mimicry.)

Figure 8.6 **Tephritid flies mimic jumping spiders to avoid predation** These bar graphs show how jumping spiders responded to the five treatment and control groups listed in Figure 8.5. The vertical axis represents the number of jumping spiders, from a total of 20, that showed each type of maximum response to each type of fly. Thus, for group A flies (unaltered *Zonosemata*), 15 of the jumping spiders retreated from their test flies [graph (a)], and five of the spiders killed their test flies [graph (c)]. For group B flies (*Zonosemata* with their wings cut and reglued), 15 of the jumping spiders retreated from their flies, two stalked and attacked but did not kill their flies, and three killed their flies. From Greene et al. (1987). Copyright © 1985, Journal of Natural History. Reprinted by permission of Taylor & Francis Ltd. http://www.tandf.co.uk

Note that most of the flies that waved marked wings (groups A and B) survived their encounters with spiders unscathed, whereas most of the flies that lacked markings or waving or both were attacked, and many were killed. The heavy black bars at the bottom of the graphic identify treatment groups where the spider responses were statistically indistinguishable from one another. Groups A and B were indistinguishable from each other, but different from groups C, D, and E. Because each spider was presented with one fly of each type, the sample size in each treatment group was 20.

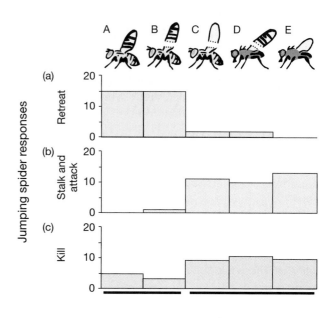

were captured and eaten. When Greene et al. placed flies before these nonsalticid predators, there was not even an appreciable difference in time-to-capture among the three treatment groups.

Comparison of Figures 8.5 and 8.6 shows that these results are consistent with hypothesis H_3, but inconsistent with hypotheses H_1 and H_2. Thus, Greene et al.'s experiment provides strong support for the hypothesis that tephritid flies mimic their own jumping-spider predators to avoid being eaten by them (see also Mather and Roitberg 1987).

In terms of experimental design, the Greene et al. study illustrates several important points:

Control groups provide a contrast to treatment groups. Control individuals may have no treatment at all, or may experience a treatment that tests the effects of experimental conditions that are predicted to have no effect on the outcome.

• Defining and testing effective control groups is critical. In Greene et al.'s study, groups A and B (Figures 8.5 and 8.6) served as controls. These individuals demonstrated that the wing surgery itself had no effect on the behavior of the flies or the spiders. Thus, when the *Zonosemata* in group C were attacked and eaten by jumping spiders, Green and colleagues could be sure that this was because the flies no longer had markings on their wings, not simply because their wings had been cut and glued.

• All of the treatments (controls and experimentals) must be handled exactly alike. It was critical that Greene et al. used the same test arena, the same time interval, and the same definitions of predator response in each test. Using standard conditions allows a researcher to avoid bias and increase the precision of the data (Figure 8.7). Think about the problems that could arise if a different test arena were used for each of the five treatment groups.

• Randomization is an important technique for equalizing other, miscellaneous effects among control and experimental groups. In essence, it is another way to avoid bias. For example, Greene and colleagues presented the different kinds of test flies to the spiders and other predators in random order. What problems could arise if they had presented the five types of flies in the same sequence to each spider?

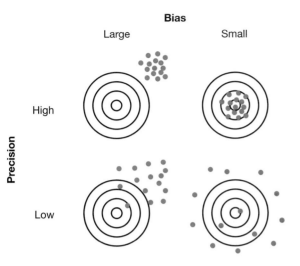

Figure 8.7 **Bias and precision** When designing an experiment or set of observations with the goal of estimating some quantity, it is important to minimize bias and maximize precision. In this cartoon the quantity being estimated in a study is represented by the bull's-eye, and the data points collected are represented by red dots. Techniques like standardizing the experimental conditions and randomizing other factors help to minimize bias and increase precision. Note that our ability to measure precision depends on having a large number of data points.

- Repeating the test on many individuals is essential. It is almost universally true in experimental (and observational) work that larger sample sizes are better. This is because the goal of an experiment is to estimate a quantity. In this case, the quantity was the likelihood that a fly will be attacked by jumping spiders as a function of its ability to wave marked wings.

Replicated experiments or observations do two things:

- They reduce the amount of distortion in the estimate caused by unusual individuals or circumstances. For example, four of the ten *Zonosemata* with marked wings that were attacked were pounced on and killed before they even had a chance to display (groups A and B in Figures 8.5 and 8.6). Because Greene and colleagues were using standardized conditions, it was not acceptable to simply throw out these data points, even though they might represent bad luck. If events like this really do represent bad luck they will be rare, and will not bias the result as long as the sample size is large.
- Replicated experiments allow researchers to understand how precise their estimate is by measuring the amount of variation in the data. Knowing how precise the data are allows the use of statistical tests. Statistical tests, in turn, allow us to quantify the probability that the result we observed was simply due to chance (see Box 8.1).

In sum, Greene et al.'s experimental design was successful because it allowed independent tests of the effect that predator type, wing type, and wing display have on the ability of *Zonosemata* flies to escape predation. Experiments are the most powerful means of testing hypotheses about adaptation. In the next section, we consider how careful observational studies can sometimes be nearly as good as experiments.

8.3 Observational Studies

Some hypotheses about adaptations are difficult or impossible to test with experiments. It is hard to imagine, for example, how we could do a controlled experiment to test alternative hypotheses about why giraffes have long necks. To do so,

Large sample sizes are critical to the success of experiments, but researchers have to trade off the costs and benefits of collecting ever-larger data sets.

When an experiment is impractical, a careful observational study may be the next best method for evaluating a hypothesis. A good observational study seeks to find circumstances in nature that resemble an experiment.

Box 8.1   A primer on statistical testing

The fundamental goal of many experimental and observational studies is to estimate the value of some quantity in two groups, such as a treatment group and a control group, and to determine if there is a difference in the quantity between the groups. In the studies we have reviewed thus far, researchers have estimated quantities like the depth of finch beaks (Chapter 3) and how frequently flies are attacked by spiders (this chapter). The groups we want to compare in these examples consist of finches before and after the drought, and flies with marked or unmarked wings.

As our example, we will focus on a comparison between the flies in groups B and C of Greene et al.'s experiment (Figures 8.5 and 8.6). To simplify the discussion, we will lump together the outcomes "stalked and attacked" and "killed" to form a single category, "attacked." When the researchers presented a group B fly to each of 20 jumping spiders, 15 of the spiders retreated and 5 attacked the fly. In contrast, when the researchers presented to each of the same 20 spiders a group C fly, 1 of the spiders retreated and 19 attacked the fly. It certainly looks as though jumping spiders are less aggressive toward flies waving marked wings (group B) than toward flies waving unmarked wings (group C).

Once we have measured the quantity in each group and have observed a difference between groups, the statistical question becomes, Is the difference real, or could it simply be due to random variation? It is conceivable that if we tested a much larger population of spiders and flies, we would discover that, in fact, spiders respond the same way to flies waving unmarked wings as they do to flies waving marked wings. Under this scenario, the apparent difference we observed in the experiment was just a chance result.

An analogy can be found in tossing coins. Imagine you have two fair coins. It is conceivable that you could toss the first coin 20 times and get 15 heads and 5 tails and then toss the second coin 20 times and get 1 head and 19 tails. It would appear that the coins are different, but the truth is that if you tossed both coins a very large number of times, you would discover that the coins are the same. (Note that this analogy is imperfect: The true rate at which a fair coin gives heads is 0.5, whereas the true rate at which spiders attack flies might be anywhere from 0 to 1.) What we want to know is, What is the probability that we could get a difference in spider behavior as large as the one we observed if the truth were that spiders actually respond the same way to both kinds of fly?

Answering this question requires a statistical test. The first step in a statistical test is to specify the null hypothesis. This is the hypothesis that there is actually no difference between the groups. In our example, the null hypothesis is that the presence or absence of wing markings does not affect the way jumping spiders respond to flies. According to this hypothesis, the true frequency of attack is the same for flies with markings on their wings as for flies without markings on their wings.

The second step is to calculate a value called a test statistic. A test statistic is a number that characterizes the magnitude of the difference between the groups. More than one test statistic might be appropriate for Greene et al.'s data. Greene and colleagues chose a test statistic that compares the actual rates of retreat and attack observed in the experiment to the rates of retreat and attack that would have been expected if the null hypothesis were true.

The third step is to determine the probability that chance alone could have made the test statistic as large as it is. In other words, if the null hypothesis were true, and we did the same experiment many times, how often would we get a value for the test statistic that is larger than the one we actually got? The answer comes from a reference distribution. This is a mathematical function that specifies the probability, under the null hypothesis, of each of all the possible values of the test statistic. Often, it is possible to look up the answer in a statistical table in a book, or to have a computer calculate it. For Greene et al.'s data and test statistic, the probability that chance alone would have made the test statistic this large is considerably less than 0.01. In other words, if the null hypothesis were true, and if Greene et al. repeated their experiment many times, they would have gotten a value of the test statistic larger than the one they actually got in fewer than one in 100 experiments. This means that the null hypothesis is probably wrong, and that there *is* a real difference in how jumping spiders respond to flies waving marked versus unmarked wings.

Box 8.1   Continued

The fourth and final step is to decide whether to consider the outcome of the experiment statistically significant. By convention, scientists generally consider the value of a test statistic significant if its probability under the null hypothesis is less than one in 20, or 0.05. By this criterion, Greene et al.'s result is significant with room to spare. In other words, when Greene and colleagues claimed to have demonstrated that flies must have markings on their wings to deter jumping-spider attacks, the researchers were taking a chance of less than 1 in 100 of later being proved wrong by someone else who might repeat their experiment. That chance was low enough for them to claim that their result is statistically significant. If there is more than a 5% probability that the difference observed is due to chance, the convention is to accept the null hypothesis of no real difference between groups.

In scientific papers the probability of finding the observed differences by chance is reported as a P value, where the P stands for probability. In the original paper published by Greene et al., for example, the caption to the figure that is our Figure 8.6 includes the phrase "$P < 0.01$."

Statistical tests are based on explicit models of the processes that produced the data and of the design of the experiment. Many different types of random processes are modeled statistically; when analyzing data, it is essential to know enough about statistics to be able to choose a model appropriate to the data collected in a particular study.

we would have to be able to make giraffes that are identical in all respects except the lengths of their necks. Experiments may also be inappropriate when a hypothesis makes predictions about how organisms will behave in nature. When experiments are impractical or inappropriate, careful observations can sometimes yield sufficient information to evaluate a hypothesis.

Behavioral Thermoregulation

The vast majority of organisms are ectothermic, which means that their body temperatures are determined by the temperatures of their environments. As Figure 8.8 demonstrates for desert iguanas (*Dipsosaurus dorsalis*), body temperature has a profound effect on an ectotherm's physiological performance. Desert iguanas can survive short exposures to body temperatures as low as 0°C and as high as about 47°C, but they can function only between about 15°C and 45°C. Within this narrower range, cold iguanas run and digest slowly, tire quickly, and hear poorly. As they get warmer, they run and digest more quickly, tire more slowly, and

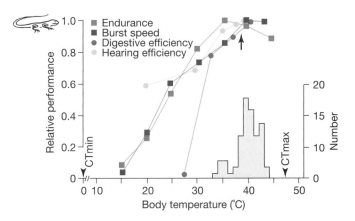

Figure 8.8   Physiological abilities of the desert iguana (*Dipsosaurus dorsalis*) as a function of body temperature The colored squares and circles show locomotor endurance, burst (sprint) speed, digestive efficiency, and hearing efficiency of iguanas as a function of body temperature. The shaded region is a histogram showing the distribution of body temperatures for active iguanas captured in nature. The black arrow indicates the body temperature chosen by iguanas in the lab. CTmax is the critical thermal maximum, that is, the upper lethal temperature. CTmin is the critical thermal minimum. Reprinted from Huey and Kingsolver (1989). Copyright © 1989, Elsevier Science. Reprinted by permission of Elsevier Science.

hear more keenly. The iguanas' physiological capacities reach a plateau in the mid-to-high 30s. Above about 45°C, the iguanas are too hot and collapse.

The relationship between physiological performance and temperature is called a thermal performance curve. The shape of the desert iguana's thermal performance curves is typical of those of a variety of physiological processes in a diversity of organisms. Given the sensitivity of physiological function to temperature, we can predict that ectotherms will exhibit behavioral thermoregulation. That is, ectotherms should move around in the environment so as to maintain themselves at or near the temperature at which they perform the best.

As the temperature of their environment changes, for example, desert iguanas do not just passively accept the consequences. Instead, they regulate their body temperature by moving into the sun to warm up or into the shade to cool off. The iguanas prefer to maintain themselves at body temperatures in the high 30s (Figure 8.8, arrow). This is the center of the range of temperatures at which the iguanas perform best. The iguana's temperature preferences are not surprising. After all, an iguana never knows when it will want to chase something to eat, or need to run away from a predator. In nature, of course, iguanas may not always have a sufficient range of environmental temperatures to move among to maintain themselves at exactly their preferred temperature. As the shaded histogram of field body temperatures in Figure 8.8 shows, however, desert iguanas do reasonably well.

Note that although we have asserted that desert iguanas thermoregulate, the mere fact that iguanas captured in nature are usually at or near their optimal body temperature does not, by itself, prove that they are active in maintaining those temperatures. It could be that the environments in which they live are always in the mid-to-high 30s. To prove behavioral thermoregulation, we must show (1) that the animal in question is choosing particular temperatures more often than it would encounter those temperatures if it simply moved at random through its environment, and (2) that its choice of temperatures is adaptive.

Do Garter Snakes Make Adaptive Choices When Looking for a Nighttime Retreat?

Ray Huey and colleagues (1989b) made a detailed study of the thermoregulatory behavior of the garter snake (*Thamnophis elegans*) at Eagle Lake, California. Garter snakes are affected by temperature in the same way as desert iguanas, except that for garter snakes, the optimal temperature, preferred temperature, and maximum survivable temperature are all a few degrees lower than the corresponding temperatures for iguanas. Huey et al. surgically implanted several snakes with miniature radio transmitters. Each implanted transmitter emits a beeping signal that allows a biologist with a handheld receiver and directional antenna to find the implanted snake, even when the snake is hiding under a rock or in a burrow. In addition, the transmitter reports the snake's temperature by changing the rate at which it beeps.

Garter snakes in the lab prefer to stay at temperatures between 28°C and 32°C. Huey and colleagues found that snakes in nature do a remarkable job of thermoregulating in the same range. Figure 8.9 shows the body temperatures of two of the implanted snakes, each over the course of a 24-hour day. Both snakes kept their temperature within or near the preferred range. How do the garter snakes manage to thermoregulate so well? The two snakes shown in Figure 8.9 spent the

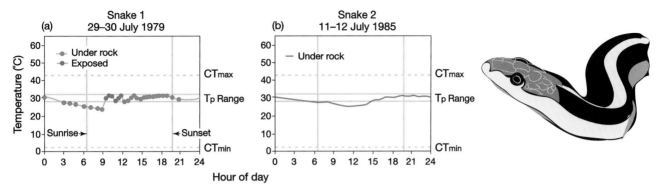

Figure 8.9 Body temperatures of garter snakes in nature (a) Snake 1 spent part of the day under a rock (red dots and line), and part of the day in the sun (blue dots and line). (b) Snake 2 spent the entire day under a rock. Tp is the preferred temperature range, measured in the lab. CTmax and CTmin are defined in Figure 8.8. Both snakes kept their temperature near 30°C for the entire day. From Huey et al. (1989b). Reprinted by permission of the Ecological Society of America.

day under or near rocks. Other options include moving up and down a burrow, and staying on the surface of the ground while moving back and forth from sunshine to shade.

Huey and colleagues compared the relative merits of each of these thermoregulatory strategies by monitoring the environmental temperature under rocks of various sizes, and at various depths in a burrow, and by monitoring the temperature of a model snake left on the surface in the sun or shade (Figure 8.10). For a snake under a rock, the thickness of the rock proves critical. A snake under a thin rock (Figure 8.10a) would not only get dangerously cold at night, but would overheat in the daytime. (As Huey says, "The snake would die by 11 a.m., and remain dead until at least 6 p.m.") A snake under a thick rock (Figure 8.10b) would remain safe all day, but would never reach its preferred temperature. Rocks of medium thickness are just right (Figure 8.10c). By moving around under the rock, a snake under a rock of medium thickness can stay close to or within its preferred temperature range for the entire day. A snake moving up and down a burrow could do reasonably well (Figure 8.10d), but would get colder at night than a snake under a medium-sized rock. Finally, a snake on the surface could thermoregulate effectively in the daytime by moving between sun and shade, but would get dangerously cold at night (Figure 8.10e). Putting these observations together, it appears that snakes have many options for thermoregulation during the daytime, as long as they avoid thin rocks or direct sun in the afternoon. At night, however, it appears that the best place for a snake to be is under a rock of medium thickness.

Most garter snakes do, in fact, retreat under rocks at night. Under the hypothesis of behavioral thermoregulation, Huey and colleagues predicted that snakes would choose their nighttime retreats adaptively. That is, the researchers predicted that snakes would preferentially select rocks of medium thickness. Huey et al. tested their prediction by comparing the availability of rocks of different sizes at Eagle Lake to the sizes of the rocks actually chosen as nighttime retreats by radio-implanted snakes (Table 8.2). Thin, medium, and thick rocks are equally available, so if the snakes chose their nocturnal retreats at random, they should be found equally often under rocks of each size. In fact, however, the garter snakes are almost always found under medium rocks or thick rocks. The fact that snakes avoid thin rocks is good evidence that the snakes are active behavioral thermoregulators.

By monitoring temperatures of potential retreats, researchers studying thermoregulation in garter snakes showed that the best place for a snake to spend the night is under a rock of medium thickness.

(a) Temperatures under a thin rock

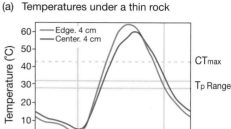

(b) Temperatures under a thick rock

(c) Temperatures under a medium rock

(d) Temperatures in a burrow

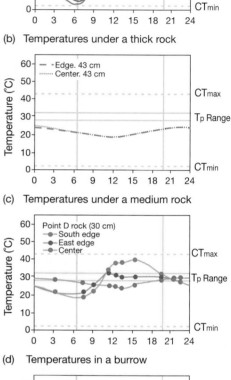

(e) Temperatures for a model on the surface

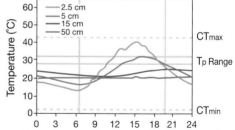

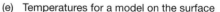

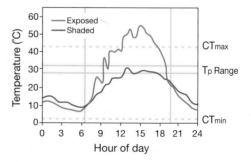

Hour of day

Figure 8.10 **Environmental temperatures available to garter snakes at Eagle Lake** The graphs show the daily cycle of temperatures in various places a snake might go. (a) Under a thin (4 cm) rock, it is cold at night and hot during the day. (b) Under a thick rock (43 cm), it is cool all the time. (c) Under a rock of medium thickness (30 cm), there is virtually always a spot within the range of temperatures preferred by snakes. (d) In a burrow, it is cool at night and cool to warm in the daytime, with the exact temperature depending on depth within the burrow. (e) On the surface, it is cold at night and just right to hot in the daytime, depending on whether a snake is in shade or direct sunlight. From Huey et al. (1989b). Reprinted by permission of the Ecological Society of America.

Table 8.2   Distributions of rocks available to snakes versus rocks chosen by snakes

Thin, medium, and thick rocks are equally abundant at Eagle Lake, but garter snakes retreating under rocks at night show a strong preference for rocks of medium thickness ($P < 0.05$; chi-square test with thin and thick rocks combined because of small expected values).

	Thin (<20 cm)	Medium (20–40 cm)	Thick (>40 cm)
Rocks available to snakes	32.4%	34.6%	33%
Rocks chosen by snakes	7.7%	61.5%	30.8%

Source: From Table 1 in Huey et al. (1989b). Copyright © 1989, Ecological Society of America.

What made the observational study by Huey and colleagues effective in testing the hypothesis that garter snakes thermoregulate is the care with which the researchers monitored the snakes' environment. By determining the options available to snakes, and measuring the frequency of each option in the environment, the researchers were able to show that the snakes they observed were not simply picking their retreats at random, but were instead making an adaptive choice. In the next section, we consider a kind of observational study that looks at adaptations on a broader scale. Biologists using the comparative method evaluate hypotheses by looking at patterns of evolution among species.

The researchers then noted the availability of thin, medium, and thick rocks, and observed that snakes choose medium rocks more often than expected under the null hypothesis that snakes select their retreats at random.

8.4 The Comparative Method

In Sections 8.2 and 8.3 we considered how experiments and observations on individuals within populations can be used to test hypotheses about adaptation. Here we examine how comparisons among species can be used to study the evolution of form and function. Our example comes from a group of bats called the Megachiroptera, which includes the fruit bats and the flying foxes (Figure 8.11a).

The comparative method seeks to evaluate hypotheses by testing for patterns across species, such as correlations among traits, or correlations between traits and features of the environment.

Why Do Some Bats Have Bigger Testes than Others?

Males in some of these bat species have larger testes for their body size than others. Based on work on a variety of other animals, David Hosken (1998) hypothesized that large testes are an adaptation for sperm competition. Sperm competition occurs when a female mates with two or more males during a single estrus cycle, and the sperm from the males are in a race to the egg. One way a male can increase his reproductive success in the face of sperm competition is to produce larger ejaculates. By entering more sperm into the race, he increases his odds of winning. And the way to produce larger ejaculates is to have larger testes.

To evaluate the sperm competition hypothesis, Hosken needed to use it to develop a testable prediction. Hosken knew that fruit bats and flying foxes roost in groups, and that the size of a typical group varies dramatically among species, from two or three individuals to tens of thousands. Hosken reasoned that females living in larger groups would have more opportunities for multiple matings, and that males living in larger groups would thus experience greater sperm competition.

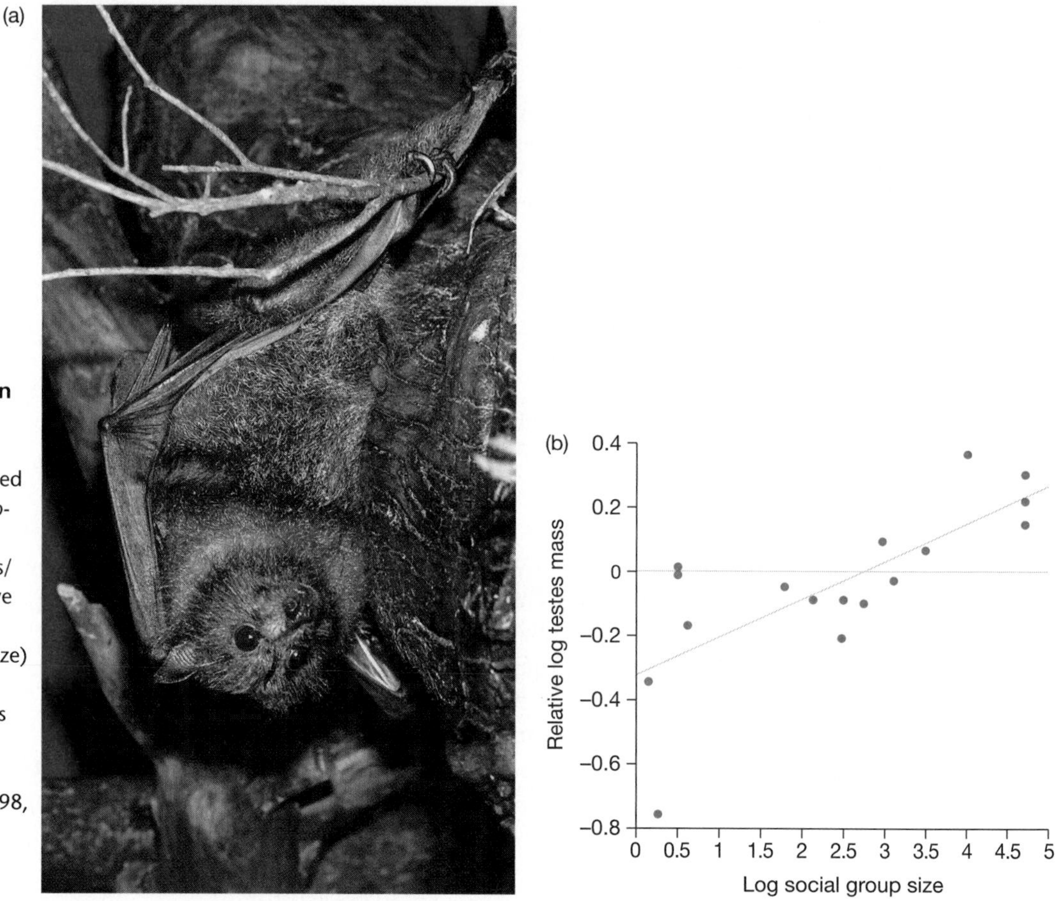

Figure 8.11 Variation in testis size among fruit bats and flying foxes (a) A grey-headed flying fox (*Pteropus poliocephalus*). (Fritz Prenzel/Animals Animals/ Earth Scenes) (b) Relative testis size (that is, testis size adjusted for body size) as a function of roost group size for 17 species of fruit bats and flying foxes. From Hosken (1998). Copyright © 1998, Springer-Verlag GmbH & Co. KG. Reprinted by permission.

Hosken predicted that whenever a bat species evolves larger roosting group sizes, its males will also evolve larger testes for their body size.

The simplest way to test this hypothesis is to gather data for a variety of species, and prepare a scatterplot showing relative testis size as a function of roost group size. When Hosken did this, he found that the two variables are strongly correlated (Figure 8.11b). Bat species that live in larger groups have larger testes for their body size. As Hosken knew, however, there may be less evidence in this graph than meets the eye.

Proper application of the comparative method requires knowledge of the evolutionary relationships among the species under study.

Figure 8.12 illustrates why. Imagine, for simplicity, that we have plotted a graph for only six species. We will call these species A, B, C, D, E, and F. Figure 8.12a shows a scatterplot for relative testis size versus group size. Like the real scatterplot in Figure 8.11, this graph shows a positive correlation between the two traits. Now imagine that the evolutionary relationships among our six species are as shown in the phylogeny in Figure 8.12b. Species A, B, and C are all closely related to each other, as are species D, E, and F. It may be that species A, B, and C all inherited their small group sizes and their small testes from their common ancestor (green arrow). Likewise, it may be that species D, E, and F all inherited their large group sizes and large testes from their common ancestor (orange arrow). The possibility that our six species inherited their traits from just two common ancestors deflates the strength of our evidence considerably.

When we prepare a scatterplot and use it as the basis for claims about nature, we want all the data points to be independent of each other. If they are independent, then each data point makes a separate statement for or against our claim. Furthermore, independence of the data points is a requirement for traditional statis-tical tests. To make sure our scatterplot accurately reflects the nature of the evidence, we should therefore replace the points for species A, B, and C with a single point representing their common ancestor, and we should do the same with the points for species D, E, and F.

The graph in Figure 8.12c shows the result. It may, in fact, be true that group size and testis size evolve together, and it may be true that sperm competition is the reason. However, a scatterplot with only two data points is weak evidence on which to base such a claim.

Joe Felsenstein (1985) developed a better way to evaluate cross–species corre-lations among traits. What we look at in Felsenstein's method are patterns of di-vergence as sister species evolve independently away from their common ancestors. Figure 8.13 shows a graphical interpretation of the method's basic approach.

The first thing we need is a phylogeny for the species we are studying. Figure 8.13a shows a phylogeny for five extant species. We will call these species A through E. The phylogeny also includes the common ancestors that lived at all the nodes on the tree. These are species F, G, H, and I. Note that there are four places on this phylogeny where sister species diverged from a common ancestor; each is indicated by a dif-ferent color. For example, A and B are sister species that diverged from common an-cestor G. Likewise, G and C are sister species that diverged from common ancestor H. What we want to know is this: When species diverge from a common ancestor, does the species that evolves larger group sizes also evolve larger testes?

We can answer this question by first plotting all of the pairs of sister species on a scatterplot, with lines connecting their data points (Figure 8.13b). We then grab each pair of points by the point closest to the vertical axis, and drag that point to the origin (Figure 8.13c). Finally, we can erase the points at the origin and the con-necting lines. We are left with a scatterplot with four data points (Figure 8.13d). Each data point represents the divergence, or contrast, that arose between a pair

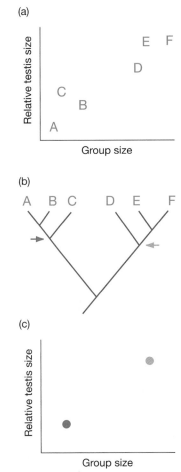

Figure 8.12   A simple scatter-plot may provide only weak evidence that two traits evolve in tandem   See text for explanation. After Fig. A in Lauder et al. (1995). Reprinted by permission of American Institute of Biological Sciences and the author.

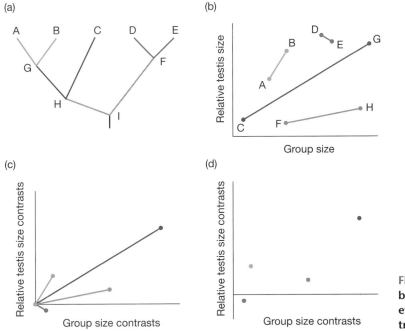

Figure 8.13   A graphical interpretation of the basic procedure in Felsenstein's method for evaluating phylogenetically independent con-trasts   See text for explanation.

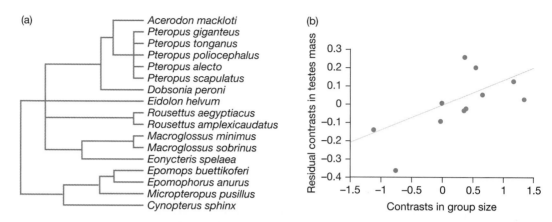

Figure 8.14   Correlated evolution of group size and testis size in fruit bats and flying foxes   (a) A phylogeny for 17 species of bats. (b) Independent contrasts for relative testis size versus group size. The points on this graph show that when a bat species evolved larger (or smaller) group sizes than its sister species, it also tended to evolve larger (or smaller) testes ($P = 0.027$). From Hosken (1998). Copyright © 1998, Springer-Verlag GmbH & Co. Reprinted by permission.

of sister species as they evolved away from their common ancestor. If the contrasts are correlated with each other, then we can conclude that when a species evolved a larger group size than its sister species, it also tended to evolve larger testes. In practice, we must make adjustments to the data before we can do statistical tests to evaluate the strength of any patterns. These adjustments are described in Box 8.2.

Hosken (1998) repeated his analysis of testis size and group size in bats using Felsenstein's method, which is called the method of phylogenetically independent contrasts. Figure 8.14a shows a phylogeny of the 17 bat species whose data were plotted in Figure 8.11b. Figure 8.14b shows a plot of the contrasts in relative testis size versus the contrasts in group size. There is a significant positive correlation among the contrasts. In other words, the data show that when a bat species evolved larger roosting group sizes than its sister species, it also tended to evolve larger testes for its body size. Hosken concluded that the evidence from flying foxes and fruit bats is consistent with the hypothesis that large testes are an adaptation to sperm competition.

When formulating and testing hypotheses about adaptation, biologists must keep in mind that organisms, and the lives they live, are complex.

We have now considered three methods evolutionary biologists use to evaluate hypotheses about adapation. In the remainder of the chapter, we turn to complexities in organismal form and function that are active areas of current research. In the examples we discuss, researchers use experiments, observational studies, and the comparative method to investigate hypotheses about phenotypic plasticity (Section 8.5), the evolutionary origin of adaptive traits (Section 8.6), and trade-offs and constraints on adaptation (Section 8.7).

8.5   Phenotypic Plasticity

Throughout much of this book, we treat phenotypes as though they were determined solely and immutably by genotypes. We know, however, that phenotypes are often strongly influenced by the environment as well. Chapter 7 included a section on estimating how much of the phenotypic variation among individuals is due to variation in genotypes and how much is due to variation in enviromnments. Here, we focus on the interplay between genotype, environment, and phenotype.

Box 8.2   Calculating phylogenetically independent contrasts

Here we use an example from Garland and Adoph (1994) to illustrate the calculation of independent contrasts from a phylogeny (see also: Felsenstein 1985; Martins and Garland 1991; Garland et al. 1999). Figure 8.15 shows the phylogeny we will use. It shows the relationships among polar bears, grizzly bears, and black bears, and gives the body mass and home range of each. We will calculate independent contrasts for both traits among the bears. The steps are as follows:

1. Calculate the contrasts for pairs of sibling species at the tips of the phylogeny. In our three-species tree, there is just one pair of sibling species in which both species reside at the tips: polar bears and grizzly bears. The polar bear–grizzly bear contrast for body mass is
$$265 - 251 = 14$$
The polar bear–grizzly bear contrast for home range is:
$$116 - 83 = 33$$

2. Prune each contrasted pair from the tree, and estimate the trait values for their common ancestor by taking the weighted average of the descendants' phenotypes. When calculating the weighted average, weight each species by the reciprocal of the branch length leading to it from the common ancestor. In our example, we are pruning polar bears and grizzlies from the tree, and estimating the body mass and home range of common ancestor A. The branch lengths from A to its decendants are both two units long. Thus, the weighted average for body mass is
$$\text{Body mass of species A} = \frac{\left(\frac{1}{2}\right)265 + \left(\frac{1}{2}\right)251}{\left(\frac{1}{2}\right) + \left(\frac{1}{2}\right)} = 258$$
The weighted average for home range is
$$\text{Home range of species A} = \frac{\left(\frac{1}{2}\right)116 + \left(\frac{1}{2}\right)83}{\left(\frac{1}{2}\right) + \left(\frac{1}{2}\right)} = 99.5$$

3. Lengthen the branch leading to the common ancestor of each pruned pair by adding to it the product of the branch lengths from the common ancestor to its descendants, divided by their sum. In our example, we are lengthening the branch leading to species A. The new branch length is
$$3 + \frac{2 \times 2}{2 + 2} = 4$$

4. Continue down the tree calculating contrasts, estimating the phenotypes of the common ancestors, and lengthening the branches leading to the common ancestors. In our example, the only remaining contrast is between species A and black bears. We do not need to estimate the phenotype of species B, or lengthen the branch leading to it, because species B is at the root of our tree. The species A–black bear contrast for body mass is
$$258 - 93 = 165$$
The species A–black bear contrast for home range is
$$99.5 - 57 = 42.5$$

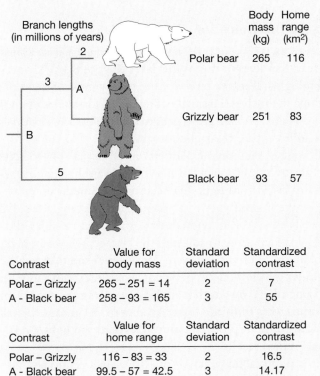

Branch lengths (in millions of years)

	Body mass (kg)	Home range (km^2)
Polar bear	265	116
Grizzly bear	251	83
Black bear	93	57

Contrast	Value for body mass	Standard deviation	Standardized contrast
Polar – Grizzly	265 – 251 = 14	2	7
A - Black bear	258 – 93 = 165	3	55

Contrast	Value for home range	Standard deviation	Standardized contrast
Polar – Grizzly	116 – 83 = 33	2	16.5
A - Black bear	99.5 – 57 = 42.5	3	14.17

Figure 8.15   An example showing how the data are adjusted when calculating phylogenetically independent contrasts   From Garland and Adolph (1994). Copyright © 1994, Physiological Zoology. Reprinted by permission of The University of Chicago Press.

Box 8.2 Continued

5. Divide each contrast by its standard deviation to yield the standardized contrasts. The standard deviation for a contrast is the square root of the sum of its (adjusted) branch lengths. The standard deviation for the polar bear–grizzly bear contrast is

$$\sqrt{2 + 2} = 2$$

The standard deviation for the species A–black bear contrast is

$$\sqrt{4 + 5} = 3$$

The standardized contrasts for our example are given in Figure 8.15.

Once we have calculated the standardized contrasts, we can use them to prepare a scatterplot and to perform traditional statistical tests.

Genetically identical individuals reared in different environments may be different in form, physiology, or behavior. Such individuals demonstrate phenotypic plasticity.

Another way to say that an individual's phenotype is influenced by its environment is to say that its phenotype is plastic. When phenotypes are plastic, individuals with identical genotypes may have different phenotypes if they live in different environments. Phenotypic plasticity is itself a trait that can evolve, and it may or may not be adaptive. As with the other traits we have discussed, to demonstrate that an example of phenotypic plasticity is adaptive, we must first determine what it is for, then show that individuals that have it achieve higher fitness than individuals that lack it.

Phenotypic Plasticity in the Behavior of Water Fleas

To illustrate phenotypic plasticity, we present the water flea, *Daphnia magna. Daphnia magna* is a tiny crustacean that lives in freshwater lakes (Figure 8.16). Conveniently for evolutionary biologists, *Daphnia* reproduce asexually most of the time. In other words, *Daphnia* clone themselves. This makes them ideal for studies of phenotypic plasticity, because researchers can grow genetically identical individuals in different environments and compare their phenotypes.

Luc De Meester (1996) studied phenotypic plasticity in *D. magna*'s phototactic behavior. An individual is positively phototactic if it swims toward light, and negatively phototactic if it swims away from light. De Meester measured the phototactic behavior typical of different genotypes of *D. magna*. In each single test, De Meester placed 10 genetically identical individuals in a graduated cylinder, illuminated them from above, gave them time to adjust to the change in environment, then watched to see where in the column they swam. De Meester summarized the results by calculating an index of phototactic behavior. The index can range in value from -1 to $+1$. A value of -1 means that all the *Daphnia* in the test swam to the bottom of the column, away from the light. A value of $+1$ means that all the *Daphnia* in the test swam to the top of the column, toward the light. An intermediate value indicates a mixed result.

De Meester measured the phototactic behavior of 10 *Daphnia* genotypes (also called clones) from each of three lakes. The results are indicated by the blue dots in Figure 8.17. The population in each lake harbors considerable genetic variation in phototactic behavior.

De Meester also measured the phototactic behavior of the same 30 *Daphnia* genotypes in water that had been previously occupied by fish. The results are indicated by the red squares in Figure 8.17. *Daphnia magna*'s phototactic behavior is

Figure 8.16   A water flea, *Daphnia sp.* The large branched appendages are antennae; the water flea uses them like oars for swimming. The dark object near the antennae is an eyespot. Also visible through this individual's transparent carapace are the intestine and other internal organs. This photograph is enlarged about 10 times. (Omikron/Photo Researchers, Inc.)

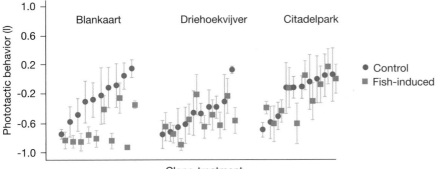

Figure 8.17 **Variation in phototactic behavior in *Daphnia magna*** Blankaart, Driehoekvijver, and Citadelpark are three lakes in Belgium. Each blue dot represents the average result from three to five tests of the phototactic behavior of a single genotype (described in main text). The error bars indicate ± 2 standard errors. Genotypes (clones) whose error bars do not overlap are significantly different. Above or below each blue dot is a red square. This red square represents the average result from three or four tests of the phototactic behavior of the same genotype. The difference is that this time the *Daphnia* were tested in water that had previously been occupied by fish. Lake Blankaart is home to many fish; Lake Driehoekvijver has few fish; Lake Citadelpark has no fish. From De Meester (1996). Copyright © 1996, Evolution. Reprinted by permission of Evolution.

phenotypically plastic. In Lake Blankaart, in particular, most *Daphnia* genotypes score considerably lower on the phototactic index when tested in the presence of chemicals released by fish.

Finally, and most importantly, De Meester's results demonstrate that phenotypic plasticity is a trait that can evolve. Recall that a trait can evolve in a population only if the population contains genetic variation for the trait. Each of the *Daphnia* populations De Meester studied contains genetic variation for phenotypic plasticity. That is, some genotypes in each population alter their behavior more than others in the presence versus the absence of fish (Figure 8.17). Genetic variation for phenotypic plasticity is called **genotype-by-environment interaction**.

Has phenotypic plasticity in fact evolved in the *Daphnia* populations De Meester studied? It has. The average genotype in Lake Blankaart shows considerably more phenotypic plasticity than the average genotype in either of the other lakes. Blankaart is the only one of the lakes with a sizeable population of fish. Fish are visual predators, and they eat *Daphnia*. A reasonable interpretation is that fish select in favor of *Daphnia* that avoid well-lit areas when fish are present.

8.6 Every Adaptive Trait Evolves from Something Else

When studying organismal form and function, it is useful to keep in mind that every adaptive trait evolves from something else. Our example of research into the origin of an adaptive structure comes from the mammalian ear.

How Did the Mammalian Ear Evolve?

One of the mammalian ear's outstanding features is a group of three bones called the ear ossicles. They are named the malleus (hammer), incus (anvil), and stapes (stirrup). Their function is to transmit wave energy from the eardrum (tympanic membrane) in the outer ear to the oval window of the cochlea in the inner ear (Figure 8.18). Why

When there is genetic variation for the degree or pattern of phenotypic plasticity, plasticity itself can evolve. Plasticity is adaptive when it allows individuals to adjust their phenotype so as to increase their fitness in the particular environment in which they find themselves.

Every adaptive trait evolves from something else. This is one reason organs and organisms sometimes appear to be rigged from an assortment of spare parts, none of which is ideally suited to the job.

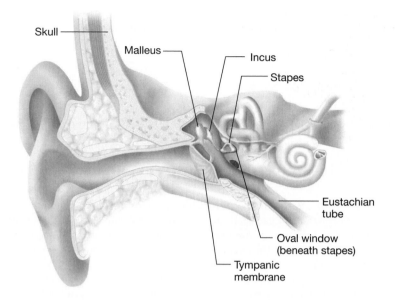

Figure 8.18   The mammalian middle ear   This drawing of a human ear shows the location of the three ear ossicles: the malleus, incus, and stapes.

have three bones instead of just one? Their lever action amplifies the force transmitted, increasing the sensitivity of hearing. The nearly thirtyfold reduction in transmission area, from the large tympanic membrane to the small oval window, also serves to amplify the signal. These ossicles are a major reason why mammals hear so well.

But where did these three little bones come from in the first place? To know, we need to

1. Establish the ancestral condition, and
2. Understand the transformational sequence, or how and why the characters changed through time.

If we can accomplish these steps, we will understand both the proximate and ultimate mechanisms of evolutionary change that produced a manifestly adaptive organ.

The logical place to start our analysis is with an animal called *Acanthostega gunnari* (Figure 8.19). This creature is among the oldest known tetrapods, or four-limbed vertebrates. A partially air-breathing swamp dweller, *Acanthostega* was one of the first animals to encounter the problem of hearing airborne sounds. *Acanthostega* lived about 360 million years ago. Its ancestor belonged to a group of fish

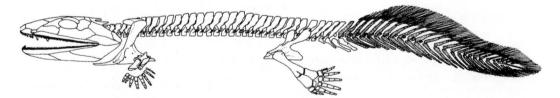
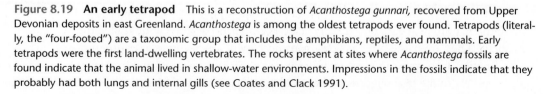

Figure 8.19   An early tetrapod   This is a reconstruction of *Acanthostega gunnari,* recovered from Upper Devonian deposits in east Greenland. *Acanthostega* is among the oldest tetrapods ever found. Tetrapods (literally, the "four-footed") are a taxonomic group that includes the amphibians, reptiles, and mammals. Early tetrapods were the first land-dwelling vertebrates. The rocks present at sites where *Acanthostega* fossils are found indicate that the animal lived in shallow-water environments. Impressions in the fossils indicate that they probably had both lungs and internal gills (see Coates and Clack 1991).

called the crossopterygians (Carroll 1988). Although the crossopterygian fish had none of the three ear ossicles, *Acanthostega* had a stapes. It was the first of the middle ear bones to appear in the fossil record (Clack 1989), and may have functioned in detecting airborne sounds.

What is the evidence for this claim? Fossilized skulls show that one end of the *Acanthostega* stapes fits into a hole in the side of the braincase that connects to the inner ear, while the other end fits into a notch in the skull near an opening called the spiracle (Clack 1994). In later fossil tetrapods and in some extant amphibians, this skull notch holds the tympanic membrane, or eardrum (Figure 8.20a). Because the form of the stapes is homologous in *Acanthostega* and later groups, we can argue that its function—transmitting airborne sound—is homologous too (Lombard and Bolt 1979; Clack 1983).

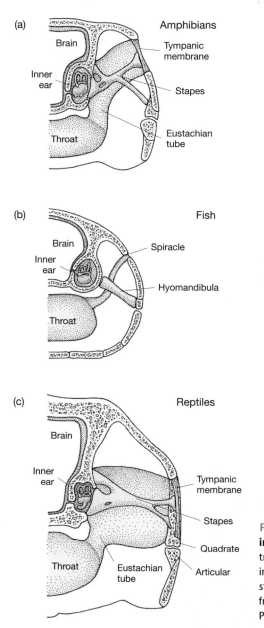

Figure 8.20 **The location of the stapes in various vertebrates**   Each of these illustrations is a cross section of the head, showing the location and form of bones and other structures discussed in the text. Modified from Romer, A.S. 1995. *The Vertebrate Body*. Philadelphia: W.B. Saunders.

But did the stapes appear out of nowhere? How can we determine the ancestral state of this evolutionary innovation? This is a general problem in paleontological analyses. Recall that in Chapter 3 we reviewed the claim that the "thumb" of the Giant Panda evolved from an elongated wrist bone. What is the evidence for assertions like these?

The key to understanding the origin of a derived trait is to establish homology with a trait present in the ancestor. For example, in form and in position, the stapes of *Acanthostega* is homologous to a structure in crossopterygian and other fish called the hyomandibula. The hyomandibula is a bone that functions as a brace between the jaw and the braincase (Figure 8.20b). Muscles attached to this bone pump the cheeks. This pumping action, in turn, ventilates the gills and opens and closes the spiracle. In modern lungfishes, the pumping action ventilates the lungs with air. Because fossils show that *Acanthostega* had both lungs and internal gills, it is reasonable to infer that muscles attached to its stapes were involved in respiration (Clack 1989; Coates and Clack 1991; Clack 1994).

All of these facts make *Acanthostega* a classic transitional form. That is, it is an intermediate between the fish and the tetrapods (amphibians, reptiles, and mammals). Its stapes was a modification of the hyomandibula that still functioned in respiration. But it was also a bone that happened to be in the right place (anatomically) at the right time (when vertebrates first ventured out onto land) to be co-opted for use as a sound transmitter. The hyomandibula was a **preadaptation** for hearing.

In addition to the similarity in form and placement of the two bones in *Acanthostega* and fish, there is a second line of evidence that the stapes is homologous to the hyomandibula. In 1837, long before Darwin established an evolutionary interpretation of homology, the German anatomist C.B. Reichert examined mammal embryos and determined that during development, the stapes originates from the second gill arch (see Gould 1993, essay 6). In fish, this same embryonic structure becomes the hyomandibula. These developmental homologies are exactly what we would predict if the two structures represent ancestral and descendant states. Combined with the morphological data in adults, we have a strong case that the two structures share a common ancestry.

Thus far, we have been able to establish the embryological origin and the ancestral condition of the stapes. But what of the other ear ossicles, the malleus and incus? *Acanthostega* does not have them. Nor do reptiles or amphibians, including the extinct forms ancestral to the mammals (Allin 1975). All of these groups transmit sound from the tympanic membrane to the inner ear directly through the stapes (Figures 8.20a and c).

The malleus and incus first appear in fossil mammals. But from where? Where were the malleus and incus found in the ancestors of mammals? In position, they are homologous with two jawbones—called the articular and quadrate—found in reptiles, amphibians, and early mammals. In fact, the malleus, incus, and stapes of modern mammals still develop, as embryonic tissues, in exactly the same positions in which they are found in adult fossils from the group ancestral to mammals.

Developmentally, the malleus and incus also originate as part of the jaw structure. In early-stage mammal embryos, the cells destined to become the malleus and incus derive from the first gill arch. In fish that have jaws, this gill arch produces the upper and lower mandible. In mammal embryos, these cells create a structure called Meckel's cartilage, which forms the lower jaw. The malleus forms at the posterior end of Meckel's cartilage and then detaches; the incus forms from a nearby structure (Allin 1975). Again we have strong evidence for homology.

This leads us to the second point in our analysis: Examining how the malleus and incus changed through time. How were jawbones transformed into ear ossicles? It is logical to start where we left off: with the ancestors of the mammals. This is a group called the cynodonts. In cynodonts (as in modern amphibians, reptiles, and birds), the jaw joint is formed by the quadrate and articular; the stapes is the only bone directly associated with hearing. The cynodont stapes happened to lie right next to the articular, however (Allin 1975). Examining cynodont fossils through time shows the following changes in the jaw:

- The part of the lower jaw near the hinge became larger, and one of the major muscles responsible for closing the jaw changed its area of attachment from the angular bone to the lower jaw.
- A later cynodont genus, called *Diarthrognathus,* has the ancestral jaw articulation involving the quadrate and articular bones and a derived articulation between the lower and upper jawbones (Figure 8.21a). *Diarthrognathus,* like *Acanthostega,* is an intermediate form. It shows both ancestral and derived states of a character (Colbert and Morales 1991).
- Early mammals have only the lower jaw–upper jaw articulation. The quadrate and articular bones are no longer involved. This is a key step, because the two bones are now free to assume a new function or disappear, depending on the direction of natural selection.
- In later mammals, the quadrate and articular bones articulate with the stapes. Consequently, they are renamed the incus and malleus. They are now located away from the jaw joint and function only in the transmission of sound (Figure 8.21b).

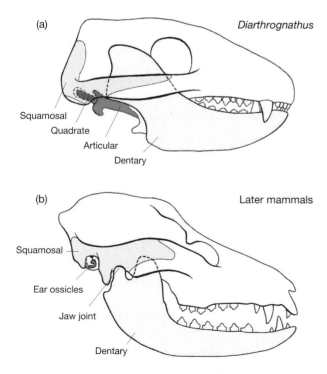

Figure 8.21   Change in mammal jaw articulation through time
The jaw articulation of an early mammal, called *Diarthrognathus,* is shown in (a). The diagram shows that the joint involves both the quadrate and articular and the dentary and squamosal bones. In later mammals, (b), the jaw articulation is made only by the dentary and squamosal. The ear ossicles have moved back and away from the jaw.

What aspects of natural selection help account for all this change? Though traditional views emphasize the importance of improved efficiency in biting and chewing during the remodeling of the jaw (see Manley 1972), Allin argues that improved hearing played a role in favoring mutants with ear ossicles detached from the jaw. His hypothesis is based on the idea that the survival of small, insectivorous, and probably nocturnal early mammals may have depended on an ability to hear high-pitched insect sounds. Because it is hard to hear while chewing, ear ossicles detached from the jaw worked better (Allin 1975).

To summarize, early in mammal evolution three bones changed function, reduced in size, and moved away from an articulation with the jaw. Currently, research on ear evolution is focused on understanding the developmental and genetic mechanisms of these changes (see Rowe 1996; Smith et al. 1997).

The fact that everything evolves from something else is just one reason why an organism's organs, even when clearly adaptive, are often imperfect. Other reasons why adaptations may be imperfect are the subject of the next section.

As a new adaptive trait evolves, some of its components may take on entirely new functions.

8.7 Trade-Offs and Constraints

It is impossible to build a perfect organism. Organismal design reflects a compromise among competing demands.

It is impossible for any population of organisms to evolve optimal solutions to all selective challenges at once. We have mentioned examples of trade-offs in passing. The giraffe's long neck, for example, may be useful for males attempting to obtain a mate, but appears to be inconvenient for drinking. In this section, we explore additional examples of research into factors that limit adaptive evolution. These factors include trade-offs, functional constraints, and lack of genetic variation.

Female Flower Size in a Begonia: A Trade-Off

The tropical plant *Begonia involucrata* is **monoecious**—that is, there are separate male and female flowers on the same plant. The flowers are pollinated by bees. As the bees travel among male flowers gathering pollen, they sometimes also transfer pollen from male flowers to female flowers. The male flowers offer the bees a reward, in the form of the pollen itself. The female flowers offer no reward; instead they get pollinated by deceit (Ågren and Schemske 1991). Not surprisingly, bees make more and longer visits to male flowers than to female flowers.

The female flowers resemble the male flowers in color, shape, and size (Figure 8.22a). This resemblance is presumably adaptive. Given that bees avoid female flowers in favor of male flowers, the rate at which female flowers are visited should depend on the degree to which they resemble male flowers. The ability to attract pollinators should, in turn, influence fitness through female function, because seed set is limited by pollen availability. Doug Schemske and Jon Ågren (1995) sought to distinguish between two hypotheses about the mode of selection imposed by bees on female flower size:

Hypothesis 1: The more closely female flowers resemble male flowers, the more often they are visited by bees. Selection on female flowers is stabilizing, with best phenotype for females identical to the mean phenotype of males (Figure 8.23a, left).

Hypothesis 2: The more closely female flowers resemble the most rewarding male flowers, the more often they will be visited by bees. If larger male flow-

Figure 8.22 *Begonia involucrata* (a) Male (left) and female (right) flowers. The flowers lack true petals. Instead, each has a pair of petaloid sepals. The sepals are white or pinkish. In the center of each flower is a cluster of yellow anthers or stigmas. The stigmas of female flowers resemble the anthers of males. (b) An inflorescence, or stalk bearing many flowers. Each inflorescence makes both male and female flowers. Typically, the male flowers open first, and the female flowers open later. The infloresence shown is unusual in having flowers of both sexes open at once. (Doug W. Schemske, University of Washington, Seattle)

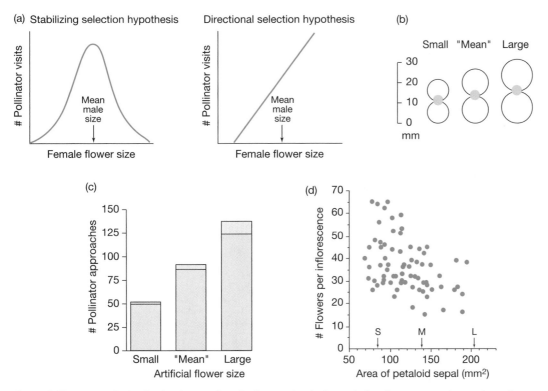

Figure 8.23 **An analysis of selection on female flower size in *Begonia involucrata*** (a) The two hypotheses investgated by Schemske and Ågren (1995). See text for more details. (b) Schemske and Ågren's three size classes of artificial flowers. The "mean" size class is the same size as the mean size of natural male flowers. (c) Pollinator preference as a function of flower size. The blue bars represent the number of bees that approached the artificial flowers; the orange bars represent the number of pollinators that actually visited the artificial flowers. Schemske and Ågren placed equal numbers of each size flower in the forest, but larger flowers attracted significantly more approaches and significantly more visits from bees. (d) Number of female flowers per inflorescence as a function of flower size. There is a statistically significant trade-off between flower size and flower number. From Schemske and Ågren (1995). Copyright © 1995, Evolution. Reprinted by permission of Evolution.

ers offer bigger rewards, then selection on female flowers is directional, with bigger flowers always favored over smaller flowers (Figure 8.23a, right).

Schemske and Ågren made artificial flowers of three different sizes (Figure 8.23b), arrayed equal numbers of each in the forest, and watched to see how often bees approached and visited them. The results were clear: The larger the flower, the more bee approaches and visits it attracted (Figure 8.23c). Selection by bees on female flowers is strongly directional.

Taken at face value, Schemske and Ågren's results suggest that female flower size in *Begonia involucrata* is maladaptive. Selection by bees favors larger flowers, yet the female flowers are no bigger than the male flowers. One solution to this paradox is that *B. involucrata* simply lacks genetic variation for female flowers that are substantially larger than male flowers. Schemske and Ågren have no direct evidence on this suggestion; *B. involucrata* is a perennial that takes a long time to reach sexual maturity, so quantitative genetic experiments are difficult to do.

> *Resources devoted to one body part or function may be resources stolen from another part or function.*

Another solution to the paradox is that focusing on individual female flowers gives us too narrow a view of selection. Schemske and Ågren expanded their focus from individual flowers to inflorescences (Figure 8.22b). The researchers measured the size and number of the female flowers on 74 inflorescences. They discovered a trade-off: The larger the female flowers on an inflorescence, the fewer flowers there are (Figure 8.23d). Such a trade-off makes intuitive sense. If an individual plant has a finite supply of energy and nutrients to invest in flowers, it can slice this pie into a few large pieces or many small pieces, but not into many large pieces. Inflorescences with more flowers may be favored by selection for two reasons. First, bees may be more attracted to inflorescences with more flowers. Second, more female flowers means greater potential seed production. Schemske and Ågren hypothesize that female flower size in *B. involucrata* has been determined, at least in part, by two opposing forces: directional selection for larger flowers, and the trade-off between flower size and number.

Flower Color Change in a Fuchsia: A Constraint

Fuchsia excorticata is a bird-pollinated tree endemic to New Zealand (Delph and Lively 1989). Its flowers hang downward like bells (Figure 8.24a and photo on page 251). The ovary is at the top of the bell. The body of the bell consists of the hypanthium,

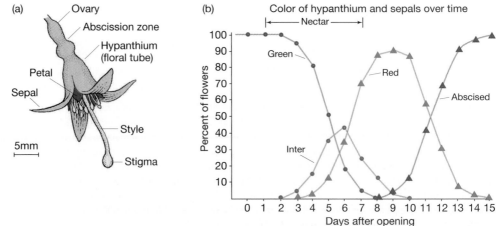

Figure 8.24 **Flower color change in *Fuchsia excorticata*** (a) A *Fuchsia excorticata* flower. (b) The horizontal axis shows flower age, in days after opening. The vertical axis and graph lines show the percentage of flowers that are in each color phase at each age. From Delph and Lively (1989). Copyright © 1989, Evolution. Reprinted by permission of Evolution.

or floral tube, and the sepals. The style resembles an elongated clapper. It is surrounded by shorter stamens and a set of reduced petals.

The hypanthium and sepals are the most conspicously showy parts of the flower.
They remain green for about 5.5 days after the flower opens, then begin to turn
red (Figure 8.24b). The transition from green to red lasts about 1.5 days, at the end
of which the hypanthium and sepals are fully red. The red flowers remain on the
tree for about 5 days. The red flowers then separate from the ovary at the abscission zone and drop from the tree.

Pollination occurs during the green phase and into the intermediate phase, but
it is complete by the time the flowers are fully red. The flowers produce nectar on
days 1 through 7 (Figure 8.24b). Most flowers have exported more than 90% of
their pollen by the end of that time. The stigmas are receptive to pollen at least until
the second day of the fully red phase, but rarely does pollen arriving after the first
day of the red phase fertilize eggs. Not surprisingly, bellbirds and other avian pollinators strongly prefer green flowers, and virtually ignore nectarless red flowers
(Delph and Lively 1985).

Why do the flowers of this tree change color? A general answer, supported by
research in a variety of plants, is that color change serves as a cue to pollinators,
alerting them that the flowers are no longer offering a reward (see Delph and
Lively [1989] for a review). By paying attention to this cue, pollinators can increase
their foraging efficiency; they do not waste time looking for nonexistent rewards.
Individual plants benefit in return, because when pollinators forage efficiently
they also transfer pollen efficiently. They do not deposit viable pollen on unreceptive stigmas, and they do not deposit nonviable pollen on receptive stigmas.

This answer is only partially satisfying, however. Why does *F. excorticata* not just
drop its flowers immediately after pollination is complete? Dropping the flowers
would give an unambiguous signal to pollinators that a reward is no longer being
offered, and it would be metabolically much cheaper than maintaining the red
flowers for several days. Retention of the flowers beyond the time of pollination
seems maladaptive.

Lynda Delph and Curtis Lively (1989) consider two hypotheses for why *F. excorticata* keeps its flowers (and changes them to red) instead of just dropping them.
The first is that red flowers may still attract pollinators to the tree displaying them,
if not to the red flowers themselves. Once drawn to the tree, pollinators could
then forage on the green flowers still present. Thus, retention of the red flowers
could increase the overall pollination efficiency of the individual tree retaining
them. If this hypothesis is correct, then green flowers surrounded by red flowers
should receive more pollen than green flowers not surrounded. Delph and Lively tested this prediction by removing red flowers from some trees but not from others, and from some branches within trees but not from others. The researchers
then compared the amount of pollen deposited on green flowers in red-free trees
and branches versus red-retaining trees and branches. They found no significant
differences. The pollinator-attraction hypothesis does not explain the retention of
the red flowers in *F. excorticata*.

The second hypothesis Delph and Lively consider is that a physiological constraint prevents *F. excorticata* from dropping its flowers any sooner than it does. This
physiological constraint is the growth of pollen tubes. After a pollen grain lands
on a stigma, the pollen germinates. The germinated pollen grain grows a tube
down through the style to the ovary. The pollen grain's two sperm travel through

Table 8.3 **Pollen tube growth in *Fuchsia excorticata***				
Days since pollination	**1**	**2**	**3**	**4**
Percentage of 10 flowers with pollen tubes in ovary	0	20%	100%	100%

Source: After Delph and Lively (1989).

Traits or behaviors that would appear to be adaptive may be physiologically or mechanically impossible.

this tube to the ovary, where one of the sperm fertilizes an egg. The growth of pollen tubes takes time, especially in a plant like *F. excorticata,* which has long styles. If the plant were to drop its flowers before the pollen tubes had time to reach the ovaries, the result would be the same as if the flowers had never been pollinated at all. Delph and Lively pollinated 40 flowers by hand. After 24 hours, they plucked 10 of the flowers, dissected them, and examined them under a microscope to see whether the pollen tubes had reached the ovary. After 48 hours, they plucked and dissected 10 more flowers, and so on. The results appear in Table 8.3. It takes about three days for the pollen tubes to reach the ovary.

This result is consistent with the physiological constraint hypothesis. *F. excorticata* cannot start the process of dropping a flower until about three days after the flower is finished receiving pollen. Dropping a flower involves forming a structure called an abscission zone between the ovary and the flower (Figure 8.24a). The abscission zone consists of several layers of cells that form a division between the ovary and the flower. In *F. excorticata,* the growth of the abscission layer takes at least 1.5 days. The plant is therefore constrained to retain its flowers for at least 4.5 days after pollination ends. In fact, the plant retains its flowers for about 5 days. Delph and Lively suggest that flower color change in *F. excorticata* is an adaptation that evolved to compensate for the physiological constraints that necessitate flower retention. Given that the plant had to retain its flowers, selection favored individuals offering cues that allow their pollinators to distinguish the receptive versus unreceptive flowers on their branches. The pollinators deposit the incoming pollen onto receptive stigmas only, and they carry away only outgoing pollen that is viable.

Host Shifts in an Herbivorous Beetle: Constrained by Lack of Genetic Variation?

In several previous chapters, we have made the point that genetic variation is the raw material for evolution by natural selection. Because natural selection is the process that produces adaptations, genetic variation is also the raw material from which adaptations are molded. Conversely, populations of organisms may be prevented from evolving particular adaptations simply because they lack the necessary genetic variation to do so.

Here is an extreme example: Pigs have not evolved the ability to fly. We can imagine that flying might well be adaptive for pigs. It would enable them to escape from predators, and to travel farther in search of their favorite foods. Pigs do not fly, however, because the vertebrate developmental program lacks genetic variation for the growth of both a trotter and a wing from the same shoulder. Other vertebrates have evolved the ability to fly, of course. But in bats and in birds, the developmental program has been modified to convert the entire forelimb from a leg to a wing; in neither group does an entirely new limb sprout from the body. Too bad for pigs.

Pig flight makes a vivid example, but in the end it is a trivial one. The wished-for adaptation is too unrealistic. Douglas Futuyma and colleagues have sought to determine whether lack of genetic variation has constrained adaptation in a more realistic and meaningful example (Funk et al. 1995; Futuyma et al. 1995; references therein). Futuyma and colleagues studied host plant use by herbivorous leaf beetles in the genus *Ophraella*. Among these small beetles, each species feeds, as larvae and adults, on the leaves of one or a few closely related species of composites (plants in the sunflower family, the Asteraceae). Each species of host plant makes a unique mixture of toxic chemicals that serve as defenses against herbivores. For the beetles, the ability to live on a particular species of host plant is a complex adaptation that includes the ability to recognize the plant as an appropriate place to feed and lay eggs, as well as the ability to detoxify the plant's chemical defenses.

An estimate of the phylogeny for 12 species of leaf beetle appears in Figure 8.25. The figure also lists the host plant for each beetle species. The evolutionary history of the beetle genus has included several shifts from one host plant to another. Four of the host shifts were among relatively distantly related plant species: They involved switches from a plant in one tribe of the Asteraceae to a plant in another tribe. These shifts are indicated in the figure by changes in the shading of the phylogeny. Other shifts involved movement to a new host in the same genus as the ancestral host, or in a genus closely related to that of the ancestral host.

Each combination of a beetle species and the host plant used by one of its relatives represents a plausible evolutionary scenario for a host shift that might have happened, but did not. For example, the beetle *Ophraella arctica* might have switched to the host *Iva axillaris*. Futuyma and colleagues have attempted to elucidate why some host shifts have actually happened while others have remained hypothetical. Here are two hypotheses:

Hypothesis 1: All host shifts are genetically possible. That is, every beetle species harbors sufficient genetic varation in its feeding and detoxifying mechanisms to allow at least some individuals to feed and survive on every potential host

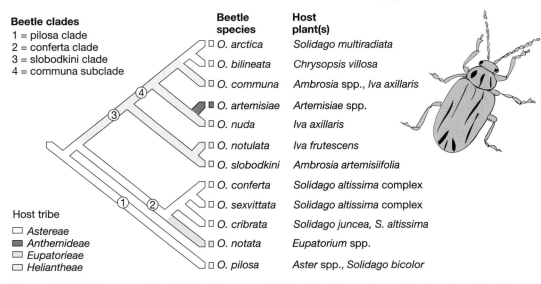

Figure 8.25   Phylogeny of the leaf beetles, genus *Ophraella*   The numbers on the branches define the major branches (clades) of the beetle evolutionary tree. The shading of branches indicates the tribes of host species. The evolutionary history of the beetle genus has included four host shifts across tribes. From Futuyma et al. (1995). Copyright © 1995, Evolution. Reprinted by permission of Evolution.

species. If a few individuals can feed and survive, they can be the founders for a new population of beetles that will evolve to become well-adapted to the new host. Because all host shifts are genetically possible, the pattern of actual host shifts has been determined by ecological factors and by chance. Ecological factors might include the abundance of the various host species within the geographic ranges of the beetle species, and the predators and competitors associated with each host species.

Hypothesis 2: Most host shifts are genetically impossible. That is, most beetle species lack sufficient genetic variation in their feeding and detoxifying mechanisms to allow any individuals to feed and survive on any but a few of the potential host species. The pattern of actual host shifts has been largely determined by what was genetically possible. Genetically possible host shifts have happened; genetically impossible host shifts have not.

We have presented these hypotheses as mutually exclusive. In fact, the truth is almost certainly that the actual pattern of host shifts has resulted from a mixture of genetic constraints, ecological factors, and chance. What Futuyma and colleagues were looking for was concrete evidence that genetic constraints have been at least part of the picture.

Futuyma and colleagues used a quantitative genetic approach (see Chapter 7) to determine how much genetic variation the beetles harbor for feeding and surviving on other potential hosts. The researchers examined various combinations of four of the beetle species listed in Figure 8.25 with six of the host plants. Their tests revealed that there is little genetic variation in most beetle species for feed-

Table 8.4 Summary of tests for genetic variation in larval or adult feeding on potential host plants

(a) Tests for genetic variation in larval or adult feeding, by relationship among host plants

Beetle tested for feeding on a plant that is . . .	Genetic variation?	
	Yes	No
. . . in the same tribe as the beetle's actual host	7	1
. . . in a different tribe than the beetle's actual host	14	17

Conclusion: Genetic variation for feeding is more likely to be found when a beetle is tested on a potential host that is closely related to its actual host

(b) Tests for genetic variation in larval or adult feeding, by relationship among beetles

Beetle tested for feeding on a plant that is . . .	Genetic variation?	
	Yes	No
. . . the host of a beetle in the same major clade	12	4
. . . the host of a beetle in a different major clade	9	14

Conclusion: Genetic variation for feeding is more likely to be found when a beetle is tested on a potential host that is the actual host of a closely related beetle

Source: From Table 7 in Futuyma et al. (1995). Copyright © 1995, Evolution. Reprinted by permission of Evolution.

ing and surviving on most potential host species. In 18 of 39 tests of whether larvae or adults of a beetle species would recognize and feed on a potential host plant, the researchers found no evidence of genetic variation for feeding. In 14 of 16 tests of whether larvae could survive on a potential host plant, the researchers found no evidence of genetic variation for survival.

These results suggest that hypothesis 2 is at least partially correct. Many otherwise-plausible host shifts appear to be genetically impossible. Futuyma and colleagues performed an additional test of hypothesis 2 by looking for patterns in their data on genetic variation for larval and adult feeding. If hypothesis 2 is correct, then a beetle species is more likely to show genetic variation for feeding on a potential new host if the new host is a close relative of the beetle's present host. Futuyma et al.'s data confirm this prediction (Table 8.4a). Likewise, if hypothesis 2 is correct, then a beetle species is more likely to show genetic variation for feeding on a potential new host if the new host is the actual host of one of the beetles' close relatives. Futuyma et al.'s data also confirm this prediction (Table 8.4b). Futuyma and colleagues conclude that hypothesis 2 is at least partially correct. The history of host shifts in the beetle genus *Ophraella* has been constrained by the availability of genetic variation for evolutionary change.

Populations sometimes lack the genetic variation that would provide the raw material to evolve particular adaptations.

8.8 Strategies for Asking Interesting Questions

We began this chapter with a review of approaches evolutionary biologists use when testing hypotheses about organismal form and function. However, testing a hypothesis is the second half of a good research project. The first half is formulating a hypothesis in the first place. Formulating hypotheses worthy of testing means asking interesting questions, then making educated guesses about the answers. We close the chapter with a brief list of strategies for asking good questions about evolution:

- Study natural history. Descriptive studies can lead to the discovery of new patterns that need explanation. Some things in nature just leap out and demand explanation, like the neck of the giraffe, or the wing-waving display of *Zonosemata*. Some of the most compelling science happens when a researcher simply picks an organism and sets out to learn about it.
- Question conventional wisdom. It is often untested. What makes Simmons and Scheeper's work on giraffes so captivating that it undermines an adaptive scenario long accepted as fact.
- Question the assumptions underlying a popular hypothesis or research technique. Felsenstein's development of an improved method of comparative analysis grew out of the recognition that the traditional approach to comparative research was violating its own assumptions.
- Draw analogies that transfer questions from field to field, or taxon to taxon. If fruit bats and flying foxes that evolve larger group sizes also evolve larger testes for their body size, might not the same be true of other kinds of animals?
- Ask why not. The studies reviewed on trade-offs and constraints were motivated by researchers who thought their study organisms were failing to do something that might be adaptive.

Learning how to ask good questions is as imporant as learning how to answer them.

Summary

Among the major activities of evolutionary biology is analyzing the form and function of organisms to determine whether and why particular traits are adaptive. To establish that a trait is adaptive, researchers must formulate hypotheses about how the trait is used and why individuals possessing the trait have higher fitness than individuals lacking it. Then, because no hypothesis should be accepted simply because it is plausible, researchers must put their hypotheses to test. Researchers test hypotheses by using them to make predictions, then collecting data to determine whether the predictions are correct.

Researchers use a variety of approaches in collecting data to test hypotheses. The most powerful method is the controlled experiment. Controlled experiments involve groups of organisms that are identical but for a single variable of interest. The experimental variable can then be confidently identified as the cause of any differences in survival and reproductive success among the groups. When experiments are impractical, careful observational studies can yield data valuable for testing hypotheses. Finally, comparisons among species can be used to confirm or refute predictions, so long as researchers take into account the shared evolutionary history of the species under study.

When analyzing adaptations, we do well to keep in mind that organisms are complicated. Individuals may be phenotypically plastic, so that genetically identical individuals reared in different environments have different phenotypes. The function of a particular trait may change over evolutionary time, and it may reflect a compromise among competing environmental or physiological demands. Finally, populations may simply lack the genetic variation required to become perfectly adapted to their environments. These and other complications are the subjects of current research by evolutionary biologists.

Questions

1. Why was it important that Greene and colleagues tested tephritid flies whose wings had been cut off and then glued back on?

2. In Huey et al.'s experiment, snakes often chose thick rocks despite the associated risk of being too cool. Outline two hypotheses for why snakes sometimes choose thick rocks. Are your hypotheses testable? Do both hypotheses assume that the behavioral trait of choosing thick rocks is adaptive?

3. Geckos are unusual lizards in that they are active at night instead of during the day. Describe the difficulties a gecko would face in trying to use its behavior to regulate its temperature at night. Would you predict that geckos have an optimal temperature for sprinting that is the same, higher, or lower than that of a typical diurnal lizard? Huey et al. (1989a) found that the geckos they studied had optimal temperatures that are the same as those of typical diurnal lizards (a finding in conflict with the researchers' own hypothesis). Can you think of an explanation?

4. Suppose that fish were introduced into Lake Citadelpark, one of the lakes De Meester stdied. What do you predict will happen to phenotypic plasticity in the *Daphnia* of Lake Citadelpark? Outline the observations you would need to make to test your prediction.

5. In reconstructing the history of the mammalian ear, we described both fossil evidence and embryological evidence. In many cases, we do not have the luxury of having both sets of evidence. Examples are the evolution of bat wings, for which there is embryological data but few fossils, and the evolution of *Triceratops* neck frills, for which there is fossil data but little embryology. In your opinion, is it necessary to have both kinds of data to determine the evolutionary history of a trait? Would fossil data alone (or embryological data alone) be enough to convince you that mammal middle-ear bones evolved from the hyomandibula, quadrate, and articular of early vertebrates? What would you think if the two data sets had contradicted each other?

6. Early tetrapods apparently did not yet have tympanic membranes, for there is no obvious place on their skulls where a tympanic membrane would have fit. In addition, the stapes in these animals was a relatively heavy bone and thus was unlikely to transmit high-frequency vibrations well. Yet the stapes was clearly in a good position to transmit vibrations to the inner ear. Early tetrapods are thought to have had a sprawled body posture, with their large heads often resting directly on the ground. With this in mind, speculate on what sort of "hearing" abilities *Acanthostega* might have had. Are your ideas testable?

7. Think about the costs and benefits of being a certain body size. For example, a mouse can easily survive a 30-foot

fall. A human falling 30 feet would probably be injured, and an elephant falling 30 feet would probably be killed. Finally, a recent study of the bone strength of *Tyrannosaurus rex* revealed that if a fast-running *T. rex* ever tripped, it would probably die (Farlow, Smith, and Robinson 1995). Given these costs, why has large body size ever evolved? Can you think of some costs of small body size? How would you test your ideas?

8. Imagine you are an explorer who has just discovered two previously unknown large islands. Each island has a population of a species of shrub unknown elsewhere. On Island A, the shrubs have high concentrations of certain poisonous chemicals in their leaves. On Island B, the shrubs have nonpoisonous, edible leaves. The islands differ in many ways—for instance, Island A has less rainfall and a colder winter than Island B, and it has some plant-eating insects that are not found on Island B. Island A also has a large population of muntjacs, a small tropical deer that loves to eat shrubs. You suspect the muntjacs have been the selective force that has caused evolution of leaf toxins. How could you test this hypothesis? What alternative hypotheses can you think of? What data would disprove your hypothesis, and what data would disprove the other hypotheses?

9. A common challenge in paleontology is to identify the selective challenge that caused a new trait to spread through a lineage, many millions of years ago. One classic example is the evolution of high-crowned teeth in grazing mammals. These teeth provide extra protection against abrasive particles in the diet, such as the silica particles in grass blades. High-crowned teeth have evolved independently in dozens of terrestrial mammals since the Miocene, including antelope, kangaroos, horses, and rabbits. Here are two hypotheses for the evolution of high-crowned teeth:

   a. It might be linked to the evolution of grasses.
   b. It might be linked to an increased amount of grit in the diet.

   Are these two hypotheses testable? What evidence would help distinguish them? (For more information on this example, see MacFadden 1997, MacFadden et al. 1999.)

10. Consider skin color in humans. Does this trait show genetic variation? Phenotypic plasticity? Genotype-by-environment interaction? Give examples documenting each phenomenon. Could phenotypic plasticity for skin color evolve in human populations? How?

11. The example on *Begonias* (Section 8.7) illustrated that organisms are frequently caught between opposing forces of selection. Each of the following examples also illustrates a tug-of-war between several forces of natural selection. For each example, hypothesize about what selective forces may maintain the trait described, and what selective forces may oppose it.

    A male moose grows new antlers, made of bone, each year.
    Douglas-fir trees often grow to over 60 feet tall.
    A termite's gut is full of cellulose-digesting microorganisms.
    Maple trees lose all of their leaves in the autumn.
    A male moth has huge antennae, which can detect female pheromones.
    A barnacle attaches itself permanently to a rock when it matures.

12. Schemske and Ågren (1995) used artificial flowers instead of real flowers in their experiment (Section 8.7). What were the advantages of using artificial flowers? (There are at least two important ones.) What were the disadvantages?

13. The popular media often presents evolution as being a predictable process with a definite goal. For instance, in one "Star Trek: Voyager" episode, the captain instructs the ship's computer to extrapolate the "probable course" of evolution of hadrosaurs (a bipedal dinosaur), if hadrosaurs had been removed from Earth before the K–T extinction and allowed to evolve on another planet. What information about the hadrosaurs' new environment would have been useful for developing the best possible prediction? Given what you know about various factors that can impose natural selection on an organism, and what you know about mutation, migration and genetic drift, do you think it is theoretically possible to accurately predict the long-term course of evolution?

14. An exercise used in some graduate programs is to have students list 20 questions that they would like to answer. Groups of students then discuss the questions and help each other sort out which would be the most interesting to pursue. There are many criteria for deciding that a question is interesting. Is it new? Does it address a large or otherwise important issue? Would pursuing it lead to other questions? Is it feasible, or would development of a new technique make it feasible? Try this exercise yourself.

Exploring the Literature

15. An important aspect of evaluating scientific papers is to consider other explanations for the data that the authors might have overlooked. See if you can think of alternate explanations for the data presented in the following papers:

Benkman, C. W., and A. K. Lindholm. 1991. The advantages and evolution of a morphological novelty. *Nature* 349: 519–520.

Soler, M., and Møller, A. P. 1990. Duration of sympatry and coevolution between the great spotted cuckoo and its magpie host. *Nature* 343: 748–750.

Finally, see the following review of Soler and Møller's work for an example of how scientific criticism can result in better science by all involved.

Lotem, A., and Rothstein, S. I. 1995. Cuckoo-host coevolution: From snapshots of an arms race to the documentation of microevolution. *Trends in Ecology and Evolution* 10: 436–437.

16. The fossil record has traditionally been the major source of data for asking questions about evolutionary history. Unfortunately, the fossil record is incomplete, and cannot answer all questions. Newer techniques of molecular and genetic analysis are sometimes seen as more useful. As a result, paleontologists have recently had to defend the "usefulness" of the fossil record. Read the following papers to explore this issue more thoroughly:

Foote, M. 1996. Perspective: Evolutionary patterns in the fossil record. *Evolution* 50: 1–11.

Benton, M. J., and G. W. Storrs. 1994. Testing the quality of the fossil record: Paleontological knowledge is improving. *Geology* 22: 111–114.

Norell, M. A., and M. J. Novacek. 1992. The fossil record and evolution: Comparing cladistic and paleontologic evidence for vertebrate history. *Science* 255: 1690–1693.

17. Male sticklebacks sometimes steal eggs from other males' nests to rear as their own. Sievert Rohwer suggested that egg stealing is a courtship strategy. Males often eat eggs out of their own nests, in effect robbing the reproductive investment made by their mates and using the proceeds to fund their own reproductive activities. Females, in consequence, should prefer to lay eggs in nests already containing the eggs of other females, thereby reducing the risk to their own. A female preference for nests already containing eggs would mean that males without eggs in their nests could increase their attractiveness by stealing eggs from other males. See

Rohwer, S. 1978. Parent cannibalism of offspring and egg raiding as a courtship strategy. *American Naturalist* 112: 429–440.

For a related phenomenon in birds, see

Gori, D. F., S. Rohwer, and J. Caselle. 1996. Accepting unrelated broods helps replacement male yellow-headed blackbirds attract females. *Behavioral Ecology* 7: 49–54.

18. For a dramatic example of phenotypic plasticity in which an herbivorous insect uses the chemical defenses of its host as a cue for the development of defenses against its predators, see

Greene, Erick. 1989. A diet-induced developmental polymorphism in a caterpillar. *Science* 243: 643–646.

19. Among the challenges faced by parasites is moving from one host to another. This challenge is a particularly potent agent of selection for parasites in which every individual must spend different parts of its life cycle in different hosts. What adaptations might you expect to find in parasites to facilitate dispersal from host to host? For dramatic examples in which parasites manipulate their hosts' behavior or appearance, see

Tierney, J. F., F. A. Huntingford, and D. W. T. Crompton. 1993. The relationship between infectivity of *Schistocephalus solidus* (Cestoda) and antipredator behavior of its intermediate host, the three-spined stickleback, *Gasterosteus aculeatus*. *Animal Behavior* 46: 603–605.

Lafferty, K. D., and A. K. Morris. 1996. Altered behavior of parasitized killifish increases susceptibility to predation by bird final hosts. *Ecology* 77: 1390–1397.

Bakker, T. C. M., D. Mazzi, and S. Zala. 1997. Parasite-induced changes in behavior and color make *Gammarus pulex* more prone to fish predation. *Ecology* 78: 1098–1104.

20. Brown-headed cowbirds (*Molothrus ater*) lay their eggs in other birds' nests, a behavior called nest parasitism. When this strategy succeeds, the host birds accept the cowbird egg as one of their own and rear the cowbird chick. When the strategy fails, the host birds recognize the cowbird egg as an imposter and eject it from the nest. Why do any host species accept cowbird eggs in their nests? Given the obvious costs involved in rearing a chick of another species, acceptance seems maladaptive. Evolutionary biologists have proposed two competing hypotheses to explain why some host species accept cowbird eggs. The evolutionary lag hypothesis posits that species that accept cowbird eggs do so simply because they have not yet evolved ejection behavior. Either the host species lack genetic variation that would allow them to evolve ejection behavior, or the host species have been exposed to cowbird nest parasitism only recently and therefore have not had sufficient time for such behavior to evolve. The evolutionary equilibrium hypothesis posits that host species that accept cowbird eggs do so because they face a fundamental mechanical constraint: Their bills are too small to allow them to grasp a cowbird egg, and if they tried to puncture the cowbird egg they would destroy too many of their own eggs in the process. Given this constraint, host species have evolved a strategy that makes the best of a bad situation. Think about how you would test each of the competing hypotheses. Then, see

Rohwer, S., and C. D. Spaw. 1988. Evolutionary lag versus bill-size constraints: A comparative study of the acceptance of cowbird eggs by old hosts. *Evolutionary Ecology* 1988: 27–36.

Rohwer, S., C. D. Spaw, and E. Røskaft. 1989. Costs to

northern orioles of puncture-ejecting parasitic cow-bird eggs from their nests. *Auk* 106: 734–738.

Røskaft, E., S. Rohwer, and C. D. Spaw. 1993. Cost of puncture ejection compared with costs of rearing cowbird chicks for northern orioles. *Ornis Scandinavica* 24: 28–32.

Sealy, S. G. 1996. Evolution of host defenses against brood parasitism: Implications of puncture-ejection by a small passerine. *Auk* 113: 346–355.

Given that some host species eject cowbird eggs by first puncturing the egg, then lifting it out of the nest, what adapatations would you expect to find in cowbird eggs? Would these adaptations carry any costs? See

Spaw, C. D., and S. Rohwer. 1987. A comparative study of eggshell thickness in cowbirds and other passerines. *Condor* 89: 307–318.

Picman, J. 1997. Are cowbird eggs unusually strong from the inside? *Auk* 114: 66–73.

21. For additional recent examples in which evolutionary biologists used a comparative approach employing independent contrasts to address interesting questions, see

Blumstein, D. T., and K. B. Armitage. 1998. Life history consequences of social complexity: A comparative study of ground-dwelling sciurids. *Behavioral Ecology* 9: 8–19.

Iwaniuk, A. N., S. M. Pellis, and I. Q. Whishaw. 1999. Brain size is not correlated with forelimb dexterity in fissiped carnivores (Carnivora): A comparative test of the principle of proper mass. *Brain, Behavior, and Evolution* 54: 167–180.

Citations

Ågren, J., and D. W. Schemske. 1991. Pollination by deceit in a Neotropical monoecious herb, *Begonia involucrata. Biotropica* 23: 235–241.

Allin, E. F. 1975. Evolution of the mammalian middle ear. *Journal of Morphology* 147: 403–438.

Carroll, R. L. 1988. *Vertebrate Paleontology and Evolution.* New York: W.H. Freeman.

Clack, J. A. 1983. The stapes of the Coal Measures embolomere *Pholiderpeton scutigerum* Huxley (Amphibia: Anthracosauria) and otic evolution in early tetrapods. *Zoological Journal of the Linnean Society* 79: 121–148.

Clack, J. A. 1989. Discovery of the earliest-known tetrapod stapes. *Nature* 342: 425–427.

Clack, J. A. 1994. Earliest known tetrapod braincase and the evolution of the stapes and fenestra ovalis. *Nature* 369: 392–394.

Coates, M. I., and J. A. Clack. 1991. Fish-like gills and breathing in the earliest known tetrapod. *Nature* 352: 234–236.

Colbert, E. H., and M. Morales. 1991. *Evolution of the Vertebrates.* New York: Wiley-Liss.

Delph, L. F., and C. M. Lively. 1985. Pollinator visits to floral colour phases of *Fuchsia excorticata. New Zealand Journal of Zoology* 12: 599–603.

Delph, L. F., and C. M. Lively. 1989. The evolution of floral color change: Pollinator attraction versus physiological constraints in *Fuchsia excorticata. Evolution* 43: 1252–1262.

De Meester, L. 1996. Evolutionary potential and local genetic differentiation in a phenotypically plastic trait of a cyclical parthenogen, *Daphnia magna. Evolution* 50: 1293–1298.

Farlow, J. D., M. B. Smith, and J. M. Robinson. 1995. Body mass, bone "strength indicator," and cursorial potential of *Tyrannosaurus rex. Journal of Vertebrate Paleontology* 15: 713–725.

Felsenstein, J. 1985. Phylogenies and the comparative method. *The American Naturalist* 125: 1–15.

Funk, D. J., D. J. Futuyma, G. Ortí, and A. Meyer. 1995. A history of host associations and evolutionary diversification for *Ophraella* (Coleoptera: Chrysomelidae): New evidence from mitochondrial DNA. *Evolution* 49: 1008–1017.

Futuyma, D. J., M. C. Keese, and D. J. Funk. 1995. Genetic constraints on macroevolution: The evolution of host affiliation in the leaf beetle genus *Ophraella. Evolution* 49: 797–809.

Garland, T., Jr., and S. C. Adolph. 1994. Why not do two-species comparative studies: Limitations on inferring adaptation. *Physiological Zoology* 67: 797–828.

Garland, T., Jr., P. E. Midford, and A. R. Ives. 1999. An introduction to phylogenetically based statistical methods, with a new method for confidence intervals on ancestral values. *American Zoologist* 39: 374–388.

Gilbert, J. J. 1966. Rotifer ecology and embryological induction. *Science* 151: 1234–1237.

Gilbert, J. J. 1980. Further observations on developmental polymorphism and its evolution in the rotifer *Brachinonus calyciflorus. Freshwater Biology* 10: 281–294.

Gould, S. J. 1993. *Eight Little Piggies.* New York: W. W. Norton.

Gould, S. J., and R. C. Lewontin. 1979. The spandrels of San Marco and the Panglossian paradigm: A critique of the adaptationist programme. *Proceedings of the Royal Society of London,* Series B 205: 581–598.

Greene, E., L. J. Orsak, and D. W. Whitman. 1987. A tephritid fly mimics the territorial displays of its jumping spider predators. *Science* 236: 310–312.

Hosken, D. J. 1998. Testes mass in megachiropteran bats varies in accordance with sperm competition theory. *Behavioral Ecology and Sociobiology* 44: 169–177.

Huey, R. B., and J. G. Kingsolver. 1989. Evolution of thermal sensitivity of ectotherm performance. *Trends in Ecology and Evolution* 4: 131–135.

Huey, R. B., P. H. Niewiarowski, J. Kaufmann, and J. C. Herron. 1989a. Thermal biology of nocturnal ectotherms: Is sprint performance of geckos maximal at low body temperatures? *Physiological Zoology* 62: 488–504.

Huey, R. B., C. R. Peterson, S. J. Arnold, and W. P. Porter. 1989b. Hot rocks and not-so-hot rocks: Retreat-site selection by garter snakes and its thermal consequences. *Ecology* 70: 931–944.

Johnson, T. P., and A. F. Bennett. 1995. The thermal acclimation of burst escape performance in fish: An integrated study of molecular and cellular physiology and organismal performance. *Journal of Experimental Biology* 198: 2165–2175.

Lauder, G. V., R. B. Huey, et al. 1995. Systematics and the study of organismal form and function. *BioScience* 45: 696–704.

Lombard, R. E., and J. R. Bolt. 1979. Evolution of the tetrapod ear: An analysis and reinterpretation. *Biological Journal of the Linnean Society* 11: 19–76.

Manley, G. A. 1972. A review of some current concepts of the functional evolution of the ear in terrestrial vertebrates. *Evolution* 26: 608–621.

Martins, E. P., and T. Garland, Jr. 1991. Phylogenetic analyses of the correlated evolution of continuous characters: A simulation study. *Evolution* 45: 534–557.

Mather, M. H., and B. D. Roitberg. 1987. A sheep in wolf's clothing: Tephritid flies mimic spider predators. *Science* 236: 308–310.

Mayr, E. 1983. How to carry out the adaptationist program? *American Naturalist* 121: 324–334.

MacFadden, B. J. 1997. Origin and evolution of the grazing guild in New World terrestrial mammals. *Trends in Ecology and Evolution* 12: 182–187.

MacFadden, B. J., N. Solounias, and T. E. Cerling. 1999. Ancient diets, ecology, and extinction of 5-million-year-old horses from Florida. *Science* 283: 824–827.

Pratt, D. M., and V. H. Anderson. 1985. Giraffe social behavior. *Journal of Natural History* 19: 771–781.

Rowe, T. 1996. Coevolution of the mammalian middle ear and neocortex. *Science* 273: 651–654.

Schemske, D. W., and J. Ågren. 1995. Deceit pollination and selection on female flower size in *Begonia involucrata:* An experimental approach. *Evolution* 49: 207–214.

Simmons, R. E., and L. Scheepers. 1996. Winning by a neck: Sexual selection in the evolution of the giraffe. *American Naturalist* 148: 771–786.

Smith, K. K., A. F. H. van Nievelt, and T. Rowe. 1997. Comparative rates of development in *Monodelphis* and *Didelphis. Science* 275: 683–684.

Young, T. P., and L. A. Isbell. 1991. Sex differences in giraffe feeding ecology: Energetic and social constraints. *Ethology* 87: 79–89.

CHAPTER 9

Sexual Selection

Marine iguanas of the Galápagos. The red iguana is a large adult male. He is surrounded by smaller males and females. (Martin Wikelski, University of Illinois at Urbana-Champaign)

MALES AND FEMALES ARE OFTEN STRIKINGLY DIFFERENT IN SIZE, APPEARANCE, and behavior. In marine iguanas, for example, the sexes differ in body mass by a factor of two. The males are larger, and become intensely territorial during the breeding season. The females are smaller, and remain gregarious throughout the year. In long-tailed widow birds, adults of opposite sex have plumages so distinct that it would be easy to mistake them for different species. The males are jet black, carry tailfeathers several times the length of their own bodies, and have red and yellow shoulder patches. The females are colored a cryptic brown, with short tail feathers and no shoulder patches. In gray tree frogs, the males have dark throats and produce a melodious call. The females have white throats and are silent. In stalk-eyed flies, both sexes wear their eyes on the ends of long thin stalks, but males have longer eyestalks than females. In some species of pipefish, the females have blue stripes and skin folds on their bellies, which the males lack. The photos of males and females in Figure 9.1 provide additional examples.

In humans, too, females and males are conspicuously different. The differences include not just the obvious and essential ones like our genitalia and reproductive organs, but also facial structure, vocal tone, distribution of body fat and body hair, and body size. The difference in body size between women and men is documented in Figure 9.2.

(a) Red deer

(b) Guppies

(c) Golden toads

Figure 9.1 **The differences between males and females (the sexual dimorphism) in red deer (*Cervus elaphus*), guppies (*Poecilia reticulata*), and golden toads (*Bufo periglenes*)** In (a), the male is on the left; in (b) and (c), the male on the top. ([a] M. Hamblin/Animals Animals/Earth Scenes; [b] Michael Gunther/PA; [c] M. Fogden/Animals Animals/Earth Scenes)

A difference between the sexes is called a sexual dimorphism.

A difference between the males and females of a species is called a **sexual dimorphism**. In this chapter, we ask why sexual dimorphism occurs in such a great variety of organisms. It is a question Charles Darwin (1871) wrote half a book about, and it has captivated evolutionary biologists ever since.

In previous chapters, we have explained many features of organisms with the theory of evolution by natural selection. The goal has been to discover whether the features in question are adaptive, and if so, then how they improve the survival or fecundity of the individuals that possess them. But differences between the

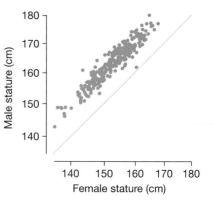

Figure 9.2 **Women and men differ in height** For each of more than 200 human societies, the average height of the men is plotted against the average height of the women. The diagonal line shows where the points would fall if men and women were of equal height. People vary widely in height from society to society: In the shortest society, the average man is about 143 cm tall (about 4 feet, 8 inches) and the average woman about 135 cm (~4′5″); in the tallest society, the average man is about 180 cm tall (~5′11″) and the average woman about 165 cm (~5′5″). But in every society the average man is taller than the average woman, usually by about 10%. From Rogers and Mukherjee (1992).

sexes of a species are often hard to explain with the theory of evolution by natural selection.

For example, compare the evolution of beak size in Darwin's finches (Chapter 3) to the evolution of long tail feathers in male long-tailed widow birds (Figure 9.3). Two problems arise. First, when the finches get hit by a drought and small soft seeds become rare, big beaks are as useful to the females as to the males. But if long tail feathers can improve the survival or fecundity of a widow bird, then why do only the males have them? Second, big beaks help the finches survive droughts by allowing the finches to open large hard seeds and get more to eat. But how could those enormously long tail feathers improve the survival, or the fecundity, of widow birds? Long tail feathers probably make male widow birds easier for predators to find and catch. Furthermore, growing long tail feathers requires considerable energy. Any energy spent on feathers is energy that cannot be spent on making offspring. It appears that the theory of evolution by natural selection can explain neither why male and female widow birds are different, nor why the unusual trait, long tail feathers, exists at all.

Sexual dimorphism is a puzzle, because natural selection cannot explain it.

As Darwin himself was the first to recognize, sex provides a solution to the puzzle of sexual dimorphism. To see why, consider life without sex. For organisms that reproduce without sex (see Chapter 7), getting genes into the next generation is fairly straightforward. The two main challenges are surviving long enough to reproduce, and then reproducing. Sex complicates life by adding a third major challenge: finding a member of the opposite sex and persuading him or her to cooperate.

Charles Darwin recognized that individuals vary not only in their success at surviving and reproducing, but also in their success at persuading members of the opposite sex to mate. About birds, for example, Darwin wrote, "Inasmuch as the act of courtship appears to be with many birds a prolonged and tedious affair, so it occasionally happens that certain males and females do not succeed during the proper season, in exciting each other's love, and consequently do not pair" (1871, page 107). In its evolutionary consequences, failing to mate is equivalent to dying young: The victim makes no genetic contribution to future generations. Darwin had already applied the label natural selection to differences among individuals in survival and reproduction. Differences among individuals in success at getting mates he called **sexual selection**. We can develop a theory of evolution by sexual selection that is logically equivalent to the theory of evolution by natural selection. If there is heritable variation in a trait that affects the ability to obtain mates, then variants conducive to success will become more common over time.

Our goal in this chapter is to explore how the theory of evolution by sexual selection explains the frequent existence of conspicuous differences between females and males, particularly when those differences involve traits that seem likely to impair survival. We first review classical work that elucidated the precise mechanism through which sexual reproduction creates different selection pressures

Figure 9.3 The sexual dimorphism in long-tailed widow birds (*Euplectes progne*) The male is black with long tail feathers and red and yellow shoulder patches; the female is brown and cryptic.

for females versus males. Then we consider recent research on the evolutionary consequences of these differing selection pressures in different species.

9.1 Asymmetries in Sexual Reproduction

In this section we argue that sexual reproduction creates different selection pressures for females versus males. The logic we develop to support this conclusion was clearly articulated by A. J. Bateman (1948) and refined by Robert Trivers (1972). It hinges on a crucial fact: Eggs (or pregnancies) are more expensive than ejaculates. In more general terms, females typically make a larger parental investment in each offspring than males. By parental investment we mean energy and time expended both in constructing an offspring and in caring for it. Ultimately, parental investment is measured in fitness. Parental investment increases the reproductive success of the offspring receiving it. At the same time, it decreases the remaining reproductive success that the investing parent may achieve in the future by way of additional offspring.

Consider the parental investments made by male and female orangutans. Adult orangutans of opposite sex tolerate each other's company only for the purpose of mating (Nowak 1991). After a brief tryst, including a copulation that lasts about 15 minutes, the male and female go their separate ways. If a pregnancy results, then the mother, who weighs about 40 kilograms, will carry the fetus for 8 months, give birth to a 1-kilogram baby, nurse it for about 3 years, and continue to protect it until it reaches the age of 7 or 8. For the father, who weighs about 70 kilograms, the beginning and end of parental investment is a few grams of semen, which he can replace in a matter of hours or days. In their pattern of parental investment, orangutans are typical mammals. Females provide substantial parental care and males provide none whatsoever in more than 90% of mammal species (Woodroffe and Vincent 1994).

Because female mammals provide such intensive parental care, mammals present a somewhat extreme example of disparity in parental investment. In most animal species, neither parent cares for the young: Mated pairs just make eggs, fertilize them, and leave them. But in these species, too, females usually make a larger investment in each offspring than males. Eggs are typically large and yolky, with a big supply of stored energy and nutrients. Think of a sea turtle's eggs, some of which are as large as a hen's eggs. Most sperm, on the other hand, are little more than DNA with a propeller. Even when a single ejaculate delivers hundreds of millions of sperm, the ejaculate seldom represents more than a fraction of the investment contained in a clutch of eggs.

The key to explaining sexual dimorphism is in recognizing that sexual reproduction imposes different selection pressures on females versus males.

Recognizing that eggs are more expensive than ejaculates allows us to predict that there will be a profound difference in the factors that limit the lifetime reproductive success of females versus males. A female's potential reproductive success is relatively small, and her realized reproductive success is likely to be limited more by the number of eggs she can make (or pregnancies she can carry) than by the number of males she can convince to mate with her. In contrast, a male's potential reproductive success is relatively large, and his realized reproductive success is likely to be limited more by the number of females he can convince to mate with him than by the number of ejaculates he can make. In other words, we predict this fundamental asymmetry: Access to females will be a limiting resource for males, but access to males will not be a limiting resource for females.

Asymmetric Limits on Reproductive Success in Fruit Flies

A. J. Bateman (1948) tested this prediction in laboratory populations of the fruit fly, *Drosophila melanogaster*. Bateman set up small populations of flies in bottles. Each population consisted of three virgin males and three virgin females. Each fly was heterozygous for a unique dominant genetic mutation: one that gave the fly curly wings, for example, or hairless patches on parts of its body. Bateman let the flies mate with each other, then raised their offspring to adulthood. He was able to identify about half of the offspring of each parent by looking at the mutations the offspring inherited. In this way, Bateman could figure out which females had mated with which males, and vice versa. He could also calculate the relative number of offspring each parent produced. Note that the males and females living together in a bottle all had exactly three potential mates. Their number of actual mates varied from zero to three. If the factors that limit reproductive success are different for the two sexes, then the data for the two sexes should show different relationships between reproductive success and number of actual mates.

The combined results from two dozen experiments appear in Figure 9.4. These results confirm the prediction that the two sexes differ in the factors that limit reproductive success:

Because females typically invest more in each offspring than males, a female's reproductive success is limited by the number of eggs she can make. In contrast, a male's reproductive success is limited by the number of females he can mate with.

- For males, reproductive success increased in direct proportion with number of mates (Figure 9.4a). Males also showed considerable variation in number of mates, with many males in the extreme groups with zero mates and three mates (Figure 9.4b). These two patterns combined to produce considerable variation

(a) Reproductive success versus number of mates:

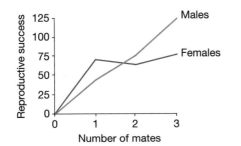

(b) Variation in number of mates:

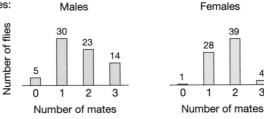

(c) Variance in reproductive success in four groups of experiments:

	Male Variance	Female Variance	Male Variance / Female Variance
Group 1	1604	985	1.63
Group 2	1700	209	8.14
Group 3	2798	993	2.82
Group 4	1098	277	3.97

Figure 9.4 The combined results from Bateman's experiments In (c), variation in reproductive success is characterized by the variance among individuals in number of offspring. To calculate the variance of a list of numbers, first calculate the mean. Then take the difference between each number and the mean. These differences are called the deviations from the mean. Now take each deviation and square it. The mean of the squared deviations is the variance. The larger the variation among the numbers in a list, the larger the variance. After Bateman (1948).

among males in reproductive success (Figure 9.4c): Males with three mates were big winners; males with no mates were big losers.

- For females, reproductive success did not substantially increase with more than one mate (Figure 9.4a). Further, females showed less variation in number of mates, with most females having one or two (Figure 9.4b). These two patterns combined to produce relatively little variation among females in reproductive success (Figure 9.4c): There were few big winners and few big losers.

In Bateman's experimental fly populations, access to mates indeed proved to be a limiting resource for males, but not for females. Sexual selection, or variation in fitness due to variation in success at getting mates, was much stronger in males than in females. Evolutionary biologists believe that this pattern is quite general (although not universal). And it has important consequences for how members of each sex approach mating.

Behavioral Consequences of Asymmetric Limits on Fitness

The asymmetry in the factors that limit reproductive success for females versus males allows us to predict differences in the mating behavior of the two sexes:

- Males should be competitive. If the fitness of males is limited by access to females, then we predict that males will compete among themselves over opportunities to mate.
- Females should be choosy. If the fitness of females is not limited by opportunities to mate, but any given mating may involve the commitment by the female to a large investment in offspring, then we predict that females will be selective about whom they mate with.

Sexual selection theory predicts that males will compete with each other over access to mates, and that females will be choosy.

Male–male competition for mates and female choosiness can play out in two ways. First, in species in which males can directly monopolize access to females, males typically fight with each other over such monopolies. The females then mate with the winners. This form of sexual selection is called **intrasexual selection**, because the key event that determines reproductive success involves interactions among the members of a single sex (the males fight). Second, in species in which males cannot directly control access to females, the males advertise for mates. The females then choose among the advertisers. This form of sexual selection is called **intersexual selection**, because the key event that determines reproductive success involves an interaction between members of the two sexes (females choose males).

Many readers will have noticed that our treatment of asymmetries in sexual reproduction has been full of crass generalizations. We want to emphasize that expectations of male–male competition and female choosiness are not based on anything inherent to maleness or femaleness per se, but on the observation that females commonly invest much more per offspring than males. This investment pattern is broken in a great variety of species. When fathers care for young, for example, male parental investment per offspring may be comparable to, or even greater than, female parental investment. Species with male parental care include humans, many fish, about 5% of frogs, and over 90% of birds. When males actually do invest more per offspring than females, access to mates will be a limiting resource for females. When access to mates is limiting for females instead of males, we predict that *females* will compete with each other over access to males, and that *males* will be choosy. Toward the end of this chapter, we will return to "sex-role reversed" species to see if they are exceptions that can prove the rules of sexual selection. For now, however, we will focus on sexual selection by male–male competition and female choice.

9.2 Male–Male Competition: Intrasexual Selection

Sexual selection by male–male competition occurs when individual males can monopolize access to females. Males may monopolize females through direct control of the females themselves, or through control of some resource important to females. In this section, we consider examples of research into three forms of male–male competition: outright combat, sperm competition, and infanticide.

Combat

Outright combat is the most obvious form of male–male competition for mates. Intrasexual selection involving male–male combat over access to mates can favor morphological traits including large body size, weaponry, and armor. Male–male combat also selects for effective tactics.

Male–male competition can take the form of combat over access to females.

Our example comes from the marine iguanas (*Amblyrhynchus cristatus*) of the Galápagos Islands (Figure 9.5). Marine iguanas have a life-style unique among the lizards. They make their living grazing on algae in the intertidal zone. Between bouts of grazing, they bask on rocks at the water's edge. Basking warms the iguanas, which aids digestion and prepares them for their next foray into the cold water. Marine iguanas grow to different sizes on different islands, but, as we mentioned earlier, on any given island the males get larger than the females (Figure 9.6a).

The sexual size dimorphism in marine iguanas is an excellent example for the study of sexual selection, because we know a great deal about how marine iguana size is affected by natural selection (Wikelski et al. 1997; Wikelski and Trillmich 1997). Martin Wikelski and Fritz Trillmich documented natural selection on iguana body size by monitoring the survival of marked individuals on two islands over one to two years. Natural selection was much harsher on Genovesa than on Santa Fé, but it was clearly at work on both islands. Moreover, selection was stabilizing. Medium-sized iguanas survived at higher rates than either small iguanas or large iguanas (Figure 9.6b).

Potential agents of this natural selection on body size are few. Marine iguanas do not compete with other species for food and have virtually no predators. Other

Figure 9.5   A Galápagos marine iguana foraging on algae in the intertidal zone   (Martin Wikelski, University of Illinois at Urbana-Champaign)

Figure 9.6 Natural selection on body size in marine iguanas (a) Histograms showing the size distributions of male and female marine iguanas on two different Galápagos Islands, Genovesa and Santa Fé. The asterisks mark the maximum sizes at which iguanas were able to maintain their weight in two different years (1991–1992 and 1992–1993). From Wikelski et al. (1997). (b) Survival rates of marked individuals of different sizes (snout–vent length, mm) from March 1991 to March 1992 on Genovesa, and from February 1990 to February 1992 on Santa Fé. The sample sizes, or number of individuals in each group, are given by *n*. From Wikelski and Trillmich (1997).

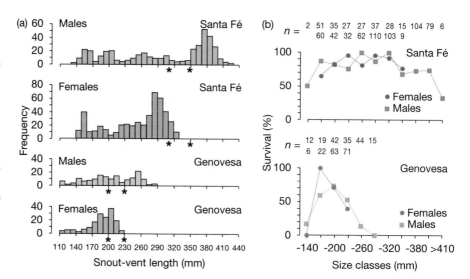

than reproduction, about all the iguanas have to contend with is competition for food among themselves.

Larger iguanas can harvest more algae, and thus gather more energy, but they also expend more energy on metabolism. Wikelski and colleagues (1997) found during two different years that small iguanas ran a net energy surplus, but large iguanas ran a net energy deficit. Consistent with the hypothesis that the availability of food limits body size, the largest iguanas on Santa Fé and Genovesa lost weight during both 1991–1992, a bad year for algae, and 1992–1993, a fairly good year (see also Wikelski and Thom 2000). The largest sizes at which iguanas were able to maintain their weight are indicated by the asterisks in Figure 9.6a.

Now compare Figure 9.6a with Figure 9.6b. The maximum sizes at which iguanas could sustain their weight are close to the optimal sizes for survival. Furthermore, the largest females in each population are near the optimal size for survival, but the largest males are much larger than the optimal size. The large body size of male marine iguanas is thus an evolutionary puzzle: We cannot explain it by natural selection, because Wikelski and Trillmich have shown that natural selection acts against it. It is exactly the kind of puzzle for which Darwin invoked sexual selection.

As we discussed earlier, a crucial issue in sexual selection is the relative parental investment per offspring made by females versus males. In marine iguanas, the parental investment by females is much larger. Each female digs a nest on a beach away from the basking and feeding areas, buries her eggs, guards the nest for a few days, and then abandons it (Rauch 1988). Males provide no parental care at all. So parental investment by females consists mostly of producing eggs, and parental investment by males consists entirely of producing ejaculates. Females lay a single clutch of one to six eggs each year, into which they put about 20% of their body mass (Rauch 1985; Rauch 1988; Wikelski and Trillmich 1997). Compared to the female investment, the cost of the single ejaculate needed to fertilize all the eggs in a clutch is paltry. This difference in investment suggests that the maximum potential reproductive success of males is much higher than that of females. Number of mates will limit the lifetime reproductive success of males, but not females.

The iguanas' mating behavior is consistent with these inferences. Females copulate only once each reproductive season. Martin Wikelski, Silke Bäurle, and their

Figure 9.7   Male marine iguanas in combat   (Martin Wikelski, University of Illinois at Urbana-Champaign)

field assistants followed several dozen marked females on Genovesa. The researchers watched the females from dawn to dusk every day during the entire month-long mating season in 1992–1993 and 1993–1994 (Wikelski and Bäurle 1996). They also watched the marked females from dawn to dusk every day during the subsequent nesting seasons. Every marked female that dug a nest and laid eggs had been seen copulating, but no marked female had been seen copulating more than once. Male iguanas, in contrast, attempt to copulate many times with many different females. But the opportunity to copulate with females is a privilege a male iguana has to fight for (Figure 9.7).

Prior to the mating season each year, male iguanas stake out territories on the rocks where females bask between feeding bouts. In these small, densely packed territories (Figure 9.8a), males attempt to claim and hold ground by ousting male interlopers. Confrontations begin with head-bobbing threats, and escalate to chases and head pushing. If neither male backs down, fights can end with bites leav-

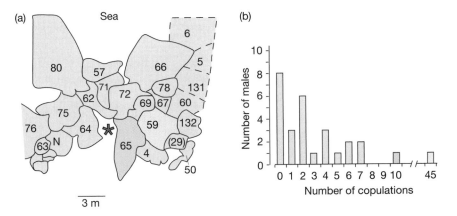

Figure 9.8   Mating success in male marine iguanas   (a) A cluster of iguana mating territories on Camaaño Islet, Galápagos. Lines show boundaries of mating territories on 16 January 1978; numbers identify territory owners. As the scale bar for this map shows, mating territories are only a few square meters in size. The dark blue asterisk indicates where Krisztina Trillmich sat to watch the iguanas. (Camaaño Islet has only 880m of shoreline and supports a population of nearly 2000 iguanas.) From Trillmich (1983). (b) Histogram showing variation in number of copulations obtained by male iguanas on mating territories shown in (a). Note the break in the horizontal scale; the most successful male, iguana 59, got more than four times as many copulations as any of his rivals. The histogram includes only males that claimed a territory for at least a short time during the mating season. From Trillmich (1983).

ing serious injuries on the head, neck, flanks, and legs (Trillmich 1983). While a male holds a territory, he has a more or less exclusive right to mate with any receptive females that happen to be there, typically females that use the territory as their basking site (Rauch 1985). Because only some males manage to claim territories, and some males manage to maintain their claims for a longer period than others, there is extreme variation among males in the number of copulations obtained (Figure 9.8b).

Male marine iguanas fight over territories where females congregate. Large iguanas win more fights, claim better territories, and thus get to copulate with more females. This pattern of sexual selection has led to the evolution of large body size in males.

Because claiming and holding a territory involves combat with other males, bigger males tend to win. In the iguana colony that Krisztina Trillmich (1983) studied on Camaaño Islet, the male that got 45 copulations (Figure 9.8b), far more than any other male, was iguana 59 (his territory is shown in Figure 9.8a). His neighbor, iguana 65, was the second most successful with 10 copulations. Both of their territories were females' favorite early-morning and late-afternoon basking places. Trillmich reported that iguana 59 was the largest male in the colony; that to claim his territory, he had to eject four other males who tried to take it; and that during his tenure, he lost parts of it to four neighboring males who were pushing their territories in from the sides. Wikelski and co-workers studied iguana colonies on Genovesa and Santa Fé (Wikelski et al. 1996; Wikelski and Trillmich 1997). Consistent with Krisztina Trillmich's observations, these researchers found that the mean size of males that got to copulate was significantly larger than the mean size of all males that tried to copulate (Table 9.1).

If we assume that body size is heritable in marine iguanas, then we have variation, heritability, and differential mating success. These are the elements of evolution by sexual selection. We thus have an explanation for why male marine iguanas get so much bigger than the optimal size for survival. Male iguanas get big because bigger males get more mates and pass on more of their big-male genes.

Male–male combat, analogous to that in marine iguanas, happens in a great variety of species, including the red deer shown in Figure 9.1. When mating opportunities are a limiting resource for males, and when males can monopolize either the females themselves, or some resource that is vital to the females and

Table 9.1 Sexual selection differentials for male body size in marine iguana colonies on Santa Fé and Genovesa

Body size is given as snout–vent length (SVL). The standardized selection differential (see Chapter 7) is the difference between the average body size of all males that copulated at least once and the average body size of all males that tried to copulate, expressed in standard deviations of the distribution of body sizes of all males that tried to copulate. (The standard deviation is the square root of the variance.) Both standardized selection differentials are positive ($P < 0.05$), indicating that males that got to copulate were larger on average than males that tried to copulate. From Wikelski and Trillmich (1997).

	N	Average size (SVL)	Standard deviation	Standardized selection differential
Santa Fé				
Males that copulated	253	401	13	0.42
All males that tried to	343	390	26	
Genovesa				
Males that copulated	25	243	26	0.77
All males that tried to	147	227	21	

thus sure to attract them, males fight among themselves for access to the females or the resource. In addition to large body size, this kind of sexual selection leads to the evolution of other traits that are assets in combat, such as weaponry and armor. Male–male combat can also lead to the evolution of alternative male mating strategies (see Box 9.1).

Sperm Competition

Male–male competition does not necessarily stop when copulation is over. The real determinant of a male's mating success is not whether he copulates, but whether his sperm fertilize eggs. If an animal has internal fertilization, and if a female mates with two or more different males within a short period, then the sperm from the males will be in a race to the eggs. Indeed, females may produce litters or clutches in which different offspring are fathered by different males. Batches of offspring with multiple fathers have been documented in a variety of animals, including squirrels (Boellstorff et al. 1994), bears (Schenk and Kovacs 1995), birds (Gibbs et al. 1990), lizards (Olsson et al. 1994), and spiders (Watson 1991). It happens in humans too; Smith (1984) reviews reports of twins with different fathers.

Male–male competition can take the form of sperm competition.

Given sperm competition, what traits contribute to victory? One useful trait might simply be the production of large ejaculates containing many sperm. If sperm competition is something of a lottery, then the more tickets a male buys, the better his chances of winning. This hypothesis has been tested by Matthew Gage (1991) with the Mediterranean fruit fly, *Ceratitis capitata*. Gage's experiment was based on the observation that, although ejaculates are cheaper than eggs, they are not free (see, for example, Nakatsuru and Kramer 1982). Gage reasoned that if male Mediterranean fruit flies are subject to any constraints on sperm production, they might benefit from conserving their sperm, using during each copulation only the minimum number necessary to ensure complete fertilization of the female's eggs. But if larger ejaculates contribute to victory in sperm competition, males whose sperm are at risk of competition should release more sperm during copulation than males whose sperm are not at risk. If the number of sperm released is unimportant to the outcome of competition, then males should release the same number of sperm regardless of the risk of competition.

Gage raised and mated male medflies under two sets of conditions. One group of 20 he raised by themselves and allowed to mate in private; the other group of 20 he raised in the company of another male and allowed to mate in the presence of that second male. Immediately after each mating, Gage dissected the females and counted the number of sperm the males had released. Males raised and mated in the presence of a potential rival ejaculated more than $2\frac{1}{2}$ times as many sperm (average $\pm$ standard error = 3520 $\pm$ 417) as males raised and mated in isolation (1379 $\pm$ 241), a highly significant difference ($P < 0.0001$). Gage's interpretation was that large ejaculates do contribute to victory in sperm competition, and that male medflies dispense their sperm to balance the twin priorities of ensuring successful fertilization and conserving sperm.

In addition to large ejaculates, sperm competition has apparently led to various other adaptations. Males may prolong copulation, deposit a copulatory plug, or apply pheromones that reduce the female's attractiveness (Gilbert 1976; Sillén-Tullberg 1981; Thornhill and Alcock 1983). During copulation in many species of damselflies, the male uses special structures on his penis to scoop out sperm left by the female's previous mates (Figure 9.9; Waage 1984, 1986). R. E. Hooper and

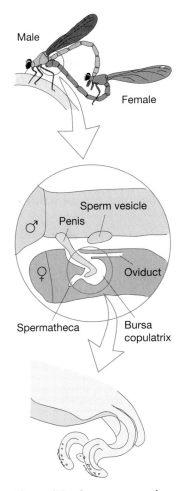

Figure 9.9 Sperm competition in damselflies During copulation (top), the male uses the barbed horns on his penis (bottom) to remove sperm left by the female's previous mates. Redrawn from Waage (1984).

Box 9.1 Alternative male mating strategies

Victory in male–male combat typically goes to the large, strong, and well armed. But what about the smaller males? Is their only chance at fitness to survive until they are large enough to win fights? Often small males attempt to mate by employing alternative strategies. Sometimes they succeed.

In marine iguanas, small adult males are ousted from the mating territories on the basking grounds. But many do not give up; they continue trying to get females to copulate with them. The small males are not terribly successful, but they do get about 5% of the matings in the colony (Wikelski et al. 1996). Needless to say, small males attempting to mate with females are often harassed by other males. This happens to large territorial males too, but it happens more often to small males. Furthermore, copulations by small males are more likely to be disrupted before the male has time to ejaculate (Figure 9.10).

The small males solve the problem of disrupted copulations by ejaculating ahead of time (Wikelski and Bäurle 1996). They use the stimulation of an attempted copulation, or even of seeing a female pass by, to induce ejaculation. The males then store the ejaculate in their cloacal pouches. If he gets a chance to mate, a small male transfers his stored ejaculate to the female at the beginning of copulation. Wikelski and Bäurle examined the cloacae of a dozen females caught immediately after copulations that had lasted less than three minutes. None of these females had

copulated earlier that mating season, but 10 of the 12 females had old ejaculates in their cloacae that must have been transferred during the short copulation.

The sperm in these old ejaculates were viable. From dawn to dusk every day for about a month, Wikelski and Bäurle watched five of the females until they laid their eggs. None of the five copulated again, but all laid fertilized eggs.

Prior ejaculation appears to be a strategy practiced more often by small nonterritorial males than by large territorial males. Wikelski and Bäurle caught 13 nonterritorial and 13 territorial males at random; 85% of the nonterritorial males had stored ejaculates in their cloacal pouches, versus only 38% of the territorial males ($P < 0.05$). This difference is unlikely to result from more frequent copulation by territorial males, because even territorial males copulate only about once every six days (Wikelski et al. 1996).

Alternative, or sneaky male, mating strategies have also evolved in a variety of other species. In coho salmon, *Oncorhynchus kisutch*, for example, males return from the sea to spawn at two different ages (Gross 1984; Gross 1985; Gross 1991). One group, called hooknoses, returns at 18 months. They are large, armed with enlarged hooked jaws, and armored with cartilaginous deposits along their backs. The other group, called jacks, returns at 6 months. They are small, poorly armed, and poorly armored.

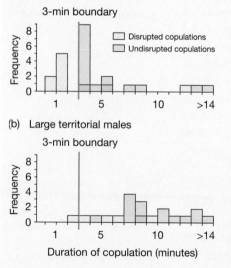

(a) Small nonterritorial males

(b) Large territorial males

Duration of copulation (minutes)

Figure 9.10 Duration of copulations by male marine iguanas Histograms showing the distribution of copulation durations for (a) 24 small nonterritorial males, and (b) 20 large territorial males. Orange areas indicate copulations that were disrupted by other males. The red vertical line at 3 minutes marks the approximate amount of time a male must copulate before he can ejaculate. Large territorial males had copulations that were significantly longer and less likely to be disrupted before the 3-minute boundary. From Wikelski and Bäurle (1996). Copyright © 1996, The Royal Society. Reprinted by permission of The Royal Society and the author.

Box 9.1 Continued

When a female coho is ready to mate, she digs a nest and then lays her eggs in it. As she prepares the nest, males congregate. The males use one of two strategies in trying to fertilize the female's eggs (Figure 9.11). Some males fight for a position close to the

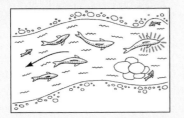

Figure 9.11 Alternative mating strategies in coho salmon This figure shows a coho mating group. The large fish at the right (upstream) is a female that has built a nest and is ready to lay her eggs. Downstream from the female are five males that have opted for the fighting strategy. These males, four hooknoses and a jack, have sorted themselves out by size. Two other jacks have opted for the sneaky strategy. They have found hiding places near the female, one behind a rock, the other in a shallow. After Gross (1991). Copyright © 1991, Ecological Society of America. Reprinted by permission.

female. These fighters quickly sort themselves out by size. When the female lays her eggs, the males spawn over them in order. The first male to spawn fertilizes the most eggs. Other males do not fight for position, but instead look for a hiding place near the female. When the female lays her eggs, these sneakers attempt to dart out and spawn over the eggs.

Among hooknoses, those that adopt the fighting strategy are more successful. Among jacks, those that adopt the sneaky strategy are more successful. The relative fitness of hooknoses versus jacks depends, in part, on the frequency of each type of male in the breeding population.

There is an important distinction between the iguana example and the coho example. In marine iguanas, the small nonterritorial males appear to be making the best of a bad situation while they grow to a large enough size to successfully fight for a territory. In coho, a male irreversibly becomes either a hooknose or a jack. Which strategy a male coho pursues depends on a mixture of environmental and genetic factors.

M. T. Siva-Jothy (1996) used genetic paternity tests to show that this strategy is highly effective. In the damselfly species they studied, the second male to mate with a female fertilized nearly all of the eggs produced during her first postcopulatory bout of oviposition.

Infanticide

In some species of mammals, competition between males continues even beyond conception. One example, discovered by B. C. R. Bertram (1975) and also studied by Craig Packer and Anne Pusey (reviewed in Packer et al. 1988), happens in lions. The basic social unit of lions is the pride. The core of a pride is a group of closely related females—mothers, daughters, sisters, nieces, aunts, and so on—and their cubs. Also in the pride is a small group of adult males; two or three is a typical number. The males are usually related to each other, but not to the adult females. This system is maintained because females reaching sexual maturity stay in the pride they were born into, whereas newly mature males move to another pride.

The move for young adult males from one pride to another is no stroll in the park. The adult males already resident in the new pride resist the invaders. That is why males stay with their other male kin: Each group, the residents and the newcomers, forms a coalition. The residents fight the newcomers, sometimes violently, over the right to live in the pride. If the residents win, they stay in the pride and the newcomers search for a different pride to take over. If the residents lose, they are evicted, and the newcomers have exclusive access to the pride's females—

Male–male competition can take the form of infanticide.

Figure 9.12 Lion infanticide In this photo by George B. Schaller, a male lion has just killed another male's cub, which it now carries in its mouth.

By killing other males' cubs, male lions gain more opportunities to mate.

exclusive, that is, until another coalition of younger, stronger, or more numerous males comes along and kicks them out. Pusey and Packer found that the average time a coalition of males holds a pride is a little over two years. Because residence in a pride is the key to reproductive success in lions, males in a victorious coalition quickly begin trying to father cubs. One impediment to quick fatherhood, however, is the presence of still-nursing cubs fathered by males of the previous coalition. That is because females do not return to breeding condition until after their cubs are weaned.

How can the males overcome this problem? They frequently employ the obvious, if grisly, solution: They kill any cubs in the pride that are not weaned (Figure 9.12). Packer and Pusey have shown that this strategy causes the cubs' mothers to return to breeding condition an average of eight months earlier than they otherwise would. Infanticide by males is the cause of about 25% of all cub deaths in the first year of life, and over 10% of all lion mortality.

Infanticide improves the males' reproductive prospects, but is obviously detrimental to the reproductive success of the females. The females have two options for making the best of their own interests in this bad situation (Packer and Pusey 1983). One is to defend their cubs from infanticidal males, which females often do, occasionally at the cost of their own lives. Nonetheless, Packer and Pusey report that young cubs rarely survive more than two months in the presence of a new coalition of males. The females' other tactic is to spontaneously abort any pregnancies in progress when a new coalition gains residence in the pride. This cuts the females' losses: They do not waste energy and time on cubs that would be killed anyway shortly after birth. With this shift in focus to female reproductive strategy, we leave the subject of male–male conflict and move to the other side of sexual selection: female choice.

9.3 Female Choice

There is a great variety of species in which male reproductive success is limited by opportunities to mate, but in which males are unable to monopolize either females themselves or any resource vital to females. In many such species, the males advertise for mates. Females typically inspect advertisements of several males before they choose a mate. Sexual selection by female choice leads to the evolution of elaborate courtship displays by males.

Charles Darwin first asserted that female choice is an important mechanism of selection in 1871, in *The Descent of Man, and Selection in Relation to Sex*. Although widely accepted today, the notion that females actively discriminate among individual males was controversial for several decades. Most evolutionary biologists thought that female discrimination was limited to choosing a male of the right species (see Trivers 1985). Beyond allowing females to identify a mate of the right species, male courtship displays were thought to function primarily in overcoming a general female reluctance to mate. Once ready to mate, a female would accept any male at hand.

We begin this section by describing two experiments that demonstrate that females are in fact highly selective, actively choosing particular males from among the many available. We then consider the functions of female choosiness. Potential benefits to a choosy female include the acquisition of good genes for her offspring, and the aquisition of resources offered by males. Alternatively, females may prefer male displays that exploit preexisting sensory biases built into the females' nervous systems.

> When males cannot monopolize access to females, they often compete by advertising for mates. Although biologists were long skeptical that females discriminate among the advertising males, female choice is now well established.

Female Choice in Barn Swallows

Our first experiment demonstrating active female choice comes from the work of Anders Møller on barn swallows. Barn swallows, *Hirundo rustica*, are small insect-eating birds that breed in colonies of up to 80 individuals. The swallows Møller (1988) studied breed in Denmark during spring and summer, after spending the European winter in Africa. Upon arriving in a Danish breeding colony, each male swallow sets up a territory a few square meters in size. He then tries to attract a mate by displaying his tail while perching and flying. Each female visits several males, then chooses one to pair with. Once paired, the male and female together build a mud nest in the male's territory. In this nest the pair raises one or, if they have time before summer's end, two clutches of young. The female incubates the eggs by herself, but both parents feed the chicks.

At first glance, barn swallows may not seem promising subjects for a study of sexual selection. The fact that the males help care for the young should tend to equalize parental investment by the two sexes, and the fact that the swallows appear to mate monogamously suggests that neither sex should be in short supply for the other. Barn swallows are, however, sexually dimorphic. The males are more brightly colored than the females, and they tend to be slightly larger (Figure 9.13a). The biggest difference is that the outermost tail feathers, which are elongated in both sexes, are about 15% longer in males than in females (Figure 9.13b).

(a)

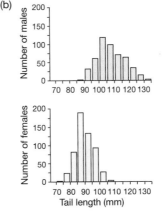

(b)

Figure 9.13 Sexual dimorphism in barn swallows (a) A female (left) and a male (right). Note that the male has richer colors, and longer outer tailfeathers. (Leonard Lee Rue III/Visuals Unlimited) (b) The distribution of tail streamer length in male and female barn swallows. The distributions overlap, but the average male has longer tail feathers than the average female. From Møller (1991).

Three additional factors suggest that sexual selection may be at work in barn swallows. First, even in a monogamous species in which both sexes care for the young, males and females may vary in quality, both as parents and as donors of genes to offspring. The members of both sexes should thus benefit, in higher reproductive success, by trying to identify and attract the best mate possible. Second, although barn swallows appear to be monogamous, many, in fact, are not. Males sometimes solicit copulations with females other than their pair-mates. When these offers of extra-pair copulation are accepted by the females, as occasionally they are, the successful male may benefit by fathering offspring that some other male will help to raise. Third, the trait that differs most prominently between the sexes—the length of the long outer tail feathers—is precisely the trait that males show off when advertising for mates. Møller hypothesized that sexual selection does indeed occur in barn swallows, through the mechanism of female choice, and that females prefer to mate with males displaying longer tail feathers.

Møller captured and color-banded 44 males that had established territories, but not yet attracted mates. He divided them at random into four groups of 11, and altered the tail feathers of each group as follows:

- **Shortened tail feathers.** Møller clipped about 2 centimeters out of the middle of each outer tail feather, then reattached the feather tips to the bases with superglue.
- **Mock-altered (control I).** Møller clipped the tail feathers, then glued them back together. This did not change the length of the feathers, but it otherwise subjected the birds to the same handling, clipping, and gluing as the shortened and lengthened groups.
- **Unaltered (control II).** Møller captured and banded these birds, but did nothing to their tail feathers.
- **Elongated tail feathers.** Møller added, by clipping and gluing, the 2 cm of feather removed from the shortened group into the middles of the tail feathers of these birds.

Møller then released the birds back into their colony. He predicted that if females prefer males with longer tail feathers, then the males with elongated tails would attract mates sooner, and fledge more young, than either control I males or control II males. The control males should, in turn, be more successful than the shortened males. If, on the other hand, the females have no preferences based on tail feathers, then there should be no differences in success among the groups.

Female barn swallows prefer to mate with males whose tail feathers are longest.

The results appear in Figure 9.14. The elongated males, on average, attracted mates more quickly than the control males, and control males attracted mates more quickly than shortened males (Figure 9.14a). Among the advantages of attracting a mate quickly is that the male and his partner can get an earlier start on rearing a clutch of chicks. An earlier start means that the parents are more likely to have time to raise a second clutch before the summer ends (Figure 9.14b). Finally, the advantage of raising two clutches is that the parents fledge more chicks over the course of the summer (Figure 9.14c).

The female barn swallows had additional opportunities to be choosy whenever they were solicited by males for extra-pair copulations. Møller was able to watch the birds closely enough to estimate the rates at which the male swallows at-

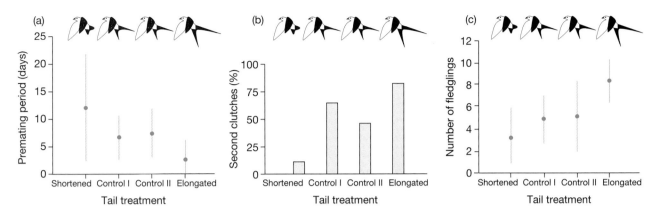

Figure 9.14 **Anders Møller's data on reproductive success and tail length in male barn swallows** (a) Length of time required by males to attract a mate. Red dots represent the average for each group; the vertical lines represent the standard deviation (a statistical measure of the amount of variation within each group). There were significant differences among groups ($P < 0.01$). (b) Differences in time required to attract a mate carried over into which males had time to raise a second clutch of chicks. Bar height indicates the percentage of males in each group who, with their mates, raised two clutches. Again, there was significant variation among groups ($P < 0.02$). (c) Differences in second-clutch success carried over into reproductive success—the number of chicks each male had fledged by summer's end. The meanings of dots and lines are the same as in (a). There was significant variation among groups ($P < 0.001$). Elongated males fledged more chicks than control males, who in turn fledged more chicks than shortened males. From Møller (1988).

tempted to copulate with females other than their pair-mates, the rates at which the females they solicited accepted these copulations, and the rates at which the study males' pair-mates copulated with other males. The males in the four study groups showed no differences in the rates at which they attempted to gain extra-pair copulations. Nor did their pair-mates differ in the rates at which they were solicited. The males in the groups did differ, however, in the rates at which the extra-pair females they propositioned actually accepted them for copulation (Table 9.2). Furthermore, the males' pair-mates differed in the rates at which they accepted extra-pair copulations with other males (Table 9.2). Apparently, females who had to settle for less desirable short-tailed males attempted to compensate by copulating out-of-pair with more desirable long-tailed males. Thanks to these females, the long-tailed males won again, at the expense of the short-tails. (For more on extra-pair copulations in birds, see Box 9.2.)

Table 9.2 Extra-pair copulations in Møller's barn swallow experiment

The numbers reported are rates, measured as extra-pair copulations per hour. *P* values give statistical significance of variation among groups. From Møller (1988).

	Male Tail Treatment				
Extra-pair copulations	**Shortened tails**	**Control I**	**Control II**	**Lengthened tails**	***P***
By males	0	0	0	0.040	< 0.001
By their social pair-mates	0.036	0.014	0.017	0	< 0.01

Box 9.2 Extra-pair copulations

Barn swallows are somewhat unusual in that a careful observer can see enough copulations to estimate the rate of extra-pair copulations directly. In recent years, however, biologists have developed methods of genetic analysis that enable them to indirectly estimate the rate of extra-pair copulations. Figure 9.15 shows two such tests performed on red-winged blackbirds.

Figure 9.15a shows paternity analysis using a restriction-fragment length polymorphism. The photo in the figure depicts an electrophoresis gel (see Box 4.1). Each lane in this electrophoresis gel contains DNA that has been extracted from an individual bird, cut with a restriction enzyme, and labeled with a probe that recognizes a sequence of DNA that occurs at a single locus. Bands on the gel are inherited as simple Mendelian alleles. Individual M_1 (center lane) is an adult male red-wing who had two mates on his territory, F_1 and F_2. M_2 and M_3 are adult male neighbors of M_1, F_1, and F_2. The numbers 1, 2, and 3 represent chicks in the nest shared by M_1 and F_1. Chick 1 has a band in its lane (arrow) that is present in neither its mother (F_1) nor its social father (M_1). This band is present, however, in M_2. We can infer that F_1 had an extra-pair copulation with M_2 (or with an unknown male with the same genotype). The numbers 4, 5, and 6 represent chicks in the nest shared by M_1 and F_2. Chick 6 has a band in its lane (arrow) that is present in neither its mother (F_2) nor its social father (M_1). This band is present, however, in M_3. We can infer that F_2 had an extra-pair copulation with M_3.

Figure 9.15b shows a paternity analysis of the same families using DNA fingerprints. Each lane contains DNA that has been extracted from an individual bird, cut with a restriction enzyme, and labeled with a probe that recognizes a sequence of DNA that occurs at many loci. Bands on the gel are inherited as simple Mendelian alleles, although we do not know which band corresponds to an allele at which locus. The

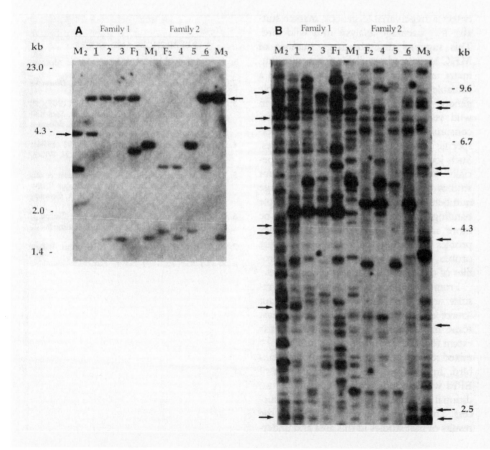

Figure 9.15 Genetic analyses demonstrating extra-pair copulations in red-winged blackbirds (a) A paternity analysis using a traditional restriction-fragment length polymorphism. (b) A paternity analysis of the same families using DNA fingerprints. Reprinted with permission from H. L. Gibbs et al., 1990. Realized reproductive success of polygynous red-winged blackbirds revealed by DNA markers. *Science* 250: 1394–96, Dec. 7, 1990, p. 1395, Fig. 1. Copyright © 1990. American Association for the Advancement of Science.

Box 9.2   **Continued**

DNA fingerprints confirm the same cases of extra-pair copulation we inferred from the gel in Figure 9.15a.

How common is extra-pair copulation in birds? Elizabeth Gray (1997) used DNA fingerprints to assess the frequency of extra-pair copulation in a population of red-winged blackbirds. She estimated that in a given breeding season between 50 and 64% of all nests contained at least one chick sired by a male other than its social parent. Red-winged black birds are not unusual. By using genetic paternity tests, biologists have discovered that many socially monogamous birds engage in frequent extra-pair copulations.

Møller's experiment demonstrates that female barn swallows are choosy. As Møller predicted, females prefer pair-mates with longer tail feathers. And for males, being one of the more desirable mates results in higher reproductive success.

Female Choice in Gray Tree Frogs

Our second experiment demonstrating active female choice comes from the work of H. Carl Gerhardt and colleagues (1996) on gray tree frogs. Gray tree frogs, *Hyla versicolor*, live in woodlands in the eastern United States. During the breeding season, males produce a melodious mating call attractive to females. Each call consists of a series of pulses, or trills. Some males give long calls, consisting of many pulses, while other males give short calls. In addition, some males are fast callers, giving many calls per minute, while other males are slow callers. Gerhardt and colleagues suspected, for at least two reasons, that female gray tree frogs discriminate among potential mates on the basis of their calls. First, when an individual male hears that many other males are calling too, he sometimes increases both the length and speed of his calls. Second, several times in the field the researchers had seen females approach and mate with the more distant of two calling males. The researchers hypothesized that females prefer to mate with longer- and faster-calling males.

Gerhardt and colleagues captured female gray tree frogs in nature and tested their preferences in the laboratory. In one series of experiments, the researchers released females between a pair of loudspeakers (Figure 9.16a). Each speaker played a computer-synthesized mating call. To make their experiment conservative, the researchers made the call they expected to be less attractive louder, either by increasing the volume on that speaker, or by releasing the female closer to it. Then they waited to see which speaker the female would approach. They found that 30 of 40 females (75%) preferred long calls to short calls, even when the short calls were louder, and that 35 of 51 females (69%) preferred fast calls to slow calls, even when the slow calls were louder.

In another experiment, Gerhardt and colleagues released female frogs facing two loudspeakers (Figure 9.16b). The closer speaker played short calls, while the more distant speaker played long calls. The researchers found that 38 of 53 females (72%) went past the short-calling speaker to approach the long-calling speaker.

The experiments of Gerhardt et al. show that female gray tree frogs are choosy. As the researchers predicted, females prefer males giving longer and faster calls.

Female choice, as illustrated by barn swallows and gray tree frogs, is thought to be the selective force responsible for the evolution of a great variety of male advertisement displays—from the gaudy tail feathers of the peacock, to the chirping

Female gray tree frogs prefer to mate with males that give longer calls, and males that repeat their calls more rapidly.

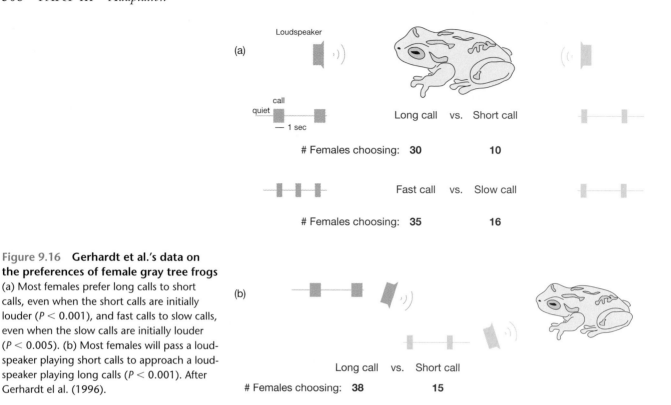

Figure 9.16 Gerhardt et al.'s data on the preferences of female gray tree frogs (a) Most females prefer long calls to short calls, even when the short calls are initially louder ($P < 0.001$), and fast calls to slow calls, even when the slow calls are initially louder ($P < 0.005$). (b) Most females will pass a loudspeaker playing short calls to approach a loudspeaker playing long calls ($P < 0.001$). After Gerhardt el al. (1996).

of crickets, to the chemical attractant of silk moths. Some male displays, like those of peacocks, are loud and clear; others, like those of barn swallows, are more subtle. It is curious that an extra two centimeters added to two tail feathers should make a male barn swallow so much more attractive to females as to dramatically improve his reproductive success. Why should the females care about such a small difference? And for that matter, why should females care about any of the advertisements, even the loud ones, that males use to attract mates? We will consider three explanations.

Choosy Females May Get Better Genes for Their Offspring

A variety of factors have been suggested to explain female preferences.

One possibility is that the displays given by males are indicators of genetic quality. If males giving more attractive displays are genetically superior to males giving less attractive displays, then choosy females will secure better genes for their offspring (Fisher 1915, Williams 1966, Zahavi 1975).

Allison Welch and colleagues (1998) used an elegant experiment to investigate whether male gray tree frogs giving long calls are genetically superior to males giving short calls (Figure 9.17). During two breeding seasons, the researchers collected unfertilized eggs from wild females. They divided each female's clutch into separate batches of eggs, then fertilized one batch of eggs with sperm from a long-calling male, and the other batch of eggs with sperm from a short-calling male. They reared some of the tadpoles from each batch of eggs on a generous diet, and the others on a restricted diet.

This experimental design allowed Welch and colleagues to compare the fitness of tadpoles that were maternal half-siblings—that is, tadpoles with the same mother, but different fathers. When comparing tadpoles fathered by long-calling males versus short-calling males, the researchers did not have to worry about uncontrolled differences in the genetic contribution of the mothers, because the mothers were the same.

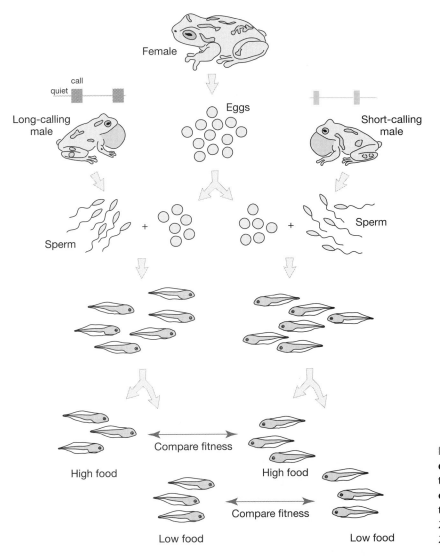

Figure 9.17 Welch et al.'s experiment to determine whether male gray tree frogs that give long calls are genetically superior to males that give short calls Overall, the experiment included batches of eggs from 20 different females fertilized with sperm from 25 different pairs of males.

Welch and colleagues measured five aspects of offspring performance related to fitness: larval growth rate (faster is better); time to metamorphosis (shorter is better); mass at metamorphosis (bigger is better); larval survival; and post-metamorphic growth (faster is better). The results of their comparisons appear in Table 9.3. In 18 comparisons between the offspring of long-calling males versus short-calling males, there was either no significant difference, or better performance by the offspring of long callers. The offspring of short callers never did better. Overall, the data indicate that the offspring of long-calling males have significantly higher fitness. This result is consistent with the good genes hypothesis. The exact nature of the genetic difference between long-calling frogs versus short-calling frogs is a subject for future research.

Choosy female gray tree frogs get better genes for their offspring.

Choosy Females May Benefit Directly Through the Acquisition of Resources

In many species the males provide food, parental care, or some other resource that is beneficial to the female and her young. If it is possible to distinguish good providers from poor ones, then choosy females reap a direct benefit in the form

Table 9.3   Fitness of the offspring of long-calling male frogs vs. short-calling male frogs

NSD = no significant difference; LC better = offspring of long-calling males performed better than offspring of short-calling males; — = no data taken. The overall result: Offspring fathered by long-calling males had significantly higher fitness than their maternal half-sibs fathered by short-calling males ($P < 0.0008$).

	1995		1996	
Fitness measure	**High food**	**Low food**	**High food**	**Low food**
Larval growth	NSD	LC better	LC better	LC better
Time to metamorphosis	LC better	NSD	LC better	NSD
Mass at metamorphosis	NSD	LC better	NSD	NSD
Larval survival	LC better	NSD	NSD	NSD
Postmetamorphic growth	—	—	NSD	LC better

of the resource provided. Such is the case in the hangingfly (*Bittacus apicalis*), studied by Randy Thornhill (1976). Hangingflies live in the woods of eastern North America, where they hunt for other insects. After a male catches an insect, he hangs from a twig and releases a pheromone to attract females. When a female approaches, the male presents his prey. If she accepts it, the pair copulates while she eats (Figure 9.18a). The larger the prey, the longer it takes her to eat it, and the longer the pair copulates (Figure 9.18b). The longer the pair copulates, the more

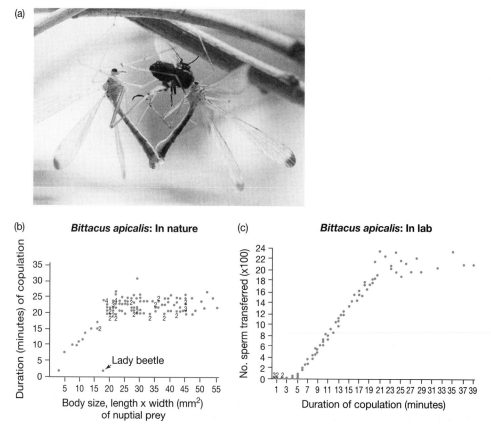

Figure 9.18   Courtship and mating in hangingflies   (a) A female (right) copulates with a male while eating a blowfly he has captured and presented to her. (Randy Thornhill, University of New Mexico). (b) The larger the gift the male presents to the female, the longer the pair copulates. Copulation ends after about 20 minutes, even if the female is still eating. A lady beetle presented by one male is an exception to the general pattern. Even though the beetle was fairly large, the female hangingfly rejected it and broke off copulation almost immediately. (c) The longer a pair copulates, the more sperm the female allows the male to transfer. The male must present a gift that takes at least five minutes to eat, or the female accepts no sperm. From Thornhill (1976). Copyright © 1997, American Naturalist. Reprinted by permission of The University of Chicago Press.

sperm the female accepts from the male (Figure 9.18c). If she finishes her meal in less than 20 minutes, the female breaks off the copulation and flies away looking for another male and another meal. The female's preference for males bearing large gifts benefits her in two ways: (1) It provides her with more nutrients, allowing her to lay more eggs; and (2) it saves her from the need to hunt for herself. Hunting is dangerous, and males die in spider webs at more than twice the rate of females. The males behave in accord with the same kind of economic analysis: If the female is still eating after accepting all the sperm she can, the male grabs his gift back and flies off to look for a second female to share it with.

Choosy female hanging flies get food from their mates.

Choosy Females May Have Preexisting Sensory Biases

Females use their sensory organs and nervous systems for many other purposes than just discriminating among potential mates. It is possible that selection for such abilities as avoiding predators, finding food, and identifying members of the same species may result in sensory biases that make females particularly responsive to certain cues (see Enquist and Arak 1993). This may in turn select on males to display those cues, even if the cues would otherwise have no relation to mating or fitness. In other words, the preexisting bias, or sensory exploitation, hypothesis holds that female preferences evolve first and that male mating displays follow.

Research by Heather Proctor on the water mite *Neumania papillator* illustrates sensory exploitation (1991; 1992). Members of this species are small freshwater animals that live amid aquatic plants, and make their living by ambushing copepods. Water mites have simple eyes that can detect light, but cannot form images. Instead of vision, water mites rely heavily on smell and touch. Both males and females hunt copepods by adopting a posture Proctor calls *net-stance*. The hunting mite stands on an aquatic plant with its four hind legs, rears up, and spreads its four front legs to form a sort of net. The mite waits until it detects vibrations in the water that might be produced by a swimming copepod, then turns toward the source of the vibrations and clutches at it.

Mating in *Neumania papillator* does not involve copulation. Instead, the male attaches sperm-bearing structures called spermatophores to an aquatic plant, then attempts to induce the female to accept them. He does this by fanning water across the spermatophores toward the female. The moving water carries to the female pheromones released by the spermatophores. When the female smells the pheromones, she may pick up the spermatophores.

Male water mites search for females by moving about on aquatic vegetation. When a male smells a female, he walks in a circle while lifting and trembling his front legs (Figure 9.19a). If the male has detected a female that is still there, not just the scent of one that has recently left, the female typically turns toward the trembling male. Often she also clutches at him. At this point, the male deposits his spermatophores and begins to fan (Figure 9.19b).

Proctor suspected that male leg-trembling during courtship evolved in *N. papillator* because it mimics the vibrations produced by copepods, and thereby elicits predatory behavior from the female. She tested this hypothesis with a series of experiments in which she observed the behavior of water mites under a microscope. First, Proctor measured the frequency of vibrations produced by trembling males, and compared it to the frequency of vibrations produced by copepods. Water mites tremble their legs at frequencies of 10 to 23 cycles per second, well within the

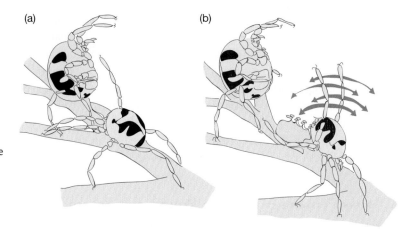

Figure 9.19 Courtship in the water mite, *Neumania papillator* (a) The female (on the left) is in net-stance, waiting to ambush a copepod; the male has found her and is now trembling his legs. (b) The female has turned toward the male in response to the trembling. The male has deposited spermatophores and is now fanning water across them. The sausage-shaped objects on top of the spermatophores are sperm packets. Redrawn from Proctor (1991) by permission of Academic Press.

Choosy female water mites may simply be responding to courting males as though the males were prey.

copepod range of 8 to 45 cycles per second. Second, Proctor observed the behavior during net-stance of female water mites when they were alone, when they were with copepods, and when they were with males. Females in net-stance rarely turned and never clutched unless copepods or males were present, and the behavior of females toward males was similar to their behavior toward copepods. Third, Proctor observed the responses to male mites of hungry females versus well-fed females. Hungry females turned toward males, and clutched them, significantly more often than well-fed females. All of these results are consistent with the hypothesis that male courtship trembling evolved to exploit the predatory behavior of females.

Males employing leg trembling during courtship probably benefit in several ways. First, males appear to use the female response to trembling to determine whether a female is actually present. Proctor observed that a male that has initiated courtship by trembling is much more likely to deposit spermatophores if the female clutches him than if she does not. Second, trembling appears to allow males to distinguish between receptive females versus unreceptive ones. Proctor observed that a male has a strong tendency to deposit spermatophores for the first female he encounters that remains in place after he initiates courtship, but that virgin females are more likely to remain in place than are nonvirgins. Third, males appear to use the female response to trembling to determine which direction the female is facing. Proctor observed that males deposit their spermatophores in front of the female more often than would be expected by chance. These benefits mean that a male that trembles should get more of his spermatophores picked up by females than would a hypothetical male that does not tremble. In other words, a male that trembles would enjoy higher mating success. This is consistent with the hypothesis that trembling evolved by sexual selection.

A key prediction of the sensory exploitation hypothesis is that net-stance evolved before male trembling. Proctor tested this hypothesis by using a suite of morphological characters to estimate the phylogeny of *Neumania papillator* and several related water mites. (Methods for estimating phylogenies will be covered in Chapter 13.) She noted which species have net stance, and which species have male courtship trembling. She then inferred the places on the phylogeny at which net-stance and courtship trembling are most likely to have evolved, based on the assumption that simpler evolutionary scenarios are more probable. Proctor concluded that one of two evolutionary scenarios is most likely to be correct (Figure 9.20).

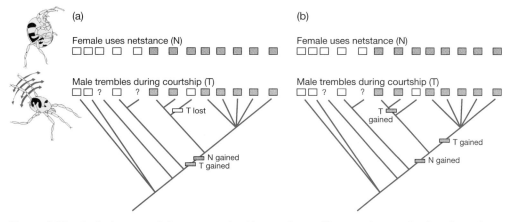

Figure 9.20   A phylogeny of the water mite *Neumania papillator* and several related species
The boxes above the tips of the branches indicate which species have net-stance, and which species have male courtship trembling. A colored box indicates the trait is present; an open box indicates the trait is absent. The two versions of the phylogeny show the two most likely scenarios for the evolution of these two traits. Redrawn from Proctor (1992) by permission of Academic Press.

In the first scenario, net-stance and courtship trembling both evolved at the base of the branch that includes all species with either or both of these traits, and trembling was subsequently lost once (Figure 9.20a). In the second scenario, net-stance evolved at the base of the branch, and courtship trembling subsequently evolved twice. The first scenario supplies insufficient evidence to test the prediction that net-stance evolved before trembling. We simply cannot tell, under this scenario, whether one trait evolved before the other, or the two traits evolved simultaneously. The second scenario is consistent with the prediction that net-stance evolved first. Which scenario is closest to the truth remains unkown. However, given the phylogenetic evidence in combination with the data from her obervations of water mite behavior, Proctor concludes that sensory exploitation is the best explanation for the evolution of courtship trembling.

Other Explanations for Female Choice

We have considered three explanations for female choice that provide good reasons why females might prefer some male displays over others. It is also possible, however, that female preferences are essentially arbitrary. One version of this idea is the sexy-son hypothesis. According to this hypothesis, once a particular male advertisement display is favored by a majority of females, selection on females will automatically reinforce a preference for the fashionable trait. The reason is that females choosing fashionable mates will have more fashionable sons, and therefore more grandchildren, than females choosing unfashionable mates. A more technically detailed version of the idea that selection may reinforce arbitrary preferences, called the runaway selection hypothesis, is discussed in Box 9.3.

We should note that all the explanations for female choice that we have discussed are mutually compatible. It is possible, at least in principle, for a single courtship display to indicate genetic quality, to predict direct benefits the female is likely to receive, to exploit biases in the female's sensory system, and to be reinforced by selection in favor of sexy sons. Much contemporary research on sexual selection is focussed on determining the relative importance of these different factors in the evolution of female preferences. One approach is to test several alternative hypotheses in a single

Theory also suggests that female preferences can be completely arbitrary.

Box 9.3 Runaway sexual selection in stalk-eyed flies?

Runaway selection is an idea first elaborated by Ronald Fisher in 1915, and perhaps traceable to a remark made by T.H. Morgan in 1903 (reviewed in Anderson 1994). The idea is worth explaining in some detail, because it illustrates useful concepts in evolutionary genetics. We will discuss runaway selection in the context of research by Gerald Wilkinson and colleagues on the stalk-eyed flies of Southeast Asia.

Stalk-eyed flies carry their eyes on the ends of long thin appendages. In both sexes bigger flies have longer eyestalks, but males have longer stalks for their size than females. By day the flies are solitary and forage for rotting plants. In the evening, the flies congregate beneath overhanging stream banks, where they cling in small groups to exposed root hairs and spend the night (Figure 9.21). At dawn and dusk, the flies roosting together on a root hair often mate with each other. Neither sex cares for the young, so a female's investment in each offspring is larger than a male's. Not surprisingly, males attempt to evict each other from the root-hair roosts in order to be the only male in the group at mating time. In male–male confrontations, the male with longer eyestalks typically wins, so male–male competition may partially explain the evolution of eyestalks (Burkhardt and de la Motte 1983; Burkhardt and de la Motte 1987; Panhuis and Wilkinson 1999). As we will see, however, there is evidence that female choice has also played a role.

To see how the runaway selection hypothesis works, imagine a population of stalk-eyed flies in which both males and females are variable, males in the lengths of their eyestalks and females in their mat-

ing preferences for stalk length. These two patterns of variation should combine to produce assortative mating; that is, the females that prefer the longest stalks will mate with the longest-stalked males, and the females that prefer the shortest stalks will mate with the shortest-stalked males (Figure 9.22a). Assume, furthermore, that in both sexes the variation is heritable—that is, that at least part of the variation in stalk length, and part of the variation in preference, is due to variation in genes (see Chapter 7). Under these assumptions, offspring that receive from their fathers genes for long eyestalks tend to also receive from their mothers genes for a preference for long-stalked males. In other words, if the assortative mating persists for some generations, then it will establish a genetic correlation (linkage disequilibrium) between the stalk-length genes and the preference genes (see Chapter 7). If we were to take a group of males, mate each with a number of randomly chosen females, and then examine their sons and daughters, we would find that sons with long eyestalks tend to have sisters with a preference for long-stalked-males (Figure 9.22b). This association means that if we conduct artificial selection on the stalk lengths of the males (and only on the males), we should get a correlated evolutionary response in the preferences of the females (Figure 9.22c).

Wilkinson and Paul Reillo (1994) tested the prediction that selection on male stalk length will produce a correlated response in female preferences. Wilkinson and Reillo collected stalk-eyed flies (*Cyrtodiopsis dalmanni*) in Malaysia and used them to establish three laboratory populations. In each population, the researchers separated the males and females immediately after the adult flies emerged from their pupae, and kept them apart for two to three months. Wilkinson and Reillo then chose breeders for the next generation of each population as follows. For the control line, they used 10 males and 25 females picked at random. For the long-selected line, they used the 10 males with longest eyestalks from a pool of 50 males picked at random, and 25 females picked at random. For the short-selected line, they used the 10 males with shortest eyestalks from a pool of 50 males picked at random, and 25 females picked at random. After 13 genera-

Figure 9.21 A group of Malaysian stalk-eyed flies (*Cyrtodiopsis whitei*) gathered on a root hair to spend the night The largest fly is a male; the others are females (Gerald Wilkinson, University of Maryland).

Box 9.3   Continued

tions the populations had diverged substantially in eye-stalk length. Wilkinson and Reillo then performed paired-choice tests to assay female preferences in each population.

In each test, Wilkinson and Reillo placed two males in a cage with five females. The males had the same body size, but one was from the long-selected line, and thus had eyestalks that were long, and the other was from the short-selected line, and thus had eyestalks that were short (but still longer than the eyestalks of females from any line). The two males were separated by a clear plastic barrier, and each had his own artificial root hair on which to roost. In the center of the plastic barrier was a hole, just large enough to allow the females to pass back and forth, but too small for the males, with their longer eyestalks, to fit through. Wilkinson and Reillo then watched to see

which male attracted more females. The scientists performed 15 to 25 tests for each of the three lines. In both the control and the long-selected lines, more females chose to roost for the night with the long-stalked male. In the short-selected line, however, more females chose to roost for the night with the short-stalked male (Figure 9.23). Artificial selection for short eyestalks in males had changed the mating preferences of females.

This result neatly accomplishes several things at once:

- It demonstrates that female stalk-eyed flies are choosy.
- It demonstrates that both male eyestalk length and female preference are heritable.
- It illustrates that selection on one trait can produce an evolutionary response in another trait (see Chapter 7).

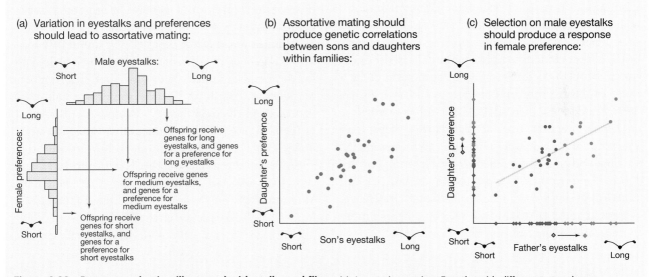

Figure 9.22   Runaway selection illustrated with stalk-eyed flies   (a) Assortative mating. Females with different tastes choose among males with different eyestalk lengths. If both traits are heritable, then offspring receiving genes for long eyestalks also tend to receive genes for "long" tastes. (b) A genetic correlation between male stalk length and female preference. Each point represents the average value of the offspring of a male mated with each of a number of randomly chosen females. Males whose sons have long eyestalks also tend to have daughters that seek long eyestalks in their mates. After Figure 4.6 in Arnold (1983). (c) Female preference evolves as a correlated response to selection on male stalk length. Each circle represents the stalklength and preference, respectively, of a father-daughter pair. The fathers are also represented by diamonds on the horizontal axis, and the daughters by diamonds on the vertical axis. If we select the longest-stalked males as breeders (red diamonds on the horizontal axis and red circles), we should see a response in the daughters. The arrows indicate the selection differential and predicted response (see Chapter 7). The gray diamond below the horizontal axis marks the average of all fathers in the population, and the red diamond marks the average of selected fathers. The gray diamond to the left of the vertical axis marks the average of all daughters, and the red diamond marks the average of daughters of selected males. After Falconer (1989).

Box 9.3   Continued

- It is consistent with Fisher's 80-year-old prediction that sexual selection by female choice produces genetic correlations between male traits and female preferences.

The scenario for the evolution of long eyestalks by runaway selection is as follows. At some time in the past, eyestalks were much shorter than they are now. At some point, a situation arose in which a majority of the females preferred longer-than-average eyestalks. Perhaps a preference for long stalks was favorable for females, because males with longer stalks were genetically superior, or perhaps the female preference was the result of genetic drift. Whatever the cause of the initial female preference, the consequence was that the males with long eyestalks left more offspring. As Fisher first noted, this can create a positive feedback loop, because as Wilkinson and Reillo's experiment showed, selection on males for longer eyestalks produces a correlated response in female preferences. Each generation's males have longer eyestalks than their fathers had, but each generation's females prefer longer stalks than their mothers did. Under the right circumstances, this positive-feedback loop can result in the automatic, or runaway, evolution of ever-longer eyestalks (see Fisher 1958; Lande 1981; Arnold 1983). In other words, it is at least theoretically possible that females prefer long eyestalks not because this preference carries any intrinsic fitness advantage for females or their young, or because of sensory biases built into the females' nervous systems, but simply because a small arbitrary preference, once established, led to runaway selection for ever more extreme preferences and ever longer eyestalks.

Is runaway selection the sole mechanism responsible for female preferences in stalk-eyed flies? We mentioned in the main text that the theories of female choice we have discussed are mutually compatible. Wilkinson and various colleagues have continued their research on stalk-eyed flies, looking for evidence of other mechanisms selecting on female preferences. Because males provide no parental care and offer no gifts to females, it seems unlikely that choosy females receive direct benefits—unless long-stalked males lay claim to better nighttime roosts. And Wilkinson, Heidi Kahler, and Richard Baker (1998) found no evidence that females had pre-existing sensory biases favoring long eyestalks before long eyestalks evolved in males. There is evidence, however, that choosy females get better genes for their offspring.

Choosy female stalk-eyed flies appear to get better genes for their offspring for at least two reasons. First, stalk-length in males is correlated with a trait biologists refer to as condition (David et al. 1998). Roughly speaking, condition is general health and vigor as demonstrated by the ability to gather and store energy. Wilkinson and Mark Taper (1999) found that condition is genetically variable. By choosing a male with long eyestalks, a female fly can give her offspring better genes for condition. Second, some male stalk-eyed flies carry an allele on their X-chromosome that causes them to have more daughters than sons, and some males carry an allele on their Y-chromosome that counteracts the allele on the X-chromosome, and causes the males to have more sons than daughters. Wilkinson, Daven Presgraves, and Lili Crymes (1998) found that the frequency of the X-chromosome allele does not vary across males with different stalk lengths,

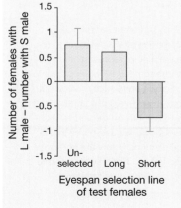

Figure 9.23   The results of Wilkinson and Reillo's paired choice tests for female preference   The height of each bar represents the average value (± standard error), for a number of trials, of the difference between the number of females that preferred the long-stalked male and the number that preferred the short-stalked male. Positive values indicate that more females preferred long-stalked males. Unselected females preferred long-stalked males ($P = 0.0033$), as did females from the long-selected line ($P = 0.005$). Females from the short-selected line preferred short-stalked males ($P = 0.023$). From Wilkinson and Reillo (1994). Copyright © 1994, The Royal Society.

Box 9.3   **Continued**

but the frequency of the Y-chromosome allele does. It is higher among males with long eyestalks. This matters because, in wild populations of stalk-eyed flies, females outnumber males and females can expect more grandoffspring through their sons than through their daughters. By choosing a mate with long eyestalks, a female fly can increase her chances of producing many sons.

In summary, long eyestalks appear to have evolved in stalk-eyed flies in response to a combination of male–male competition, a female preference for mates with good genes, and possibly a female preference reinforced by runaway selection. This combination of forces favoring long eyestalks raises a new question: Why are the males' eyestalks not even longer than they are now—for example, twice the length of the flies' bodies, or three times? One hypothesis is that if the eyestalks were any longer they would be a serious impediment to survival. As far as we know, this hypothesis has not been tested.

---

species (see Box 9.3 for an example). Another approach is to use the alternative hypotheses to develop testable predictions about how sexually selected traits will vary among closely related species on an evolutionary tree (see Prum 1997 for an example).

9.4 Diversity in Sex Roles

In all of the examples we have presented so far, the crucial fact explaining the roles taken by each sex is that access to mates limits the reproductive success of males more than it limits the reproductive success of females. This pattern is widespread, but it is by no means universal (Arnold and Duvall 1994). The pipefish species *Nerophis ophidion* and *Syngnathus typhle*, studied by Gunilla Rosenqvist, Anders Berglund, and their colleagues, provide a counterexample. These pipefish, which live in eelgrass beds, are relatives of the seahorses (Figure 9.24). As in seahorses, males provide all the parental care. In *N. ophidion*, the male has a brood patch on his belly; in *S. typhle*, the male has a brood pouch. In both species, the female lays her eggs directly onto or into the male's brood structure. The male supplies the eggs with oxygen and nutrients until they hatch.

Species in which males invest more in each offspring, and are thus a limiting resource for females, are the exceptions that can prove the rules of sexual selection.

Although the extensive parental care provided by male pipefish requires energy, the pivotal currency for pipefish reproduction is not energy but time (Berglund et al. 1989). Females of both *N. ophidion* and *S. typhle* can make eggs faster than males can rear them to hatching. As a result, access to male brood space limits female reproductive success. If the theory of sexual selection we have developed is correct, then in these pipefish the females should compete with each other over access to males, and the males should be choosy.

In *N. ophidion,* the females are larger than the males and have two traits the males lack: dark blue stripes and skin folds on their bellies. These traits appear to function primarily as advertisements for attracting mates. For example, females develop skin folds during the breeding season and lose them after, and in captivity females develop skin folds only when males are present (Rosenqvist 1990). In paired-choice tests (Figure 9.24a), *N. ophidion* males are choosy, preferring larger females (Figure 9.24b) and females with larger skin folds (Figure 9.24c). Females, in contrast, appear to be less choosy. In paired-choice tests, females showed no tendency to discriminate between males of different sizes (Berglund and Rosenqvist 1993).

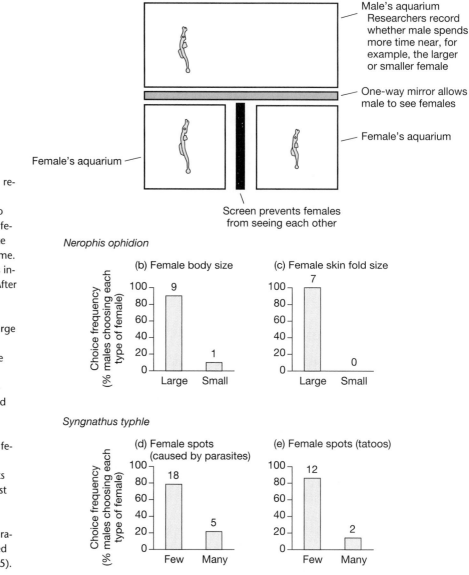

(a) Experimental design for paired choice tests

Male's aquarium Researchers record whether male spends more time near, for example, the larger or smaller female

One-way mirror allows male to see females

Female's aquarium

Female's aquarium

Screen prevents females from seeing each other

Nerophis ophidion

(b) Female body size

(c) Female skin fold size

Syngnathus typhle

(d) Female spots (caused by parasites)

(e) Female spots (tatoos)

Figure 9.24 Male choice in pipefish (a) In paired-choice tests, researchers place a male pipefish in an aquarium from which he can see two females. The researchers infer which female the male would prefer as a mate from where he spends more of his time. In (b–d) the numbers above the bars indicate the number of males tested. After Rosenqvist and Johansson (1995). (b) Given a choice between large or small females, male pipefish prefer large females ($P = 0.022$). Replotted from Rosenqvist (1990). (c) Given a choice between females with large or small skin folds, male pipefish prefer large-folded females ($P = 0.016$). Replotted from Rosenqvist (1990). (d) Given a choice between females with many black spots (caused by a parasite) or females with few black spots, male pipefish prefer females with few spots ($P < 0.05$). Replotted from Rosenqvist and Johansson (1995). (e) Males still prefer females with few spots, even when the spots are tattooed onto parasite-free females ($P < 0.01$). Replotted from Rosenqvist and Johansson (1995).

Male pipefish brood their young, and access to males is a limiting resource for females. As predicted, females fight among themselves, and males are choosy.

In *S. typhle*, the males and females are similar in size and appearance. Females, however, can change their color to intensify the zigzag pattern on their sides (Berglund et al. 1997; Bernet et al. 1998). The females compete with each other over access to males (Berglund 1991), and while doing so display their dark colors. Females initiate courtship, and mate more readily than males (Berglund and Rosenqvist 1993). Males are choosy (Rosenqvist and Johansson 1995). In paired-choice tests (Figure 9.24a), male *S. typhle* prefer females showing fewer of the black spots that indicate infection with a parasitic worm, whether the black spots were actually caused by parasites (Figure 9.24d) or were tattooed onto the females (Figure 9.24e). This choosiness benefits the males directly, because females with fewer parasites lay more eggs for the males to fertilize and rear.

The mating behavior of pipefish males and females is consistent with the theory of sexual selection. Other examples of "sex-role reversed" species whose behavior appears to support the theory include moorhens (Petrie 1983), spotted sandpipers (Oring et al. 1991a,b, 1994), giant waterbugs (see Anderson 1994), and some species of katydids (Gwynne 1981; Gwynne and Simmons 1990).

9.5 Sexual Selection in Plants

Plants are often sexually dimorphic. Orchids in the genus *Catasetum* provide the most dramatic example. So different are the flowers of the two sexes that early or-chid systematists placed the males (Figure 9.25a) in one genus and the females (Figure 9.25b) in another. The herb *Wurmbea dioica*, from Australia, provides a more typical example. The males make larger flowers than the females (Figure 9.25c). We have seen that sexual selection can explain sexual dimorphism in ani-mals. Can it also explain sexual dimorphism in plants?

Many of the ideas we have developed about sexual selection in the context of animal mating can, in fact, be applied to plants (Bateman 1948; Willson 1979). In plants, mating involves the movement of pollen from one individual to another. The recipient of the pollen, the seed parent, must produce a fruit. As a result, the

Sexual selection theory can be applied to plants as well as animals.

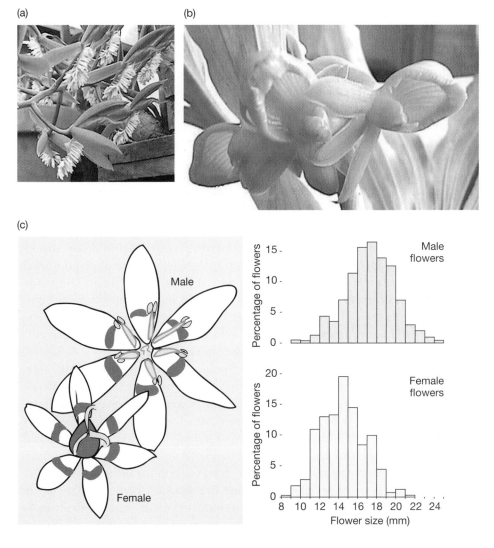

Figure 9.25 Sexual dimor-phism in plants (a) In the orcid *Catasetum barbatum*, males (a) and females (b) produce strikingly dif-ferent flowers. (Sharon Dahl) In the herb *Wurmbea dioica* (c), males make larger flowers. Redrawn from Vaughton and Ramsey (1998). Copyright © 1998, Springer Verlag. Reprinted by permission.

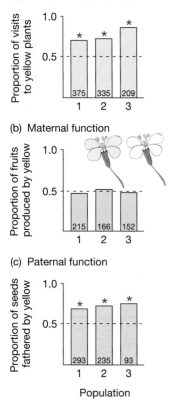

(a) Pollinator discrimination

(b) Maternal function

(c) Paternal function

Population

Figure 9.26 Reproductive success through pollen donation is more stongly affected by the number of pollinator visits than is reproductive success through the production of seeds The numbers inside each bar represent the number of pollinator visits (a), the number of fruits (b), and the number of seeds (c) examined. Bars marked with an asterisk have heights significantly different from 0.5 ($P < 0.0001$). (a) In populations with equal numbers of white and yellow flowers, the yellow-flowered plants got most of the pollinator visits. (b) In spite of the inequality in pollinator visits, white- and yellow-flowered plants produced equal numbers of fruits. (c) The majority of the seeds, however, were fathered by yellow plants. From Stanton et al. (1986). Copyright © 1986, Association for the Advancement of Science.

seed parent may make a larger reproductive investment per seed than the pollen donor, which must only make pollen. When pollen is transported from individual to individual by animals, a plant's access to mates is a function of its access to pollinators. Based on the principles of sexual selection in animals, we can hypothesize that access to pollinators limits the reproductive success of pollen donors to a greater extent than it limits the reproductive success of seed parents.

Maureen Stanton and colleagues (1986) tested this hypothesis in wild radish (*Raphanus raphanistrum*). Wild radish is a self-incompatible annual herb that is pollinated by a variety of insects, including honeybees, bumblebees, and butterflies. Many natural populations of wild radish contain a mixture of white-flowered and yellow-flowered individuals. Flower color is determined by a single locus: White (*W*) is dominant to yellow (*w*). Stanton and colleagues set up a study population with eight homozygous white plants (*WW*) and eight yellow plants (*ww*). The scientists monitored the number of pollinator visits to plants of each color, then measured reproductive success through female and male function.

Measuring reproductive success through female function was easy: The researchers just counted the number of fruits produced by each plant of each color. Measuring reproductive success through male function was harder; in fact, it was not possible at the level of individual plants. Note, however, that a yellow seed parent (*ww*) will produce yellow offspring (*ww*) if it mated with a yellow pollen donor (*ww*), but white offspring (*Ww*) if it mated with a white pollen donor (*WW*). Thus by rearing the seeds produced by the yellow seed parents and noting the color of their flowers, Stanton and colleagues could compare the population-level reproductive success of white versus yellow pollen donors. The relative reproductive success of pollen donors through yellow seed parents should be a reasonable estimate of the pollen donors' relative reproductive success through seed parents of both colors. The scientists repeated their experiment three times.

As Stanton and colleagues expected from previous research, the yellow-flowered plants got about 3\4 of the pollinator visits (Figure 9.26a). If reproductive success is limited by the number of pollinator visits, then the yellow-flowered plants should also have gotten about 3\4 of the reproductive success. This was true for reproductive success through pollen donation (Figure 9.26c), but not for reproductive success through seed production (Figure 9.26b). Reproductive success through seed production was simply proportional to the number of plants of each type. These results are consistent with the typical pattern in animals: The reproductive success of males is more limited by access to mates than is the reproductive success of females. The results also suggest that the evolution of showy flowers that attract pollinators has been driven more by their effect on male reproductive success than on female reproductive success (Stanton et al. 1986).

If it is true in general that the number of pollinator visits is more important to male reproductive success than to female reproductive success, then in animal-pollinated plant species with separate male and female flowers, the flowers should be dimorphic and the male flowers should be more attractive. Lynda Delph (1996) and colleagues tested this hypothesis with a survey of animal- and wind-pollinated plants, including both **dioecious** species (separate male and female individuals) and **monoecious** species (separate male and female flowers on the same individual).

Delph and her coauthors first noted that the showiest parts of a flower, the petals and sepals that together form the perianth, serve not only to attract pollinators, but also to protect the reproductive structures when the flower is developing in the bud. If protection were the only function of the perianth, then the sex that has the bigger reproductive parts should always have the bigger perianth. This was indeed the case in all 11 wind-pollinated species Delph and colleagues measured (Figure 9.27a, right). If, however, pollinator attraction is also important, and more important to males than to females, then there should be a substantial number of species in which the female flowers have bigger reproductive parts, but the male flowers have bigger perianths. This was the case in 29% of the 42 animal-pollinated plants Delph and colleagues measured (Figure 9.27a, left). Furthermore, in species that are sexually dimorphic, male function tends to draw a greater investment in number of flowers per inflorescence and in strength of floral odor, although not in quantity of nectar (Figure 9.27b). These results are consistent with the hypothesis that sexual selection, via pollinator attraction, is often stronger for male flowers than for female flowers.

Can sexual selection explain the particular examples of sexual dimorphism we introduced at the beginning of this section? Recall that in the herb *Wurmbea dioica*, males make larger flowers than females. This plant is pollinated by bees, butterflies, and flies. Glenda Vaughton and Mike Ramsey (1998) found that bees and butterflies visit larger flowers at higher rates than smaller flowers. As a result, pollen is removed from large flowers more quickly than from small flowers. Males with large flowers may benefit from exporting their pollen more quickly if a head start allows their pollen to beat the pollen of other males in the race to females' ovules. In addition, larger male flowers make more pollen, giving the pollen donor more chances to win. For females, larger flowers probably do not confer any benefit. Female flowers typically receive more than four times the pollen needed to fertilize all their ovules, and seed production is therefore not limited by pollen. These patterns are consistent with the hypothesis that sexual selection on males is responsible for the sexual dimorphism in flower size.

When male reproductive success is limited by access to pollinators, but female reproductive success is not, male flowers may evolve showier displays than female flowers.

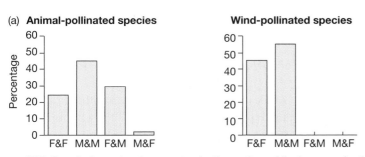

(a) **Animal-pollinated species**   **Wind-pollinated species**

F&F: Female flower has larger reproductive parts and the larger perianth.
M&M: Male flower has larger reproductive parts and the larger perianth.
F&M: Female flower has larger reproductive parts, but male has the larger perianth.
M&F: Male flower has larger reproductive parts, but female has the larger perianth.

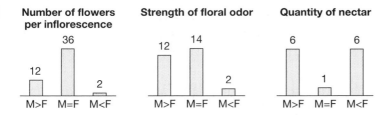

M>F: Male investment is greater than female investment.
M=F: Male investment and female investment are equal.
M<F: Male investment is less than female investment.

Figure 9.27   Patterns of sexual dimorphism in plants with separate male and female flowers (a) In all 11 wind-pollinated species measured (right), the sex with larger reproductive parts has the larger perianth. In 29% of the 42 animal-pollinated species measured (left), the female has larger reproductive parts, but the male has the larger perianth, a pattern significantly different from that for wind-pollinated species ($P < 0.001$). From Delph et al. (1996). (b) When animal-pollinated plants have flowers that are sexually dimorphic for investment in pollinator attraction, investment by males tends to be larger for two of the three traits studied ($P < 0.01$; $P = 0.01$; N.S.). Drawn from data in Delph et al. (1996). Copyright © 1996, American Naturalist. Reprinted by permission of The University of Chicago Press.

Orchids in the genus *Catasetum*, the plants with the dramatically dimorphic flowers, have an unusual pollination system. They are pollinated exclusively by male euglossine bees. The orchids attract the bees with fragrant chemicals, such as cineole, which the bees collect and use in their own attempts to attract mates. The flowers of male *Catasetum* orchids are loaded with a single pollen-bearing structure, called a pollinarium. When a bee trips the trigger in a male flower, the flower shoots the pollinarium at the bee like a rubber band off a finger. The pollinarium sticks to the back of the bee with an adhesive that makes it impossible for the bee to remove. When the bee later visits a female flower, one of the pollen masses on the pollinarium lodges in the receptive structure on the female flower, called the stigmatic cleft, and is torn from the pollinarium. The stigmatic cleft quickly swells shut. This means that a female flower typically receives pollen from just one male.

Gustavo Romero and Craig Nelson (1986) observed bees pollinating the flowers of *Catasetum ochraceum*. The researchers found that after being shot with a pollinarium by one male flower, the bees avoided visiting other male flowers but continued to forage in female flowers. Romero and Nelson offer the following scenario to account for the sexual dimorphism in the flowers of *Catasetum ochraceum*. At any given time, there are many more male flowers blooming than female flowers. This, in combination with the fact that female flowers accept pollen from only one male, means that there is competition among male flowers over opportunities to mate. The competition is further intensified by the fact that a second pollinarium attached to a bee would probably interfere with the first. A male flower that has attracted a bee and loaded it with a pollinarium would be at a selective advantage if it could prevent the bee from visiting another male flower. It is therefore adaptive for male flowers to train bees to avoid other male flowers, so long as they do not also train the bees to avoid female flowers. If this scenario is correct, forcible attachment of the pollinarium to the bee and sexually dimorphic flowers make sense together, and both are due to competition for mates—that is, to sexual selection.

9.6 Sexual Dimorphism in Body Size in Humans

One of the examples of sexual dimorphism that we cited at the beginning of this chapter was body size in humans (Figure 9.2). We now ask whether the sexual dimorphism in human size is the result of sexual selection. It is a difficult question to answer, because sexual selection concerns mating behavior. The evolutionary significance of human behavior is hard to study for at least two reasons:

- Human behavior is driven by a complex combination of culture and biology. Studies based on the behavior of people in any one culture provide no means of disentangling these two influences. Cross-cultural studies can identify universal traits or broad patterns of behavior, either of which may warrant biological explanations. Cultural diversity is rapidly declining, however, and some biologists feel that it is no longer possible to do a genuine cross-cultural study.

- Ethical and practical considerations prohibit most of the kinds of experiments we might conduct on individuals of other species. This means that most studies of human behavior are observational. Observational studies can identify correlations between variables, but they offer little evidence of cause and effect.

Human behavior is inherently fascinating, however, and we therefore proceed, with caution, to briefly consider the question of sexual selection and body size in humans.

The most basic knowledge of human reproductive biology indicates that the opportunity for sexual selection is greater in men than in women. Data from a single culture will suffice to illustrate this point. Research by Monique Borgerhoff Mulder (1988) on the Kipsigis people of southwestern Kenya revealed that the men with highest reproductive success had upwards of 25 children, while the most prolific women rarely had more than 10 (Figure 9.28). In Kipsigis culture, it appears that the reproductive success of men was limited by mating opportunities to a greater extent than was the reproductive success of women. But is there any evidence that reproductive competition, either via male–male interactions or female choice, selects for larger body size in men?

The most obvious kind of sexual selection to look at is male–male competition, because it drives the evolution of large male size in a great variety of other species. Men do, on occasion, compete among themselves over access to mates, but so do women. Do men compete more intensely? On the reasoning that homicide is an unambiguous indication of conflict, and that virtually all homicides are reported to the police, Martin Daly and Margo Wilson (1988) assembled data on rates of same-sex homicide from a variety of modern and traditional cultures. In all of these cultures, men kill men at much higher rates than women kill women. In the culture with the most *balanced* rates of male–male vs. female-female killings, men committed 85% of the same-sex homicides. In several cultures, men committed all of the same-sex homicides. Data from the United States and Canada show that the majority of perpetrators, and victims, of male–male homicides are in their late teens, twenties, and early thirties. On these and other grounds, Daly and Wilson interpret much male–male homicide as a manifestation of sexually selected competition among men.

If Daly and Wilson's interpretation is correct, then men who are more successful in male–male combat should have higher mating success and higher fitness, at least in pre-modern cultures without formal police and criminal justice systems. Napoleon Chagnon (1988) reported data on the Yąnomamö that confirm this prediction, at least for one culture. The Yąnomamö are a pre-modern people that live in the Amazon rain forest in Venezuela and Brazil. They take pride in their ferocity. Roughly 40% of the adult men in Chagnon's sample had participated in a homicide, and roughly 25% of the mortality among adult men was due to homicide. The Yąnomamö refer to men who have killed as *unokais*. Chagnon's data show that *unokais* have singificantly more wives, and significantly more children, than non-*unokais* (Figure 9.29).

The Yąnomamö fight with clubs, arrows, spears, machetes, and axes. It would be reasonable to predict that *unokais* are larger that non-*unokais*. Chagnon (1988) reports, however, that "Personal, long-term familiarity with all the adult males in this study does not encourage me to conclude at this point that they could easily be sorted into two distinct groups on the basis of obvious biometric characters, nor have detailed anthropometric studies of large numbers of Yąnomamö males suggested this as a very likely possibility." Data on the relationship between male–male competition, body size, and mating success in other cultures are scarce.

B. Pawlowski and colleagues (2000) investigated the hypothesis that the sexual dimorphism in human body size is a result of female choice. The researchers gathered data from the medical records of 3201 Polish men. They used statistical techniques to remove the effects of a variety of confounding variables, including residence in cities versus rural areas, age, and education. Pawlowski and colleagues then com-

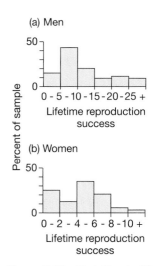

Figure 9.28 Variation in lifetime reproductive success among Kipsigis men and women (a) For the men, the height of each bar represents the percentage of men who had 0 to 5 children, 6 to 10 children, and so on. (b) For the women, the height of each bar represents the percentage of women who had 0 to 2 children, 3 or 4 children, 5 or 6 children, and so on. Some of the men had more than 25 children; few of the women had more than 10. From Borgerhoff Mulder (1988). Copyright © 1988, University of Chicago Press. Reprinted by permission.

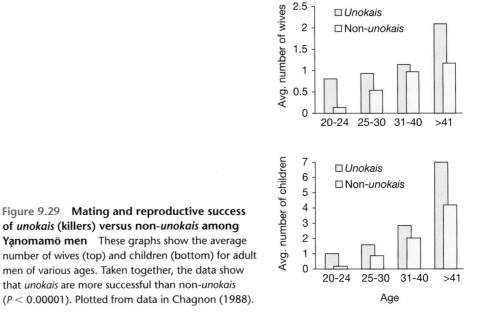

Figure 9.29 **Mating and reproductive success of *unokais* (killers) versus non-*unokais* among Yanomamö men**   These graphs show the average number of wives (top) and children (bottom) for adult men of various ages. Taken together, the data show that *unokais* are more successful than non-*unokais* ($P < 0.00001$). Plotted from data in Chagnon (1988).

pared bachelors to married men. The married men were taller by a small but statistically significant margin. In addition, men with one or more children were significantly taller than childless men. The exception to this pattern was the group of men in their fifties, within which there was no difference in height between fathers versus childless men. Pawlowski and colleagues note that the men in their fifties reached marrying age shortly after World War II, when the ratio of women to men in Poland was unusually high. The researchers speculate that the men in their fifties had experienced less intense sexual selection, via female choice, than is the norm.

The evolutionary significance of sexual size dimorphism in humans is unresolved. The observational studies we have reviewed provide mixed results, and evidence that is suggestive at best. It is also possible that we humans simply inherited our sexual size dimorphism from our ancestors, who were more sexually dimorphic in size than we are (McHenry 1992). What is really needed to settle the issue is data from a larger number of cultures on the relationship between body size, number of mates, survival, and reproductive success for both women and men. Preferably, the data would come from hunter-gatherer cultures, whose members live the life-style ancestral for our species. The most technically challenging factor to measure accurately is the reproductive success of men. Modern techniques for genetic analysis have made it feasible, in principle, to collect such data (Figure 9.15). However, the research remains to be done.

It is unclear whether sexual selection has played a role in maintaining the sexual dimorphism in body size in humans. Males compete for mates, but larger males do not necessarily win. Females are choosy, and limited data suggest a slight preference for taller men. More studies must be done on a greater variety of cultures.

Summary

Sexual dimorphism, a difference in form or behavior between females and males, is common. The difference often involves traits, like the enormous tail feathers of the peacock, that appear to be opposed by natural se-

lection. To explain these puzzling traits, Darwin invoked sexual selection. Sexual selection is differential reproductive success resulting from variation in mating success.

Mating success is often a more important determinant of fitness for one sex than for the other. Usually, but by no means always, it is males whose reproductive success is limited by mating opportunities, and females whose reproductive success is limited by resources rather than matings.

The members of the sex experiencing strong sexual selection typically compete among themselves over access to mates. This competition may involve direct combat, gamete competition, infanticide, or advertisement.

The members of the sex whose reproductive success is limited by resources rather than matings are typically choosy. This choosiness may provide the chooser with direct or indirect benefits, such as food or better genes for its offspring, or it may be the result of a pre-existing sensory bias.

The theory of sexual selection was developed to explain sexual dimorphism in animals, but it applies to plants as well. In plants, access to pollinators is often more limiting to reproductive success via pollen donation than to reproductive success via seed production. This can lead to the evolution of sexual dimorphism in which male flowers are showier than female flowers.

Questions

1. The graphs in Figure 9.30 show the variation in lifetime reproductive success of male vs. female elephant seals (Le Bouef and Reiter 1988). Note that the scales on the horizontal axes are different. Why is the variation in reproductive success so much more extreme in males than females? Draw a graph showing your hypothesis for the relationship between number of mates and reproductive success for male and female elephant seals. Why do you think male elephant seals are four times larger than female elephant seals? Why aren't males even bigger?

2. What sex is the sage grouse pictured in Figure 9.31? What is it doing and why? Do you think this individual provides parental care? What else can you guess about the social system of this species?

3. We used long-tailed widowbirds as an example of a dimorphic species at the beginning of this chapter. Suggest two hypotheses for why male widowbirds have such long tails. Design an experiment to test your ideas. (*Hint:* Anders Møller's experiments on barn swallows were inspired by Malte Anderson's research on widowbirds.) Look up Anderson's paper on widowbirds (Anderson 1982) and see if he designed the same experiment that you did.

4. Male butterflies and moths commonly drink from puddles, a behavior known as puddling. Scott Smedley and Thomas Eisner (1996) report a detailed physiological analysis of puddling in the moth *Gluphisia septentrionis*. A male *G. septentrionis* may puddle for hours at a time. He rapidly processes huge amounts of water, extracting the sodium and expelling the excess liquid in anal jets (see Smedley and Eisner's paper for a dramatic photo). The male moth will later give his harvest of sodium to a female during mating. The female will then put much of the sodium into her eggs. Speculate on the role this gift plays in the moth's mating ritual, and in the courtship roles taken by the male and the female. How would you test your ideas?

5. Males in many species often attempt to mate with strikingly inappropriate partners. Ryan (1985), for example, describes male túngara frogs clasping other males. Some orchids mimic female wasps and are pollinated by amorous male wasps—who have to be fooled twice for the strategy to work. Would a female túngara or a female wasp make

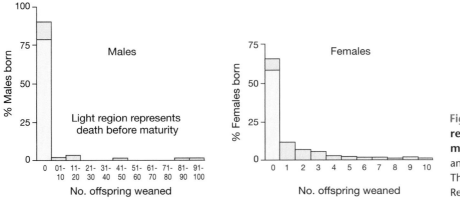

Figure 9.30 **Distributions of lifetime reproductive success in male and female elephant seals** From Le Boeuf and Reiter (1988). Copyright © 1988, The University of Chicago Press. Reprinted by permission.

Figure 9.31 Sage grouse

the same mistake? Why or why not? (More general answers—applicable to a wider range of species—are better.)

6. Do you think there is any association in humans between infection with parasites and physical appearance? The scatterplot in Figure 9.32 shows the relationship between the importance of attractiveness in mate choice (as reported by subjects responding to a questionnaire) and the prevalence of six species of parasites (including leprosy, malaria, and filaria) in 29 cultures (Gangestad 1993; Gangestad & Buss 1993). (Note that statistical techniques have been used to remove the effects of latitude, geographic region, and mean income). What is the pattern in the graph? Does this pattern make sense from an evolutionary perspective?

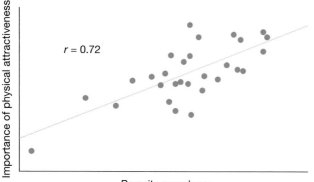

Figure 9.32 Importance of physical attractiveness in mate choice versus parasite prevalence in 29 human cultures

One of the parasitic diseases is schistosomiasis. We presented evidence in Chapter 4 that resistance to schistosomiasis is heritable. What do women gain (evolutionarily) by choosing an attractive mate? What do men gain (evolutionarily) by choosing an attractive mate? Can you offer a cultural explanation that could also account for this pattern?

7. In many katydids, the male delivers his sperm to the female in a large spermatophore which contains nutrients the female eats (for a photo, see Gwynne 1981). The female uses these nutrients in the production of eggs. Darryl Gwynne and L.W. Simmons (1990) studied the behavior of caged populations of an Australian katydid under low food (control) and high food (extra) conditions. Some of their results are graphed in Figure 9.33. (The graph shows the results from four sets of replicate cages; calling males = average number of males calling at any given time; matings/female = average number of times each female mated; % reject by M = fraction of the time a female approached a male for mating and was rejected; % reject by F = fraction of the time a female approached a male but then rejected him before copulating; % with F–F comp = fraction of matings in which one or more females were seen fighting over the male.) When were the females choosy and the males competitive? When where the males choosy and the females competitive? Why?

8. In some species of deep sea anglerfish, the male lives as a symbiont permanently attached to the female (see Gould 1983, essay 1). The male is tiny compared to the female. Many of the male's organs, including the eyes, are reduced, though the testes remain large. Others, such as the jaws and teeth, are modified for attachment to the female. The circulatory systems of the two sexes are fused, and the male receives all of his nutrition from the female via the shared bloodstream. Often, two or more males are attached to a single female. What are the costs and benefits of the male's symbiotic habit for the male? For the female? What limits the lifetime reproductive

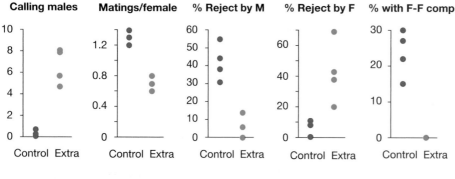

Figure 9.33 Behavior of male and female katydids under control versus extra-food conditions

success of each sex—the ability to gather resources, or the ability to find mates? Do you think that the male's symbiotic habit evolved as a result of sexual selection, or natural selection? (It may be helpful to break the male symbiotic syndrome into separate features, such as staying with a single female for life, physical attachment to the female, reduction in body size, and nutritional dependence on the female.)

Exploring the Literature

9. If a single insemination provides all the sperm necessary to fertilize an entire clutch of eggs, then what do females gain by engaging in extra-pair copulations? For hypotheses and tests, see

   Kempenaers, B., G. R. Verheyn, M. van den Broeck, T. Burke, C. van Broeckhoven, and A. A. Dhondt. 1992. Extra-pair paternity results from female preference for high-quality males in the blue tit. *Nature* 357: 494–496.

   Madsen, T., R. Shine, J. Loman, and T. Hakansson. 1992. Why do female adders copulate so frequently? *Nature* 355: 440–441.

   Gray, E.M. 1996. Female control of offspring paternity in a western population of red-winged blackbirds (*Agelaius phoeniceus*). *Behavioral Ecology and Sociobiology* 38: 267–278.

   Gray, E. M. 1997. Do female red-winged blackbirds benefit genetically from seeking extra-pair copulations? *Animal Behaviour* 53: 605–623.

   Gray, E. M. 1997. Female red-winged blackbirds accrue material benefits from copulating with extra-pair males. *Animal Behaviour* 53: 625–629.

10. Why do females sometimes copulate more than once with the same male?

    Petrie, M. 1992. Copulation frequency in birds: Why do females copulate more than once with the same male? *Animal Behaviour* 44: 790–792.

11. The testes in many mammals, including humans, are positioned in a scrotum outside the abdominal cavity. This dangerous arrangement has long defied adequate evolutionary explanation. One hypothesis is that the evolution of scrotal testes was driven by sperm competition. See

    Freeman, S. 1990. The evolution of the scrotum: A new hypothesis. *Journal of Theoretical Biology* 145: 429–445.

12. In plants, the analog of sperm competition is a race among pollen tubes to reach the ovules. For an exploration of whether the pollen of some donors is consistently superior in competition with the pollen from other donors, see

    Snow, A. A., and T. P. Spira. 1991. Pollen vigor and the potential for sexual selection in plants. *Nature* 352: 796–797.

    Snow, A. A., and T. P. Spira. 1996. Pollen-tube competition and male fitness in *Hibiscus moscheutos*. *Evolution* 50: 1866–1870.

13. Peacocks are among the most famous animals with an elaborate male mating display. For research on sexual selection in peacocks, see

    Manning, J. T., and M. A. Hartley. 1991. Symmetry and ornamentation are correlated in the peacock's train. *Animal Behaviour* 42: 1020–1021.

    Petrie, M. 1992. Peacocks with low mating success are more likely to suffer predation. *Animal Behaviour* 44: 585–586.

    Petrie, M., T. Halliday, and C. Sanders. 1991. Peahens prefer peacocks with elaborate trains. *Animal Behaviour* 41: 323–331.

Citations

Anderson, M. 1982. Female choice selects for extreme tail length in a widowbird. *Nature* 299: 818–820.

Anderson, M. 1994. *Sexual Selection*. Princeton, NJ: Princeton University Press.

Arnold, S. J. 1983. Sexual selection: The interface of theory and empiricism. In P. Bateson, ed. *Mate Choice*. Cambridge: Cambridge University Press, 67–107.

Arnold, S. J., and D. Duvall. 1994. Animal mating systems: A synthesis based on selection theory. *American Naturalist* 143: 317–348.

Bateman, A. J. 1948. Intra-sexual selection in Drosophila. *Heredity* 2: 349–368.

Berglund, A. 1991. Egg competition in a sex-role reversed pipefish: Subdominant females trade reproduction for growth. *Evolution* 45: 770–774.

Berglund, A., and G. Rosenqvist. 1993. Selective males and ardent females in pipefish. *Behavioral Ecology and Sociobiology* 32: 331–336.

Berglund, A., G. Rosenqvist, and P. Bernet. 1997. Ornamentation predicts reproductive success in female pipefish. *Behavioral Ecology and Sociobiology* 40: 145–150.

Berglund, A., G. Rosenqvist, and I. Svensson. 1989. Reproductive success of females limited by males in two pipefish species. *American Naturalist* 133: 506–516.

Bernet, P., G. Rosenqvist, and A. Berglund. 1998. Female-female competition affects female ornamentation in the sex-role reversed pipefish *Syngnathus typhle*. *Behaviour* 135: 535–550.

Bertram, C. R. 1975. Social factors influencing reproduction in wild lions. *Journal of Zoology* 177: 463–482.

Boellstorff, D. E., D. H. Owings, M. C. T. Penedo, and M. J. Hersek. 1994. Reproductive behavior and multiple paternity of California ground squirrels. *Animal Behaviour* 47: 1057–1064.

Borgerhoff Mulder, M. 1988. Reproductive success in three Kipsigis co-horts. In T.H. Clutton-Brock, ed. *Reproductive Success.* Chicago: University of Chicago Press, 419–435.

Burkhardt, D., and I. de la Motte. 1983. How stalk-eyed flies eye stalk-eyed flies: Observations and measurements of the eyes of *Cyrtodiopsis whitei* (Dopsidae, Diptera). *Journal of Comparative Physiology* 151: 407–421.

Burkhardt, D., and I. de la Motte. 1987. Physiological, behavioural, and morphometric data elucidate the evolutive significance of stalked eyes in Diopsidae (Diptera). *Entomologia Generalis* 12: 221–233.

Burkhardt, D., and I. de la Motte. 1988. Big "antlers" are favored: Female choice in stalk-eyed flies (Diptera, Insecta), field collected harems and laboratory experiments. *Journal of Comparative Physiology* A 162: 649–652.

Chagnon, N. A. 1988. Life histories, blood revenge, and warfare in a tribal population. *Science* 239: 985–992.

Clutton-Brock, T. H. 1985. Reproductive Success in Red Deer. *Scientific American* 252(February): 86–92.

Daly, M., and M. Wilson. 1988. *Homicide.* New York: Aldine de Gruyter.

Darwin, C. 1871. *The Descent of Man, and Selection in Relation to Sex.* London: John Murray.

David, P., A. Hingle, et al. 1998. Male sexual ornament size but not asymmetry reflects condition in stalk-eyed flies. *Proceedings of the Royal Society London,* Series B 265: 2211–2216.

Delph, L. F., L. F. Galloway, and M. L. Stanton. 1996. Sexual dimorphism in flower size. *American Naturalist* 148: 299–320.

Enquist, M., and A. Arak. 1993. Selection of exaggerated male traits by female aesthetic senses. *Nature* 361: 446–448.

Falconer, D. S. 1989. *Introduction to Quantitative Genetics.* New York: John Wiley & Sons.

Fisher, R. A. 1915. The evolution of sexual preference. *Eugenics Review* 7: 184–192.

Fisher, R. A. 1958. *The Genetical Theory of Natural Selection,* 2nd ed. New York: Dover.

Gage, M. J. G. 1991. Risk of sperm competition directly affects ejaculate size in the Mediterranean fruit fly. *Animal Behaviour* 42: 1036–1037.

Gangestad, S. W. 1993. Sexual selection and physical attractiveness: Implications for mating dynamics. *Human Nature* 4: 205–235.

Gangestad, S. W., and D. M. Buss. 1993. Pathogen prevalence and human mate preferences. *Ethology and Sociobiology* 14: 89–96.

Gerhardt, H. C., M. L. Dyson, and S. D. Tanner. 1996. Dynamic properties of the advertisement calls of gray tree frogs: Patterns of variability and female choice. *Behavioral Ecology* 7: 7–18.

Gibbs, H. L., P. J. Weatherhead, P. T. Boag, B. N. White, L. M. Tabak, and D. J. Hoysak. 1990. Realized reproductive success of polygynous red-winged blackbirds revealed by DNA markers. *Science* 250: 1394–1397.

Gilbert, L. E. 1976. Postmating female odor in *Heliconius* butterflies: A male-contributed anti-aphrodisiac? *Science* 193: 419–420.

Gould, S. J. 1983. *Hen's Teeth and Horse's Toes.* New York: W.W. Norton & Company.

Gray, E. M. 1997. Do female red-winged blackbirds benefit genetically from seeking extra-pair copulations? *Animal Behaviour* 53: 605–623.

Gross, M. R. 1984. Sunfish, salmon, and the evolution of alternative reproductive strategies and tactics in fishes. In G. W. Potts and R. J. Wootton, eds. *Fish Reproduction: Strategies and Tactics.* London: Academic Press, 55–75.

Gross, M. R. 1985. Disruptive selection for alternative live histories in salmon. *Nature* 313: 47–48.

Gross, M. R. 1991. Salmon breeding behavior and life history evolution in changing environments. *Ecology* 72: 1180–1186.

Gwynne, D. T. 1981. Sexual difference theory: Mormon crickets show role reversal in mate choice. *Science* 213: 779.

Gwynne, D. T., and L. W. Simmons. 1990. Experimental reversal of courtship roles in an insect. *Nature* 346: 172–174.

Hooper, R. E., and M. T. Siva-Jothy. 1996. Last male sperm precendence in a damselfly demonstrated by RAPD profiling. *Molecular Ecology* 5: 449–452.

Lande, R. 1981. Models of speciation by sexual selection on polygenic traits. *Proceedings of the National Academy of Sciences, USA* 78: 3721–3725.

Le Boeuf, B. J., and J. Reiter. 1988. Lifetime reproductive success in northern elephant seals. In T.H. Cutton-Brock, ed. *Reproductive Success.* Chicago: University of Chicago Press, 344–362.

McHenry, H. M. 1992. Body size and proportions in early hominids. *American Journal of Physical Anthropology* 87: (4) 407–431 Apr 1992

Møller, A. P. 1988. Female choice selects for male sexual tail ornaments in the monogamous swallow. *Nature* 332: 640–642.

Møller, A. P. 1991. Sexual selection in the monogamous barn swallow (*Hirundo rustica*). I. Determinants of tail ornament size. *Evolution* 45: 1823–1836.

Nakatsuru, K., and D. L. Kramer. 1982. Is sperm cheap? Limited male fertility and female choice in the lemon tetra (Pisces, Characidae). *Science* 216: 753–755.

Nowak, R. M. 1991. *Walker's Mammals of the World.* Baltimore: Johns Hopkins University Press.

Olsson, M., A. Gullberg, and H. Tegelstrom. 1994. Sperm competition in the sand lizard, *Lacerta agilis. Animal Behaviour* 48: 193–200.

Oring, L. W., M. A. Colwell, J. M. Reed. 1991a. Lifetime reproductive success in the spotted sandpiper (*Actitis macularia*)—sex-differences and variance-components. *Behavioral Ecology and Sociobiology* 28: 425–432.

Oring L. W., J. M. Reed, et al. 1991b. Factors regulating annual mating success and reproductive success in spotted sandpipers (*Actitis macularia*). *Behavioral Ecology and Sociobiology* 28: 433–442.

Oring, L. W., J. M. Reed, and S. J. Maxson. 1994. Copulation patterns and mate guarding in the sex-role reversed, polyandrous spotted sandpiper, *Actitis macularia. Animal Behaviour* 47: (5) 1065–1072.

Packer, C., L. Herbst, A. E. Pusey, J. D. Bygott, J. P. Hanby, S. J. Cairns, and M. Borgerhoff Mulder. 1988. Reproductive success of lions. In T. H. Clutton-Brock, ed. *Reproductive Success: Studies of Individual Variation in Contrasting Breeding Systems.* Chicago: University of Chicago Press, 263–283.

Packer, C., and A. E. Pusey. 1983. Adaptations of female lions to infanticide by incoming males. *American Naturalist* 121: 716–728.

Panhuis, T. M., and G. S. Wilkinson. 1999. Exaggerated male eye span influences contest outcome in stalk-eyed flies (Diopsidae). *Behavioral Ecology and Sociobiology* 46: 221–227.

Pawlowski, B., R. I. M. Dunbar, and A. Lipowicz. 2000. Tall men have more reproductive success. *Nature* 403: 156.

Petrie, M. 1983. Female moorhens compete for small fat males. *Science* 220: 413–415.

Proctor, H. C. 1991. Courtship in the water mite, *Neumania papillator.* Males capitalize on female adaptations for predation. *Animal Behaviour* 42: 589–598.

Proctor, H. C. 1992. Sensory exploitation and the evolution of male mating behaviour: a cladistic test using water mites (Acari: Parasitengona). *Animal Behaviour* 44: 745–752.

Prum, Richard O. 1997. Phylogenetic tests of alternative intersexual selection mechanisms: Trait macroevolution in a polygynous clade (*Aves: Pipridae*). *American Naturalist* 149: 668–692.

Rauch, N. 1985. Female habitat choice as a determinant of the reproductive success of the territorial male marine iguana (*Amblyrhynchus cristatus*). *Behavioral Ecology and Sociobiology* 16: 125–134.

Rauch, N. 1988. Competition of marine iguana females *Amblyrhynchus cristatus* for egg-laying sites. *Behavior* 107: 91–106.

Rogers, A. R., and A. Mukherjee. 1992. Quantitative genetics of sexual dimorphism in human body size. *Evolution* 46: 226–234.

Romero, G. A., and C. E. Nelson. 1986. Sexual dimorphism in *Catasetum* orchids: Forcible pollen emplacement and male flower competition. *Science* 232: 1538–1540.

Rosenqvist, G. 1990. Male mate choice and female-female competition for mates in the pipefish *Nerophis ophidion. Animal Behaviour* 39: 1110–1115.

Rosenqvist, G., and K. Johansson. 1995. Male avoidance of parasitized females explained by direct benefits in a pipefish. *Animal Behaviour* 49: 1039–1045.

Ryan, M. J. 1985. *The Túngara Frog: A Study in Sexual Selection and Communication.* Chicago: University of Chicago Press.

Schenk, A., and K. M. Kovacs. 1995. Multiple mating between black bears revealed by DNA fingerprinting. *Animal Behaviour* 50: 1483–1490.

Sillén-Tullberg, B. 1981. Prolonged copulation: A male "postcopulatory" strategy in a promiscuous species, *Lygaeus equestris* (Heteroptera: Lygaeidae). *Behavioral Ecology and Sociobiology* 9: 283–289.

Smedley, S. R., and T. Eisner. 1996. Sodium: A male moth's gift to its offspring. *Proceedings of the National Academy of Sciences, USA* 93: 809–813.

Smith, R. L. 1984. Human sperm competition. In R. L. Smith, ed. *Sperm Competition and the Evolution of Animal Mating Systems.* Orlando: Academic Press, 601–659.

Stanton, M. L., A. A. Snow, and S. N. Handel. 1986. Floral evolution: Attractiveness to pollinators increases male fitness. *Science* 232: 1625–1627.

Thornhill, R. 1976. Sexual selection and nuptial feeding behavior in *Bittacus apicalis* (Insecta: Mecoptera). *American Naturalist* 110: 529–548.

Thornhill, R., and J. Alcock. 1983. *The Evolution of Insect Mating Systems.* Cambridge, MA: Harvard University Press.

Trillmich, K. G. K. 1983. The mating system of the marine iguana (*Amblyrhynchus cristatus*). *Zeitschrift für Tierpsychologie* 63: 141–172.

Trivers, R. L. 1972. Parental investment and sexual selection. In B. Campbell, ed. *Sexual Selection and the Descent of Man 1871–1971.* Chicago: Aldine, 136–179.

Trivers, R. 1985. *Social Evolution.* Menlo Park, CA: Benjamin/Cummings.

Vaughton, G. and M. Ramsey. 1998. Floral display, pollinator visitation, and reproductive success in the dioecious perennial herb *Wurmbea dioica* (Liliaceae). *Oecologia* 115: 93–101.

Waage, J. K. 1984. Sperm competition and the evolution of Odonate mating systems. In R. L. Smith, ed. *Sperm Competition and the Evolution of Animal Mating Systems.* Orlando: Academic Press, 251–290.

Waage, J. K. 1986. Evidence for widespread sperm displacement ability among Zygoptera (Odonata) and the means for predicting is presence. *Biological Journal of the Linnean Society* 28: 285–300.

Watson, P. J. 1991. Multiple paternity as genetic bet-hedging in female sierra dome spiders, *Linyphia litigiosa* (Linyphiidae). *Animal Behaviour* 41: 343–360.

Welch, Allison, R. D. Semlitsch, and H. Carl Gerhardt. 1998. Call duration as an indicator of genetic quality in male gray tree frogs. *Science* 280: 1928–1930.

Wikelski, M., and S. Bäurle. 1996. Precopulatory ejaculation solves time constraints during copulations in marine iguanas. *Proceedings of the Royal Society of London,* Series B 263: 439–444.

Wikelski, M., C. Carbone, and F. Trillmich. 1996. Lekking in marine iguanas: Female grouping and male reproductive strategies. *Animal Behaviour* 52: 581–596.

Wikelski, M., V. Carrillo, and F. Trillmich. 1997. Energy limits to body size in a grazing reptile, the Galapagos marine iguana. *Ecology* 78: 2204–2217.

Wikelski, M., and C. Thom. 2000. Marine iguanas shrink to survive El Niño. *Nature* 403: 37.

Wikelski, M., and F. Trillmich. 1997. Body size and sexual size dimorphism in marine iguanas fluctuate as result of opposing natural and sexual selection: An island comparison. *Evolution* 51: 922–936.

Wilkinson, G. S., H. Kahler, and R. H. Baker. 1998. Evolution of female mating preferences in stalk-eyed flies. *Behavioral Ecology* 9: 525–533.

Wilkinson, G. S., D. C. Presgraves, and L. Crymes. 1998. Male eye span in stalk-eyed flies indicates genetic quality by meiotic drive suppression. *Nature* 391: 276–279.

Wilkinson, G. S., and P. R. Reillo. 1994. Female choice response to artificial selection on an exaggerated male trait in a stalk-eyed fly. *Proceedings of the Royal Society of London,* Series B 255: 1–6.

Wilkinson, G. S., and M. Taper. 1999. Evolution of genetic variation for condition-dependent traits in stalk-eyed flies. *Proceedings of the Royal Society London,* Series B 266: 1685–1690.

Williams, G. C. 1966. *Adaptation and Natural Selection: A Critique of Some Current Evolutionary Thought.* Princeton University Press, Princeton, NJ.

Willson, M. F. 1979. Sexual selection in plants. *American Naturalist* 113: 777–790.

Woodroffe, R., and A. Vincent. 1994. Mother's little helpers: Patterns of male care in mammals. *Trends in Ecology and Evolution* 9: 294–297.

Zahavi, A. 1975. Mate selection—A selection for a handicap. *Journal of Theoretical Biology* 53: 205–214.

CHAPTER 10

Kin Selection and Social Behavior

One of these crows has a fish; the other two appear to want it. In many cases, the outcome of a social interaction like this depends on the genetic relationship of the participants. (Gregory K. Scott/Photo Researchers, Inc.)

SOCIAL INTERACTIONS CREATE THE POSSIBILITY FOR CONFLICT AND COOPERATION. Consider two American crows (*Corvus brachyrhynchos*) patrolling the edge of their adjacent nesting territories. If one moves across the established boundary, its action may trigger aggressive calls, a flight chase, or even physical combat. But if a hawk flies by, the two antagonists will cooperate in chasing the predator away. Later in the day, these same individuals may spend considerable time and effort feeding the young birds in their respective nests, even though the nestlings are siblings or half-siblings and not their own offspring.

When and why do these individuals cooperate with each other, and why do they help their parents raise their siblings instead of raising their own offspring? What conditions lead to conflicts with each other and with their parents, and how are these conflicts resolved? These are the types of questions addressed in this chapter.

In fitness terms, an interaction between individuals has four possible outcomes (Table 10.1). Cooperation (or **mutualism**) is the term for actions that result in fitness gains for both participants. **Altruism** represents cases in which the individual instigating the action pays a fitness cost and the individual on the receiving end benefits. **Selfishness** is the opposite: The actor gains and the recipient loses. **Spite** is the term for behavior that results in fitness losses for both participants.

> **Table 10.1 Types of social interactions**
>
> The "actor" in any social interaction affects the recipient of the action as well as itself. The costs and benefits of interactions are measured in units of surviving offspring (fitness).
>
> | | Actor benefits | Actor is harmed |
> |---|---|---|
> | **Recipient benefits** | Cooperative | Altruistic |
> | **Recipient is harmed** | Selfish | Spiteful |

Understanding the evolution of these four interactions is made simpler because there are no clear-cut examples of spite in nature (Keller et al. 1994). It is straightforward to understand why spite has not evolved: An allele that results in fitness losses for both actor and recipient would quickly be eliminated by natural selection. But altruism would seem equally difficult to explain, because one of the participants suffers a fitness loss. Altruistic behavior appears to be common, however. Examples range from the crows that help at their parents' nests to a human who dives into a river and saves a drowning child. This is the first question we need to address: Why does altruism exist in nature?

10.1   Kin Selection and the Evolution of Altruism

Explaining altruistic behavior is a challenge for the Theory of Evolution by Natural Selection.

Altruism is a central paradox of Darwinism. It would seem impossible for natural selection to favor an allele that results in behavior benefiting other individuals at the expense of the individual bearing the allele. For Darwin (1859: 236), the apparent existence of altruism presented a "special difficulty, which at first appeared to me insuperable, and actually fatal to my whole theory." Fortunately he was able to hint at a resolution to the paradox: Selection could favor traits that result in decreased personal fitness if they increase the survival and reproductive success of close relatives. Over a hundred years passed, however, before this result was formalized and widely applied.

Inclusive Fitness

In 1964, William Hamilton developed a genetic model showing that an allele that favors altruistic behavior could spread under certain conditions. The key parameter in Hamilton's formulation is the **coefficient of relationship**, r. This is the probability that the homologous alleles in two individuals are identical by descent (Box 10.1). The parameter is closely related to F, the coefficient of inbreeding, which we introduced in Chapter 6. F is the probability that homologous alleles in the same individual are identical by descent.

Given r, the coefficient of relatedness between the actor and the recipient, **Hamilton's rule** states that an allele for altruistic behavior will spread if

$$Br - C > 0$$

where B is the benefit to the recipient and C is the cost to the actor. Both B and C are measured in units of surviving offspring. This simple law means that altruism is more likely to spread when the benefits to the recipient are great, the cost to the actor is low, and the participants are closely related.

BOX 10.1   Calculating coefficients of relatedness

Calculating r, the coefficient of relatedness, requires a pedigree that includes the actor (the individual dispensing the behavior) and the recipient (the individual receiving the behavior). The researcher then performs a path analysis. Starting with the actor, all paths of descent are traced through the pedigree to the recipient. For example, half-siblings share one parent and have two genealogical connections, as indicated in Figure 10.1a. Parents contribute half their genes to each offspring, so the probability that genes are identical by descent (ibd) in each step in the path is 1/2. Put another way, the probability that a particular allele was transmitted from parent to actor is 1/2. The probability that the same allele was transmitted from parent to recipient is 1/2. The probability that this same allele was transmitted to both the actor and the recipient (meaning that the alleles in actor and recipient are ibd) is the product of these two independent probabilities, or 1/4.

Full siblings, on the other hand, share genes inherited from both parents. To calculate r when actor and recipient are full-sibs, we have to add the probabilities that genes are ibd through each path in the pedigree. In this case, we add the probability that genes are ibd through the mother to the probability that they are ibd through the father (see Figure 10.1b). This is $\frac{1}{4} + \frac{1}{4} = \frac{1}{2}$.

Using this protocol results in the following coefficients:

- First cousins, $\frac{1}{8}$ (Figure 10.1c)
- Parent to offspring, $\frac{1}{2}$
- Grandparent to grandchild, $\frac{1}{4}$
- Aunt or uncle to niece or nephew, $\frac{1}{4}$

The analyses we have just performed work for autosomal loci in sexual organisms, and assume that no inbreeding has occurred. If the population is inbred then coefficients will be higher. But when studying populations in the field, investigators usually have no data on inbreeding and have to assume that individuals are completely outbred. On this basis, coefficients of relationship that are reported in the literature should be considered minimal estimates. Another uncertainty in calculating r's comes in assigning paternity in pedigrees. As we indicated in Chapter 9, extra-pair copulations are common in many species. If paternity is assigned on the basis of male-female pairing relationships and extra-pair copulations go undetected, estimates of r may be inflated.

When constructing genealogies is impractical, coefficients of relatedness can be estimated directly from genetic data (Queller and Goodnight 1989). The microsatellite loci introduced in Chapter 18 are proving to be extremely useful markers for calculating r in a wide variety of social insects (e.g., Peters et al. 1999).

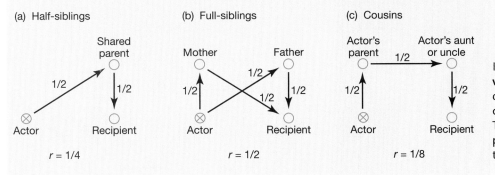

(a) Half-siblings   (b) Full-siblings   (c) Cousins

$r = 1/4$ $r = 1/2$ $r = 1/8$

Figure 10.1   Path analysis with pedigrees The arrows describe paths by which genes can be identical by descent. The text explains how these paths are used to calculate r, the coefficient of relatedness.

To generalize this result, Hamilton created the concept of **inclusive fitness**. He pointed out that an individual's fitness can be partitioned into two components, which he called direct and indirect. **Direct fitness** results from personal reproduction. **Indirect fitness** results from additional reproduction by relatives that is made possible by an individual's actions. Indirect fitness accrues when relatives achieve reproductive success above and beyond what they would have achieved on their own—meaning, without aid. When natural selection favors the spread of alleles that increase

Inclusive fitness consists of direct fitness due to personal reproduction and indirect fitness due to additional reproduction by relatives. Behavior that results in indirect fitness gains is favored by kin selection.

the indirect component of fitness, **kin selection** occurs. As we will see, most instances of altruism in nature are the result of kin selection.

Robert Trivers (1985:47) called Hamilton's rule and the concept of inclusive fitness "the most important advance in evolutionary theory since the work of Charles Darwin and Gregor Mendel." To see why, we will apply the theory by venturing to the Sierra Nevada of California and observing an intensely social mammal: Belding's ground squirrel (*Spermophilus beldingi*).

Alarm Calling in Belding's Ground Squirrels

Explaining alarm calling in birds and mammals is a classical application of inclusive fitness theory. When flocks or herds are stalked by a predator, prey individuals that notice the intruder sometimes give loud, high-pitched calls. These warnings alert nearby individuals and allow them to flee or dive for cover. They may also expose the calling individual to danger. In Belding's ground squirrels, 13% of callers are stalked or chased by predators while only 5% of non-calling individuals are (Sherman 1977).

Paul Sherman (1977, 1980) studied patterns in alarm calls given by Belding's ground squirrels to determine why such seemingly altruistic behavior evolved. Belding's ground squirrels are rodents that breed in colonies established in alpine meadows. Males disperse far from the natal burrow, while female offspring tend to remain and breed close by. As a result, females in proximity tend to be closely related. Because Sherman had individually marked many individuals over the course of the study, he was able to construct pedigrees and calculate coefficients of relationship for most members of the colony.

When Sherman compiled data on which squirrels called at the approach of a weasel, coyote, or badger, two outstanding patterns emerged: Females were much more likely to call than males (Figure 10.2), and females were much more likely

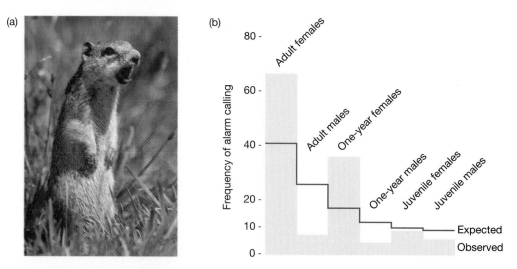

Figure 10.2 In ground squirrels, most alarm calling is done by females (a) This female Belding's ground squirrel is giving an alarm call. (Richard R. Hansen/Photo Researchers, Inc.) (b) This bar chart reports the observed and expected frequencies of alarm calling by different sex and age classes of Belding's ground squirrels, based on 102 encounters with predatory mammals. The expected values are indicated by the blue line and are calculated by assuming that individuals call randomly—that is, in proportion to the number of times they are present when the predator approaches. The observed and expected values are significantly different ($P = 0.001$). Reprinted with permission from Sherman (1977). Copyright © 1977, American Association for the Advancement of Science.

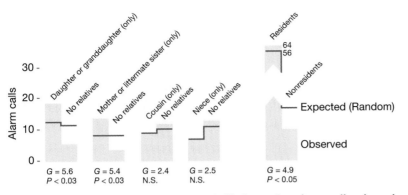

Figure 10.3 **Female ground squirrels are more likely to give alarm calls when close kin are nearby** This bar chart summarizes data from 119 cases in which a mammalian predator approached a ground squirrel colony. Each paired comparison in the figure represents the occasions when at least one female in each category was present, but no others. The expected values are computed as in Figure 10.2. The G values reported here represent a modified form of the χ^2 statistic introduced in Chapter 5. "N.S." stands for not significant. From Sherman (1980).

to call when they had close relatives within earshot (Figure 10.3). These data strongly support the hypothesis that kin selection is responsible for the evolution of alarm calling. Sherman (1981) has also been able to show that mothers, daughters, and sisters are much more likely to cooperate when chasing trespassing squirrels off their territory than are more distant kin or nonrelatives (Figure 10.4).

The data show that altruistic behavior is not dispensed randomly. It is nepotistic. Self-sacrificing behavior is directed at close relatives and should result in indirect fitness gains.

Individuals are more likely to give alarm calls when close relatives are nearby.

White-Fronted Bee-Eaters

Another classical system for studying kin selection in vertebrates is helping behavior in birds (see Brown 1987; Stacey and Koenig 1990). In species from a wide variety of bird families, young that are old enough to breed on their own will instead remain and help their parents rear their brothers, sisters, or half-siblings. Helpers

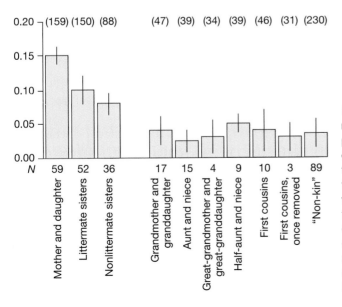

Figure 10.4 **Closely related female ground squirrels are more likely to cooperate than distant kin** This bar chart reports how frequently territory-owning females were joined by different categories of relatives and nonrelatives in chasing away trespassing ground squirrels. The number in parentheses indicates the number of chases that occurred when both types of individuals were present. N gives the number of different dyads of each kind that were observed. The horizontal bars give the standard deviations of the frequencies. There is a significant difference in the frequency of cooperation between the three categories on the left and the seven categories on the right ($P < 0.01$), but no significant difference among the seven kinship categories on the right ($P > 0.09$). From Sherman (1981).

Figure 10.5 White-fronted bee-eaters The individual in the middle is performing a wing-waving display, and may be soliciting a feeding. (Gerard Lacz/Animals Animals/Earth Scenes)

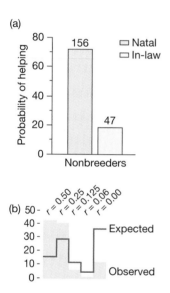

Figure 10.6 In bee-eaters, helpers assist close relatives
(a) Bee-eater clans often contain nonbreeders that have paired with members of the clan. Their *r* with the offspring being raised that season is 0. This bar chart shows that they are much less likely to help than are clan members (*P* < 0.01). (b) In this bar chart, the expected probability of helping is calculated by assuming that helpers assist clan members randomly, in proportion to the *r*'s of nestlings in clan nests. A *G*-test rejects the null hypothesis that helping is directed randomly with respect to kinship (*P* < 0.01). From Emlen and Wrege (1988).

assist with nest building, nest defense, or food delivery to incubating parents and the nestlings.

Helping at the nest is usually found in species where breeding opportunities are extremely restricted, either because habitats are saturated with established breeding pairs or because suitable nest sites are difficult to obtain. In these cases, gaining direct fitness is almost impossible for young adults. Gaining indirect fitness by helping becomes a best-of-a-bad-job strategy.

Steve Emlen, Peter Wrege, and Natalie Demong have completed an intensive study of helping behavior in the white-fronted bee-eater (*Merops bullockoides*). This colonial species, native to East and central Africa, breeds in nesting chambers excavated in sandy riverbanks (Figure 10.5). The 40–450 individuals in a colony are subdivided into groups of 3 to 17, each of which defends a feeding territory up to seven kilometers away. These clans may include several sets of parents and off-spring.

Many year-old bee-eaters stay to help at the nest during what would otherwise be their first breeding season. Clan members are related, so helpers usually have a choice of nestlings with different degrees of kinship as recipients of their helping behavior (Hegner et al. 1982; Emlen et al. 1995). This choice is a key point: Because kinship varies among the potential recipients of altruistic behavior, white-fronted bee-eaters are an excellent species for researchers to use in testing theories about kin selection.

After marking large numbers of individuals and working out genealogies over an eight-year study period, Emlen and Wrege (1988, 1991) found that bee-eaters conform to predictions made by Hamilton's rule. They determined, for example, that the coefficient of relatedness with recipients has a strong effect on whether a nonbreeding member of the clan helps (Figure 10.6a). Further, nonbreeders actively decide to help the most closely related individuals available (Figure 10.6b). That is, when young with different coefficients of relationship are being reared within their clan, helpers almost always chose to help those with the highest *r* (Box 10.2). Their assistance is an enormous benefit to parents. More than half of bee-eater young die of starvation before leaving the nest. On average, the presence of each helper results in an additional 0.47 offspring being reared to fledging (Figure 10.7). For young birds, helping at the nest results in clear benefits for inclusive fitness.

BOX 10.2 Kin recognition

The data in Figures 10.2–6 suggest that individuals have accurate mechanisms for assessing their degree of kinship with members of their own species, or conspecifics. This phenomenon, called kin recognition, has been divided into two broad categories: direct and indirect (Pfennig and Sherman 1995). Indirect kin recognition is based on cues like the timing or location of interactions. Many species of adult birds rely on indirect kin recognition when their chicks are young, and will feed any young bird that appears in their nest. Direct kin recognition, in contrast, is based on specific chemical, vocal, or other cues.

There is currently a great deal of interest in determining whether loci in the major histocompatibility complex (MHC) function in direct kin recognition. In Chapter 5, we introduced the MHC and its role in self-nonself-recognition by cells in the immune system. Although loci in the MHC clearly evolved to function in disease prevention, polymorphism is so extensive that non–kin share very few alleles. As a result, these genes can serve as reliable markers of kinship (see Brown and Eklund 1994). Could similarity in MHC provide a reliable cue of kinship and offer a criterion for dispensing altruistic behavior?

Jo Manning and colleagues (1992) addressed this question in a population of house mice (*Mus musculus domesticus*). A kin recognition system requires three components: production of the signal, recognition of the signal by conspecifics, and action based on that recognition. Previous work had shown that glycoproteins coded for by MHC loci are released in the urine of mice, and that mice can distinguish these molecules by smell. Mice are, for example, able to distinguish full siblings from half-siblings on the basis of their MHC genotypes. But do mice dispense altruistic behavior accordingly?

House mice form communal nests and nurse each others' pups. Because individuals could take advantage of this cooperative system by contributing less than their fair share of milk, Manning et al. predicted that mothers would prefer to place their young in nests containing close relatives. The logic here is that close kin should be less likely to cheat on one another because of the cost to their indirect fitness. Through a program of controlled breeding in which wild-caught mice were crossed with laboratory strains, Manning et al. created a population of mice with known MHC genotypes. This population was allowed to establish itself in a large barn. The researchers then recorded where mothers in this population placed their newborn pups after birth. The null hypothesis was that mothers would choose to rear their offspring randomly with respect to the MHC genotypes present in the communal nests available at the time. Contrary to the null expectation, mothers showed a strong preference for rearing their young in nests containing offspring with similar MHC genotypes. This result confirms a role for MHC as a signal used in direct kin recognition, and shows that mice are capable of dispensing altruistic behavior on the basis of MHC genotypes.

10.2 Evolution of Eusociality

Darwin (1859) recognized that social insects represent the epitome of altruism, and thus a special challenge to the theory of evolution by natural selection. Many worker ants and bees, for example, do not reproduce at all. They are helpers at the nests of their parents, for life. This is an extreme form of reproductive altruism.

Eusociality (true sociality) is used to describe social systems with three characteristics (Michener 1969; Wilson 1971; Alexander et al. 1991): (1) overlap in generations between parents and their offspring, (2) cooperative brood care, and (3) specialized castes of nonreproductive individuals. Eusocial species are found in a variety of insect orders (Table 10.2), snapping shrimp (Duffy 1996), and one family of rodents (the mammal family Bathyergidae, or mole-rats).

As an entree to the extensive literature on eusociality, in this section we consider how reproductive altruism evolved in two very different groups: the Hymenoptera (ants, bees, and wasps) and mole-rats.

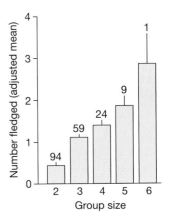

Figure 10.7 Fitness gains due to helping From Emlen and Wrege (1991).

Haplodiploidy and Eusocial Hymenoptera

The Hymenoptera represent the pinnacle of social evolution. A single ant colony may number millions of individuals, each appearing to function more like a cell in a superorganism than an individual pursuing its own reproductive interests. Worker, soldier, and reproductive castes, which seem analogous to tissues in a body, can be identified on the basis of their morphology and the tasks they perform. But unlike cells and tissues, individuals in the colony are not genetically identical. What factors laid the groundwork for such extensive altruism? Why is eusociality so widespread in Hymenoptera?

William Hamilton (1972) proposed that the unique genetic system of ants, wasps, and bees predisposes them to eusociality. Hymenoptera have an unusual form of sex determination: Males are haploid and females are diploid. Males develop from unfertilized eggs; females develop from fertilized eggs. As a result of this system, called **haplodiploidy**, female ants, bees, and wasps are more closely related to their sisters than they are to their own offspring. This follows because sisters share all of the genes they inherited from their father, which is half their genome, and half the genes they inherited from their mother (the colony's queen), which is the other half of their genome. Thus, the probability that homologous alleles in hymenopteran sisters are identical by descent is $(1 \times 1/2) + (1/2 \times 1/2) = 3/4$. To their own offspring, however, females are related by the usual r of $1/2$. This unique system favors the production of reproductive sisters over daughters, sons, or brothers. (Females are related to their brothers by $r = 1/4$; see Figure 10.8). Thus, females will maximize their inclusive fitness by acting as workers rather than as reproductives (Hamilton 1972). Specifically, their alleles will increase in the population faster when they invest in the production of sisters rather than producing their own offspring. This is the haplodiploidy hypothesis for the evolution of eusociality in Hymenoptera.

In haplodiploid species, females are more closely related to their sisters than they are to their own offspring.

Testing the Haplodiploidy Hypothesis

In addition to offering an explanation for why workers prefer to invest in sisters rather than their own offspring, the haplodiploidy hypothesis predicts that workers prefer to invest in sisters over brothers. Because their r with sisters is $3/4$ and only $1/4$ with brothers, workers should favor a 3:1 female-biased sex ratio in reproductive offspring (meaning, offspring that are not destined to become sterile workers or soldiers; Trivers and Hare 1976). Queens, in contrast, are equally related to their sons and daughters and should favor a 1:1 sex ratio in the reproductives produced (see Box 10.3). The fitness interests of workers and queens are not the same. The question is: Who wins the conflict? Do queens or workers control the sex ratio of reproductive offspring?

Liselotte Sundström and coworkers (1996) set out to answer this question by determining the sex ratio of reproductive offspring in wood ant (*Formica exsecta*) colonies. They found that queens laid a roughly equal number of male and female eggs, but that sex ratios were heavily female-biased at hatching. To make sense of this result, the researchers hypothesize that workers are able to determine the sex of eggs and that they selectively destroy male offspring.

Based on results from similar studies, most researchers acknowledge that female-biased sex allocation is widespread among eusocial hymenoptera. (For recent reviews, see Bourke and Franks 1995; Crozier and Pamilo 1996.) In the tug-of-war over the fitness interests, workers appear to have the upper hand over queens.

Table 10.2   Sociality in insects

This table summarizes the taxonomic distribution of eusociality in insects. Species are called "primitively eusocial" if queens are not morphologically differentiated from other individuals.

Order	Family	Subfamily	Eusocial species
Hymenoptera	Anthophoridae (carpenter bees)		In seven genera
	Apidae	Apinae (honeybees)	Six highly eusocial species
		Bombinae (bumble bees)	300 primitively eusocial species
		Euglossinae (orchid bees)	None
		Meliponinae (stingless bees)	200 eusocial species
	Halictidae (sweat bees)		In six genera
	Sphecidae (sphecoid wasps)		In one genus
	Vespidae (paper wasps, yellow jackets)	Polistinae	Over 500 species, all eusocial
		Stenogastrinae	Some primitively eusocial species
		Vespinae	Ca. 80 species, all eusocial
	Formicidae (ants)	11 subfamilies	Over 8,800 described species, all eusocial or descended from eusocial species
	Many other families		None
Isoptera (termites)	Nine families		All species (over 2,288) are eusocial
Homoptera (plant bugs)	Pemphigidae		Sterile soldiers found in six genera
Coleoptera (beetles)	Curculionidae		*Austroplatypus incompertus*
Thysanoptera (thrips)	Phlaeothripidae		Subfertile soldiers are found in *Oncothrips*

Source: From Crozier and Pamilo (1996).

Perhaps the more important general message of this work, however, is that colonies of ants, bees, and wasps are not harmonious "superorganisms." The asymmetry in relationship between queens and offspring versus workers and offspring produces a sharp conflict of interest.

Does the Haplodiploidy Hypothesis Explain Eusociality?

The prediction and affirmation of 3:1 sex ratios in reproductive offspring, at least in some hymenopterans, confirms that the haplodiploid system of sex determination has a strong effect on how workers behave. But is haplodiploidy the reason that so many hymenopteran species are eusocial? Most researchers are concluding that the answer is no. There are several reasons for this.

First, the prediction that workers favor the production of sisters over the production of their own offspring is based on an important assumption: that all of the female workers in the colony have the same father. In many species this is not true. Multiple mating is common in certain groups of eusocial Hymenoptera. Honeybee queens, for example, mate an average of 17.25 times before founding a colony (Page and Metcalf 1982). As a result, it is common to find that the average coefficient of relatedness among honeybee workers is under 1/3 (Oldroyd et

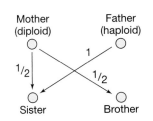

Figure 10.8   Haplodiploidy produces unusual coefficients of relationship   The arrows describe the paths by which genes can be identical by descent in hymenopterans. Note that there is no path of shared descent between sisters and brothers through their father, because males have no father.

BOX 10.3 The evolution of the sex ratio

In many species, the sex ratio at hatching, germination, or birth is 1:1. Ronald Fisher (1930) explained why this should be so. Fisher pointed out that if one sex is in short supply in a population, then an allele that leads to the production of the rarer sex will be favored. This is because individuals of the rarer sex will have more than one mate on average when they mature, simply because in a sexual species every individual has one mother and one father. Members of the rarer sex will experience increased reproductive success relative to individuals of the more common sex. Indeed, whenever the sex ratio varies from 1:1, selection favoring the rarer sex will exist until the ratio returns to unity. Fisher's explanation is a classic example of frequency-dependent selection—a concept we introduced in Chapter 5 with the example of left- and right-handed scale-eating fish.

Fisher's argument is based on an important assumption: that parents invest equally in each sex. When one sex is more costly than the other, parents should adjust the sex ratio to even out the investment in each. For this reason evolutionary biologists distinguish the numerical sex ratio from the investment sex ratio and speak, in general terms, about the issue of sex allocation (Charnov 1982).

Robert Trivers and Dan Willard (1973) came up with an important extension to Fisher's model. They suggested that when females are in good physiological condition and are better able to care for their young, and when differences in the condition of young are sustained into adulthood, then they should preferentially invest in male offspring. This is because differences in condition affect male reproductive success RS more than female RS (see Chapter 9). This prediction, called condition-dependent sex allocation, has been confirmed in a wide variety of mammals, including humans. (For examples, see Clutton-Brock et al. 1984; Betzig and Turke 1986.)

A third prominent result in sex-ratio theory is due to William Hamilton (1967). In insects that lay their eggs in fruit or other insects, the young often hatch, develop, and mate inside the host. Frequently, hosts are parasitized by a single female. Given this situation, Hamilton realized that selection should favor females that produce only enough males to ensure fertilization of their sisters, resulting in a sex ratio with a strong female bias. This phenomenon, known as local mate competition, has been observed in a variety of parasitic insects. (For examples, see Hamilton 1967; Werren 1984.)

al. 1997, 1998). In these colonies, workers are *not* more closely related to their sisters than they are to their own offspring.

Second, in many species more than one queen is active in founding the nest. If they have neither parent in common, then workers in these colonies have a coefficient of relatedness of 0.

Haplodiploidy affects the behavior of eusocial species, but it is not the most important factor leading to the evolution of eusociality.

Third, many eusocial species are not haplodiploid, and many haplodiploid species are not eusocial. Nonreproductive castes are found in all termite species, for example, even though termites are diploid and have a normal chromosomal system of sex determination (Thorne 1997). And although eusociality is common among hymenopterans, it is by no means universal. Reviewing recent work on the phylogeny of the hymenoptera will help drive this last point home.

Using Phylogenies to Analyze Social Evolution

To understand which traits are most closely associated with the evolution of eusociality in hymenoptera, James Hunt (1999) analyzed the evolutionary trees shown in Figure 10.9. Because all hymenopterans are haplodiploid, Hunt could infer that this system of sex determination evolved early in the evolution of the group, at the point marked A on the tree. Eusociality is found in just a few families of hymenopterans, however. Because these families are scattered around the tree, it is

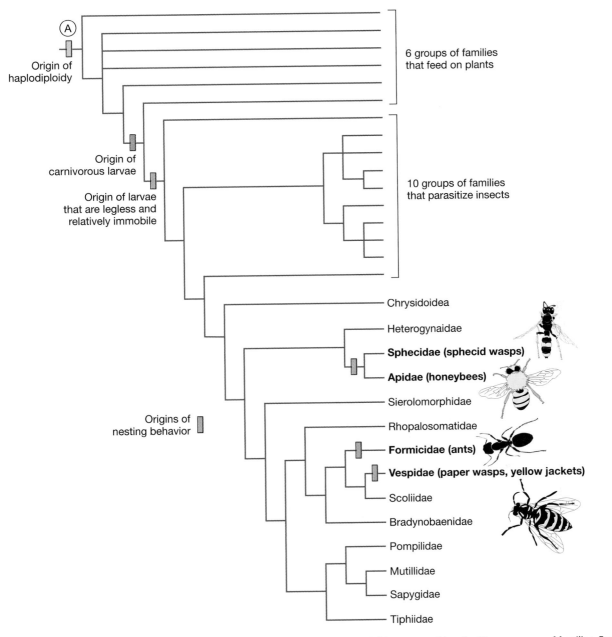

Figure 10.9 A phylogeny of the hymenoptera The taxa at the tips of this tree are either families or groups of families. Families that include eusocial species are indicated in bold type. (Not all of the species in these families are eusocial, however.) The arrows indicate points where certain key traits evolved. Modified from Hunt (1999). Copyright © 1999 Evolution. Reprinted by permission of Evolution.

likely that eusociality evolved not just once, but several times independently. Most importantly, Hunt noted that eusociality only evolved in groups that build complex nests and that care for their larvae for extended periods.

The association between nest building, care of larvae, and eusociality is important because it suggests that the primary agent favoring reproductive altruism in insects is ecological in nature—not genetic as proposed by the haplodiploidy hypothesis. The logic here is similar to the "best-of-a-bad-job" explanation for helping behavior in birds reviewed in Section 10.1. Nest building and the need to supply larvae with a continuous supply of food make it difficult or impossible for a female to breed on her own (see Alexander et al. 1991). Also, when predation rates are high but young are dependent on parental care for a long period, then individuals who breed alone are unlikely to survive long enough to bring their young to adulthood (Queller 1989; Queller and Strassmann 1998). In short, to explain the evolution of eusociality we clearly need to consider ecological factors that affect B and C as well as genetic factors that dictate r.

Facultative Strategies in Paper Wasps

Paper wasps in the genus *Polistes* have been an especially productive group for research into the costs and benefits of reproductive altruism. Unlike workers and soldiers in ants and termites, paper wasp workers are not sterile. Instead of being obligate helpers, *Polistes* females are capable of reproducing on their own. This contrast is important. To achieve reproductive success, worker and soldier ants and termites have no choice but to assist relatives—the nutrition they received as larvae guarantees that they are sterile. But in paper wasps, females have the option of helping relatives or breeding on their own.

In paper wasps, reproductive altruism is facultative. Females can choose between helping at a nest or breeding on their own.

In *Polistes dominulus,* Peter Nonacs and Hudson Reeve (1995) found that females pursue one of three distinct strategies: They either initiate their own nest, join a nest as a helper, or wait for a breeding opportunity. Each option is associated with costs and benefits.

What are the costs and benefits of founding a nest? In the population that Nonacs and Reeve studied, nests were founded by single females or by multifemale groups. Earlier studies had shown that single foundresses are at a distinct disadvantage compared to multifemale coalitions. Adult mortality is high, and nests with multiple foundresses are less likely to fail because surviving females keep the nest going. Nonacs and Reeve also found that multifoundress coalitions are more likely to renest after a nest is destroyed. When they analyzed 106 instances of nest failure due to predation or experimental removal, they determined that only 5 of 54 single foundresses rebuilt while 21 of 51 multifoundress groups did.

Although the success rate of multifemale nests is high when compared to single-foundress nests, multifemale coalitions are not free of conflicts. Fights between wasps are decided by body size. As the graph in Figure 10.10 shows, multifoundress nests grew fastest when there was a large difference in the body size of the dominant female and her subordinate helpers. To interpret this result, Nonacs and Reeve suggest that productivity is low in coalitions where body size is similar be-

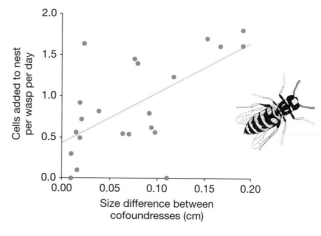

Figure 10.10   In paper wasps, the success of female coalitions varies This graph plots the growth rate of *Polistes dominulus* nests as a function of the size difference between cofoundresses. Nests grow fastest when the founding females are markedly different in size. Modified from Nonacs and Reeve (1995). Copyright © 1995, Ecological Society of America. Reprinted with permission.

cause subordinate females frequently challenge for control of the nest and the right to lay most of the eggs.

Why would females join a coalition and help rear offspring that are not their own? Subordinates gain indirect fitness benefits because they are usually closely related to the dominant female. They may also gain direct fitness benefits if the dominant individual dies and they are able to take over the nest. Thus, the costs and benefits of helping depend on a female's body size and her coefficient of relatedness with other members of the coalition.

Nonacs and Reeve also found that a "sit-and-wait" strategy could make sense in fitness terms, because females that do not participate in nest initiation are able to adopt orphaned nests (after all of the adult females have died) or usurp small nests later in the season by defeating the attending female(s) in combat. In *Polistes fuscatus,* Reeve and colleagues (1998) found that some females pursue an interesting twist on these sit-and-wait tactics: They leave the nest early in the spring, enter a dormant state in a sheltered location, and wait until the following breeding season before attempting to nest.

The fundamental message of these studies is that reproductive altruism is facultative. It is an adaptive response to environmental conditions. In the case of the population studied by Nonacs and Reeve, the conditions that are relevant to a female are its body size relative to its competitors, its coefficient of relatedness to members of a nesting coalition, and the availability of other nests or nest-sites. Genetic, social, and ecological factors have also been invoked to explain the evolution of eusociality in naked mole-rats.

For female paper wasps, the decision to join an existing nest or to breed independently hinges on a series of costs of benefits. These costs and benefits are dicated by environmental and social conditions, and may change through time.

Naked Mole-Rats

Naked mole-rats (*Heterocephalus glaber*) are one of the great oddities of the class Mammalia (Figure 10.11). They are neither moles nor rats, but are members of the family Bathyergidae, native to desert regions in the Horn of Africa. They eat tubers, live underground in colonies of 70–80 members, and construct tunnel systems up to two miles long by digging cooperatively in the fashion of a bucket brigade. Mole-rats are nearly hairless and ectothermic ("cold-blooded") and, like termites, can digest cellulose with the aid of specialized microorganisms in their intestines.

Figure 10.11 **Naked mole-rats** This photo shows a naked mole-rat queen threatening a worker. For superb introductions to the biology of naked mole-rats, see Honeycutt (1992) and Sherman et al. (1992). (Raymond A. Mendez/Animals Animals/Earth Scenes)

In naked mole-rats, helpers gain indirect fitness benefits because they are very closely related to the queen's offspring.

Naked mole-rats are also eusocial. All young are produced by a single queen and all fertilizations are performed by a group of 2–3 reproductive males. As other members of the colony grow older and increase in size, their tasks change from tending young and working in the tunnels to specializing in colony defense. For unknown reasons, there is a slight male bias in the colony sex ratio, of 1.4:1. Naked mole-rats are diploid and have an XY-chromosome system of sex determination.

The leading hypothesis to explain why naked mole-rats are eusocial centers on inbreeding. Analyses of microsatellite loci confirm that colonies are highly inbred (Reeve et al. 1990). Researchers studying colonies established in the lab have determined that approximately 85% of all matings are between parents and their offspring or between full siblings, and that the average coefficient of relationship among colony members is 0.81 (Sherman et al. 1992). These are among the highest coefficients ever recorded in animals.

Even extensive inbreeding does not mean that the reproductive interests of workers and reproductives are identical, however. Conflicts exist because workers are still more closely related to their own offspring than they are to their siblings and half-siblings. Queens are able to maintain control, however, through physical dominance. If nonreproductives slow their pace of work, mole-rat queens push them. These head-to-head shoves are aggressive and can move a worker more than a meter backward through a tunnel. Shoves are directed preferentially toward non-relatives and toward relatives more distant than offspring and siblings of the queen (Figure 10.12). Workers respond to shoves by nearly doubling their work rate (Table 10.3). These data suggest that queens impose their reproductive interests on subordinates through intimidation.

Inbreeding has also been hypothesized as a key factor predisposing termites to eusociality (Bartz 1979; but see Pamilo 1984; Roisin 1994). Not all inbred species are eusocial, however. This means that inbreeding is just one of several factors that contribute to eusociality in naked mole-rats and termites. Ecological factors such as extended parental care, group defense against predation, and severely constrained breeding opportunities are also important in explaining the evolution of reproductive altruism.

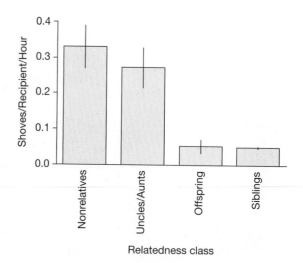

Figure 10.12 Naked mole-rat queens preferentially shove nonrelatives   These data were collected from a captive colony of naked mole-rats. The bars indicate the average and standard errors of shoves given to different kin classes by a queen called three-bars. There is a statistically significant difference between shoving rates for nonrelatives and uncles/aunts versus the two closer kin classes. From Reeve and Sherman (1991). Copyright © 1991, Princeton University Press. Reprinted by permission of Princeton University Press.

Table 10.3   Shoves from naked mole-rat queens motivate workers

Working in captive colonies, Hudson Reeve (1992) performed regular scans and recorded the activity of all individuals in the colony. The work level reported in this table represents the proportion of scans during which individuals were performing work, with standard errors given in parentheses. Shoving events by queens that took place in the nestbox or tunnels were also recorded. There is a statistically significant difference between work rates before and after shoves, for both types of shoves ($P < 0.01$).

Shove recipient's work level	Before shove	After shove
All shoves	0.14 (0.03)	0.25 (0.06)
Tunnel shoves only	0.34 (0.05)	0.58 (0.07)

10.3   Parent–Offspring Conflict

The theory of kin selection has been remarkably successful in explaining the structure and dynamics of social groups such as bee-eater clans and wasp colonies. Now, we consider how the theory might inform questions about a more fundamental social unit: parents and offspring.

Parental care is a special case of providing fitness benefits for close relatives. Although kin selection can lead to close cooperation between related individuals such as parents and offspring, even close kin can be involved in conflicts when the costs and benefits of altruism change or when degrees of relatedness are not symmetrical. Robert Trivers (1974) was the first to point out that parents and offspring are *expected* to disagree about each other's fitness interests. Because parental care is so extensive in birds and mammals, conflicts over the amount of parental investment should be especially sharp.

Weaning Conflict

Weaning conflict is a well-documented example of parent–offspring strife. Aggressive and avoidance behaviors are common toward the end of nursing in a wide variety of mammals. Mothers will ignore or actively push young away when they attempt to nurse, and offspring will retaliate by screaming or by attacking their mothers (Figure 10.13).

The key to explaining weaning conflict is to recognize that the fitness interests of parents and offspring are not symmetrical. Offspring are related to themselves with $r = 1$, but parents are related to their offspring with $r = 1/2$. Further, parents are equally related to all their offspring and are expected to equalize their investment in each. Siblings, in contrast, are related by 1 to themselves but 1/2 to each other. The theory of evolution by natural selection predicts that each offspring will demand an unequal amount of parental investment for itself.

When these asymmetries are applied to nursing, conflicts arise. At the start of nursing, the benefit to the offspring is high relative to the cost to the parent (Figure 10.14a). As nursing proceeds, however, this ratio declines. Young grow and demand more milk, which increases the cost of care. At the same time, they are increasingly able to find their own food, which decreases the benefit. Natural selection should favor mothers who stop providing milk when the benefit-to-cost

Parents maximize their fitness by investing in all of their offspring equally. Offspring, in contrast, maximize their fitness by receiving more parental investment than their siblings.

(a)

(b)

Figure 10.13   Weaning conflict   (a) The infant langur monkey on the left has just attempted to nurse from its mother, at the right. The mother refused to nurse. In response, the infant is screaming at her. Reproduced by permission from Trivers, *Social Evolution,* page 147, Fig. 7.2a. Menlo Park, CA: Benjamin Cummings Publishing Co. (1985). (b) The infant then dashes across the branch and slaps its mother. (Sarah Blaffer Hrdy/Anthro-Photo File)

ratio reaches 1 (this is time *P* in Figure 10.14a). From the mother's perspective, this is when weaning should occur. Offspring, on the other hand, devalue their mother's cost of providing care. They do this because the "savings" that a parent achieves through weaning will be invested in brothers or sisters with $r = 1/2$ instead of in themselves with $r = 1.0$. Natural selection should favor offspring who try to coerce continued parental investment until the benefit-cost ratio is $1/2$ (this is time *O* on Figure 10.14a). The period between time *P* and time *O* defines the interval of weaning conflict. Avoidance and aggressive behavior should be observed

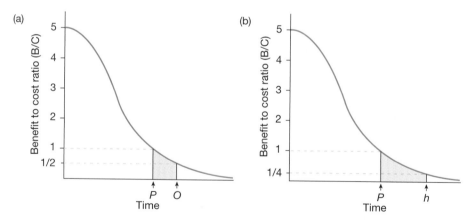

Figure 10.14   Parent–offspring conflict results from changes in the costs and benefits of parental care and asymmetries in relationship   (a) This graph illustrates why parent–offspring conflict occurs. The *y*-axis plots the benefit-to-cost ratio (*B/C*) for an act of parental care such as providing milk. Benefit is measured in terms of increased survival by the offspring receiving the care, while cost is measured as decreased production of additional offspring by the parent. Time is plotted along the *x*-axis. The curve drawn here is hypothetical; its shape will vary from species to species. See the text for an explanation of how this curve is interpreted. (b) This is the same graph as in (a), modified to illustrate how the period of parent–offspring conflict is extended when parents produce half-siblings instead of full siblings. From Trivers (1985).

throughout this period. If mating systems are such that mothers routinely remate and produce half-siblings, the period of weaning conflict will extend to time h, when the ratio of benefit to cost $(B/C) = r = 1/4$ (Figure 10.14b). For field studies that confirm weaning conflict, see Trivers (1985, Chapter 7); for a theoretical treatment of other types of parent–offspring strife, see Godfrey (1995).

Harassment in White-Fronted Bee-Eaters

Another dramatic example of parent–offspring conflict occurs in the white-fronted bee-eaters introduced earlier. Steve Emlen and Peter Wrege (1992) have collected data suggesting that fathers occasionally coerce sons into helping to raise their siblings. They do this by harassing sons who are trying to raise their own young.

A variety of harassment behaviors are observed at bee-eater colonies. Individuals chase resident birds off their territory, physically prevent the transfer of food during courtship feeding, or repeatedly visit nests that are not their own before egg laying or hatching. During the course of their study, Emlen and Wrege observed 47 cases of harassment. Over 90% of the instigators were male and over 70% were older than the targeted individual. In 58% of the episodes, the instigator and victim were close genetic kin. In fact, statistical tests show that harassment behavior is not targeted randomly, but is preferentially directed at close kin ($P < 0.01$; χ^2 test).

Emlen and Wrege interpret this behavior by proposing that instigators are actively trying to break up the nesting attempts of close kin. Further, they suggest that instigators do this to recruit the targeted individuals as helpers at their own (the instigator's) nest.

What evidence do Emlen and Wrege present to support this hypothesis? In 16 of the 47 harassment episodes observed, the behavior actually resulted in recruitment: The harassed individuals abandoned their own nesting attempts and helped at the nest of the instigator. Of these successful events, 69% involved a parent and offspring and 62% involved a father and son. The risk of being recruited is clearly highest for younger males and for males with close genetic relatives breeding within their clan (Figure 10.15).

These data raise the question of why sons do not resist harassment more effectively. Emlen and Wrege suggest that harassment can be successful because sons are equally related to their own offspring and to their siblings. Parents, in contrast, are motivated to harass because they are more closely related to their own offspring ($r = 1/2$) than they are to their grandchildren ($r = 1/4$). We have already mentioned that on average, each helper is responsible for an additional 0.47 offspring being raised. In comparison, each parent at a nest unaided by relatives is able to raise 0.51 offspring. This means that for a first-time breeder, the fitness payoff from breeding on its own is only slightly greater than the fitness payoff from helping. The payoffs are close enough to suggest that parents can change the bottom line of the fitness accounting. Perhaps harassing a son tips the balance by increasing his cost of rearing young. Then helping becomes a more favorable strategy for the son than raising his own young. Emlen and Wrege's data imply that bee-eater fathers recognize sons and coerce them into serving the father's reproductive interests.

Siblicide

In certain species of birds and mammals, it is common for young siblings to kill each other while parents look on passively. How can this behavior be adaptive, given that the parents and siblings are related by an r of 1/2?

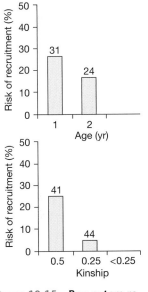

Figure 10.15 Bee-eaters recruit helpers who are younger and closely related Emlen and Wrege (1992) considered each paired male in the colony who had a same-age or older male present in its clan, and calculated the percentage of these individuals who were recruited as helpers after experiencing harassment. This probability is plotted on the y-axis of these two bar charts. Different age and kinship categories of targeted individuals are plotted on the x-axis. Here kinship represents the r-value between the targeted individual and the offspring in the instigator's nest.

Lynn Lougheed and David Anderson (1999) took an experimental approach to answering this question. Their research subjects were species of seabirds called the blue-footed booby and the masked booby (Figure 10.16). In both taxa, females normally lay a two-egg clutch. Because the eggs are laid 2–10 days apart, one chick hatches before the other. In the masked booby, the older offspring pushes its younger sibling from the nest within a day or two of hatching. There the smaller chick quickly dies of exposure or is taken by a predator.

Siblicide is more complex in the blue-footed booby, however. The older nestling does not always kill its younger sibling right after it hatches. During short-term food shortages, Anderson and Robert Ricklefs (1995) actually found that older chicks reduce their food intake. By doing so, they help their younger siblings survive. But if food shortages continue, the older chick attacks and kills its sibling. Presumably, this enhances the older chick's chance of survival by removing competition for food.

Lougheed and Anderson (1999) wanted to understand whether parents play a role in these events. In both masked boobies and blue-footed boobies, siblicide is sensible in light of the relatedness asymmetry between individuals (where $r = 1$) and their siblings (where $r = 1/2$). But parents are equally related to each chick and would be expected to intervene and prevent attacks.

To explore whether parental behavior differs between masked boobies and blue-footed boobies, Lougheed and Anderson performed a reciprocal transplant experiment. They placed newly hatched masked booby chicks in blue-footed booby nests, and vice versa. As controls, they also monitored the fate of masked booby chicks transferred to other masked booby nests and blue-footed booby broods transferred to other blue-footed booby nests.

Siblicide may increase the fitness of parents as well as the siblicidal offspring if the offspring that is killed is likely to die anyway.

As the data in Table 10.4 show, the fate of the chicks varied dramatically in the four treatments. Chicks were much more likely to die if they had a masked booby nestmate. This is consistent with the observation that siblicide is virtually universal in this species. But nestlings were also much more likely to die if they had masked booby parents. To explain this result, Lougheed and Anderson proposed that masked booby parents tolerate siblicidal chicks, while blue-footed booby parents attempt to intervene and prevent the death of their younger offspring. Their data set is the first to indicate that in some siblicidal species, parents act to defend their reproductive interests.

Why do masked booby and blue-footed boobies differ so strongly in their response to parent–offspring conflict? The leading hypothesis is that food shortages

Figure 10.16 Blue-footed and masked boobies These photos show masked boobies (left) and blue-footed boobies (right). The two species nest in adjacent colonies at the study site established by Lougheed and Anderson in the Galápagos islands off the northwest coast of South America. (Left: D. Cavagnaro/ Visuals Unlimited; right: Tui De Roy/Minden Pictures)

Table 10.4 **In boobies, the probability of siblicide varies with parent species and nestmate species**

These data show that siblicide is much more common in nests with masked booby parents or masked booby nestlings than in nests with blue-footed booby nestlings or blue-footed booby parents. To explain these patterns, Lougheed and Anderson hypothesize that masked booby nestlings are more siblicidal than blue-footed booby nestlings, and that blue-footed booby parents attempt to intervene and prevent siblicide.

Treatment	No siblicide	Siblicide
Masked booby nestlings with masked booby parents	0	25
Blue-footed booby nestlings with masked booby parents	12	8
Masked booby nestlings with blue-footed booby parents	4	16
Blue-footed booby nestlings with blue-footed booby parents	17	0

Source: Lougheed and Anderson (1999).

are much more likely in masked boobies, and that second chicks almost always starve to death even without siblicide. This hypothesis is still untested, however. Research into the dynamics of siblicide and other forms of parent–offspring conflict continues (see Mock and Parker 1997).

10.4 Reciprocal Altruism

Inclusive fitness theory has been remarkably successful in explaining a wide range of phenomena in social evolution. In many cases, altruistic acts can be understood in light of Hamilton's rule, and conflicts can be understood by analyzing asymmetries in coefficients of relatedness and differences in fitness payoffs. But the theory and data we have reviewed thus far are only relevant to interactions among kin. What about the frequent occurrence of cooperation among unrelated individuals?

Reciprocal altruism provides one theoretical framework for studying cooperation among non-kin. Robert Trivers (1971) proposed that individuals can be selected to dispense altruistic acts if equally valuable favors are later returned by the beneficiaries. According to Trivers, natural selection can favor altruistic behavior if the recipients reciprocate.

Two important conditions must be met for reciprocal altruism to evolve. First, selection can favor altruistic acts only if the cost to the actor is smaller than or equal to the benefit to the recipient. Alleles that lead to high-cost, low-benefit behavior cannot increase in the population even if this type of act is reciprocated. (As with kin selection, the costs and benefits of altruistic acts are measured by the numbers of surviving offspring.) Second, individuals that fail to reciprocate must be punished in some way. If they are not, then altruistic individuals suffer fitness losses with no subsequent return. Alleles that lead to cheating behavior would increase in the population and altruists would quickly be eliminated. As a result, the theory predicts that altruists will be selected to detect and punish cheaters by physically assaulting them or by withholding future benefits (Box 10.4).

BOX 10.4   Prisoner's dilemma: analyzing cooperation and conflict using game theory

Robert Trivers (1971) recognized that a classical problem from the branch of mathematics called game theory closely simulates the problems faced by nonrelatives in making decisions about their interactions. The central idea in game theory is that the consequences of any move in a game are contingent: The result or payoff from an action depends on the move made by the opponent. When players in a game pursue contrasting strategies, game theory provides a way to quantify the outcomes and decide which strategy works best.

Game theory was invented in the 1940s to analyze contrasting strategies in games like poker and chess. Later, the approach was applied by economists to a variety of problems in market economics and business competition. John Maynard Smith (1974, 1982) pioneered the use of game theory in evolutionary biology in analyses of animal contests. His work inspired a series of productive studies on the evolution of display behavior and combat (e.g., Sigurjónsdóttir and Parker 1981; Hammerstein and Reichert 1988).

The game that Trivers employed to analyze cooperation is called Prisoner's Dilemma. Prisoner's Dilemma models the following situation: Two prisoners who have been charged as accomplices in a crime are locked in separate cells. The punishment they suffer depends on whether they cooperate with one another in maintaining their innocence or implicate the other in the crime. Each prisoner has to choose his strategy without knowing the other prisoner's choice. The payoffs to Player A in this game are as follows:

		Player B's action	
		C Cooperation	**D** Defection
Player A's action	**C** Cooperation	R (reward for cooperation—both receive light sentences)	S (sucker gets longer sentence if partner defects)
	D Defection	T (temptation—reduced sentence for defector)	P (punishment for mutual defection—both receive intermediate sentences)

where $T > R > P > S$ and $R > (S + T)/2$. The highest payoff in the game comes when Players A and B cooperate by maintaining silence, but Player A does best when A defects and B cooperates. When Players A and B interact just once, the best strategy for each player is to defect.

What happens when the two players interact repeatedly? Robert Axelrod and William Hamilton (1981) performed a widely cited analysis of an iterated Prisoner's Dilemma. They invited game theorists from all over the world to submit strategies for players competing in a computerized simulation of the game. Each round in this tournament had the following payoffs: $R = 3, T = 5, S = 0$, and $P = 1$. Axelrod and Hamilton let each strategy play against all of the other strategies submitted, and computed the outcome of every one-on-one game over many interactions. Most of the theorists submitted complicated decision-making algorithms, but the winner was always the simplest strategy of all, called tit for tat, TFT. An individual playing TFT starts by cooperating, then simply does whatever the opponent did in the previous round. This strategy has three prominent features: (1) It is never the first to defect, (2) it is provoked to immediate retaliation by defection, and (3) it is willing to cooperate again after just one act of retaliation for a defection.

In analyzing the outcome of games like this, researchers use the concept of an evolutionarily stable strategy, ESS. A strategy is an ESS if a population of individuals using it cannot be invaded by a rare mutant adopting a different strategy. Axelrod and Hamilton's tournament showed that TFT is an evolutionarily stable strategy with respect to other strategies employed in the tournament. Their result offers an explanation for the evolution of cooperative behavior in unrelated individuals. Laboratory experiments with guppies and sticklebacks suggest that animals may actually play TFT when they interact (see Milinski 1996; Dugatkin 1998).

Trivers pointed out that reciprocal altruism is most likely to evolve when

- each individual repeatedly interacts with the same set of individuals (groups are stable);
- many opportunities for altruism occur in an individual's lifetime;
- individuals have good memories; and
- potential altruists interact in symmetrical situations.

Reciprocal altruism can evolve only under a restricted set of conditions.

This means that interacting individuals are able to dispense roughly equivalent benefits at roughly equivalent costs.

Accordingly, we expect reciprocal altruism to be characteristic of long-lived, intelligent, social species with small group size, low rates of dispersal from the group, and a high degree of mutual dependence in group defense, foraging, or other activities. Reciprocal altruism should be less likely to evolve in species where strong dominance hierarchies are the rule. In these social systems, subordinate individuals are rarely able to provide benefits in return for altruistic acts dispensed by dominant individuals.

Based on these characteristics, Trivers (1971, 1985; see also Packer 1977) has suggested that reciprocal altruism is responsible for much of the cooperative behavior observed in primates like baboons, chimpanzees, and humans. Indeed, Trivers has proposed that human emotions like moralistic aggression, gratitude, guilt, and trust are adaptations that have evolved in response to selection for reciprocal altruism. He suggests that these emotions function as "scorekeeping" mechanisms useful in moderating transactions among reciprocal altruists.

Studying reciprocal altruism in natural populations is exceptionally difficult, however. For example, it is likely that kin selection and reciprocal altruism interact and are mutually reinforcing in many social groups. This makes it difficult for researchers to disentangle the effect that each type of selection has independently of the other. Also, the fitness effects of some altruistic actions can be difficult to quantify. When a young male baboon supports an older, unrelated male in his fight over access to a female, or when a young lioness participates in group defense of the territory, how do we quantify the fitness costs and benefits for each participant? What are our chances of observing the return behavior, and quantifying its costs and benefits as well? Finally, it can be difficult to distinguish reciprocal altruism from what biologists call by-product mutualism. This is cooperative behavior that benefits both individuals more or less equally. The critical difference between reciprocity and mutualism is that there is a time lag between the exchange of benefits in reciprocal altruism.

For these reasons, it has been difficult for evolutionary biologists to document reciprocal altruism—difficult enough, in fact, to make many biologists suspect that this type of natural selection is rare. Here we will examine one of the most robust studies done to date: food sharing in vampire bats. In this system, the altruistic act is regurgitating a blood meal. The cost of altruism can be measured as an increase in the risk of starvation, and the benefit as a lowered risk of starvation.

Blood-Sharing in Vampire Bats

Gerald Wilkinson (1984) worked on a population of about 200 vampire bats (*Desmodus rotundus*) at a study site in Costa Rica (Figure 10.17). The basic social unit in this species consists of 8–12 adult females and their dependent offspring.

Figure 10.17 **Vampire bats**
This photo shows a group of vampire bats roosting in a hollow tree. (Gerald Wilkinson, University of Maryland)

Members of these groups frequently roost together in hollow trees during the day, although subgroups often move from tree to tree. (There was a total of 14 roosting trees at Wilkinson's study site.) As a result of this social structure, many individuals in the population associate with one another daily. The degree of association between individuals varies widely, however. Wilkinson quantified the degree of association between each pair of bats in the population by counting the number of times they were seen together at a roost and dividing by the total number of times they were observed roosting.

Wilkinson was able to capture and individually mark almost all of the individuals at his study site over a period of four and a half years, and estimate coefficients of relatedness through path analysis of pedigrees. In the female group for which he had the most complete data, the average *r* between individuals was 0.11. (Recall that cousins have an *r* of 0.125.)

The combination of variability in association and relatedness raises interesting questions about the evolution of altruism. Vampire bats dispense altruistic behavior by regurgitating blood meals to one another. This food sharing is important because blood meals are difficult to obtain. The bats leave their roosts at night to search for large mammals—primarily horses and cattle—that can provide a meal. Prey are wary, however, and 33% of young bats and 7% of adults fail to feed on any given night. By studying weight loss in captive bats when food was withheld, Wilkinson was able to show that bats who go three consecutive nights without a meal are likely to starve to death.

Vampire bats reciprocate by sharing blood meals. They usually share with close relatives or nonrelatives who are nestmates and may later reciprocate.

Because the degree of relatedness and degree of association varied among individuals in the population, either kin selection or reciprocal altruism could operate in this system. Wilkinson was able to show that both occur. Over the course of the study, he witnessed 110 episodes of regurgitation. Seventy-seven of these were between mother and child and are simply examples of parental care. In 21 of the remaining 33 cases, Wilkinson knew both the *r* and the degree of association between the actor and beneficiary and could examine the effect of both variables. Wilkinson discovered that both degree of relatedness and degree of association have a statistically significant effect in predicting the probability of regurgitation (Figure 10.18). Bats do not regurgitate blood meals to one another randomly. They are much more likely to regurgitate to relatives and to nonrelatives who are frequent roostmates.

To confirm that bats actually do reciprocate, Wilkinson held nine individuals in captivity, withheld food from a different individual each night for several weeks,

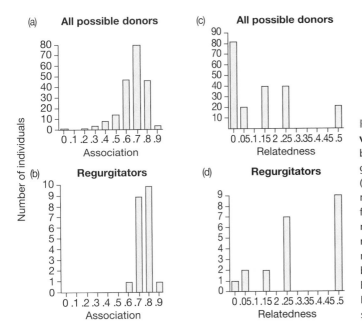

(a) **All possible donors**

Number of individuals / Association

(b) **Regurgitators**

Association

(c) **All possible donors**

Relatedness

(d) **Regurgitators**

Relatedness

Figure 10.18 Association, relatedness, and altruism in vampire bats These histograms plot the total number of bat–bat pairs at Gerald Wilkinson's study site versus their (a) degree of association, for all potential blood donors in the roost, (b) degree of association, for blood-sharing pairs who are not related as mother–offspring, (c) coefficient of relatedness, for all potential blood donors in the roost, and (d) coefficient of relatedness, for blood-sharing pairs who were not related as mother–offspring. The visual impression in these figures is that regurgitators are more likely to be related and more likely to be roostmates than the general population. This is confirmed by a statistical procedure called a stepwise logistic regression. Both relatedness and association affect the probability of blood-sharing. From Wilkinson (1984).

and recorded who regurgitated to whom over the course of the experiment. Statistical tests rejected the null hypothesis that hungry individuals received blood randomly from cagemates. Instead, hungry individuals were much more likely to receive blood from an individual they had fed before. This confirms that vampire bats are reciprocal altruists.

Territory Defense in Lions

Lions (*Panthera leo;* Figure 10.19) are the only species of cat that lives in social groups. These groups, called prides, consist of 3–6 related females, their offspring, and a coalition of males. Males are often related to one another but are unrelated to pride females (Packer and Pusey 1982; Packer et al. 1991). Males cooperate in defending themselves against other coalitions of males that attack and try to take over the pride. If a pride takeover occurs, the incoming males often kill the young cubs present (see Section 9.2).

Females cooperate in defending their young against infanticidal attacks, in nursing young, in hunting prey that are difficult to capture (females do the vast majority

Figure 10.19 African lions When female lions hear roars made by unfamiliar females, they move across their territory toward the source. Females that lead the movement frequently glance back at female pride members that are lagging behind. (Mitsuaki Iwago/Minden Pictures)

of hunting in lions), and in defending the pride's territory against incursions by females in neighboring prides (Packer and Pusey 1983, 1997). Battles with intruding females are dangerous, especially if pride females do not cooperate in defense. Solitary lions are often killed in same-sex encounters.

To study cooperation during incursions by strange females who threaten the territory, Robert Heinsohn and Craig Packer (1995) placed speakers near the edge of pride territories and played tape recordings of roars given by unfamiliar females. Pride females respond to these roars by approaching the speakers; they even attack if a stuffed female lion is placed near the speaker. They also do not habituate to this stimulus—they continue responding when the experiment is repeated. As a result, Heinsohn and Packer were able to quantify how females responded to threats over multiple trials. In addition to recording how long it took an individual to reach the midpoint between the pride's original position and the speaker, they also calculated the difference between each individual's time-to-midpoint and the leader's time, documented the order within the group that each female reached the midpoint, and counted the number of glances a female made back at lagging pride members. They collected data on female responses in eight different prides.

Statistical analyses showed significant differences in the strength of responses among females within prides. Some females in each pride always led; others always lagged behind. Other females adopted conditional strategies: They would assist the lead female(s) more frequently when Heinsohn and Packer played tapes of more than one female roaring (that is, when the threat to the pride was greater). Still others would lag behind more when multiple roars were played. These differences were not correlated with the age or body size of females or with their coefficient of relationship with other pride members.

The dynamics of group defense in lions have yet to be explained by kin selection or reciprocal altruism.

These results are paradoxical in the context of both inclusive fitness and reciprocity theory. Why do the leaders tolerate the laggards? According to the theory we developed earlier, they should be punished. But leaders were never observed to threaten laggards or to withhold benefits. For example, they did not stop leading until laggards caught up. They glanced back and appeared to recognize that laggards were lagging, but continued approaching the speaker.

The simple answer is that we do not know why this combination of altruist and non-reciprocator strategies exists among lions. The leading hypothesis, which is still to be tested, is that laggard females make fitness benefits up to the leaders in other ways: through exceptional hunting prowess or milk production, for example. But clearly, social interactions among lions are substantially more complex than current theory would suggest.

Summary

When individuals interact, four outcomes are possible with respect to fitness: cooperation, altruism, selfishness, and spite. The evolution of altruism was one of the great paradoxes of evolutionary biology until it was resolved by two important advances:

1. William Hamilton showed mathematically that a gene for altruism will spread when $Br - C > 0$

where B is the benefit to the recipient in units of surviving offspring, r is the coefficient of relatedness between actor and recipient, and C is the cost to the actor. When Hamilton's rule holds, kin selection results in altruistic behavior.

2. Robert Trivers developed the theory of reciprocal altruism. Altruism among unrelated individuals can evolve if the benefits of an altruistic act to the re-

cipient are large and the cost to the actor is small, and the benefits are later returned to the actor by the recipient.

Kin selection explains phenomena such as alarm calling in ground squirrels and helping behavior in birds. In Belding's ground squirrels, individuals are more willing to risk giving an alarm call if close relatives are nearby. In white-fronted bee-eaters, 1-year-old individuals preferentially help at the nests of closely related individuals (often their parents).

Kin selection has also been important in explaining the evolution of eusociality in Hymenoptera and in naked mole-rats. Phylogenetic analyses show that nest-building and extensive care of young was a precondition for the evolution of reproductive altruism in ants, wasps, and bees. In these groups, individuals are unlikely to reproduce on their own successfully, so reproductive altruism is favored by kin selection. In obligately eusocial groups like ants and bees, asymmetries in relatedness between egg-laying queens and nonreproductive workers result from the haplodiploid system of sex determination and produce sharp conflicts over the sex ratio of offspring. Helping behavior is facultative in *Polistes* wasps, however. In this group, females may or

may not help at the nest depending on their body size and their coefficient of relatedness relative to nestmates. In naked mole-rats, eusociality is supported by the queen's physical dominance and by extensive inbreeding that leads to high coefficients of relatedness among individuals.

Parent–offspring conflict and sibling conflict occur when the fitness interests of individuals within families clash. Parents are equally related to each of their offspring, but offspring are more closely related to themselves than to their siblings. Weaning conflict occurs because young mammals demand more resources for themselves than for their siblings; siblicide occurs if perpetrators gain enough direct fitness benefits by removing competition for food to outweigh the indirect fitness cost of killing a sibling. Parents may acquiesce to siblicide if it increases the probability that at least one offspring will survive.

Reciprocal altruism has been successful as an explanation for food-sharing between vampire bats, and may be involved in interactions among females in lion prides. It is often difficult, however, to distinguish when cooperation is due to reciprocal altruism and when it results from kin selection or simple mutualism.

Questions

1. Suppose adult bee-eaters could raise only 0.3 more offspring with a helper than without a helper. Would you still expect male bee-eaters to "give in" to the harassment of their fathers, or would male bee-eaters tend to fight off their fathers? Explain your reasoning.

2. When a Thomson's gazelle detects a nearby stalking cheetah, the gazelle often begins bouncing up and down with a stiff-legged gait called "stotting" (see Figure 10.20). One

hypothesis is that stotting has evolved because it may help alert the gazelle's kin to the presence of a predator, analogous to the alarm calls of ground squirrels. Caro (1986) reports that stotting does not seem to increase the gazelle's risk of being attacked. In fact, once a gazelle begins to stott, the cheetah often gives up the hunt. How is C (the cost of stotting) different for a gazelle, compared to C (the cost of alarm calls) for a ground squirrel? Do you

Figure 10.20 A Thomson's gazelle stotting. (R. D'Estes/ Photo Researchers, Inc.)

think it is likely that stotting is an altruistic behavior? With this in mind, make a prediction about whether a gazelle will stott when there are no other gazelles around, and then look up Caro's papers to see if you are right.

Caro, T. M. 1986. The function of stotting in Thomson's gazelles: Some tests of the hypotheses. *Animal Behaviour.* 34: 663–684.

Caro, T. M. 1994. Ungulate antipredator behaviour: Preliminary and comparative data from African bovids. *Behaviour* 128: 189–228.

3. The cubs of spotted hyenas often begin fighting within moments of birth, and often one hyena cub dies. The mother hyena does not interfere. How could such a behavior have evolved? For instance, from the winning sibling's point of view, what must B (benefit of siblicide) be, relative to C (cost of siblicide), to favor the evolution of siblicide? From the parent's point of view, what must B be, relative to C, for the parent to watch calmly rather than to interfere? [See Frank (1997) for more about the unusual social system of spotted hyenas, and Golla et al. (1999) for new information from studies of wild hyenas.]

Frank, Laurence G. 1997. Evolution of genital masculinization: Why do female hyaenas have such a large 'penis'? *Trends in Ecology and Evolution* 12: 58–62.

Golla, W., H. Hofer, and M. L. East. 1999. Within-litter sibling aggression in spotted hyaenas: Effect of maternal nursing, sex and age. *Animal Behaviour* 58: 715–726.

4. Blue jays (*Cyanocitta cristata*) seem to be better than American robins (*Turdus migratorius*) at recognizing individuals. In one study, blue jays raised with American robins could distinguish strange from familiar robins better than the robins themselves (Schimmel and Wasserman 1994). Do you think these species differ in occurrence of kin selection or reciprocal altruism (or both)? Why?

Schimmel, K. L. and F. E. Wasserman. 1994. Individual and species preference in two passerine birds: Auditory and visual cues. *Auk* 111: 634–642.

5. The first paragraph of this chapter refers to crows who cooperate in chasing a predator away (a behavior known as mobbing) and who help their parents raise their siblings. Which behavior is mutualistic, and which behavior is favored by kin selection? Suggest a hypothesis to explain why young crows in some populations might help at the nest, while in other populations they nest on their own.

6. The biologist J. B. S. Haldane was once explaining kin selection to some friends in a pub. As the story goes, he scribbled some calculations on an envelope and announced that he would be willing to die for two brothers or eight cousins. Explain his reasoning.

7. Look at Figure 10.14 on parent–offspring conflict. Explain, in general terms, why the behavior of females should evolve so that mothers start weaning when B/C falls below 1. (*Hint:* Consider the reproductive success of mothers who wean very early, and of mothers who wean very late.) If a mother could have only one litter of young in her lifetime, how would the period of weaning conflict change?

8. The text claims that eusociality has evolved several times independently within the hymenoptera. What is the evidence for this statement? If it is true, in what sense is eusociality in ants, bees, and wasps an example of convergent evolution (see Chapter 9)?

9. How would you go about testing the hypothesis that female lions who do not participate in territory defense reciprocate by providing milk to the offspring of territory defenders? List the predictions made by the hypothesis and the types of data you would have to collect.

10. House sparrows often produce two successive broods of young. Males feed their first brood only briefly, but feed their second brood for much longer [see Hegner & Wingfield (1986) for further information]. Why do males feed first broods less than second broods? (*Hint:* Consider how C, the cost of feeding the current brood, changes). How could you test your hypothesis? How is this situation analogous to weaning conflict in mammals?

Hegner, R. E. and J. C. Wingfield. 1986. Behavioral and endocrine correlates of multiple brooding in the semicolonial house sparrow *Passer domesticus*. I. Males. *Hormones and Behavior* 20: 294–312.

11. Which is more common in human cultures—eusociality (look back at the three requirements of eusociality; can you think of any human cultures that fit?), or a helper-at-the-nest social system? Which do you think is generally more common in social animals? Why?

12. Human siblings often show intense sibling rivalry that typically declines during the teenage years. Suggest an evolutionary explanation for this pattern.

Exploring the Literature

13. Throughout this chapter, we concentrated on the fitness consequences of social interactions and paid little attention to the issue of why organisms live in groups in the first place. To learn how social living has been favored by

factors such as a requirement for group defense against predators, benefits from group foraging, and a need for long-term care of dependent young, see

Alexander, R. D. 1974. The evolution of social behavior. *Annual Review of Ecology and Systematics* 5: 325–383.

Packer, C., D. Scheel, and A. E. Pusey. 1990. Why lions form groups: Food is not enough. *American Naturalist* 136: 1–19.

14. Because this chapter emphasizes theories that explain why cooperative behavior can evolve when a fitness cost is involved, we spent little time on the evolution of mutualism. For examples of cooperative behavior that are not caused by kin selection or reciprocal altruism, see

McDonald, D. B., and W. K. Potts. 1994. Cooperative display and relatedness among males in a lek-breeding bird. *Science* 266: 1030–1032.

Watts, D.P. 1998. Coalitionary mate guarding by male chimpanzees at Ngogo, Kibale National Park, Uganda. *Behavioral Ecology and Sociobiology* 44: 43–55.

15. A variety of models and experiments (some using college students as experimental subjects) have shown that certain variations on the tit-for-tat strategy are extremely successful in interactions among individuals. To begin reviewing this literature, see

Wedekind, C., and M. Milinski. 1996. Human cooperation in the simultaneous and the alternating Prisoner's Dilemma: Pavlov versus generous tit-for-tat. *Proceedings of the National Academy of Sciences, USA* 93: 2686–2689.

Roberts, G., and T. N. Sherratt. 1998. Development of cooperative relationships through increasing investment. *Nature* 394: 175–179.

Citations

Alexander, R. D., K. M. Noonan, and B.J. Crespi. 1991. The evolution of eusociality. In P.W. Sherman, J.U.M. Jarvis, and R.D. Alexander. *The Biology of the Naked Mole Rat.* Princeton, NJ: Princeton University Press, 3–44.

Anderson, D. J., and R. E. Ricklefs. 1995. Evidence of kin-selected tolerance by nestlings in a siblicidal bird. *Behavioral Ecology and Sociobiology* 37: 163–168.

Axelrod, R., and W. D. Hamilton. 1981. The evolution of cooperation. *Science* 211:1390–1396.

Bartz, S. H. 1979. Evolution of eusociality in termites. *Proceedings of the National Academy of Sciences, USA* 76:5764–5768.

Betzig, L. L., and P. W. Turke. 1986. Parental investment by sex on Ifaluk. *Ethology and Sociobiology* 7:29–37.

Bourke, A. F. G., and N. R. Franks. 1995. *Social Evolution in Ants.* Princeton, NJ: Princeton University Press.

Brown, J. L. 1987. *Helping and Communal Breeding in Birds.* Princeton, NJ: Princeton University Press.

Brown, J. L., and A. Eklund. 1994. Kin recognition and the major histocompatibility complex: An integrative review. *American Naturalist* 143: 435–461.

Charnov, E. L. 1982. *The Theory of Sex Allocation.* Princeton, NJ: Princeton University Press.

Clutton-Brock, T. H., S. D. Albon, and F. E. Guinness. 1984. Maternal dominance, breeding success and birth sex ratios in red deer. *Nature* 308: 358–360.

Crozier, R. H., and P. Pamilo. 1996. *Evolution of Social Insect Colonies.* Oxford: Oxford University Press.

Darwin, C. 1859. *The Origin of Species.* London: John Murray.

Duffy, J. E. 1996. Eusociality in a coral-reef shrimp. *Nature* 381: 512–514.

Dugatkin, A. L. 1998. *Cooperation among animals.* Oxford: Oxford University Press.

Emlen, S. T., and P. H. Wrege. 1988. The role of kinship in helping decisions among white-fronted bee-eaters. *Behavioral Ecology and Sociobiology* 23: 305–315.

Emlen, S. T., and P. H. Wrege. 1991. Breeding biology of white-fronted bee-eaters at Nakuru: The influence of helpers on breeder fitness. *Journal of Animal Ecology* 60: 309–326.

Emlen, S. T., and P. H. Wrege. 1992. Parent–offspring conflict and the recruitment of helpers among bee-eaters. *Nature* 356:331-333.

Emlen, S. T., P. H. Wrege, and N.J. Demong. 1995. Making decisions in the family: An evolutionary perspective. *American Scientist* 83: 148–157.

Fisher, R. A. 1930. *The Genetical Theory of Natural Selection.* Oxford: Clarendon Press.

Godfrey, H. C. J. 1995. Evolutionary theory of parent–offspring conflict. *Nature* 376: 133–138.

Hamilton, W. D. 1964. The genetical evolution of social behaviour. I. *Journal of Theoretical Biology* 7: 1–16.

Hamilton, W. D. 1964. The genetical evolution of social behaviour. II. *Journal of Theoretical Biology* 7: 17–52.

Hamilton, W. D. 1967. Extraordinary sex ratios. *Science* 156: 477–488.

Hamilton, W. D. 1972. Altruism and related phenomena, mainly in the social insects. *Annual Review of Ecology and Systematics* 3: 193–232.

Hammerstein, P., and S. E. Riechert. 1988. Payoffs and strategies in territorial contests: ESS analyses of two ecotypes of the spider *Agelenopsis aperta. Evolutionary Ecology* 2: 115–138.

Hegner, R. E., S.T. Emlen, and N. J. Demong. 1982. Spatial organization of the white-fronted bee-eater. *Nature* 298: 264–266.

Heinsohn, R., and C. Packer. 1995. Complex cooperative strategies in group-territorial African lions. *Science* 269: 1260–1262.

Honeycutt, R. L. 1992. Naked mole-rats. *American Scientist* 80: 43–53.

Hunt, J. H. 1999. Trait mapping and salience in the evolution of eusocial vespid wasps. *Evolution* 53: 225–237.

Keller, L., M. Milinski, M. Frischknecht, N. Perrin, H. Richner, and F. Tripet. 1994. Spiteful animals still to be discovered. *Trends in Ecology and Evolution* 9: 103.

Lougheed, L.W., and D. J. Anderson. 1999. Parent blue-footed boobies suppress siblicidal behavior of offspring. *Behavioral Ecology and Sociobiology* 45: 11–18.

Manning, C. J., E. K. Wakeland, and W. K. Potts. 1992. Communal nesting patterns in mice implicate MHC genes in kin recognition. *Nature* 360: 581–583.

Maynard Smith, J. 1974. The theory of games and the evolution of animal conflicts. *Journal of Theoretical Biology* 47: 209–221.

Maynard Smith, J. 1982. *Evolution and the Theory of Games.* Cambridge: Cambridge University Press.

Michener, C. D. 1969. Comparative social behavior of bees. *Annual Review of Entomology* 14: 299–342.

Milinski, M. 1996. By-product mutualism, tit-for-tat reciprocity and cooperative predator inspection: A reply to Connor. *Animal Behaviour* 51: 458–461.

Mock, D. W., and G. A. Parker. 1997. *The evolution of sibling rivalry*. Oxford: Oxford University Press.

Nonacs, P., and H. K. Reeve. 1995. The ecology of cooperation in wasps: Causes and consequences of alternative reproductive decisions. *Ecology* 76: 953–967.

Oldroyd, B. P., M. J. Clifton, S. Wongsiri, T. E. Rinderer, H. A. Sylvester, and R. H. Crozier. 1997. Polyandry in the genus *Apis,* particulary *Apis andreniformis*. *Behavioral Ecology and Sociobiology* 40: 17–26.

Oldroyd, B. P., M. J. Clifton, K. Parker, S. Wongsiri, T. E. Rinderer, and R. H. Crozier. 1998. Evolution of mating behavior in the genus *Apis* and an estimate of mating frequency in *Apis cerana* (Hymenoptera: Apidae). *Annals of the Entomological Society of America* 91: 700–709.

Packer, C. 1977. Reciprocal altruism in *Papio anubis. Nature* 265: 441–443.

Packer, C., and A. E. Pusey. 1982. Cooperation and competition within coalitions of male lions: Kin selection or game theory? *Nature* 296: 740–742.

Packer, C., and A. E. Pusey. 1983. Adaptations of female lions to infanticide by incoming males. *American Naturalist* 121: 716–728.

Packer, C., and A. E. Pusey. 1997. Divided we fall: Cooperation among lions. *Scientific American* 276 (May): 52–59.

Packer, C., D. A. Gilbert, A. E. Pusey, and S. J. O'Brien. 1991. A molecular genetic analysis of kinship and cooperation in African lions. *Nature* 351: 562–565.

Page, R. E. Jr., and R. A. Metcalf. 1982. Multiple mating, sperm utilization, and social evolution. *American Naturalist* 119: 263–281.

Pamilo, P. 1984. Genetic relatedness and evolution of insect sociality. *Behavioral Ecology and Sociobiology* 15: 241–248.

Peters, J. M., D. C. Queller, V. L. Imperatriz-Fonseca, D. W. Roubik, and J. E. Strassmann. 1999. Mate number, kin selection and social conflicts in stingless bees and honeybees. *Proceedings of the Royal Society of London,* Series B 266: 379–384.

Pfennig, D. W., and P. W. Sherman. 1995. Kin recognition. *Scientific American* 272: 98–103.

Queller, D. C. 1989. The evolution of eusociality: Reproductive head starts of workers. *Proceedings of the National Academy of Sciences, USA* 86: 3224–3226.

Queller, D. C., and K. F. Goodnight. 1989. Estimating relatedness using genetic markers. *Evolution* 43: 258–275.

Queller, D. C., and J. E. Strassmann. 1998. Kin selection and social insects. *BioScience* 48: 165–175.

Reeve, H. K. 1992. Queen activation of lazy workers in colonies of the eusocial naked mole-rat. *Nature* 358: 147–149.

Reeve, H. K., D. F. Westneat, W. A. Noon, P. W. Sherman, and C. F. Aquadro. 1990. DNA "fingerprinting" reveals high levels of inbreeding in colonies of the eusocial naked mole-rat. *Proceedings of the National Academy of Sciences, USA* 87: 2496–2500.

Reeve, H. K., J. M. Peters, P. Nonacs, and P. T. Starks. 1998. Dispersal of first "workers" in social wasps: Causes and implications of an alternative reproductive strategy. *Proceedings of the National Academy of Sciences, USA* 95: 13737–13742.

Reeve, H. K., and P. W. Sherman. 1991. Intracolonial aggression and nepotism by the breeding female naked mole-rat. In P. W. Sherman, J. U. M. Jarvis, and R. D. Alexander, eds. *The Biology of the Naked Mole Rat.* Princeton, NJ: Princeton University Press, 337–357.

Roisin, Y. 1994. Intragroup conflicts and the evolution of sterile castes in termites. *American Naturalist* 143: 751–765.

Sherman, P. W. 1977. Nepotism and the evolution of alarm calls. *Science* 197:1246–1253.

Sherman, P. W. 1980. The limits of ground squirrel nepotism. In George W. Barlow and J. Silverberg, eds. *Sociobiology: Beyond Nature/Nurture?* Washington, DC: AAAS; and Boulder, CO: Westview Press, 505–544.

Sherman, P. W. 1981. Kinship, demography, and Belding's ground squirrel nepotism. *Behavioral Ecology and Sociobiology* 8:251–259.

Sherman, P. W., J. U. M. Jarvis, and S. H. Braude. 1992. Naked mole rats. *Scientific American* 267: 72–78.

Sigurjónsdóttir, H., and G. A. Parker. 1981. Dung fly struggles: Evidence for assessment strategy. *Behavioral Ecology and Sociobiology* 8: 219–230.

Stacey, P. B., and W. D. Koenig, eds. 1990. *Cooperative Breeding in Birds: Longterm Studies of Ecology and Behaviour.* Cambridge: Cambridge University Press.

Sundström, L., M. Chapuisat, and L. Keller. 1996. Conditional manipulation of sex ratios by ant workers: A test of kin selection theory. *Science* 274: 993–995.

Thorne, B. L. 1997. Evolution of eusociality in termites. *Annual Review of Ecology and Systematics* 28: 27–54.

Trivers, R. L. 1971. The evolution of reciprocal altruism. *Quarterly Review of Biology* 46: 35–57.

Trivers, R. L. 1974. Parent–offspring conflict. *American Zoologist* 14: 249–264.

Trivers, R. L. 1985. *Social Evolution.* Menlo Park, CA: Benjamin Cummings.

Trivers, R. L., and D. E. Willard. 1973. Natural selection of parental ability to vary the sex ratio of offspring. *Science* 179: 90–92.

Trivers, R. L., and H. Hare. 1976. Haplodiploidy and the evolution of the social insects. *Science* 191: 249–263.

Werren, J. H. 1984. A model for sex ratio selection in parasitic wasps: Local mate competition and host quality effects. *Netherlands Journal of Zoology* 34: 81–96.

Wilkinson, G. S. 1984. Reciprocal food sharing in the vampire bat. *Nature* 308: 181–184.

Wilson, E. O. 1971. *The Insect Societies.* Cambridge, MA: Harvard University Press.

CHAPTER 11

Aging and Other Life History Characters

A female seed beetle (*Stator limbatus*) looking for a place to lay her eggs on the seeds of catclaw acacia (*Acacia greggii*) and blue palo verde (*Cercidium floridum*). (Timothy A. Mousseau, University of South Carolina)

Evolution by natural selection has engineered all organisms to perform the same single ultimate task: to reproduce. How organisms go about the business of reproducing, however, is enormously variable. A few examples illustrate the diversity:

- Some mammals mature early and reproduce quickly, whereas others mature late and reproduce slowly. For example, female deer mice (*Peromyscus maniculatus*) mature at about seven weeks and have three or four litters of pups each year, whereas female black bears (*Ursus americanus*) mature at four or five years and produce cubs only once every two years (Nowak 1991).

- Plants have a wide range of reproductive life spans. Some, like the California poppy (*Eschscholtzia californica*), live and flower for just a single season. Others, like the black cherry (*Prunus serotina*), flower yearly for decades.

- Some bivalves produce enormous numbers of tiny eggs, whereas others produce small numbers of large eggs (Strathmann 1987). The oyster *Crassostrea gigas,* for example, releases 10 to 50 million eggs in a single spawn, each 50 to 55 micrometers in diameter. The clam *Lasaea subviridis,* in contrast, broods fewer than 100 eggs at a time, each some 300 micrometers in diameter.

The branch of evolutionary biology that attempts to make sense of the diversity in reproductive strategies is called life history analysis.

An organism truly perfected for reproduction would mature at birth, continuously produce high-quality offspring in large numbers, and live forever. Richard Law (1979) called such an organism a Darwinian demon: It would bedevil all other organisms and eventually monopolize life on the planet. No such organism exists, even after 3.8 billion years of evolution by natural selection. The reason is that such an organism is impossible. Some actual organisms come close to realizing one or another of the traits of an ideal reproducer, but all such organisms fall strikingly short by one or more of the remaining measures. For example, the female thrips egg mite (*Adactylidium* sp.) is mature at birth. Furthermore, she is already inseminated, having hatched inside her mother's body and mated with her brother (Elbadry and Tawfik 1966; see also Gould 1980, essay 6). But she produces just one clutch of offspring and her life is brief: She dies at the age of just four days, when her own offspring eat her alive from the inside (Figure 11.1a). Another example, the brown kiwi (*Apteryx australis mantelli*), produces high-quality offspring (Taborsky and Taborsky 1993). Female kiwis weigh about six pounds and lay eggs that weigh one pound (Figure 11.1b). The chicks that hatch from these huge eggs become largely self-reliant within a week. However, kiwi parents cannot produce these chicks continuously, and they cannot produce them in large numbers. It takes the female over a month to make each of the eggs in a typical two-egg clutch. The male has to incubate the eggs for about three months, during which time he loses some 20% of his body weight.

Organisms face fundamental trade-offs in their use of energy and time.

As the egg mite and the kiwi suggest, the laws of physics and biology impose fundamental trade-offs. The amount of energy an organism can harvest is finite, and biological processes take time. Energy and time devoted to one activity are energy and time that cannot be devoted to another. For example, an individual can allocate energy to growth for a long time, which may enable it to reach a larger

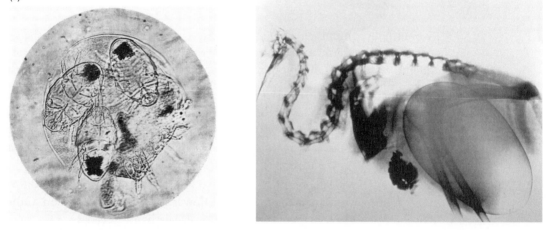

(a) (b)

Figure 11.1 Extreme reproductive strategies (a) Having devoured their mother from the inside, three thrips egg mites (*Adactylidium* sp.) prepare to depart her empty cuticle. The mother's legs are visible at lower right (180×). Reproduced by permission from E.A. Elbadry and M.S.F. Tawfik, 1966. Life cycle of the mite *Adactylidium* sp. (Acarina: Pyemotidae), a predator of thrips eggs in the United Arab Republic. *Annals of the Entomological Society of America* 59(3): 458–61, May 1966, p. 460, Fig. 6. (b) An x-ray of a female brown kiwi (*Apteryx australis mantelli*) ready to lay an egg. (Otorohanga Kiwi House, New Zealand)

size and ultimately enable it to produce more offspring. This benefit of large size, however, is balanced by a cost. The time required to grow to a large size is time during which predators, diseases, or accidents may strike. An individual that takes the time to grow to a large size thus incurs a greater risk of dying without ever having reproduced at all. We introduced the concept of trade-offs in Chapter 8, and we discussed how trade-offs constrain the evolution of adaptations. Whenever there is a trade-off between different components of fitness, we expect natural selection to favor individuals that allocate energy and time with an optimal balance between benefits and costs, thereby maximizing lifetime reproductive success. Different balances are optimal in different environments. Environmental variation is undoubtedly the source of much of the life history variation seen among living organisms.

In exploring the evolution of life histories, we analyze costs and benefits, and fitness trade-offs, as they apply to the following questions:

- Why do organisms age and die?
- How many offspring should an individual produce in a given year?
- How big should each offspring be?

These questions focus on the balance among aspects of fitness for individual organisms. One emerging new area of life history studies is the analysis of conflicts of interest between organisms of the same species and the evolution of strategies for managing these conflicts. We briefly discuss two of these conflicts between the interests of male and female parents. In the final section, we place life history analysis in a broader evolutionary context by considering the maintenance of genetic variation and evolutionary transitions in life history.

11.1 Basic Issues in Life History Analysis

An example of a life history, one that we will return to near the end of Section 11.2, appears in Figure 11.2. The figure follows the career of a hypothetical female Virginia opossum (*Didelphis virginiana*). As a baby, this female nursed for a little more than three months, was then weaned, and became independent. She continued to grow for another several months, reaching sexual maturity at an age of about 10 months. Shortly thereafter the female had her first litter consisting of eight

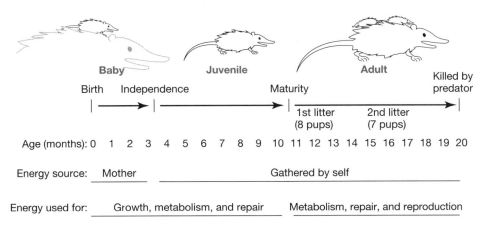

Figure 11.2 The life history of a hypothetical female Virginia opossum (*Didelphis virginiana*) This hypothetical female has a life history typical of female Virginia opossums living in the mainland United States (Austad 1988; 1993). Figure designed after Charnov and Berrigan (1993).

offspring. A few months later, she had a second litter, this time with seven offspring. At the age of 20 months, the female was killed by a predator.

The figure also indicates where the female opossum got her energy from at different stages of her life, and the functions to which she allocated that finite energy supply. Before she became sexually mature, the female used her energy for growth, metabolic functions like thermoregulation, and the repair of damaged tissues. After she became sexually mature, the female stopped growing, thereafter using her energy for metabolism, repair, and reproduction.

Changes in life history are caused by changes in the allocation of energy.

Fundamentally, differences among life histories concern differences in the allocation of energy. For example, a different female opossum than the one shown in Figure 11.2 might stop allocating energy to growth at an earlier age, thereby reaching sexual maturity more quickly. This strategy involves a trade-off: The female also matures at a smaller size, which means that she has to produce smaller litters. Still another female might, after reaching sexual maturity, allocate less energy to reproduction and more to repair, thereby keeping her tissues in better condition. Again there is a trade-off: Allocating less energy to reproduction means having smaller litters. Natural selection acts on life histories to adjust energy allocation in a way that maximizes the total lifetime production of offspring.

11.2 Why Do Organisms Age and Die?

Aging should be opposed by natural selection.

Aging, or **senescence**, is a late-life decline in an individual's fertility and probability of survival (Partridge and Barton 1993). Figure 11.3 documents aging in three animal species: a bird, a mammal, and an insect. All three show declines in both fertility and survival. All else being equal, aging reduces an individual's fitness. Aging should therefore be opposed by natural selection.

We consider two theories on why aging persists. The first, called the rate-of-living theory, invokes an evolutionary constraint (see Chapter 8); it posits that populations lack the genetic variation to respond any further to selection against aging. The second, called the evolutionary theory, invokes, in part, a trade-off between the allocation of energy to reproduction versus repair.

The Rate-of-Living Theory of Aging

The rate-of-living theory of senescence holds that aging is caused by the accumulation of irreparable damage to cells and tissues (reviewed in Austad and Fischer 1991). Damage to cells and tissues is caused by errors during replication, transcription, and translation, and by the accumulation of poisonous metabolic by-products. Under the rate-of-living theory, all organisms have been selected to resist and repair cell and tissue damage to the maximum extent physiologically possible. They have reached the limit of biologically possible repair. In other words, populations lack the genetic variation that would enable them to evolve more effective repair mechanisms than they already have.

One theory holds that aging is a function of metabolic rate . . .

The rate-of-living theory makes two predictions: (1) because cell and tissue damage is caused in part by the by-products of metabolism, the aging rate should be correlated with the metabolic rate; and (2) because organisms have been selected to resist and repair damage to the maximum extent possible, species should not be able to evolve longer life spans, whether subjected to natural or artificial selection.

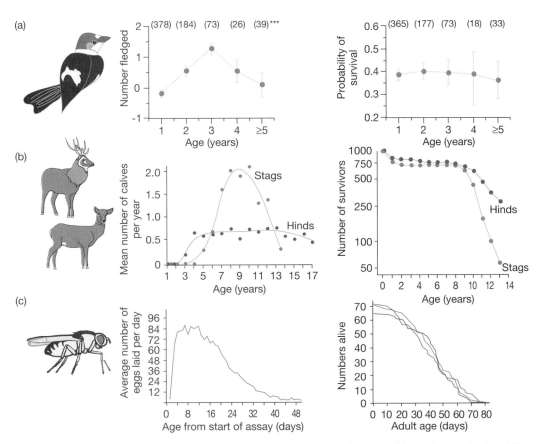

Figure 11.3   Aging in three animals   (a) Aging in a bird. For females in a wild population of collared fly-catchers (*Ficedula albicollis*), the number of young fledged each year declines after age three (age-related differences in number fledged are significant at *P* < 0.001). The probability of survival from one year to the next declines slightly, but not significantly, after age two. Sample sizes in parentheses. From Gustafsson and Pärt (1990). (b) Aging in a mammal. For males in a wild population of red deer (*Cervus elaphus*), the number of calves fathered each year declines sharply after a peak at about age nine; for females, the number of calves produced each year declines gradually starting at about age 13. For both sexes, the probability of surviving from one year to the next is nearly 100% from age two to age nine. After age nine, the probability of surviving plummets. From Clutton-Brock et al. (1988). (c) Aging in an insect. For females in a laboratory population of fruit flies (*Drosophila melanogaster*), the average number of eggs laid per day declines after the age of about 12 days. In three laboratory populations, the probability of surviving from one day to the next falls at the age of about 20 days. Modified from Rose (1984).

Steven Austad and Kathleen Fischer (1991) tested the first prediction, that aging rate will be correlated with metabolic rate, using comparative data on a diversity of mammals. With data from the literature, Austad and Fischer calculated the amount of energy expended per gram of tissue per lifetime for 164 mammal species in 14 orders. According to the rate–of–living theory, all species should expend about the same amount of energy per gram per lifetime, whether they burn it slowly over a long lifetime or rapidly over a short lifetime. In fact, there is wide variation among mammal species (Figure 11.4). Across all 164 species Austad and Fischer surveyed, energy expenditure ranges from 39 kilocalories per gram per lifetime in an elephant shrew to 1102 kcal/g/lifetime in a bat. Even within orders, energy expenditure varies greatly. The range for bat species runs from 325 to 1102 kcal/g/lifetime. As a group, bats have metabolic rates that are similar to those of other mammals of the same size, but life spans that average nearly three times

. . . but data on variation in metabolic rate and aging among mammals refute this theory.

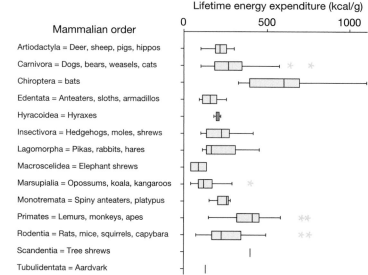

Lifetime energy expenditure (kcal/g)

Mammalian order

Artiodactyla = Deer, sheep, pigs, hippos
Carnivora = Dogs, bears, weasels, cats
Chiroptera = bats
Edentata = Anteaters, sloths, armadillos
Hyracoidea = Hyraxes
Insectivora = Hedgehogs, moles, shrews
Lagomorpha = Pikas, rabbits, hares
Macroscelidea = Elephant shrews
Marsupialia = Opossums, koala, kangaroos
Monotremata = Spiny anteaters, platypus
Primates = Lemurs, monkeys, apes
Rodentia = Rats, mice, squirrels, capybara
Scandentia = Tree shrews
Tubulidentata = Aardvark

Figure 11.4 Variation among mammals in lifetime energy expenditure This box plot represents the range of lifetime energy expenditure within each of 14 orders of mammals. The vertical line dividing each box represents the median value for that order. The right and left ends of each box represent the 75th and 25th percentiles. The horizontal lines extending to the right and left of each box represent the range of values; the asterisks represent statistical outliers. From Austad and Fischer (1991). Copyright © 1991, Gerontological Society of America. Reprinted with permission.

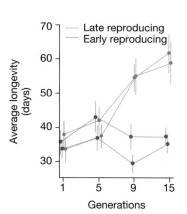

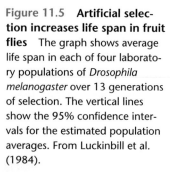

Figure 11.5 Artificial selection increases life span in fruit flies The graph shows average life span in each of four laboratory populations of *Drosophila melanogaster* over 13 generations of selection. The vertical lines show the 95% confidence intervals for the estimated population averages. From Luckinbill et al. (1984).

longer. Energy expenditure for marsupial species ranges from 43 to 406 kcal/g/lifetime. As a group, marsupials have metabolic rates that are significantly lower than those of other mammals of the same size ($P = 0.001$), but life spans that are significantly shorter ($P < 0.001$). These patterns are not consistent with the rate-of-living theory of aging.

Leo Luckinbill and colleagues (1984) tested the second prediction, that species cannot evolve longer life spans, by artificially selecting for longevity in laboratory populations of fruit flies (*Drosophila melanogaster*). Luckinbill et al. collected wild flies and used them to establish four laboratory populations. In two populations, the researchers selected for early reproduction by collecting eggs from young adults (two to six days after eclosion) and using the individuals that hatched from these eggs as the next generation's breeders. Longevity in these populations did not change during 13 generations of selection (Figure 11.5). In the other two populations, Luckinbill and colleagues selected for late reproduction by collecting eggs from old adults. "Old" meant 22 days after eclosion at the beginning of the experiment, and 58 days after eclosion by the end. Longevity in these populations increased dramatically during 13 generations of selection (Figure 11.5). At the beginning of the experiment, the average life span of the flies in these populations was about 35 days; by the end, the average life span was about 60 days. Other researchers conducting similar experiments have confirmed that average life span increases in *Drosophila* populations in response to selection for late-life reproduction (Rose 1984; Partridge 1987; Partridge and Fowler 1992; Roper et al. 1993). These results are consistent with the rate-of-living theory of aging only if the long-lived populations have evolved lower metabolic rates. Phillip Service (1987) found that fruit flies selected for long life span indeed had lower metabolic rates than controls, but only in the first 15 days of life. It is not clear that an evolved difference in metabolic rate can explain an evolved difference in life span as large as that obtained by Luckinbill and colleagues.

These experiments and observations seem to falsify the predictions of the rate-of-living theory. However, the general idea that organisms live fast and die young has persisted, perhaps in part because cellular and genetic mechanisms seem to link these two aspects of a life history. One mechanism in animals is based not on the rate of energy expenditure by whole individuals but rather on the rate of division of their cells and chromosomes. Normal animal cells are capable of some limited number of divisions (and duplications of their chromosomes), after which the cells cease dividing and eventually die. This pattern occurs in all cells except germ line cells, cancer cells, and some embryonic and blood stem cells, and may be caused by damage to chromosomes (Campisi 1996; Reddel 1998). Each end—or telomere—of a eukaryotic chromosome consists of many copies of a repetitive DNA sequence (in humans the repeat is TTAGGG) that is tagged onto the end of the chromosome by a DNA polymerase enzyme called telomerase. Telomerase is strongly expressed in cancers and germ line cells but not in most other cells. A portion of this telomere is lost with each cycle of DNA replication and cell division (Harley et al. 1990). The progressive loss of part of the telomere with each cell division is associated with senescence and death of the cell (Harley et al. 1990; Reddel 1998).

This observation suggests a simple answer to the question, Why do organisms age and die? They die, in part, because their telomeres are lost and their chromosomes become too damaged to function. But how is this relevant to life history evolution? The answer depends on the association between senescence of individual cells and the longevity of whole organisms. Dan Röhme (1981) showed that the life span of mammal species (measured in years) is correlated with life spans of their skin and blood cells (measured in days) (Figure 11.6). These correlations hold up both for cells cultured in the laboratory and for measures of cell life span in whole organisms (de Haan and Van Zant 1999), and it suggests that organisms live longer if their cells are capable of more cell divisions. But is this correlation based on a causal connection between the destruction of telomeres and the senescence of the whole organism? Probably so: By forcing laboratory cultures of skin cells to express the gene for telomerase, Bodnar et al. (1998) could prevent telomere loss in these laboratory cell lines and increase the life span of these cells by at least an extra 20 cell divisions. If such an experiment could be done in whole living organisms, the method could be used to test directly the association between telomere loss, cell life span, and the death of organisms.

Increased telomerase activity increases the life span of cells.

These results seem to be consistent with the rate-of-living theory of senescence. However, they also include a fundamental contradiction: The rate of living theory predicts that organisms cannot evolve longer lives, yet it appears that organisms could extend their lives by evolving higher telomerase expression in many of their cells. Why do they not do so? The solution could involve trade-offs between extending the lives of cells (through telomerase activity) and preventing the uncontrolled proliferation of cells (leading to cancer). Trade-offs are at the heart of the alternative theory that we will discuss next.

The Evolutionary Theory of Aging

The results discussed so far present a paradox. Fruit fly populations can evolve longer life spans, bat species apparently have evolved longer life spans than other eutherian mammals, and genetic engineers can artificially increase the life span of cells by boosting telomerase expression. If natural selection can lead to longer life spans, and the physiological mechanisms for longer life already exist, why has natural selection not produced this result in all species? The evolutionary theory of

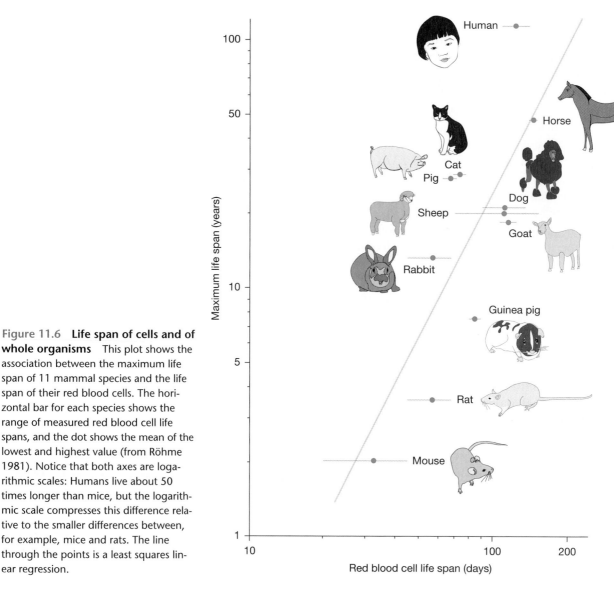

Figure 11.6   Life span of cells and of whole organisms   This plot shows the association between the maximum life span of 11 mammal species and the life span of their red blood cells. The horizontal bar for each species shows the range of measured red blood cell life spans, and the dot shows the mean of the lowest and highest value (from Röhme 1981). Notice that both axes are logarithmic scales: Humans live about 50 times longer than mice, but the logarithmic scale compresses this difference relative to the smaller differences between, for example, mice and rats. The line through the points is a least squares linear regression.

senescence offers two related mechanisms to resolve this paradox (Medawar 1952; Williams 1957; Hamilton 1966; Partridge and Barton 1993; Neese and Williams 1995). Under the evolutionary theory, aging is caused not so much by cell and tissue damage itself as by the failure of organisms to completely repair such damage. George C. Williams argues that complete repair ought to be physiologically possible (Williams 1957; Neese and Williams 1995). Given that organisms are capable of constructing themselves via a complex process of development, they should, in principle be capable of the easier job of maintaining the organs and tissues thus formed. Indeed, organisms do have remarkable abilities to replace or repair damaged parts. Yet in many organisms repair is incomplete. Under the evolutionary theory of senescence, the failure to completely repair damage is ultimately caused by either (1) deleterious mutations, or (2) trade-offs between repair and reproduction.

Figure 11.7 uses a simple genetic and demographic model of a hypothetical population to show how deleterious mutations or trade-offs can lead to the evolution

(a) Wild type matures at age 3 and dies at age 16; prior to age 16 annual rate of survival = 0.8

Age	Fraction of Zygotes Surviving	RS of Survivors	Expected RS for Zygotes
0	1.000	0	0.000
1	0.800	0	0.000
2	0.640	0	0.000
3	0.512	1	0.512
4	0.410	1	0.410
5	0.328	1	0.328
6	0.262	1	0.262
7	0.210	1	0.210
8	0.168	1	0.168
9	0.134	1	0.134
10	0.107	1	0.107
11	0.086	1	0.086
12	0.069	1	0.069
13	0.055	1	0.055
14	0.044	1	0.044
15	0.035	1	0.035

Expected lifetime RS: 2.419

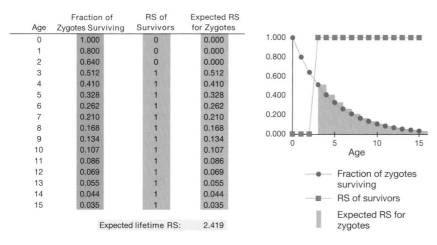

(b) Mutation that causes death at age 14; prior to age 14 annual rate of survival = 0.8

Age	Fraction of Zygotes Surviving	RS of Survivors	Expected RS for Zygotes
0	1.000	0	0.000
1	0.800	0	0.000
2	0.640	0	0.000
3	0.512	1	0.512
4	0.410	1	0.410
5	0.328	1	0.328
6	0.262	1	0.262
7	0.210	1	0.210
8	0.168	1	0.168
9	0.134	1	0.134
10	0.107	1	0.107
11	0.086	1	0.086
12	0.069	1	0.069
13	0.055	1	0.055
14	0.000	1	0.000
15	0.000	1	0.000

Expected lifetime RS: 2.340

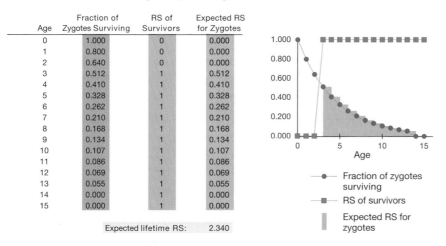

(c) Mutation that causes maturation at age 2 and death at age 10; prior to age 10 annual rate of survival = 0.8

Age	Fraction of Zygotes Surviving	RS of Survivors	Expected RS for Zygotes
0	1.000	0	0.000
1	0.800	0	0.000
2	0.640	1	0.640
3	0.512	1	0.512
4	0.410	1	0.410
5	0.328	1	0.328
6	0.262	1	0.262
7	0.210	1	0.210
8	0.168	1	0.168
9	0.134	1	0.134
10	0.000	1	0.000
11	0.000	1	0.000
12	0.000	1	0.000
13	0.000	1	0.000
14	0.000	1	0.000
15	0.000	1	0.000

Expected lifetime RS: 2.663

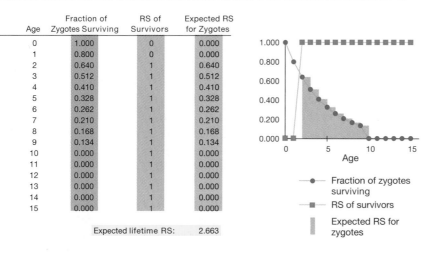

Figure 11.7 A simple genetic model reveals two mechanisms through which aging can evolve (a) The life history of individuals with the wild-type genotype. The sum of the values in column 4 gives the expected lifetime reproductive success of wild-type zygotes: 2.419. (b) The life history of individuals with a mutation that causes death at age 14. The expected lifetime reproductive success of zygotes with this mutation is 2.340. The mutation is deleterious, but not extremely so. (c) The life history of individuals with a mutation that causes maturation at age 2 and death at age 10. The expected lifetime reproductive success of zygotes with this mutation is 2.663. The mutation is advantageous. See text for more details.

of senescence. The figure follows the life histories of individuals in the population from birth until death. Individuals in the population are always at risk of death due to accidents, predators, and diseases. Except where noted, the probability that an individual will survive from one year to the next is 0.8. This leads to an exponential decline over time in the fraction of individuals that are still alive.

Figure 11.7a tracks the life histories of individuals with the wild-type genotype. These individuals mature at age 3 and die at age 16. The columns of the table are as follows:

- The first column lists ages.
- The second column indicates the fraction of all wild-type zygotes born that are still alive at each age. From age 1 onward, each number in the second column is simply the number immediately above it multiplied by 0.8. Few individuals survive to the age of 15. To keep the size of the table reasonable, we assume that all individuals that survive until their 16th birthday die before reproducing that year (this assumption affects only about 3% of the population).
- The third column shows that wild-type individuals reach reproductive maturity at age 3. Once the organisms start to reproduce at age 3, they each have one offspring every year (for as long as they survive).
- The fourth column shows the expected reproductive success at each age for wild-type zygotes. The expected reproductive success at age 5, for example, is simply the fraction of zygotes that will survive to age 5 multiplied by the number of offspring each survivor will have at age 5. The sum of the numbers in this column gives the expected lifetime reproductive success of wild-type zygotes.

The numbers in the table are plotted in the graph; the expected lifetime reproductive success of the wild-type zygotes is equal to the area of the shaded region. The expected lifetime reproductive success of wild-type individuals is about 2.42.

We now consider two mutations that change the life histories of the individuals that carry them. If we imagine that these mutations are dominant in their effects, then our consideration covers both homozygotes and heterozygotes. If we imagine that the mutations are recessive, then our consideration covers only homozygotes.

Deleterious Mutations and Aging: The Mutation Accumulation Hypothesis

Figure 11.7b depicts a mutation that causes death at age 14. In other words, the mutation causes premature senescence. All other aspects of life history are unchanged. The mutation is obviously deleterious, but how strongly will it be selected against? As shown in the table and graph, the expected lifetime reproductive success of zygotes with the mutation is about 2.34. This is over 96% of the fitness of wild-type zygotes. Because few zygotes survive to age 14 anyway, zygotes carrying the mutation causing death at 14 do not, on average, suffer much of a penalty. The mutation is not selected against very strongly.

At first glance, it is a bit surprising that a mutation causing death is only mildly deleterious. Many mutations that cause death are, in fact, highly deleterious. A mutation causing death at age 2, for example, would be selected against strongly. Zygotes carrying such a mutation would have an expected lifetime reproductive success of zero. But mutations causing death after reproduction has begun are selected against less strongly. The later in life that such mutations exert their deleterious effects, the more weakly they are selected against. Mutations that are selected against only weakly can persist in mutation-selection balance (see Chapter 5). The accumulation in populations of deleterious mutations whose effects occur only late in life is one evolutionary explanation for aging (Medawar 1952).

What kind of mutation could cause death, but only at an advanced age? One possibility is a mutation that reduces an organism's ability to maintain itself in good repair. Humans provide an example. Among the kinds of cellular damage that humans (and other organisms) must repair are DNA mismatch errors. Mismatched nucleotide pairs can be created by mistakes during DNA replication, or they can be induced by chemical damage to DNA (Vani and Rao 1996). Repair of these errors is performed by a suite of special enzymes. Germ-line mutations in the genes that code for these enzymes can result in the accumulation of mismatch errors, which in turn can result in cancer.

Germ-line mutations in DNA mismatch repair genes cause a form of cancer in humans called hereditary nonpolyposis colon cancer (Eshleman and Markowitz 1996; Fishel and Wilson 1997). In one study, the age at which individuals were diagnosed with hereditary nonpolyposis colon cancer ranged from 17 to 92. The median age of diagnosis was 48 (Rodriguez Bigas et al. 1996). Thus, most people carrying mutations in the genes for DNA mismatch repair enzymes do not suffer the deleterious consequences of the mutations until well after the age at which reproduction begins. In an evolutionary sense, hereditary nonpolyposis colon cancer is a manifestation of senescence that is caused by deleterious mutations. These deleterious mutations persist in populations because they reduce survival only late in life.

Hereditary nonpolyposis colon cancer is caused by a mutation that acts late in life—usually long after reproduction begins.

Such deleterious mutations may accumulate rapidly. Several experiments on fruit flies have measured the decline in fitness of captive populations that are not subject to selection (which would remove such mutations from the population). One recent innovative approach is to collect flies from the wild, mate them, and then freeze some of their embryos cryogenically while allowing others to develop into a laboratory population in which deleterious mutations are allowed to accumulate (Shabalina et al. 1997). Then, the frozen flies can be reanimated and their fitness compared to flies that experienced mutation accumulation. The crucial point of such an experiment is that the fitness of flies before and after mutation accumulation can be measured at the same time under the same conditions, a feat that would be impossible without cryopreservation. Shabalina et al. (1997) found that fitness (measured as the number of surviving offspring per female) was reduced on average by 0.2% to 2.0% every generation in just 30 generations in such a population. This is a potentially disastrous decline, caused almost entirely by the accumulation of mutations. The important difference between this artificially selected population and a natural population (of fruit flies or other organisms) is that, in a natural population, mutations affecting survival and reproduction early in life will be effectively removed from the population by selection, while mutations affecting survival and reproduction late in life will accumulate. Like hereditary nonpolyposis colon cancer, these mutations have little influence on lifetime reproductive success and so can become abundant, resulting in senescence and death.

Trade-Offs and Aging: The Antagonistic Pleiotropy Hypothesis

Figure 11.7c depicts a mutation that affects two different life history characters. That is, the mutation is pleiotropic. The mutation causes reproductive maturation at age 2 instead of age 3, and the mutation causes death at age 10. In other words, the mutation involves a trade-off between reproduction early in life and survival late in life; its pleiotropic effects are antagonistic.

As shown in the table and graph, the expected lifetime reproductive success of zygotes with the mutation is about 2.66. This is 1.1 times the expected lifetime reproductive success of wild-type zygotes. Most of the zygotes born with the mutation will live long enough to reap the benefit of earlier reproduction, but few will survive long enough to pay the cost of early aging. This mutation that causes both early maturation and early senescence is therefore favored by selection. Selection for alleles with pleiotropic effects that are advantageous early in life and deleterious late in life is a second evolutionary explanation for aging (Williams 1957; Rose 1991).

What kind of mutation could increase reproduction early in life at the same time it reduced reproduction or survival late in life? Perhaps a mutation that causes less energy to be allocated to repair early in life, and more energy to be allocated to reproduction (see Figure 11.2). Until very recently, specific genes with this kind of pleiotropic action had not been identified. However, at least one candidate gene has now been found: the gene encoding the 70 kilodalton heatshock protein called hsp70.

The heatshock proteins are a class of molecules called chaperones: The products of these genes act as stress reducers, binding to other macromolecules within the cell in order to prevent damage such as denaturation of folded proteins caused by high temperature or other sources of environmental stress (Morimoto et al. 1994). However, this protective binding of heat-shock proteins to other components of the cell interferes with some normal cellular functions, so that the heat-shock protein genes are only expressed after environmental stress is detected (such as an increase in temperature). The heat-shock proteins are rapidly removed from the cytoplasm after the stress has passed. The transient expression of one of these genes, *hsp70,* increases the life span of *Drosophila melanogaster* in the laboratory by reducing mortality about twofold during the 2–3 weeks following heat shock (Tatar et al. 1997; Tatar 1999; Figure 11.8a). Now Rebecca Silbermann and Marc Tatar (2000) have shown that this increased life span comes at the expense of lower offspring production early in life, and that this trade-off between early fecundity and late survival is directly mediated by *hsp70* expression. Silberman and Tatar used a clever genetic technique to more than double the number of *hsp70* genes in these flies, then exposed both the transgenic flies and control flies (with the normal complement of *hsp70* genes) to a mild and brief temperature increase. When Silberman and Tatar measured the amount of hsp70 protein produced in the fly tissues, the transgenic flies manufactured increased amounts of hsp70 (though both groups experienced the same thermal environment). Beginning about two days after heat shock, the increased amount of hsp70 in the transgenic flies caused a reduction of about 30% in the number of eggs hatching from each female (Figure 11.8b). The reduction is probably caused by the cellular effects of hsp70 on egg formation.

In short, the expression of *hsp70* in *Drosophila* cells mediates a trade-off between early reproduction and late survival. Mutations in *hsp70* that increase or decrease its expression affect both reproductive success and the life span of the mother. As a result, we would expect the expression of this protein to reflect a balance between these competing life history interests. If *hsp70* expression is a common mechanism by which this trade-off evolves in fruit fly populations, then many of the experi-

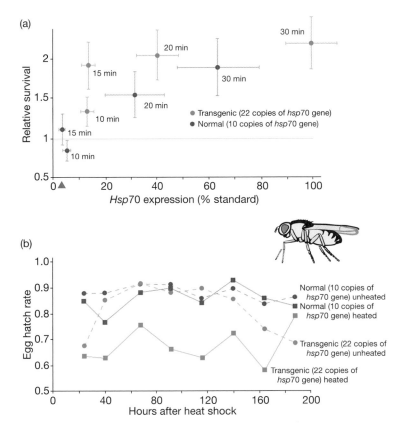

(a)

(b)

Figure 11.8 Expression of a heat-shock protein in transgenic fruit flies produces antagonistic pleiotropy (a) Transgenic flies have 12 extra copies of the *hsp70* gene (compared to 10 copies in the control, or normal, flies). The numbers beside each symbol indicate the duration of a heat shock—an increase in temperature from 24 to 36°C. To read this graph, compare the level of *hsp70* expression and survival for transgenic and normal flies heated for the same length of time. In every case, transgenic individuals produce more hsp70 protein and have higher survival rates. The error bars show ± one standard error. The arrow on the x-axis shows the relative amount of *hsp70* expression in unheated flies. After Tatar et al. (1997). Reprinted by permission from Nature. © 1997, Macmillan Magazines Ltd. (b) Increased hsp70 production lowers the proportion of eggs that hatch. The difference persists up to a week after heat shock. Unheated flies (dashed lines) have egg hatch rates similar to the normal flies. From Silbermann and Tatar (2000). Reprinted by permission of M. Tatar.

mental laboratory *Drosophila* populations that have been subject to selection on life span or fecundity might differ in their *hsp70* expression patterns.

In addition to identifying specific genetic mechanisms for pleiotropy, such as *hsp 70* expression, evolutionary biologists have found widespread evidence that there is indeed a trade-off between reproduction early in life and reproduction or survival late in life. Most of this evidence comes from the analysis of quantitative genetic or phenotypic trade-offs between traits. In the paragraphs that follow, we review two examples.

Lars Gustafsson and Tomas Pärt (1990) studied trade-offs in a bird, the collared flycatcher (*Ficedula albicollis*). Working for 10 years with a flycatcher population on the Swedish island of Gotland, Gustafsson and Pärt followed the life histories of individual birds from hatching to death. The researchers found that some female flycatchers begin breeding at age 1, whereas others wait until age 2. The females that breed at age 1 have smaller clutch sizes throughout life (Figure 11.9a), indicating that there is a cost later in life to breeding early. To further investigate this interpretation, Gustafsson and Pärt manipulated early reproductive effort by giving some first-year breeders extra eggs. The females given extra eggs had progressively smaller clutch sizes in subsequent years, whereas control females did not begin to show reproductive senescence until age 4 (Figure 11.9b). Gustafsson and Pärt conclude that there is a trade-off in collared flycatchers between early-life and late-life reproduction. The researchers note that despite this trade-off, first-year breeders had higher lifetime reproductive success than second-year breeders (1.24 ± 0.08 versus 0.90 ± 0.14 offspring surviving to adulthood $P < 0.05$).

Researchers have documented a trade-off between reproductive effort early in life and reproductive success late in life.

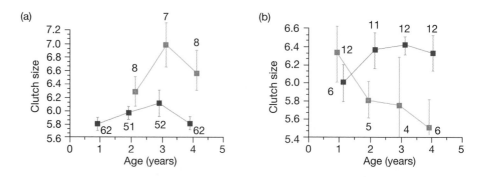

Figure 11.9 **In collared flycatchers (*Ficedula albicollis*), natural variation and experimental manipulations demonstrate a trade-off between reproduction early in life versus reproduction late in life** (a) Females that first breed at age 1 (blue boxes) have smaller clutches at ages 2, 3, and 4 than females that do not breed until age 2 (red boxes). The error bars indicate the standard errors of the estimated mean clutch sizes; the numbers above the symbols give the sample sizes. (b) Females given extra eggs at age 1 (red boxes) have progressively smaller clutches each year at ages 2, 3, and 4. In contrast, control females (blue boxes) do not begin to show a decline in clutch size until age 4. From Gustafsson and Pärt (1990).

Truman Young (1990) studied trade-offs in plants. Young reviewed data from the literature on the energy allocated to reproduction by closely related pairs of annuals versus perennials (Table 11.1). Annuals, which reproduce once and die, always allocate more energy to their sole bout of reproduction than perennials allocate to any given bout. This indicates that there is a trade-off in plants between reproduction and survival. Annual plants enjoy enhanced reproduction in their first reproductive season at the expense of drastically accelerated senescence.

A Natural Experiment on the Evolution of Aging

Steven Austad took advantage of a natural experiment in order to compare populations that had been historically exposed to different rates of mortality caused by extrinsic factors such as predators, diseases, and accidents. We will call this kind of mortality "ecological mortality," in contrast to mortality caused by processes intrinsic to the organism, like the wearing out of body parts. (Mortality caused by intrinsic processes might be called physiological mortality.)

The evolutionary theory of senescence predicts that populations with lower rates of ecological mortality will evolve delayed senescence (Austad 1993). What is the logic behind this prediction? Both of the evolutionary mechanisms that lead to senescence have reduced effectiveness in populations with lower ecological mortality rates. In the case of late-acting deleterious mutations (Figure 11.7b), lower ecological mortality means that a higher fraction of zygotes will live long enough to experience the deleterious effects. Late-acting deleterious mutations are thus more strongly selected against, and will be held at lower frequency in mutation-selection balance. In the case of mutations with pleiotropic effects (Figure 11.7c), lower ecological mortality means that a higher fraction of zygotes will live long enough to experience both the early-life benefits and the late-life costs. The change in the fraction of zygotes experiencing the benefits and costs is more pronounced, however, for the costs. Thus mutations with pleiotropic effects are less

Table 11.1 **Reproductive allocation in annual versus perennial plants**

Each line in the table gives the amount of energy allocated to reproduction by annual plants during their only bout of reproduction, divided by the amount of energy allocated to reproduction by closely related perennial plants during any single bout of reproduction. All the values are larger than one, indicating that annuals always allocate more per reproductive episode. The comparisons for *Oryza perennis* and *Ipomopsis aggregata* are within species; all other comparisons are between species. From Young (1990); see this source for citations to individual studies.

Species	Allocation by annuals/allocation by perennials
Oryza perennis	2.9
Oryza perennis	5.3
Gentiana spp.	2.2–3.5
Lupinus spp.	2.2–3.2
Helianthus spp.	1.7–4.0
Temperate herbs	2.8–2.9
Old field herbs	1.7
Ipomopsis aggregata	1.5–2.3
Sesbania spp.	2.1–2.3
Hypochoeris spp.	2.4–3.7

strongly favored by selection. All else being equal, then, individuals in populations with lower ecological mortality should show later senescence.

Austad (1993) studied the Virginia opossum (Figure 11.2). He compared a population living in the mainland southeastern United States to a population living on Sapelo Island, located off the coast of Georgia. In the mainland population, opossums have high ecological mortality rates. In one study reviewed by Austad, more than half of all naturally occurring opossum deaths were caused by predators. When identifiable, two-thirds of the predators were mammals, including bobcats and feral dogs. Mammalian predators are absent on Sapelo Island, however. Sapelo Island supports an opossum population that has been isolated from the mainland population for four to five thousand years. Other than the difference in mammalian predators, Sapelo Island differs little from Austad's mainland study site at Savanna River, South Carolina. The two habitats are similar in temperature, rainfall, opossum ectoparasite loads, and food available per opossum. The evolutionary theory of senescence predicts that the Sapelo Island opossums will show delayed senescence relative to the mainland opossums.

To test this prediction, Austad put radio collars on 34 island females and 37 mainland females and followed their life histories from birth until death. By three different measures, island females indeed show delayed senescence:

In populations where mortality rates are high, individuals tend to breed earlier in life.

- Island females show delayed senescence in month-to-month probability of survival (Figure 11.10a). The probability of surviving from one month to the next falls with age in both populations, a manifestation of senescence. Monthly survival falls more slowly, however, for island females than for mainland females. As a result, the average life span of island females is significantly longer than the average life span of mainland females (24.6 versus 20.0; $P < 0.02$).

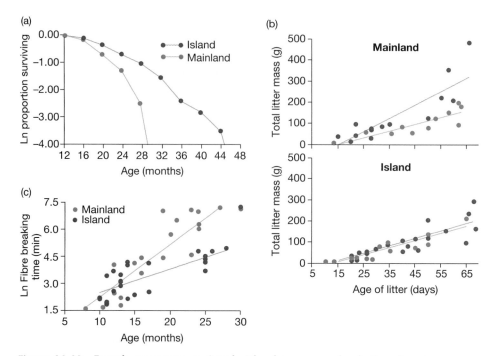

Figure 11.10   Female opossums on Sapelo Island age more slowly than female opossums on the mainland   (a) The rate of survival from one month to the next falls more rapidly with age for mainland opossums than for island opossums. (b) Each graph shows total litter mass as a function of litter age for females in their first year of reproduction (blue) and for females in their second year of reproduction (red). In mainland females (top graph), offspring in second-year litters grow more slowly than offspring in first-year litters ($P < 0.001$). In island females, second-year litters grow just as fast as first-year litters. (c) Tail collagen fiber-breaking time increases more slowly with age in island females than in mainland females ($P < 0.001$). From Austad (1993).

- Island females show delayed senescence in reproductive performance (Figure 11.10b). Austad measured reproductive performance by monitoring the growth rates of litters of young. For mainland females, litters produced in the mother's second year of reproduction grew more slowly than litters produced in the mother's first year of reproduction. This difference indicates that second-year mothers are less efficient at nourishing their young. Island females show no such decline in performance with age.

- Island females show delayed senescence in connective tissue physiology (Figure 11.10c). As mammals age, the collagen fibers in their tendons develop cross-links between protein molecules. These cross-links reduce the flexibility of the tendons. The amount of cross-linking in a tendon can be determined by measuring how long it takes for collagen fibers from the tendon to break. In both island and mainland opossums, breaking time in tail-tendon collagen increases with age, but breaking time increases less rapidly with age in island opossums. In other words, island opossums have slower rates of physiological aging.

These results are all consistent with the evolutionary theory of senescence.

While they support the conclusion that ecological mortality is an important factor in the evolution of senescence, Austad's results do not allow us to determine which of the evolutionary theory's two hypotheses is more important. Is the more rapid aging of mainland opossums due to late-acting deleterious mutations, or to trade-offs between early reproduction and late reproduction and survival? At least part of the difference in rates of senescence appears to be due to trade-offs. Island opossums have, on average, significantly smaller litters (5.66 versus 7.61; $P < 0.001$). This suggests that mainland opossums are, physiologically and evolutionarily, trading increased early reproduction for decreased later reproduction and survival.

In summary, the evolutionary theory of senescence hinges on the observation that the force of natural selection declines late in life. This is because most individuals die—due to predators, diseases, or accidents—before reaching late life. Two mechanisms can lead to the evolution of senescence: (1) Deleterious mutations whose effects occur late in life can accumulate in populations, and (2) when there are trade-offs between reproduction and maintenance, selection may favor investing in early reproduction even at the expense of maintaining cells and tissues in good repair. The evolutionary theory of senescence has been successful in explaining variation in life history among populations and species. Among the questions that remain are these: What is the relative importance of deleterious mutations and trade-offs in the evolution of senescence (Partridge and Barton 1993)? Can evolutionary theory explain unusual reproductive life histories, such as menopause in human females (Box 11.1)?

11.3 How Many Offspring Should an Individual Produce in a Given Year?

Section 11.2 dealt, in part, with the optimal allocation of energy to reproduction versus repair over an organism's entire life history. In this section, we turn to the related issue of how much an organism should invest in any single episode of reproduction. Again we are concerned with trade-offs. Primary among them is this intuitively straightforward constraint: The more offspring a parent (or pair of parents) attempts to raise at once, the less time and energy the parent can devote to caring for each one.

Questions about the optimal number of offspring have been addressed most thoroughly by biologists studying clutch size in birds. The reason is probably that it is easy to count the eggs in a nest, and it is easy to manipulate clutch size by adding or removing eggs. Assuming that the size of individual eggs is fixed, how many eggs should a bird lay in a single clutch?

Clutch Size in Birds

The simplest hypothesis for the evolution of clutch size, first articulated by David Lack (1947), is that selection will favor the clutch size that produces the most surviving offspring. Figure 11.12 (page 378) shows a simple mathematical version of this hypothesis (for a more detailed mathematical treatment, see Stearns 1992). The model assumes a fundamental trade-off in which the probability that any individual offspring will survive decreases with increasing clutch size. Many researchers have tested this assumed trade-off by adding eggs to nests; in the majority of cases they have found that adding eggs indeed reduces the survival rate for individual chicks (Stearns 1992). One explanation could be that the ability of the parents to feed any individual offspring declines as the number of offspring increases. In Figure 11.12a, we assume that the decline in offspring survival is a linear function of clutch size, but the model depends only on survival being a decreasing function. Given a

BOX 11.1   **Is there an evolutionary explanation for menopause?**

In humans, reproductive capacity declines earlier and more rapidly in women than in men (Figure 11.11a). The early decline in the reproductive capacity of women is puzzling, especially given that other measures of women's physiological capacity decline much more slowly (Figure 11.11b). Why should women's reproductive systems shut down by age 50, while the rest of their organs and tissues are still in good repair?

We consider two hypotheses. One hypothesis suggests that menopause is a nonadaptive artifact of our modern lifestyle (see Austad 1994). The other hypothesis suggests that menopause is a life history adaptation associated with the contribution grandmothers make to feeding their grandchildren (Hawkes at al. 1989).

Advocates of the artifact hypothesis point out that archaeologists reconstructing the demography of ancient peoples have often concluded that in premodern cultures, virtually all adults died by age 50 or 55 (see Hill and Hurtado 1991). If death by age 50 or 55 was the rule for our hunter–gatherer ancestors, then the modern situation, in which individuals often live into their 80s and 90s, is unprecedented in our evolutionary history. Menopause cannot be an adaptation, because our hunter–gatherer ancestors never lived long enough to experience it. When other mammals are kept in captivity and given modern medical care, they too live far longer than individuals do in nature. In captive mammals, females in at least some species show a decline in reproductive capacity well in advance of the decline in male reproductive capacity, and long before death (Figure 11.11c). Thus, menopause may need no other explanation than that our modern lifestyle has extended our life span beyond that experienced by our ancestors.

Critics of the artifact hypothesis point out that in contemporary hunter–gatherer societies, many individuals live into their 60s and 70s (Figure 11.11d). These data may be more reliable indicators of the demography of our hunter–gatherer ancestors than are archaeological reconstructions (Hill and Hurtado 1991; see also Austad 1994). If a substantial fraction of our female hunter–gatherer ancestors lived long enough to experience menopause, then menopause needs an evolutionary explanation.

Advocates of the grandmother hypothesis note that human children are dependent on their mothers for food for several years after weaning. This is true in contemporary hunter–gatherer cultures, particularly when mothers harvest foods that yield a high return for adults, but are difficult for children to process (Hawkes et al. 1989). Thus, a woman's ability to have additional children may be substantially limited by her need to provision her older, still-dependent children. Furthermore, as a woman gets older, several relevant trends are likely to occur: (1) The probability that she will live long enough to be able to nurture another baby from birth to independence declines, (2) the risks associated with pregnancy and childbirth rise, and (3) her own daughters will themselves start to have children. The grandmother hypothesis suggests that older women may reach a point at which they can get more additional copies of their genes into future generations by ceasing to reproduce themselves, and instead helping to provision their weaned grandchildren so that their daughters can have more babies. In other words, grandmothers face a trade-off between investment in children and investment in grandchildren.

Kristen Hawkes and colleagues (1989; 1997) studied postmenopausal women in the Hadza, a contemporary hunter–gatherer society in East Africa. If the grandmother hypothesis is correct, then women in their 50s, 60s, and 70s should continue to work hard at gathering food. If the grandmother hypothesis is wrong, then we might expect older women (who no longer have dependent children) to relax. In fact, older Hadza women work harder at foraging than any other group (Figure 11.11e). Furthermore, for at least some crops at some times of year, older women are the most effective foragers (Figure 11.11f). Older women do with their extra food exactly what the grandmother hypothesis predicts: They share it with young relatives, thereby improving the childrens' nutritional status.

These data are consistent with the grandmother hypothesis in a large number of ways, as summarized by Hawkes et al. (1998), but the data do not provide a definitive test. As Austad (1994) points out, the crucial issue is whether the daughters of helpful grandmothers are actually able to have more children, and whether the grandmothers thereby achieve higher inclusive fitness (see Chapter 16) than they would by attempting to have more children of their own. Kim

BOX 11.1   Continued

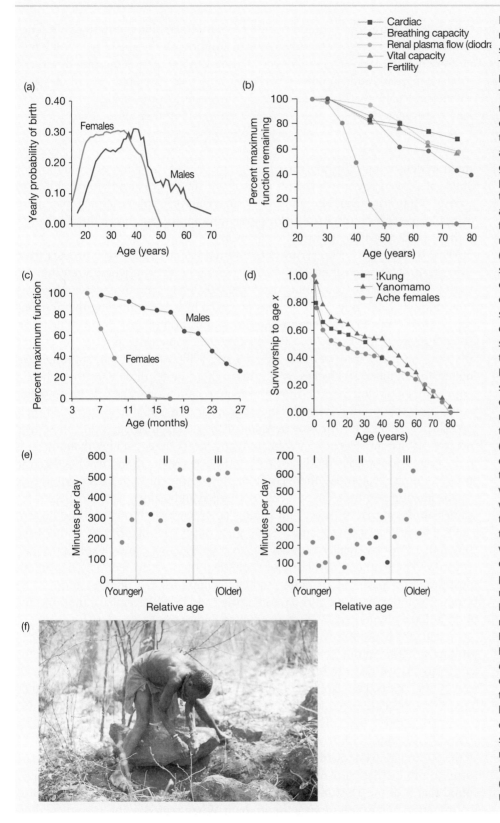

Figure 11.11   **Data on menopause**   (a) Reproductive senescence in women and men. These data are from the Ache, hunter–gatherers of Paraguay. The graph shows the probability that women and men will have a child born during the year that they are any given age. From Hill and Hurtado (1996). (b) This graph shows the functional capacity of various physiological systems in women as a function of age. Capacity is calculated as the fraction of youthful capacity still remaining. From Hill and Hurtado (1991); see citations therein for sources of data. (c) Reproductive capacity as a function of age in captive rats. From Austad (1994); see citations therein for sources of data. (d) Fraction of individuals surviving as a function of age in three hunter–gatherer cultures. From Hill and Hurtado (1991); see citations therein for sources of data. (a, b, d) Reprinted with permission. Copyright © 1996, Walter de Gruyter, Inc., New York. (e) Time spent foraging each day during the wet season (left) and the dry season (right) by Hadza women of different ages: I = women who have reached puberty, but not yet married or begun to have children; II = women who are pregnant or have young children; III = women who are past childbearing age and have no children younger than 15. Blue circles represent women nursing young children. From Hawkes et al. (1989). (f) This Hadza woman is approximately 65 years old. She is using a digging stick and muscle power to dig tubers from underneath large rocks. Digging tubers requires knowledge, skill, patience, strength, and experience, making Hadza grandmothers the most productive foragers. (Copyright James F. O'Connell, University of Utah)

BOX 11.1   Continued

Hill and Magdalena Hurtado (1991; 1996) addressed this issue with data on the Ache hunter–gatherers of Paraguay. Hill and Hurtado's data show that the average 50-year-old woman has 1.7 surviving sons and 1.1 surviving daughters. The researchers calculate that by helping these children reproduce, the average Ache grandmother can gain the inclusive fitness equivalent of only 5% of an additional offspring of her own. This conclusion offers little support for the grandmother hypothesis. More complete data on more cultures are needed for definitive evaluation of both the artifact hypothesis and the grandmother hypothesis.

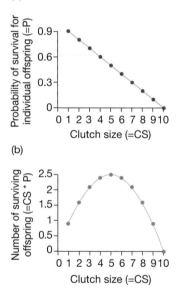

Figure 11.12   A mathematical treatment of Lack's hypothesis on the evolution of clutch size   (a) The probability that any individual offspring will survive declines with increasing clutch size. Here, the probability of survival starts at 0.9 for the single offspring in a one-egg clutch, then declines by 0.1 for each one-egg increment in clutch size. (b) The number of surviving offspring per clutch is the number of eggs multiplied by the probability that any individual offspring will survive. The optimal clutch size is the one that produces the maximum number of survivors. Here, the optimal clutch size is five.

function describing offspring survival, the number of surviving offspring from a clutch of a given size is just the product of the clutch size and the probability of survival (Figure 11.12b). The number of surviving offspring reaches a maximum at an intermediate clutch size. It is this most-productive clutch size that Lack's hypothesis predicts will be favored by selection.

Mark Boyce and C. M. Perrins (1987) tested Lack's hypothesis with data from a long-term study of great tits (*Parus major*) nesting in Wytham Wood, a research site near Oxford, England. Combining data for 4489 clutches monitored over the years 1960 through 1982, Boyce and Perrins plotted a histogram showing the distribution of clutch sizes in the Wytham Wood tit population (Figure 11.13). The mean clutch size was 8.53. Boyce and Perrins also determined the average number of surviving offspring from clutches of each size (Figure 11.13). This number was highest for clutches of 12 eggs. When researchers added three eggs to each of a large number of clutches, the most productive clutch size was still 12. In other words, birds that produced smaller clutches apparently could have increased their reproductive success for the year by laying 12 eggs. Taken at face value, these data indicate that natural selection in Wytham Wood favors larger clutches than the birds in the population actually produce. Because the average clutch size was less than the most productive clutch size, these results are not consistent with Lack's hypothesis.

The literature on Lack's hypothesis is extensive, and many researchers have done studies similar to that of Boyce and Perrins (see reviews in Roff 1992 and Stearns 1992). The results of Boyce and Perrins are typical: The majority of studies have shown that birds lay smaller clutches than predicted. How can we explain this discrepancy? The mathematical logic of Lack's hypothesis is correct. The hypothesis must therefore make one or more implicit assumptions that often turn out to be wrong. Evolutionary biologists have identified and tested several assumptions implicit in Lack's hypothesis. We discuss two of them now.

First, Lack's hypothesis assumes that there is no trade-off between a parent's reproductive effort in one year and its survival or reproductive performance in future years. As we discussed in Section 11.2, however, reproduction often entails exactly such costs. The data in Figure 11.9b demonstrated that when female collared flycatchers are given an extra egg in their first year, their clutch size in future years is lower than that of control females. In a review of the literature on reproductive costs in birds, Mats Lindén and Anders Møller (1989) found that 26 of the 60 studies that looked for trade-offs between current reproductive effort and future reproductive performance found them. In addition, Lindén and Møller found that 4 of the 16 studies that looked for trade-offs between current reproductive effort and future survival found them. When reproduction is costly and se-

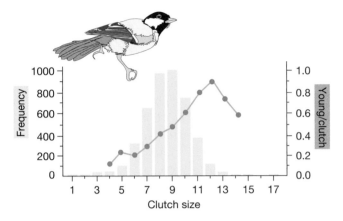

Figure 11.13 **Lack's hypothesis tested with data on great tits (*Parus major*)** The bars and left vertical axis form a histogram showing the variation in clutch size. The average clutch size for 4489 clutches was 8.53. The red dots connected by line segments and the right vertical axis show the number of surviving young per clutch as a function of clutch size. The number of surviving young per clutch was highest for clutches of 12 eggs. From Roff (1992); redrawn from Boyce and Perrins (1987).

lection favors withholding some reproductive effort for the future, the optimal clutch size may be less than the most productive clutch size.

Second, Lack's hypothesis assumes that the only effect of clutch size on offspring is in determining whether the offspring survive. Being part of a large clutch may, however, impose other costs on individual offspring than just reducing their probability of survival. Dolph Schluter and Lars Gustafsson (1993) added or removed eggs from the nests of collared flycatchers, put leg bands on the chicks that hatched from the nests, and then monitored the chicks' subsequent life histories. When the female chicks matured and built nests of their own, there was a strong relationship between the size of the clutches they produced and how much the clutch they were reared in had been manipulated (Figure 11.14). Females reared in nests from which eggs had been removed produced larger clutches, whereas females reared in nests to which eggs had been added produced smaller clutches. This indicates that clutch size affects not only offspring survival, but also offspring reproductive performance. These data suggest that there is a trade-off between the quality and quantity of offspring produced. When larger clutches mean lower offspring reproductive success, the optimal clutch size will be smaller than the most numerically productive clutch size.

Note that we have been assuming that clutch size is fixed for any given genotype. In fact, clutch size is often phenotypically plastic (see Chapter 8). When clutch size is plastic, and when birds can predict whether they are going to have a good year or a bad year, then we would predict that individuals will adjust their clutch size to the optimum value for each kind of year (for example, see Sanz and Moreno 1995).

Lack's Hypothesis Applied to Parasitoid Wasps

Although Lack's hypothesis often proves to be too simple to accurately predict clutch size, the examples we have reviewed demonstrate that it offers a useful null model. By explicitly specifying what we should expect to observe under minimal assumptions, Lack's hypothesis alerts us to interesting patterns we might not otherwise have noticed. This application of Lack's hypothesis is not limited to birds.

Eric Charnov and Samuel Skinner (1985) used Lack's hypothesis to explore the evolution of clutch size in parasitoid wasps. Parasitoid wasps use a stinger-like ovipositor to inject their eggs into the eggs or body cavity of a host insect. When the larval parasitoids hatch, they eat the host alive from the inside. The larvae then

Efforts to identify which of Lack's assumptions are violated have led to the discovery of additional trade-offs, and improved estimates of lifetime fitness.

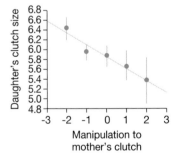

Figure 11.14 **Does clutch size affect offspring reproductive performance?** The graph shows the relationship between daughters' clutch sizes and the number of eggs added to or removed from the mothers' nests in which the daughters were reared. From Schluter and Gustafsson (1993).

pupate inside the empty cuticle of the host, finally emerging as adults to mate and repeat the life cycle.

For a parasitoid, a host is analogous to a nest. A female parasitoid can lay a clutch of one or more eggs in a single host. The larvae compete among themselves for food, so there is a trade-off between clutch size and the survival of individual larvae. An added twist with insects is that adult size is highly flexible. In addition to reducing offspring survival, competition for food may result in larvae simply becoming smaller adults. The maternal fitness associated with a given clutch size must therefore be calculated as the product of the clutch size, the probability of survival of individual larvae, and the lifetime egg production of offspring of the size that emerge.

Lack's hypothesis is a useful null model for other organisms as well.

Charnov and Skinner used this modified version of Lack's hypothesis to analyze the oviposition behavior of female parasitoid wasps in the species *Trichogramma embryophagum*. This wasp deposits its eggs in the eggs of a variety of host insects. Using data from the literature, Charnov and Skinner calculated maternal fitness as a function of clutch size for three different host species (Figure 11.15). Table 11.2 gives the most productive clutch sizes and the actual clutch sizes female wasps lay in each species of host egg. The data indicate that female wasps shift their behavior in a manner appropriate to different hosts. Females lay fewer eggs in the relatively poor hosts, and more eggs in the relatively good hosts. As with many birds, however, female wasps lay smaller clutches than those predicted by Lack's hypothesis.

Why do female wasps lay clutches smaller than the predicted sizes? Charnov and Skinner consider three reasons. Two of the three are similar to factors we discussed for birds: Larger clutch sizes may reduce offspring fitness in ways that Charnov and Skinner did not include in their calculations; and there may be trade-offs between a female's investment in a particular clutch and her own future survival or reproductive performance. Charnov and Skinner's third hypothesis is novel to parasitoid wasps. Unlike birds, female parasitoid wasps may produce more than one clutch in rapid succession. Soon after she has laid one clutch, a female wasp may begin looking for another host to parasitize. The appropriate measure of a wasp's fitness with regard to clutch size may not be the discrete fitness she gains from a single clutch. Instead, it may be the rate at which her fitness rises as she searches for hosts and lays eggs in them. Readers familiar with behavioral ecology may recognize this as an optimal foraging problem.

Figure 11.16 presents a graphical analysis of a female's rate of increase in fitness over time. The figure follows the female from the time she sets out to find a host

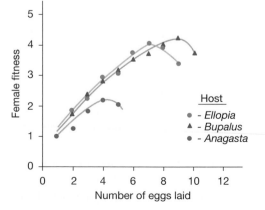

Figure 11.15 Lack's hypothesis applied to a parasitoid wasp This graph shows the fitness of female parasitoid wasps (*Trichogramma embryophagum*) as a function of clutch size. Each curve represents wasp clutches deposited in a different species of host egg. From Charnov and Skinner (1985).

Table 11.2	Parasitoid wasp (*Trichogramma embryophagum*) clutch sizes predicted by Lack's hypothesis versus actual clutch sizes for three host species	
Host species	**Lack's optimal clutch size**	**Actual clutch size**
Anagasta	4	1–2
Ellopia	7	5–8
Bupalus	9	5–8

Source: From Charnov and Skinner (1985).

egg until she leaves that host egg to look for another. While she is searching, the female gains no fitness. Once she finds a host and begins to lay eggs in it, however, her fitness begins to rise. The fitness she gets from a clutch of any given size is determined by a parabolic function, as in our original depiction of Lack's hypothesis (Figure 11.12). In this example, if a female leaves to look for a new host after laying just one egg, her total fitness gain from the first host is 0.9. Her average rate of fitness gain from the time she set out looking for the first host to the time she leaves to look for a second is given by 0.9 divided by the total elapsed time. This rate of fitness gain is equal to the slope of the diagonal line from the origin to the point representing a clutch size of one. Likewise, if the female stays to lay five eggs, her average rate of fitness gain for the whole trip is given by the slope of the upper diagonal line. In this example, the female would get the highest rate of fitness gain from this host if she left after laying four eggs. This is one egg less than the most productive clutch size. Thus, if female parasitoids are selected to maximize their rate of fitness increase, they may produce smaller clutches than those predicted by Lack's hypothesis.

To summarize, Lack's hypothesis is a useful starting point for the evolutionary analysis of clutch size. Assuming only that there is a trade-off between the number of offspring in a clutch and the survival of individual offspring, Lack's hypothesis predicts that parents will produce clutches of the size that maximizes the number of surviving offspring. This prediction is often violated, with actual clutches typically smaller than expected. These violations indicate the possible presence of other trade-offs. Current parental reproductive effort may be negatively correlated with

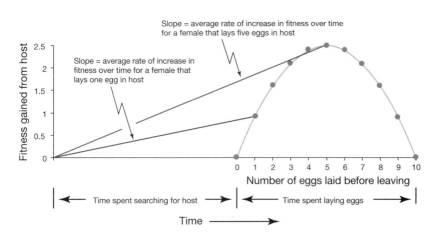

Figure 11.16 Rate of increase in parasitoid maternal fitness with time spent searching for hosts and laying eggs The horizontal axis represents the time spent by a female parasitoid in searching for a host egg and depositing a clutch. The vertical axis represents female fitness, in units of surviving offspring. The red dots show the relationship between number of surviving offspring and clutch size, as in Lack's hypothesis. After Charnov and Skinner (1985).

future parental survival or reproductive performance, or clutch size may be negatively correlated with offspring reproductive performance. Alternatively, a violation of predicted clutch size may indicate that we have chosen the wrong measure of parental fitness.

11.4 How Big Should Each Offspring Be?

In Section 11.3, we assumed that the size of individual offspring was fixed. We now relax that assumption. Given that an organism will invest a particular amount of energy in an episode of reproduction, we can ask whether that energy should be invested in many small offspring or a few large offspring.

Organisms face a trade-off between making many low-quality offspring or a few high-quality offspring.

A trade-off between the size and number of offspring should be fundamental. A pie can be sliced into many small pieces or a few large pieces, but it cannot be sliced into many large pieces. Biologists have found empirical evidence for a size–versus–number trade-off in a variety of taxa. Mark Elgar (1990), for example, analyzed data from the literature on 26 families of fish. Elgar found a clear negative correlation between clutch size and egg size: Fish that produce larger eggs produce fewer eggs per clutch (Figure 11.17a). David Berrigan (1991) performed a similar analysis of variation in egg size and number among species in three orders of insects. In each case, Berrigan found a clear negative correlation between egg number and egg size (Figure 11.17b).

Selection on Offspring Size

If selection on parents is forced by a fundamental constraint to strike a balance between the size and number of offspring, what is the optimal compromise? Christopher Smith and Stephen Fretwell (1974) offered a mathematical analysis of this

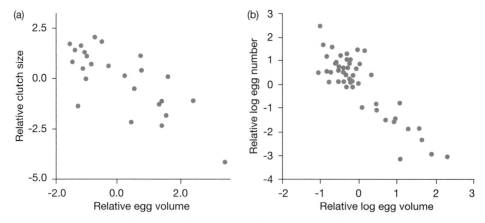

Figure 11.17 Trade-offs across taxa in size and number of offspring (a) A trade-off across 26 fish families. Larger fish produce bigger clutches, so Elgar had to use statistical techniques to remove the effect of variation among fish families in body size. The vertical axis shows relative clutch size, or the number of eggs per clutch adjusted statistically for differences in body size among families. The horizontal axis shows relative egg volume, or egg size adjusted for differences in body size among families. The negative correlation between size and number of eggs is statistically significant ($P < 0.001$). From Elgar (1990). (b) A trade-off across fruit fly species. Larger fruit flies produce more and larger eggs, so Berrigan also had to use statistical techniques to remove the effect of variation in body size. The vertical axis shows relative egg number; the horizontal axis shows relative egg volume. The negative correlation is statistically significant ($P < 0.001$). Berrigan found similar patterns in wasps and beetles. Provided by David Berrigan using data analyzed in Berrigan (1991).

question. Smith and Fretwell's analysis is based on two assumptions. The first assumption is the trade-off between size and number of offspring (Figure 11.18a). The second assumption is that individual offspring will have a better chance of surviving if they are larger (Figure 11.18b). There must be a minimum size below which offspring have no chance of survival. As offspring get larger, their probability of surviving rises. If survival probability approaches one, it must do so in a saturating fashion, because survival probability cannot exceed one. Given the two assumptions, the analysis is simple: The expected fitness of a parent producing offspring of a particular size is the number of such offspring the parent can make, multiplied by the probability that any individual offspring will survive. A plot of expected parental fitness versus offspring size (Figure 11.18c) reveals the size of offspring that gives the highest parental fitness.

The optimal offspring size depends on the shapes of the relationships for offspring number versus size and offspring survival versus size. In many cases, as in Figure 11.18, the optimal offspring size is intermediate. The important point here is that selection on parental fitness often favors offspring smaller than the size favored by selection on offspring fitness. This identification of a potential conflict of interest between parents and offspring is the primary contribution of Smith and Fretwell's model.

The shape of the offspring survival curve is particularly important (Figure 11.18b). In Smith and Fretwell's model, survival probability increases with offspring size, but the rate of increase declines: that is, increasingly large offspring gain a progressively smaller survival benefit. This leads directly to the prediction of an optimal offspring size and gives the highest parental fitness (Figure 11.18c). If the offspring survival curve were a linear relationship instead of a concave curve (Vance 1973), the model would predict selection favoring extremes of offspring size: the smallest offspring capable of development, or the largest offspring that a female could manufacture, rather than some optimal intermediate offspring size (see Levitan 1993, 1996; Podolsky and Strathmann 1996).

It is possible to test Smith and Fretwell's analysis empirically only if there is substantial variation in offspring size among parents within a population. Variation in offspring size is relatively small in most species (Stearns 1992). We review two recent studies that have confirmed both the assumptions and the conclusion of Smith and Fretwell's analysis. In one study, researchers generated extra variation in egg size in a lizard by surgically manipulating females. In the other study, researchers took advantage of phenotypic plasticity in egg size in a beetle.

Selection on parents favors a compromise between the quality and quantity of offspring, but selection on individual offspring favors high quality.

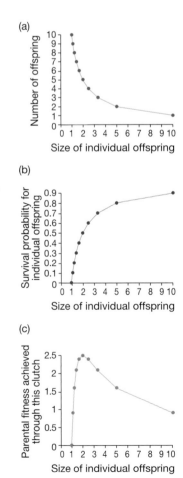

(a)

(b)

(c)

Figure 11.18   The optimal compromise between size and number of offspring   (a) Assumption 1: There is a trade-off between size and number of offspring. The units we have used are arbitrary. The shape of the curve may vary from species to species. Here we have used the equation: Number = 10/Size. (b) Assumption 2: Above a minimum size, the probability that any individual offspring will survive is an increasing function of its size. Again, we have used arbitrary units. The shape of the curve may vary from species to species. Here we have used the equation: Survival = 1 − (1/size). (c) Analysis: The parental fitness gained from a single clutch of offspring of a given size is the number of offspring in the clutch multiplied by the probability that any individual offspring will survive. For example, given the equations and units used here, if a parent makes offspring of size five it can make two of them, and each has a probability of survival of 0.8. Thus, the expected fitness gained by the parent from this clutch is 2 × 0.8 = 1.6. For some (but not all) combinations of the trade-off function (a) and the survival function (b), parents achieve maximum fitness through offspring of intermediate size (as in c). After Smith and Fretwell (1974).

Selection on Offspring Size in a Lizard

Barry Sinervo and colleagues (1992) studied the side-blotched lizard (*Uta stansburiana*), which lives in the deserts of western North America. Females lay clutches of one to nine eggs (Sinervo and Licht 1991). Wild populations of side-blotched lizards show heritable variation in egg size (Sinervo and Doughty 1996). Sinervo and colleagues (1992) further expanded the range of egg sizes by surgically manipulating gravid females. The researchers induced near-term females to lay unusually small eggs by withdrawing a fraction of the yolk from each egg. The researchers induced early-term females to lay unusually large eggs by destroying all but two or three of the developing follicles. This caused the females to allocate to a few remaining eggs yolk that they would otherwise have allocated to several. The graphs in Figure 11.19a show the variation in egg size in two populations of lizards in each of two years.

Sinervo and colleagues (1992) tested the first assumption of Smith and Fretwell at the level of individual females. Each plot in Figure 11.19b shows an estimated relationship between clutch size and egg size for females in a population in 1989 or 1990. In both populations and both years, there was a trade-off between size and number of eggs.

Sinervo and colleagues tested the second assumption of Smith and Fretwell by following the fates of individual hatchlings. The researchers marked several hundred hatchling lizards and released them into the wild. One month later, the researchers censused the hatchlings to see which had survived. Each plot in Figure 11.19c shows an estimate of the probability of hatchling survival as a function of egg mass. In 1989, selection favored larger offspring in both populations—exactly as Smith and Fretwell assumed it would. In 1990, selection favored medium-sized offspring—a surprising result.

How did the egg size versus egg number trade-off and selection on hatchlings combine to select on the mothers? Each plot in Figure 11.19d shows an estimate of the relationship between maternal fitness and egg size. In all cases, selection on mothers favored eggs of intermediate size. In 1989, selection on the mothers favored smaller eggs than did selection on offspring. This is the conflict of interest that Smith and Fretwell predicted would sometimes occur. In 1990, selection on mothers favored eggs almost exactly the same size as did selection on offspring.

Phenotypic Plasticity in Egg Size in a Beetle

Charles Fox and colleagues (1997) studied the seed beetle *Stator limbatus* (see photo on page 359). The females of this small beetle lay their eggs directly onto the surface of host seeds. The larvae hatch and burrow into the seed. Inside, the larvae feed, grow, and pupate. They emerge from the seed as adults. *S. limbatus* is a generalist seed predator; it has been reared on the seeds of over 50 different host plants.

Fox and colleagues studied *S. limbatus* on two of its natural hosts: an acacia (*Acacia greggii*) and a palo verde (*Cercidium floridum*). The acacia is a good host; most larvae living in its seeds survive to adulthood. The palo verde is a poor host; fewer than half the larvae living in its seeds survive. We can easily add hosts of different quality to the Smith and Fretwell analysis (Figure 11.20). When we do so, we get a clear prediction: Females should lay larger eggs on the poor host than on the good host. Recall from Chapter 8 that when selection favors different phenotypes at different times or in different places, organisms sometimes evolve phenotypic plasticity. The

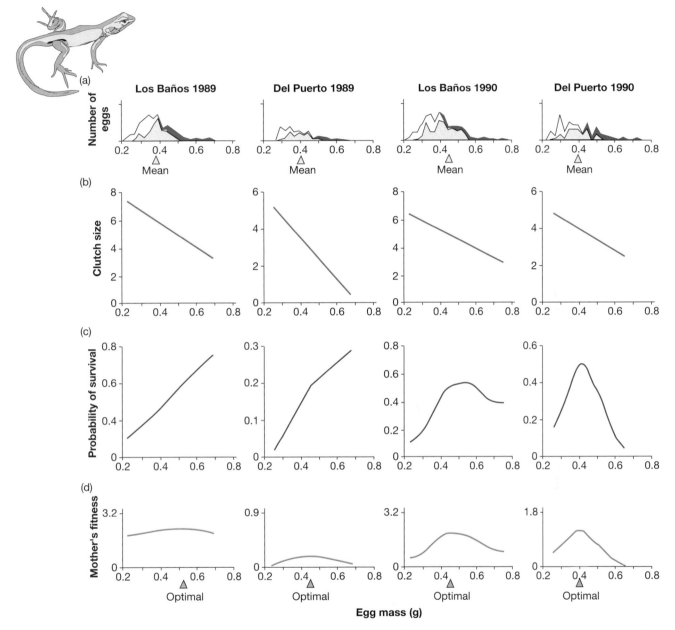

Figure 11.19 **Selection on egg size in the side-blotched lizard (*Uta stansburiana*)** (a) These graphs show the distribution of egg sizes in the populations. The tan region represents natural variation; the white region represents surgically induced small eggs; the dark blue portion represents surgically induced large eggs. The mean egg size for natural eggs in each population is indicated by the tan triangle. (b) Each plot shows the estimated relationship between maternal clutch size and egg size. Clutch size was a decreasing function of egg size. This indicates a trade-off, confirming the first assumption of Smith and Fretwell's analysis. (c) Each plot shows the estimated relationship in nature between egg size and the probability of hatchling survival to the age of one month. In 1989, offspring survival increased with egg size, confirming the second assumption of Smith and Fretwell's analysis. In 1990, offspring survival was highest for medium-sized hatchlings. (d) Each plot shows the estimated relationship between maternal fitness (clutch size × probability that an individual offspring will survive) and egg size. In both populations in both years the optimal egg size for mothers (red triangle) was an intermediate value. In 1989, this value was smaller than the optimal egg size for offspring. Reprinted with permission from Sinervo et al. (1992). Copyright ©1992, American Association for the Advancement of Science.

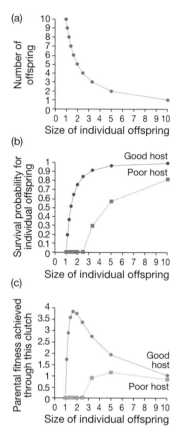

Figure 11.20 The optimal compromise between size and number of offspring on a good host and a poor host (a) As in Figure 11.18, we assume a trade-off between size and number of offspring. (b) As in Figure 11.18, we assume that there is a minimum size below which individual offspring do not survive; above this minimum size, the probability that any individual offspring will survive is an increasing function of offspring size. The minimum size for offspring survival is smaller on the good host. Furthermore, survival is higher on the good host at all sizes above the minimum. (c) Analysis: The parental fitness gained from a single clutch of offspring of a given size is the number of offspring in the clutch multiplied by the probability that any individual offspring will survive. The optimal offspring size (for the mother) is bigger on the poor host than on the good host. After Smith and Fretwell (1974).

analysis in Figure 11.20 predicts that *S. limbatus* should exhibit phenotypic plasticity in egg size.

Fox and colleagues found that, as predicted, female *S. limbatus* adjust the size of the eggs that they lay to the host on which they deposit them. When the researchers took newly emerged females from the same population and gave them only one kind of seed, females given palo verde seeds (the poor host) laid significantly larger eggs than females given acacia seeds (Figure 11.21a). Confirming assumption 1 of Smith and Fretwell, these larger eggs came at the cost of fewer eggs produced over a lifetime (Figure 11.21b).

For females laying on palo verde seeds, the production of large eggs is adaptive. Fox et al. manipulated females into laying small eggs on palo verde seeds by keeping the females on acacia seeds until they laid their first egg, then moving them to palo verde seeds. Only 0.3% of the larvae hatching from small eggs on palo verde seeds survived to adulthood, whereas 24% of the larvae hatching from large eggs on palo verde seeds survived ($P < 0.0001$). Confirming assumption 2 of Smith and Fretwell, even among the large eggs on palo verde seeds, the probability of survival from egg to adult was positively correlated with egg size. For females laying on acacia seeds, the production of small eggs is adaptive. Given that nearly all larvae hatching on acacia seeds survive, females producing more and smaller eggs have higher lifetime reproductive success.

Fox and colleagues even showed that individual females that had started to lay size-appropriate eggs on one host could readjust their egg size when switched to the other host (Figure 11.21c). Control females left on one kind of seed consistently produced large or small seeds for life (Figure 11.21d).

In summary, selection on offspring size often involves a conflict of interest between parents and offspring. Because making larger offspring also means making fewer offspring, selection on parents can favor smaller offspring sizes than are optimal for offspring survival. The exact balance between size versus number of offspring depends on the relationship between offspring size and survival. Poor environments pose a greater obstacle to offspring survival and thus favor larger offspring.

11.5 Conflicts of Interest Between Life Histories

Analyzing trade-offs among life history traits has helped to explain many aspects of the extraordinary life history diversity known among living organisms. However, this view can sometimes obscure the fact that the life history of each organism unfolds in an ecological context that includes other individuals. For example, the opposum life history in Figure 11.2 shows reproduction by a hypothetical female, but does not show the males with which she mated to produce offspring. This simplification might imply that the males are mere sperm providers, and that their interests in the production of offspring are the same as the interests of the female shown in the figure. In fact, the reproductive interests of males and females will often be different. In this section, we discuss two such conflicts of interest and their evolutionary consequences.

Genetic Conflict Between Mates: Genomic Imprinting

Opposums and other mammals that nourish their offspring through a placenta offer a surprising opportunity for conflict between the reproductive interests of females (which brood the offspring) and males (which do not). Consider the copies

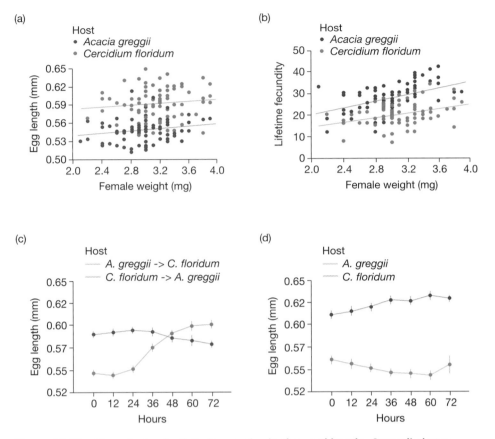

Figure 11.21 Phenotypic plasticity in egg size in the seed beetle, *Stator limbatus*
(a) Larger females lay slightly larger eggs, but females laying on *Cercidium floridum* lay larger eggs for their size ($P < 0.001$). (b) Larger females have slightly higher fecundity (number of eggs), but females laying on *A. greggii* have higher fecundity for their size ($P < 0.001$) (c) In a host switch experiment, Fox and colleagues placed newly emerged females on each kind of seed, let them lay their first egg, and then switched them to the other kind of seed. The graph shows egg size as a function of time for the subsequent 72 hours. Females that laid their first egg on acacia (*A. greggi,* the good host) produced small eggs at first, but gradually switched to large eggs. Females that laid their first egg on palo verde (*C. floridum,* the poor host) produced large eggs at first, but gradually switched to smaller eggs. (d) In a control experiment for (c), Fox and colleagues left females on each kind of seed for life. Each group laid large or small eggs for life. From Fox et al. (1997). Copyright © 1997, American Naturalist. Reprinted by permission of The University of Chicago Press.

of a mammalian gene inherited from the father versus the mother. Why might these alleles be in conflict? In most mammals, females carry offspring from many different males in the course of a lifetime. Indeed, offspring with different fathers are frequently found in the same litter. Because the mother is related to each of these offspring equally, natural selection should act to equalize her physiological investment in each. Natural selection should, on the other hand, favor a father that can coerce the mother into investing more heavily in his offspring at the expense of offspring from other males.

Consistent with this prediction, at least some loci are biochemically marked (or imprinted) in mammals, to distinguish paternal and maternal alleles (Barlow 1995). This marking of alleles occurs in the testis and ovary during production of gametes. Imprinting affects the subsequent transcription of the marked genes within cells of

When different males father offspring within the same litter or clutch, the reproductive interests of the fathers and the mother conflict.

Genomic imprinting occurs when male and female alleles contain distinct chemical markers and are transcribed differently.

the embryo after fertilization. The paternal allele of a hormone called insulin-like growth factor II (IGF-II), for example, is widely expressed in mice, while the maternal copy is hardly transcribed. This is a surprising pattern of gene expression for a diploid organism, because natural selection should favor the equal expression of both alleles. Equal expression protects the offspring against the effects of deleterious recessive mutations that interfere with the function of one allele (Hurst 1999). Why should a mother imprint her IGF-II alleles to reduce transcription of this gene in her offspring, especially when the paternal allele of the same gene is actively transcribed?

The answer hinges on the function of IGF-II and its interaction with other molecules. This hormone is a general stimulant to cell division, and acts through a cell surface protein called the type-1 IGF-II receptor. However, it happens that another abundant cell surface protein in mice, called the cation-independent mannose-6-phosphate receptor (CI-MPR), also has a binding site for IGF-II (this alternative binding site is called the type-2 receptor). CI-MPR's function is completely unrelated to growth, and in mouse embryos it is transcribed only from the maternal allele.

David Haig and colleagues have proposed that this bizarre arrangement of hormones, receptors, and transcription patterns results from a tug-of-war between the interests of maternal and paternal alleles within the uterus. According to this interpretation, the paternally transcribed IGF-II is selected to maximize rates of cell division in the developing embryo. This increases growth rates and monopolizes the flow of maternal resources to the embryo through the placenta. The maternal IGF-II allele is turned off in order to conserve resources for future reproduction. In contrast, the maternally transcribed type-2 receptor is selected to bind excess paternal hormone, mitigate the effects of IGF-II overtranscription, and equalize the flow of resources to different embryos, while the paternal type-2 receptor allele is turned off in order to maximize the influence of the paternal IGF-II hormone on the mother (Haig and Graham 1991; Moore and Haig 1991).

Consistent with this interpretation, CI-MPR does not bind IGF-II in chickens and frogs; their embryos are provisioned before fertilization. Chicken and frog fathers have no opportunity to manipulate the distribution of maternal resources among offspring. This is a hint that the type-2 receptor of mammals evolved after the advent of the placenta, in response to selection that favored equalization of maternal resources among all offspring. Genomic imprinting has also been confirmed in flowering plants, and may have been important in the evolution of the nutritive tissue called endosperm (see Haig and Westoby 1989, 1991).

The qualitative predictions of Haig's hypothesis for genomic imprinting have generally been confirmed, though there is some debate over whether quantitative variation in imprinting occurs, and whether multiple paternity is necessary for imprinting to arise (Haig 1999; Hurst 1999; Spencer et al. 1999). For example, alleles could vary in the amount of transcription rather than being turned "on" or "off." Imprinting is known to be widespread in mammalian genomes (see references in Spencer et al. 1999), and the details of the imprinting mechanism and interactions among imprinted alleles could vary among genes and species.

Finally, it is important to note that mammals are not the only animals that have evolved placental development. For example, lizards (Guillette and Jones 1985), sharks (Wourms 1993), and numerous marine invertebrate groups (Strathmann

1987) have evolved structures like a placenta that transfer materials between the maternal body and the internally brooded offspring. Haig's hypothesis predicts that imprinted genes should be found in these groups, and that these will be genes that moderate the conflict among offspring within a brood as they compete for maternal resources. The hypothesis has not yet been tested in these groups, however (Spencer et al. 1999).

Physiological Conflict Between Mates: Sexual Coevolution

In Chapter 7, we introduced the idea of adaptations arising in competing species, such as hosts and pathogens, that counteract each other's effects so that neither lineage shows a net gain in fitness. In these circumstances, fitness evolves around a kind of dynamic equilibrium in which the environment, and thus the nature of selection acting on a population of organisms, is largely determined by interactions with other organisms and their adaptations (Van Valen 1973).

This idea can be extended to life history adaptations arising within species as well. Experiments by William Rice and colleagues show that, where the reproductive interests of male and female fruit flies differ, sexual selection may favor adaptations that arise in one sex, but are actually detrimental to the other sex. One of these adaptations involves the biochemistry of male seminal fluid, which has evolved to influence female behavior, such as egg laying rate or the tendency to remate with another male (Fowler and Partridge 1989). These effects are beneficial to a male if his mate is likely to have multiple partners because these adaptations will tend to increase the number of eggs that are fertilized by his sperm. Such seminal fluid is toxic and increases mortality of females, however (Fowler and Partridge 1989). Toxic effects favor the subsequent evolution of resistance among females, followed by more extreme adaptations among males to overcome female resistance. This iterated process has been called chase-away sexual selection (Rice 1987; Rice and Holland 1997; Holland and Rice 1998).

Direct evidence for this kind of antagonistic sexual adaptation comes from a series of experiments conducted by Rice (1996). In these experiments, male flies competed with each other for matings with females. Females, in turn, were able to mate with multiple partners. The competition among males selected for traits such as high rate of remating with the same female and highly toxic seminal fluid. However, only male offspring were retained from each generation of experimental mating. After each round of selection, the selected males were mated to females from a control group in which competition for mates was not occurring. In this way, Rice kept the female response to male sexual adaptations static while the selected males competed with one another to overcome female defenses.

The results of 31–41 generations of such selection are shown in Figure 11.22. Compared to males in the control group, selected males had higher fitness (more sons born per male, shown as the "Net fitness" assays in Figure 11.22a). The data on the right-hand side of Figure 11.22a suggests that two traits contributed to the higher fitness of selected males. They were more likely to remate with the same female, and they fertilized a much higher proportion of eggs when the female was remated to another male ('Defense' assays, Fig. 11.22a). These benefits to male reproductive success came at a cost to females, however: After 41 generations of selection, the mortality rate for females mated to selected males was about 50% higher than the mortality rate for females

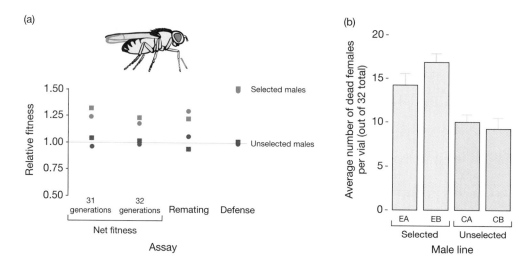

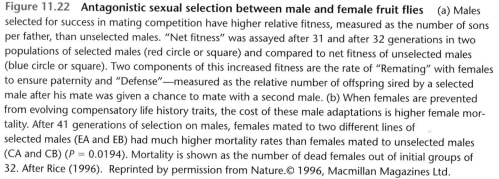

Figure 11.22   Antagonistic sexual selection between male and female fruit flies   (a) Males selected for success in mating competition have higher relative fitness, measured as the number of sons per father, than unselected males. "Net fitness" was assayed after 31 and after 32 generations in two populations of selected males (red circle or square) and compared to net fitness of unselected males (blue circle or square). Two components of this increased fitness are the rate of "Remating" with females to ensure paternity and "Defense"—measured as the relative number of offspring sired by a selected male after his mate was given a chance to mate with a second male. (b) When females are prevented from evolving compensatory life history traits, the cost of these male adaptations is higher female mortality. After 41 generations of selection on males, females mated to two different lines of selected males (EA and EB) had much higher mortality rates than females mated to unselected males (CA and CB) ($P = 0.0194$). Mortality is shown as the number of dead females out of initial groups of 32. After Rice (1996). Reprinted by permission from Nature.© 1996, Macmillan Magazines Ltd.

mated to unselected males (Figure 11.22b). The experiment suggests that males and females are engaged in a reproductive "arms race," which males win if female countermeasures are prevented.

It is important to recognize, however, that this result is based on a particular mating system. By enforcing monogamous mating upon flies from the same source population for many generations, Holland and Rice (1999) showed that the effects of antagonistic sexual adaptation could actually be reversed: Monogamous lines of flies evolved lower sperm toxicity in males and lower resistance to sperm toxicity in females. These results make sense in light of the new relationship between male and female fitness: Males with only one lifetime mate depend on the fitness of that female alone to produce offspring, and they have no fear of being cuckolded by another male. These males should evolve less harmful life history traits in order to increase their own fitness. The defenses evolved by females against toxic male seminal fluid (or other male life history traits) become less beneficial as males evolve more benign traits of their own. If female resistance is expensive in time or energy, then resistance traits should be selected against (Holland and Rice 1999).

11.6   Life Histories in a Broader Evolutionary Context

In this final section of the chapter, we place life histories in a broader evolutionary context. We briefly consider examples of research addressing two general questions: What forces maintain genetic variation in populations? How do novel traits evolve?

The Maintenance of Genetic Variation

Natural selection on a trait should reduce the genetic variation for the trait (Fisher 1930). A simple example illustrates why (Roff 1992). Imagine a series of loci, each with two alleles, that affect a single trait correlated with fitness. At each locus, one allele ("0") contributes to the trait in such a way as to add zero units to the fitness of individuals, whereas the other allele ("1") contributes one unit to individual fitness. The genotype with the highest fitness is homozygous for allele "1" at all loci. Over time, selection should lead to the fixation of the "1" allele at each locus, and there will no longer be genetic variation in the trait.

Life history traits, because of their intimate connection with reproduction, should be more closely correlated with fitness than other kinds of traits, including behavioral, physiological, or morphological traits (Mousseau and Roff 1987). Consequently, life history traits should show less genetic variation—lower heritability—than other kinds of traits (for a discussion of heritability, see Chapter 7). From the literature, Mousseau and Roff (1987) assembled a sample of 1120 estimates of the heritability of various traits. They found that life history traits indeed tend to have the lowest heritabilities (Figure 11.23). This result is consistent with the expectation from our simple theoretical treatment (for an alternative interpretation, see Price and Schluter 1991).

Life history traits are closely correlated with fitness and have relatively low heritabilities.

Nonetheless, Mousseau and Roff's review documents that life history traits typically have substantial genetic variation. What evolutionary forces maintain genetic variation in populations? The list of possibilities includes mutation, heterozygote advantage, frequency-dependent selection, and genotype-by-environment interaction in which different genotypes have higher fitness in different environments or at different times (see Chapters 4-7).

Richard Grosberg (1988) studied the maintenance of genetic variation for life history traits in a population of the sea squirt *Botryllus schlosseri*. *Botryllus schlosseri* is a colonial animal that lives attached to hard surfaces in shallow marine waters of the temperate zone. Colonies consist of a number of identical modules. The modules in a colony are physiologically connected, and their life histories are synchronous. The population Grosberg studied contains two distinct life history morphs. One morph is semelparous: Upon reaching sexual maturity, the modules in a colony reproduce once and die. The other morph is iteroparous: Colonies have at least three episodes of sexual reproduction before they die. In a series of experiments in which he grew sea squirts in a common environment and bred the

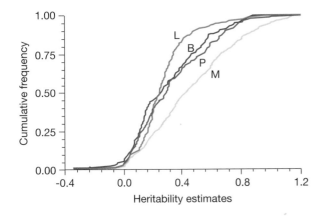

Figure 11.23   Life history traits have lower heritabilities than other kinds of traits   This plot shows four cumulative frequency distributions. A cumulative frequency distribution is a running sum, moving across a histogram, of the heights of the bars. The more rapidly the curve in a cumulative frequency distribution rises to 1, the lower the mean of the histogram. The line marked L is the cumulative frequency distribution for estimates of the heritability of life history traits; B = behavioral traits; P = physiological traits; M = morphological traits. Life history traits tend to have the lowest heritabilities. From Mousseau and Roff (1987).

morphs with each other, Grosberg demonstrated that the two morphs are genetically determined.

What maintains genetic variation for life history morphs in this sea squirt population? Grosberg tracked the seasonal frequency of the two morphs over two years (Figure 11.24). In both years, the semelparous morph dominated the population in the spring and early summer, whereas the iteroparous morph dominat-

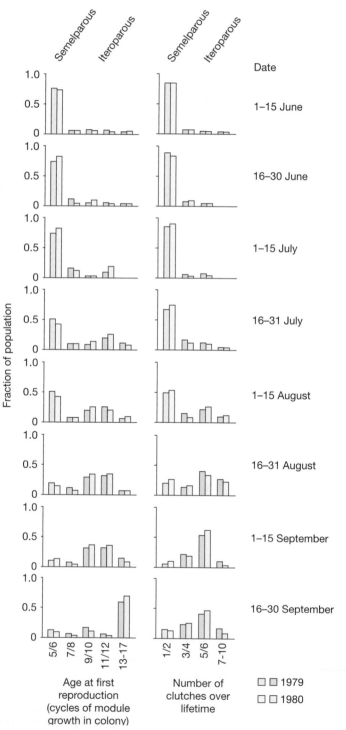

Figure 11.24   Annual cycles in the frequencies of two life history morphs in a population of sea squirts   The photo shows a colony of the sea squirt, *Botryllus schlosseri*. (Richard K. Grosberg, University of California at Davis) Each of the bar graphs shows a frequency distribution for the population during a two-week period. Colonies of the semelparous morph (dark and light orange bars) reproduce at an early age (age cycles of module growth in the colony) and produce only a single clutch of offspring. Colonies of the iteroparous morph (dark and light blue bars) reproduce at a late age and produce at least three clutches of offspring. From Grosberg (1988).

ed in the late summer. This indicates that the two morphs are maintained in the population by seasonal variation in selection. One important selective factor may be competitive interactions with another sea squirt (*Botryllus leachi*). This competitor, which becomes more abundant late in the summer, overgrows colonies of the semelparous *B. schlosseri* morph but not the iteroparous morph—a genotype-by-environment interaction.

In some species, genetic variation for life history traits is maintained by selection that varies through time.

The Evolution of Novel Traits

The evolution of novel traits represents a challenge for the theory of evolution by natural selection (see Chapter 3). Providing an example that is the focus of current research, closely related species of sea urchins can have strikingly different larval forms. Figure 11.25a shows the larvae of two urchins in the same genus: one is a pluteus; the other is a schmoo. Pluteus larvae hatch from small eggs. Before metamorphosis to the adult form, pluteus larvae live, feed, and grow in the plankton. Schmoo larvae hatch from large eggs. Schmoo larvae undergo metamorphosis earlier than do pluteus larvae, and they do not feed.

The process of development in pluteus larvae versus schmoo larvae is as strikingly different as their morphology (Wray and Bely 1994; Wray 1995; 1996; Raff 1996).

(a)

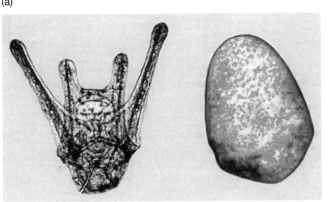

(b)

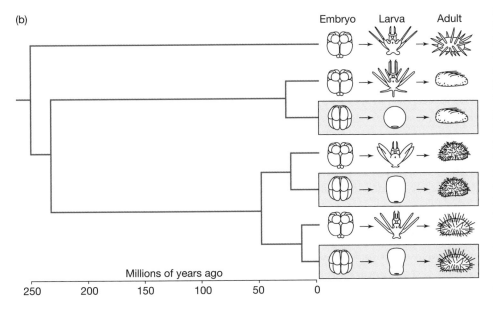

Millions of years ago

Figure 11.25 The evolution of nonfeeding larvae in sea urchins (a) The larva on the left is a form called a pluteus. The larva on the right is a form called a schmoo. These two larvae belong to closely related species of sea urchin: The pluteus is from *Heliocidaris tuberculata;* the schmoo is from *H. erythrogramma.* Both larvae in this photo are about three days old. Photo by R. A. Raff, reproduced by permission from G.A. Wray, 1995. Punctuated evolution of embryos. *Science* 267: 1115–6, Feb. 24, 1995. © American Association for the Advancement of Science. (b) The pluteus larval form is ancestral and ancient in sea urchins. In the urchin phylogeny shown here, the pluteus form is at least 250 million years old. More extensive phylogenies have shown the pluteus form to be about 500 million years old. Extant species with schmoo larvae appear to be relatively recently derived from ancestors with pluteus larvae. As the diagrams of early embryos included in the figure show, the pattern of the earliest cell divisions is different in species with pluteus versus schmoo larvae. Reprinted with permission from Wray (1995). Copyright © 1995, American Association for the Advancement of Science.

The two larval forms differ in the pattern of the earliest cell divisions (Figure 11.25b). Likewise, the larval forms differ in the expression of a variety of genes.

Although the schmoo form is simpler than the pluteus, the pluteus form is probably ancestral (Strathmann 1978). Strathmann argued that the pluteus form is so complex, and is shared among such distantly related sea urchin species, that the same larval form is unlikely to have evolved twice. Strathmann argued instead that pluteus larvae evolve into schmoo larvae, but this change is not reversible. From their morphological features, schmoo larvae appear to have evolved by simplification and loss of the complex features used by a pluteus to feed and swim in the plankton.

How can this hypothesis be tested? The alternative hypothesis is that the larval forms of sea urchins can switch from pluteus to schmoo and back again. One approach to compare these hypotheses is to build a phylogeny for sea urchin groups, and ask whether such reversibility seems likely from the distribution of larval forms on the phylogeny (Figure 11.25b). We introduced this idea of using phylogenies to reconstruct evolutionary history in Chapter 1, and we analyzed numerous examples of this approach in Chapter 13. In the case of sea urchins, reconstructed phylogenies indicate that the pluteus form is hundreds of millions of years old. In contrast, extant species with schmoo larvae appear to have been derived within the last 50 million years from ancestors with pluteus larvae. In addition, Jurassic-aged fossils of pluteus skeletons also support an ancient origin of the pluteus form (see Wray 1996). If a pluteus is the ancestral form for the sea urchins in Figure 11.25b, then a schmoo evolved three times independently. On the other hand, if a schmoo can reversibly give rise to a pluteus, then four different origins of a pluteus could have occurred among these species. By the parsimony method that we introduced in Chapter 1, we would conclude that three changes are more likely to have happened than four changes, and by this method we would conclude that the transition from pluteus larvae to schmoo larvae appears to be one-way. Larger phylogenies for more species indicate that schmoo larvae have evolved from pluteus ancestors at least 20 times in the sea urchins (Wray and Bely 1994; Wray 1995).

Figure 11.26 shows a hypothesis for how the schmoo larval form evolves from a pluteus ancestor. In the first step in this scheme, selection favors the production of larger, more energy-rich eggs. This makes feeding optional for the derived larva. In the second step, the derived larva loses the ability to feed. Finally, selection for earlier metamorphosis causes the loss of all feeding structures, ultimately resulting in the most derived larval form, the schmoo. Michael Hart (1996) studied the larval form of one sea urchin (*Brisaster latifrons*) that appears to represent the first transitional step in this scenario. *Brisaster latifrons* larvae hatch from large eggs. Although they can feed, *B. latifrons* larvae do not have to feed in order to complete metamorphosis to the adult form. If the scenario outlined in Figure 11.26 is both correct and general, then the evolution of the novel developmental and morphological features of the schmoo form has been repeatedly initiated by selection for a simple life history trait: larger eggs.

Suboptimal Life Histories

A crucial assumption behind many of the models and experiments reviewed in this chapter is that populations have had both the time and the additive genetic variation in life history traits to enable evolution toward an optimum. We end this

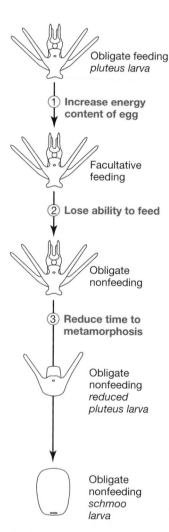

Obligate feeding
pluteus larva

① **Increase energy content of egg**

Facultative feeding

② **Lose ability to feed**

Obligate nonfeeding

③ **Reduce time to metamorphosis**

Obligate nonfeeding
reduced pluteus larva

Obligate nonfeeding
schmoo larva

Figure 11.26   Hypothesized steps in the transition from pluteus larva to schmoo From Wray (1996).

discussion with an important caveat: Life histories (like other kinds of adaptations) are not perfect, and they need not be optimal. Suboptimal life histories may be found because organisms lack the time or variation to evolve toward an optimum, or there may be fundamental limits on the ability of populations to evolve an optimal life history (other than the limits imposed by time and energy trade-offs).

Douglas Gill (1989) described one kind of suboptimal life history in a population of pink lady's slipper orchids, *Cypripedium acaule*. These orchids can live more than 20 years, have high adult survival rates, and produce large, showy flowers to attract pollinating insects (usually bumblebees in the genus *Bombus*). The flowers do not self-fertilize, so they require pollinators to deliver pollen from another *C. acaule*. However, the orchids deceive the pollinator because they offer no nectar reward in return for pollination. As a result, only naïve pollinators (newly emerged queen bees) visit the orchids, and even these pollinators do so only once.

Gill measured reproductive success of hundreds of these orchids in one study plot in the mountains of Virginia. In 10 years, this population produced 895 flowers, of which only 20 produced a natural fruit capsule. Pollinators are common at the study site, and this very low rate of fruiting is typical of other *C. acaule* populations. Gill concluded that such a life history is suboptimal for these orchids because this population could rapidly be invaded by several kinds of life history mutations that would increase fruiting success. For example,

- The flowers are self-fertile. When Gill self-pollinated flowers by hand they produced abundant fruits. A self-pollinating mutant would be an improved life history in this population.

- Many other orchids do offer nectar rewards. A mutant that provides nectar to attract repeated visits by bees would also be an improved life history.

Not all life history traits are optimized.

- New queen bees are more abundant later in the summer. A mutant that flowered later, even without a nectar reward, would be more likely to attract repeated visits (especially by bees carrying pollen from another *C. acaule*) and would be an improved life history.

Given that such life histories can and do evolve in other orchids, why does this suboptimal life history persist in *C. acaule* and evolve repeatedly among other orchids? Gill could not explain this, and the life history of this plant remains an enigma.

Richard Strathmann and colleagues (1981) described another suboptimal life history that gives some clues to potential constraints on the evolution of an optimal life history. They studied the reproduction and dispersal of the larvae of an intertidal barnacle, *Balanus glandula*. Adult barnacles release their larvae into the plankton, where they must develop for about two weeks before they are able to settle and become benthic juveniles. The adults tend to live in the low rocky intertidal zone, and settling larvae preferentially look for areas low in the intertidal zone, probably by evaluating the algae and other organisms growing on the rocks (Strathmann et al. 1981). Unfortunately for the barnacles, this preference sometimes misleads them to settle in low intertidal areas that are unsuitable for later adult life. In some parts of the geographic range of this species, the low intertidal is an ideal place to live, while in other parts of the range settlement in the low intertidal zone results in massive mortality of young barnacles (Strathmann and Branscomb 1979). In this case, the limitation on a more optimal life history for *B. glandula* is the length of the early larval life. Barnacles in the low intertidal zone of a good habitat might benefit by retaining their larvae in that habitat, but these barnacles have an obligate period of about two weeks as a planktonic larva. In those two weeks, larvae are so widely dispersed that it is impossible for a single larva to predict whether the low intertidal habitat it has selected for settlement will be a good or poor habitat as an adult. This obligate dispersal of the larva prevents natural selection from optimizing the life history to keep offspring living in the same neighborhood as their parents.

Summary

Organisms face fundamental trade-offs. The amount of energy available is finite, and energy devoted to one function—such as growth or tissue repair—cannot be devoted to other functions—such as reproduction. Furthermore, biological processes take time. An individual growing to a large size before maturing risks dying from disease or predation before ever reproducing. Fundamental trade-offs involving energy and time mean that every organism's life history is an evolutionary compromise.

Senescence evolves because natural selection is weaker late in life. Late-acting deleterious mutations can persist in populations under mutation–selection balance. Selection may favor increased investment in reproduction early in life at the expense of repair. Both mechanisms can result in a decline in reproductive performance and survival with age.

A trade-off between the number of offspring in a clutch and the survival of individual offspring constrains the evolution of clutch size. Additional constraints may involve trade-offs between present parental reproductive effort and future reproductive performance or survival, as well as trade-offs between clutch size and offspring reproductive performance.

A trade-off between the size and number of offspring constrains the evolution of offspring size. Selection on parents may favor smaller offspring than does selection on the offspring themselves.

Life history traits may reflect conflicts of interest between individuals. These conflicts have led to the evolution of differential gene expression (imprinting) and sexually antagonistic traits in males and females.

Theory predicts that life history traits should have low heritability because they are closely related to fitness. Life history traits do tend to have lower heritability than other kinds of traits, but nonetheless they typically show substantial genetic variation. One mechanism demonstrated to maintain genetic variation in life history traits is temporally varying selection.

Selection on life history traits can have dramatic consequences for other aspects of an organism's biology. Selection for larger eggs in echinoderms appears to have initiated, in numerous independent lineages, dramatic rearrangements in larval form and development. However, not all life history traits have evolved toward a stable optimum, and there are some examples of obviously suboptimal life histories.

Questions

1. Listed below are four possible causes of aging that were discussed in the text. As a review, name the theory that is associated with each cause, and describe whether or not selection for a longer life span is possible under each theory. What predictions does each theory make about the effect of ecological mortality (death due to external causes—predators, starvation, etc.) on aging rate?
   - "Wearing out" due to metabolic activity
   - Reduction in size of telomeres with each cell division
   - Mutations that have negative effects late in life
   - Mutations that have positive effects early and negative effects late in life

2. How is the replicative senescence theory related to the antagonistic pleiotropy theory? (*Hint:* Consider the possible cost of mutations that increase the level of telomerase and thus enable cells to continue dividing.)

3. Most domestic female rabbits will get uterine cancer if they are not spayed. The uterine cancer usually appears sometime after the age of two years. Describe a hypothesis for why rabbits have not evolved better defenses against uterine cancer. What do you think the average life span of a wild female rabbit might be? What do you think is a typical cause of death in wild rabbits? Why?

4. We have seen how aging can evolve due to two different phenomena: First, aging may evolve due to mutations that have deleterious effects only late in life. As a review, explain how such mutations could ever become common in a population. Second, aging may evolve due to mutations with pleiotropic effects that cause "trade-offs"—positive effects early and negative effects late. What would happen if a mutation arose with a reverse trade-off, that is, a mutation with negative effects early and positive effects late in life? Could such a mutation ever be selected for?

5. Look again at Figure 11.7, which shows life history trade-offs for a hypothetical species. Suppose you have a large captive population of these (hypothetical) organisms, and you notice a new mutation that causes its carriers to have two offspring per year instead of just one. The new mutation does not alter the age of maturation, which still occurs at three years. Your preliminary observations indicate that the new mutation may cause an early death, but you are not certain exactly how early the deaths occur. You do notice, however, that the new mutation is increasing in frequency and the wild-type allele is decreasing. Make a prediction about the minimum possible age of death of organisms that carry this mutation, and explain your reasoning.

6. Assuming for the moment that the grandmother hypothesis of menopause is correct, speculate on what aspects of a species' behavior and sociality may make menopause likely to evolve. For instance, is it important whether or not the species is highly social, or whether or not the species lives in kin groups? Might the age of independence of the young be important? Could menopause ever evolve in a species without parental care, such as aphids or willow trees? As fuel for thought, consider the likelihood of evolution of menopause in (1) orangutans, who live in small groups consisting simply of a female and her dependent young; (2) lions, in which females are very social and remain with their female kin for most of their lives; and (3) Arabian oryx, who live in small family groups in the desert, and must sometimes find distant waterholes known only to the older oryx.

7. The examples of the side-blotched lizards and seed beetles indicate that females probably cannot produce many large eggs; instead, they must choose between producing many small eggs or producing a few large eggs (and sometimes, in unfortunate cases, just a few small eggs). Explain, then, how it is possible for a queen honeybee to produce a very large number of relatively large eggs. (*Hint:* Consider what the other bees are doing.) Does this suggest a general way in which a female can escape from the size–number trade-off?

8. The 1998 movie "Godzilla" and the 1986 movie "Aliens" each depict a fictional large female carnivore. The "Godzilla" female lives off a large prey population of humans and fishes, but has no assistance from others of her kind. In a few days she produces hundreds of seven-foot-tall eggs, enough to fill Madison Square Garden. The "Aliens" female lives off a small prey population of a few dozen humans, is assisted by nonreproducing workers, and produces hundreds of large eggs in a few weeks. Comment on what is realistic and unrealistic about the life history traits and egg production abilities of each of these fictional animals. If they were real, would they have long or short life spans? Why?

9. Dairy farmers are sometimes frustrated in their attempts to breed a better milk cow because heritability values for milk production and reproductive traits are low—generally below 0.10. In addition, those cows that produce the most milk tend to have longer intervals between birth of successive calves, and require more breedings to a bull before the cow will conceive. Do these patterns make sense in light of evolutionary life history theory? Explain.

   Reference: T. E. Aitchison, "Genetic Improvement" page at the University of Maryland's "Dairy Cattle Genetics and Selection" website: http://www.inform. umd.edu/ EdRes/Topic/AgrEnv/ndd/genetics/

Exploring the Literature

10. A restricted diet prolongs life span in mammals. One hypothesis for the mechanism behind this phenomenon is that a restricted diet reduces the metabolic rate. If this is true, then the effect of a restricted diet would be consistent with the rate-of-living theory of aging. For an experimental evaluation of the effect of diet on metabolic rate, see

    McCarter, R., E. J. Masoro, and B. P. Yu. 1985. Does food restriction retard aging by reducing the metabolic rate? *American Journal of Physiology* 248: E488–E490.

11. Genes that might produce antagonistic pleiotropic effects on life history traits are difficult to identify. One approach is to predict which cellular or physiological functions are important for organisms, and study the genes known to regulate those functions (we discussed such genes in Section 11.2). An alternative approach is to perform massive mutation experiments and analyze the mutant strains for different life history traits. For an example of this approach (and a candidate aging gene in *Drosophila*), see

Lin, Y. J., L. Seroude, and S. Benzer. 1998. Extended life span and stress resistance in the *Drosophila* mutant *methuselah*. *Science* 282: 943–946.

How would you determine whether the trade-off between life span and early fecundity in *Drosophila* is influenced by the *methuselah* mutation?

12. Graham Bell distinguished between the rate-of-living versus evolutionary theories of aging by comparing invertebrates that lay eggs (and thus have a distinct soma

and germ line) with invertebrates that reproduce by fission (and thus have no soma/germ line division). According to the rate-of-living theory, both kinds of organisms will inevitably accumulate irreparable damage. According to the evolutionary theory, selection would allow genes responsible for senescence to accumulate only in organisms with a disposable soma. See

Bell, G. 1984. Evolutionary and nonevolutionary theories of senescence. *American Naturalist* 124: 600–603.

Citations

Austad, S. N. 1988. The adaptable opossum. *Scientific American* (February): 98–104.

Austad, S. N. 1993. Retarded senescence in an insular population of Virginia opossums (*Didelphis virginiana*). *Journal of Zoology, London* 229: 695–708.

Austad, S. N. 1994. Menopause: An evolutionary perspective. *Experimental Gerontology* 29: 255–263.

Austad, S. N., and K. E. Fischer. 1991. Mammalian aging, metabolism, and ecology: Evidence from the bats and marsupials. *Journal of Gerontology* 46: B47–53.

Barlow, D. P. 1995. Gametic imprinting in mammals. *Science* 270: 1610–1613.

Berrigan, D. 1991. The allometry of egg size and number in insects. *Oikos* 60: 313–321.

Bodnar, A. G., M. Ouellette, M. Frolkis, S. E. Holt, C.-P. Chiu, G. B. Morin, C. B. Harley, J. W. Shay, S. Lichtsteiner, and W. E. Wright. 1998. Extension of life span by introduction of telomerase into normal human cells. *Science* 279: 349–352.

Boyce, M. S., and C. M. Perrins. 1987. Optimizing great tit clutch size in a fluctuating environment. *Ecology* 68: 142–153.

Campisi, J. 1996. Replicative senescence: an old lives' tale? *Cell* 84: 497–500.

Charnov, E. L., and D. Berrigan. 1993. Why do female primates have such long life spans and so few babies? or Life in the slow lane. *Evolutionary Anthropology* 1: 191–194.

Charnov, E. L., and S. W. Skinner. 1985. Complementary approaches to the understanding of parasitoid oviposition decisions. *Environmental Entomology* 14: 383–391.

Clutton-Brock, T. H., S. D. Albon, and F. E. Guinness. 1988. Reproductive success in male and female red deer. In T. H. Clutton-Brock, ed. *Reproductive Success*. Chicago: University of Chicago Press, 325–343.

De Haan, G., and G. Van Zant. 1999. Genetic analysis of hemopoietic cell cycling in mice suggests its involvement in organismal life span. *The FASEB Journal* 13: 707–713.

Elbadry, E. A., and M. S. F. Tawfik. 1966. Life cycle of the mite *Adactylidium* sp. (Acarina: Pyemotidae), a predator of thrips eggs in the United Arab Republic. *Annals of the Entomological Society of America* 59: 458–461.

Elgar, M. A. 1990. Evolutionary compromise between a few large and many small eggs: Comparative evidence in teleost fish. *Oikos* 59: 283–287.

Eshleman, J. R., and S. D. Markowitz. 1996. Mismatch repair defects in human carcinogenesis. *Human Molecular Genetics* 5: 1489–1494.

Fishel, R., and T. Wilson. 1997. *MutS* homologs in mammalian cells. *Current Opinion in Genetics and Development* 7: 105–133.

Fisher, R. A. 1930. *The Genetical Theory of Natural Selection*. Oxford: Clarendon Press.

Fowler, K., and L. Partridge. 1989. A cost of mating in female fruit flies. *Nature* 338: 760-761.

Fox, C. W., M. S. Thakar, and T. A. Mousseau. 1997. Egg size plasticity in a seed beetle: An adaptive maternal effect. *American Naturalist* 149: 149–163.

Gill, D. E. 1989. Fruiting failure, pollinator inefficiency, and speciation in orchids. In D. Otte and J. A. Endler, eds. *Speciation and Its Consequences*. Sunderland, MA: Sinauer Associates.

Gould, S. J. 1980. *The Panda's Thumb: More Reflections in Natural History*. New York: W. W. Norton.

Grosberg, R. K. 1988. Life-history variation within a population of the colonial ascidian *Botryllus schlosseri*. I. The genetic and environmental control of seasonal variation. *Evolution* 42: 900–920.

Guillette, L. J. Jr., and R. E. Jones. 1985. Ovarian, oviductal and placental morphology of the reproductively bimodal lizard, *Sceloporus aeneus*. *Journal of Morphology* 184: 85-98.

Gustafsson, L., and T. Pärt. 1990. Acceleration of senescence in the collared flycatcher (*Ficedula albicollis*) by reproductive costs. *Nature* 347: 279–281.

Haig, D. 1999. Multiple paternity and genomic imprinting. *Genetics* 151: 1229-1231.

Haig, D., and C. Graham. 1991. Genomic imprinting and the strange case of the insulin-like growth factor II receptor. *Cell* 64: 1045–1046.

Haig, D. and M. Westoby. 1989. Parent-specific gene expression and the triploid endosperm. *American Naturalist* 134: 147–155.

Haig, D., and M. Westoby. 1991. Genomic imprinting in endosperm: Its effect on seed development in crosses between species, and between different ploidies of the same species, and its implications for the evolution of apomixis. *Philosophical Transactions of the Royal Society of London,* Series B 333: 1–13.

Hamilton, W. D. 1966. The moulding of senescence by natural selection. *Journal of Theoretical Biology* 12: 12–45.

Harley, C. B., A. B. Futcher, and C. W. Greider. 1990. Telomeres shorten during ageing of human fibroblasts. *Nature* 345: 458-460.

Hart, M. W. 1996. Evolutionary loss of larval feeding: Development, form, and function in a facultatively feeding larva, *Brisaster latifrons*. *Evolution* 50: 174–187.

Hawkes, K., J. F. O'Connell, and N. G. Blurton Jones. 1989. Hardworking Hadza Grandmothers. In V. Standen and R. A. Foley, eds. *Comparative Socioecology*. Oxford: Blackwell Scientific Publications, 341–366.

Hawkes, K., J. F. O'Connell, and N. G. Blurton Jones. 1997. Hadza women's time allocation, offspring provisioning, and the evolution of long postmenopausal life spans. *Current Anthropology* 38: 551–577.

Hawkes, K., J. F. O'Connell, N. G. Blurton Jones, H. Alvarez, and E. L. Charnov. 1998. Grandmothering, menopause, and the evolution of human life histories. *Proceedings of the National Academy of Sciences, USA* 95: 1336-1339.

Hill, K., and A. M. Hurtado. 1991. The evolution of premature reproductive senescence and menopause in human females: An evaluation of the "grandmother hypothesis." *Human Nature* 2: 313–351.

Hill, K., and A. M. Hurtado. 1996. *Ache Life History: The Ecology and Demography of a Foraging People.* New York: Aldine de Gruyter.

Holland, B., and W. R. Rice. 1998. Perspective: Chase-away sexual selection: antagonistic seduction versus resistance. *Evolution* 52: 1–7.

Holland, B., and W. R. Rice. 1999. Experimental removal of sexual selection reverses intersexual antagonistic coevolution and removes a reproductive load. *Proceedings of the National Academy of Sciences, USA* 96: 5083–5088.

Hurst, L. D. 1999. Is multiple paternity necessary for the evolution of genomic imprinting? *Genetics* 153: 509–512.

Lack, D. 1947. The significance of clutch size. *Ibis* 89: 302–352.

Law, R. 1979. Ecological determinants in the evolution of life histories. In R. M. Anderson, B. D. Turner, and L.R. Taylor, eds. *Population Dynamics.* Blackwell Scientific, Oxford, 81–103.

Levitan, D. R. 1993. The importance of sperm limitation to the evolution of egg size in marine invertebrates. *American Naturalist* 141: 517–536.

Levitan, D. R. 1996. Predicting optimal and unique egg sizes in free-spawning marine invertebrates. *American Naturalist* 148: 174–188.

Lindén, M., and A. P. Møller. 1989. Cost of reproduction and covariation of life history traits in birds. *Trends in Ecology and Evolution* 4: 367–371.

Luckinbill, L. S., R. Arking, M. J. Clare, W. C. Cirocco, and S. A. Buck. 1984. Selection for delayed senescence in *Drosophila melanogaster. Evolution* 38: 996–1003.

Medawar, P. B. 1952. *An Unsolved Problem in Biology.* London: H. K. Lewis.

Moore, T., and D. Haig. 1991. Genomic imprinting in mammalian development—a parental tug-of-war. *Trends In Genetics* 7: 45–49.

Morimoto, R. I., A. Tissieres, and C. Georgopoulos, eds. 1994. *The Biology of Heat Shock Proteins and Molecular Chaperones.* Cold Spring Harbor, NY: Cold Spring Harbor Press.

Mousseau, T. A., and D. A. Roff. 1987. Natural selection and the heritability of fitness components. *Heredity* 1987: 181–197.

Neese, R. M., and G. C. Williams. 1995. *Why We Get Sick: The New Science of Darwinian Medicine.* New York: Vintage Books.

Nowak, R. M. 1991. *Walker's Mammals of the World,* 5th ed. Baltimore: The Johns Hopkins University Press.

Partridge, L. 1987. Is accelerated senescence a cost of reproduction? *Functional Ecology* 1: 317–320.

Partridge, L., and N. H. Barton. 1993. Optimality, mutation, and the evolution of ageing. *Nature* 362: 305–311.

Partridge, L., and K. Fowler. 1992. Direct and correlated responses to selection on age at reproduction in *Drosophila melanogaster. Evolution* 46: 76–91.

Podolsky, R. D., and R. R. Strathmann. 1996. Evolution of egg size in free-spawners: Consequences of the fertilization-fecundity trade-off. *American Naturalist* 148: 160–173.

Pough, F. H., J. B. Heiser, and W. N. McFarland. 1996. *Vertebrate Life.* Upper Saddle River, NJ: Prentice Hall.

Price, T., and D. Schluter. 1991. On the low heritability of life-history traits. *Evolution* 45: 853–861.

Raff, R. A. 1996. *The Shape of Life : Genes, Development, and the Evolution of Animal Form.* Chicago: University of Chicago Press.

Reddel, R. R. 1998. A reassessment of the telomere hypothesis of senescence. *BioEssays* 20: 977–984.

Rice, W. R. 1987. The accumulation of sexually antagonistic genes as a selective agent promoting the evolution of reduced recombination between primitive sex chromosomes. *Evolution* 41: 911–914.

Rice, W. R. 1996. Sexually antagonistic male adaptation triggered by experimental arrest of female evolution. *Nature* 381: 232–234.

Rice, W. R., and B. Holland. 1997. The enemies within: Intergenomic conflict, interlocus contest evolution (ICE), and the intraspecific Red Queen. *Behavioral Ecology and Sociobiology* 41: 1–10.

Rodriguez Bigas, M. A., P. H. U. Lee, L. O'Malley, T. K. Weber, O. Suh, G. R. Anderson, and N. J. Petrelli. 1996. Establishment of a hereditary nonpolyposis colorectal cancer registry. *Diseases of the Colon & Rectum* 39: 649–653.

Roff, D. A. 1992. *The Evolution of Life Histories.* New York: Chapman & Hall.

Röhme, D. 1981. Evidence for a relationship between longevity of mamamlian species and life spans of normal fibroblasts *in vitro* and erythrocytes *in vivo. Proceedings of the National Academy of Sciences, USA* 78: 5009–5013.

Roper, C., P. Pignatelli, and L. Partridge. 1993. Evolutionary effects of selection on age at reproduction in larval and adult *Drosophila melanogaster. Evolution* 47: 445–455.

Rose, M. R. 1984. Laboratory evolution of postponed senescence in *Drosophila melanogaster. Evolution* 38: 1004–1010.

Rose, M. R. 1991. *Evolutionary Biology of Aging.* New York: Oxford University Press.

Sanz, J. J., and J. Moreno. 1995. Experimentally induced clutch size enlargements affect reproductive success in the Pied Flycatcher. *Oecologia* 103: 358–364.

Schluter, D., and L. Gustafsson. 1993. Maternal inheritance of condition and clutch size in the collared flycatcher. *Evolution* 47: 658–667.

Service, P. 1987. Physiological mechanisms of increased stress resistance in *Drosophila melanogaster* selected for postponed senescence. *Physiological Zoology* 60: 321–326.

Shabalina, S. A., L.Y. Yampolsky, and A.S. Kondrashov. 1997. Rapid decline of fitness in panmictic populations of *Drosophila melanogaster* maintained under relaxed natural selection. *Proceedings of the National Academy of Sciences, USA* 94: 13034–13039.

Silbermann, R., and M. Tatar. 2000. In C.W. Fox, D. A. Roff, and D. J. Fairbain, eds., *Evolutionary Ecology: Perspectives and Synthesis.* Oxford: Oxford University Press.

Sinervo, B., and P. Doughty. 1996. Interactive effects of offspring size and timing of reproduction on offspring reproduction: Experimental, maternal, and quantitative genetic aspects. *Evolution* 50: 1314–1327.

Sinervo, B., P. Doughty, R. B. Huey, and K. Zamudio. 1992. Allometric engineering: A causal analysis of natural selection on offspring size. *Science* 258: 1927–1930.

Sinervo, B., and P. Licht. 1991. Proximate constraints on the evolution of egg size, number, and total clutch mass in lizards. *Science* 252: 1300–1302.

Smith, C. C., and S. D. Fretwell. 1974. The optimal balance between size and number of offspring. *American Naturalist* 108: 499–506.

Spencer, H. G., A. G. Clark., and M. W. Feldman. 1999. Genetic conflicts and the evolutionary origin of genomic imprinting. *Trends in Ecology and Evolution* 14: 197–201.

Stearns, S. C. 1992. *The Evolution of Life Histories.* Oxford: Oxford University Press.

Strathmann, M. F. 1987. *Reproduction and Development of Marine Invertebrates of the Northern Pacific Coast.* Seattle, WA: University of Washington Press.

Strathmann, R. R. 1978. The evolution and loss of feeding larval stages of marine invertebrates. *Evolution* 32: 894–906.

Strathmann, R. R., and E. S. Branscomb. 1979. Adequacy of cues to favorable sites used by settling larvae of two intertidal barnacles. In S. E. Stancyk, ed. *Reproductive Ecology of Marine Invertebrates.* Columbia, SC: University of South Carolina Press.

Strathmann, R. R., E. S. Branscomb, and K. Vedder. 1981. Fatal errors in set as a cost of dispersal and the influence of intertidal flora on set of barnacles. *Oecologia* 48: 13–18.

Taborsky, M., and B. Taborsky. 1993. The kiwi's parental burden. *Natural History* 1993: 50–56.

Tatar, M. 1999. Transgenes in the analysis of life span and fitness. *American Naturalist* 154: S67–S81.

Tatar, M., A. A. Kazaeli, and J.W. Curtsinger. 1997. Chaperoning extended life. *Nature* 390: 30.

Vani, R. G., and M. R. S. Rao. 1996. Mismatch repair genes of eukaryotes. *Journal of Genetics* 75: 181–192.

Van Valen, L. 1973. A new evolutionary law. *Evolutionary Theory* 1: 1–30.

Vance, R. R. 1973. On reproductive strategies in marine benthic invertebrates. *American Naturalist* 107: 339–352.

Williams, G. C. 1957. Pleiotropy, natural selection, and the evolution of senescence. *Evolution* 11: 398–411.

Wourms, J. P. 1993. Maximization of evolutionary trends for placental viviparity in the spadenose shark, *Scoliodon laticaudus. Environmental Biology of Fish* 38: 269–294.

Wray, G. A. 1995. Punctuated evolution of embryos. *Science* 267: 1115–1116.

Wray, G. A. 1996. Parallel evolution of nonfeeding larvae in echinoids. *Systematic Biology* 45: 308–322.

Wray, G. A., and A. E. Bely. 1994. The evolution of echinoderm development is driven by several distinct factors. *Development 1994 Supplement* 97–106.

Young, T. P. 1990. Evolution of semelparity in Mount Kenya lobelias. *Evolutionary Ecology* 4: 157–171.

PART IV

According to the best data currently available, the Earth formed about 4.6 billion years ago. Life has existed for about 3.85 billion years. (NASA Headquarters)

THE HISTORY OF LIFE

PART II INTRODUCED THE MECHANISMS RESPONSIBLE FOR EVOLUTIONARY CHANGE, AND Part III probed the result of one of these processes—natural selection—in depth. Now, our focus shifts dramatically. We again consider all four evolutionary processes, and ask about their consequences over the sweep of Earth's history. Instead of studying evolutionary change within populations, we focus on how and why lineages diversified through time.

When did life on Earth begin, and what did the first organism look like? When did multicellularity evolve? What species is the closest relative of *Homo sapiens,* and how are modern humans related to fossil hominids like the Neanderthals?

Before we can address these questions, we need to explore how diversification occurs. Chapter 10 analyzes how populations diverge to become distinct species, and features studies on populations in which divergence is underway. Chapter 11 reviews how we can estimate the historical relationships among species, and shows how we can use phylogenetic trees to answer questions about the rate and pattern of change through time.

The rest of Part IV focuses on major events in the history of life. Chapter 14 introduces the earliest life forms and the branching events that led to the origin of life and the three main branches on the tree of life: the Bacteria, the Archaea, and the Eucarya. Chapter 15 investigates the evolution of multicellular life, with an emphasis on the initial diversification of animals and the catastrophic events called mass extinctions. The unit closes with a look at the evolution of our own species.

CHAPTER 12

Mechanisms of Speciation

These flowers are from different species of monkeyflowers and their hybrid offspring. Researchers produced the hybrids to investigate the loci and alleles responsible for the differences between monkeyflower species. (Douglas W. Schemske, University of Washington, Seattle)

ALL ORGANISMS ALIVE TODAY TRACE THEIR ANCESTRY BACK THROUGH TIME TO the origin of life some 3.8 billion years ago. Between then and now, millions—if not billions—of branching events have occurred as populations split and diverged to become separate species. In this chapter we examine how these branching events happened. In Chapters 5-7 we investigated how mutation, natural selection, migration, and drift act to change allele frequencies within populations; now we ask how these four processes can lead to genetic differences between populations.

In addition to providing a foundation for studying the history of life, studying speciation has important practical applications. Much of the material we explore focuses on the extent and causes of gene flow (or lack of gene flow) among differentiated groups of organisms. Understanding these topics is fundamental to creating effective strategies for preserving biodiversity, and to managing genetically engineered organisms that are released into the environment.

Along with considering these applied issues and the general problem of how evolutionary processes can isolate populations, we need to examine two additional questions: What happens when recently diverged populations come into contact and interbreed? What genetic changes take place during differentiation? To begin our analysis we start with the field's most fundamental question: What is a species?

12.1 Species Concepts

All human cultures recognize different types of organisms in nature and name them. These taxonomic systems are based on judgments about the degrees of similarity among organisms. People intuitively group like with like. The challenge to biologists has been to move beyond these informal judgments to a definition of species that is mechanistic and testable, and to a classification system that accurately reflects the evolutionary history of organisms. This has been difficult to do. In the past 30 years alone, there have been at least half a dozen species concepts proposed, recurrent controversies about which definition is best, and even philosophical debates about whether the unit we call the species actually exists in nature or whether it is merely a linguistic and cultural construct (for example see Donoghue 1985; Templeton 1989).

Species consist of interbreeding populations that evolve independently of other populations.

In this section, we describe the pros and cons of three major species concepts and review how they are applied. Although they differ in detail, these definitions agree that species share a distinguishing characteristic, which is evolutionary independence. Evolutionary independence occurs when mutation, selection, migration, and drift operate on each species separately. This means that species form a boundary for the spread of alleles. Consequently, different species follow independent evolutionary trajectories.

The differences among species concepts center on the problem of establishing practical criteria for identifying evolutionary independence. This is a challenge because the data available to define species vary between sexual and asexual organisms and between fossil and extant groups. The following three examples illustrate this point.

The Biological Species Concept

Under the biological species concept (BSC), the criterion for identifying evolutionary independence is reproductive isolation. Specifically, if populations of organisms do not hybridize, or fail to produce fertile offspring when they do, then they are reproductively isolated and considered good species. The BSC has been the textbook definition of a species since Ernst Mayr proposed it in 1942. It is used in practice by many zoologists and is the legal definition employed in the Endangered Species Act, which is the flagship biodiversity legislation in the United States.

Reproductive isolation is clearly an appropriate criterion for identifying species because it confirms lack of gene flow. This is the litmus test of evolutionary independence. But although the BSC is compelling in concept and useful in some situations, it is often difficult to apply. For example, if nearby populations do not actually overlap, we have no way of knowing whether they are reproductively isolated. Instead, biologists have to make subjective judgments to the effect that, "If these populations were to meet in the future, we believe that they are divergent enough already that they would not interbreed, so we will name them different species." In these cases, species designations cannot be tested with data. Further, the biological species concept can never be tested in fossil forms, is irrelevant to asexual populations (see Box 12.1), and is difficult to apply in the many plant groups where hybridization between strongly divergent populations is routine.

The Phylogenetic Species Concept

Systematists are the people responsible for classifying the diversity of life, and a growing number are promoting an alternative to the BSC called the phylogenetic or evolutionary species concept. This approach focuses on a criterion called

BOX 12.1 Species concepts in bacteria

Much of the research reviewed in this chapter focuses on evolutionary processes that lead to reproductive isolation. Indeed, the biological species concept treats this property as *the* criterion of speciation. But in bacteria and many other asexual forms, reproduction takes place via mitosis—without an exchange of genetic material. When gene exchange does occur between bacteria, it is limited to small segments of the genome and is unidirectional (Figure 12.1). Just as important, recombination can occur between members of widely diverged taxa. In bacteria, gene flow can occur between cells whose genomes have diverged up to 16% (Cohan 1995). Bacteria that are classified as members of different phyla can and do exchange genes, such as the alleles responsible for antibiotic resistance, via the extra-chromosomal loops of DNA called plasmids (Cohan 1994). In contrast, genetic exchange between eukaryotes is generally limited to organisms whose genomes have diverged a total of 2% or less. As a result, eukaryotic species that hybridize are almost always classified in the same genus.

In short, what most of us consider "normal" sex—meaning meiosis followed by the reciprocal exchange of homologous halves of genomes, among members of the same species—is unheard of in enormous numbers of organisms. As a result, gene flow in bacteria plays a relatively minor role in homogenizing allele frequencies among populations (Cohan 1994, 1995). The primary consequence of gene exchange in bacteria is that certain populations acquire alleles with high fitness advantages, such as sequences that confer antibiotic resistance.

Based on these data, Lawrence and Ochman (1998) have proposed that acquiring novel alleles through gene exchange is the primary mechanism for speciation in bacteria. Their hypothesis is that gene flow triggers divergence among bacteria populations, even though it prevents divergence among eukaryotes.

Work continues on the task of creating a workable species concept for bacteria and on measuring the rate of recombination in natural populations of these organisms. Exploring the mechanisms of speciation in bacteria is an exciting frontier in speciation research.

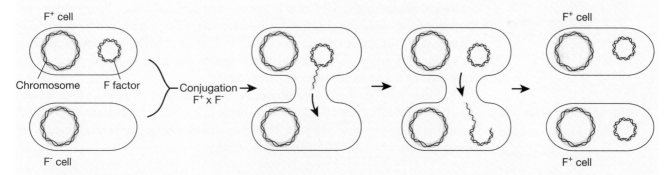

FIGURE 12.1 Genetic recombination in bacteria *Escherichia coli* cells with an extrachromosomal loop of DNA called an F (for fertility) factor can engage in recombination. The process starts when cells with F factors form conjugation tubes with F⁻ cells. A copy of the F factor migrates through the conjugation tube, converting the recipient cell from F⁻ to F⁺. Occasionally F factors will integrate into the chromosome. These integrated sequences can later leave the chromosome. When they do, they frequently take chromosomal sequences (that is, new genes) with them. In this way, F factors can transfer alleles between bacterial chromosomes.

monophyly. Monophyletic groups are defined as taxa or suites of taxa that contain all of the known descendants of a single common ancestor (Figure 12.2).

Under the phylogenetic species concept (PSC), species are identified by estimating the phylogeny of closely related populations and finding the smallest monophyletic groups. On a tree like this, species form the tips. For example, the taxa labeled A–J in Figure 12.2 are the smallest monophyletic groups on the tree and represent distinct species.

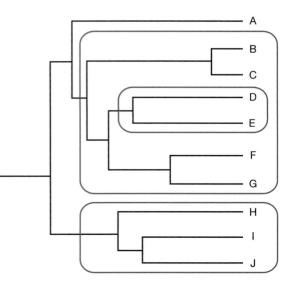

Figure 12.2 **Monophyletic groups** The taxa labeled A–J on the tips of this phylogeny may represent any taxonomic level ranging from populations to phyla. The groups that are circled on this phylogenetic tree are all monophyletic. Note that monophyletic groups can be nested—one inside another—and that there are many other monophyletic subgroups in this clade besides the ones that are circled.

The rationale behind the PSC is that traits can only distinguish populations on a phylogeny if the populations have diverged from one another in isolation. Put another way, to be called separate species under the PSC, populations must have been evolutionarily independent long enough for diagnostic traits to emerge. The appeal of this approach is that it is testable: Species are named on the basis of statistically significant differences in the traits used to estimate the phylogeny.

The problem comes with putting this criterion into practice. Carefully constructed phylogenies are available for only a handful of groups thus far. Also, many biologists object to the idea that a "species-specific trait" may be anything that distinguishes populations in a phylogenetic context. Such traits can be as trivial as a single substitution in DNA that is fixed in one population but not in another, or a slight but measurable and statistically significant increase in hairiness on the underside of leaves in different populations.

Each species concept has advantages and disadvantages.

Estimates vary, but the general opinion is that instituting the phylogenetic species concept could easily double the number of named species. Proponents of the PSC are not bothered by that prospect. They respond by saying that if this increase did occur, it would merely reflect biological reality.

The Morphospecies Concept

Paleontologists define species on the basis of morphological differences among fossils. When rigorous tests of reproductive isolation or well-estimated phylogenies are lacking, as they usually are, botanists and zoologists working on extant species do the same. The great advantage of the morphospecies concept is that it is so widely applicable. But when it is not applied carefully, species definitions can become arbitrary and idiosyncratic. In the worst-case scenario, species designations made by different investigators are not comparable.

Paleontologists have to work around other restrictions when identifying species. Fossil species that differed in color or the anatomy of soft tissues cannot be distinguished. Neither can populations that are similar in morphology but were strongly divergent in traits like songs, temperature or drought tolerance, habitat use, or courtship displays. These are called **cryptic species**.

Given these limitations, is the morphospecies concept still useful? Specifically, are the species we identify in the fossil record analogous to those we recognize today, using much larger suites of characters?

Jeremy Jackson and Alan Cheetham (1990, 1994) have performed the most careful analysis to date of the morphological species concept. Jackson and Cheetham study speciation in fossil forms of the colonial, marine-dwelling invertebrates called cheilostome Bryozoa. (The Cheilostomata is an order in the animal phylum Bryozoa.) Accordingly, they set out to check a critical assumption in their work: that the fossil morphospecies they have named conform to genetically differentiated species of bryozoans living today (Figure 12.3).

Jackson and Cheetham's first task was to establish that the skeletal measurements used to distinguish fossil morphospecies have a genetic basis. It is possible that these skeletal characteristics vary among bryozoan species simply because of environmental differences. To test this hypothesis, the researchers collected embryos from several extant cheilostomes and raised them in the same environment: a shallow-water habitat in the Caribbean. When they measured skeletal characters in the full-grown individuals, all but nine of the 507 offspring they raised were assigned to the correct morphospecies (that of their parents). This confirmed that the species-specific characters have a genetic basis. Then, in the critical experiment, Jackson and Cheetham surveyed variation in proteins isolated from eight species in three different genera. They found unique types of proteins in each of the named morphospecies. This result indicates that strong genetic divergence has occurred, and that no gene flow is occurring between the populations. Jackson and Cheetham interpreted these experiments as evidence that morphospecies, at least in Bryozoa, correspond to independent evolutionary units.

(a)   (b)   (c)

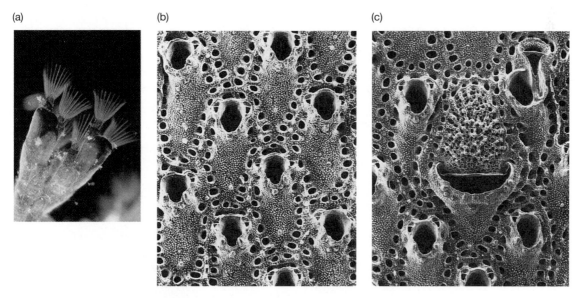

FIGURE 12.3   Morphological characters used to distinguish morphospecies in cheilostome Bryozoa   Living colony members are shown in photo (a). (Kjell B. Sandved/Butterfly Alphabet, Inc.) The enlarged photos show the skeletons of *Stylopoma spongites* (b) and *Metrarabdotos tenue* (c). Each colony member occupies one of the long chambers that have an orifice near the top. To distinguish morphospecies in these genera, Jackson and Cheetham (1994) measured traits like orifice length and width and the number of pores present per 0.2mm square. (b, c: Smithsonian Institution Photo Services)

Applying Species Concepts: The Case of the Red Wolf

Although it is probably unrealistic to insist on a single, all-purpose criterion for identifying species (Endler 1989), the major species concepts that have been proposed are productive when applied in appropriate situations. Consider, for example, how different species definitions have informed the controversy over the red wolf (Wayne and Gittleman 1995).

The red wolf, *Canis rufus,* is a member of the dog family native to the southeastern United States (Figure 12.4). Due to widespread hunting and clearing of forest habitats, its population had dwindled to a mere handful of individuals by the early 1970s. Many of the animals that were left showed characteristics typical of coyotes (*Canis latrans*), which started to become abundant in the wolf's range in the 1930s. This morphological similarity suggested that wolves and coyotes were hybridizing extensively. Fortunately, biologists from the U.S. Fish and Wildlife Service were able to capture 14 red wolves that apparently had no coyote traits before the population went extinct in the wild. These animals bred readily in captivity, and now several hundred red wolves await reintroduction to protected habitats in the wild.

Under the BSC, however, the extensive hybridization with coyotes made the species status of the red wolf questionable. Are red wolves reproductively isolated, evolutionarily independent units? Or are they actually a population of hybrids between the gray wolf (*Canis lupus*) and coyote? Ronald Nowak (1979, 1992) studied a large series of skull and dental characters and showed that red wolves collected before 1930 were a clearly identifiable morphospecies. These individuals had characteristics intermediate between gray wolves and coyotes. Because his data showed that animals collected after 1930 were much more similar to coyotes, Nowak suggested that hybridization was a recent phenomenon caused by the expansion of the coyote population, and the inability of wolves to find mates when their population dwindled. His conclusion was that red wolves clearly qualify as a species.

Genetic studies using DNA extracted from the captive population and from wolf pelts collected before 1930 told a different story, however. Surveys of mitochondrial DNA variation and alleles at 10 microsatellite loci showed no diagnostic (that is, no species-specific) differences between red wolves and coyotes (Wayne and Jenks 1991; Roy et al. 1994; Reich et al. 1999). Instead, the genetic data strongly supported the

Figure 12.4 The red wolf (Barbara von Hoffman/ Animals Animals/Earth Scenes)

hypothesis that red wolves are a hybrid between gray wolves and coyotes, and have no unique genetic characteristics of their own. This analysis implies that morphological data are simply not informative in this case, and that the red wolf's intermediate characteristics are the result of hybridization and not independent evolution. Under the PSC as well as the BSC, red wolves are not a distinct species.

Reliable criteria for identifying species are essential for preserving biodiversity.

The fate of the red wolf remains to be decided. The genetic research suggests that other species may be a higher priority for public monies committed to preserving biodiversity, and that the reintroduction program's greatest value may lie in bringing a top predator back into the ecosystem of North America's southern woodlands. Employing several criteria for identifying species can be a productive approach in clarifying conservation and evolutionary issues.

12.2 Mechanisms of Isolation

Having explored different criteria for identifying species, we now consider how species form. Speciation can be analyzed as a three-stage process: an initial step that isolates populations, a second step that results in divergence in traits such as mating tactics or habitat use, and a final step that produces reproductive isolation. In this section, we consider how physical separation or changes in chromosome complements can reduce gene flow. Once gene flow is dramatically reduced or ceases, evolutionary isolation occurs and speciation is initiated. In Section 12.3, we ask how genetic drift and natural selection, in combination with mutation, can cause populations to diverge; in Section 12.4, we consider how natural selection can complete the speciation process by causing reproductive isolation.

Physical Isolation as a Barrier to Gene Flow

In Chapter 6, we introduced migration as gene flow between populations and developed models showing that migration tends to homogenize gene frequencies and reduce the differentiation of populations. Using the example of water snakes from mainland and island habitats in Lake Erie, we also introduced the idea of a balance between migration and natural selection. Recall that experiments had shown a selective advantage for unbanded snakes on island habitats. But because migration of banded forms from the mainland occurs regularly, and because banded and unbanded forms subsequently interbreed, the island populations did not completely diverge from mainland forms. Migration continually introduced alleles for bandedness, even though selection tended to eliminate them from the island populations.

Now consider a thought experiment: What would happen if changes in shoreline habitats or lake currents effectively stopped the migration of banded forms from the mainland to the islands? The island populations would then be isolated from the mainland population. Gene flow would stop and the migration–selection balance would tip. The island population would be free to differentiate as a consequence of mutation, natural selection, and drift. These forces would act on them independently of the forces acting on mainland forms.

The speciation process begins when gene flow is disrupted and populations become genetically isolated.

This scenario illustrates a classical theory for how speciation begins, called the **allopatric model**. This theory was developed by Ernst Mayr (1942, 1963). One of Mayr's most important hypotheses is that speciation is especially likely to occur in small populations that become isolated on the periphery of a species' range, such as the island forms of water snakes. Population genetic models have shown

Geographic isolation produces genetic isolation.

that speciation in peripheral populations can occur rapidly when selection for divergence is strong and gene flow is low (García-Ramos and Kirkpatrick 1997).

Physical isolation is obviously an effective barrier to gene flow, and undoubtedly has been an important trigger for the second stage in the speciation process: genetic and ecological divergence. Geographic isolation can come about through dispersal and colonization of new habitats or through vicariance events, where an existing range is split by a new physical barrier (Figure 12.5).

Geographic Isolation Through Dispersal and Colonization

One of the most spectacular radiations in the class Insecta is also a superb example of geographic isolation through dispersal. The Hawaiian drosophilids, close relatives of the fruit flies we have encountered before, number over 500 named species in two genera and an estimated 350 species yet to be formally described and named.

The ecological diversification in this group is also unprecedented. Hawaiian flies can be found from sea level to montane habitats and from dry scrub to rainforests. Food sources, especially the plant material used as the medium for egg laying and larval development, vary widely among species. One of the Hawaiian flies even lays its eggs in spiders, while another has aquatic larvae. In addition, many species have elaborate traits, such as patterning on their wings or modified head shapes, that are used in combat or courtship displays (Figure 12.6).

Populations can become geographically isolated when individuals colonize a new habitat.

How did this enormous diversity come to be? The leading explanation is called the **founder hypothesis**. Many of the Hawaiian flies are island endemics, meaning that their range is restricted to a single island in the archipelago. The founder hypothesis maintains that this endemism results when small populations of flies, or perhaps even single gravid females, disperse to new habitats or islands. As a result, the colonists found new populations that are cut off from the ancestral species. Divergence begins after the founding event, resulting from drift and selection on the genes involved in courtship displays and habitat use.

The logic of the founder hypothesis is compelling, but do we have evidence, other than endemism, that these events actually occurred? Because the geology of

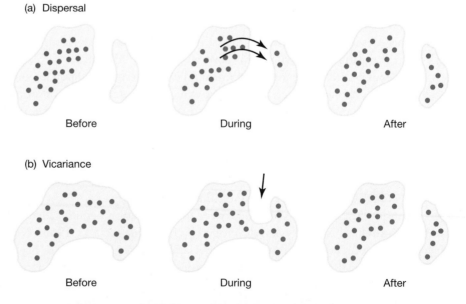

(a) Dispersal

Before   During   After

(b) Vicariance

Before   During   After

Figure 12.5   Isolation by dispersal and vicariance   In the diagram of dispersal (a), the arrows indicate movement of individuals. In the diagram of vicariance (b), the arrows indicate an encroaching physical feature such as a river, glacier, lava flow, or new habitat.

Figure 12.6 **Hawaiian *Drosophila*** As these photos of *Drosophila nigribasis*, *D. macrothrix*, and *D. suzukii* (left to right) show, the *Drosophila* found in Hawaii are remarkably diverse in body size, wing coloration, and other traits. (Kenneth Y. Kaneshiro, University of Hawaii)

the Hawaiian islands is well known, the hypothesis makes a strong prediction about speciation patterns in flies. The Hawaiian islands are produced by a volcanic hot spot under the Pacific Ocean. The hot spot is stationary, but periodically spews magma up and out onto the Pacific plate. After islands form, continental drift carries them to the north and west (see Figure 12.7a). As time passes, the volcanic cones gradually erode down to atolls and submarine mountains.

The founder hypothesis makes two predictions based on these facts: (1) Closely related species should almost always be found on adjacent islands, and (2) at

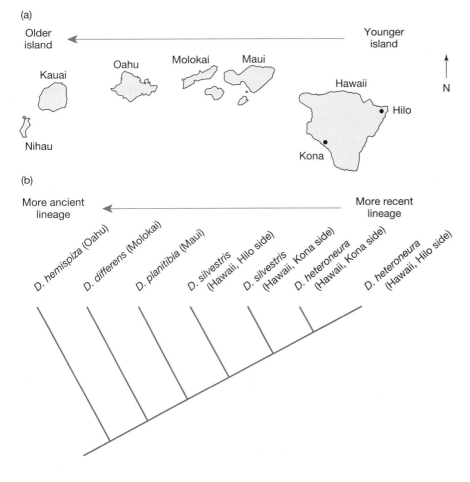

Figure 12.7 **Evidence for speciation by dispersal and colonization events** (a) The Hawaiian islands are part of an archipelago that stretches from the island of Hawaii to the Emperor Seamounts near Siberia. The youngest landform in the chain is the island of Hawaii, which still has active volcanoes. (b) *Drosophila silvestris*, *D. heteroneura*, *D. planitibia*, and *D. differens* are a closely related group. Note that the older-to-younger sequence of branches on the phylogeny shown here corresponds to the older-younger sequence of island formation shown in part (a). This pattern is consistent with the hypothesis that at least some of the speciation events in this group were the result of island hopping. (The phylogeny was estimated from data on sequence divergence in mitochondrial DNA; see De-Salle and Giddings 1986.)

least some sequences of branching events should correspond to the sequence in which islands were formed. Rob DeSalle and Val Giddings (1986) used sequence differences in mitochondrial DNA to estimate the phylogeny of four closely related species, and found exactly these patterns. The most recent species are found on the youngest islands, and several of the branching events correspond to the order of island formation (Figure 12.7b). This is strong evidence that dispersal to new habitats can trigger speciation.

As a mechanism for producing physical isolation and triggering speciation, the founder hypothesis is relevant to a wide variety of habitats in addition to oceanic islands. Hot springs, deep sea vents, fens, bogs, caves, mountaintops, and lakes or ponds with restricted drainage also represent habitat islands. Dispersal to novel environments is a general mechanism for initiating speciation.

Geographic Isolation Through Vicariance

Populations can also become geographically isolated when a species' former range is split into two or more distinct areas.

Vicariance events split a species' distribution into two or more isolated ranges and discourage or prevent gene flow between them. There are many possible mechanisms of vicariance, ranging from slow processes such as the rise of a mountain range or a long-term drying trend that fragments forests, to rapid events such as a mile-wide lava flow that bisects a snail population.

Nancy Knowlton and colleagues have been studying a classical vicariance event: the recent separation of marine organisms on either side of Central America. We know from geological evidence that the Isthmus of Panama closed (and the land bridge between South and North America opened) about 3 million years ago. Is this enough time for speciation to occur?

Knowlton et al. (1993) recently looked at a series of snapping shrimp (*Altheus*) populations from either side of the isthmus (Figure 12.8a). The populations they sampled appeared to represent seven pairs of closely related morphospecies, with one member of each pair found on each side of the land bridge. The phylogeny of these shrimp, estimated from differences in their mitochondrial DNA sequences, confirms this hypothesis (Figure 12.8b). The species pairs from either side of the isth-

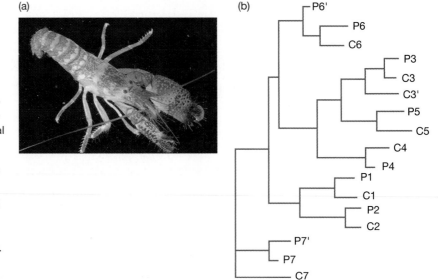

Figure 12.8 Snapping shrimp (a) This is *Alpheus malleator,* found on the Pacific side of the Panamanian isthmus. (Carl C. Hansen/Nancy Knowlton/Smithsonian Institution Photo Services) (b) This tree was estimated from sequence divergence in mitochondrial DNA. Reprinted with permission from Knowlton et al. (1993). Copyright © 1993, American Association for the Advancement of Science. The P or C designations with each number refer to whether that species is found in the Pacific or Caribbean. The prime (') marks after some letters indicate cryptic species distinguished by sequence differences. In every case, the putative sister species are indeed each others' closest relative.

mus, reputed to be sisters on the basis of morphology, are indeed each others' closest relatives. This is consistent with the prediction from the vicariance hypothesis.

Further, when Knowlton et al. put males and females of various species pairs together in aquaria and watched for aggressive or pairing interactions, the researchers found a strong correlation between the degree of genetic distance between species pairs and how interested the shrimp were in courting. Males and females from species with greater genetic divergence, indicative of longer isolation times, were less interested in one another. Finally, almost none of the pairs that formed during the courtship experiments produced fertile clutches. This last observation confirms that the Pacific and Caribbean populations are indeed separate species under all three of the species concepts we have reviewed.

One of the most interesting aspects of the study, though, was that the data contradicted a prediction made by the vicariance hypothesis. If the land bridge had formed rapidly, we would expect that genetic distances and degrees of reproductive isolation would be identical in all seven species pairs. This is not the case. For example, DNA sequence divergence between species pairs varied from about 6.5% to over 19%.

What is going on? Because it is unlikely that the land bridge popped up all at once, a prediction of identical divergence would be naïve. Instead, as the land rose and the ocean gradually split and retreated on either side, different shrimp populations would become isolated in a staggered fashion, depending on the depth of water each species occupied and how efficiently their larvae dispersed. The ranges of deeper-water species, or those with less-motile larvae, would be cut in two first. Consistent with this idea, the species numbered 6 and 7 in Figure 12.8b inhabit deep water, while the species numbered 1–4 live in shallower water (see Knowlton and Weigt 1998). Note also that the degree of genetic divergence in the species pairs numbered 1–4 is very similar. These are the lowest values observed, perhaps indicating "the final break" between the two oceans.

Changes in Chromosomes as a Barrier to Gene Flow

If a mutation results in a large-scale change in an individual's chromosomes, the event may cause rapid or even instantaneous isolation between the individual's descendants and the parental population. In Chapter 4, for example, we pointed out that polyploidization can result in instant reproductive isolation because of incompatibilities between gametes with different chromosome numbers. Purely on the basis of its distribution in plants, polyploidization has to be considered an important mechanism for initiating speciation.

Changes in chromosome number isolate populations genetically.

We need to introduce a distinction, however, between two types of polyploidy. Polyploids whose chromosomes originate from the same ancestral species are called autopolyploids. Polyploids that result from hybridization events between different species are called allopolyploids. Although no exact figures are available, allopolyploids are thought to be significantly more common than autopolyploids (Soltis and Soltis 1993). We examine allopolyploids in more detail in Section 12.4, when we consider gene flow between closely related populations and the third stage in the speciation process.

How important is polyploidization as a mechanism of speciation's first stage? Unfortunately, estimates for the frequency of auto- and allopolyploid species are rarely done in a phylogenetic context, where we can infer which ploidy is ancestral and get precise counts for the number of times that polyploidization has occurred. But under the assumption that any plant with a haploid number of chromosomes (symbolized n) greater than 14 is polyploid, Verne Grant (1981) estimated that 43% of species in the flowering plant class Dicotyledonae and 58% of species in the flowering plant class Monocotyledonae are the descendants of ancestors that underwent polyploidy. Jane Masterson (1994) came up with an even higher estimate. She showed that cell size correlates strongly with chromosome number in extant plants, and then measured the size of cells in plant fossils that lived early in the radiation of flowering plants. Her data support the hypothesis that $n = 7$ or 9 or lower is the ancestral haploid number in angiosperms. If so, then 70% of flowering plants have polyploidy in their history. Polyploidy is also common in mosses and almost the rule in ferns, where 95% of species occur in clades with an ancestral polyploidization event. Grant goes so far as to say that polyploidy "is a characteristic of the plant kingdom" (1981: 289).

Speciation triggered by changes in chromosome number has been especially important in plants.

Changes in chromosome number less drastic than polyploidization may also be important in speciation. For example, Oliver Ryder and colleagues (1989) studied chromosome complements in a series of small African antelopes—called dik-diks—that were being displayed in North American zoos. Although zookeepers traditionally recognized just two species, Ryder's team distinguished three on the basis of chromosome number and form. Further, they were able to show that hybrid offspring between unlike karyotypes are infertile. Their research revealed a cryptic species. Two of the dik-dik species have apparently differentiated on the basis of karyotype.

It is extremely common to find small-scale chromosomal changes like these when the karyotypes of closely related species are compared. Although these mutations could be important in causing genetic divergence between populations (White 1978), much of the extensive work on chromosomal differentiation done to date is merely correlative. That is, many studies have measured chromosome differences in related species and claimed that chromosomal incompatibilities are responsible for genetic isolation. But in many cases, it is likely that the chromosome differences arose after speciation had occurred due to other causes (Patton and Sherwood 1983). Until more causative links are established, we have to be cautious in interpreting the importance of small-scale karyotype differences in speciation.

12.3 Mechanisms of Divergence

Polyploidization, dispersal, and vicariance only create the conditions for speciation. For the event to continue, genetic drift and natural selection have to act on mutations in a way that creates divergence in the isolated populations. In this section we review how drift and selection act on closely related populations once gene flow between them has been reduced or eliminated.

Genetic Drift

In Chapter 6, we introduced population genetic models that quantified the major effects of genetic drift within populations: random fixation of alleles and random loss of alleles. We also reviewed data from a founder event observed by Peter Grant and Rosemary Grant during a study of Darwin's finches in the Galápagos. Their measurements of body size in a flock of colonizing birds confirmed that the founding population was a nonrandom sample of the source population.

Because genetic drift is a sampling process, its effects are most pronounced in small populations. This is important because most species are thought to have originated with low population sizes. Normally, only tiny numbers of individuals are involved in colonization events, and peripheral populations tend to be small. As a consequence, genetic drift has long been hypothesized as the key to speciation's second stage.

Drift can produce rapid genetic divergence in small, isolated populations.

A variety of genetic models have examined how drift might lead to rapid genetic differentiation in small populations (for a recent review, see Templeton 1996). The general message of these models is that small populations that become isolated start out as a nonrandom sample of the ancestral population. Subsequently, other effects of drift—including random loss of alleles and random fixation of existing and new alleles—encourage rapid divergence in the isolated population. In general, though, the overall genetic diversity of the founding populations is not necessarily lower than the source population. Lande (1980, 1981) has shown that when a population is reduced to a small size for a short period of time—the phenomenon known as **bottlenecking**—only very rare alleles tend to be lost due to drift. For genetic diversity to be reduced dramatically, the founding population has to be extremely small.

The role of drift in speciation events is controversial, however. Peter Grant and Rosemary Grant (1996) point out that hundreds of small populations have been introduced to new habitats around the world in the last 150 years due to the action of humans, but that few, if any, dramatic changes in genotypes have resulted because of genetic drift. Although genetic drift once dominated discussions of speciation mechanisms, most evolutionary biologists now take a much more balanced view. Natural selection has also been shown to be an important force promoting the divergence of isolated populations.

Natural Selection

Marked genetic differences have to emerge between closely related populations for speciation to proceed beyond the first stage. Drift almost always plays a role when at least one of the populations is small. But natural selection can also lead to divergence if one of the populations occupies a novel environment.

Selection's role in the second stage of speciation is clearly illustrated by recent research on apple and hawthorn flies. These closely related insect populations are diverging because of natural selection on preferences for a crucial resource: food.

The apple maggot fly, *Rhagoletis pomonella,* is found throughout the northeastern and north central United States (Figure 12.9). The species is a major agricultural pest, causing millions of dollars of damage to apple crops each year. The flies also parasitize the fruits of trees in the hawthorn group (species of *Crataegus*), which are closely related to apples.

Male and female *Rhagoletis* identify their host trees by sight, touch, and smell. Courtship and mating occur on or near the fruits. Females lay eggs in the fruit while it is still on the tree; the eggs hatch within two days and then develop through three larval stages in the same fruit. This takes about a month. After the fruit falls to the ground, the larvae leave and burrow a few inches into the soil. There they pupate after 3–4 days and spend the winter in a resting state called diapause. Most leave this resting stage and emerge as adults the next summer, starting the cycle anew.

Figure 12.9 Apple and hawthorn maggot flies
Apple and hawthorn races of *Rhagoletis pomonella* are indistinguishable to the eye. This is an apple fly. (Guy Bush, Michigan State University, and Jeff Feder, University of Notre Dame)

Apple trees clearly represent a novel food source for *Rhagoletis.* Hawthorn trees and *Rhagoletis* are native to North America, but apple trees were introduced from Europe less than 300 years ago. Our question is this: Are the flies that parasitize apple fruits and hawthorn fruits distinct populations? This hypothesis implies that natural selection, based on a preference for different food sources, has created two distinct races of flies. The contrasting hypothesis is that flies parasitizing hawthorns and apples are members of the same population. This hypothesis predicts that flies on hawthorns and apples interbreed freely, and that selection for exploiting different hosts has not occurred.

The hypothesis of no differentiation actually appears much more likely, because the first stage in speciation has not occurred. The two host trees and fly populations occur together throughout their ranges. Far from being isolated, at some sites hawthorn and apple trees are almost in physical contact. Marked flies have been captured over a mile from the site where they were originally captured, proving that individuals do search widely for appropriate fruit to parasitize. Thus, flies from the same population might simply switch from apple to hawthorn trees and back, based on fruit availability.

To test these two hypotheses, Jeff Feder and colleagues looked at the genetic makeup of flies collected from hawthorns versus apples, using the technique called protein electrophoresis introduced in Chapter 4. Remarkably, they found a clear distinction in the two samples: Flies collected from hawthorns versus apples have statistically significant differences in the frequencies of alleles for six different enzymes (Feder et al. 1988, 1990). This is strong support for the hypothesis that hawthorn and apple flies have diverged and now form distinct populations. Even though the two races look indistinguishable, they are easily differentiated on the basis of their genotypes.

How could this have occurred? Have these flies skipped the first stage in speciation (Box 12.2)? The key is that instead of being isolated by geography or by chromosomal incompatibilities, apple and hawthorn flies are isolated on different host species. In experiments where individuals are given a choice of host plants, apple and hawthorn flies show a strong preference for their own fruit type (Prokopy et al. 1988). Because mating takes place on the fruit, this habitat preference should result in strong nonrandom mating. Feder and colleagues (1994) confirmed this prediction by following marked individuals in the field. They found that matings between hawthorn and apple flies accounted for just 6% of the total observed.

Natural selection can cause populations to diverge even when a small amount of gene flow occurs.

Although host plant fidelity serves as an important barrier to mating, the two populations continue to exchange alleles. To cause the genetic divergence that has been observed, natural selection must overwhelm this gene flow in some way. Feder and co-workers (1997) hypothesized that natural selection for divergence is triggered by a marked difference in when apple and hawthorn fruits ripen. Because hawthorn fruits ripen about three weeks after apples, Feder and colleagues suggested that hawthorn fly larvae are selected to develop rapidly so they can pupate and enter diapause before the ground freezes. In contrast, they predicted that apple fly larvae are selected to develop slowly, so that they avoid emerging from diapause, as adults, before the onset of winter.

To test this hypothesis the researchers collected a large number of pupae from hawthorn fruits, split them into groups, and exposed each sample to 1–5 weeks of warm weather. In each group of pupae, the warm period was followed by a period of cold to simulate winter and then a period of warm temperatures to simulate

BOX 12.2   Parapatric and sympatric speciation

An enduring debate in speciation research has been whether physical isolation is an absolute requirement for populations to diverge, or whether natural selection for divergence can overwhelm gene flow and trigger speciation. Careful genetic modeling suggests that it is possible for populations to diverge even when they remain in physical contact. Joseph Felsenstein (1981) and William Rice (1987) were the first to show that populations can diverge even with low to moderate degrees of gene flow if two important conditions are met: Selection for divergence must be strong, and mate choice must be correlated with the factor that is promoting divergence. Exactly these conditions are met by apple and hawthorn flies, because mating takes place on different host plants.

Parapatric speciation is the other major mode of divergence that does not require physical isolation of populations. In parapatric speciation, strong selection for divergence causes gene frequencies in a continuous population to diverge along a gradient (Figure 12.10). Although theory demonstrates that parapatric divergence is possible (Endler 1977), we still lack well-documented examples in natural populations.

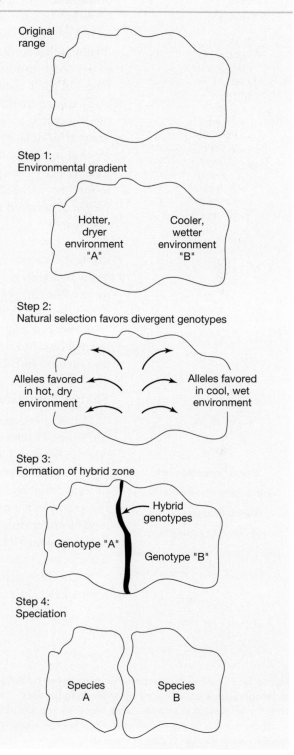

Figure 12.10   Parapatric speciation   This diagram illustrates the series of steps involved in a mode of divergence called parapatric ("beside place") speciation.

spring. The last step in the experiment was to collect the individuals that emerged as adults in the "spring" and assay the frequencies of the six allozymes that differ between the hawthorn and apple races. The results for the allele called *Acon-2 95*, shown in Figure 12.11, are typical. This graph shows that the hawthorn race individuals that were exposed to a month of warm days as pupae, and that survived to develop into adults during the spring, had allozyme frequencies similar to those found in apple flies.

This is a remarkable result. It confirms the hypothesis that hawthorn flies making the switch to apples have to develop slowly to stay in diapause through the warm days of late fall and emerge at the correct time in the spring. The experiment also suggests that the six alleles surveyed, or closely linked loci, are responsible for the change in development time. In a single generation, then, the researchers succeeded in replicating the selection events that have produced divergence between the apple race and the hawthorn race in nature over the past 300 years.

The diapause experiment demonstrates that natural selection is responsible for strong divergence between *R. pomonella* populations even in the face of gene flow. Many biologists now consider hawthorn and apple flies to be incipient species. This means that the populations have clearly diverged and are largely, but not completely, isolated in terms of gene flow.

Finally, it is important to recognize that apple maggot flies are by no means an isolated case. As Table 12.1 shows, many other examples of divergence due to selection on food or habitat choice, primarily in insects and fish, are being documented.

Sexual Selection

Sexual selection acts on characters involved in mate choice. Changes in sexual selection may isolate populations and cause rapid divergence.

Sexual selection results from differences among individuals in their ability to obtain mates. This is considered a form of selection distinct from natural selection and was the focus of Chapter 9. Population genetic models have shown that changes in the way that a population of sexual organisms chooses or acquires mates can lead to rapid differentiation from ancestral populations (Fisher 1958; Lande 1981, 1982). For example, if a new mutation led females in a certain population of barn swallows to prefer males with iridescent feathers instead of preferring males with long tails, then sexual selection would trigger rapid divergence. The key point is that sexual selection promotes divergence efficiently because it affects gene flow directly.

In the Hawaiian *Drosophila*, for example, sexual selection is thought to have been a key factor in promoting divergence among isolated populations. Many of

Figure 12.11 Allele frequency changes caused by differences in temperatures experienced by apple maggot flies This graph plots the frequencies of an allele called *Acon-2 95* in populations of hawthorn maggot fly pupae that survived to emerge as adults, as a function of the number of days of warm temperatures the pupae experienced. In the population that experienced an extended period of warm days—similar to the regime experienced by apple flies in nature—allele frequencies approximated those observed in natural populations of apple flies. From Feder et al. (1997). Copyright © 1997, National Academy of Sciences, USA.

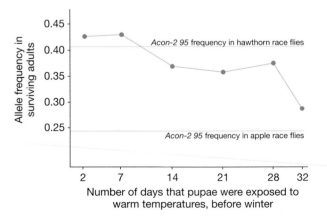

Table 12.1   Speciation in action

The right-most column in this table furnishes citations for recent studies on populations that are diverging due to differences in habitat use or resource use.

Species	Type of divergence that is underway	Citation
Lake whitefish	Populations of dwarf versus normal-sized individuals	Lu, G. and L. Bernatchez. 1999. *Evolution* 53: 1491–1505.
Three-spine sticklebacks (freshwater fish)	Benthic (bottom-dwelling) versus limnetic (open-water dwelling) populations	Hatfield, T. and D. Schluter. 1999. *Evolution* 53: 866–873.
Sockeye salmon	Sea-run versus lake-dwelling populations	Wood, C. C., and C. J. Foote. 1996. *Evolution* 50: 1265–1279.
Pea aphids	Different host plants	Via, S. 1999. *Evolution* 53: 1446-1457.
Army worms	Different host plants	Pashley, D. P. 1988. *Evolution* 42: 93–102.
Soapberry bugs	Different host plants	Carroll, S., H. Dingle, and S. P. Klassen. 1997. *Evolution* 51: 1182–1188.
Goldenrod ball gallmakers	Different host plants	Brown, J. M., W. G. Abrahamson, and P. A. Way. 1996. *Evolution* 50: 777–786.
Blueberry and apple maggot flies	Different host plants	Feder, J. L., C. A. Chilcote, and G.L. Bush. 1989. *Entomological Experiments and Applications* 51: 113–123.
Heliconius butterflies	Different habitats and types of warning coloration	McMillan, W. O., C. D. Jiggins, and J. Mallet. 1997. *Proceedings of the National Academy of Sciences, USA* 94: 8628–8633.

these flies court and copulate in aggregations called leks. In this mating system, males fight for small display territories and dance or sing for females, who visit the lek to select mates. Lek breeding systems are often associated with elaborate male characters, which vary widely among Hawaiian flies. Does this imply that sexual selection has been important in speciation?

The evidence in favor of the hypothesis is tantalizing, though not yet conclusive. For example, males of *Drosophila heteroneura* have wide, hammer-shaped heads (Figure 12.12a). Because males butt heads when fighting to stake out a courting arena on the lek, the unusual head shape appears to be a product of sexual selection (Kaneshiro and Boake, 1987). In contrast, males and females of *D. heteroneura's* closest relative, *Drosophila silvestris,* have heads that are similar in size and shape to female *D. heteroneura* (Figure 12.12b). Instead of head-butting, *D. silvestris* males fight on the lek by rearing up and grappling with one another. Both species are endemic to the island of Hawaii.

These facts are consistent with the following scenario:

1. In the ancestor to *silvestris* and *heteroneura,* males had normal heads, courted on leks, and fought for display territories by rearing up and grappling. Females chose the males who were most successful in combat.

2. A mutation occurred in an isolated subpopulation, which led to males with a new fighting behavior: head-butting.

3. The mutant males were more efficient in combat on leks and experienced increased reproductive success, because females still preferred to mate with males who won the most contests.

(a)   (b)

Figure 12.12 **Contrasting head shapes and fighting strategies in Hawaiian *Drosophila*** (a) Male *Drosophila heteroneura* have wide heads. As the photo at the bottom left shows, they butt heads to establish display territories on a lek. (b) Male *Drosophila silvestris* have normally shaped heads. They fight over display territories by rearing up and grappling with one another. (Kenneth Y. Kaneshiro, University of Hawaii)

4. The mutation increased to fixation, and additional mutations led to the elaboration of the trait over time. For example, it is possible that mutations leading to widely spaced eyes were favored because they made them less prone to damage during head-butting fights.

As a result of this sequence of events, strong divergence among the populations would occur due to sexual selection. The differentiation between populations would be based on the strategies and weaponry employed in male–male combat and female choice.

Formulating this type of plausible sequence is a productive way to generate testable hypotheses, but it does not substitute for genetic models, experiments, or other types of evidence. For example, consider recent work by Christine Boake and associates (1997) that tested two assumptions of the sexual selection scenario outlined above. These researchers staged a series of tests in the laboratory to assess whether female *D. heteroneura* prefer to mate with especially wide-headed males. They also staged male–male contests, to test the prediction that males with wider heads are more likely to win fights. As Figure 12.13 shows, both patterns were strongly supported by their data. The results increase our confidence that sexual selection has been a prominent cause of divergence in these populations.

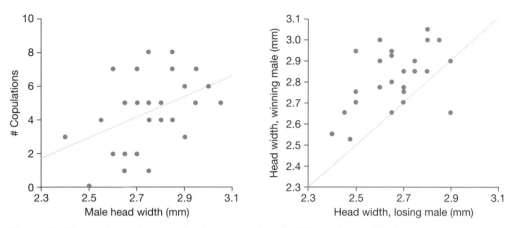

Figure 12.13 Evidence for sexual selection on head width in *Drosophila heteroneura*
The graph on the left shows the number of copulations achieved by male *Drosophila heteroneura* when paired with a series of different females, as a function of their head width. The best-fit line indicates that there is a positive relationship between copulation success and head width. The graph on the right compares the head width of winning versus losing males in staged contests. The straight line divides the plot into sections indicating that the wider-headed male won (upper left half) or the narrower-headed male won (lower right half). From Boake et al. (1997). Copyright © 1997, National Academy of Sciences, USA.

12.4 Secondary Contact

We have portrayed speciation as a three-step process that begins with the isolation of populations and continues when selection, mutation, and drift create divergence. A third step may occur if recently diverged populations come back into contact and have the opportunity to interbreed. Hybridization events between recently diverged species are especially common in plants. For example, over 700 of the plant species that have been introduced to the British Isles in the recent past have hybridized with native species at least occasionally, and about half of these native/non-native matings produce fertile offspring (Abbott 1992). Ten percent of all bird species also hybridize regularly and produce fertile offspring (Grant and Grant 1992).

Hybridization occurs when recently diverged populations interbreed.

In at least some cases, the fate of these hybrid offspring determines the outcome of the speciation event. Will the hybrids thrive, interbreed with each of the parental populations, and eventually erase the divergence between them? Or will hybrids have new characteristics and create a distinct population of their own? And what happens if hybrid offspring have reduced fitness relative to the parental populations?

Reinforcement

The geneticist Theodosius Dobzhansky (1937) formulated an important hypothesis about speciation's third stage. Dobzhanzky reasoned that if populations have diverged sufficiently in allopatry, their hybrid offspring should have markedly reduced fitness relative to individuals in the parental population. Because producing hybrid offspring reduces a parent's fitness in this case, there should be strong natural selection favoring assortative mating. That is, selection should favor individuals that choose mates only from the same population. Selection that reduces the frequency of hybrids in this way is called **reinforcement**. If reinforcement occurs, it would finalize the speciation process by producing complete reproductive isolation.

The reinforcement hypothesis predicts that some sort of mechanism of pre-mating isolation will evolve in closely related species that come into contact and hybridize. Selection might favor mutations that alter aspects of mate choice, genetic compatibility, or life history (such as the timing of breeding). Divergence in these traits prevents fertilization from occurring and results in **prezygotic isolation** of the two species. But populations can also remain genetically isolated in the absence of reinforcement if hybrid offspring are sterile or infertile. This possibility is known as **postzygotic isolation**.

Reinforcement is a type of selection that leads to assortative mating and the prezygotic isolation of populations.

Some of the best data on prezygotic isolation and the reinforcement hypothesis have been assembled and analyzed by Jerry Coyne and Allen Orr (1997). Coyne and Orr examined data from a large series of sister-species pairs in the genus *Drosophila*. Some of these species pairs live in allopatry and others in sympatry. Coyne and Orr's data set included estimates of genetic distance, calculated from differences in allozyme frequencies, along with measurements of the degree of pre- and postzygotic isolation. When they plotted the degree of prezygotic isolation against genetic distance, which they assumed correlates at least roughly with time of divergence, they found a striking result: Prezygotic isolation evolves much faster in sympatric species pairs than it does in allopatric species pairs (Figure 12.14). This is exactly the prediction made by the reinforcement hypothesis.

These laboratory experiments with *Drosophila* are some of the best evidence we have in support of the reinforcement hypothesis. In contrast, field studies that have looked for evidence of reinforcement in hybridizing populations have produced mixed results. Genetic models exploring reinforcement's effectiveness have generally been unconvincing as well (see Butlin 1987). Although Dobzhansky considered reinforcement as a universal stage in speciation, this view is probably overstated. A new consensus is emerging, based on a series of recent studies like

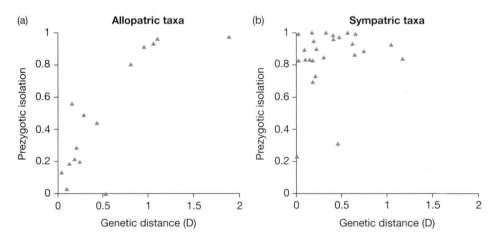

Figure 12.14   Prezygotic isolation in allopatric versus sympatric species pairs of *Drosophila*   These graphs (Coyne and Orr 1997) plot degree of prezygotic isolation versus genetic distance in a variety of sister-species pairs from the genus *Drosophila*. Prezygotic isolation is estimated from mate-choice tests performed in the laboratory. A value of 0 indicates that different populations freely interbreed (0% prezygotic isolation) and 1 indicates no interbreeding (100% prezygotic isolation). Genetic distance is estimated from differences in allele frequencies found in allozyme surveys. Sibling species with the same degree of overall genetic divergence show much more prezygotic isolation if they live in sympatry.

Coyne and Orr's: Reinforcement can and does occur, but it is not essential to completing speciation (Butlin 1995; Noor 1995).

Hybridization

Reinforcement should occur when hybrid offspring have reduced fitness. But what happens to hybrid offspring that survive and reproduce well? Their fate has important consequences for speciation, and practical applications as well. For example, because several crop plants have close relatives that are serious weeds, evolutionary biologists have expressed concern about the release of genetically engineered crop plants to the wild. What benefit comes from introducing a gene for herbicide resistance into a crop species if hybridization quickly carries the allele into a closely related weed population?

Using an experimental system featuring crop sorghum (*Sorghum bicolor*) and johnsongrass (*S. halepense*), Paul Arriola and Norman Ellstrand (1996) recently confirmed that this scenario could occur. Sorghum is one of the world's most important crops; johnsongrass is a serious weed. Arriola and Ellstrand sowed a field with crop sorghum seeds that carried a distinctive allozyme marker and planted seedlings of johnsongrass at various distances nearby (Figure 12.15a). After the johnsongrass plants had set seed, Arriola and Ellstrand harvested them and raised the progeny. Gel electrophoresis of proteins isolated from these first-generation (F₁) plants confirmed that the allozyme marker had been carried by crop sorghum

Biologists are concerned about gene flow between genetically modified crops and closely related weeds.

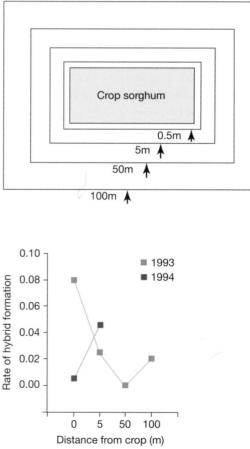

Figure 12.15 Gene flow from crops to weeds through hybridization The photographs show the flowering heads of crop sorghum (top) and johnsongrass (bottom). (a) This diagram shows the planting scheme employed by Arriola and Ellstrand (1996). Potted seedlings of johnsongrass were placed along each of four rectangles, as shown in the diagram. (The diagram is not to scale.) (b) In this graph, the percentage of johnsongrass plants that produced at least one hybrid offspring is plotted versus the distance that johnsongrass seedlings were growing from the edge of the crop. [Photos: (top) Michael P. Gadomski/Photo Researchers, Inc.; (bottom) Nigel Cattlin/Holt Studios International/Photo Researchers, Inc.]

pollen to johnsongrass plants. Significant gene flow had taken place in the crop-to-weed direction (Figure 12.15b).

Creation of New Species Through Hybridization

The concern about sorghum–johnsongrass hybrids illustrates a general point: Hybrid offspring do not necessarily have lower fitness than their parents. If crop sorghum plants were transformed with a gene for herbicide resistance and these individuals hybridized with johnsongrass, it is conceivable that a few of the offspring would end up with a favorable combination of traits from each parent. The hybrids might receive the allele for herbicide resistance from the crop parent, but a suite of genes that cause rapid growth from the weedy parent. This is the basis for concern about "superweeds." And this concern is not idle. Both classical and recent studies have confirmed that interspecific hybridization is a major source of evolutionary novelty in plants (Stebbins 1950; Grant 1981; Abbott 1992; Rieseberg 1997). In newly colonized environments or in certain novel habitats, hybrid offspring may have higher fitness than either of the parental species. Will these hybrid populations occupy the new environment and become distinct species?

A recent experimental study of plant hybridization, conducted in Loren Rieseberg's lab (1996), has actually duplicated a natural hybridization event that led to speciation. These researchers worked with three annual sunflower species native to the American southwest: *Helianthus annuus, H. petiolaris,* and *H. anomalous.* Based on morphological and chromosome studies, it had long been thought that *H. anomalous* originated in a hybridization event between *H. annuus* and *H. petiolaris.*

To test this hypothesis rigorously, Rieseberg and co-workers crossed individuals of *H. annuus* and *H. petiolaris* to produce three lines of F_1 hybrids. They then either mated each line back to *H. annuus* (this is called a backcross) or sib-mated the individuals for four additional generations. As a result, each experiment line underwent a different sequence and combination of backcrossing and sib-mating. This protocol simulated different types of matings that may have occurred when populations of *H. annuus* and *H. petiolaris* hybridized naturally.

At the end of the experiment, Rieseberg and colleagues surveyed the three hybrid populations genetically. Their goal was to determine how similar the hybrids were to each other and to *H. anomalous* individuals. To make this comparison possible, the researchers mapped a large series of species-specific DNA sequences, called randomly amplified polymorphic DNA (RAPD) markers, in the two parental species. These markers allowed the researchers to determine which alleles from the two parental species were present in the three experimental hybrid populations and compare them to alleles present in *H. anomalous.* The results were striking: The three independently derived experimental hybrids and the natural hybrid shared an overwhelming majority of markers. The experimental and natural hybrids were almost identical.

Researchers have experimentally recreated a speciation event that occurred naturally via hybridization.

To interpret this result, Rieseberg and co-workers contend that only certain alleles from *H. petiolis* and *H. anuus* work in combination, and that other hybrid types are inviable or have reduced fitness. The genetic composition of the hybrids quickly sorted out into a very similar, favorable combination. Even more remarkable, this combination of alleles was nearly the same as that produced by a natural hybridization event that occurred thousands of years ago. This puts an interesting twist on speciation's third stage: Secondary contact and gene flow between recently diverged species can result in the formation of a new, third, species.

Hybrid Zones

A **hybrid zone** is a region where interbreeding between diverged populations occurs and hybrid offspring are frequent. Hybrid zones are produced by two distinct situations (Hewitt 1988): (1) After secondary contact between species that have diverged in allopatry, a hybrid zone can form where they meet and interbreed, and (2) during parapatric speciation, a hybrid zone can develop between populations that are diverging (see Box 12.2).

We have already seen that it is possible for hybrid offspring to have lower or higher fitness than purebred offspring, with very different consequences (reinforcement of parental forms or the formation of a new species). Research on hybrid zones has confirmed that a third outcome is also possible. Frequently no measurable differences can be found between the fitnesses of hybrid and pure offspring. The following three possibilities dictate the size, shape, and longevity of hybrid zones (Endler 1977; Barton and Hewitt 1985; see Table 12.2):

- When hybrid and parental forms are equally fit, the hybrid zone is wide. Individuals with hybrid traits are found at high frequency at the center of the zone and progressively lower frequencies with increasing distance. In this type of hybrid zone, the dynamics of gene-frequency change are dominated by drift. The width of the zone is a function of two factors: how far individuals from each population disperse each generation, and how long the zone has existed. The farther the individuals move each generation and the longer the populations are in contact, the wider the zone.

- When hybrids are less fit than purebred individuals, the fate of the hybrid zone depends on the strength of selection against them. If selection is very strong and reinforcement occurs, then the hybrid zone is narrow and short lived. If selection is weak, then the region of hybridization is wider and longer lived. These types of hybrid zones are an example of a selection–migration balance, analogous to the situation with water snakes in Lake Erie.

- When hybrids are more fit than purebreds, the fate of the hybrid zone depends on the extent of environments in which hybrids have an advantage. If hybrids achieve higher fitness in environments outside the ranges of the parental species, then a new species may form (as reviewed earlier). If hybrids have an advantage at the boundary of each parental population's range, then a stable hybrid zone may form. For example, many hybrid zones are found in regions called ecotones, where markedly different plant and animal communities meet. In this case, two

Table 12.2 Outcomes of secondary contact and hybridization

When populations hybridize after diverging in allopatry, several different outcomes are possible. The type of hybrid zone formed and the eventual outcome depend on the relative fitness of hybrid individuals.

Fitness of hybrids	Hybrid zone	Eventual outcome
Lower than parental forms	Relatively narrow and short lived	Reinforcement (differentiation between parental populations increases)
Equal to parental forms	Relatively wide and long lived	Parental populations coalesce (differentiation between parental populations decreases)
Higher than parental forms	Depends on whether fitness advantage occurs in ecotone or new habitat	Stable hybrid zone or formation of new species

closely related species or populations (often called subspecies or varieties) are found on either side of the ecotone, with a hybrid zone between them. To explain this pattern, researchers hypothesize that hybrid individuals with intermediate characteristics have a fitness advantage in these transitional habitats.

To illustrate how biologists go about distinguishing between these possibilities, we review recent work on what may be the most widespread and economically important plant in the American west: big sagebrush (*Artemesia tridentata*). A total of four subspecies of big sagebrush have been described, including two that hybridize in the Wasatch Mountains of Utah (Freeman et al. 1995). Basin big sagebrush (*A.t. tridentata*) is found at low elevations in river flats, while mountain sagebrush (*A.t. vaseyana*) grows at higher elevations in upland habitats. The two subspecies hybridize where they make contact at intermediate elevations.

The first task in analyzing a hybrid zone is to describe the distribution and morphology of hybrids relative to parental populations. Previous work had shown that hybrid zones between sagebrush populations are narrow—often less than the length of a football field—and Carl Freeman and colleagues (1991) found that hybrids are intermediate in form between basin and mountain subspecies (Figure 12.16). Historical records indicate that the size and distribution of hybrid zones have been stable in extent for at least 2–3 sagebrush generations.

To assess the relative fitness of hybrid offspring versus pure forms, John Graham et al. (1995) compared a variety of fitness components in individuals sampled along an elevation gradient. These fitness components included seed and flower production, seed germination, and extent of browsing by mule deer and grasshoppers. Table 12.3 shows some data representative of their results. In general, hybrids show equal or even superior production of flowers and seeds, and resistance to herbivores that is equal to the mountain forms. Thus, hybrid offspring do not appear to be less fit than offspring of the parental populations.

In intermediate or transitional habitats, hybrid populations may be more fit than either parental population.

This leaves two possibilities: The hybrid zone could be maintained by selection–mutation balance, or by positive selection on hybrids in the ecotone. To test these alternative hypotheses, the research group studied growth rates and other components of fitness in basin, hybrid, and mountain seedlings that were transplanted to habitats at low, intermediate, and high elevation (Wang et al. 1997). These reciprocal transplant experiments showed that each form performed best in its native habitat (Figure 12.17). These data suggest that the hybrid zone is maintained because hybrid offspring have superior fitness in a transitional habitat.

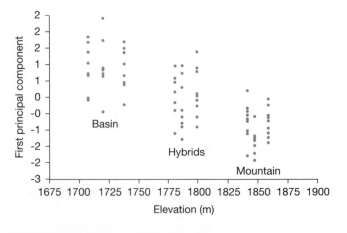

Figure 12.16 Hybrid sagebrush are intermediate in form between parental subspecies On this graph, a quantity called the principal component score is plotted against the elevation where sagebrush plants were sampled. Principal component analysis (PCA) is a statistical procedure for distilling information from many correlated variables into one or two quantities that summarize the variation measured among individuals in the study. In this case, Carl Freeman and colleagues (1991) measured a large series of morphological traits in sagebrush such as height, circumference, crown diameter, and branch length. The PCA was performed to combine these many variables into a single quantity, the PCA score, that summarizes overall size and shape. Each data point represents an individual.

Table 12.3 Fitness of sagebrush hybrids

The rows in this table, listing basin through mountain populations, represent big sagebrush plants sampled along a lower-to-higher elevational gradient. *N* is the sample size and the numbers in parentheses are standard deviations—a measure of variation around the average value. Differences among these populations in the number of inflorescences (flowering stalks) are not statistically significant. But hybrids have significantly more flowering heads per unit inflorescence length—a measure of flower density ($P < 0.05$).

Population	*N*	Number of inflorescences		Number of flowering heads	
Basin	25	19.92	(6.16)	175.1	(124.9)
Near basin	25	17.72	(6.59)	174.4	(92.5)
Hybrid	27	20.11	(6.75)	372.7	(375.9)
Near mountain	25	17.04	(6.50)	153.7	(75.2)
Mountain	25	16.80	(6.34)	102.0	(59.4)

Source: Adapted from Table 2 in Graham et al. (1995).

Having reviewed isolation, divergence, and secondary contact, we can move on to consider the genetic mechanisms responsible for these events. Understanding the genetic basis of speciation is our focus in Section 12.5.

12.5 The Genetics of Differentiation and Isolation

What degree of genetic differentiation is required to isolate populations and produce new species? The traditional view was that some sort of radical reorganization of the genome, called a genetic revolution, was necessary (Mayr 1963). This hypothesis was inspired by a strict interpretation of the biological species concept. The logic went as follows: Under the biological species concept (BSC), species are reproductively isolated if and only if hybrids are inviable or experience dramatic reductions in fitness. For this to happen, sister species would have to be genetically incompatible. Combining their alleles would produce dysfunctional development, morphology, or behavior.

Genetic models have shown that these types of large-scale changes in the genome are not only unlikely, but unnecessary for divergence and speciation to occur (Lande 1980; Barton and Charlesworth 1984). These theoretical results have

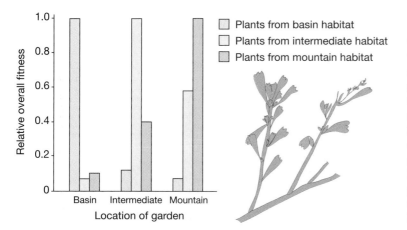

Figure 12.17 Relative fitness of big sagebrush taxa The vertical axis on this graph plots an overall measure of fitness that combines data on survivorship, flowering, seed production, and seed germination rate. The data are expressed as relative fitness by assigning a value of 1.0 to the group that had the highest fitness in each of the three experimental gardens, and then expressing the fitness of the other groups as a percentage of that group's fitness. The horizontal axis indicates whether the data come from gardens at the basin, intermediate, or mountain elevations. From Wang et al. (1997). Copyright © 1997 Evolution. Reprinted by permission of Evolution.

been verified by the work we reviewed in Section 12.3, which demonstrates that marked differentiation can occur between populations of sunflowers and sagebrush that still produce fertile hybrid offspring. As a result, the questions that motivate current research in the genetics of speciation are focused on the number, location, and nature of genes that distinguish closely related species.

Insights from Classical Genetics

When experimenters make crosses between sister species, it is not unusual to find that one sex in the F_1 generation is absent or sterile. Although premating isolation does not occur in these cases, postzygotic isolation is pronounced. The offspring that survive are incapable of mating further among themselves. This postzygotic isolation confirms that the populations are reproductively isolated and that the speciation process is complete.

Postzygotic isolation occurs when populations have diverged so thoroughly that hybrid offspring are inviable or sterile.

An important idea here is that populations that are physically isolated from one another can diverge so thoroughly in their genetic makeup over time that hybrid offspring simply do not develop normally and are sterile or inviable as a result. Allen Orr and Lynne Orr captured this point when they pointed out that reproductive isolation by postzygotic incompatibilities "is almost certainly an 'epiphenomenon' of divergence by ordinary population genetic processes ..." (Orr and Orr 1996: 1748).

Identifying the genes responsible for postzygotic isolation might tell us something interesting about this aspect of speciation. One of our best clues in this gene hunt comes from a striking observation made in experimental crosses between recently diverged taxa: If one sex in the hybrid offspring is sterile or inviable, it is almost always the one heterozygous for the sex chromosomes (Coyne and Orr 1989). This is referred to as the heterogametic sex. In species with X and Y sex chromosomes, such as insects and mammals, males are the heterogametic sex. But in species with Z and W sex chromosomes (birds and butterflies), females are the heterogametic sex. The pattern of sterility or inviability in the heterogametic sex holds, irrespective of how sex is specified genetically (Table 12.4). The generalization is so pervasive that it has been called Haldane's rule in honor of population geneticist J.B.S. Haldane, who first published on the phenomenon in 1922.

Table 12.4   Haldane's rule

In this table, the "Hybridizations with asymmetry" column indicates the number of closely related populations that have been crossed and that exhibit asymmetries in the fertility or viability of males and females in the F_1 hybrids. The "Number obeying Haldane's rule" column reports how many of these crosses showed greater fertility or viability loss in the heterogametic sex.

Group	Trait	Hybridizations with asymmetry	Number obeying Haldane's rule
Mammals	Fertility	20	19
Birds	Fertility	43	40
	Viability	18	18
Drosophila	Fertility and viability	145	141

Source: Coyne and Orr 1989.

The question is why? What is it about having one of each sex chromosome that contributes to sterility or inviability when hybrids are formed between divergent populations? Recently, a consensus has emerged that a hypothesis put forth by H. J. Muller in the early 1940s is probably correct (see Orr 1997). Muller started out by considering an autosomal locus, *A*, and an X-linked locus, *B*, in a species where males are the heterogametic sex. Then he supposed that individuals from one species are fixed for alleles A_1 and B_1, while individuals from a sister species are fixed for alleles A_2 and B_2. Further, he supposed that alleles A_1 and B_2 interact to cause inviability. Muller pointed out that if females from the first species mate with males of the second, male hybrid offspring have the genotype $A_1A_2B_1$. As a result, they are inviable. Female hybrid offspring, in contrast, have the genotype $A_1A_2B_1B_2$. They are viable because they have copies of A_2 and B_1 which interact to give a normal phenotype.

By mapping loci associated with hybrid incompatibility in crosses between *Drosophila simulans* and *D. seychellia*, Jerry Coyne and Marty Kreitman (1986) confirmed that at least one—and possibly two—loci on the X chromosome are involved in postzygotic isolation, as Muller's hypothesis predicts. More recent studies of offspring from crosses between sister species of fruit flies have shown that in hybrid males, pronounced defects occur in mitosis or sperm formation (Orr et al. 1997; Kulathinal and Singh 1998). The loci responsible for postzygotic isolation appear to be associated with basic cell functions.

In Drosophila, the genes responsible for postzygotic isolation are located on the X chromosome.

Analyzing Quantitative Trait Loci

An innovative experimental approach called quantitative trait loci (QTL) mapping is offering an additional way to locate genes involved in divergence and measure their effects. As Chapter 7 showed, many or most of the morphological and behavioral differences we observe between closely related species are quantitative traits. Characteristics like body size, body shape, songs, and flower shape result from the combined effects of many loci. QTL mapping is a technology for locating genes with small, but significant, effects on these types of traits.

H. D. Bradshaw and co-workers (1995) recently used QTL mapping to investigate divergence between sister species of monkeyflowers called *Mimulus cardinalis* and *M. lewisii* (Figure 12.18a). These two species hybridize readily in the lab and produce fertile offspring. They also have overlapping ranges in the Sierra Nevada of California. But no hybrids have ever been found in the field. The reason is that they attract different pollinators. *M. cardinalis* is hummingbird pollinated and *M. lewisii* is bee pollinated.

Flower morphology in these species correlates with the differences in the pollinators. Bees do not see well in the red part of the visible spectrum and need a platform to land on before walking into a flower and foraging. Hummingbirds, in contrast, see red well, have long, narrow beaks, and hover while harvesting nectar. *M. lewisii* has a prominent landing spot, while *M. cardinalis* has an elongated tube with a nectar reward at the end. The flower characteristics of *M. lewisii* and *M. cardinalis* conform to classical bee- and bird-pollinated colors and shapes.

Although a formal phylogenetic analysis has not been done, most species in the genus *Mimulus* are bee pollinated. This implies that the flatter, purple, bee pollinated flower is ancestral and the more tubular, reddish, hummingbird-pollinated form derived. The question is, what genes are responsible for the radical make

(a)

(b)

FIGURE 12.18 Contrasting floral traits in monkeyflowers (a) *Mimulus cardinalis* (left) is hummingbird pollinated while *M. lewisii* (right) is bee pollinated. The table below summarizes contrasts in eight floral traits. (Courtesy of Toby Bradshaw and Douglas Schemske, University of Washington. Photo by Jordan Rehm.) (b) These individuals are the products of self-fertilization of F_1 interspecific hybrids, and show a wide variety of floral characters. (Douglas W. Schemske, University of Washington, Seattle)

Characteristic	*M. cardinalis*	*M. lewisii*
Purple pigment (anthocyanins) in petals	high	low
Yellow pigment (carotenoids) in petals	low	high
Corolla width	low	high
Petal width	low	high
Nectar volume	high	low
Nectar concentration	low	high
Stamen (male structure) length	high	low
Pistil (female structure) length	high	low

Notes:

• The yellow pigment in *M. lewisii* petals is arranged in stripes called nectar guides, which are interpreted as a "runway" for bees as they land on the wide petals.

• Nectar volume and concentration are thought to contrast in bird- and bee-flowers simply because of the enormous difference in body size (hummingbirds can drink a lot more).

• The difference in stamen and pistil length is important: In *M. cardinalis* these structures extend beyond the flower and make contact with the hummingbird's forehead as it feeds.

over in flower shape? Can we pinpoint the loci that have responded to selection for hummingbird pollination? QTL mapping gives researchers a way to answer these questions.

In a QTL study, the hunt for the targeted genes takes place in hybrids between the two populations or species that have diverged. For example, Bradshaw et al. first created a large population of hybrids between *M. cardinalis* and *lewisii*. These F_1 individuals tend to be intermediate in flower type and not especially variable. This is because they are all heterozygous for flower size, shape, and color alleles from *M. cardinalis* and *M. lewisii*. But when these F_1 individuals are self-pollinated, their alleles for flower morphology segregate into many different combinations. Accordingly, the F_2 offspring have an astonishing variety of flower colors, shapes, and nectar rewards (Figure 12.18b).

In QTL mapping, researchers try to find correlations between the size or color or shape of these F_2 individuals and genetic markers that they carry (Box 12.3). If a statistically significant association is found between the trait and the marker that has been mapped, it implies that a QTL near that marker contributes to that trait. In the study done by Bradshaw et al., the researchers found that each of the eight floral traits listed in Figure 12.18a was associated with at least one QTL. In fact, in every case there was one QTL that accounted for 25% or more of the total variation in the trait. Because this is high for a polygenic trait, Bradshaw et al. refer to these loci as genes with major effects.

To confirm that these were the loci that underwent selection during the diversification of the two species, Douglas Schemske and Bradshaw (1999) reared a large series of F_2 individuals in the greenhouse and recorded the amount of purple pigment, yellow pigment, and nectar in their flowers, along with overall flower size. Then they planted the individuals out in a habitat where both species coexist and recorded which pollinators visited which flowers. Their data showed that there was a strong trend for bees to prefer large flowers and avoid flowers with a high concentration of yellow pigments. Hummingbirds, in contrast, tended to visit the most nectar-rich flowers and those with the highest amounts of purple pigment.

QTL studies can map the loci responsible for divergence between closely related species.

BOX 12.3 QTL mapping

The goal of QTL mapping is to find statistical associations between genetic markers that are unique to each parental species and the value of the trait in the hybrid offspring. Whenever an association between a marker and a trait is confirmed, we know that a locus at the marker, or tightly linked to it, contributes to the trait in question. By quantifying the degree of association between the QTL and the phenotype, we can also estimate how much of the observed variation is due to the alleles at each locus. That is, we can distinguish between genes that have major or minor effects on the traits we are interested in. (For a superb overview of QTL mapping, see Tanksley 1993.)

There are several tricks that increase the efficiency of a QTL analysis. First, it helps to have a very large series of species-specific genetic markers scattered among all of the chromosomes in the species being studied. In the monkeyflower study, Bradshaw et al. (1995) mapped 153 RAPD markers, five allozymes, and one visible trait (a change in the distribution of yellow color due to the *yup* locus). Second, it is easiest to identify statistically significant associations between the markers and the trait in question if the trait varies dramatically. As Figure 12.18b shows, the phenotypic variation that Bradshaw et al. could measure in F_2 hybrids was remarkable. Third, it is crucial to measure these effects in a very large number of hybrid offspring in order to find statistically significant associations. Bradshaw et al. measured flower phenotypes and mapped genetic markers in 93 hybrid offspring.

By collecting tissues from each F_2 individual planted in the field and determining which QTL markers they contained, the researchers were able to calculate that an allele associated with increased concentration of yellow pigments reduced bee visitation by 80%, while an allele responsible for increasing nectar production doubled hummingbird visitation. It is reasonable to conclude that changes in the frequencies of these alleles, driven by differential success in attracting hummingbirds as pollinators, was the mechanism behind speciation. Studies like these promise to identify and characterize "speciation genes" in a diverse array of organisms.

Summary

Although a wide variety of species concepts have been proposed, all agree that the distinguishing characteristic of a species is evolutionary independence. The various species concepts differ in the criteria that are employed for recognizing evolutionary independence.

Speciation can be analyzed as a three-step process: (1) Isolation of populations caused by dispersal, vicariance, or large-scale chromosome changes such as polyploidization, (2) divergence based on drift or selection, and (3) completion or elimination of divergence upon secondary contact. There are numerous exceptions to this sequence, however. In some cases, selection for divergence is strong enough that populations can differentiate without physical isolation, as research on apple and hawthorn maggot flies illustrates. Further, a variety of outcomes are possible after secondary contact. These include formation of stable hybrid zones and creation of a new species containing genes from each of the parental forms.

The primary strategy employed in genetic analyses of speciation is to look for correlations between mapped phenotypic or molecular markers and the distribution of traits in F_2 offspring of recently diverged species. These strategies have confirmed that loci on the X chromosome are particularly important in postzygotic isolation in *Drosophila* and have identified several loci responsible for changes in flower shape during speciation in monkeyflowers.

Speciation is the event that produces new branches on the tree of life. Understanding how biologists estimate the size, shape, and growth of this tree is the focus of Chapter 13.

Questions

1. Different species definitions tend to be advocated by people working in different disciplines. Consider the needs of a botanist studying the ecology of oak trees, a zoologist studying bird distributions, a conservation biologist studying endangered marine turtle populations, and a paleontologist studying extinction events using fossils from planktonic organisms. Which species definition would be most useful to each type of scientist?

2. Global travel and the shipping industry are causing widespread interchange of some "weedy" species across oceans and continents, while many "nonweedy" species (less tolerant of human activity) are losing many small fringe populations and subspecies. What are the implications for speciation rates for weedy and nonweedy species, and for the current global speciation rate?

3. When the Panama land bridge between North and South America was uncovered, some North American mammal lineages crossed to South America and underwent dramatic radiations. For terrestrial species, did the completion of the land bridge represent a vicariance or dispersal event? Does the recent building of the Panama Canal represent a vicariance or dispersal event for terrestrial organisms? For marine organisms? Briefly outline one experiment that would test whether the Panama Canal is affecting speciation in a terrestrial or marine species.

4. Would the recent glaciation events in northern Europe and North America have created vicariance events? If so, how? Which organisms might have been affected? For example, consider the different effects glaciation might have on small mammals, migratory birds, and trees.

5. Apple maggot flies are not the only insects that may be evolving to exploit new host plants. Within the past 50 years, soapberry bug populations in the U.S. have diversified into host race populations distinguished by markedly different beak lengths (Carroll and Boyd 1992).

These bugs eat the seeds at the center of soapberry fruits (Family Sapindaceae). Native and introduced varieties of soapberries differ greatly in fruit size. Describe the experiments or observations you would make to launch an in-depth study of host race formation in these bugs. What data would tell you whether they are separate populations evolving independently, or a single interbreeding population? Many museums contain insect specimens from decades ago. What would you examine in these old specimens? What information about the host plants would be useful?

6. Red crossbills are small finches specialized for eating seeds out of the cones of conifer trees. They fly thousands of kilometers each year in search of productive cone crops. Despite their mobility, crossbills have diverged into several "types" that differ in bill shape, body size, and vocalizations. Each type prefers to feed on a different species of conifer, and each species of conifer is only found in certain forests. Bill size and shape affects how efficiently a bird can open cones of a certain conifer species.

   Explain how a highly mobile animal such as the red crossbill could have diverged into different types in the absence of any geographic barrier. How could you test your ideas? If crossbills could not fly, do you think speciation would occur more quickly or more slowly? If conifer species were not patchily distributed (i.e. in dif-ferent forests), do you think crossbill speciation would occur more quickly or more slowly? In general, how do habitat patchiness and dispersal ability interact to affect divergence?

7. The monkeyflowers studied by Bradshaw et al. tend to be found at different altitudes. The hummingbird-pollinated *M. cardinalis* occurs at higher elevations, and the bee-pollinated *M. lewisii* at lower elevations. In addition to the QTLs responsible for the differences in flower color and shape, it is likely that loci affecting physiological traits, such as the ability to photosynthesize and grow at colder temperatures, have also diverged between the two species. Write a protocol describing how you would go about mapping these traits.

8. Ellen Censky and coworkers (1998) recently documented the arrival of a small group of iguanas to the Caribbean island of Anguilla, which previously had no iguanas. The animals were carried there on a raft of fallen trees and other debris during a hurricane. Outline a long-term study that would document whether this newly isolated population diverges from iguanas on nearby islands to form a new species.

9. Despite intensive fieldwork by several hundred evolutionary biologists, very few speciation events have been observed first-hand. Comment on why this is so.

Exploring the Literature

10. Biologists frequently use the word spectacular to describe the species numbers and ecological and morphological diversity of cichlid fish found in east Africa's Lakes Malawi, Tanganyika, and Victoria. Males are brightly colored in many of the 1,000 species and recent work has shown that sexual selection may be intense. Phylogenies and geologic data indicate that the 300 species found in Lake Victoria are derived from a single founding population that arrived just 12,000 years ago. Further, each lake contains species that eat fish, mollusks, insect larvae, algae, zooplankton, or phytoplankton. To learn more about this dramatic adaptive radiation, see

Galis, R. and J. A. J. Metz. 1998. Why are there so many cichlid species? *Trends in Ecology and Evolution* 13: 1–2.

Knight, M. E., G. F. Turner, C. Rico, M. J. H. van Oppen, and G. M. Hewitt. 1998. Microsatellite paternity analysis on captive Lake Malawi cichlids supports reproductive isolation by direct mate choice. *Molecular Ecology* 7: 1605–1610.

Johnson, T. C., C. A. Scholz, M. R. Talbot, K. Kelts, R. D. Ricketts, G. Ngobi, K. Beuning, I. Ssemmanda, and J. W. McGill. 1996. Late Pleistocene desiccation of Lake Victoria and rapid evolution of cichlid fishes. *Science* 273:1091–1093.

Citations

Abbott, R. J. 1992. Plant invasions, interspecific hybridization and the evolution of new plant taxa. *Trends in Ecology and Evolution* 7: 401–405.

Arriola, P. E., and N. C. Ellstrand. 1996. Crop-to-weed gene flow in the genus *Sorghum* (Poaceae): Spontaneous interspecific hybridization between johnsongrass, *Sorghum halepense,* and crop sorguhm, *S. bicolor. American Journal of Botany* 83: 1153–1160.

Barton, N. H., and B. Charlesworth. 1984. Genetic revolutions, founder effects, and speciation. *Annual Review of Ecology and Systematics* 15: 133–164.

Barton, N. H., and G. M. Hewitt. 1985. Analysis of hybrid zones. *Annual Review of Ecology and Systematics* 15: 133–164.

Boake, C. R. B., M. P. DeAngelis, and D. K. Andreadis. 1997. Is sexual selection and species recognition a continuum? Mating behavior of the stalk-eyed fly *Drosophila heteroneura*. *Proceedings of the National Academy of Sciences, USA* 94: 12442-12445.

Bradshaw, H. D. Jr., S. M. Wilbert, K. G. Otto, and D.W. Schemske. 1995. Genetic mapping of floral traits associated with reproductive isolation in monkeyflowers (*Mimulus*). *Nature* 376: 762–765.

Butlin, R. 1987. Speciation by reinforcement. *Trends in Ecology and Evolution* 2: 8–13.

Butlin, R. 1995. Reinforcement: An idea evolving. *Trends in Ecology and Evolution* 10: 432–434.

Carroll, S. P., and C. Boyd. 1992. Host race radiation in the soapberry bug: Natural history with the history. *Evolution* 46: 1052–1069.

Censky, E. J., K. Hodge, and J. Dudley. 1998. Over-water dispersal of lizards due to hurricanes. *Nature* 395: 556.

Cohan, F. M. 1994. Genetic exhange and evolutionary divergence in prokaryotes. *Trends in Ecology and Evolution* 9: 175–180.

Cohan, F. M. 1995. Does recombination constrain neutral divergence among bacterial taxa? *Evolution* 49:164–175.

Coyne, J. A., and M. Kreitman. 1986. Evolutionary genetics of two sibling species, *Drosophila simulans* and *D. sechellia*. *Evolution* 40:673–691.

Coyne, J. A., and H. A. Orr. 1989. Two rules of speciation. In D. Otte and J. A. Endler, eds. *Speciation and Its Consequences*. Sunderland, MA: Sinauer, 180–207.

Coyne, J. A., and H. A. Orr. 1997. "Patterns of speciation in *Drosophila*" revisited. *Evolution* 51: 295–303.

DeSalle, R., and L. V. Giddings. 1986. Discordance of nuclear and mitochondrial DNA phylogenies in Hawaiian *Drosophila*. *Proceedings of the National Academy of Sciences, USA* 83: 6902–6906.

Dobzhansky, T. 1937. *Genetics and the Origin of Species*. New York: Columbia University Press.

Donoghue, M. J. 1985. A critique of the biological species concept and recommendations for a phylogenetic alternative. *Bryologist* 88: 172–181.

Endler, J. A. 1977. *Geographic Variation, Speciation, and Clines*. Princeton, NJ: Princeton University Press.

Endler, J. A. 1989. Conceptual and other problems in speciation. In D. Otte and J. A. Endler, eds. *Speciation and Its Consequences*. Sunderland, MA: Sinauer, 625–648.

Feder, J. L., C. A. Chilcote, and G. L. Bush. 1988. Genetic differentiation between sympatric host races of the apple maggot fly *Rhagoletis pomonella*. *Nature* 336: 61–64.

Feder, J. L., C. A. Chilcote, and G. L. Bush. 1990. The geographic pattern of genetic differentiation beween host associated populations of *Rhagoletis pomonella* (Diptera: Tephritidae) in the eastern United States and Canada. *Evolution* 44: 570–594.

Feder, J. L., S. B. Opp, B. Wlazlo, K. Reynolds, W. Go, and S. Spisak. 1994. Host fidelity is an effective premating barrier between sympatric races of the apple maggot fly, *Rhagoletis pomonella*. *Proceedings of the National Academy of Sciences, USA* 91: 7990–7994.

Feder, J. L., J. B. Roethele, B. Wlazlo, and S. H. Berlocher. 1997. Selective maintenance of allozyme differences among sympatric host races of the apple maggot fly. *Proceedings of the National Academy of Sciences, USA* 94: 11417–11421.

Felsenstein, J. 1981. Skepticism towards Santa Rosalia, or why are there so few kinds of animals? *Evolution* 35:124–138.

Fisher, R. A. 1958. *The Genetical Theory of Natural Selection*. New York: Dover.

Freeman, D. C., W. A. Turner, E. D. McArthur, and J. H. Graham. 1991. Characterization of a narrow hybrid zone between two subspecies of big sagebrush (*Artemisia tridentata*: Asteraceae). *American Journal of Botany* 78:805–815.

Freeman, D. C., J. H. Graham, D. W. Byrd, E. D. McArthur, and W. A. Turner. 1995. Narrow hybrid zone between two subspecies of big sagebrush *Artemisia tridentata* (Asteraceae). III. Developmental instability. *American Journal of Botany* 82: 1144–1152.

García-Ramos, G., and M. Kirkpatrick. 1997. Genetic models of adaptation and gene flow in peripheral populations. *Evolution* 51: 21–28.

Graham, J. H., D. C. Freeman, and E. D. McArthur. 1995. Narrow hybrid zone between two subspecies of big sagebrush. II. Selection gradients and hybrid fitness. *American Journal of Botany* 82: 709–716.

Grant, P. R., and B. R. Grant. 1992. Hybridization of bird species. *Science* 256: 193–197.

Grant, P. R., and B. R. Grant. 1996. Speciation and hybridization in island birds. *Philosophical Transactions of the Royal Society of London*, Series B 351: 765–772.

Grant, V. 1981. *Plant Speciation*. New York: Columbia University Press.

Hewitt, G. M. 1988. Hybrid zones—natural laboratories for evolutionary studies. *Trends in Ecology and Evolution* 3: 158–166.

Jackson, J. B. C., and A. H. Cheetham. 1990. Evolutionary significance of morphospecies: A test with cheilostome Bryozoa. *Science* 248: 579–583.

Jackson, J. B. C., and A. H. Cheetham. 1994. Phylogeny reconstruction and the tempo of speciation in the cheilostome Bryozoa. *Paleobiology* 20: 407–423.

Kaneshiro, K. Y., and C. R. B. Boake. 1987. Sexual selection and speciation: Issues raised by Hawaiian *Drosophila*. *Trends in Ecology and Evolution* 2: 207–213.

Knowlton, N., L. A. Weigt, L. A. Solórzano, D. K. Mills, and E. Bermingham. 1993. Divergence in proteins, mitochondrial DNA, and reproductive incompatibility across the isthmus of Panama. *Science* 260: 1629–1632.

Knowlton, N. and L. A. Weigt. 1998. New dates and new rates for divergence across the Isthmus of Panama. *Proceedings of the Royal Society of London*, Series B 265: 2257–2263.

Kulathinal, R. and R. S. Singh. 1998. Cytological characterization of premeiotic versus postmeiotic defects producing hybrid male sterility among sibling species of the *Drosophila melanogaster* complex. *Evolution* 52: 1067–1079.

Lande, R. 1980. Genetic variation and phenotypic evolution during allopatric speciation. *American Naturalist* 116: 463–479.

Lande, R. 1981. Models of speciation by sexual selection on polygenic traits. *Proceedings of the National Academy of Sciences, USA* 78: 3721–3725.

Lande, R. 1982. Rapid origin of sexual isolation and character divergence in a cline. *Evolution* 36: 213–223.

Lawrence, J. G. and H. Ochman. 1998. Molecular archaeology of the *Escherichia coli* genome. *Proceedings of the National Academy of Sciences, USA* 95: 9413–9417.

Masterson, J. 1994. Stomatal size in fossil plants: Evidence for polyploidy in majority of angiosperms. *Science* 264:421–424.

Mayr, E. 1942. *Systematics and the Origin of Species*. New York: Columbia University Press.

Mayr, E. 1963. *Animal Species and Evolution*. Cambridge, MA: Harvard University Press.

Noor, A. M. 1995. Speciation driven by natural selection in *Drosophila*. *Nature* 375:674–675.

Nowak, R. M. 1979. North American Quaternary *Canis*. Monograph 6, Museum of Natural History, University of Kansas, Lawrence.

Nowak, R. M. 1992. The red wolf is not a hybrid. *Conservation Biology* 6:593–595.

Orr, H. A. 1997. Haldane's rule. *Annual Review of Ecology and Systematics* 28: 195–218.

Orr, H. A. and L. H. Orr. 1996. Waiting for speciation: The effect of population subdivision on the time to speciation. *Evolution* 50: 1742–1749.

Orr, H. A., L. D. Madden, J. A. Coyne, R. Goodwin, and R. S. Hawley. 1997. The developmental genetics of hybrid inviability: A mitotic defect in *Drosophila* hybrids. *Genetics* 145: 1031–1040.

Patton, J. L., and S. W. Sherwood. 1983. Chromosome evolution and speciation in rodents. *Annual Review of Ecology and Systematics* 14: 139–158.

Prokopy, R. J., S. R. Diehl, and S. S. Cooley. 1988. Behavioral evidence for host races in *Rhagoletis pomonella* flies. *Oecologia* 76: 138–147.

Reich, D. E., R. K. Wayne, and D. B. Goldstein. 1999. Genetic evidence for a recent origin by hybridization of red wolves. *Molecular Ecology* 8: 139-144.

Rice, W. R. 1987. Speciation via habitat specialization: The evolution of reproductive isolation as a correlated character. *Evolutionary Ecology* 1: 301–314.

Rieseberg, L. H. 1997. Hybrid origins of plant species. *Annual Review of Ecology and Systematics* 28: 359–389.

Rieseberg, L. H., B. Sinervo, C. R. Linder, M. C. Ungerer, and D. M. Arias. 1996. Role of gene interactions in hybrid speciation: Evidence from ancient and experimental hybrids. *Science* 272: 741–745.

Roy, M. S., E. Geffen, D. Smith, E. A. Ostrander, and R. K. Wayne. 1994. Patterns of differentiation and hybridization in North American wolflike canids, revealed by analysis of microsatellite loci. *Molecule Biology and Evolution* 11: 553–570.

Ryder, O. A., A. T. Kumamoto, B. S. Durrant, and K. Benirschke. 1989. Chromosomal divergence and reproductive isolation in dik-diks. In D. Otte and J. A. Endler. *Speciation and its Consequences.* Sunderland, MA: Sinauer, 180–207.

Schemske, D. W. and H. D. Bradshaw, Jr. 1999. Pollinator preference and the evolution of floral traits in monkeyflowers (*Mimulus*). *Proceedings of the National Academy of Sciences, USA* 96: 11910–11915.

Soltis, D. E., and P. S. Soltis. 1993. Molecular data and the dynamic nature of polyploidy. *Critical Reviews in the Plant Sciences* 12: 243–273.

Stebbins, G. L. 1950. *Variation and Evolution in Plants.* New York: Columbia University Press.

Tanksley, S. D. 1993. Mapping polygenes. *Annual Review of Genetics* 27: 205–233.

Templeton, A. R. 1989. The meaning of species and speciation: A genetic perspective. In D. Otte and J. A. Endler. *Speciation and its Consequences.* Sunderland, MA: Sinauer, 3–27.

Templeton, A. R. 1996. Experimental evidence for the genetic-transilience model of speciation. *Evolution* 50: 909–915.

Wang, H., E. D. McArthur, S. C. Sanderson, J. H. Graham, and D. C. Freeman. 1997. Narrow hybrid zone between two subspecies of big sagebrush (*Artemisia tridentata:* Asteraceae). IV. Reciprocal transplant experiments. *Evolution* 51: 95–102.

Wayne, R. K., and J. L. Gittleman. 1995. The problematic red wolf. *Scientific American* 273: 36–39.

Wayne, R. K., and S. M. Jenks. 1991. Mitochondrial DNA analysis implying extensive hybridizaton of the endangered red wolf *Canis rufus. Nature* 351: 565–568.

White, M. J. D. 1978. *Modes of Speciation.* San Francisco: W. H. Freeman.

CHAPTER 13

Reconstructing Evolutionary Trees

This chapter uses research on the dolphins, porpoises, and whales as an example of how evolutionary biologists estimate phylogenies. (Flip Nicklin/Minden Pictures)

THE EVOLUTIONARY HISTORY OF A GROUP IS CALLED ITS **PHYLOGENY**. A **phylogenetic tree** is a graphical summary of this history. The tree describes the pattern, and in some cases the timing, of branching events. It records the sequence of speciation and documents which taxa are more closely or distantly related.

In Chapter 1, we introduced the use of phylogenetic thinking by using a parsimony analysis to determine whether HIV was transmitted from humans to monkeys or from monkeys to humans. In Chapter 2, we introduced the idea of tree thinking more formally and explored how to use a phylogenetic tree to determine whether the swim bladders found in fish evolved from lungs, or vice versa. But in these and other instances, we have been using trees without asking how they were put together.

The task before us now is to understand how phylogenies are estimated, so we can interpret them critically and make independent judgments about their quality. The goal of this chapter is to explain and illustrate the logic of inferring a phylogeny from data.

Because we do not have direct knowledge of evolutionary history, a phylogeny must be inferred indirectly from data. In the cases we have already discussed, many different trees could possibly represent the evolutionary history of a group. In any particular case, what data should we use, and how do we know that we have inferred the true tree?

We illustrate these issues with a controversial and surprising example: the phylogenetic relationships between whales and dolphins (the order Cetacea) and other mammals. We use this case study to show how the principle of parsimony is used in phylogeny estimation, how researchers choose data for a phylogenetic problem, and how the reliability of a particular phylogeny can be tested. This presentation parallels the sequence of decisions that face researchers when estimating phylogenies (Swofford et al. 1996).

Finally, we explore the various ways that phylogenies can be used to answer questions about evolutionary patterns and processes. The chapter closes with examples illustrating how phylogenetic thinking is applied in contemporary research. Topics in this last section range widely, from how we should classify the diversity of life to the evolution of ants that farm fungi.

13.1　Parsimony and Phylogeny

At its most basic level, the logic of estimating evolutionary relationships is simple: The most closely related taxa should have the most traits in common. Naively, we would say that any traits or characters that are independent of one another, heritable, and variable among the taxa in the study can help us reconstruct who evolved from whom. These characters might be things like DNA sequences, the presence or absence of certain skeletal elements or flower parts, or the mode of embryonic or larval development. The only requirements are that different characters be independent of each other and that each one can be scored as a homologous character state in all of the species to be studied. (We reviewed the criteria for making a statement about homology in Chapter 2.)

In choosing characters to study, however, it is essential to realize that evolutionary relationships are only revealed by traits that are similar because they were derived from a common ancestor. (Box 13.1 explains why this is so.) These shared

---

BOX 13.1　Cladistic methods

Techniques that identify monophyletic groups on the basis of shared derived characters are called cladistic methods. Traits that are shared because they were modified in a common ancestor are called synapomorphies. Mammals, for example, share fur and lactation as traits that were derived from their common ancestor (a population from a long-extinct group called the synapsids). Synapomorphies can be identified at whatever taxonomic level a researcher might be interested in: populations, species, genera, families, orders, or classes.

Two ideas are key to understanding why this approach to phylogeny inference is valid. The first is that synapomorphies identify evolutionary branch points. Why? After a species splits into two lineages that begin evolving independently, some of their homologous traits undergo changes due to mutation, selection, and drift. These changed traits are synapomorphies that identify

populations in the two independent lineages. The second key idea is that synapomorphies are nested. That is, as you move back in time and trace an evolutionary tree backward from its tips to its root, each branching event adds one or more synapomorphies. As a result, the hierarchy described by synapomorphies replicates the hierarchy of branching events. The source of these insights is the German entomologist Willi Hennig, who began writing on cladistic approaches in the 1950s.

The cornerstone of tree building with cladistic methods is to identify these shared, derived traits accurately. Identifying and evaluating synapomorphies requires judgments about the homology of traits and the direction, or polarity, of change through time. There are several ways to determine which states of a trait are ancestral and which are derived. One of the most basic and reliable ways is called outgroup analysis. In outgroup analy-

BOX 13.1 Continued

sis, the character state in the group of interest (the in-group) is compared to the state in a close relative that clearly branched off earlier (the outgroup). The outgroup is assumed to represent the ancestral state. Finding an appropriate outgroup, in turn, involves borrowing conclusions from other phylogenetic analyses or confirming an earlier appearance in the fossil record. Most investigators prefer to examine several outgroups so they can corroborate that the characters used in the analysis were not present in the "derived" state earlier in evolution.

When no convergence, parallelism, or reversal occurs, all the similarities observed in a data set are due to modifications that took place in common ancestors. A researcher would say that all synapomorphies identified are congruent. In this case, inferring the phylogeny is straightforward: The investigator simply groups the taxa according to their synapomorphies. Each branch on the tree corresponds to one or more

synapomorphies that distinguish the derived groups. A phylogenetic tree inferred in this manner is called a cladogram. By convention, the location of these synapomorphies is indicated on published trees with bars across the branches. The synapomorphies are described in a key or labels that accompany the diagram. Figure 13.1 presents an example cladogram.

When homoplasy is present, synapomorphies are not completely congruent and the strategy for inferring phylogenies changes slightly. The goal now is to find the branching pattern that minimizes homoplasy. Many or all of the possible branching patterns have to be examined and evaluated, and a best tree chosen. How the evaluation and selection are done is the subject of Sections 13.2 through 13.5.

For further background on cladistic approaches, see Hennig (1979), Eldredge and Cracraft (1980), Nelson and Platnick (1981), and Mayden and Wiley (1992).

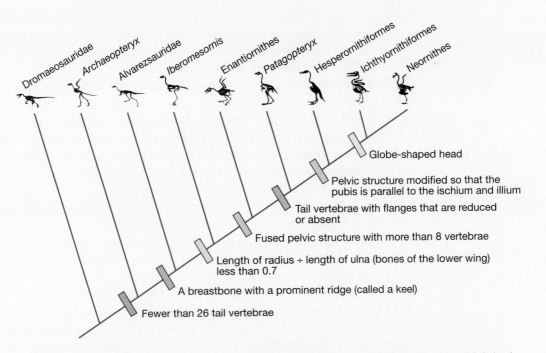

Figure 13.1 Phylogeny of bird groups from the Mesozoic era Luis Chiappe (1995) used skeletal characters to infer the phylogeny of early bird lineages. *Archaeopteryx, Iberomesornis,* and *Patagopteryx* are genera; all the rest of the taxa indicated are families or higher-level groupings. The synapomorphies listed on this cladogram identify trends in changes to the forelimbs, hindlimbs, breastbone, tail, and pelvis of birds. The fusions, reductions in bone number, and other modifications were undoubtedly favored as adaptations for flight.

Shared traits that are derived from a common ancestor are called synapomorphies.

derived character states are similarities that arose due to a change from the character state in the common ancestor of all members of the group. For example, Figure 13.1 shows that a series of these shared derived character states, or **synapomorphies**, defines phylogenetic relationships among groups of birds that lived during the Mesozoic Era. These synapomorphies identify trends in changes to the forelimbs, hindlimbs, breastbone, tail, and pelvis of birds. The fusions, reductions in bone number, and other modifications were undoubtedly favored as adaptations for flight.

If similarities among organisms occurred only because of descent from a common ancestor, then inferring a phylogeny for any group of species would be a simple problem of combining progressively larger sets of species together based on their synapomorphies, as in Figure 13.1.

Unfortunately, things are not nearly this simple in practice. For example, species can have similar traits because those traits evolved independently in each of the lineages leading to those species. Flippers are a derived trait found in both penguins and seals, but they did not evolve through modification in the common ancestor of penguins and seals. Instead, the similarity we observe in the forelimbs of penguins and seals results from natural selection that favored the same type of structure in ancestors from very different lineages (birds and mammals). This is called **convergent evolution**. We can be fairly sure that this similarity is due to convergence because many other traits (other than the shape of the forelimbs) unite penguins with other birds and seals with other mammals. From this point of view, the phylogenetic information from the forelimb characteristics of penguins and seals disagrees with information from many other characters.

In addition, derived traits can also revert to the ancestral form due to mutation or selection. Events like these remove similarity caused by descent from a common ancestor. A **reversal** wipes out the phylogenetic signal and restricts our ability to estimate evolutionary relationships.

Convergence and reversal are lumped under the term **homoplasy**. Homoplasy represents noise in the data sets used to reconstruct phylogenies, and it will arise whenever some characters give conflicting information about relationships among a group of species. Phylogeneticists try to minimize the confusing influence of homoplasy by choosing characters that evolve slowly relative to the age of the groups involved and that show few instances of apparent convergence or reversal. But given that such noise will be present in data used to infer a phylogeny, how can the true phylogeny still be estimated, and how do we decide the strength of support for a particular phylogenetic conclusion?

Parsimony provides one approach for identifying which branching pattern, among the many that are possible, most accurately reflects evolutionary history. Under parsimony, the preferred tree is the one that minimizes the total amount of evolutionary change that has occurred. The rationale for invoking parsimony is simple and compelling. In many instances, we can assume that convergence and reversal will be rare relative to similarity that is due to modification from a common ancestor (but see Felsenstein 1978, 1983). In the terminology of Box 13.1, synapomorphies are usually more common than convergence and reversals. It makes sense, then, that the most parsimonious tree will minimize the amount of homoplasy inferred for the data, and this will be the best estimate of the true phylogenetic relationships among the species being studied. In the following sections, we use recent analyses of whale phylogeny to illustrate the process of inferring this best phylogeny by parsimony.

13.2 The Phylogeny of Whales

The whales, dolphins, and porpoises (called Cetacea) share a set of features that are unusual among the mammals, such as loss of the posterior limbs. The relationships between the cetaceans and other mammal groups are difficult to discern because whales are highly evolved for aquatic life. Several studies, however, have placed cetaceans as a relative of the ungulates (including horses, rhinoceroses, deer, pigs, antelope, and camels) as in Figure 13.2a (Flower 1883; Simpson 1945; Novacek 1993). The oldest fossils that can be recognized as whales come from Eocene rocks in the Himalayas that are about 55 million years old. These archaeocete whales had some terrestrial traits, such as hind limbs (Thewissen and Hussain 1993; Bajpai and Gingerich 1998), and resembled an extinct group of amphibious mammals called the mesonychians (Thewissen et al. 1994).

Recent analyses of DNA sequences and other molecular characters have challenged this view of whale relationships. These molecular studies strongly suggest that whales are not merely related to the ungulates but are, in fact, close relatives of one particular group of ungulates: the hippopotamuses (Figure 13.2b). Notice that the two trees in Figure 13.2 differ only in the location of the whale branch, and that otherwise the relationships between the artiodactyl groups are not different.

This conclusion would upset our traditional understanding of mammalian evolution and classification. For example, it would suggest that some traits of hippos and whales, thought to be convergent adaptations for aquatic life, are perhaps shared

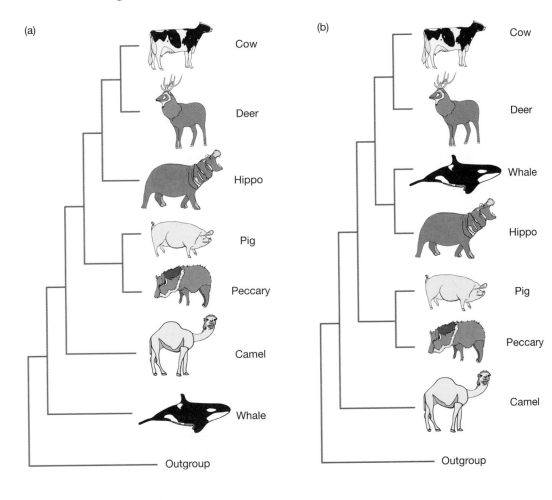

(a)

Cow

Deer

Hippo

Pig

Peccary

Camel

Whale

Outgroup

(b)

Cow

Deer

Whale

Hippo

Pig

Peccary

Camel

Outgroup

Figure 13.2 Phylogenetic hypotheses for whales and other mammals The tree in (a) shows the Artiodactyla hypothesis: Whales and dolphins are related to the ungulates, possibly as the sister group to the artiodactyls (represented by cows, deer, hippos, pigs, peccaries, and camels). The outgroup to these species is from the ungulate group called Perissodactyla (horses and rhinos). The tree in (b) shows the whale+hippo hypothesis. It is identical to (a) with one exception: the branch leading to whales is moved so that whales are the sister group to the hippos.

derived characters. How do we test these competing ideas about whale phylogenetic relationships?

Choosing Characters: Morphology and Molecules

The first task of any phylogenetic analysis is choosing which characters to use as data. Like many other phylogenetic problems, the phylogeny of whales has been studied using two very different types of characters: morphological traits (especially those of the skeleton) and molecular traits (including allozymes, DNA restriction sites, immunological similarity, and especially DNA sequences). Morphological traits are essential in the case of extinct species found only in museum collections or as fossils. To reduce homoplasy in the data, the homology of morphological similarities can often be assessed by looking at the embryological origins of similar structures. A disadvantage of morphological characters is that scoring a single morphological trait for a group of species may require slow, painstaking work by a highly trained taxonomic expert.

Morphological and molecular traits each have advantages and disadvantages when used to infer evolutionary relationships.

Molecular characters (especially DNA sequences) have other advantages and disadvantages. Nucleotides may be scored rapidly in nearly limitless numbers in different genes, and molecular biologists have developed sophisticated models to predict how sequences change through time. However, homoplasy in molecular similarities can be difficult to identify, and DNA sequence characters are limited to just four character states (A, C, G, and T). These different features often lead phylogeneticists to use both morphological and molecular characters whenever possible to analyze relationships.

Parsimony with a Single Morphological Character

The ungulates are traditionally divided into two monophyletic taxa: hippos, cows, deer, pigs, giraffes, antelope, and camels (called Artiodactyla) and the horses and rhinoceroses (called Perissodactyla). The Artiodactyla are grouped together on the basis of some skull and dental characteristics, but most notably by the form of a bone in the ankle called the astragalus (Prothero et al. 1988; Milinkovich and Thewissen 1997). In all artiodactyls, the astragalus has an unusual shape. Both ends of the bone are smooth and pulley-shaped (Figure 13.3). This shape allows the foot

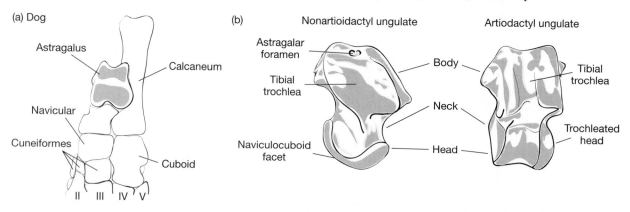

Figure 13.3 **The astragalus defines artiodactyl mammals as a taxon** (a) The astragalus (in this case, of a dog) is the highest bone in the ankle, around which the foot rotates to extend forward or backward. Numbers II–V indicate the four digits. From Thewissen and Madar (1999). (b) The astragalus of an artiodactyl (right) and a nonartiodactyl (left) ungulate. In the artiodactyl, both ends of the astragalus are pulley-shaped. From Schaeffer (1948). Copyright © 1948 Evolution. Reprinted by permission of Evolution.

to rotate in a wide arc around the end of the ankle, and contributes to the long stride and excellent running ability of most artiodactyls.

This shared derived character state is one reason some morphologists have rejected the idea that hippos and whales could be sister groups (Luckett and Hong 1998). The straightforward logic is illustrated in Figure 13.4. If hippos and other Artiodactyla form a monophyletic group, then the pulley-shaped astragalus evolved just once without subsequent changes (a single change in character state), as shown in Figure 13.4a. On the other hand, if whales are a sister group to the hippos, then the origin of the pulley-shaped astragalus (one evolutionary event or step) was followed by the loss of this synapomorphy in the lineage leading to whales (a second step), as seen in Figure 13.4b. Based on this character, the whale + hippo hypothesis (Figure 13.2b) is less parsimonious than the Artiodactyla hypothesis (Figure 13.2a) because it implies one extra step in ankle bone evolution. This kind of inference is the heart of a phylogenetic analysis based on parsimony: comparing possible alternative trees, and concluding that the tree implying the fewest evolutionary changes is the tree most likely given the available data.

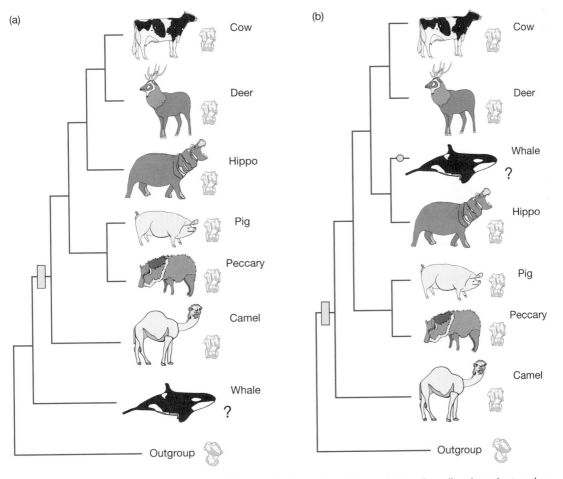

Figure 13.4   Using the astragalus as a phylogenetic character   The evolution of a pulley-shaped astragalus can be reconstructed by finding the smallest number of changes implied by each possible phylogenetic tree. In the Artiodactyla hypothesis (a), the fewest changes are implied if the pulley-shaped astragalus evolved once, at the point marked by the pink bar. In the whale + hippo hypothesis (b), the simplest reconstruction is that the pulley-shaped astragalus evolved once and was lost in the lineage leading to modern whales (marked by the blue dot), implying at least two changes in ankle morphology. There is a question mark by the whales because they do not have ankles.

Living whales have no ankles, so the shape of the whale astragalus as a possible artiodactyl trait cannot be assessed. However, some fossil whales have hindlimbs (Gingerich et al. 1990). Johannes Thewissen and Sandra Madar (1999) found fossil ankle bones in the same deposits containing the oldest archaeocete whales and compared these to the ankles of living and extinct artiodactyls. They concluded that some features of the pulley-shaped astragalus are, in fact, found in the archaeocete whales. This conclusion confirms that the whales are descended from an artiodactyl ancestor. However, this conclusion is vigorously disputed by some morphological systematists (Luckett and Hong 1998; O'Leary and Geisler 1999), who question the origin of these bones and suggest that they may belong to some other artiodactyl and not to a whale. In general, morphological traits tend to support the Artiodactyla hypothesis (Luckett and Hong 1998; O'Leary and Geisler 1999).

Parsimony with Multiple Molecular Characters

How do we extend this parsimony method to the analysis of multiple characters, especially when homoplasy is likely to appear among the data? The answer is deceptively simple. Each character is treated independently, mapped onto each of the possible trees, and the most parsimonious pattern of character change is noted for each character on each tree. Then, the number of changes is summed across characters for each tree, and the total number of changes is assessed. Under parsimony, the best tree is the one that implies the fewest character state changes across all characters.

John Gatesy and his colleagues (1999) summarized the available molecular evidence on whale relationships to other mammals. They assembled all of the DNA sequence data relevant to this question into a single data set for four whales and eight artiodactyls, including a hippo, and an outgroup from the Perissodactyla. Their analyses of this data set strongly supported the whale + hippo hypothesis, and they dubbed this collection of molecular characters the WHIPPO-1 data set (for WHale-hIPPO).

Sixty of those characters for a subset of the taxa in the WHIPPO-1 data set are shown in Figure 13.5. They correspond to nucleotide sites 141-200 in Gatesy's alignment of DNA sequences for a milk protein gene called beta-casein. We use these eight taxa and 60 characters in the following sections, in order to illustrate the way that multiple characters (in this case, nucleotides) are used to infer a phylogeny.

Of the 60 characters in Figure 13.5, 15 group 2 or more taxa. As a result, these sites contain phylogenetic information. The others are invariant (like site 142, for which all taxa have G) or variable but uninformative (like site 192, for which all taxa, including the outgroup, have C, except the camel, which has G).

The parsimony informative characters provide support for clades among the whales and artiodactyls. For example, site 162 provides a synapomorphy for a clade including the cow, deer, whale, and hippo sequences (all with T, a shared difference from the outgroup and all other artiodactyl sequences which have C). Site 166 provides a synapomorphy (the only one in this example) for a clade consisting of hippos and whales (the only sequences with C at this site). However, not all informative characters support the same groupings. Site 177 provides a synapomorphy for a clade of whales, hippos, pigs, and peccaries, excluding the cow and deer sequences. This synapomorphy conflicts with site 162, and indicates that reversal or convergence has resulted in homoplasy at one of these sites that does not reflect the evolutionary history of artiodactyls. In most cases, this is the only way that

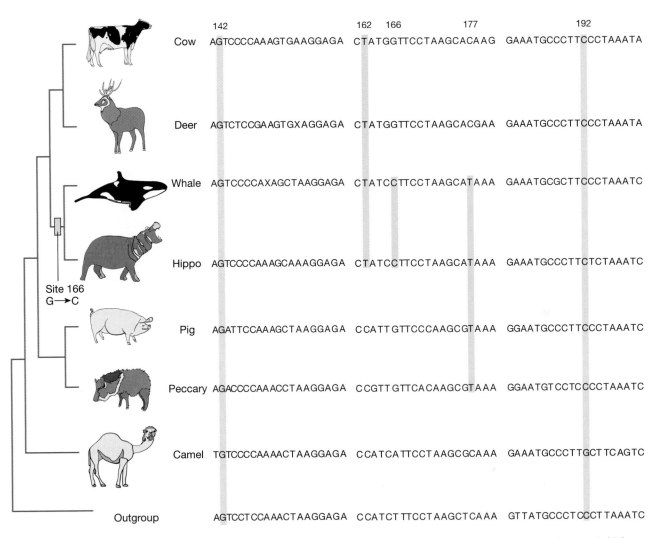

Figure 13.5 Sequence data for parsimony analysis Sixty nucleotides of aligned sequence from the beta-casein gene (which encodes a milk protein) of six artiodactyls, a whale (the dolphin *Lagenorhynchus obscurus*), and a perissodactyl. A, C, G, and T denote the nucleotides at homologous sites from 141 to 200 in the alignment of Gatesy et al. (1999). X indicates an ambiguously identified nucleotide. Some sites are invariant or uninformative (shaded blue), while others provide synapomorphies that define clades (shaded red). The phylogeny is based on a parsimony analysis of these nucleotide synapomorphies. One synapomorphy (at site 166) defines a clade of whales and hippos. Notice that not all synapomorphies agree with each other, indicating that some homoplasy has occurred in the evolution of these nucleotides.

homoplasy in molecular characters can be discovered: by comparison to other characters and the phylogeny that is implied by them.

We use the information from these characters to choose between alternative phylogenies by, first, finding the most parsimonious reconstruction of character change for each character on each tree. Then we add the total number of character changes. For the Artiodactyla hypothesis (Figure 13.2a), 47 nucleotide changes are implied by this tree for the data shown in Figure 13.5; for the same data, the whale + hippo hypothesis implies only 41 changes. This difference between the two trees can be traced back to six characters in Figure 13.5: nucleotide sites 151, 162, 166, 176, 177, and 194. For each of these characters, the whale + hippo hypothesis implies one fewer change than does the Artiodactyla hypothesis. As an exercise, try to find the most parsimonious reconstruction for each of these six characters on both trees (as we did for the astragalus in Figure 13.4), and satisfy yourself that Figure 13.2a is, in fact, six steps longer than Figure 13.2b. For these 60 characters and eight taxa, the whale +hippo hypothesis is more parsimonious than the Artiodactyla hypothesis and is, therefore, the preferred tree.

Searching Among Possible Trees

So far, we have restricted ourselves to comparing two specific phylogenetic hypotheses. However, the true number of possible trees is actually much larger, and the task is to evaluate the many possible tree topologies and find the best one. This is difficult, because the number of tree topologies that are possible becomes astonishingly large, even in a moderately sized study. When four species are included, only three different branching patterns are possible. Adding a fifth species to the data set makes the number of possible topologies jump from 3 to 15. A sixth taxon leads to 105, a seventh to 945, and for the eight taxa in Figure 13.2, there are 10,395 possible trees. However, it is fairly routine now for studies to include 50 taxa. In this case, an incomprehensibly large number of different trees is possible. Obviously, so many alternative trees cannot be evaluated manually as we have done so far. Fast computers are required to automate the task. Three different approaches can be used to search among all of these possible trees in order to find the most parsimonious one.

When the number of taxa in a study is relatively low—typically fewer than 11—it is feasible to employ computer programs that evaluate all of the possible trees. This strategy is called an exhaustive search. Because it guarantees that the optimal tree implied by a particular data set will be found, the approach is called an exact method. An exhaustive search of all 10,395 trees for the data in Figure 13.5 produced a single shortest phylogeny of 41 steps identical to the tree in Figure 13.2b. This is the same branching pattern among these eight sequences as that found by Gatesy et al. (1999), and it supports the whale + hippo hypothesis.

However, the original WHIPPO-1 data set consists of 13 taxa (including three other whale sequences) and about 8,000 molecular characters. Exhaustive searching of this data set is prohibitively slow, so Gatesy et al. (1999) used two other methods for searching among the possible trees. These methods take advantage of some logical or computational shortcuts. They search only some parts of the landscape of all possible trees, while maximizing the probability of finding the most parsimonious tree.

Analyses of DNA sequence data support a close relationship between whales and hippos.

In this case, both methods produced two most-parsimonious trees, both of which are consistent with the phylogeny in Figure 13.2b. This consistency among search methods indicates that the close relationship between whales and hippos is not an artifact of incomplete searching or of the inability of search methods to find the most parsimonious solutions.

Evaluating the Trees

Having compared several (or all) possible trees, we now need to ask: How good is the most parsimonious tree? In particular, how sure can we be that the node joining whales with hippos is reliably supported by the data? Is this tree really substantially better than one in which whales are not a sister group to the hippos?

Many investigators examine the topologies of trees close to the optimal tree visually, and make an informed judgment about how different they are from it. If the nearly optimal trees differ from the optimal tree in only minor ways, we can be more confident in using the most optimal tree to draw conclusions about the group's evolution. In addition, computer programs can evaluate multiple trees and create a consensus representing the topology supported by all of the nearly optimal trees.

The reliability of most-parsimonious trees can also be evaluated statistically (see Bremer 1994; Swofford et al. 1996; Huelsenbeck and Rannala 1997). One approach to this problem is called **bootstrapping** (Felsenstein 1985). In bootstrapping, a computer creates a new data set from the existing one by repeated sampling. For example, if there are 300 base pairs of sequence in a study, the computer begins the bootstrapping process by randomly selecting one of the sites and using it as the first entry in a new data set. Then, it randomly selects another site, which becomes the second data point in the new data set. (There is a 1/300 chance that this second point will be the same site as the first.) The computer keeps resampling like this until the new data set has 300 base-pairs of data, representing a random selection of the original data. This new data set is then used to estimate the phylogeny. By repeating this process many times, the investigator can say that particular branches occur in 50%, 80%, or 100% of the trees estimated from the resampled data sets. The more times a branch occurs in the bootstrapped estimates, the more confidence we have that the branch actually exists. If bootstrap support for a particular branch is low, say under 50%, an investigator will usually conclude that she cannot determine the branching pattern in that part of the tree, and will therefore collapse that particular branch into a polytomy (a point of uncertainty) in the published tree.

Resampling the data in Figure 13.5 through bootstrapping indicates strong support for the whale-hippo clade. Of 1000 resampled data sets, 71% included a whale-hippo clade (more than any other branching arrangement). When all taxa and all molecular characters in the WHIPPO-1 data set were analyzed, bootstrap support for this node approached 100% (Hillis 1999). In some analyses of known phylogenies (where investigators bred laboratory organisms and split their populations to create lineages of known relatedness), bootstrap support of around 70% or greater was usually associated with the identification of the true phylogeny (Hillis and Bull 1993). From this perspective as well, the close relationship between whales and hippos seems sound.

A second approach is to use a phylogenetic method other than simple parsimony. Two such alternatives, called **maximum likelihood** and **genetic distances**, are described in Box 13.2. These methods use different assumptions about how characters evolve and different methods for joining similar taxa together in the search for the best phylogeny. When these very different methods agree on the same best phylogeny, this gives some degree of confidence that the best tree has actually been found (Huelsenbeck and Hillis 1993; Hillis et al. 1994). When they conflict, we are left to make a judgment about which criterion for choosing among trees is most valid in the case being analyzed. We face a similar situation when comparing trees inferred from entirely different data sets. Trees produced by different studies are treated as competing estimates. When they conflict, researchers have more confidence in trees estimated with larger data sets and from characters thought to be less subject to homoplasy.

Bootstrapping is a technique for evaluating which branches on a particular tree are more well-supported than others.

BOX 13.2   Alternatives to parsimony: Maximum likelihood and genetic distances

One drawback of parsimony methods is that they do not use all of the information available to an investigator. Recall, from the whale example, that only 15 of 60 nucleotide sites were useful in a parsi-

mony analysis. To use information more efficiently and to create the possibility of using statistical tests to choose the best tree among the many possible trees, Joseph Felsenstein (1981) introduced an inno-

	Cow	Deer	Whale	Hippo	Pig	Peccary	Camel
Deer	0.073						
Whale	0.150	0.197					
Hippo	0.148	0.197	0.053				
Pig	0.264	0.270	0.197	0.217			
Peccary	0.340	0.412	0.266	0.287	0.129		
Camel	0.284	0.347	0.216	0.236	0.291	0.340	
Outgroup	0.306	0.340	0.241	0.261	0.311	0.306	0.210

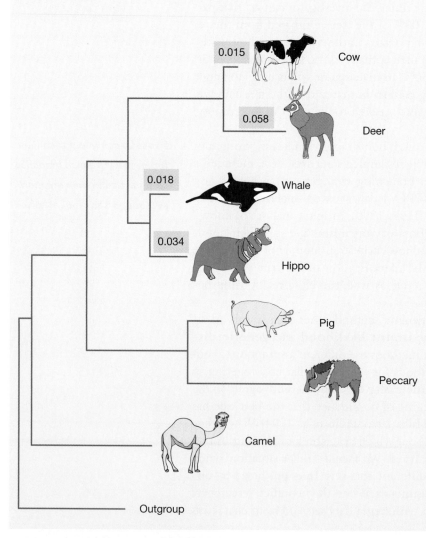

Figure 13.6   Genetic distances for cluster analysis   Each entry in this table is a genetic distance between a pair of taxa, calculated from the sequence data in Figure 13.5. The phylogeny below was produced by a clustering analysis of these genetic distances. Notice that pairs of taxa, such as the cow and deer (blue), or the whale and hippo (red), with low genetic distances are grouped as sister taxa. The lengths of the branches are proportional to the expected proportion of nucleotide differences between groups (also shown numerically for several branches).

BOX 13.2   Continued

vative and powerful approach called maximum likelihood estimation.

The essence of the likelihood approach for molecular phylogenetics is to ask the question: Given a mathematical formula that describes the probability that different types of nucleotide substitution will occur, and given a particular phylogenetic tree with known branch lengths, how likely am I to obtain this particular set of DNA sequences? To implement this strategy, a computer program evaluates each tree topology and computes the probability of producing the observed data, given the specified model of character change. The sum of the probabilities of getting each branch represents the probability of getting the observed data, if the tree is correct. This probability is reported as the tree's likelihood. The criterion for accepting or rejecting competing tree topologies, then, is to choose the one with the highest likelihood. Unfortunately, likelihood methods are computationally slow, and very large data sets cannot be analyzed as completely by this technique as they can using much faster parsimony methods.

A radically different approach is to convert discrete character data (such as the presence or absence of a morphological trait, or the identity of a nucleotide at a homologous location in a gene) into a distance value (Swofford et al. 1996). For example, the percentage of nucleotide sites that differ between two taxa can be computed (a 10% difference means that an average of 10 nucleotides have changed per 100 bases). This conversion of discrete characters to a single distance measure results in loss of specific information, but it can capture the overall degree of similarity between taxa.

As in likelihood methods, distance analyses require the investigator to assume a model of character evolution in order to convert information from multiple characters into one composite measure of the distance between two taxa. One commonly used formula for DNA sequences corrects for multiple substitutions at the same site and for differences in the frequency of transition and transversion substitutions (Kimura 1980; Wakely 1996).

To estimate a phylogeny from distance data, computer programs are used to cluster taxa, so that the most similar forms are found close to one another on the resulting tree. This general strategy, based on clustering taxa according to their similarities, is called a phenetic approach (Sneath and Sokal 1973). The preferred tree is the one that minimizes the total distance among taxa. Several different clustering algorithms are commonly used that make more or less restrictive assumptions about the nature of the distances analyzed.

An example of this approach is shown in Figure 13.6. The top part of the figure shows the matrix of pairwise genetic distances among the sequences in Figure 13.5, calculated using a formula devised to account for multiple substitutions at the same sites (Kimura 1980). Small genetic distances should indicate recent divergence from a common ancestor and close phylogenetic relationships. Notice that the smallest genetic distances are those between the whale and hippo sequences and between the deer and cow sequences. Note also that genetic distances among these four sequences are all relatively small. The bottom part of the figure shows that a clustering analysis groups whales and hippos together as sister taxa based on these genetic distances, with cows and deer as their next nearest relatives.

Resolving Character Conflict

We began by contrasting the results of morphological and molecular character analysis for whale phylogenies. The two sets of results appear to be in strong conflict. Gatesy et al. (1999) attempted to include both morphological and molecular data in one analysis, but they could not include the wealth of morphological characters from fossils without scoring all of the molecular characters as missing for the fossil taxa. What are we to do in such a situation? The most prudent act is to wait for more data. For example, better fossil sampling may reveal character states in extinct artiodactyls and whales that would support grouping whales with hippos (or other artiodactyl lineages). On the other hand, molecular data for a larger number of artiodactyl groups might change the inferred order of character state changes and thereby support removing whales from within the artiodactyls.

A 'Perfect' Phylogenetic Character?

One way in which the passage of time tends to help resolve phylogenetic issues such as the whale-artiodactyl problem is the discovery of new kinds of phylogenetic characters. Recently, a new and potentially ideal molecular character has been applied to the problem of whale phylogeny (Milinkovitch and Thewissen 1997). SINEs and LINEs (for Short or Long INterspersed Elements) are parasitic DNA sequences that occasionally insert themselves into new locations within a host genome. The presence or absence of a particular SINE or LINE at a homologous location in two different host genomes can be used as a trait in phylogeny inference. David Hillis (1999) outlined the potential advantages of such a molecular phylogenetic character. Transposition events (in which the sequence inserts itself in a new location in the host genome) are relatively rare, so that two homologous SINEs in two independent host lineages are unlikely to insert into exactly the same place in their host genomes. This kind of convergence is possible, but it is probably extremely unusual. As a result, convergence in SINE or LINE characters should be rare. Reversal to the ancestral condition is also unlikely, because the deletion of a SINE or LINE can be detected as the loss of parts of the host genome along with the loss of the interpersed element. These features of SINE and LINE genetics allow a phylogeneticist to tell the difference between the ancestral absence of a SINE or LINE and the secondary loss of one of these elements. If convergence and reversal are rare or can be identified, then homoplasy is an unlikely source of SINE or LINE similarity. Without homoplasy, SINEs and LINEs should give excellent (though perhaps not perfect) insight into some phylogenetic relationships.

Such is the case for whales. Interspersed elements in the genomes of whales and artiodactyls were used by Masato Nikaido and colleagues (1999) in a test of the whale + hippo phylogenetic hypothesis. They analyzed 20 different SINE or LINE insertions and found no homoplasy at all! Look carefully at the matrix of SINE or LINE presence/absence data shown in Figure 13.7 for representatives of the same taxa analyzed using DNA sequences. Notice that variation in presence or absence of an element at each locus corresponds to exactly one clade in the phylogeny. This phylogeny is identical to the tree produced by sequence analysis. Four different insertions are shared by whales and hippos alone. Notice also that—unlike the DNA sequence characters in Figure 13.5—none of the SINE or LINE characters are in conflict with each other. This is similar to the pattern of synapomorphies used to define Mesozoic bird clades by Chiappe (Figure 13.1). The shared presence of SINEs or LINEs provides a perfect window into whale phylogeny and strongly corroborates the conclusion from sequence studies indicating that whales and hippos are close relatives.

13.3 Using Phylogenies to Answer Questions

The first two sections of this chapter focused on the methods of phylogeny inference. The fundamental message of this analysis is that estimating evolutionary relationships requires a series of careful decisions about which data are appropriate for the task and how they should be analyzed. Now we turn to the issue of how evolutionary trees can be used to answer interesting questions.

Locus	1	2	3	4	5	6	7	8	9	10	11	12	13	14	15	16	17	18	19	20
Cow	0	0	0	0	0	0	0	1	1	1	1	1	1	1	1	1	1	1	0	0
Deer	0	0	0	0	0	0	0	1	?	1	1	1	1	1	1	?	1	1	0	0
Whale	1	1	1	1	1	1	1	0	?	1	0	1	1	0	0	0	?	1	0	0
Hippo	0	?	0	1	1	1	1	0	1	1	0	1	1	0	0	0	?	1	0	0
Pig	0	0	0	?	0	0	0	0	?	0	0	0	?	?	0	0	0	1	1	1
Peccary	?	?	?	?	?	?	?	?	?	?	?	?	?	?	?	?	?	?	1	1
Camel	0	0	0	0	0	0	0	0	0	0	0	0	0	0	0	0	0	0	0	0

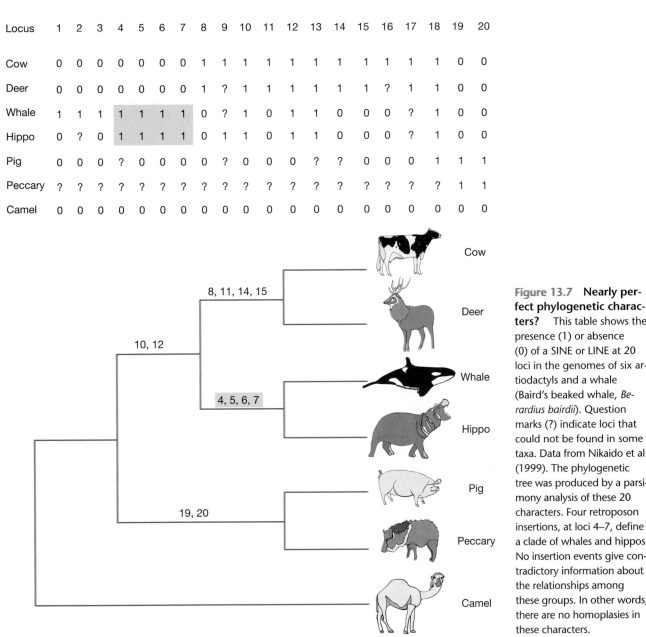

Figure 13.7 Nearly perfect phylogenetic characters? This table shows the presence (1) or absence (0) of a SINE or LINE at 20 loci in the genomes of six artiodactyls and a whale (Baird's beaked whale, *Berardius bairdii*). Question marks (?) indicate loci that could not be found in some taxa. Data from Nikaido et al. (1999). The phylogenetic tree was produced by a parsimony analysis of these 20 characters. Four retroposon insertions, at loci 4–7, define a clade of whales and hippos. No insertion events give contradictory information about the relationships among these groups. In other words, there are no homoplasies in these characters.

Research programs based on estimating and interpreting evolutionary trees are growing rapidly (Hillis 1997). In the remainder of this chapter, our goal is to sample the diversity of applications for phylogenetic thinking. This discussion will set the stage for reviewing the history of life in Chapters 14 through 16.

Rates of Change: Radiation of Hawaiian Fruit Flies

Some of our most basic questions about the history of life concern when major events occurred, and how rapidly. In Chapter 15, we introduce events that have been dated from the fossil record. Here, we look at how data on rates of divergence in a protein were used to time a major episode in evolution.

The 500-plus species of fruit flies endemic to the Hawaiian archipelago rank as one of the most spectacular of all sequences of speciation and divergence events known. The oldest island, Kauai, is 5–6 million years old, implying that speciation was extraordinarily rapid. But divergence times estimated from molecular data suggest a very different view. (Box 13.3 explains how divergence times can be estimated from molecular data—a phenomenon known as the **molecular clock**.)

To determine when flies first colonized the Hawaiian archipelago, Stephen Beverly and Allan Wilson (1985) estimated how extensively larval hemolymph

---

BOX 13.3   The molecular clock

Many of the uses of a phylogeny (and many questions about the evolution of organisms) have inspired evolutionary biologists to ask about the age of particular evolutionary events or taxonomic groups. Fossils give direct minimum estimates of the ages of clades in a phylogeny. However, if the characters used for phylogenetic analysis could themselves be shown to evolve with clock-like regularity, then ages of nodes in a tree could be estimated from the number or amount of character change without direct observations from the fossil record. This hypothesis, called the molecular clock, originated with Emile Zuckerkandl and Linus Pauling (1962) and was promoted and elaborated throughout the 1970s and 1980s in a series of papers from Allan Wilson's laboratory. (See Wilson 1985.) The neutral theory of molecular evolution (see Chapter 18) later provided a theoretical basis for expecting that certain nucleotide changes should accumulate at a rate equal to the mutation rate. If this mutation rate does not change much over time, and if generation times are similar, then the number of neutral molecular differences between two taxa should be proportional to the age of their most recent common ancestor. The prospect of estimating the age of nodes in a molecular phylogeny from the molecular data themselves has tremendous appeal. The idea was to calculate the rate of molecular divergence in groups where the time of divergence was known from the fossil record and then use this calibration to date divergence times in groups with no fossil record.

How fast does a particular molecular clock run? Eldredge Bermingham and Harilaos Lessios (1993), for example, estimated mitochondrial DNA (mtDNA) sequence divergence in sister species of sea urchins found on either side of the Isthmus of Panama. Because the land bridge between North and South America separated these populations about 3 million years ago, the authors were able to calibrate the amount of sequence divergence for absolute time. The urchins provide an estimate of 1.8% to 2.2% divergence per million years for a protein-coding gene—remarkably close to the rate of mtDNA-sequence divergence of mammals (Brown et al. 1979). Similar mtDNA calibrations have been made in butterflies (Brower 1994) and geese (Shields and Wilson 1987).

Many such clocks have eventually proven to be unreliable. Nancy Knowlton and her colleagues (Knowlton et al. 1993; Knowlton and Weigt 1998) examined similar mtDNA sequence data, as well as allozyme data for pairs of sister species of snapping shrimp (genus *Alpheus*). These shrimp had also been separated from each other by the Panamian land bridge. We discussed this same example as a study of speciation through a vicariance event in Chapter 12. Knowlton and her colleagues found that pairs of sister species varied widely in their genetic divergence. Because the genetic differences for mtDNA were correlated with allozyme differences within pairs of *Alpheus* species, the variation among pairs of species probably arose because some pairs were isolated on either side of the rising land bridge up to 15 Ma before the final closure of the seaway. Knowlton concluded that many other studies using the Panamian land bridge or other biogeographic events may provide poor estimates of divergence dates based on sequence differences. This is because the ages of the biogeographic events do not correspond closely to the actual times when lineages of organisms became separate from each other and stopped interbreeding. Other departures from clock-like molecular evolution may arise from variation in mutation rate (Martin et al. 1992; Martin and Palumbi 1993) or generation time (Martin and Palumbi 1993; Hillis et al. 1996). As a result of such problems, researchers are cautious about applying molecular clocks to interpret phylogenies (Hillis et al. 1996).

protein (LHP) has diverged in a series of Hawaiian flies and outgroup species. They used these data to estimate the phylogeny of the Hawaiian flies and outgroup species (Figure 13.8). Beverly and Wilson (1984) could also estimate the *rate* of evolution of LHP, from comparisons of LHP differences between other fly groups in which the time of divergence was already known. In other words, they could calibrate a molecular clock for fly LHP (see Box 13.3). This allowed them to date each of the major nodes on the phylogenetic tree for Hawaiian flies. The analysis pointed to a startling conclusion: The amount of LHP difference among Hawaiian flies (compared to LHP differences among flies with known age of divergence) implied that Hawaiian flies first colonized the archipelago some 42 million years ago. This was long before the existing Hawaiian islands were formed.

How can this be? Recall, from our discussion of fly speciation in Chapter 12, that the Hawaiian islands originated from a hot spot under the Pacific plate. These volcanic islands drift to the north and west after they are formed and gradually erode, eventually becoming underwater mountains called seamounts. Many of the landforms in the chain have been dated using the radiometric dating techniques introduced in Chapter 2, confirming the old-to-new sequence. In Figure 13.8, this picture of island formation is matched to the fly phylogeny, the LHP divergence data, and the dates implied by the molecular clock.

Taken together, the geographical, phylogenetic, and clock data offer a logical resolution to the paradox of "young islands, old flies." *Drosophila* first colonized the Hawaiian chain when the islands that are now in the Koko seamount were over the hot spot. As these older islands eroded and new islands were formed, the flies hopped from the old landforms to new. This "ancient origin" hypothesis for Hawaiian flies has now been confirmed by investigators using different gene sequences and independent clock calibrations (see DeSalle and Hunt 1987). Hawaiian crickets, in contrast, have radiated solely on the current group of islands (Shaw 1996).

The Age of Clades

When the origin and duration of a lineage is well-documented in the fossil record, branching points on a phylogeny can be directly aligned with an absolute time scale. This combination of phylogenetic and geological information can lead to some surprising results. Luis Chiappe (1995) was able to use dates from the fossil record to scale his cladogram of Mesozoic era bird groups for time (Figure 13.9). Adding information on time enriches the interpretation of this phylogeny. It is clear that several of the major bird lineages disappeared 65 million years ago, during the mass extinction event at the Cretaceous–Tertiary boundary. Even more interesting is that the branching patterns and the chronology do not align closely in every case. Look again at the tree in Figure 13.9. Note that the early dates documented for *Archaeopteryx* and the group called Enantiornithes dictate that branches 1–4 must be placed deep in time. But fossil material that old had not been found from the derived groups called Dromaeosauridae, Alvarezsauridae, and the genus *Iberomesornis*. The placement of branches 5–7 in the early Cretaceous is also startling, because fossil material that old has not yet been found. The point is that many of the branches on this tree have to be much longer than the current fossil record indicates (that is, they have to go further back in time.) The cladogram thus makes a strong prediction: Older representatives of these three fossil taxa will eventually be

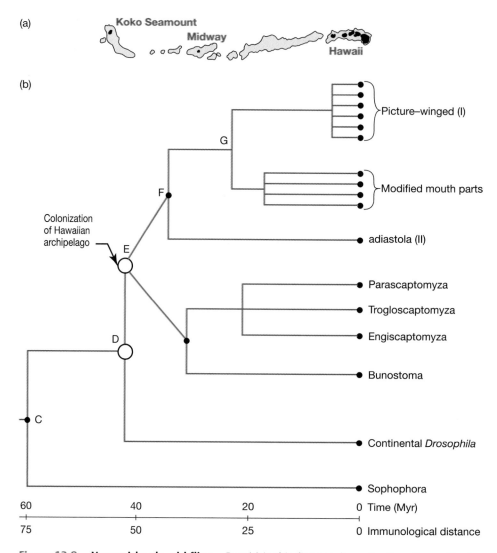

(a)

Koko Seamount
Midway
Hawaii

(b)

Picture–winged (I)

G

Modified mouth parts

F

adiastola (II)

Colonization
of Hawaiian
archipelago

E

Parascaptomyza

Trogloscaptomyza

Engiscaptomyza

D

Bunostoma

C

Continental *Drosophila*

Sophophora

60		40		20		0 Time (Myr)
75		50		25		0 Immunological distance

Figure 13.8   Young islands, old flies   Part (a) in this diagram shows the Hawaiian archipelago, scaled so that the age of the landforms corresponds to the absolute time scale at the bottom of the figure. Islands are shown in black and the extent of undersea structures is outlined in beige. The phylogeny of fruit flies, part (b), is based on clustering of genetic distance between larval hemolymph proteins (LHP). The branch lengths are scaled so that a molecular-clock calibration for divergence in LHP corresponds to the absolute time scale at the bottom of the figure. The "Picture-winged" through "Bunostoma" designations on the right refer to subgroups of Hawaiian fruit flies. The outgroup called Sophophora is represented by the common household fruit fly, *Drosophila melanogaster*. Node E, then, represents the ancestral population that first colonized the Hawaiian islands. From Beverly and Wilson (1985).

Note that several nodes are reported as polytomies (among species within the picture-winged group, for example). These branch points cannot be distinguished on the basis of LHP distance. Note also that the picture-winged group has radiated within the last 5 million years, solely within the current group of Hawaiian islands.

found. In at least one case, this prediction may have been confirmed. Xing Xu and colleagues (1999) discovered and described an early Cretaceous dromaeosaurid from China. Their discovery extends the fossil record for dromaeosaurids to near the predicted age of this group, based on Chiappe's estimate for the phylogeny.

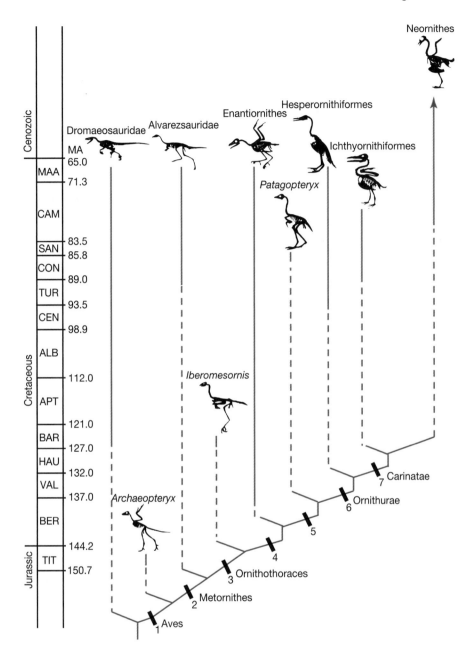

Figure 13.9   A phylogeny of fossil birds, scaled for time   This is the same cladogram given in Figure 13.1, modified to reflect additional information on times of divergence and the duration of branches. The scale at the left of this diagram gives the era or period, stage, and absolute age of the events depicted in this phylogeny. The solid vertical branches leading to each taxon represent intervals where fossil representatives of a lineage have been found; the dashed bars represent intervals where fossil forms are yet to be discovered. From Chiappe (1995).

Classifying the Diversity of Life

The phylogenetic species concept, introduced in Chapter 12, considers every population that occupies a tip on an evolutionary tree as a different species. What does phylogenetic thinking have to say about how we should organize higher-order taxa such as genera, families, orders, and classes?

Traditional classifications group organisms according to similar features. This is called a **phenetic** approach. Phylogenetic schemes for classifying organisms, in contrast, are based on evolutionary relationships. Phylogenetic systematics argues that classification systems should be tree-based, with names and categories that reflect the actual sequence of branching events. According to this point of view, only

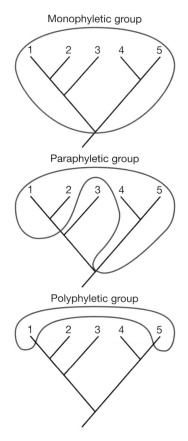

Figure 13.10 Monophyletic, paraphyletic, and polyphyletic groups Taxonomic groups can be monophyletic, paraphyletic, or polyphyletic. Monophyletic groups, or clades, contain all the descendants of a common ancestor. Paraphyletic groups contain some, but not all, the descendants of a common ancestor. A polyphyletic grouping does not contain the most recent common ancestor of all the taxa.

In a cladistic classification, only monophyletic groups are named. In contrast, traditional taxonomic schemes occasionally assign names to polyphyletic and paraphyletic groups.

monophyletic groups, which include all descendants of a common ancestor, are named (Figure 13.10).

Although phenetic and phylogenetic naming schemes frequently lead to the same conclusions, they can produce conflicts (see Schwenk 1994). The relationship between whales and ungulates is one of many recent examples. A cladistic classification would nest whales as a subgroup within the Artiodactyla (Figure 13.5), which in turn are a subgroup of the Mammalia. The important break from traditional mammalian classification in this case would be the reduction of the Cetacea from a high-level taxon (an order) to a low-level group (perhaps a subfamily related to hippos).

Phylogenetic systematics represents an important philosophical break from classical approaches to taxonomy (de Queiroz and Gauthier 1992; Simpson and Cracraft 1995). An increasing number of taxonomists and systematists are calling for a complete overhaul of the traditional phenetic scheme, the goal being to create a phylogenetic classification (see Pennisi 1996). A phylogenetic naming scheme is already being implemented in prominent forums like GenBank, the online database of DNA sequences. However, some organismal biologists disagree. For example, whale specialists might object that a strictly phylogenetic scheme for naming and ranking groups would ignore the many functionally and ecologically important differences between whales and other artiodactyls. Ungulate specialists might also object that classifying whales within the Artiodactyla detracts from the unity of form found among the rest of the artiodactyls (apart from whales).

Coevolution

Coevolution is an umbrella term for interactions among species that result in reciprocal adaptation. We introduced this field of research in Chapter 8, when we discussed predation, mutualism, and parasitism. Here, we introduce how phylogenetic thinking is used in coevolutionary studies. Our focal system is ants that farm fungi.

The 200 species of ants in the tribe Attini are the dominant herbivores in the New World tropics. The group includes the leaf-cutting ants, which dissect pieces from leaves and carry them to their nests. There, they use the leaf material as a substrate for growing fungi in underground gardens. The fungi are harvested and serve as the primary food source for the ant colony. This symbiosis, or mutually beneficial relationship, between ants and fungi is thought to be 50 million years old. This time estimate is based on the first appearance of attine ants in the fossil record. The relationship is also obligate: To the best of our knowledge, none of the symbiotic ant or fungal species can live without the other.

We would like to answer the following question about this association: If the symbiosis originated just once, did the ants and fungi subsequently evolve in tandem? That is, did the two groups "cospeciate"? If so, we would expect the phylogenies of the two groups to be congruent (match up branch for branch).

To answer this question, Gregory Hinkle and colleagues (1994) collected fungi from the nests of five different species of attine ants and sequenced over 1800 base pairs from the coding region for the small subunit of their ribosomal RNA. They used these data to estimate the phylogeny of the attine fungi and several outgroups, using a parsimony criterion. When they matched the fungal phylogeny to a cladogram for the attine ants, they found a close correspondence: Branching patterns were identical in four of the five species compared (Figure 13.11). This is strong support for the idea that the ants and fungi have cospeciated. The fifth fungus-farming ant species, however, posed a dilemma. The sequence data were unable to resolve the relationship of the fungus used by the ant *Apterostigma collare* relative to free-living and symbiotic forms. If this fungus branched earlier than the free-living forms in the polytomy shown in Figure 13.11, it breaks the pattern of cospeciation. If so, it means that ants acquired fungal symbionts more than once, or that some fungi have escaped from the farms to resume a free-living lifestyle.

A study performed by Ignacio Chapela et al. (1994) has helped clarify the history of mutualism. These researchers sequenced the 28S rRNA gene in 37 species of ant-associated and free-living fungi and matched the branching pattern inferred from these data to the same ant phylogeny shown in Figure 13.11. Their analysis clearly shows that the *Apterostigma* fungus split earlier than free-living forms in the polytomy we just examined (Figure 13.12). This confirms that strict cospeciation has not been universal in the coevolution of the attine ants and fungi. Early in the evolution of ant–fungi symbiosis, some ants that farmed fungi switched species, picking up new, free-living fungi to "domesticate." Cospeciation has occurred only in the most recently evolved forms.

It is noteworthy that, in at least some of these highly derived ant genera, like *Trachymyrmex* and *Atta*, queens that leave established nests to found new colonies carry a small ball of fungus with them, and use it to start a new garden. This behavior, which should lead to strict cospeciation, has not been found in the older lineages like *Apterostigma* and *Myrmicocrypta*.

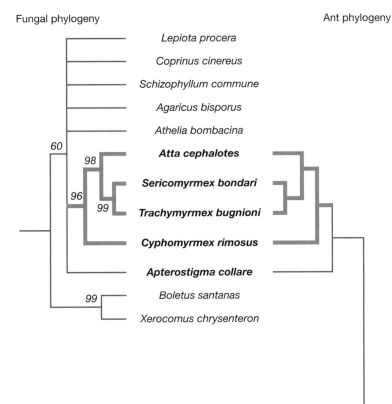

Fungal phylogeny

Ant phylogeny

Lepiota procera

Coprinus cinereus

Schizophyllum commune

Agaricus bisporus

Athelia bombacina

Atta cephalotes

Sericomyrmex bondari

Trachymyrmex bugnioni

Cyphomyrmex rimosus

Apterostigma collare

Boletus santanas

Xerocomus chrysenteron

Blepharidatta brasiliensis

Figure 13.11 Comparing phylogenies of fungi and their ant symbionts I The left side of this figure shows the phylogeny of five species of fungi cultivated by ants, along with several free-living fungi. The free-living forms are shown in light type and the symbiotic forms in bold type. Note that the free-living forms included in the study are ancestral to the species cultivated by ants. This observation confirms that symbiosis is a derived trait. The numbers indicate the percentage of bootstrapped trees that include given branches; branches that appeared in less than 50% of the bootstrapped replicates were collapsed into polytomies.

The right side of the figure shows the phylogeny of some fungal-farming ants. The topology is based on synapomorphies identified in the morphology of prepupal worker larva. The names in bold type are the names of ants that farm fungi; *Blepharidatta* serves as an outgroup.

The thick red lines highlight the congruence between ants and the fungi they farm. From Hinkle et al. (1994).

Fungal phylogeny   Ant phylogeny

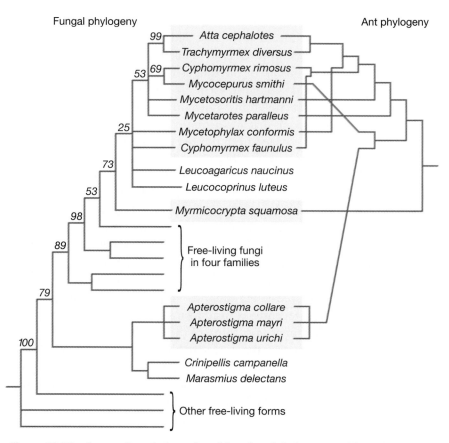

Figure 13.12 Comparing phylogenies of fungi and their ant symbionts II The left side of this figure (Chapela et al. 1994) shows the phylogeny of fungi cultivated by ants, along with free-living fungi. The free-living forms are shown on a light background and the symbiotic forms in a shaded background. This is a strict consensus tree of the 22 equally most-parsimonious trees (meaning that only branches found in all 22 best trees are shown). The numbers indicate the percentage of bootstrapped phylogenies that include given branches. The right side of the figure shows the phylogeny of fungal-farming ants, based on the same cladistic data used in Figure 13.16.

To interpret this (very complicated) diagram, begin by tracing the ant tree up from the base. Note that the first lineage to branch off leads to *Myrmicocrypta squamosa*. If cospeciation were occurring, the next branch off should have led to *Leucocoprinus* or *Leucoagaricus*. Instead it leads to *Apterostigma*. This means that ants "domesticated" free-living fungal species more than once. Now read from the fungal side of the paired phylogenies. Note that *Leucocoprinus* and *Leucoagaricus* are part of an ant-associated clade. This fact suggests that these fungal species escaped domestication and again became free-living.

The Spread of AIDS

In Chapter 1, we used the biology of HIV to illustrate the kinds of questions and answers that occupy evolutionary biologists. Phylogeny inference methods are being used to understand the origins of emerging viruses like HIV and how AIDS and other diseases spread. In the early 1990s, an HIV–positive dentist in Florida was suspected of transmitting the virus to one of his patients. In response, many of his patients were tested for the presence of the virus. Several tested positive. Did they get the virus from the dentist? This was unclear, because some had other risk factors for transmission. To answer the question, investigators from the Centers for Disease Control sequenced the *gp120* gene from viruses in infected patients and

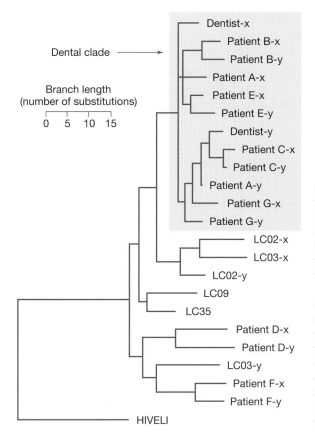

Dental clade →

Branch length
(number of substitutions)

0 5 10 15

Dentist-x
Patient B-x
Patient B-y
Patient A-x
Patient E-x
Patient E-y
Dentist-y
Patient C-x
Patient C-y
Patient A-y
Patient G-x
Patient G-y
LC02-x
LC03-x
LC02-y
LC09
LC35
Patient D-x
Patient D-y
LC03-y
Patient F-x
Patient F-y
HIVELI

Figure 13.13 Using a phylogeny to determine the history of HIV transmission This tree shows the evolutionary relationships among human immunodeficiency viruses collected from a Florida dentist, a series of his patients, and HIV-positive individuals from the same geographic area (denoted as LC) in the early 1990s. The *x* and *y* designations refer to different HIV clones cultured from the same individual, at different times. The outgroup, called HIVELI, is from an African HIV-1 sequence.

Reading up the tree, the first branch-off identifies a clade that includes a local individual (LC03-*y*) who was *not* a patient of the dentist, along with patients D and F. The existence of this branch shows that these patients did not get the virus from the dentist; it happens that these individuals had other risk factors for transmission. The clade enclosed by the tan box, however, suggests that the virus that infected the dentist in the late 1980s is ancestral to those found in patients A, B, C, E, and G (and in the dentist) in the early 1990s. These five patients did not have other risk factors for the disease, suggesting that they contracted the virus while being treated by the dentist. From Hillis et al. (1994).

infected nonpatients, and estimated the viruses' evolutionary relationships (Ou et al. 1992). The resulting tree clearly demarcated a "dental clade" (Figure 13.13). This defined a group of individuals in whom the *gp120* sequences were closely related to the *gp120* sequences of the dentist, leading to the conclusion that these patients (but not others) had acquired the AIDS virus from the dentist. This conclusion was immediately controversial (DeBry et al. 1993) but has been confirmed (Hillis et al. 1994). The fact that transmission was confirmed helped promulgate extensive changes in the practice of dentistry *and* medicine. The changes were designed to decrease the risk of transmission from health care providers to patients and vice versa.

Summary

Recent conceptual and technological advances have revolutionized our ability to estimate phylogenies accurately. Research in systematics is advancing rapidly, and phylogenetic thinking is beginning to pervade evolutionary biology.

The first step in estimating a phylogeny is to select and measure characters that can be phylogenetically informative. The molecular or morphological characters employed in phylogeny inference have to be independent, homologous, variable among the taxa in the study, and resistant to homoplasy. The second step in phylogeny inference is to decide whether a parsimony, maximum likelihood, or distance method is most appropriate for analyzing the data in hand. Parsimony approaches are implemented by reconstructing the pattern of change in each character that implies the smallest number of character state changes. This process is repeated for each character on each candidate tree. Several different computer algorithms can be employed to search among the very large number of trees that are possible and evaluate them. Under parsimony, for example, the best tree is the one on which the smallest number of changes (across all characters) is implied for the data available. Then, a variety of statistical techniques can be used to evaluate the degree of support for the whole tree or for specific clades.

Phylogenetic thinking has been applied to a wide variety of problems in evolutionary biology, from the transmission of HIV to systems for classifying the diversity of life. Informative uses of phylogenies include dating events that are poorly documented in the fossil record, studying the rate and pattern of evolution in characters other than those used in constructing the phylogeny, and studying coevolution.

Questions

1. Mammals with high-crowned teeth, which are well suited for grazing, include some rodents (Rodentia), rabbits and hares (Lagomorpha), most cloven-hoofed animals (Artiodactyla), horses (Perissodactyla), and elephants (Proboscidea). Examine Figure 13.14, which shows the relationships of these and other mammalian orders. Are high-crowned teeth a synapomorphy (e.g., did the last common ancestor of mammals have high-crowned teeth) or have they arisen via convergent evolution?

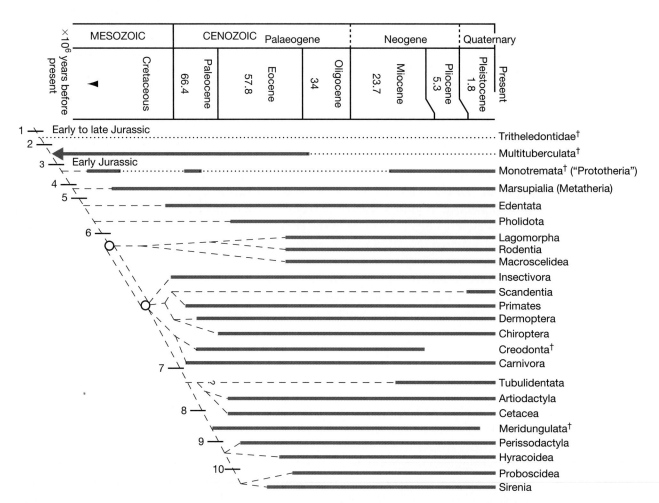

Figure 13.14 Phylogeny of the mammals This cladogram was modifed by Pough et al. (1996) based on work by Michael Novacek (1992, 1993). The names given along the right of the tree refer to mammalian orders. The numbers across branches refer to synapomorphies listed in Pough et al. The dark branches indicate the presence of fossil forms. The dashed lines indicate when the presence of an order has to be inferred.

2. The *gp120* gene of HIV evolves so quickly that clones of the virus from the HIV-positive Florida dentist (labelled Dentist-*x* and Dentist-*y* in Figure 13.13) evolved many nucleotide differences between the time that patients were infected in the late 1980s and the time that viral samples were collected from the dentist and his patients in the early 1990s. As a result, the two *gp120* sequences from the dentist are not very closely related to each other. Imagine, instead, that Ou et al. (1992) had found the Dentist-*x* and Dentist-*y* viral clones to be sister groups arising from the basal node of the dental clade (Figure 13.15 compared to Figure 13.13). How would this affect your hypothesis about the history and source of the infections of patients A, B, C, E, and G (and the dentist himself)? Would you still conclude that the den-

tist had transmitted the virus to each of these patients independently? What more parsimonious explanation could be found if this had been the result observed by Ou et al.?

3. In what sense do the HIV-positive local individuals included in Figure 10.3 serve as a control?

4. Examine the three primate phylogenies shown in Figure 13.16. Figure 13.16a shows a detailed phylogeny for most Old World primates, Figure 13.16b emphasizes relationships of great apes and humans, and Figure 13.16c emphasizes relationships of Old World monkeys. Do the three phylogenies agree with each other? (That is, do they show the same relationships and the same order of branching?) Do they give different impressions of whether there was a "goal" of primate evolution, or of what the "highest" primate is? If so, what aspects of the figures give the impression of a goal, and is this truly reflected in the complete data set? (*Sources:* data for great apes from various sources listed in Chapter 16; data for macaques from Hayasaka, K., Fujii, K., and Horai, S. 1996. Molecular phylogeny of macaques: Implications of nucleotide sequences from an 896-base pair region of mitochondrial DNA. *Molecular Biology and Evolution* 13:1044–1053.)

5. For several centuries, biologists have used the Linnaean system of classification, in which all species are assigned a spot in a hierarchy with seven major tiers (kingdom, phylum, class, order, family, genus, species). As explained in the text, some biologists have called for the abolition of the Linnaean system and the institution of new names, such as "Tetrapoda" and "Certartiodactyla," for each major evolutionary branching event. In this new system,

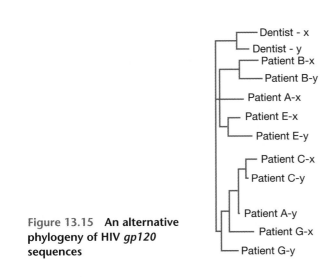

Figure 13.15 **An alternative phylogeny of HIV *gp120* sequences**

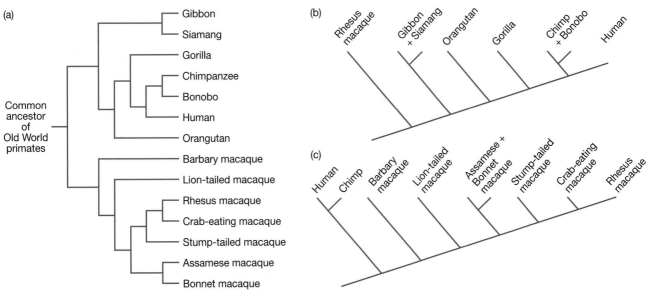

Figure 13.16 **Phylogenies showing relationships of several Old World primates** (Branch lengths are not scaled.)

what would determine the number of tiers required to classify a species? Would different species have the same number of tiers? The Linnaean system gives extra weight to those evolutionary branching events that produced noticeable morphological innovations, such as the evolution of birds (see text). Do you think it is worthwhile to distinguish between those branching events that produced noticeable morphological innovation and those that did not?

6. In Figure 13.5, site 192 was considered "variable but uninformative." Why is it uninformative?

7. Why does a fossil provide only a minimum age for its taxon?

8. Suppose it were discovered that aquatic mammals show convergence in milk proteins—e.g., suppose a certain form of milk protein, with certain amino acids in key positions, is particularly beneficial for aquatic mammals. How would that affect your interpretation of the milk casein data shown in Figure 13.5?

9. Do base pairs from different positions in one gene represent independent information? Why do evolutionary geneticists try to include as many genes in their analysis as their computers can handle?

10. In the late 1990s, hundreds of fossils of primitive birds were discovered in northeastern China, including birds with various combinations of traits 4, 5, and 6 of Figure 13.9 (see, for example, Martin & Zhou 1997; Hou et al. 1996, 1999). Early reports dated these fossil beds as late Jurassic. Does this match the predictions of Figure 10.8? The late Jurassic date was later shown to be incorrect—do you think the corrected date is likely to be earlier or later? Look up Swisher et al.'s 1999 paper to see if you are right.

Exploring the Literature

11. Leading phylogenetic systematists are now maintaining World Wide Web sites containing finished trees for all types of life forms and/or copies of the raw data sets used to estimate phylogenies. One major effort is called The Tree of Life (http://www.phylogeny.arizona.edu/tree/phylogeny.html). For background on this effort and the addresses of some key sites, see Morell (1996) and the Web site associated with this text.

12. Phylogenetic studies of coevolution are a booming research area. For entrees to the literature on relationships between flowering plants and herbivorous insects, fungal–algal relationships in lichen, and the evolution of warning coloration in butterflies, see

Brower, A. V. Z. 1996. Parallel race formation and the evolution of mimicry in *Heliconius* butterflies: A phylogenetic hypothesis from mitochondrial DNA sequences. *Evolution* 50: 195–221.

Farrell, B. D. 1998. Inordinate fondness explained: Why are there so many beetles? *Science* 281: 555–559.

Gargas, A., P. T. DePriest, M. Grube, and A. Tehler. 1995. Multiple origins of lichen symbioses in fungi suggested by SSU rDNA phylogeny. *Science* 268: 1492–1495.

13. Mapping traits onto phylogenies has confirmed or challenged several traditional views of how traits evolved. The following papers offer a test of how social behavior evolved in wasps, insight into the evolution of endothermy in fish, an analysis of the evolution of self-fertilization in a family of flowering plants, and a general review of the field.

Block, B. A., J. R. Finnerty, A. F. R. Stewart, and J. Kidd. 1993. Evolution of endothermy in fish: Mapping physiological traits on a molecular phylogeny. *Science* 260: 210–214.

Carpenter, J. M. 1989. Testing scenarios: Wasp social behavior. *Cladistics* 5: 131–144.

Kohn, J. R., S. W. Graham, B. Morton, J. J. Doyle, and S. C. H. Barrett. 1996. Reconstruction of the evolution of reproductive characters in Pontederiaceae using phylogenetic evidence from chloroplast DNA restriction-site variation. *Evolution* 50: 1454–1469.

Maddison, D. R. 1994. Phylogenetic methods for inferring the evolutionary history and processes of change in discretely valued characters. *Annual Review of Entomology* 39: 267–292.

Citations

Bajpai, S., and P. D. Gingerich. 1998. A new Eocene archaeocete (Mammalia, Cetacea) from India and the time of origin of whales. *Proceedings of the National Academy of Sciences, USA* 95: 15464–15468.

Bermingham, E., and H. A. Lessios. 1993. Rate variation of protein and mitochondrial DNA evolution as revealed by sea urchins separated by the Isthmus of Panama. *Proceedings of the National Academy of Sciences, USA* 90: 2734–2738.

Beverly, S. M., and A. C. Wilson. 1984. Molecular evolution in *Drosophila* and the higher Diptera II. A time scale for fly evolution. *Journal of Molecular Evolution* 21: 1–13.

Beverly, S. M., and A. C. Wilson. 1985. Ancient origin for Hawaiian Drosophilinae inferred from protein comparisons. *Proceedings of the National Academy of Sciences, USA* 82: 4753–4757.

Bremer, K. 1994. Branch support and tree stability. *Cladistics* 10: 295–304.

Brower, A. V. Z. 1994. Rapid morphological radiation and convergence among races of the butterfly *Heliconius erato*, inferred from patterns of mitochondrial DNA evolution. *Proceedings of the National Academy of Sciences, USA* 91: 6491–6495.

Brown, W. M., M. George, Jr., and A. C. Wilson. 1979. Rapid evolution of animal mitochondrial DNA. *Proceedings of the National Academy of Sciences, USA* 76: 1967–1971.

Chapela, I. H., S. A. Rehner, T. R. Schultz, and U. G. Mueller. 1994. Evolutionary history of the symbiosis between fungus-growing ants and their fungi. *Science* 266: 1691–1694.

Chiappe, L. M. 1995. The first 85 million years of avian evolution. *Nature* 378: 349–355.

DeBry, R. W., L. G. Abele, S. H. Weiss, M. D. Hill, M. Bouzas, E. Lorenzo, F. Graebnitz, and L. Resnick. 1993. Dental HIV transmission? *Nature* 361: 691.

DeSalle, R., and J. A. Hunt. 1987. Molecular evolution in Hawaiian drosphilids. *Trends in Ecology and Evolution* 2: 212–215.

Dixon, M. T., and D. M. Hillis. 1993. Ribosomal RNA structure: Compensatory mutations and implications for phylogenetic analysis. *Molecular Biology and Evolution* 10: 256–267.

Eldredge, N., and J. Cracraft. 1980. *Phylogenetic Patterns and the Evolutionary Process.* New York: Columbia University Press.

Felsenstein, J. 1978. Cases in which parsimony or compatibility methods will be positively misleading. *Systematic Zoology* 27: 401–410.

Felsenstein, J. 1981. Evolutionary trees from DNA sequences: A maximum likelihood approach. *Journal of Molecular Evolution* 17: 368–376.

Felsenstein, J. 1983. Parsimony in systematics: Biological and statistical issues. *Annual Review of Ecology and Systematics* 14: 313–333.

Felsenstein, J. 1985. Confidence limits on phylogenies: An approach using the bootstrap. *Evolution* 39: 783–791.

Flower, W. H. 1883. On whales, present and past and their probable origin. *Proceedings of the Zoological Society of London* 1883: 466–513.

Gatesy, J., M. Milinkovitch, V. Waddell, and M. Stanhope. 1999. Stability of cladistic relationships between Cetacea and higher-level artiodactyl taxa. *Systematic Biology* 48: 6–20.

Gingerich, P. D., B. H. Smith, and E. L. Simons. 1990. Hind limbs of Eocene *Basilosaurus*: Evidence of feet in whales. *Science* 249: 154–157.

Hennig, W. 1979. *Phylogenetic Systematics.* Urbana: University of Illinois Press.

Hillis, D. M. 1997. Biology recapitulates phylogeny. *Science* 276: 218–219.

Hillis, D. M. 1999. SINEs of the perfect character. *Proceedings of the National Academy of Sciences, USA* 96: 9979–9981.

Hillis, D. M., and J. J. Bull. 1993. An empirical test of bootstrapping as a method for assessing confidence in phylogenetic analysis. *Systematic Biology* 42: 182–192.

Hillis, D. M., J. P. Huelsenbeck, and C. W. Cunningham. 1994. Application and accuracy of molecular phylogenies. *Science* 264: 671–677.

Hillis, D. M., B. K. Mable, and C. Moritz. 1996. Applications of molecular systematics: The state of the field and a look to the future. In D. M. Hillis, C. Moritz, and B. K. Mable, eds. *Molecular Systematics.* Sunderland, MA: Sinauer, 515–543.

Hinkle, G., J. K. Wetterer, T. R. Schultz, and M. L. Sogin. 1994. Phylogeny of the attine ant fungi based on analysis of small subunit ribosomal RNA gene sequences. *Science* 266: 1695–1697.

Hou, L-H., L. D. Martin, Z-H. Zhou, and A. Feduccia. 1996. Early adaptive radiation of birds: Evidence from fossils from northeastern China. *Science* 274:1164–1167.

Hou, L-H., L. D Martin, Z-H. Zhou, A. Feduccia, and F. Zhang. 1999. A diapsid skull in a new species of the primitive bird *Confuciusornis*. *Nature* 399: 679–682.

Huelsenbeck, J. P., and D. M. Hillis. 1993. Success of phylogenetic methods in the four-taxon case. *Systematic Biology* 42: 247–264.

Huelsenbeck, J. P., and B. Rannala. 1997. Phylogenetic methods come of age: Testing hypotheses in an evolutionary context. *Science* 276: 227–232.

Kimura, M. 1980. A simple method for estimating evolutionary rates of base substitution through comparative studies of nucleotide sequences. *Journal of Molecular Evolution* 16: 111–120.

Knowlton, N., L. A. Weigt, L. A. Solorzano, D. K. Mills, and E. Bermingham. 1993. Divergence in proteins, mitochondrial DNA, and reproductive compatibility across the Isthmus of Panama. *Science* 260: 1629–1632.

Knowlton, N., and L. A. Weigt. 1998. New dates and new rates for divergence across the Isthmus of Panama. *Proceedings of the Royal Society of London, Series B* 265: 2257–2263.

Luckett, W. P., and N. Hong. 1998. Phylogenetic relationships between the orders Artiodactyla and Cetacea: A combined assessment of morphological and molecular evidence. *Journal of Mammalian Evolution* 5: 127–182.

Maddison, W. P., M. J. Donoghue, and D. R. Maddison. 1984. Outgroup comparison and parsimony. *Systematic Zoology* 33: 83–103.

Martin, A. P., G. J. P. Naylor, and S. R. Palumbi. 1992. Rates of mitochondrial DNA evolution in sharks are slow compared with mammals. *Nature* 357: 153–155.

Martin, A. P., and S. R. Palumbi. 1993. Body size, metabolic rate, generation time, and the molecular clock. *Proceedings of the National Academy of Sciences, USA* 90: 4087–4091.

Martin, L. D., and Z. Zhou. 1997. Archaeopteryx-like skull in enantiornithine bird. *Nature* 389:556.

Mayden, R. L., and E. O. Wiley. 1992. The fundamentals of phylogenetic systematics. In R. L. Mayden, ed. *Systematics, Historical Ecology, and North American Freshwater Fishes.* Stanford, CA: Stanford University Press, 114–185.

Milinkovitch, M. C., and J. G. M. Thewissen. 1997. Even-toed fingerprints on whale ancestry. *Nature* 388: 622–624.

Morell, V. 1996. Web-crawling up the tree of life. *Science* 273: 568–570.

Nelson, G., and N. Platnick. 1981. *Systematics and Biogeography: Cladistics and Vicariance.* New York: Columbia University Press.

Nikaido, M., A. P. Rooney, and N. Okada. 1999. Phylogenetic relationships among cetartiodactyls based on insertions of short and long interspersed elements: Hippopotamuses are the closest extant relatives of whales. *Proceedings of the National Academy of Sciences, USA* 96:10261–10266.

Novacek, M. J. 1992. Mammalian phylogeny: Shaking the tree. *Nature* 356: 121–125.

Novacek, M. J. 1993. Reflections on higher mammalian phylogenetics. *Journal of Mammalian Evolution* 1: 3–30.

O'Leary, M. A., and J. H. Geisler. 1999. The position of Cetacea within Mammalia: Phylogenetic analysis of morphological data from extinct and extant taxa. *Systematic Biology* 48: 455–490.

Ou, C.-Y., Carol A. Ciesielski, G. Myers, C. I. Bandea, C.-C. Luo, B. T. M. Korber, J. I. Mullins, G. Schochetman, R. L. Berkelman, A. N. Economou, J. J. Witte, L. J. Furman, G. A. Satten, K. A. MacInnes, J. W. Curran, H. W. Jaffe, Laboratory Investigation Group, and Epidemiologic Investigation Group. 1992. Molecular epidemiology of HIV transmission in a dental practice. *Science* 256: 1165–1171.

Pennisi, E. 1996. Evolutionary and systematic biologists converge. *Science* 273: 181–182.

Pough, F. H., J. B. Heiser, and W. N. McFarland. 1996. *Vertebrate Life.* Upper Saddle River, NJ: Prentice Hall.

Prothero, D. R., E. M. Manning, and M. Fischer. 1988. The phylogeny of the ungulates. In M. J. Benton, ed. *The phylogeny and classification of the tetrapods, Volume 2–Mammals.* Oxford: Clarendon Press, 201–234.

de Queiroz, K., and J. Gauthier. 1992. Phylogenetic taxonomy. *Annual Review of Ecology and Systematics* 23: 449–480.

Schaeffer, B. 1948. The origin of a mammalian ordinal character. *Evolution* 2: 164–175.

Schwenk, K. 1994. Comparative biology and the importance of cladistic classification: A case study from the sensory biology of squamate reptiles. *Biological Journal of the Linnaean Society* 52: 69–82.

Shaw, K. L. 1996. Sequential radiations and patterns of speciation in the Hawaiian cricket genus *Laupala* inferred from DNA sequences. *Evolution* 50: 237–255.

Shields, G. F., and A. C. Wilson. 1987. Calibration of mitochondrial DNA evolution in geese. *Journal of Molecular Evolution* 24: 212–217.

Simpson, B. B., and J. Cracraft. 1995. Systematics: The science of biodiversity. *BioScience* 45: 670–672.

Simpson, G. G. 1945. The principles of classification and a classification of mammals. *Bulletin of the American Museum of Natural History* 85: 1–350

Sneath, P. H. A., and R. R. Sokal. 1973. *Numerical Taxonomy: The Principles and Practice of Numerical Classification.* San Francisco: Freeman.

Swisher, C. C., Y-Q Wang, X-L Wang, X. Xu, and Y. Wang. 1999. Cretaceous age for the feathered dinosaurs of Liaoning, China. *Nature* 400: 58–61.

Swofford, D. L., G. J. Olsen, P .J. Waddell, and D. M. Hillis. 1996. Phylogenetic Inference. In D. M. Hillis, C. Moritz, and B. K. Mable, eds. *Molecular Systematics.* Sunderland, MA: Sinauer, 407–514.

Thewissen, J. G. M., and S. T. Hussain. 1993. Origin of underwater hearing in whales. *Nature* 361: 444–445.

Thewissen, J. G. M., S. T. Hussain, and M. Arif. 1994. Fossil evidence for the origin of aquatic locomotion in archaeocete whales. *Science* 263: 210–212.

Thewissen, J. G. M., and S. I. Madar. 1999. Ankle morphology of the earliest cetaceans and its implications for the phylogenetic relations among ungulates. *Systematic Biology* 48: 21–30.

Wakely, J. 1996. The excess of transitions among nucleotide substitutions: New methods of estimating transition bias underscore its significance. *Trends in Ecology and Evolution* 11: 158–163.

Wilson, A. C. 1985. The molecular basis of evolution. *Scientific American* 253: 164–173.

Xu, X., X.-L. Wang, and X.-C. Wu. 1999. A dromaeosaurid dinosaur with a filamentous integument from the Yixian formation of China. *Nature* 401: 262–266.

Zuckerkandl, E., and L. Pauling. 1962. Molecular disease, evolution and genic heterogeneity. In M. Kash and B. Pullman, eds. *Horizons in Biochemistry.* New York: Academic Press.

CHAPTER 14

The Origins of Life and Precambrian Evolution

Biologists, from the laboratory of Norman Pace, collecting heat-loving microbes from Obsidian Pool, a hot spring in Yellowstone National Park. (Norman R. Pace, University of California at Berkeley)

IN THE EARLY 1980S, TWO TEAMS OF SCIENTISTS INDEPENDENTLY DISCOVERED SMALL enzymes that could break and reform the chemical bonds that hold strings of nucleic acids intact. The enzymes did their job relatively poorly; compared to the hundreds of other such enzymes previously discovered, they were slow at their catalytic task and were not very versatile. Yet the discovery has been recognized as one of the most significant biological breakthroughs in the last 20 years. In 1989, the teams' leaders, Sidney Altman and Thomas Cech, shared the Nobel prize for the achievement.

Why were biologists so enthralled with these new enzymes? The answer is that the enzymes were made not of protein, but of nucleic acid—specifically RNA. Until 1982, all known enzymes were proteins. RNA was often considered to be DNA's poor cousin, relegated to the task of shuttling biological information from DNA, where the information is stored, to proteins, which carry out all the work of the cell. But Altman and Cech's unveiling of RNA enzymes, or "ribozymes," has changed how biologists view the operations of the cell. Perhaps more importantly, the existence of ribozymes has forever changed how biologists view the origin of the first cells—how they believe life originated and evolved on the early Earth.

To understand the impact that the discovery of catalytic RNA had on the study of life's origins, we must first consider the enormous challenge the origin of life presents for scientists. By our best estimate, life arose on the Earth approximately 4 billion years ago. No physical record of biological events has survived for this long. In contrast to the evolutionary processes we have investigated so far in this book, the origins of life must be reconstructed using indirect evidence alone. Consequently, biologists have turned to gathering disparate bits of information and fitting them together like pieces of a jigsaw puzzle. When more complete, this puzzle should present a clearer picture of life's origins on the Earth. Working forward in time from the origin of Earth, three main questions are now being pursued:

1. What was the Earth's environment like when life arose?
2. What is life, and how did it arise from non-life?
3. What were the first cells, and what characteristics did they have?

The first question is intimately tied to the study of the formation of the Earth itself. Rocks dating from the time of Earth's formation do not exist on the planet's surface, but radioisotopic dating of meteorites yields an estimated age for the solar system, and hence the Earth, of 4.5 to 4.6 billion years (see Badash 1989). The newborn Earth remained inhospitable for at least a few hundred million years. At first, it was simply too hot for life. The collisions of the planetesimals that formed the Earth released enough heat to melt the entire planet (Wetherill 1990). Eventually, the Earth's outer surface cooled and solidified to form the crust, and water vapor released from the planet's interior cooled and condensed to form the oceans. As we will discuss later, an active area of geochemical research involves determining the timing of the events that comprise the early history of the Earth. When conditions on Earth became permissive, the planet, at one location or perhaps many, became the "cradle of life" (Schopf 1983).

The third question takes us back to the advent of cellular life, the sort that exists today. We will focus our attention on the first living thing, or "primordial form," and one of its descendants, the most recent common ancestor of all extant organisms (Figure 14.1)—sometimes referred to as the cenancestor (Fitch and Upper 1987). Recent discoveries suggest that the cenancestor might not have been a single species, but instead a community of interbreeding forms. In either

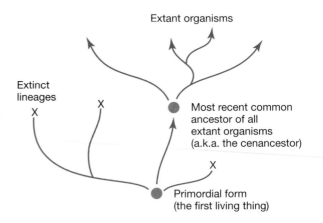

Figure 14.1 Cartoon of the tree of life Presumably, the first living organism had several descendant lineages, all but one of which eventually died out. The most recent common ancestor of all living things is the organism whose immediate descendants diverged to found the lineages that ultimately became all extant organisms. The whole-life phylogeny drawn here does not include the viruses, whose position on the tree is unclear.

case, the challenge is to use the same evolutionary analyses that work to reconstruct Phanerozoic relationships (see Figure 2.11) to trace events that predate even the earliest fossils.

The second question—the appearance of life itself—has been under investigation, via observation and experimentation, for almost 80 years. Biologists have made artificial cells and artificial cell membranes, and have zeroed in on some of the chemical reactions that could have made cellular material from nonliving sources. However, in the course of these studies, a quandary quickly became apparent. Which substance did life acquire first, proteins or DNA? Proteins can do all sorts of complicated biological tasks, but there is no evidence that proteins can propagate themselves; they cannot transmit the information needed to replicate. DNA, on the other hand, is perfectly suited to store and transmit information by complementary base pairing, but it was not known to be able to perform any biological work. This chicken-and-egg problem of which came first was essentially resolved with the discovery of catalytic RNA. Since RNA has a capacity for information storage and transmission *and* the ability to perform biological work, we now think that it preceded both proteins and DNA in the march toward the origins of life. Was there once a time when life was based entirely on RNA—an RNA World (Gilbert 1986)? This question is the topic of Section 14.1.

14.1 The RNA World

Before we look back on what the Earth was like before life, and forward to the first type of cellular life, let us consider the possibility that life was once a milieu of RNA molecules interacting without being confined by cell membranes. This RNA World is appealing to scientists because it can possess many of the characteristics of modern life without the need for much more than a few organic molecules in solution. The hypothesis of an RNA World is based on the realization, since the discovery of ribozymes, that RNA can simultaneously possess both a genotype and a phenotype (Joyce 1989). The genotype is the primary sequence of nucleotides along the RNA (Figure 14.2a), much like the genotype of a modern organism is the sequence of nucleotides along the DNA in the chromosome. Catalytic RNA, for example, contains between 30 and 1000 ribonucleotides that form its primary sequence, and hence its genotype. The *Tetrahymena* ribozyme discovered by Cech and colleagues (Kruger et al. 1982; Zaug and Cech 1986) is a stretch of about 400 nucleotides from head (the 5′ end) to tail (the 3′ end). However, unlike genomic DNA, which is usually double-stranded (see Figure 4.1), RNA usually exists as a single-stranded molecule that can fold back on itself many times to form a three-dimensional structure. In the case of ribozymes, this folded state can have an active site that enables the RNA molecule to catalyze a chemical reaction on a substrate, as do protein enzymes. This reactivity gives RNA its phenotype (Figure 14.2b).

The RNA World hypothesis proposes that catalytic RNA molecules were a transitional form between nonliving matter and the earliest cells.

Defining Life

All living organisms possess both a genotype and a phenotype. In fact, when we consider what life really is, and how living systems can be distinguished from nonliving ones, the ability to store and transmit information (a genotype) and the ability to express that information (a phenotype) are perhaps the most important

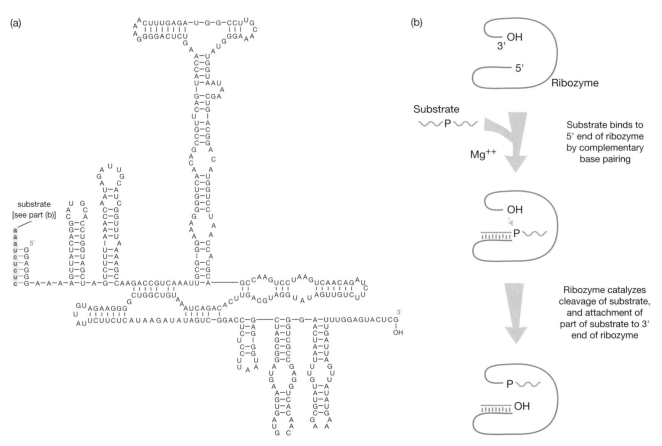

Figure 14.2 **The ribozyme from *Tetrahymena thermophila*** (a) The primary nucleotide sequence, which is the genotype of this catalytic RNA. This RNA is an intron (an intervening sequence between two genes) that separates two regions of the *Tetrahymena* genome that code for ribosomal RNA (rRNA) genes. Tom Cech and colleagues found that this sequence has the catalytic capability to splice itself out from between the two adjacent rRNAs after they have been transcribed (Kruger et al. 1982). The sequence shown here is a 413-nucleotide version, shortened from the naturally occurring form by the use of a restriction enzyme. A secondary structure of this molecule is formed when nucleotides base-pair with each other as the molecule folds back upon itself during transcription from DNA. Note that in RNA, uridine (U) replaces thymidine (T), and that three types of base pairs are common: A–U, G–U, and G–C. The secondary structure shown here is drawn such that no RNA strands cross each other, and as such does not accurately reflect how the molecule folds further into a tertiary (three-dimensional) structure. (b) A cartoon of the catalysis performed by the *Tetrahymena* ribozyme *in vitro,* which is its phenotype. A short oligonucleotide substrate (orange) binds to the 5′ end of the ribozyme (red) through complementary base-pairing (green ticks). In the presence of a divalent cation, such as Mg^{2+}, the ribozyme catalyzes the breakage of a phosphoester bond in the substrate and the ligation of the 3′ fragment to its own 3′ end. This "pick-up-the-tail" event can be used to discriminate catalytically active mutant sequences from less active or inactive mutants during *in vitro* evolution experiments (see Figure 14.4).

Here is one way we might define life: If it forms populations capable of evolving by natural selection, then it is alive.

criteria that set life apart from nonlife. Unfortunately, there is no neat list of characteristics that define life. Most biologists would include traits like growth and reproduction on such a list, but they cannot agree on what else should be used to exclude such lifelike systems as a growing salt crystal or a computer virus (if indeed these should be excluded). However, many now agree that the ability to evolve is a crucial component of any definition of life. Evolution—or change over time—requires both the ability to record and make alterations in heritable information, and some way of distinguishing valuable changes from detrimental ones. The former is carried out by a genotype, while the latter is carried out by a phenotype.

Dozens of naturally occurring ribozymes have now been discovered, and the phenotypes of all of them involve the formation and breaking of phosphoester bonds in RNA or DNA (Figure 14.2b). The chemistry of these reactions is precisely what is needed to replicate nucleic acids. This observation gives support to the idea of a primordial RNA World, where RNA would be responsible for replicating itself in order to persist. If a molecule of RNA could make a copy of itself while accommodating the possibility of mistakes, or mutations, then that molecule would exhibit many of the characteristics of modern life and could therefore be considered alive.

The Case for RNA as an Early Life Form

The RNA World hypothesis posits that an RNA-based living system evolved into one more like the life we see today, with DNA storing the biological information and proteins manifesting this information. DNA is better suited as an information repository because it is chemically more stable than RNA. Especially when double-stranded, DNA can better withstand high temperatures and spontaneous degradation by acids or bases.

What evidence do we have that RNA is ancient? The existence of catalytic RNA is certainly critical, but there are other indicators as well. A strong clue that RNA could have been involved in early life forms is its ubiquitous presence in the basic replication machinery of cells. The most highly conserved and universal component of the information-processing machinery in cells, for example, is the apparatus for translation, the ribosome. This apparatus, while it incorporates proteins, is built on a frame made of RNA (rRNA). Not only do ribosomes contain RNA themselves, but they require RNA adaptors (tRNAs) to carry out the task of protein synthesis. Furthermore, provocative evidence indicates that it is the RNA portion of ribosomes that actually carries out the catalytic steps in protein synthesis (Noller et al. 1992; Nitta et al. 1998). Another powerful argument for the antiquity of RNA is that the basic currency for biological energy is ribonucleoside triphosphates, such as ATP and GTP (Joyce 1989). These molecules are involved in almost every energy-transfer operation of all cells, and are even components of electron-transfer cofactors such as NAD (nicotinamide adenine dinucleotide), FAD (flavin adenine dinucleotide), and SAM (S-adenosyl methionine). With these ghosts of an RNA World in mind, we turn to the next important question: Can RNA evolve?

The Experimental Evolution of RNA

RNA sequences can provide the blueprint for the formation of their complementary sequences by base-pairing. Thus, like DNA, RNA has the capacity to store heritable information that can be propagated. A good example of this is the life cycle of the HIV virus we tracked in Chapter 1. In the case of HIV, the protein enzyme reverse transcriptase is used to copy an RNA strand into a DNA complement, which can then be converted into double-stranded DNA (see Figure 1.3). Any RNA, including a ribozyme, has this information-storage capacity. Researchers have begun to exploit the unique duality of *catalytic* RNA to ask if RNA sequences can demonstrate an evolutionary capacity in the laboratory. If RNA can evolve in test tubes, then the plausibility is greatly improved that an RNA World existed before the advent of distinct cells.

One way researchers have tested the RNA World hypothesis is to check whether populations of RNA molecules can evolve by natural selection.

The experimental evolution of RNA actually predates the discovery of ribozymes. Sol Spiegelman and colleagues used purified replicase protein from the bacteriophage Q_β to demonstrate the evolvability of Q_β RNA in the test tube (Mills et al. 1967). Replicase is a general term to describe any enzyme that can make a copy of another molecule. When a small amount of Q_β RNA is incubated with Q_β replicase for a few minutes, the replicase makes copies of the RNA, and copies of the copies, and so on. With each copying, there exists a small, but finite, probability that the replicase will make a mistake and miscopy a nucleotide, by putting an A instead of a G across from a C, for example. This mutability provides the raw material for evolution. After only four serial transfers of the Q_β RNA solution (removing a small amount of RNA after a prespecified time and using it to seed a fresh test tube containing fresh replicase), the phenotypic composition of the RNA population had already changed: The Q_β RNA had a dramatically reduced ability to infect bacteria (Figure 14.3).

An important aspect of test-tube evolution experiments is to demonstrate that any observed phenotypic change has an underlying genotypic basis. Although the Q_β RNA experiment was performed before technology existed to determine the nucleotide sequences of large numbers of nucleic acids, Spiegelman and colleagues were able to infer a genotypic shift in the RNA population after 74 transfers by digesting the RNA into its composite nucleotides. The researchers determined the percentages of all four nucleotides by chromatographic analysis, and found that values from the evolved RNA population differed from those of the original Q_β RNA by as much as 5%.

The experiments with Q_β RNA showed that mutations could arise and spread in a test-tube RNA population, but they were not designed to correlate a specific phenotypic trait of RNA with a particular genotype during evolution. The novel genotypes that rose to high frequency in the Spiegelman experiment were simply favored by selection for an RNA sequence that could be more quickly replicated by the Q_β replicase protein. In fact, the RNA from the 75th transfer was 83% shorter than the original Q_β RNA, and it could be replicated about 15 times faster.

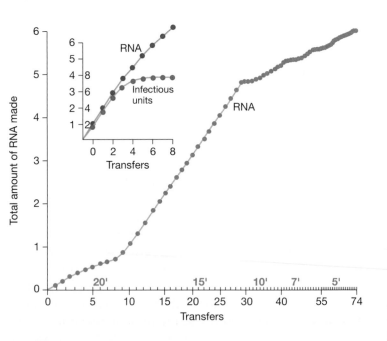

Figure 14.3   Sol Spiegelman's experiment with Q_β RNA   The graphs show the accumulation of Q_β RNA over many serial transfers from one test tube to another. In each test tube, a small amount of Q_β RNA from the previous tube is mixed with some purified Q_β replicase protein and the four ribonucleotide triphosphates, incubated for the length of time indicated on the *x*-axis, and then transferred to a new tube containing fresh replicase and nucleotides. After the fourth transfer, the RNA had evolved a new phenotype in that it could no longer infect *E. coli* bacteria (inset). The RNA from the 74th transfer had evolved a base composition that was about 5% different from the original Q_β RNA and was only 17% the length of the original Q_β RNA. From Fig. 1, p. 219, in Mills et al. (1967). Reprinted by permission of the author.

With the discovery of ribozymes came a more powerful way to perform the test-tube evolution of RNA, and with that, a more convincing argument that life may have passed through an RNA-based period. If an RNA molecule has a specific phenotype, can we apply selection to improve or modify this phenotype and observe a heritable change? Beaudry and Joyce (1992) exploited the catalytic capacity of the *Tetrahymena* ribozyme to address this question (Figure 14.4). Researchers had previously determined that a shortened form of the *Tetrahymena* ribozyme could catalyze a phosphoester transfer reaction on a short RNA substrate, called an oligonucleotide (a piece of single-stranded nucleic acid between about 5 and 30 nucleotides in length). In the reaction, the 3′ half of the substrate is broken off by the ribozyme and attached to the ribozyme's own 3′ end (Kruger et al. 1982). If this 3′ "tail" could be used as a tag, then ribozymes that performed the catalysis could be distinguished from those that did not. Beaudry and Joyce (1992) first made a large population of RNA molecules by sprinkling random mutations throughout the *Tetrahymena* ribozyme at a rate of 5% per position. Then, this mutant population was challenged with a novel task to select certain genotypes away from others. The task in this case was that the substrate oligonucleotide was provided in the form of DNA, not RNA. The naturally occurring sequence of the *Tetrahymena* ribozyme (the "wild type") used to start these experiments could cleave a DNA substrate only at a miserably slow rate. Beaudry and Joyce hoped that in the mutant pool there were variant sequences that, by chance, had an increased capacity for DNA cleavage.

The researchers incubated the mutant RNA population with a DNA substrate for an hour, and then amplified the ribozyme RNA into many more copies by adding two protein enzymes—reverse transcriptase and RNA polymerase. Because ribozymes pick up a 3′ tail as a consequence of cleaving the substrate, a DNA primer for reverse transcriptase that is complementary to the 3′ tail can be used to discriminate reactive sequences from nonreactive sequences. A 3′ tail is necessary to bind the primer, which in turn is necessary to initiate reverse transcription, which is in turn necessary to make more RNA. The RNA that results from this cycle of events can be used to seed a completely new cycle (a new generation) to continue and refine the selection process. After 10 such generations, the activity of the RNA pool towards cleaving DNA substrates had improved by a factor of 30. Importantly, this phenotypic enhancement could be traced to specific changes in the nucleotide sequence (Figure 14.5). Specific mutations at four nucleotide positions in the ribozyme's sequence were responsible for the majority of the catalytic improvement. Individual ribozymes carrying mutations at positions 94, 215, 313, and 314 proved to have a catalytic efficiency over 100 times greater than the wild-type sequence.

This experiment demonstrated that RNA molecules in solution can possess the features of living organisms that allow them to evolve. Each RNA can be ascribed a particular fitness, which is a function of both survival (substrate catalysis) and reproduction (ability to be reverse and forward transcribed). The fitness of the molecule is a reflection of its phenotype, which, in the case of ribozymes, is immediately specified by their primary sequence. Variation in an RNA population can be introduced at the outset, by the randomization of a wild-type sequence, as was the case with the Beaudry and Joyce (1992) experiment. Alternatively, an investigator can rely on the intrinsic error rates of the protein enzymes used in RNA amplification and can even alter the chemical environment to make the error rates

Populations of catalytic RNA molecules exhibit variation in nucleotide sequence. This variation is heritable when RNA is replicated. And researchers have devised experimental conditions under which sequence variation results in differences in survival.

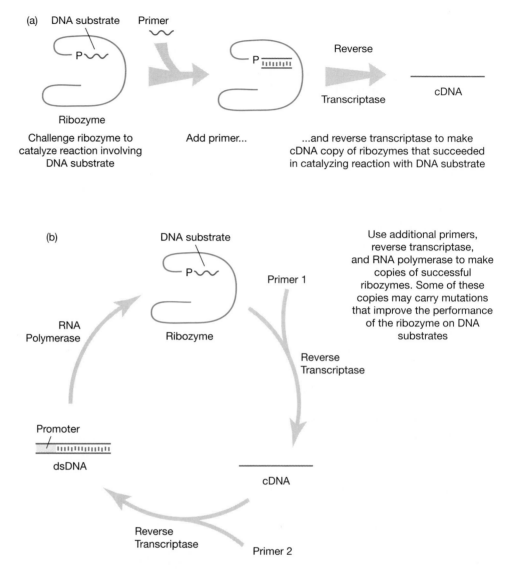

Figure 14.4 **Test-tube selection and reproduction of RNA** (a) Selection. A pool of RNA sequences (red), made by random mutagenesis of a ribozyme such as the *Tetrahymena* ribozyme (Figure 14.2a), is challenged to perform a desired chemical reaction. Only those that can perform the reaction acquire a short "tail" of DNA nucleotides attached to their 3' end (blue). This tail of nucleotides is complementary to a primer that is required for copying of the ribozyme's RNA into cDNA by reverse transcriptase. (b) Reproduction. RNA sequences that have successfully acquired a 3' tail (top) can bind primer 1 by complementary base pairing and be copied by reverse transcriptase into complementary DNA (cDNA). A second primer (primer 2) then binds to the cDNA so that the reverse transcriptase can make the DNA double-stranded. Primer 2 contains the promoter region for RNA polymerase, so that in the last step, RNA polymerase can bind to the double-stranded DNA and copy it back into RNA many times. The overall effect of this cycle is that successful RNA sequences, those that can perform the initial chemical reaction, are able to reproduce themselves into thousands of copies. Additional variation is introduced into the RNA population during this cycle because the two protein enzymes, reverse transcriptase and RNA polymerase, can make mistakes during copying and create mutations, which themselves are subject to selection. Reprinted with permission from Fig. 1, p. 636, in Beaudry and Joyce (1992). Copyright © 1992, American Association for the Advancement of Science.

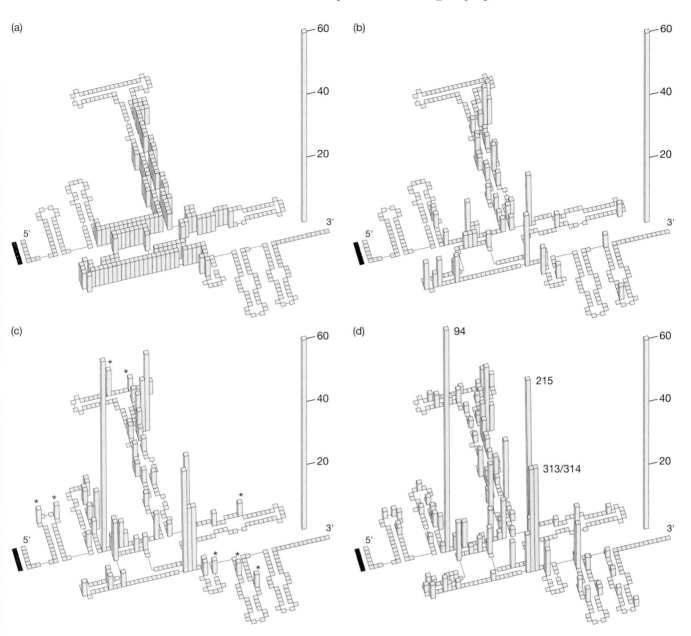

Figure 14.5 Genotypic changes in an evolving RNA population   These histograms depict the genotypic changes that occurred during the test-tube evolution experiment of Beaudry and Joyce (1992) with variants of the *Tetrahymena* ribozyme. (a) Mutations were introduced randomly throughout the middle portion of the ribozyme. A crude representation of the folded secondary structure of the 413-nucleotide ribozyme is shown as a base from which the heights of each bar depict the frequencies of mutations in the test-tube population relative to the wild-type *Tetrahymena* sequence. In the starting generation, 140 of the nucleotides were randomly mutated such that each nucleotide had a 5% chance of not being the wild-type nucleotide. This pool of variants (about 10^{13} molecules) was challenged such that only those sequences that could catalyze the cleavage of a DNA oligonucleotide substrate (black boxes) would be allowed to reproduce (Figure 14.4b). (b) The genotypic composition of the population after three rounds of selection and reproduction. Certain mutations in the original variant pool were deleterious and selected out of the population (boxes with no height), while others increased in frequency (taller boxes). (c) The genotypic composition of the population after six rounds of selection and reproduction. Some mutations are becoming very frequent, and even some not present in the original mutant pool, but introduced by on-line mutagenesis through copying mistakes of the protein enzymes during replication, are evident (asterisks). (d) The genotypic composition of the population after nine rounds of selection and reproduction. Four mutations, at nucleotides 94, 215, 313, and 314, have increased in frequency over 50% and are mainly responsible for the new phenotypic characteristics of the population. Reprinted with permission from Fig. 4, p. 638, in Beaudry and Joyce (1992). Copyright © 1992, American Association for the Advancement of Science.

higher. With such on-line mutagenesis the system becomes truly evolutionary, and selection can operate on variants of variants over many generations. Thus, it is easy to see a parallel between an evolving population of RNA in a test tube, and an evolving population of modern organisms in the natural environment (Lehman and Joyce 1993).

In test-tube experiments like these, researchers have evolved many ribozymes with either improved function or an entirely new function. The catalytic repertoire of RNA has been greatly expanded (Joyce 1998), and we now know that RNA can catalyze such reactions as phosphorylation (Lorsch and Szostak 1994), aminoacyl transfer (Illangasekare et al. 1995), peptide-bond formation (Zhang and Cech 1997), and carbon–carbon bond formation (Tarasow et al. 1997). Ribozymes have been designed that are allosteric, requiring a small-molecule cofactor to carry out catalysis (Tang and Breaker 1997). Ribozymes can be selected that can play a role in ribonucleotide synthesis (Unrau and Bartel 1998), retain activity with only three of four nucleotides (Rogers and Joyce 1999), and operate without divalent metal–ion cofactors (Geyer and Sen 1997). RNA sequences, called aptamers, can be selected to bind tightly to almost any other molecule desired (Tuerk and Gold 1990; Ellington and Szostak 1990), much like the immunoglobulin proteins of the mammalian immune system today. Together, all of these developments implicate RNA as a possible living system that existed before cellular life.

As discussed in Chapter 3, Darwin deduced that when the individuals in a population exhibit (1) variation, (2) inheritance, (3) excess reproduction, and (4) variation in survival or reproductive success, populations will evolve. When stripped of the particular characteristics of an intact, complex organism, we realize that traits (1) and (2) are about having a genotype, trait (3) is about being self-replicating, and trait (4) is about having a phenotype that makes a difference. Consequently, a self-replicating population of RNA would have the essence of life, even without the cells or organelles or tissues or leaves or fur or behavioral characteristics, and so on, that we are accustomed to seeing in living creatures.

Self-Replication

Although populations of catalytic RNA molecules have most of the properties required for evolution by natural selection, they still cannot evolve on their own without considerable help from human researchers. This is because catalytic RNA molecules cannot copy themselves.

From what we have discussed so far, it may seem like an open-and-shut case for an RNA World. However, this is far from true. We know that RNA is an incredibly diverse molecular structure, and that it can evolve under the right circumstances. But there is something conspicuously missing. In all the experiments previously described, RNA was copied by protein enzymes. These proteins, of course, were not around in the RNA World. A main premise of the RNA World hypothesis is that RNA predates the time when life used proteins to do most of the biological work. A puzzle piece that we lack for the RNA World is the demonstration that RNA can copy itself. The as-yet-undiscovered "RNA-dependent RNA autoreplicase" remains a Holy Grail for origins-of-life research (Bartel and Unrau 1999). Whether the RNA World used only one type of self-replicating RNA or a suite of interacting RNAs, an RNA with a replicase phenotype would be necessary (Bartel 1999). The acquisition of the ability to self-replicate by a collection of organic molecules, such as RNA, is arguably the point at which non-living matter came to life.

The hypothesis that an RNA molecule could replicate itself, serving as a simple proto-organism, is testable. If the hypothesis is correct, then we should be able to make a self-replicating RNA molecule in the lab. Although this has not been achieved to date, researchers have made significant advances. David Bartel and co-workers, for example, are using test-tube evolution to search for ribozymes capable of synthesizing RNA (Bartel and Szostak 1993; Ekland et al. 1995).

Figure 14.6 shows the selection scheme that Bartel and Szostak (1993) used to make ribozymes that can catalyze the formation of a phosphoester bond to link a pair of adjacent RNA nucleotides. Bartel and Szostak started with a large pool of RNA polynucleotides. Every polynucleotide had the same sequence on its 5′ and 3′ ends (represented by lines), plus a unique 220-nucleotide stretch of random sequence in the middle (represented by the box labeled "Random 220"). Figure 14.6 follows two molecules that are members of this RNA pool: Random 220 A (left column) and Random 220 B (right column).

Bartel and Szostak bound the pool RNAs to agarose beads by means of a base-pairing interaction on their 3′ ends. The scientists then bathed the pool RNAs in a solution containing many copies of a specific substrate polynucleotide (row 1). This short RNA molecule had, on its 5′ end, a sequence of nucleotides forming a tag, whose function will soon become clear. On its 3′ end, the substrate RNA had a sequence of nucleotides complementary to the free end of the pool RNA molecules. The substrate molecules quickly became bound, by base-pairing hydrogen bonds, to the pool RNAs (row 2).

This annealing brought into adjacent position the triphosphate group (PPP) on the 5′ end of the pool RNA and the hydroxyl group (OH) on the 3′ end of the substrate RNA. If, by chance, the 220-nucleotide stretch of random sequence in a pool RNA molecule had some ability to catalyze the formation of phosphoester bonds, then it catalyzed the formation of such a bond between the substrate and pool RNA molecules. In row 3, Random 220 A has catalyzed such a reaction, liberating a diphosphate molecule, whereas Random 220 B has not.

Bartel and Szostak then rinsed the pool RNAs under conditions that washed away any substrate RNAs not covalently bound (by phosphoester bonds) to pool RNAs, and liberated the pool RNAs from their agarose beads (row 4). Random 220 A still has its substrate (with tag); Random 220 B does not.

Finally, the scientists ran the pool RNAs through an affinity column (row 5). The affinity column caught hold of the tag sequence on the substrate RNA by base-pairing. The column thus captured any pool RNA whose 220-nucleotide stretch of random sequence had catalytic activity (like Random 220 A), and let pass any pool RNA whose 220-nucleotide sequence did not. This selection step is analogous to the discrimination between tailed and untailed *Tetrahymena* ribozymes in the experiment of Beaudry and Joyce (1992).

Now Bartel and Szostak released the captured pool RNAs from the affinity column, made many copies of each by using replication enzymes that allowed some mutations, and repeated the whole process again. Notice that Bartel and Szostak's protocol also gives, to the pool RNAs, all the properties necessary and sufficient for evolution by natural selection. The RNAs have reproduction with heritability (via the copying process), variation (due to mutation), and differential survival (in the affinity column). The RNAs most likely to survive from one generation to the next are the ones that are most efficient at catalyzing phosphoester bonds. After 10 rounds of selection, the RNA pool had evolved ribozymes that could catalyze the formation of phosphoester bonds at a rate several orders of magnitude faster than such bonds form without a catalyst (Figure 14.7).

Researchers are performing selective breeding experiments with populations of catalytic RNAs in an effort to develop catalytic RNAs that can replicate themselves. If the researchers succeed, then by the definition we proposed earlier, they will have created life.

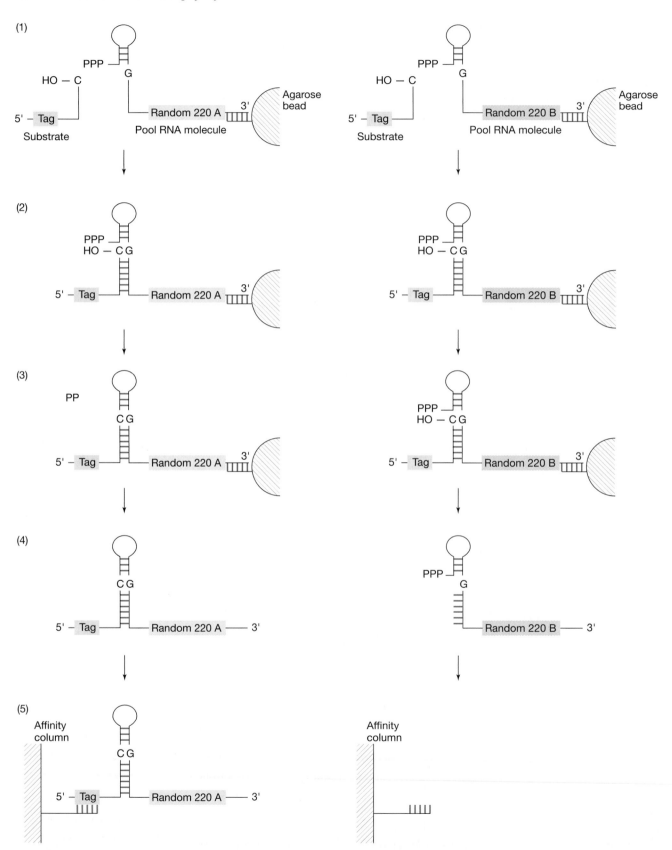

Figure 14.6   Test-tube selection scheme for identifying ribozymes that can synthesize RNA   See text for explanation. After Bartel and Szostak (1993).

More recently, Ekland and Bartel (1996) reported that they had used a similar scheme to evolve a catalytic RNA that can add up to six nucleotides to a growing RNA chain. This ribozyme uses template RNA and nucleoside triphosphates, and catalyzes RNA polymerization by the same chemical reaction promoted by the protein-based RNA polymerase enzymes used by living organisms. Laboratory-evolved ribozymes are not yet capable of self-replication, but it appears that biochemists are rapidly homing in on a sequence, or set of sequences, that are.

As Gerald Joyce (1996) put it, "Once an RNA enzyme with RNA replicase activity is in hand, the dreaming stops and the fun begins." If given the right organic molecules to feed on, a population of self-replicating RNAs should be able to evolve on its own by mutation and natural selection. The population would not require the generation-by-generation management practiced by Beaudry and Joyce or Bartel and Szostak. Would a species of self-replicating RNA evolve a DNA genome with DNA replication and transcription? Would it invent proteins and translation? Would its machinery be anything like the machinery in naturally evolved organisms? Would one of its descendants resemble the cellular life we see today? Perhaps one day soon we will have more pieces to fit into the puzzle.

14.2 Looking Further Back—How Do We Get to RNA?

The RNA World has many attractive features, and it solves the problem of having to propose the advent of proteins before DNA existed to encode them. But an RNA World comes with troubles of its own. Some critics have claimed that it simply pushes the problem of the origin of self-replication one step back, much like the Panspermia Hypothesis (Box 14.1) pushes the problem of the origins of life off the Earth to some other location. The issue here is simple: How could any RNA sequences come into being from an abiotic environment?

Chemists have studied the ways in which nucleic acids could be made without the aid of living systems. Certain aspects of the abiotic synthesis of nucleic acids turn out to be surprisingly easy, while others turn out to be dauntingly difficult. The general consensus is that the RNA World was probably not the first self-replicating system. This is because the likelihood of making RNA abiotically is too minute. Later, we will talk about the challenges of RNA synthesis, but for now we will note that RNA was almost certainly evolved from a more primitive chemical system. Regardless, in order to reconstruct the advent of any information-containing organic molecule with the properties of self-replication, the following four issues need to be addressed:

1. Information-containing biomolecules need to be made from simple inorganic compounds. Where did these compounds come from?
2. The chemical reactions that construct larger molecules from simple inorganics must be favorable and have a source of energy. What were these reactions?
3. The building blocks must be able to self-assemble into polymers such as RNA and polypeptides. How did this happen?
4. Larger biomolecules must be protected from harsh environmental conditions. How was this accomplished?

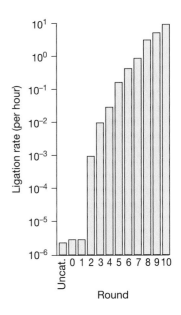

Figure 14.7 Evolution of catalytic ability in a laboratory population of ribozymes The graph shows the average rate at which the members of Bartel and Szostak's (1993) RNA pool catalyzed the formation of phosphoester bonds (ligation rate) as a function of round of selection. Note the logarithmic scale on the vertical axis. Over the course of the experiment, the catalytic activity of the molecules in the RNA pool increased by several orders of magnitude. From Bartel and Szostak (1993).

The proposition that catalytic RNAs were a transitional form between nonliving matter and cellular life leaves many gaps. We must still explain where the first RNA molecules came from, and how a population of self-replicating RNA molecules evolved into DNA- and protein-based cells.

BOX 14.1 The panspermia hypothesis

Although most contemporary specialists assume that life originated on Earth, there is an alternative: Life could have originated elsewhere and traveled to Earth. This suggestion, the panspermia hypothesis, begs the complaint that it merely shifts the problem of life's origin to some remote location where it is even harder to study. Francis Crick and Leslie Orgel (1973) pointed out, however, that such criticism is not only unfair, it could even prevent us from discovering the truth:

> For all we know there may be other types of planet on which the origin of life … is greatly more probable than on our own. For example, such a planet may possess a mineral, or compound, of crucial catalytic importance, which is rare on Earth.

One version of the panspermia hypothesis is that life originated on another planet within our own solar system. Microbes could then have been dislodged from their home world by a meteor impact, carried through space on a chunk of debris, and dropped to Earth in another meteor impact. The field of exobiology, or astrobiology, is centered around the study of life elsewhere in the solar system. Two crucial questions are (1) Do (or did) microbes exist on other planets in our solar system? and (2) Could they survive such a trip?

In an effort, among other things, to answer the first question, the United States landed two Viking spacecraft on Mars in 1976. The Viking landers carried three experiments designed to determine whether microbes were living in the Martian soil. All three involved attempts to detect gases released as by-products of metabolism. None yielded positive results. The results did not, however, rule out the possibility that life existed on Mars at other places or times.

David McKay and colleagues (1996) reported evidence suggesting that life did indeed exist on Mars some 4 billion years ago. McKay's team studied a rock from Mars that had fallen as a meteorite onto Antarctica. On freshly chipped pieces of the rock, the team found globules of carbonate ($-CO_3$) in close association with magnetite (Fe_3O_4), iron sulfide (FeS_2), and organic molecules called polycyclic aromatic hydrocarbons. All of these chemicals can be produced by either biological or nonbiological processes. Carbonate crystals, however, form in the presence of water, a requisite for life as we know it. In addition, McKay and colleagues found objects on the Martian rock that resemble tiny bacteria (Figure 14.8). McKay et al. concluded that the most plausible explanation for their findings is that the objects are fossils of biological processes that produced the minerals and chemicals. Other scientists are not convinced (see Anders et al. 1996; Kerr 1997a). As Bill Schopf noted, "This attempt failed to find life on Mars. That does not mean Mars contained no life—just that these scientists didn't find any." A new series of expeditions to Mars is in progress, and promises eventually to yield more decisive information about past or present life on the red planet.

There are other candidate locations for extraterrestrial life in the solar system. One is Jupiter's moon, Europa. Recent photographs taken by the spacecraft Galileo suggest that Europa has abundant liquid water and active volcanoes. Chyba (2000) argues that energy capable of sustaining life on this moon could be generated by charged particles provided in Jupiter's magnetosphere. Together, these data offer the possibility that life could have evolved in a liquid ocean hidden under Europa's icy surface (Belton et al. 1996; Kerr 1996, 1997b).

What about a microbe's chances of surviving a trip to Earth? For microbes on meteorites knocked loose from Mars, the trip to Earth would typically take several million years, but a lucky few would reach Earth

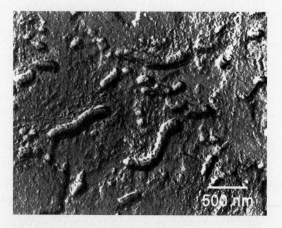

Figure 14.8 Fossils of Martian bacteria? This scanning electron microscope image of a Martian meteorite shows objects that resemble tiny bacteria. (NASA/Johnson Space Center)

14.1 Continued

in less than a year (Gladman et al. 1996; Gladman and Burns 1996). Spacefaring microbes would be exposed to cold, vacuum, and both UV and ionizing radiation. Two teams, Peter Weber and J. Mayo Greenberg (1985) and Klaus Dose and Anke Klein (1996), have measured the survival rates of spores of the bacterium *Bacillus subtilis* exposed to various combinations of these conditions; Jeff Secker and colleagues (1994) have made theoretical calculations. The consensus is that bare spores could not survive an interplanetary trip. With some sort of shield against radiation, however, spores would have some chance of making it. A shield could be provided by ice, rock, or carbon.

To our knowledge, only once has the long-term survival of microbes in space been directly tested. In November of 1969, astronauts from Apollo 12 recovered a camera from the unmanned lunar lander Surveyor 3, which had touched down on the Moon two and a half years earlier. Back on Earth, NASA scientists found that a piece of foam insulation inside the camera contained viable bacteria (*Streptococcus mitis*). The microbes had apparently stowed away on Surveyor before it left Earth, and, shielded from radiation by the camera, survived their stay in the vacuum and cold of space (Mitchell and Ellis 1972).

A second possibility under panspermia is that life originated in another solar system and traveled to Earth through interstellar space. Spores embarking on such a voyage would need a force to accelerate them to sufficient velocity to escape the gravitational field of their home star. This force can be supplied by radiation pressure (Arrhenius 1908; Secker et al. 1994). Sailing on radiation pressure severely limits the mass of the shielding a spore can carry. Secker and colleagues suggest that spores encased in a film of carbonaceous material could both achieve escape velocity and survive the radiation they would encounter on a trip between stars.

Finally, Crick and Orgel (1973) suggest a third possibility, which they call directed panspermia: Earth's founding microbes were sent here intentionally, aboard a spacecraft, by intelligent extraterrestrials bent on seeding the galaxy with life. Crick and Orgel argue that, within the foreseeable future, it will probably be possible for us to launch such a mission. Therefore, it is at least conceivable that some other civilization actually did so 4 billion years ago.

Where Did the Stuff of Life Come From?

On September 28, 1969 at about 11 o'clock in the morning, a meteor entered the Earth's atmosphere over the town of Murchison, Australia and broke up into several meteorites that scattered over a five-square-mile area of the ground (Figure 14.9). Soon after, scientists collected some of the meteorites and carefully brought them back to the laboratory for chemical analysis (Kvenvolden et al. 1970). To their astonishment, the analyses showed that organic compounds were present in the interior of the rocks. In particular, the amino acids glycine, alanine, glutamic acid, valine, and proline were found in significant concentrations (1–6 micrograms of amino acid per gram of meteorite). These amino acids are among the ones used by modern organisms to make proteins. Amino acids had been found in meteorites before, but their presence was more than likely the result of contamination from human handling. The scientists who studied the Murchison meteorites fractured them in the lab and analyzed only the interior portions. In addition, the amino acids they found in the Murchison stones were racemic; that is, they included roughly equal proportions of the D- and L-stereoisomers (mirror-image forms). By contrast, biological amino acids are almost purely of the L-form, and thus terrestrial life could not be the source of the compounds the researchers found in the meteorites.

Why were the Murchison meteorites significant? The biomolecules of life, as well as their likely precursors, all require the elements carbon, hydrogen, oxygen,

Figure 14.9 The Murchison Meteorite This photo shows one fragment of the approximately 100 kg of meteorites that fell near Murchison, Australia in 1969. Chemical analyses of these meteorites have detected dozens of amino acids. (New England Meteoritical Services)

nitrogen, sulfur, and phosphorus in large amounts, plus trace quantities of other elements such as magnesium, calcium, and potassium. Moreover, these elements must be in a chemical form that allows them to be used in the construction of biological building blocks like amino acids, sugars, and carbohydrates. If these building blocks could have been synthesized on the primitive Earth, then presumably they would have been available for condensation into larger biomolecules. But if they could not have been made on Earth, we would have to look to extraterrestrial sources, such as meteors, to account for their presence.

The problem with terrestrial sources is that, 4 billion years ago, the Earth's environment might not have been permissive for the synthesis of life's building blocks. In addition to temperature and pressure, a key feature of the environment is whether it was primarily oxidizing, with high abundances of molecular oxygen (O_2), and carbon dioxide (CO_2), or primarily reducing, with high concentrations of hydrogen (H_2), methane (CH_4), and ammonia (NH_3). Or it could have been intermediate in oxidizing activity. Which state it was in would have determined which chemical reactions were possible. The composition of the early atmosphere remains a subject of debate (Lazcano and Miller 1996), and atmospheric chemists are looking for mechanisms by which organic molecules could have been synthesized, even in rel-

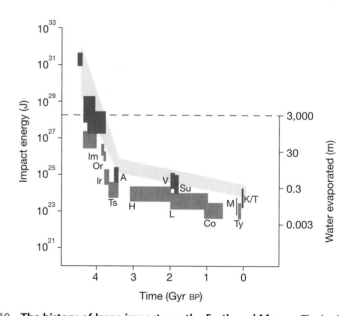

Figure 14.10 The history of large impacts on the Earth and Moon The horizontal axis represents time (billions of years before present). The left vertical axis represents the energy of impact (joules). The right vertical axis shows the depth (m) to which the oceans would be vaporized by an impact with a given energy. The dashed line represents vaporization of the entire global ocean. Each box encloses the range of times during which a particular impact is estimated to have occurred, and the range of energies the impact is estimated to have fallen within. The red boxes are for the Moon; the blue boxes are for the Earth. Boxes with labels represent impacts documented by craters or other geological evidence. The unlabeled box at the upper left represents a large impact thought to be responsible for the formation of the Moon. The other unlabeled boxes are for hypothetical impacts. The gray band shows the largest impacts likely to have hit the Earth at any given time. Lunar craters—Im = Imbrium, Or = Orientale, Ir = Iridium, Ts = Tsiolkovski, H = Hausen, L = Langrenus, Co = Copernicus, Ty = Tycho. Terrestrial craters—A = Archaean spherule beds, V = Vredevort, Su = Sudbury, M = Manicougan, K–T = Cretaceous–Tertiary impact (crater located off the Yucatan peninsula). From Sleep et al. (1989).

atively impermissive mixtures of gases (Kasting 1993). Some feel that geochemical evidence points to an atmosphere that would not be favorable for the generation of biologically important molecules, or at least would not produce them in the concentrations needed for the origins of life. Thus, many have explored an alternative hypothesis that certain critical biochemicals were made elsewhere in the solar system and delivered to the Earth in vehicles such as the Murchison meteorite. Many comets also contain a variety of organic molecules (Chyba et al. 1990; Cruikshank 1997). In fact, the young Earth experienced heavy bombardment by meteors and comets. Figure 14.10 shows the history of very large impacts on both the Earth and the Moon. Like the Murchison meteorite, several carbonaceous chondrite meteorites, believed to be fragments of asteroids, have proven to contain an abundance of organic molecules (see Chyba et al. 1990; Lazcano and Miller 1996).

There is at least one crucial difficulty with the hypothesis that life's building blocks came from space: When meteors and comets crash to Earth, friction with the atmosphere and collision with the ground generate tremendous heat (Anders 1989). This heat may destroy most or all of the organic molecules the meteors and comets carry (Chang 1999). Edward Anders (1989) notes that very small incoming particles are slowed gently enough by the atmosphere to avoid incinerating all of their organics; he suggests that cometary dust may have been the primary source of the young Earth's organic molecules. Christopher Chyba (1990) and colleagues look instead to the possibility that the early atmosphere was dense with carbon dioxide. A dense CO_2 atmosphere may have provided a soft enough landing, even for large meteors and comets, for some of their organics to survive. The Murchison meteorites certainly provide direct evidence that at least some organic molecules can survive a descent to Earth.

> The simple organic molecules from which life was built may have formed in space and then fallen to Earth. Researchers have tested this idea by looking for amino acids and other organic molecules inside meteorites.

The Oparin–Haldane Model

Originally, there was great hope that the Earth itself could provide the "right stuff" for prebiotic synthesis. In 1953, Stanley Miller, then a graduate student in Harold Urey's laboratory at the University of Chicago, reported a simple and elegant experiment. He built an apparatus that boiled water and circulated the hot vapor through an atmosphere of methane, ammonia, and hydrogen, past an electric spark, and finally through a cooling jacket that condensed the vapor and directed it back into the boiling flask. Miller let the apparatus run for a week; the water inside turned deep red and cloudy. Using paper chromatography, Miller identified the cause of the red color as a mixture of organic molecules, most notably the amino acids glycine, α-alanine, and β-alanine. Since 1953, chemists working on the prebiotic synthesis of organic molecules have, in similar experiments, documented the formation of a tremendous diversity of organic molecules, including amino acids, nucleotides, and sugars (see Fox and Dose 1972; Miller 1992).

Miller used methane, ammonia, and hydrogen as his atmosphere; in the 1950s, this highly reducing mixture was thought to model the atmosphere of the young Earth. The implication of Miller's result was that if lightning or UV radiation could have played the role that the spark did in his experiment, then the young Earth's oceans would have quickly become rich in biological building blocks. Many atmospheric chemists now believe that Earth's early atmosphere was not so reducing, being dominated by carbon dioxide rather than methane, and molecular nitrogen (N_2) rather than by ammonia (Kasting 1993). This conclusion is based on the mixture of gases

> The simple organic molecules from which life was built may also have formed on Earth. Researchers have tested this idea by trying to recreate the chemical conditions on the early Earth and replicate the chemical reactions that might have created amino acids and nucleotides.

released by contemporary volcanoes, and on improved knowledge of the chemical reactions that occur in the upper atmosphere. Reaching a consensus on the prebiotic environment is important, because an atmosphere dominated by carbon dioxide and molecular nitrogen appears to be much less conducive to the formation of certain organic molecules. However, the formation of aldehydes, especially formaldehyde (H_2CO), from carbon dioxide has been deemed plausible by several researchers, and aldehydes are necessary in the construction of the ribose sugars needed to make the nucleotides of an RNA World (Mojzsis et al. 1999).

The view that the Earth possessed all the necessary ingredients for the origins of life is perhaps the most thoroughly investigated hypothesis and holds great appeal for many scientists even today. This opinion dates back to the efforts of A. Oparin and J. B. S. Haldane, working in the first half of the 20th century, to reconstruct how life may have begun. These scientists and others (including Charles Darwin) created a lasting image of life arising in an aqueous environment brimming in high concentrations of biological building blocks. This was Darwin's "warm little pond" (Darwin 1887), the famous "prebiotic soup." There are many severe criticisms of this vision, not the least of which is whether liquid water existed on the Earth at the time of life's origin. Nonetheless, this view remains as sort of a null model against which deviations can be tested, much like the Hardy–Weinberg equilibrium principle in population genetics. This scenario is often referred to as the Oparin–Haldane model.

We can break the Oparin–Haldane model into a series of steps that occurred sequentially in the waters or moist soil of the young Earth (Figure 14.11). First,

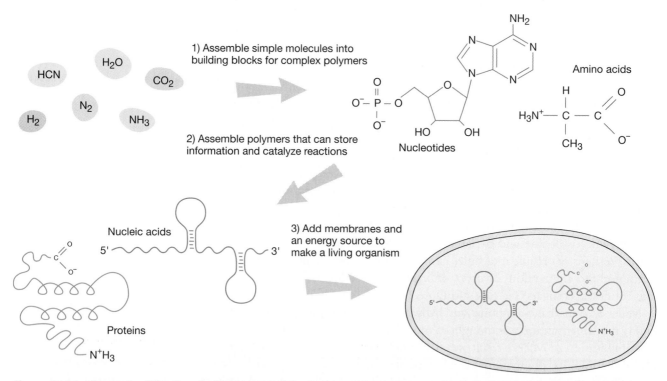

Figure 14.11 Synopsis of the Oparin–Haldane model The first stage would have been the formation of biological building blocks (nucleotides and amino acids) from existing inorganic material on the early Earth. The second stage would have been the polymerization of these building blocks to form biological macromolecules (nucleic acids and proteins). Finally, these macromolecules would have directed the formation of other biological structures, such as cell membranes.

nonbiological processes synthesized organic molecules, such as amino acids and nucleotides, that would later serve as the building blocks of life. Then, the organic building blocks in the prebiotic soup were assembled into biological polymers, such as proteins and nucleic acids. Last, some combination of biological polymers was assembled into a self-replicating organism that fed off of the existing organic molecules, much as we discussed earlier for the RNA World.

From Simple Inorganics to the Building Blocks of Life

Previously, we saw the ease with which amino acids can be made from simple inorganics like methane, ammonia, and hydrogen. What about nucleotides? A second monumental achievement in origins-of-life research was the demonstration by Juan Oró (1961) that the nitrogenous base adenine (a purine) could be readily made from a thermodynamically favorable reaction involving only ammonia and hydrogen cyanide (HCN). When these two compounds are heated in water, adenine is produced in yields as high as 0.5%, which is significant if the early atmosphere was reducing and contained large amounts of ammonia and hydrogen cyanide. Miller refers to this reaction as the "rock of faith" for terrestrial prebiotic synthesis. Other chemists have had similar results for other purine bases. Pyrimidines (C, U, and T) are slightly more difficult to construct abiotically, but chemists have had some successes (Voet and Schwartz 1982). Finally, the ribose sugars that form nucleotides can, at least under the right environmental conditions, be derived from a cascade of condensation reactions that begin only with formaldehyde.

Unfortunately, the description of several mostly independent, plausible chemical pathways that could have produced amino acids, nucleotides, and sugars leaves us a long way from fully formed building blocks that are on the verge of becoming a self-replicating system. A major obstacle that has plagued biochemists for decades is the origin of chirality, or handedness. As noted above, living systems today use only one stereoisomer, or mirror-image form, of the amino acids in their proteins, and the same is true of nucleotides. In many of the chemical syntheses described by adherents to the Oparin–Haldane model, both mirror images of the building blocks are made in roughly equal quantities, and it is difficult to devise mechanisms that produce only one or the other. Exacerbating this problem is the fact that one mirror-image form would inhibit the polymerization of the other during any type of polymer self-replication (see Joyce et al. 1987). Furthermore, as is the case with sugar formation, not only does the sugar that we see today in nucleic acids (ribose) constitute a very small percentage of all the sugar products of formaldehyde condensation, but there also exist multiple equally probable ways that the nitrogenous bases could be attached to the sugar. Each of these combinations produces a subtly, but importantly, different nucleotide isomer than that used by contemporary RNA. To make matters worse, each building block needs to be activated, or chemically charged, before it can be incorporated into a polymer. Activation requires a pre-existing source of chemical energy. Without cell membranes to concentrate this energy, it is challenging to understand how building blocks became activated in the RNA World (Orgel 1986). These problems have been described by Joyce and Orgel (1999) as turning the "molecular biologist's dream" (...once upon a time there was a prebiotic pool full of β-D-nucleotides...) into the "prebiotic chemist's nightmare."

So where does this leave the RNA World? Staggering, perhaps, but certainly still standing. Many researchers now think that the RNA World did not arise *de novo* from a warm little pond. Instead, RNA was likely to have been a later stage in an evolutionary lineage that derived from a simpler genetic system. Several non-RNA self-replicating systems have been proposed. Among them are: polymers made up not of ribonucleotides as we know them today, but of ribonucleotide analogs that have only one stereoisomer (Joyce et al. 1987); polymers made up of a hybrid between peptides and nucleic acids (Egholm et al. 1992); polymers made up of nucleotides composed of pyranose sugar instead of ribose sugar (Eschenmoser 1999); and even polymers made up of inorganic substances such as clay (Cairns-Smith et al. 1992). Christian deDuve (1991) has outlined a "Thioester World" in which information transfer is intricately linked to the metabolic turnover of thioester linkages in a complex chemical milieu. All of these scenarios are based on the presumption that another self-replicating system could arise abiotically with a higher probability than RNA. Some of these are envisioned such that RNA could develop from them; presumably, the preexistence of a self-replicator could allow for a bias in the way RNA is synthesized, such that the problem of chirality could be overcome, for example. Others are envisioned as alternatives to an RNA World, many formulated in a way that would favor the construction and use of catalysts other than RNA.

The Assembly of Biological Polymers

The second step in the Oparin–Haldane theory, the formation of biological polymers from the building blocks in the prebiotic soup, has presented other theoretical and practical challenges. The prebiotic soup would contain organic building blocks dissolved in water, and although biological polymers can readily be synthesized in water, they also break down by hydrolysis even as they are being built. This problem raises doubts that polymers sufficiently long to serve as the basis of a self-replicating primordial organism would ever have formed in a simple organic soup (Ferris et al. 1996).

James Ferris and colleagues (1996), extending a tradition that dates from the 1940s and 1950s (see Ferris 1993), demonstrated a plausible mechanism to overcome the hydrolysis problem. Ferris et al. prepared a simple prebiotic soup in the lab and added the common clay mineral montmorillonite. Montmorillonite is a naturally occurring aluminum–silicate clay to which organic molecules readily adhere. When activated nucleotides stick to montmorillonite, the clay acts as a catalyst and will join them together in a polynucleotide chain. While bound to the clay, the polynucleotides form more rapidly than they are hydrolyzed, and the researchers succeeded in encouraging the formation of polynucleotide chains containing 8–10 nucleotides in a row.

Ferris and colleagues then demonstrated that it was possible to prepare much longer polynucleotides by the daily addition of activated nucleotides to a premade oligonucleotide primer. They started with a polyadenylate primer 10 nucleotides long, and let it bind to the montmorillonite. The scientists then added to the polyadenylate/clay solution a bath of activated adenosine nucleotides. The activated nucleotides reacted with the polyadenylate primers, adding themselves to the nucleotide chains. Ferris and colleagues then used a centrifuge to spin down the clay (and its attached nucleotide chains), poured off the spent solution, and added a fresh bath of activated nucleotides. By repeating this process, adding a fresh bath of activated nucleotides once each day, Ferris et al. synthesized polyadenylates over 40 nucleotides long (Figure 14.12).

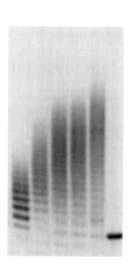

Figure 14.12 Synthesis of long nucleotide chains on clay This electrophoresis gel has separated mixtures of polyadenylates by size. The right lane contains a single band that corresponds to nucleotide chains 10 bases long; this was the starting point for Ferris et al.'s (1996) experiment. The left lane contains the mixture of polynucleotides produced when 10-nucleotide polyadenylates were allowed to bind to montmorillonite, then given two successive baths with activated adenosine nucleotides. Each successive band represents a one-nucleotide difference in length. The leftmost lane thus contains polyadenylates ranging from 11 to 20 nucleotides in length. The second lane from the left shows the results of four successive baths with activated nucleotides, and so on. Reactions run without montmorillonite failed to produce elongated nucleotide chains. (James P. Ferris, Chemistry Dept., Rensselaer Polytechnic Institute, Troy, NY 12180)

Ferris and Orgel have since used an analogous procedure to grow polypeptides up to 55 amino acids long on the minerals illite and hydroxylapatite (Ferris et al. 1996; Hill et al. 1998). The teams assert that their method models a mechanism by which biological polymers could have grown on the early Earth. Minerals in sediments that were repeatedly splashed with the pre-biotic soup, or continuously bathed by it, could have nursed the formation of polymers that were long enough to form a self-replicating primordial form. Thus, the second step of the Oparin–Haldane model appears to have experimental support. We will briefly discuss the third step in Section 14.3.

The building blocks of life may have been assembled into polymers while stuck to the surface of clay crystals. Adhering to clay helps a growing polymer avoid being broken apart by hydrolysis.

Protecting Life from the Environment

At this point, one can at least grapple with the possibility that there did exist a logical chain of events that led from simple inorganics, such as carbon dioxide, ammonia, and hydrogen cyanide, to fully formed nucleic acids. These events could have all taken place on Earth, or some of them could have taken place on extraterrestrial bodies and other ones on Earth. Some researchers have even suggested that certain chemical reactions might have occurred in the atmosphere itself, perhaps suspended in water droplets that rose and fell with the temperature. Regardless of the chemistry of the early atmosphere, early Earth probably offered many local opportunities for organic synthesis—hydrothermal environments, ocean water rich in ferrous iron, or the caldera of volcanoes, just to name a few. However the final challenge for any model of the origins of life is not whether the early Earth would have provided for the needs of life, but whether it would have been hospitable enough to allow life to evolve.

The oldest known sedimentary rocks contain evidence suggesting that life was already established on Earth by 3.85 billion years ago. The rocks, from Akilia Island and Isua, both in Greenland, are of a type known as banded iron formations (Figure 14.13a). Geological processes have exposed the rocks to high temperatures

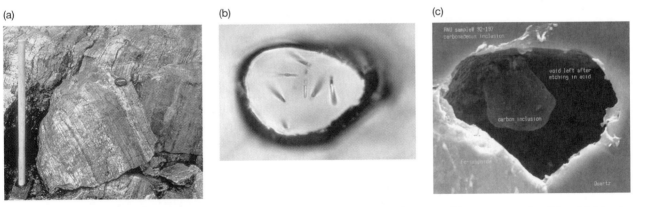

Figure 14.13 **The oldest known sedimentary rocks also contain the oldest known evidence suggesting life** (a) A boulder from the banded iron formation on Akilia Island, Greenland. This rock is at least 3.85 billion years old. The dark bands are magnetite (an iron-containing mineral); the light bands are silicates (silicon-containing minerals). The rock hammer at the left edge of the photograph is about 40 cm tall (Allen P. Nutman, Australian National University/Scripps Institution of Oceanography). (b) An apatite crystal from an Akilia Island sedimentary rock. The dark border around the crystal is made of carbonaceous material. The dark streaks within the crystal are tracks left by fission products released during the decay of radioactive elements embedded in the crystal. The crystal is a little over 20 micrometers long (Scripps Institution of Oceanography). (c) A photograph, taken with a scanning electron microscope, of the void left after an apatite crystal has been etched away with nitric acid. The void contains a speck of carbonaceous material (an inclusion) that was embedded in the apatite. The ratio of ^{12}C to ^{13}C in this carbonaceous speck suggests that the speck was produced by a living organism. (Stephen J. Mojzsis, Scripps Institution of Oceanography, University of California at San Diego)

and pressures, which have compacted the rocks and crystallized many of the minerals they contain. This transformation would have destroyed any microfossils the rocks might have originally harbored. The rocks do contain apatite crystals, however (Figure 14.13b). Apatites are calcium phosphate minerals. Apatites are produced by many microorganisms, and are also the main component of bones and teeth in vertebrates. Apatites can also be produced by nonbiological processes.

Following up on earlier work by Manfred Schidlowski (1988), Stephen Mojzsis, Gustaf Arrhenius, and their colleagues (1996) hypothesized that the apatite in the Akilia and Isua rocks was produced by organisms. They tested this hypothesis by examining specks of carbonaceous material embedded in the apatite crystals (Figure 14.13c). Carbon has two stable isotopes: ^{12}C and ^{13}C. When organisms capture and fix environmental carbon, during photosynthesis for example, they harvest ^{12}C at a slightly higher rate than ^{13}C. As a result, carbonaceous material produced by biological processes has a slightly higher ratio of ^{12}C to ^{13}C than does carbonaceous material produced by nonbiological processes. Mojzsis, Arrhenius, and colleagues used an ion microprobe to measure the ratio of ^{12}C to ^{13}C in the carbonaceous specks embedded in Akilia and Isua apatite crystals. The ion microprobe knocks individual carbon atoms out of a sample; a mass spectrometer then weighs the atoms. The scientists also examined carbonaceous specks in apatite crystals from 3.25-billion-year-old sedimentary rocks from Pilbara, Australia. In all three cases, the scientists found carbon isotope ratios characteristic of life. Other workers in the field have greeted these results with cautious enthusiasm. John M. Hayes (1996), for example, called for further evaluation of the suitability of the ion microprobe for measuring carbon isotope ratios, and notes that some as-yet-undiscovered prebiotic chemical reaction could explain the data. Hayes agrees with Mojzsis, Arrhenius, et al., however, that their evidence does suggest that life had made its appearance by 3.85 billion years ago.

Solid crust and oceans probably existed earlier than the 3.85 billion years ago demonstrated by the Akilia Island rocks, but erosion, plate tectonics, and volcanic eruptions have obliterated all direct evidence. Even if crust and oceans did exist earlier, however, continued bombardment of the planet by large meteors could have prevented life from being established earlier than 3.85 billion years ago (see Figure 14.10). Large meteor impacts generate heat, create sun-blocking dust, and produce a blanket of debris. As time passed, and the largest planetesimals got swept up by the Earth and other planets, the sizes of the largest impacts decreased. Norman Sleep and colleagues (1989) estimated that the last impact with sufficient energy to vaporize the entire global ocean, and thereby frustrate the emergence of any self-replicating system, probably happened between 4.44 and 3.8 billion years ago.

Life's development may have actually been initiated many times over, if sterilizing events allowed enough time in between for self-replication to reevolve each time. Alternatively, life may have survived some of these impacts, sequestered in protective niches in the environment, such as deep-sea hydrothermal settings. Regardless, we can estimate that the origins of life were threatened by an inhospitable environment until about 4 billion years ago. Whether the Oparin–Haldane model leading to an RNA World is correct, or whether some other scenario turns out to be more plausible, we should note one last point before we consider cellular life— the origins of life occurred in a tumultuous abiotic environment. Paradoxically, the Earth today is even more inhospitable to the origins of life. Life has become so

Although we still lack a complete scenario for how the first living things arose from nonliving matter, it appears that they did so quickly— almost as soon as the early Earth was habitable.

successful at exploiting extreme niches that there are no places left for inorganic molecules to reinvent self-replication before the first stages of these attempts would be gobbled up by extant creatures.

14.3 Looking Further Forward—When Life Went Cellular

Once self-replicating systems evolved on the Earth, at least one of them adapted to the use of DNA to store heritable information and to the use of proteins to express that information. This system eventually gave rise to all lineages of life on the planet today. We draw this conclusion because all life forms (except viruses) use DNA and proteins. In fact, all modern organisms use them in the same way; the same 20 amino acids and the same basic structure of the genetic code have been found in all creatures studied to date. Thus, we apply the principle of parsimony to infer that all organisms share a common ancestor.

What Was the Most Recent Common Ancestor of All Living Things?

Because another shared feature of all extant life is the existence of cells, we also infer that the common ancestor was a cellular form. Technically speaking, we need to say that all life has descended from a population of interbreeding cells, because if portions of the primitive genome could be readily swapped, then life today cannot trace its ancestry to a single organism. The picture that emerges of the origins and early evolution of life on the Earth can be diagrammed as in Figure 14.14. The first cellular life whose descendants ultimately survived, the cenancestor (or cenancestors), appeared at least 2 billion years ago and probably much earlier. The advantages of cellular membranes, as well as internal organellar membranes (see below) would have been enormous. Cells allow for compartmentalization. Certain chemicals can be concentrated inside the cell, while others can be pumped outside the cell. This allowed life to accumulate its necessary constituents in much higher concentrations than they are found free in solution—activated nucleotides, for example. Cells also allowed genotypes and phenotypes to be linked, even after the latter had become the domain of proteins and not the genetic material itself. It does a genotype little (evolutionary) good if the phenotype it encodes is free to diffuse to other genotypes.

It is a long way from a self-replicating RNA molecule to the cenancestors, and many questions remain. For example, how did the earliest organisms acquire cellular form? One potential answer has come from the work of Sidney Fox and colleagues, who found that mixtures of polyamino acids in water or salt solution spontaneously organize themselves into microspheres with properties reminiscent of living cells (see Fox and Dose 1972; Fox 1988; Fox 1991).

About the ancestral cellular lineage, like the first self-replicating system, we can ask what its general characteristics were, when it lived, and by what route did its descendants evolve into today's orchids, ants, mushrooms, amoebae, and bacteria? Again, these events occurred early in the Earth's history and much direct information has been lost. But if we know what questions to ask, the available data in the geological record can begin to remove the mystery of the first cellular life (Schopf 1994b).

The first place we might look in trying to identify the ancestral cells is the fossil record. In principle, a complete fossil record would allow us to trace lines of descent

Little is known about how the first self-replicating molecules evolved into cellular life forms, although researchers have shown that structures reminiscent of cell membranes form spontaneously.

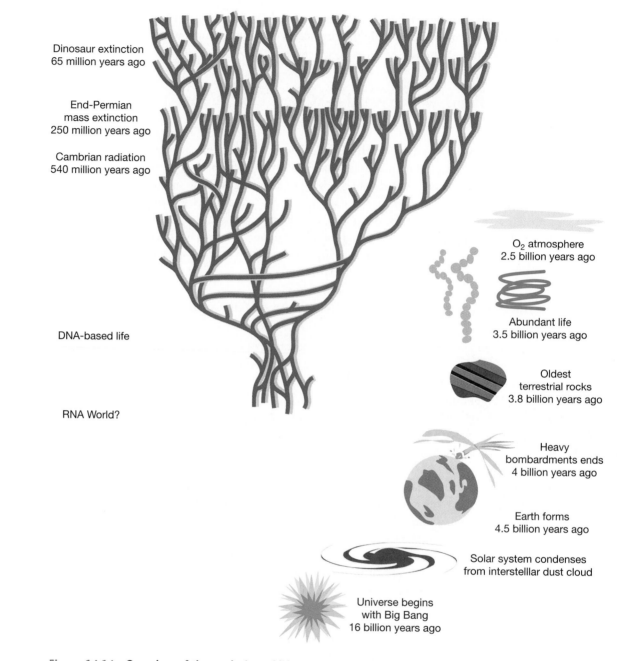

Dinosaur extinction
65 million years ago

End-Permian
mass extinction
250 million years ago

Cambrian radiation
540 million years ago

DNA-based life

RNA World?

O₂ atmosphere
2.5 billion years ago

Abundant life
3.5 billion years ago

Oldest
terrestrial rocks
3.8 billion years ago

Heavy
bombardments ends
4 billion years ago

Earth forms
4.5 billion years ago

Solar system condenses
from interstellar dust cloud

Universe begins
with Big Bang
16 billion years ago

Figure 14.14   Overview of the evolution of life   In the tree of life shown here (blue), the fusion of branches represents the acquisition of symbionts and other forms of lateral gene transfer. These phenomena will be discussed later in the chapter. After Atkins and Gesteland (© 1998) in Gesteland et al. (1999); Doolittle (2000).

from present-day organisms all the way back to the cenancestors. However, it does not appear that the fossil record so far assembled can take us that deep into the past. The oldest fossils definitively established as those of living organisms are 3.465 billion years old (Schopf 1993; see also Schopf and Packer 1987; Schopf 1992). These fossils, which come from a rock formation called the Apex chert in Western Australia, show simple cells growing in short filaments (Figure 14.15). Their discoverer, J. William Schopf (1993), believes these organisms belonged to the cyanobacteria, a group of photosynthesizing bacteria formerly known as the blue-green algae. This identification is tentative. It is based on the morphology of the

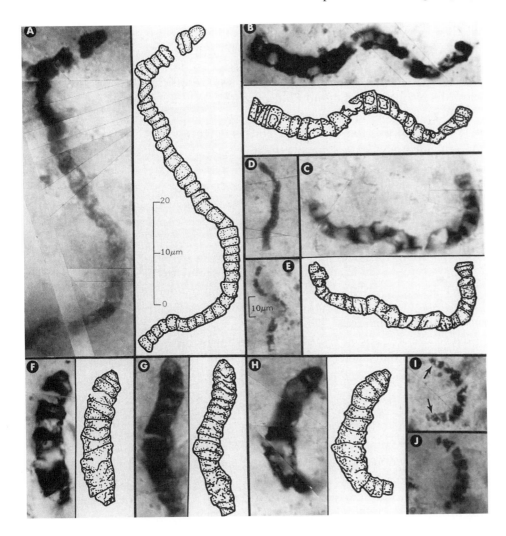

Figure 14.15 **The oldest known fossils of living organisms** Most photographs of fossils are accompanied by an interpretive drawing showing the fossil's morphology. The fossils show filaments composed of individual cells lined up like beads on a string. Specimens A, B, C, D, and E belong to the species *Primaevifilum amoenum*; F, G, H, I, and J belong to *P. conicoterminatum.* Reprinted with permission from J. William Schopf, Microfossils of the early Archean Apex chert: New evidence of the antiquity of life, *Science* 260: 640–646, 1993. Copyright © 1993 by the American Association for the Advancement of Science.

fossil cells and on analysis of the isotopic composition of the organic carbon in the rocks, neither of which is taxonomically definitive. If Schopf's identification of the Apex chert fossils is correct, then the fossils probably represent organisms that are already fairly high up the evolutionary tree, well above the most recent common ancestors (see later). Furthermore, the fossil record for times earlier than 2.5 billion years ago is too spotty to allow paleontologists to trace lines of evolutionary descent from present-day organisms back to the Apex chert fossils (Altermann and Schopf 1995). As a result, we have no direct way of knowing whether the organisms recorded in the Apex chert represent extinct or living branches of the tree of life. If we want to discover the characteristics of ancestral cell lineages, we must use methods other than examination of the fossil record.

One way to learn about the characteristics of the earliest cells is to look for their fossils.

The Phylogeny of All Living Things

Another way to study the ancestral lineage is to reconstruct the phylogeny of all living things. A universal phylogeny should allow us to infer additional characteristics of the earliest life forms (see Chapter 13), beyond just their cellular nature. The first attempts to reconstruct the phylogeny of everything were based on the morphologies of organisms (see reviews in Woese 1991; Doolittle and Brown 1994). The morphological approach was productive for biologists interested in the branches of the tree of life that contain eukaryotes. Morphology was, historically, the basis of the phylogeny of many taxonomic groups. The morphological approach led only to frustration, however, for biologists interested in the branches of the universal phylogeny containing prokaryotes. Prokaryotes lack sufficient structural diversity to allow the reconstruction of morphology-based evolutionary trees.

When biologists developed methods for reading the sequences of amino acids in proteins, and the sequences of nucleotides in DNA and RNA, a new technique for estimating phylogenies quickly became established (Zuckerkandl and Pauling 1965). Some of the details of this technique are devilish (see Chapter 13), but the basic idea is straightforward. Imagine that we have a group of species, all carrying in their genomes a particular gene. We can read the sequence of nucleotides in this gene in each of the species, then compare the sequences. If species are closely related, their sequences ought to be fairly similar. If species occupy distant branches on the evolutionary tree, then their sequences ought to be less similar. As a result, we can use the relative similarity of the sequences of species to infer their evolutionary relationships. We place species with more similar sequences on neighboring branches of the evolutionary tree, and species with less similar sequences on more distant branches.

The challenge in using sequence data to estimate the evolutionary tree for all living things is to find a gene that shows recognizable sequence similarities even between species that are as distantly related as *Escherichia coli* and *Homo sapiens* (Woese 1991). We need a gene that is present in all organisms, and that encodes a product whose function is essential and thus subject to strong stabilizing selection. Without strong stabilizing selection, billions of years of genetic drift will have obliterated any recognizable similarities in the sequences of distantly related organisms. Additionally, the function of the gene must have remained the same in all organisms. This is because when a gene product's function shifts in some species but not in others, selection on the new function can cause a rapid divergence in nucleotide sequence that makes species look more distantly related than they actually are.

One gene that meets all the criteria for use in reconstructing the universal phylogeny is the gene that codes for the small-subunit ribosomal RNA (Woese and Fox 1977; Woese 1991). All organisms have ribosomes, and in all organisms the ribosomes have a similar composition, including both rRNA and protein. All ribosomes have a similar tertiary structure, including small and large subunits. In all organisms the function of the ribosomes is the same: They are the machines responsible for translation. Translation is so vital, and organisms are under such strong natural selection to maintain it, that the ribosomal RNAs of humans and their intestinal bacteria show recognizable similarities in nucleotide sequence, even though humans and bacteria last shared a common ancestor billions of years ago. The small-subunit rRNA was the molecule chosen by Carl R. Woese, the chief pioneer of the use of molecular sequences in estimating the universal phylogeny (Fox et al. 1977; Woese and Fox 1977; see also Doolittle and Brown 1994). Though it is not a perfect solution, the small-subunit rRNA remains an informative resource for whole-life phylogenies.

An estimate of the universal phylogeny, based on sequences of the small-subunit rRNA, appears in Figure 14.16. Before attempting to use this tree to infer the characteristics of the most recent common ancestor, we need to discuss the phylogeny's general shape. The whole-life rRNA phylogeny has prompted a dramatic revision of our traditional view of the organization of life, because it reveals that

Another way to learn about the earliest cells is to estimate the phylogeny of all living things, then infer the characteristics of the common ancestors.

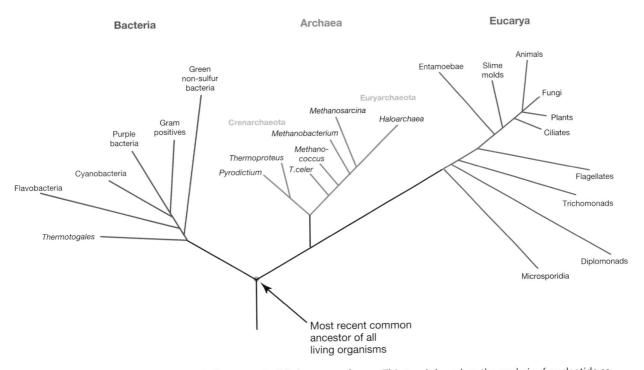

Figure 14.16 **An estimate of the phylogeny of all living organisms** This tree is based on the analysis of nucleotide sequences of small-subunit rRNAs. From Woese (1996).

the five-kingdom system of classification (Whittaker 1969) bears only a limited resemblance to actual evolutionary relationships (Woese et al. 1990; for a contrary view, see Margulis 1996).

The prokaryotes, for example, which are all grouped in the kingdom Monera in the traditional classification, occupy two of the three main branches of the rRNA phylogeny. One of these two branches, the Bacteria, includes virtually all of the well-known prokaryotes. The gram positive bacteria, for instance, include *Mycobacterium tuberculosis,* the pathogen that causes tuberculosis. The purple bacteria include *E. coli.* (The purple bacteria are so named because some of them are purple and photosynthetic, although *E. coli* is neither.) The cyanobacteria, all of which are photosynthetic, include *Nostoc,* an organism often seen in introductory biology labs.

The other prokaryote branch, the Archaea, is not as well known. Many of the Archaea live in physiologically harsh environments, are difficult to grow in culture, and were discovered only recently (see Madigan and Marrs 1997). Most of the Crenarchaeota, for example, are hyperthermophiles, living in hot springs at temperatures as high as 110°C. Many of the Euryarchaeota are anaerobic methane producers. Another group in the Euryarchaeota, the Haloarchaea, are highly salt dependent and are thus referred to as extreme halophiles.

Because of their prokaryotic cell structure, the Archaea were originally considered bacteria. When Woese and colleagues discovered that these organisms were only distantly related to the rest of the bacteria, they renamed them the archaebacteria (Fox et al. 1977; Woese and Fox 1977). Eventually biologists realized that, as the phylogeny in Figure 14.16 shows, the archaebacteria are in fact more closely related to the eukaryotes than they are to the true bacteria (see Bult et al. 1996; Olsen and Woese 1996). In recognition of this, Woese and colleagues (1990) proposed the new classification

The first whole-life phylogenies based on sequence data were estimated on the basis of small-subunit rRNA genes. These rRNA phylogenies revealed that the traditional five-kingdom system of classification offers a misleading view of evolutionary relationships.

used in Figure 14.16. Woese and colleagues dropped the *bacteria* from "archaebacteria," renaming this group the Archaea. The most inclusive taxonomic units in the new classification are three domains corresponding to the three main branches on the tree of life: the Bacteria, the Archaea, and the Eucarya. Woese and colleagues proposed that the two fundamental branches of the Archaea, the Crenarchaeota and the Euryarchaeota, be designated as kingdoms.

Woese et al. (1990) declined to offer a detailed proposal on how to divide the Eucarya into kingdoms. The Protista, a single kingdom in the traditional classification, are scattered across several fundamental limbs on the eukaryotic branch of the tree of life. The diplomonads, for example, which include the intestinal parasite *Giardia lamblia,* represent one of the deepest branches of the Eucarya. They are well separated from such other protists as the flagellates, which include *Euglena,* and the ciliates, which include *Paramecium.* If we want our kingdoms to be natural evolutionary groups, they should be monophyletic. That is, each kingdom should include all the descendants of a single common ancestor. Unless we want the kingdom Protista to include the animals, plants, and fungi, it will have to be disbanded and replaced by several new kingdoms.

The remaining three kingdoms in the traditional classification, the Animals, Plants, and Fungi, require only minor revision. To make the Fungi a natural group, for example, the cellular slime molds (such as *Dictyostelium,* a favorite of developmental biologists) will have to be removed.

The universal rRNA phylogeny demonstrates, however, that the Animals, Plants, and Fungi, the kingdoms that have absorbed most of the attention of evolutionary biologists (and represent most of the examples in this book), are mere twigs on the tip of one branch of the tree of life. The multicellular, macroscopic organisms in these three kingdoms are newcomers on the evolutionary scene, with a relatively recent common ancestor. For genes shared among all organisms, like the gene for the small-subunit rRNA, Animals, Plants, and Fungi appear to possess less than 10% of the nucleotide-level diversity observed on Earth (Olsen and Woese 1996).

An Examination of Early Cellular Life

Now that we have a universal phylogeny, what does it tell us about the earliest cellular life forms? The arrow in Figure 14.16 points to the last common ancestor of all extant organisms. According to this tree, the common ancestor's descendants diverged to become the bacteria on one side and the Archaea–Eucarya on the other. Rooting the tree of life in this way was and remains an enormous challenge, because there is no outgroup to work with. The position of the root shown in Figure 14.16 is based on the work of several groups of researchers who used different analytical tricks (see Figure 14.17), but who all came up with approximately the same answer: The Archea and Eucarya are more closely related to each other than either of them is to the Bacteria (Gogarten et al. 1989; Iwabe et al. 1989; Brown and Doolittle 1995; Baldauf et al. 1996). More recent data have yielded surprises, as we will discuss shortly. Estimating the location of the root is likely to remain an active area of research for the foreseeable future.

Assuming that we can accurately estimate a phylogeny that extends so far back in time, and that our placement of the root is reasonable, we can make inferences about when certain fundamental cellular traits evolved, and in which lineages. Recall from Chapter 13 that we can map character-state changes onto phyloge-

Whole-life phylogenies based on molecular data suggest that the most recent common ancestor of all extant life was a sophisticated organism, with a DNA genome and much of the machinery of modern cells. . . .

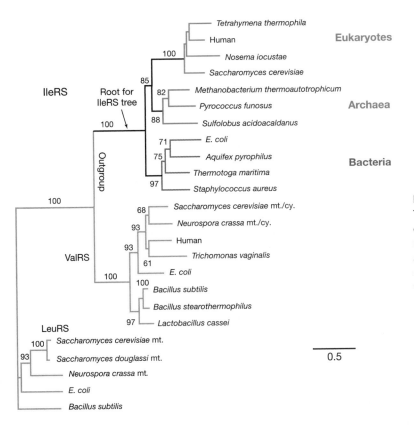

Figure 14.17   Rooting the universal phylogeny
To root the tree of life, which has no organism that can serve as an outgroup, an analytical trick must be used. Here, the observation that gene families exist as a result of ancient duplications can provide a molecular outgroup to the tree of life. The aminoacyl-tRNA synthetase gene family is an example. The top portion of this phylogeny is based on sequence similarities among the isoleucine aminoacyl-tRNA synthetase (IleRS) genes of organisms representing all three domains (the Bacteria, the Archaea, and the Eucarya). Sequences of the valine (ValRS) and leucine (LeuRS) aminoacyl-tRNA synthetase genes from a variety of bacteria and eukaryotes are the outgroups that root the tree. From Brown and Doolittle (1995).

nies using the principle of parsimony. For examples of how we might do this on the universal phylogeny look at Figure 14.18. If a trait occurs in all three domains (Figure 14.18a), or if it occurs in the Bacteria and the Archaea but not in the Eucarya (Figure 14.18b), or if it occurs in the Bacteria and the Eucarya but not in the Archaea (Figure 14.18c), we can infer that the trait was present in the common ancestor and was lost on the lineage (if there is one) that lacks it. Alternative scenarios would require that the trait arose independently two or three times, and therefore would require more evolutionary transitions.

Using the principle of parsimony, we might ask whether the common ancestor of all extant organisms was already storing its genetic information in DNA. The fact that all extant organisms use DNA suggests that the common ancestor did the same. An alternative possibility is that the common ancestor stored its genetic information in some other molecule, such as RNA, but that storage in DNA was favored so strongly by natural selection that a conversion from RNA storage to DNA storage occurred independently in more than one domain. Use of DNA by the common ancestor appears more likely than this scenario of convergent evolution. One clue is that the DNA-dependent RNA polymerases used in transcription show strong similarities across all three domains. This suggests that a DNA-dependent RNA polymerase was present in the last common ancestor. The possession of a DNA-dependent RNA polymerase implies the possession of DNA (Benner et al. 1989). Likewise, DNA polymerases found in all three domains show enough similarities to suggest that the common ancestor had a DNA polymerase too. And where there was a DNA polymerase, again, there was probably DNA.

Based on similar kinds of evidence and reasoning, many researchers have tentatively concluded that the most recent common ancestor was highly evolved and

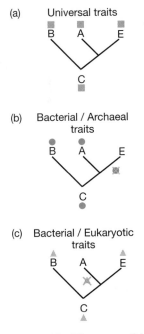

Figure 14.18   Three possible distributions of complex traits among the three domains of life   The first appearance of a symbol on a tree represents the origin of a trait. A crossed-out symbol represents the loss of a trait.

biologically sophisticated. Overall, the common ancestor appears to have been rather like a modern bacterium.

Our Picture of the Earliest Cells Will Continue to Improve

Our understanding of the earliest cells whose descendants survive today depends crucially on genetic sequence data. Such data allow us to estimate the universal phylogeny. Furthermore, sequence data provide much of the information about the traits of organisms that, when placed on the universal phylogeny, allow us to make inferences about the common ancestors. As of this writing, the amount of sequence data we have is limited, but growing rapidly. Two trends in particular promise to yield many new insights.

First, our knowledge of the Archaea is increasing dramatically. As we mentioned earlier, many archaeans live in harsh and unusual environments. *Methanococcus jannaschii,* for example, lives anaerobically in deep-sea hydrothermal vents, at temperatures near 85°C and depths of at least 2600 m (Jones et al. 1983). Not surprisingly, most archaeans are difficult or impossible to grow in culture, and thus hard to study. In 1984, a team of biologists working in the laboratory of Norman Pace pioneered a new approach to studying the environmental distribution of the Archaea. The researchers extracted DNA directly from mud and water samples collected in nature, then amplified and sequenced the DNA in the lab (Stahl et al. 1984). Following this approach, Edward DeLong and colleagues examined ribosomal RNA genes extracted from seawater collected in the Antarctic and off the coast of North America. DeLong and colleagues found many genes that were recognizable, based on their sequences, as belonging to previously unknown archaeans (DeLong 1992; DeLong et al. 1994). Susan Barns and colleagues (1994) likewise looked at rRNA genes extracted directly from mud in a hot spring in Yellowstone National Park. They also detected several rRNAs from previously unknown archaeans. Researchers in several laboratories are now pursuing similar studies (Service 1997).

The new Archaean sequences, plus additional new gene sequences from organisms in the other two domains, will improve our understanding of the universal phylogeny. Barns and colleagues (1996) used the new archaean rRNA sequences in the estimate of the whole-life phylogeny shown in Figure 14.19. This tree suggests the existence of a previously unknown kingdom of archaeans, the Korarchaeota. Also note that some of the deep branches in the Eucarya occur in a different order than in the tree in Figure 14.16.

. . . However, whole-life phylogenies based on molecular data have also yielded rude surprises. . . .

The second trend that will improve our understanding of both the universal phylogeny and the biology of the most recent common ancestor is the advent of whole-genome sequencing. As of this writing, the complete genomes of at least three eukaryotes (the fruit fly *Drosophila,* the nematode worm *Caenorhabditis,* and the yeast *Saccharomyces*), four archaeans, and 20 bacteria have been sequenced. Efforts are underway to sequence dozens more (Wade 1997). The availability of whole genome sequences gives researchers the opportunity to estimate the universal phylogeny based on information from a great variety of different genes. And this has produced some surprises.

We would expect estimates of the universal phylogeny based on different genes to be broadly congruent. In fact, however, they are sometimes radically incongruent. One example appears in Figure 14.20. This whole life phylogeny, produced by

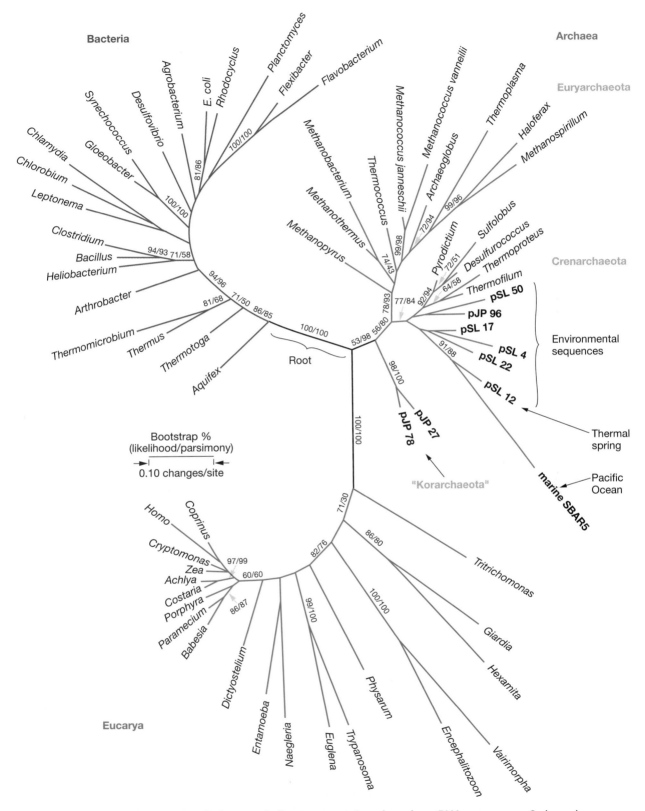

Figure 14.19 **An estimate of the phylogeny of all extant organisms based on rRNA sequences** Codes such as "pSL 22" designate organisms known only from their ribosomal RNA genes. From Barns et al. 1996.

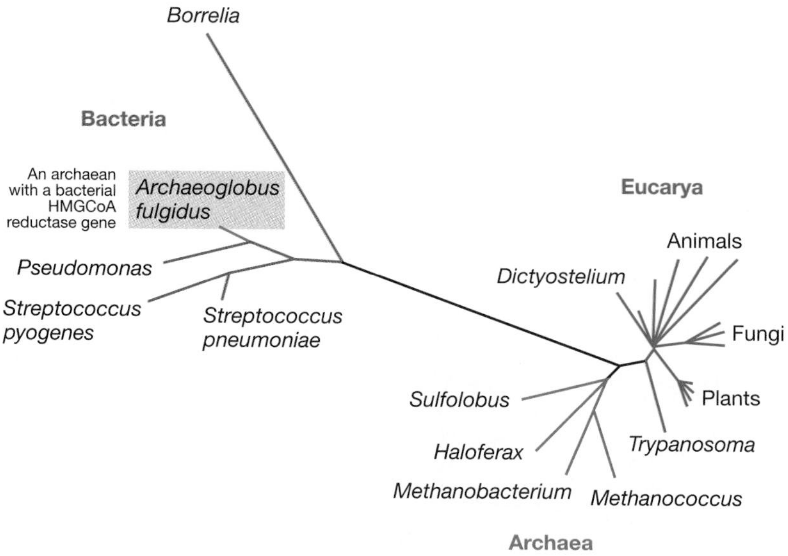

Figure 14.20 **A universal phylogeny based on the HMGCoA reductase gene reveals lateral gene transfer** The organism *Archaeoglobus fulgidus* is an archaean, yet its gene for HMGCoA reductase branches from within the bacteria. The most reasonable explanation is that *Archaeoglobus* has lost its native archaean version of the HMGCoA reductase gene and replaced it with a version of the gene acquired from a bacterium. From Doolittle and Logsdon (1998), after Doolittle (2000). Illustrated by C. Blumrich. Reprinted by permission.

Ford Doolittle and John Logsdon (1998), is based on the gene for an enzyme called HMGCoA reductase. Look at the location on the phylogeny of the HMG-CoA reductase gene from *Archaeoglobus fulgidus. Archaeoglobus fulgidus* is unambiguously an archaean. Its small-subunit rRNA gene places it in the Archaea (see the position of *Archaeoglobus* in Figure 14.19), its machinery for transcription and translation is typical of Archaeans, and it has Archaean lipids in its cell membrane. And yet its gene for HMGCoA reductase appears to be bacterial. How could this be? The likely answer is that its HMGCoA reductase *is* bacterial. In other words, *Archaeoglobus fulgidus,* or one of its ancestors, lost its native Archaean HMGCoA reductase gene and replaced it with a gene picked up from a bacterium. The movement of a gene from one taxon to another is called **lateral gene transfer**.

Researchers working on the universal phylogeny are discovering examples of lateral gene transfer at a rapid pace. Many now believe that early during life's history lateral gene transfer was rampant. So rampant, in fact, that we can no longer think of phylogenies like the ones in Figures 14.16, 14.19, and 14.20 as representing the histories of organismal lineages. Instead, these phylogenies reveal only the histories of the particular genes they are based on. Furthermore, it may be incorrect to think of the last common ancestor of all extant organisms as having been a single species. A more accurate depiction of the common ancestor may be that it was a community of interacting species that readily traded their genes (Woese 1998; Doolittle 2000).

> ... Chief among the rude surprises from whole-life phylogenies is that organisms appear to have swapped their genes much more readily than anyone suspected. This means that the phylogenies of genes may be different from the phylogenies of the organisms that harbor them.

The Latest Possible Date for the Root of the Tree of Life

The organisms at the root of Figure 14.16 could not have lived any more recently than the occurrence of any of the branch points above them on the universal phylogeny. Attempts have been made to calibrate the timing of the branch points using sequence data and molecular clocks (for example, see Doolittle et al. 1996, but also Hasegawa et al. 1996). However, the most definitive information about branching times comes from fossils. The fossils useful in this regard are those that can be confidently identified as belonging to a particular group of organisms. If we can place a fossil on the whole-life phylogeny, then we know that the first branch point is older than the fossil.

Fossils of single-celled organisms are identifiable as eukaryotes if they show sufficient structural complexity. The 590-million-year-old fossil in Figure 14.21a, for example, is clearly eukaryotic: It is about 250μm in diameter and, unlike any known Archaean or Bacterium, it is covered with spikes. The 850-to-950-million-year-old fossil in Figure 14.21b is also eukaryotic: It is about 40μm in diameter and covered with knobs. The 1.4-to-1.5-billion-year-old fossil in Figure 14.21c is about 60μm in diameter, but simple in structure. It is probably, but not certainly, eukaryotic.

The oldest known fossils that are probably those of eukaryotes are 1.85–2.1 billion years old (Figure 14.22). Found by Tsu-Ming Han and Bruce Runnegar (1992) at the Empire Mine in Michigan, these fossils show a spiral-shaped organism similar to a more recent fossil named *Grypania spiralis*. *G. spiralis* is known from fossils in Montana, China, and India that range in age from 1.1 to 1.4 billion years old (see Han and Runnegar 1992). Because of its large size and structural complexity, paleontologists believe that *Grypania* was a eukaryote—probably an alga.

Fossil cyanobacteria also suggest that the root of the universal phylogeny predates 2 billion years (Schopf 1994a). In a testament to the length of time that successful organisms can remain at least superficially unchanged, many fossil cyanobacteria are identifiable based on their structural similarity to extant forms. Each row of Figure 14.23 shows an extant species of cyanobacteria on the left, and a similar fossil form on the right. The fossils range in age from 850 million to 2 billion years old. The extant cyanobacteria occupy a limb of the universal phylogeny

(a)

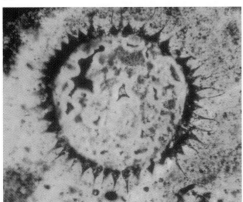

(c)

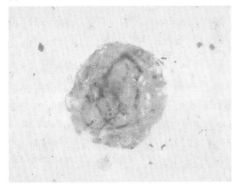

(b)

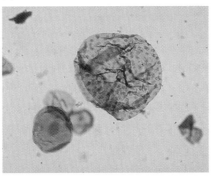

Figure 14.21 Fossils of single-celled Eucarya (a) A spiny fossil from the Doushantuo Formation, China. This 590-million-year-old fossil represents either the preserved cell wall of a single-celled eukaryote, the reproductive cyst of a multicellular alga, or the egg case of an early animal. The fossil has a diameter of about 250μm. From Knoll (1994); see also Knoll (1992). (b) A structurally complex fossil from the Miroyedicha Formation, Siberia. This 850-to-950-million-year-old fossil is clearly eukaryotic, but like that shown in (a), its exact identification is unclear. It probably represents a single-celled organism. It has a diameter of about 40μm. From Knoll (1994); see also Knoll (1992). (c) Fossil cell from the Roper Group, Australia. This cell is 1.4 to 1.5 billion years old; it probably represents a eukaryote, but it lacks sufficient complexity for a definitive identification. The fossil is about 60μm in diameter. (Andrew H. Knoll, Harvard University)

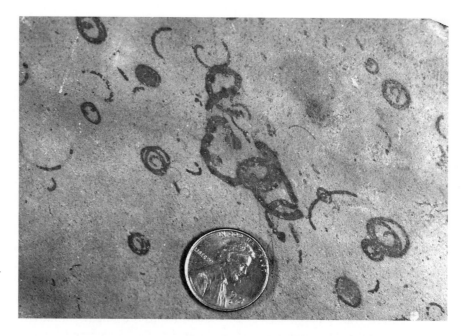

Figure 14.22 2-billion-year-old fossils from Michigan Paleontologists believe these fossils represent eukaryotic algae. The penny is 18.5 mm in diameter. (Tsu-Ming Han)

that is, like those occupied by the extant algae, several branch points above the last common ancestors (Figure 14.16). Again, we can conclude that the last common ancestors lived at least 2 billion years ago, and probably much earlier. If Schopf (1993) is correct in identifying the 3.465-billion-year-old Apex Chert fossils (Figure 14.15) as belonging to a diversity of cyanobacteria, then the organisms at the root of Figure 14.16 must have lived at least 3.5 billion years ago.

In summary, we can use the estimated universal phylogeny, along with geological and paleontological data, to bracket the time when the first branching in the universal phylogeny took place. The earliest possible date is between 4.4 and 3.85 billion years ago; the most recent possible date is between 3.5 and 2 billion years ago.

The Origin of Organelles

If we accept the reconstruction of the most recent common ancestors outlined earlier, then modern Bacteria and Archaea appear to resemble them in many respects. Evolution from the common ancestors to the extant Bacteria and Archaea appears to have been a process of gradual refinement. The Eucarya, on the other hand, are dramatically different. Among other things, eukaryotes have complex cells containing a variety of organelles, and dramatically more complicated genomes, in which the coding regions of genes are frequently interrupted by intervening sequences of noncoding DNA (introns). In addition, many eukaryotes are multicellular, with differentiated cells and tissues. We will only consider one of these problems here, the evolutionary origin of organelles, as we will discuss the evolution of multicellularity in Chapter 15.

The organelles whose evolution is best understood are mitochondria and chloroplasts. Most Eucarya have mitochondria and many also have chloroplasts, but some early-branching eukaryotes, such as *Giardia lamblia* (Figure 14.19), have neither (Sogin et al. 1989). Thus, mitochondria and chloroplasts do not appear to be defining characteristics of the Eucarya; instead, they arose within the eukaryotic branch of the tree of life (but see Palmer 1997).

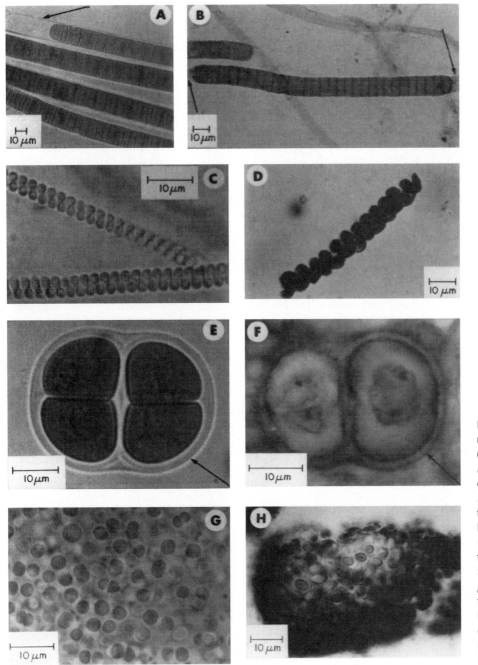

Figure 14.23 Fossil cyanobacteria and their extant relatives (a) *Lyngbya* (extant); (b) *Paleolyngbya* fossil from the 950-million-year-old Lakhanda Formation, Siberia. (c) *Spirulina* (extant); (d) *Heliconema* fossil from the 850-million-year-old Miroedikha Formation, Siberia. (e) *Gloeocapsa* (extant); (f) *Gloeodiniopsis* fossil from the 1.55-billion-year-old Satka Formation, Bashkiria. (g) *Entophysalis* (extant); (h) *Eoentophysalis* fossil from the 2-billion-year-old Belcher Group, Canada. For more details see Schopf (1994a). (J. William Schopf, University of California at Los Angeles)

Superficially, mitochondria and chloroplasts resemble simple bacteria (Schwartz and Dayhoff, 1978). With the discovery that mitochondria and chloroplasts have their own chromosomes, and that these chromosomes are small circular DNA molecules similar to those of bacteria, biologists began to take seriously an old idea that had been dismissed by all but a persistent few. This idea is that mitochondria and chloroplasts originated as bacteria that lived as internal symbionts of early eukaryotic cells (Margulis 1970; 1993). The definitive test of this hypothesis was to sequence genes from mitochondria and chloroplasts, such as the genes for the small-subunit ribosomal RNA, then determine their position on the

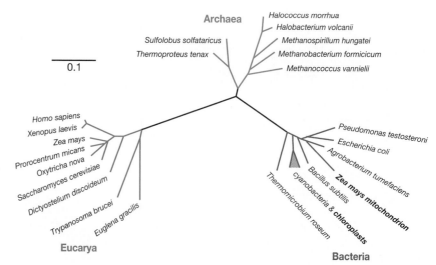

Figure 14.24 **An estimate of the universal phylogeny, showing the locations of the mitochondria and chloroplasts** The tree is based on sequences of small-subunit rRNA genes. The mitochondria are represented by that of *Zea mays* (corn). The chloroplasts are represented by a variety of species, all of which branch within the region shaded blue. The scale bar represents the number of point mutations fixed per nucleotide. From Giovannoni et al. (1988).

Whole-life molecular phylogenies have allowed researchers to test hypotheses about the origin of eukaryotic cells. For example, such phylogenies confirm that mitochondria and chloroplasts are descended from free-living bacteria.

universal phylogeny. If the organelles arose via symbiosis, then their rRNA genes should branch from within the Bacteria; if, instead, the organelles arose independently within the Eucarya, then their rRNA genes should branch from within the Eucarya. The results are in, and they prove that Lynn Margulis was correct in championing the endosymbiosis theory (Figure 14.24).

Mitochondria are closely related to the proteobacteria, previously called purple bacteria (Yang et al. 1985). The proteobacteria include well-known genera such as *Agrobacterium* and *Pseudomonas.* In an evolutionary sense, the mitochondria *are* proteobacteria. Yang et al. (1985) note that the branch of the proteobacteria that holds the mitochondria also includes a variety of other species, including the rhizobacteria, the agrobacteria, and the rickettsias, that typically form close associations with eukaryotic cells. Mitochondria are the descendants of such an association, one that became intimate to the point of mutual dependence.

Chloroplasts and other photosynthetic organelles are cyanobacteria (Giovannoni et al. 1988). Some eukaryotes that acquired photosynthetic organelles were later acquired themselves, as photosynthetic organelles. *Cryptomonas F* is an alga whose chloroplasts have two pairs of envelope membranes instead of the more typical single pair. Inside the inner membrane pair, the *Cryptomonas* chloroplast has a typical circular chloroplast chromosome; between the inner and outer membrane pairs, it also has a small nucleuslike organelle, called the nucleomorph. The nucleomorph makes a functional ribosome, which remains between the membrane pairs. Susan Douglas and colleagues (1991) sequenced the small-subunit ribosomal RNA from both the nucleomorph and the nucleus, then placed these sequences on a phylogeny of the eukaryotes. Both rRNAs are clearly of eukaryotic origin, but they are not closely related to each other. The implication is that the outer chloroplast membrane pair and the nucleomorph are vestiges of a eukaryotic ancestor. This ancestor, itself host to a chloroplast, became the endosymbiont of another eukaryotic host.

The process of evolution from the root of the universal phylogeny to modern Bacteria and Archaea was largely one of refinement. The evolution of the Eucarya, in contrast, involved major innovations. Eucarya obtained mitochondria and chloroplasts by acquiring symbiotic bacteria, clearly an important factor in eukaryotic evolution. Other innovations of the Eucarya are the subject of the next chapter.

Summary

Life arose from an abiotic environment about 4 billion years ago. As a consequence of its extreme antiquity, the reconstruction by scientists of this event involves many challenges. Life may have begun only once and spread quickly throughout the Earth. It may have arisen several times, each time only to be extinguished by the vaporization of Earth's water by the impact of meteorites. It may have evolved entirely on the Earth, or had its origins elsewhere in the solar system.

Scientists have broken down the origins of life, regardless of its particulars, into three phases. The first phase would have been the synthesis of the building blocks of life, such as amino acids, nucleotides, and simple carbohydrates, from small inorganic molecules. Many plausible scenarios for these reactions exist, but significant uncertainties remain. The second phase would be the assembly of building blocks into an information-containing and -transmitting polymer, such as RNA. Again, researchers have demonstrated that many of the details of such polymerization may be possible. And the third phase would be the advent of cellular compartmentalization, which would allow significant advances in phenotypic evolution, and lead to the community of cells from which all current life is descended—the cenancestors.

The study of life's origins is a highly collaborative venture, drawing on expertise from such diverse fields as astronomy, geology, chemistry, molecular biology, and evolutionary biology. It has forced us to consider exactly what "life" means. It is an excellent example of how science works, by formulating and testing hypotheses. It also reveals how great progress can be made in the absence, at this time, of a general-consensus viewpoint. Notably, the editors of *Chemical and Engineering News* (December 6, 1999, issue) asked prominent chemists what will be the major scientific questions for the next hundred years. Three responded that the origins of life would be one of the major topics of study. Rita R. Colwell, current director of the National Science Foundation, remarked, "Chemists also will develop self-replicating molecular systems to provide insights into the molecular origins of life." This would be a milestone achievement, and yet it would merely be another piece in an elusive puzzle.

Questions

1. The genesis of life is sometimes said to have required four things: energy, concentration, protection, and catalysis (for example, Cowen 1995). Explain why each of these four things was necessary for the generation of the primordial form.

2. It has often been thought that life could develop on Earth only because Earth is just the right distance from the sun. Any closer, and the Earth would have been too hot (like Mercury or Venus); any farther away, and there would not have been sufficient solar energy for the evolution of living things. Recently, communities of organisms have been found in deep-sea vents on Earth. These communities seem to get all of their energy from the vents rather than the sun. That is, the vent communities derive energy from the inner heat of the Earth (which is provided ultimately by radioactivity). How does this discovery inform consideration of whether life might exist on other planets or moons that are not at "the right distance" from the sun?

3. Recently, it has been proposed that the conditions in the solar system are exceedingly rare in the universe at large and that life has a very poor chance of evolving past a microbial-like state, even on other planets in other solar systems (Ward and Brownlee 2000). For one thing, our solar system lies in a region of the galaxy that experiences relatively little bombardment from solid debris and a minimum of ionizing radiation. In fact, the presence of the massive planet Jupiter helps to absorb and eject much of the potentially destructive material that enters our solar system. Using the definitions of life given in this chapter, propose a living system that is based on a chemistry that is more likely to withstand these threats.

4. Many people have attempted to calculate the chance that "advanced" extraterrestrial life exists on other planets. By "advanced," they typically mean highly intelligent lifeforms that have a technological civilization with radio communication, such as has existed on Earth during the past century. One of the fundamental uncertainties in these calculations is the probability that any life at all will evolve on a planet. On Earth, how soon after the Earth became habitable did life appear? How long did it take until eukaryotes appeared? How long until intelligent life appeared? In your opinion, do the answers indicate that the evolution of life (of any kind) on other earth-like planets

is probable or improbable? How about the evolution of intelligent life? How about advanced civilization?

5. Leslie Orgel admitted to John Horgan (1991) that he and Crick intended their directed panspermia hypothesis as "sort of a joke." In their 1973 paper, however, Crick and Orgel treat the idea seriously enough to consider biological patterns that might serve as evidence. They point out, for example, that

> It is a little surprising that organisms with somewhat different [genetic] codes do not exist. The universality of the code follows naturally from an "infective" theory of the origins of life. Life on Earth would represent a clone derived from a single extraterrestrial organism.

Since 1973, biologists have discovered that the genetic code is not universal, and that organisms with "somewhat different codes" do, in fact, exist. Our mitochondria, for example, use a code slightly different from that used by our nuclei (see Ayala and Kiger 1984). Many ciliates and other organisms also have slightly deviant codes (see Osawa et al. 1992). How strongly does the discovery that the genetic code is not universal refute the directed panspermia hypothesis? How strongly does it refute other versions of panspermia? Explain your reasoning. Can you think of other kinds of evidence that could (or do) either support or refute some version of panspermia?

6. Examine closely Figure 14.5. What fraction of the 140 randomized nucleotide positions did not mutate by the ninth round of selection and reproduction? Why did these positions show no mutations?

7. In the experiment diagrammed in Figure 14.6, why was it important for the researchers to include a tag on the 5′ end of the small substrate RNAs?

8. In the entire chain of events leading from the abiotic synthesis of biological building blocks to the evolution of eukaryotes, which transition appears to be the least characterized? Why do you think this is the case?

9. Imagine an extremely primitive organism that has very primitive ribosomes with no proteins. Would it be possible to place this organism on the phylogenies in Figure 14.16 or Figure 14.19? Why or why not? How about an organism with no ribosomes? (Can you think of such an organism?) Is it conceivable that there are some as-yet-undiscovered primitive organisms that cannot be placed on these phylogenies? How would the discovery of such organisms affect our reconstruction of the cenancestor?

10. When biologists worked out the details of DNA replication in bacteria and eukaryotes, many researchers were surprised to discover that there are several different DNA polymerases, each with a different role. The machinery for replication seemed enormously complex, and every piece seemed essential if the whole system was to function at all. Many people found it hard to imagine how such a complex system of interdependent parts could have evolved by natural selection. Does the discovery of organisms with only one DNA polymerase (such as *Methanococcus jannaschii*) offer new insight into the evolution of replication? Why or why not?

11. Suppose you are trekking through remote Greenland on a day off from your summer job at a scientific camp and you find an unusual layer of sedimentary rock that is not mapped on your geological charts. You suspect this rock might be even older than the 3.85-billion-year-old rocks from Akilia Island. What would you do to determine whether these rocks have any evidence of ancient life? What results would show that life was indeed present before 3.85 billion years ago?

Exploring the Literature

12. Many of the fields discussed in this chapter are progressing very rapidly. Search Current Contents, the Science Citation Index, or another on-line database for the most recent papers by the researchers whose work we have cited. What has changed since this writing?

13. During their first 2 billion years on earth, organisms wrought dramatic changes in the chemical composition of the atmosphere and oceans. For an introduction, see

Schopf, J. W. 1992. The oldest fossils and what they mean. In J. W. Schopf, ed. *Major Events in the History of Life*. Boston: Jones and Bartlett Publishers, 29–63.

14. The possibility that life exists on Mars has engendered a great debate, and all evidence for or against is heavily scrutinized. The argument in favor is clearly presented in

Gibson, E. K., Jr., et al. 1997. The case for relic life on Mars. *Scientific American* 277(6): 58–65.

15. Currently scientists are in the process of seeking and judging the utility of several "biomarkers" that are chemical traces that life exists or existed in a particular environment or in a fossil. For a recent discussion of the use of biomarkers to date eukaryotic fossils to at least 2.7 billion years ago, see

Brocks, J. J. et al. 1999. Archean molecular fossils and the early rise of eukaryotes. *Science* 285: 1033–1036.

16. Zhu Shixing and Chen Huineng reported 1.7-billion-year-old fossils that they interpreted as representing multicellular algae with differentiated tissues. See

   Zhu Shixing and Chen Huineng. 1995. Megascopic multicellular organisms from the 1700-million-year-old Tuanshanzi Formation in the Jixian area, North China. *Science* 270: 620–622.

Examine Zhu and Chen's photographs. How convincing is their evidence?

17. The organisms on the deepest eukaryotic branches in Figure 14.16 were long thought to lack mitochondria. For a review of recent evidence suggesting that this belief is mistaken, see

   Palmer, J. D. 1997. Organelle genomes: Going, going, gone! *Science* 275: 790–791.

Citations

Altermann, W., and J. W. Schopf. 1995. Microfossils from the Neoarchean Campbell Group, Griqualand West Sequence of the Transvaal Supergroup, and their paleoenvironmental and evolutionary implications. *Precambrian Research* 75: 65–90.

Anders, E. 1989. Prebiotic organic matter from comets and asteroids. *Nature* 342: 255–257.

Anders, E., C. K. Shearer, J. J. Papike, J. F. Bell, S. J. Clemett, R. N. Zare, D. S. McKay, K. L. Thomas-Keprta, C. S. Romanek, E. K. Gibson, Jr., and H. Vali. 1996. Evaluating the evidence for past life on Mars. *Science* 274: 2119–2125.

Arrhenius, S. 1908. *Worlds in the Making.* New York: Harper and Row.

Ayala, F. J., and J. A. Kiger, Jr. 1984. *Modern Genetics.* Menlo Park, CA: Benjamin/Cummings.

Badash, L. 1989. The Age-of-the-Earth Debate. *Scientific American* 261: 90–96.

Baldauf, S. L., J. D. Palmer, and W. F. Doolittle. 1996. The root of the universal tree and the origin of eukaryotes based on elongation factor phylogeny. *Proceedings of the National Academy of Sciences, USA* 93: 7749–7754.

Barns, S. M., R. E. Fundyga, M. W. Jeffries, and N. R. Pace. 1994. Remarkable archaeal diversity detected in a Yellowstone National Park hot spring environment. *Proceedings of the National Academy of Sciences, USA* 91: 1609–1613.

Barns, S. M., C. F. Delwiche, J. D. Palmer, and N. R. Pace. 1996. Perspectives on archaeal diversity, thermophily and monophyly from environmental rRNA sequences. *Proceedings of the National Academy of Sciences, USA* 93: 9188–9193.

Bartel, D. P. 1999. Recreating an RNA replicase. In R. F. Gesteland, T. R. Cech, and J. F. Atkins, eds. *The RNA World,* 2nd ed. Cold Spring Harbor, NY: Cold Spring Harbor Laboratory Press, 143–162.

Bartel, D. P., and J. W. Szostak. 1993. Isolation of new ribozymes from a large pool of random sequences. *Science* 261: 1411–1418.

Bartel, D. P., and P. J. Unrau. 1999. Constructing an RNA World. *Trends in Cell Biology* 9: M9–M13.

Beaudry, A. A., and G. F. Joyce. 1992. Directed evolution of an RNA enzyme. *Science* 257: 635–641.

Belton, M. J. S., et al. 1996. Galileo's first images of Jupiter and the Galilean satellites. *Science* 274: 377–385.

Benner, S. A., A. D. Ellington, and A. Tauer. 1989. Modern metabolism as a palimpsest of the RNA world. *Proceedings of the National Academy of Sciences, USA* 86: 7054–7058.

Brown, J. R., and W. F. Doolittle. 1995. Root of the universal tree of life based on ancient aminoacyl-tRNA synthetase gene duplications. *Proceedings of the National Academy of Sciences, USA* 92: 2441–2445.

Bult, C. J., et al. 1996. Complete genome sequence of the methanogenic archaeon, *Methanococcus jannaschii. Science* 273: 1058–1073.

Cairns-Smith, A. G., A. J. Hall, and M. J. Russell. 1992. Mineral theories of the origin of life and an iron sulfide example. *Origins of Life and Evolution of the Biosphere* 22: 161–180.

Chang, S. 1999. Planetary environments and the origin of life. *Biological Bulletin* 196: 308–310.

Chyba, C. F. 2000. Energy for microbial life on Europa. *Nature* 403: 381–382.

Chyba, C. F., P. J. Thomas, L. Brookshaw, and C. Sagan. 1990. Cometary delivery of organic molecules to the early Earth. *Science* 249: 366–373.

Cowen, R. 1995. *History of Life,* 2nd ed. Cambridge: Blackwell Scientific Publications.

Crick, F. H. C., and L. E. Orgel. 1973. Directed panspermia. *Icarus* 19: 341–346.

Cruikshank, D. P. 1997. Stardust memories. *Science* 275: 1895–1896.

Darwin, F. 1887. *The Life and Letters of Charles Darwin,* vol. 2. New York: Appleton.

DeDuve, C. 1991. *Blueprint for a Cell: The Nature and Origin of Life.* Burlington, NC: Patterson.

DeLong, E. F. 1992. Archaea in coastal marine environments. *Proceedings of the National Academy of Sciences, USA* 89: 5685–5689.

DeLong, E. F., K. Y. Wu, B. B. Prézelin, and R. V. M. Jovine. 1994. High abundance of Archaea in Antarctic marine picoplankton. *Nature* 371: 695–697.

Doolittle, R. F., D.-F. Feng, S. Tsang, G. Cho, and E. Little. 1996. Determining divergence times of the major kingdoms of living organisms with a protein clock. *Science* 271: 470–477.

Doolittle, W. F. 1999. Phylogenetic classification and the universal tree. *Science* 284: 2124–2128.

Doolittle, W. F. 2000. Uprooting the tree of life. *Scientific American* (February): 90–95.

Doolittle, W. F., and J. R. Brown. 1994. Tempo, mode, the progenote, and the universal root. *Proceedings of the National Academy of Sciences, USA* 91: 6721–6728.

Doolittle, W. F., and J. M. Logsdon, Jr. 1998. Archaeal genomics: Do archaea have a mixed heritage? *Current Biology* 8: R209–R211.

Douglas, S. E., C. A. Murphy, D. F. Spencer, and M. W. Gray. 1991. Cryptomonad algae are evolutionary chimaeras of two phylogenetically distinct unicellular eukaryotes. *Nature* 350: 148–151.

Dose, K., and A. Klein. 1996. Response of Bacillus subtilis spores to dehydration and UV irradiation at extremely low temperatures. *Origins of Life and Evolution of the Biosphere* 26: 47–59.

Egholm, M., O. Buchardt, L. Christensen, P. E. Nielson, and R. H. Berg. 1992. Peptide nucleic acids (PNA). Oligonucleotide analogues with an achiral peptide backbone. *Journal of the American Chemical Society* 114: 1895–1897.

Ekland, E. H., J. W. Szostak, and D. P. Bartel. 1995. Structurally complex and highly active RNA ligases derived from random RNA sequences. *Science* 269: 364–370.

Ekland, E. H., and D. P. Bartel. 1996. RNA-catalysed RNA polymerization using nucleoside triphosphates. *Nature* 382: 373–376.

Ellington, A. D., and J. W. Szostak. 1990. *In vitro* selection of RNA molecules that bind specific ligands. *Nature* 346: 818–822.

Eschenmoser A. 1999. Chemical etiology of nucleic acid structure. *Science* 284: 2118–2124.

Ferris, J. P. 1993. Catalysis and prebiotic RNA synthesis. *Origins of Life and Evolution of the Biosphere* 23: 307–315.

Ferris, J. P., A. R. Hill, Jr., R. Liu, and L. E. Orgel. 1996. Synthesis of long prebiotic oligomers on mineral surfaces. *Nature* 381: 59–61.

Fitch, W. M., and K. Upper. 1987. The phylogeny of tRNA sequences provides evidence for ambiguity reduction in the origin of the genetic code. *Cold Spring Harbor Symposia on Quantitative Biology* 52: 759–767.

Fox, G. E., L. J. Magrum, W. E. Balch, R. S. Wolfe, and C. R. Woese. 1977. Classification of methanogenic bacteria by 16S ribosomal RNA characterization. *Proceedings of the National Academy of Sciences, USA* 74: 4537–4541.

Fox, S. W. 1988. *The Emergence of Life: Darwinian Evolution from the Inside.* New York: Basic Books.

Fox, S. W. 1991. Synthesis of life in the lab? Defining a protoliving system. *Quarterly Review of Biology* 66: 181–185.

Fox, S. W., and K. Dose. 1972. *Molecular Evolution and the Origin of Life.* San Francisco: W. H. Freeman.

Gesteland, R. F., T. R. Cech, and J. F. Atkins, eds. 1999. *The RNA World,* 2nd ed. Cold Spring Harbor, NY: Cold Spring Harbor Laboratory Press.

Geyer, C. R., and D. Sen. 1997. Evidence for the metal-cofactor independence of an RNA phosphodiester-cleaving DNA enzyme. *Chemistry and Biology* 4: 579–593.

Gilbert, W. 1986. The RNA world. *Nature* 319: 618.

Giovannoni, S. J., S. Turner, G. J. Olsen, S. Barns, D. J. Lane, and N. R. Pace. 1988. Evolutionary relationships among cyanobacteria and green chloroplasts. *Journal of Bacteriology* 170: 3584–3592.

Gladman, B. J., and J. A. Burns. 1996. Mars meteorite transfer: Simulation. *Science* 274: Letters.

Gladman, B. J., J. A. Burns, M. Duncan, P. Lee, and H. F. Levison. 1996. The exchange of impact ejecta between terrestrial planets. *Science* 271: 1387–1392.

Gogarten, J. P., et al. 1989. Evolution of the vacuolar H1-ATPase: Implications for the origin of eukaryotes. *Proceedings of the National Academy of Sciences, USA* 86: 6661–6665.

Han, T.-M., and B. Runnegar. 1992. Megascopic eukaryotic algae from the 2.1-billion-year-old Negaunee Iron-Formation, Michigan. *Science* 257: 232–235.

Hasegawa, M., W. M. Fitch, J. P. Gogarten, L. Olendzenski, E. Hilario, C. Simon, K. E. Holsinger, R. F. Doolittle, D.-F. Feng, S. Tsang, G. Cho, and E. Little. 1996. Dating the cenancestor of organisms. *Science* 274: 1750–1753.

Hayes, J. M. 1996. The earliest memories of life on Earth. *Nature* 384: 21–22.

Hill, A. R., Jr., C. Böhler, and L.E. Orgel. 1998. Polymerization on the rocks: Negatively-charged α-amino acids. *Origins of Life and Evolution of the Biosphere* 28: 235–243.

Horgan, J. 1991. In the beginning ... *Scientific American* 264(2): 116–125.

Illangasekare, M., G. Sanchez, T. Nickles, and M. Yarus. 1995. Aminoacyl-RNA synthesis catalyzed by an RNA. *Science* 267: 643–647.

Iwabe, N., K.-i. Kuma, M. Hasegawa, S. Osawa, and T. Miyata. 1989. Evolutionary relationship of archaebacteria, eubacteria, and eukaryotes inferred from phylogenetic trees of duplicated genes. *Proceedings of the National Academy of Sciences, USA* 86: 9355–9359.

Jones, W. J., J. A. Leigh, F. Mayer, C. R. Woese, and R. S. Wolfe. 1983. *Methanococcus jannaschii sp. nov.*, an extremely thermophilic methanogen from a submarine hydrothermal vent. *Archives of Microbiology* 136: 254–261.

Joyce, G. F. 1989. RNA evolution and the origins of life. *Nature* 338: 217–224.

Joyce, G. F. 1996. Ribozymes: Building the RNA world. *Current Biology* 6: 965–967.

Joyce, G. F. 1998. Nucleic acid enzymes: Playing with a fuller deck. *Proceedings of the National Academy of Sciences, USA* 95: 5845–5847.

Joyce, G. F., A. Schwartz, S. L. Miller, and L. Orgel. 1987. The case for an ancestral genetic system involving simple analogues of the nucleotides. *Proceedings of the National Academy of Sciences, USA* 84: 4398–4402.

Joyce, G. F. and L. E. Orgel. 1999. Prospects for understanding the RNA world. In R. F. Gesteland, T. R. Cech, and J. F. Atkins, eds. *The RNA World,* 2nd ed. Cold Spring Harbor, NY: Cold Spring Harbor Laboratory Press, 49–77.

Kasting, J. F. 1993. Earth's early atmosphere. *Science* 259: 920–926.

Kerr, R. A. 1996. Galileo turns geology upside down on Jupiter's icy moons. *Science* 274: 341.

Kerr, R. A. 1997a. Martian "microbes" cover their tracks. *Science* 276: 30–31.

Kerr, R. A. 1997b. An ocean emerges on Europa. *Science* 276: 355.

Knoll, A. H. 1992. The early evolution of eukaryotes: A geological perspective. *Science* 256: 622–627.

Knoll, A. H. 1994. Proterozoic and Early Cambrian protists: Evidence for accelerating evolutionary tempo. *Proceedings of the National Academy of Sciences, USA* 91: 6743–6750.

Koonin, E. V., and A. R. Mushegian. 1996. Complete genome sequences of cellular life forms: Glimpses of theoretical evolutionary genomics. *Current Opinion in Genetics and Development* 6: 757–762.

Kruger, K., Grabowski, P. J., Zaug, A. J., Sands, J., Gottschling, D. E., and T. R. Cech. 1982. Self-splicing RNA: Autoexcision and autocatalyzation of the ribosomal RNA intervening sequence of *Tetrahymena. Cell* 31: 147–157.

Kvenvolden, K., et al. 1970. Evidence for extraterrestrial amino acids and hydrocarbons in the Murchison meteorite. *Nature* 228: 923–926.

Lake, J. A., M. Rivera, R. Haselkorn, W. J. Buikema, S. W. Golomb, G. J. Olsen, C. R. Woese, O. White, and J. C. Venter. 1996. Methanococcus genome. *Science* 274: 901–903.

Lazcano, A., and S. L. Miller. 1994. How long did it take for life to begin and evolve to cyanobacteria? *Journal of Molecular Evolution* 39: 546–554.

Lazcano, A., and S. L. Miller. 1996. The origin and early evolution of life: Prebiotic chemistry, the pre-RNA world, and time. *Cell* 85: 793–798.

Lehman, N., and G. F. Joyce. 1993. Evolution *in vitro* of an RNA enzyme with altered metal dependence. *Nature* 361: 182–185.

Leipe, D. D. 1996. Biodiversity, genomes, and DNA sequence databases. *Current Opinion in Genetics and Development* 6: 686–691.

Lorsch, J. R., and J. W. Szostak. 1994. *In vitro* evolution of new ribozymes with polynucleotide kinase activity. *Nature* 371: 31–36.

Madigan, M. T., and B. L. Marrs. 1997. Extremophiles. *Scientific American* 276(4): 82–87.

Margulis, L. 1970. *Origin of Eukaryotic Cells.* New Haven, CT: Yale University Press.

Margulis, L. 1993. *Symbiosis in Cell Evolution.* New York: Freeman.

Margulis, L. 1996. Archaeal-eubacterial mergers in the origin of Eukarya: Phylogenetic classification of life. *Proceedings of the National Academy of Sciences, USA* 93: 1071–1076.

McKay, D. S., et al. 1996. Search for past life on Mars: Possible relic biogenic activity in martian meteorite ALH84001. *Science* 273: 924–930.

Miller, S. L. 1953. A production of amino acids under possible primitive Earth conditions. *Science* 117: 528–529.

Miller, S. L. 1992. The prebiotic synthesis of organic compounds as a step toward the origin of life. In J. W. Schopf, ed. *Major Events in the History of Life.* Boston: Jones and Bartlett, 1–28.

Mills, D. R., R. L. Peterson, and S. Spiegelman. 1967. An extracellular Darwinian experiment with a self-duplicating nucleic acid molecule. *Proceedings of the National Academy of Sciences, USA* 58: 217–220.

Mitchell, F. J., and W. L. Ellis. 1972. Surveyor 3: Bacterium isolated from lunar-retrieved television camera. In *Analysis of Surveyor 3 Material and Photographs Returned by Apollo 12.* Washington, DC: National Aeronautics and Space Administration, 239–248.

Mojzsis, S. J., G. Arrhenius, K. D. McKeegan, T. M. Harrison, A. P. Nutman, and C. R. L. Friend. 1996. Evidence for life on Earth before 3,800 million years ago. *Nature* 384: 55–59.

Mojzsis, S. J., R. Krishnamurthy, and G. Arrhenius. 1999. Before RNA and after: Geophysical and geochemical constraints on molecular evolution. In

R. F. Gesteland, T. R. Cech, and J. F. Atkins, eds. *The RNA World,* 2nd ed. Cold Spring Harbor, NY: Cold Spring Harbor Laboratory Press. 1–47.

Nitta I., T. Ueda, and K. Watanabe. 1998. Possible involvement of *Escherichia coli* 23S ribosomal RNA in peptide bond formation. *RNA* 4: 257–267.

Noller, H. F., V. Hoffarth, and L. Zimniak. 1992. Unusual resistance of peptidyl transferase to protein extraction procedures. *Science* 256: 1416–1419.

Olsen, G. J., and C. R. Woese. 1996. Lessons from an Archaeal genome: What are we learning from *Methanococcus jannaschii*? *Trends in Genetics* 12: 377–379.

Orgel, L. E. 1986. RNA catalysis and the origins of life. *Journal of Theoretical Biology* 123: 127–149.

Oró, J. 1961. Mechanism of synthesis of adenine from hydrogen cyanide under plausible primitive earth conditions. *Nature* 191: 1193–1194.

Osawa, S., T. H. Jukes, K. Watanabe, and A. Muto. 1992. Recent evidence for evolution of the genetic code. *Microbiological Reviews* 56: 229–264.

Palmer, J. D. 1997. Organelle genomes: Going, going, gone! *Science* 275: 790–791.

Rivera, M. C., and J. A. Lake. 1992. Evidence that eukaryotes and eocyte prokaryotes are immediate relatives. *Science* 257: 74–76.

Rogers, J., and Joyce, G. F. 1999. A ribozyme that lacks cytidine. *Nature* 402: 323–325.

Schidlowski, M. 1988. A 3,800-million-year isotopic record of life from carbon in sedimentary rocks. *Nature* 333: 313–318.

Schopf, J. W., ed. 1983. *Earth's Earliest Biosphere: Its Origin and Evolution.* Princeton, NJ: Princeton University Press.

Schopf, J. W. 1992. The oldest fossils and what they mean. In J. W. Schopf, ed. *Major Events in the History of Life.* Boston: Jones and Bartlett, 29–63.

Schopf, J. W. 1993. Microfossils of the early Archean apex chert: New evidence of the antiquity of life. *Science* 260: 640–646.

Schopf, J. W. 1994a. Disparate rates, differing fates: Tempo and mode of evolution changed from the precambrian to the phanerozoic. *Proceedings of the National Academy of Sciences, USA* 91: 6735–6742.

Schopf, J. W. 1994b. The early evolution of life: Solution to Darwin's dilemma. *Trends in Ecology and Evolution* 9: 375–378.

Schopf, J. W., and B. M. Packer. 1987. Early Archean (3.3-billion to 3.5-billion-year-old) microfossils from Warrawoona Group, Australia. *Science* 237: 70–73.

Schwartz, R. M., and M. O. Dayhoff. 1978. Origins of prokaryotes, eukaryotes, mitochondria, and chloroplasts. *Science* 199: 395–403.

Secker, J., J. Lepock, and P. Wesson. 1994. Damage due to ultraviolet and ionizing radiation during the ejection of shielded micro-organisms from the vicinity of 1 M main sequence and red giant stars. *Astrophysics and Space Science* 219: 1–28.

Service, R. F. 1997. Microbiologists explore life's rich, hidden kingdoms. *Science* 275: 1740–1742.

Sleep, N. H., K. J. Zahnle, J. F. Kasting, and H. J. Morowitz. 1989. Annihilation of ecosystems by large asteroid impacts on the early Earth. *Nature* 342: 139–142.

Sogin, M. L., J. H. Gunderson, H. J. Elwood, R. A. Alonso, and D. A. Peattie. 1989. Phylogenetic meaning of the kingdom concept: An unusual ribosomal RNA from *Giardia lamblia*. *Science* 243: 75–77.

Stahl, D. A., D. J. Lane, G. J. Olsen, and N. R. Pace. 1984. Analysis of hydrothermal vent-associated symbionts by ribosomal RNA sequences. *Science* 224: 409–411.

Szathmáry, E., and I. Gladkih. 1989. Suboptimal growth and coexistence of nonenzymatically replicating templates. *Journal of Theoretical Biology* 138: 55–58.

Tang, J., and R. R. Breaker. 1997. Rational design of allosteric ribozymes. *Chemistry and Biology* 4: 453–459.

Tarasow, T. M., S. L. Tarasow, and B. E. Eaton. 1997. RNA-catalyzed carbon–carbon bond formation. *Nature* 389: 54–57.

Tuerk, C., and L. Gold. 1990. Systematic evolution of ligands by exponential enrichment: RNA ligands to bacteriophage T4 DNA polymerase. *Science* 249: 505–510.

Unrau, P. J., and D. P. Bartel. 1998. RNA-catalyzed nucleotide synthesis. *Nature* 395: 260–263.

Voet, A. B., and A. W. Schwartz. 1982. Uracil synthesis via HCN oligomerization. *Origins of Life* 12: 45–49.

Wade, N. 1997. Thinking small paying off big in gene quest. *The New York Times,* February 3: A1; A14.

Ward, P. D., and D. Brownlee. 2000. *Rare Earth: Why Complex Life Is Uncommon in the Universe.* New York: Copernicus Books.

Weber, P., and J. M. Greenberg. 1985. Can spores survive in interstellar space? *Nature* 316: 403–407.

Wetherill, G. W. 1990. Formation of the Earth. *Annual Review of Earth and Planetary Science* 18: 205–256.

Whittaker, R. H. 1969. New concepts of kingdoms of organisms. *Science* 163: 150–160.

Woese, C. R. 1991. The use of ribosomal RNA in reconstructing evolutionary relationships among bacteria. In R. K. Selander, A. G. Clark, and T. S. Whittam, eds. *Evolution at the Molecular Level.* Sunderland, MA: Sinauer, 1–24.

Woese, C. R. 1996. Phylogenetic trees: Whither microbiology? *Current Biology* 6: 1060–1063.

Woese, C. R. 1998. The universal ancestor. *Proceedings of the National Academy of Sciences, USA* 95: 6854–6859.

Woese, C. R., and G. E. Fox. 1977. Phylogenetic structure of the prokaryotic domain: The primary kingdoms. *Proceedings of the National Academy of Sciences, USA* 74: 5088–5090.

Woese, C. R., O. Kandler, and M. L. Wheelis. 1990. Towards a natural system of organisms: Proposal for the domains Archaea, Bacteria, and Eucarya. *Proceedings of the National Academy of Sciences, USA* 87: 4576–4579.

Yang, D., Y. Oyaizu, H. Oyaizu, G. J. Olsen, and C. R. Woese. 1985. Mitochondrial origins. *Proceedings of the National Academy of Sciences, USA* 82: 4443–4447.

Zaug, A. J., and T. R. Cech. 1986. The intervening sequence RNA of *Tetrahymena* is an enzyme. *Science* 231: 470–475.

Zhang, B., and T. R. Cech. 1997. Peptide bond formation by in vitro selected ribozymes. *Nature* 390: 96–100.

Zillig, W. 1991. Comparative biochemistry of Archaea and Bacteria. *Current Opinion in Genetics and Development* 1: 544–551.

Zuckerkandl, E., and L. Pauling. 1965. Molecules as documents of evolutionary history. *Journal of Theoretical Biology* 8: 357–366.

CHAPTER 15

The Cambrian Explosion and Beyond

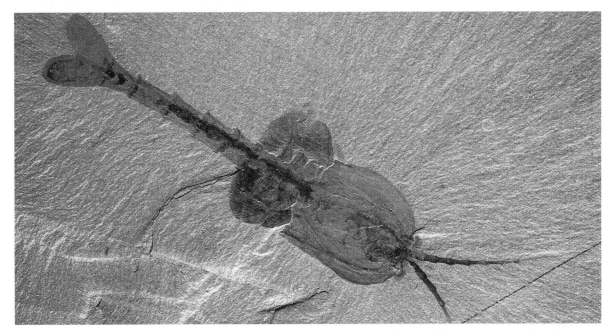

This fossil, of a crustacean called *Waptia fieldensis,* is over 515 million years old. (Chip Clark/Stuart Kenter Associates)

ONCE THE FUNDAMENTAL LIFE PROCESSES OF DNA REPLICATION, PROTEIN synthesis, respiration, and cell division had evolved, a spectacular diversification of life ensued. Breakthroughs like the evolution of photosynthesis and the nuclear membrane occurred. These events, along with others we reviewed in Chapter 14, spanned some 3.2 billion years and created the deep branches on the tree of life. During this interval all organisms, with the exception of some red, brown, and green algae, were unicellular.

The first multicellular animals do not appear in the fossil record until about 565 million years ago (Ma). But then, over a span of just 40 million years, virtually every major phylum of animals appeared. This period is called the Cambrian explosion. Although it represents a mere 0.8% of the Earth's history, it ranks as one of the great events in the history of life. In addition to the Cambrian explosion and other more recent periods of rapid species diversification, the history of organismal evolution is marked by several episodes of cataclysmic extinction. Our goal in this chapter is to introduce these and other turning points that have taken place over the past 543 million years. This interval in earth history is called the Phanerozoic ("visible life") eon.

We focus on the broad patterns of multicellular organismal evolution in the Phanerozoic. What are the major groups, and when did they appear? What patterns

507

can we detect in the history of adaptation and diversification? Why did mass extinctions occur during the Phanerozoic? We start, however, with the basics: a look at how paleontologists read the fossil record and document the history of life.

15.1 The Nature of the Fossil Record

In Chapter 2, we introduced the geological time scale that was established by paleontologists in the early 19th century. We also looked at how 20th-century geochronologists are using radioactive isotopes to estimate the absolute age of each eon, era, period, and epoch. We now elaborate on how life has changed through time by reviewing the process of fossilization, examining the strengths and weaknesses of the fossil record, and presenting a time line of major events in evolution.

How Organic Remains Fossilize

A **fossil** is any trace left by an organism that lived in the past. Fossils are enormously diverse, but four general categories can be defined by method of formation. In the list that follows, there are two important issues to focus on: Which part of the organism is preserved and available for study? What kinds of habitats produce fossils?

- Compression and impression fossils (Figure 15.1) can result when organic material is buried in water- or wind-borne sediment before it decomposes. Under the weight of sand, mud, ash, or other particles deposited above, a structure can leave an impression in the material below. The resulting fossil is analogous to the record left by footprints in mud or leaves in wet concrete.
- Permineralized fossils (Figure 15.2) can form when structures are buried in sediments and dissolved minerals precipitate in the cells. This process, which is analogous to the way a microscopist embeds tissues with resin before sectioning, can preserve details of internal structure.
- Casts and molds (Figure 15.3) originate when remains decay or when shells are dissolved after being buried in sediment. Molds consist of unfilled spaces, while casts form when new material infiltrates the space, fills it, and hardens into rock.

Figure 15.1 Compression and impression fossils   These are two-dimensional fossils, usually found by splitting sedimentary rocks along the bedding plane. (a) A compression fossil of a Paleocene leaf found near Alberta, Canada. (b) A close-up from the leaf pictured in (a), showing a stomate. From Wilson N. Stewart, *Paleobotany and the Evolution of Plants,* page 12, Fig. 2.5A and 2.5D. New York: Cambridge University Press, 1983. © 1983 Cambridge University Press. Reprinted with the permission of Cambridge University Press.

(a)   (b)

(a)

(b)

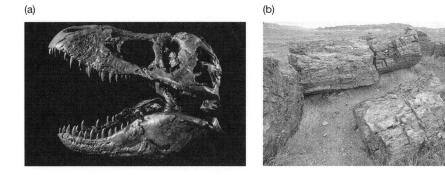

Figure 15.2 **Permineralized fossils** Permineralized fossils are usually found in rock outcrops after they have been partially exposed by natural weathering. (a) The skull of *Tyrannosaurus rex,* a predatory dinosaur (Francis Gohier/Photo Researchers, Inc.). (b) Petrified wood (John D. Cunningham/Visuals Unlimited).

This process is analogous to the lost-wax casting technique used by sculptors. Molds and casts preserve information about external and internal surfaces.

- Unaltered remains (Figure 15.4) are sometimes preserved in environments that discourage loss from weathering, consumption by scavenging animals, and decomposition by bacteria and fungi. How long can organic material remain unaltered? Human cadavers from the Iron Age, buried in the highly acidic environment of peat bogs, have been recovered with some flesh still intact. Woolly mammoths dug out of permafrost have fur and many tissues preserved. Dried but otherwise unaltered dung from giant ground sloths, which lived in the Pleistocene, can be found in protected, desiccating environments like desert caves. Viscous plant resins can harden into amber, preserving the insects trapped inside so well that wing veins are visible. Paleobotanists have driven nails into 100-million-year-old logs recovered from oil-saturated tar sands.

Figure 15.3 **Casts and molds** This is a cast of a horsetail stem from the Carboniferous. (Ted Clutter/Photo Researchers, Inc.)

Though spectacular, intact remains are so rare that they represent only a small fraction of the total fossil record. Compression, impression, casting, molding, and permineralization are far more common. All of these processes depend on three key features of the specimen: durability, burial (usually in a water-saturated sediment), and lack of oxygen. As a result, the fossil record consists primarily of hard structures left in depositional environments such as river deltas, beaches, floodplains, marshes, lakeshores, and sea floors.

It is not hard to understand why these structures predominate, while soft tissues and organisms from upland habitats are rarely preserved. Marine bivalves that burrow are automatically buried in saturated sediments after death. Because enamel is one of the highest-density substances known in nature, teeth decay slowly and have more time to fossilize. Trees that grow in floodplain forests drop their leaves and seeds onto substrates that are frequently washed over with sediment, and the saturated soils of swamps are routinely anaerobic and permit only slow decomposition. Shelled organisms in the marine plankton drift to the ocean floor after death and are likewise buried.

Strengths and Weaknesses of the Fossil Record

There are three main types of bias in the fossil record: geographic, taxonomic, and temporal. We have already mentioned the propensity for fossils to come from lowland and marine habitats. To appreciate the taxonomic bias more fully, consider this:

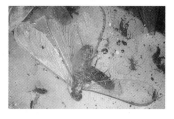

Figure 15.4 **Unaltered remains** This is a winged male termite, preserved in amber, from the Upper Cretaceous of Canada (John Koivula/Science Source/Photo Researchers, Inc.).

Marine organisms dominate the fossil record, but make up only 10% of extant species. A full two-thirds of animal phyla living today lack any sort of mineralized hard parts, such as bone or shell, that are amenable to fossilization. Critical parts of plants, including reproductive structures such as flowers, are seldom preserved. The temporal bias results because the Earth's crust is constantly being recycled. When tectonic plates subduct or mountains erode, their fossils go with them. As a result, old rocks are rarer than new rocks, and our ability to sample ancient life forms is relatively poor.

It is important to realize, however, that these types of sampling issues are by no means unique to paleontology. Advances in developmental genetics depend on the generality of a few model systems, such as *Drosophila melanogaster,* the roundworm *Caenorhabditis elegans,* the annual flower *Arabidopsis thaliana*, corn (*Zea mays),* and zebrafish. Most work in molecular genetics is done on a few phages, the filamentous fungus *Neurospora crassa*, and *E. coli.* The vast majority of research in behavioral ecology is done on birds and mammals; community ecology has historically focused on upland habitats in North America and Europe.

Like all sources of data, the fossil record has inherent strengths and limitations.

The important point is to recognize that, like any source of data, the fossil record has characteristics that limit the types of information that can be retrieved and how broadly the data can be interpreted. The goal for paleontologists is to recognize the constraints and work creatively within them.

With these caveats in place, we begin our intensive use of the fossil record with a broad look at the sequence of events during the Phanerozoic.

Life Through Time: An Overview

The geologic time scale is a hierarchy divided into eons, eras, periods, epochs, and stages. Each named interval is defined by a suite of diagnostic fossils. When the scale was first being formulated, in the early 1800s, the intervals were arranged by relative age only. It was only much later, after the discovery of radioisotopes and the development of accurate dating techniques, that absolute times were assigned to each interval. Consequently, levels in the hierarchy are not equivalent in terms of time. The Paleozoic era lasted 292 million years and the Mesozoic era 186 million, for example. Also, the geologic time scale is a work in progress: Estimates for the absolute ages in the scale are constantly improving as dating techniques become more sophisticated and more rocks are sampled.

Figure 15.5 presents time lines for the three eras that make up the Phanerozoic. The eon begins with the Cambrian explosion and ends in the present. Its three component eras are the Paleozoic (ancient life), Mesozoic (middle life), and Cenozoic (recent life). In addition to offering a compact overview of the history of multicellular life, the diagrams in Figure 15.5 should inspire questions. For example, each of the fossil "firsts" mapped along the top of the time lines represents a suite of new traits. What are these? Why did selection favor them? How long did they persist? Note, too, that many of these events led to the recognition of some organisms as new orders, classes, and phyla. Why did some of them become extinct?

Given enough time and space, we could explore these questions for any of the many events diagrammed on the time lines. However, the history of Phanerozoic evolution includes a smaller number of broad patterns. Because our goal is to introduce how research in contemporary paleontology is done and to illustrate the most important concepts in the field, we have selected a few of these patterns to review in detail.

15.2 The Cambrian Explosion

Almost all of the animal phyla currently recognized by taxonomists make their first appearance in the fossil record over a span of just 40 million years during the Cambrian period, at the beginning of the Phanerozoic. Geologically speaking, this span of time represents the blink of an eye. The unprecedented diversification of multicellular animals during this short time ranks as one of the great events in the history of life. Documenting what happened and testing hypotheses to explain the events have been major goals of paleontology, so we give this great episode particularly close attention. To appreciate its full drama, we need to review the changes that occurred. This means delving into a little classical embryology.

Diversification of Animal Body Plans

Animal phyla are distinguished by how they develop as embryos and how their adult body plan is organized. There are three key divisions (Figure 15.6, page 514):

- **Diploblasts and triploblasts**. Diploblastic animals have two embryonic tissue types. Some are radially symmetrical in whole or in part (like the cnidarians and ctenophores), while others are asymmetrical (like sponges). Triploblasts have three embryonic tissues and bilateral symmetry. The tissue types present in both groups are the ectoderm (outer skin) and endoderm (inner skin). Ectodermal cells produce the adult skin and nervous system, and endodermal cells produce the gut and associated organs. The third embryonic tissue is the mesoderm (middle skin). In the bilateral animals, these cells develop into the gonads, heart, connective tissues, and blood.

  The bilaterally symmetric animals, or Bilateria, have three embryonic tissues.

- **Coelomates, pseudocoelomates, and acoelomates**. Coelomates have a true coelom. This is a fluid-filled cavity in the body that is derived from mesoderm and lined with mesodermal cells called the peritoneum. Coeloms create a "tube within a tube" body plan that provides space for the arrangement of internal organs. Because fluid-filled cavities are required for hydrostatic skeletons to operate, coeloms also triggered improvements in swimming and crawling ability. In contrast, acoelomates have no body cavity. Their mesoderm forms a solid mass between the body wall and the gut. Pseudocoelomates have cavities in their bodies that are not derived from mesoderm and are not lined completely with mesoderm. This condition is found mainly in phyla with very small bodies and is probably derived from larger, coelomate ancestors.

  Most bilaterians have a coelom, or body cavity.

- **Protostomes and deuterostomes**. Although both of these groups have three embryonic tissue types, true coeloms, and bilaterally symmetric body plans, they differ in several important developmental processes, especially how their embryos undergo a key developmental event called gastrulation. This is the mass movement of cells that rearranges the embryonic cells after cleavage and defines the ectoderm, endoderm, and mesoderm. In protostomes, gastrulation forms the mouth region first; in deuterostomes, gastrulation forms the mouth region after the anal region forms. These differences inspired the use of the Greek roots *proto* (first), *deutero* (second), and *stoma* (mouth).

  Within the Bilateria, the protostomes and deuterostomes form distinct groups.

The Cambrian explosion produced other major morphological innovations, including the first segmented body plans, shells, external skeletons, appendages, and notochords. In short, an enormous amount of morphological diversity suddenly

(a)

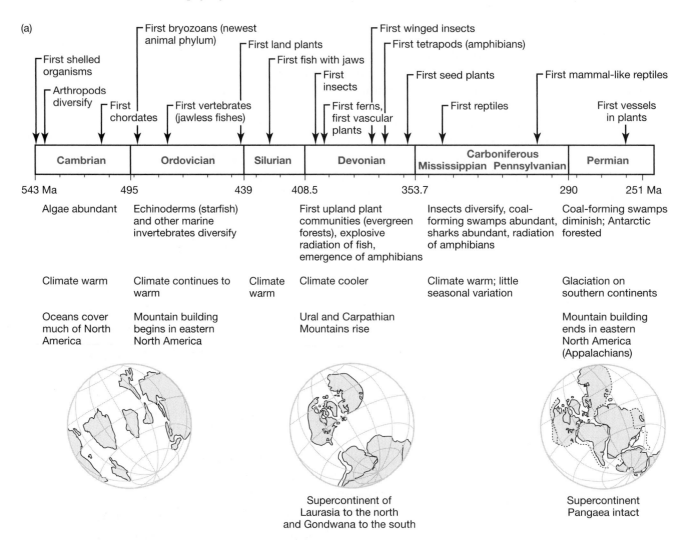

First shelled organisms
Arthropods diversify
First chordates
First bryozoans (newest animal phylum)
First vertebrates (jawless fishes)
First land plants
First fish with jaws
First insects
First ferns, first vascular plants
First winged insects
First tetrapods (amphibians)
First seed plants
First reptiles
First mammal-like reptiles
First vessels in plants

Cambrian	Ordovician	Silurian	Devonian	Carboniferous Mississippian Pennsylvanian	Permian

543 Ma   495   439   408.5   353.7   290   251 Ma

Algae abundant

Echinoderms (starfish) and other marine invertebrates diversify

First upland plant communities (evergreen forests), explosive radiation of fish, emergence of amphibians

Insects diversify, coal-forming swamps abundant, sharks abundant, radiation of amphibians

Coal-forming swamps diminish; Antarctic forested

Climate warm

Climate continues to warm

Climate warm

Climate cooler

Climate warm; little seasonal variation

Glaciation on southern continents

Oceans cover much of North America

Mountain building begins in eastern North America

Ural and Carpathian Mountains rise

Mountain building ends in eastern North America (Appalachians)

Supercontinent of Laurasia to the north and Gondwana to the south

Supercontinent Pangaea intact

Figure 15.5   The Phanerozoic eon   The diagrams show a selection of events from the three eras that make up the Phanerozoic. We have included the first appearances of life forms discussed in this and other chapters, the names of periods or epochs within the era, absolute ages assigned by radioactive dating, notes on prominent plant communities, the climate and important geological events, and the estimated positions of major land masses. References used in compiling this figure include Niklas et al. 1983; Taylor and Hickey 1990; Niklas 1994; Li et al. 1996; Taylor and Taylor 1993; Irving 1977; Harland et al. 1990; and Erwin, D. 1996, pers. comm.

(a) The Paleozoic, or "ancient life" era. The Paleozoic begins with the radiation of animals and ends with a mass extinction at the end of the Permian. The time bar is labeled with the names of periods and subperiods within the Paleozoic, but omits the names of epochs and stages.

(b) The Mesozoic, or "middle life" era. The Mesozoic, sometimes nicknamed the age of reptiles, begins after the end-Permian extinction event and ends with the extinction of the dinosaurs and other groups at the Cretaceous–Tertiary boundary. The time bar is labeled with the names of periods within the Mesozoic, but omits the names of epochs and stages.

(c) The Cenozoic, or "recent life" era. The Cenozoic is divided into the Tertiary and Quaternary periods. The Tertiary includes the Paleocene, Eocene, Oligocene, Miocene, and Pliocene epochs. The Quaternary includes the Pleistocene and Holocene epochs. The Cenozoic is sometimes nicknamed the age of mammals.

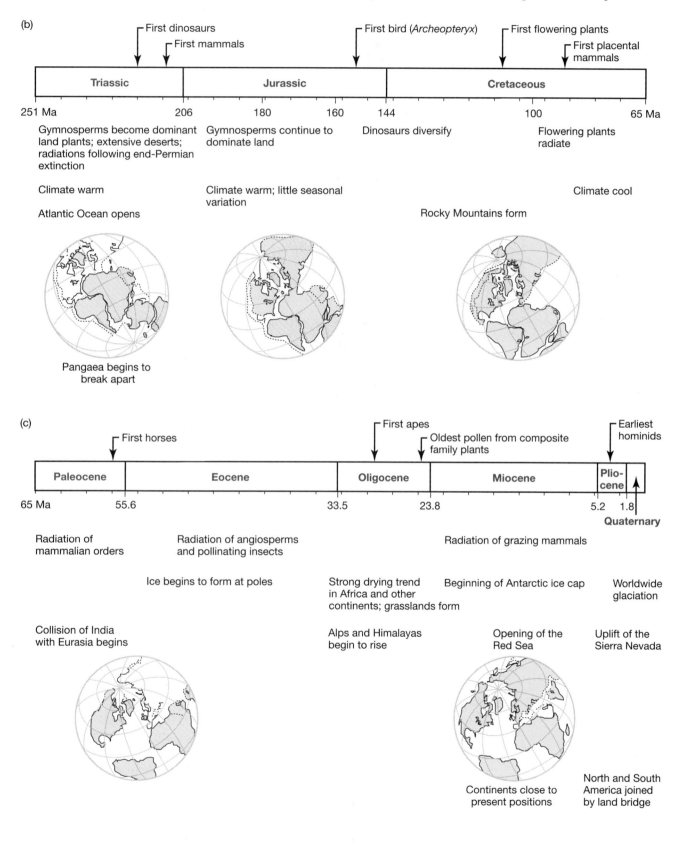

(b)

First dinosaurs
First mammals
First bird (*Archeopteryx*)
First flowering plants
First placental mammals

Triassic	Jurassic	Cretaceous

251 Ma   206   180   160   144   100   65 Ma

Gymnosperms become dominant land plants; extensive deserts; radiations following end-Permian extinction

Gymnosperms continue to dominate land

Dinosaurs diversify

Flowering plants radiate

Climate warm

Climate warm; little seasonal variation

Climate cool

Atlantic Ocean opens

Rocky Mountains form

Pangaea begins to break apart

(c)

First apes
First horses
Oldest pollen from composite family plants
Earliest hominids

Paleocene	Eocene	Oligocene	Miocene	Plio-cene	

65 Ma   55.6   33.5   23.8   5.2   1.8

Quaternary

Radiation of mammalian orders

Radiation of angiosperms and pollinating insects

Radiation of grazing mammals

Ice begins to form at poles

Strong drying trend in Africa and other continents; grasslands form

Beginning of Antarctic ice cap

Worldwide glaciation

Collision of India with Eurasia begins

Alps and Himalayas begin to rise

Opening of the Red Sea

Uplift of the Sierra Nevada

Continents close to present positions

North and South America joined by land bridge

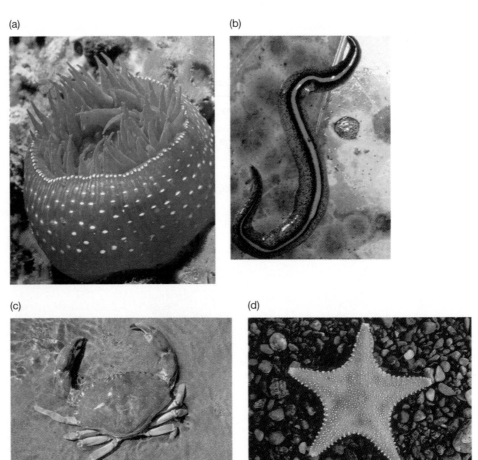

Figure 15.6 Animal body plans (a) Diploblastic animals like this sea anemone (phylum Cnidaria) have just two embryonic tissue layers: ectoderm and endoderm. They are radially symmetric, meaning that their body parts are arranged around a central axis. (Bruce Watkins/Animals Animals/Earth Scenes) (b–d) Bilaterian animals have left–right symmetry and are triploblastic (with mesoderm). (b) Flatworms (Platyhelminthes) and other acoelomates lack a fluid-filled body cavity or coelom. (Michael Fogden/Oxford Scientific Films/Animals Animals/Earth Scenes) (c) Some coelomates like this crab (Arthropoda), and other protostomes, develop the mouth first during gastrulation (Tom McHugh/Photo Researchers, Inc.), (d) while others, like this sea star (Echinodermata) and other deuterostomes, develop the mouth secondarily after gastrulation is complete (Andrew J. Martinez/Photo Researchers, Inc.).

appeared. How do we know all this? Snapshots offered by two beautifully preserved faunas, dated about 40 million years apart, tell the story.

The Ediacaran and Burgess Shale Faunas

The first unequivocal evidence for multicellular animals in the fossil record comes from the Ediacaran faunas (Conway Morris 1989; Weiguo 1994). These specimens were first found in the 1940s in the Ediacara Hills of south Australia; similar fossils have now been found at some 20 sites around the world. The earliest members of this fauna are dated at or around 565 Ma, placing them at the start of the Vendian period in the Proterozoic (early life) era, and the youngest at 544 Ma. These latest Precambrian faunas were entirely soft bodied and are preserved primarily as compression and impression fossils (Figure 15.7). The partial- or full-body specimens that are present are often difficult to identify, but most experts now agree that sponges, jellyfish, and comb jellies (diploblastic phyla) are represented (Conway Morris 1989; Weiguo 1994; Conway Morris 1998b). Many of the specimens are traces, meaning that they are the remnants of burrows, fecal pellets, or tracks. Trace fossils can be difficult to interpret, but at least some must have been made by bilaterally symmetric organisms (that is, triploblastic phyla), possibly including arthropods and molluscs or their close relatives (Waggoner 1998).

The second, and slightly younger, snapshot is provided by the Burgess Shale fauna (Figure 15.8), discovered early in this century near the town of Field, British

(a)

(b)

Figure 15.7 **Ediacaran faunas** (a) Jellyish, like this *Brachina delicata*, are radially symmetric. (Simon Conway Morris, University of Cambridge, Cambridge, United Kingdom) (b) Polychaete worms (this is *Spriggina floundersi*) are bilaterally symmetric. (Ken Lucas/Visuals Unlimited)

Columbia. The Burgess Shale, dated to 520–515 Ma, and the Chengjiang biota from Yunnan Province in China (525–520 Ma) are arguably the most spectacular fossil deposits ever found (Gould 1989; Conway Morris 1989, 1993, 1998b; Briggs 1991; Weiguo 1994; Erwin et al. 1997). They are primarily impression and compression fossils, but are extraordinary for the detail they preserve and the story they tell. The specimens include a wide diversity of unusual arthropods, including trilobites, but there are also segmented worms, wormlike priapulids, molluscs, and several chordates, including at least two species of jawless vertebrates that resemble living hagfishes and lampreys (Shu et al. 1999). There is little overlap between species found in the Ediacaran and Burgess Shale deposits, but at least a few organisms, such as large, colonial cnidarians (similar to modern sea pens), are present in both (Conway Morris 1998b).

In addition, the Burgess Shale and Chengjiang faunas include some organisms whose body plans are so unusual that they have been difficult to assign to extant phyla. Taxonomists have traditionally grouped these animals in a miscellaneous bin called the problematica (Figure 15.9). If we imposed the criteria we use today to classify these organisms, at least some might be assigned to new phyla represented by single species (Conway Morris 1989; Briggs et al. 1992). However, it now appears that

All but one of the animal phyla living today are present in the Burgess Shale deposits.

(a)

(b)

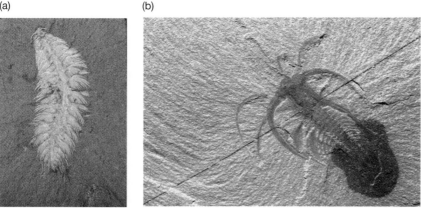

Figure 15.8 **Burgess shale faunas** The Burgess Shale and Chengjiang faunas are extraordinary not only for the diversity of arthropods and other hard-shelled animals, but for the number of soft-bodied organisms that are preserved. In other localities from the same period, only the shelled organisms fossilized. But shelled organisms represent a mere 20% of the diversity recorded in the Burgess Shale and Chengjiang deposits. Note that animals with well-developed segmentation, heads, and limbs are present. (Photos from Chip Clark)

Figure 15.9 Problematica? Some 60% of the Burgess Shale fauna are so unusual that they were initially left out of the classification of known living phyla (Conway Morris 1989). Many of these problematica are now more reliably grouped with living phyla, or at least with early Phanerozoic fossils of known affinity. (a) *Opabinia* is related to arthropods (or may belong within the arthropods); (b) *Wiwaxia* is most likely a polychaete worm. (Photos from A. J. Copley/Visuals Unlimited)

(a) (b)

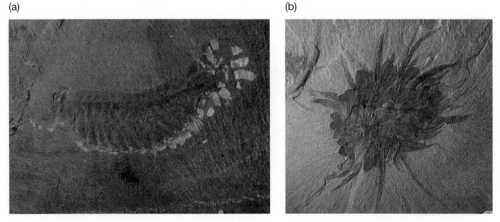

many of the problematica are in fact members or close relatives of living phyla. *Opabinia regalis* (Figure 15.9a), for example, was first described as an elongate fossil with serially repeated lateral plates, five dorsal eyes, and an elongate nozzle on the anterior end. This description left *Opabinia* out of the classification of living phyla. More recently described specimens, though, appear to have legs with a terminal claw (Budd 1996). This redescription suggests that *Opabinia* is in fact a close relative or a member of the arthropods (Conway Morris 1998b). Similarly, *Wiwaxia corrugata* (Figure 15.9b), was formerly given phylum status as a problematic fossil consisting of spines and plates, but is almost certainly a polychaete worm (Butterfield 1990) or a member of the stem group that gave rise to the annelids (Conway Morris 1998b). In short, the number of phyla that existed during the Cambrian was probably not much greater than later in the Phanerozoic (Briggs et al. 1992). Nevertheless, something unusual and compelling did happen in the Cambrian explosion: The earliest members of the major animal lineages appeared suddenly in the fossil record, at the same time, in geographically distant parts of the globe, without obvious precursors older than the Ediacaran fauna. What happened, and how can this remarkable pattern be explained?

Was the Cambrian Explosion Really Explosive?

Figure 15.10 shows a phylogeny of the living animal phyla that was estimated from DNA sequence data. The blue bars to the right of the tree indicate which lineages appear in the Ediacara and Burgess Shale fossil deposits. Here we ask: When did the branching events on this tree take place? Did they all occur between 565–525 million years ago?

To answer this question, evolutionary biologists have used molecular clocks to estimate when the deep branches on the animal phylogeny occurred. As Chapter 13 explained, in addition to revealing the order of branching events, DNA or protein sequences can also be used to estimate when they took place. For example, Bruce Runnegar (1982) used differences among hemoglobin amino acid sequences to measure the genetic differences between the vertebrates and various invertebrate phyla. Runnegar turned these genetic distances into times of divergence by using the genetic distances among vertebrate groups with known fossil ages as a calibration. He concluded that the earliest branches in Figure 15.10 occurred about 900 Ma—long before the Cambrian explosion. Greg Wray, Jeff Levinton, and Leo Shapiro (1996) came to a similar conclusion. Using a different data set, they estimated that

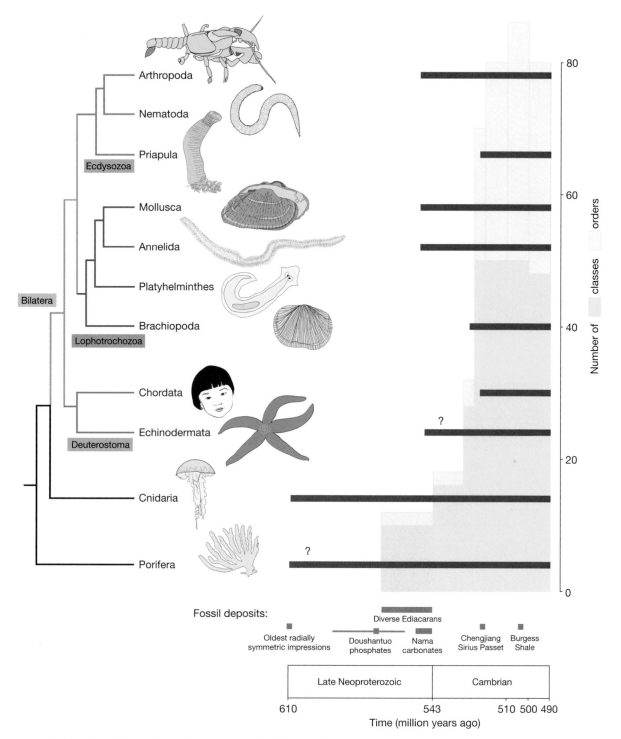

Figure 15.10   The Cambrian explosion: taxa, fossils, and lineages.   The phylogeny on the left shows the relationships among some of the major taxa represented in the earliest fossil faunas. The phylogeny is based on molecular data from living members of these phyla (Aguinaldo et al. 1997; de Rosa et al. 1999). The blue bars to the right of the tree indicate which phyla are represented in the fossil faunas identified by the red bars. (A question mark along a blue bar indicates that the presence of a group in a certain fossil fauna is controversial.) The yellow and tan histogram plots the total number of animal classes and orders that are present through time. Reprinted with permission from A. Knoll and S. Carroll, 1999, *Science* 284: 2129–2137. Copyright © 1999, American Association for the Advancement of Science.

chordates and echinoderms diverged about 1000 Ma, while protostomes and deuterostomes diverged about 1200 Ma. This study used more genes and more taxa, and used the same vertebrate fossil record for calibration as the Runnegar study. Both of these analyses admitted some uncertainty about the exact ages of the divergences among the bilateraly symmetric organisms, but concluded that these divergences occurred hundreds of millions of years before their first appearance in the fossil record.

These conclusions have generated a storm of controversy, because such early divergence dates imply a long history of animal evolution that has not been discovered in the fossil record. If the dates are correct, then older Proterozoic rocks should eventually yield fossils of bilaterally symmetric animals. A few such animals are now coming to light, mostly from fossil deposits in China, and we discuss two exciting examples.

The earliest confirmed chordates are similar to the Burgess Shale animal *Pikaia gracilens*. This animal had segmented trunk muscles and a skeletal rod called a notochord. The main group of chordates, the vertebrates that have a vertebral column replacing the notochord, were thought to have evolved much later. But D.-G. Shu and colleagues (1999) recently described two species of jawless fishes from the Chengjiang fauna that are about 530 million years old (Figure 15.11). The discovery is exciting because a longer fossil history for vertebrates suggests that chordates must be much older than the current fossil record indicates (see Janvier 1999; Zimmer 1999). The find is consistent with a longer and undiscovered period of Precambrian divergences among chordate lineages.

(a)

(b)

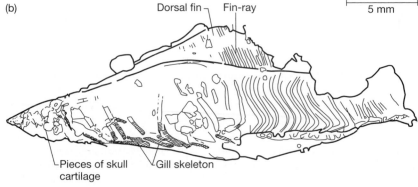

Figure 15.11   A lower Cambrian vertebrate: *Haikouichthys ercaicunensis.* The photograph in part (a) shows the whole specimen. (D. G. Shu, People's Republic of China) The interpretive drawing below (b) shows some distinctive vertebrate characters, including a cartilaginous skull, gill skeleton, and fin rays. From Shu et al. (1999). Reprinted by permission from *Nature* 402: 42–46, © 1999, Macmillan Magazines Ltd.

Because the diversification of animal forms included major changes in their embryonic development, finding fossil embryos or larvae might improve our understanding of when and in what order the lineages split from each other. Amazingly, some Cambrian and Precambrian embryos are now being discovered. They come from the Proterozoic Doushantuo formation of southern China, in which phosphate minerals replaced soft tissue to leave behind fossils that show even the finest details of animal anatomy. One research group (Li et al. 1998) has described beautifully preserved sponge specimens with larvae similar to the forms found in living sponges. Another group (Xiao et al. 1998) has found microfossils that they interpret as cleavage-stage animal embryos. As Figure 15.12 shows, the specimens are about half a millimeter in diameter and consist of two, four, eight, or more round parts interpreted as blastomeres in a cleaving embryo, possibly that of an arthropod (Bengtson 1998). Xiao et al. (1998) interpret their fossils cautiously, and some researchers wonder whether these are really animal embryos (Conway Morris 1998a), but "the Doushantuo fossils are most probably bilaterian, and they may document cladogenesis within the Bilateria before the appearance of diverse Ediacaran assemblages" (Xiao et al. 1998: 557).

As predicted, paleontologists are beginning to find bilaterian fossils in Precambrian deposits.

What is the General Message of These Recent Fossil Finds?

Recall that the literal interpretation of the Cambrian animal explosion—that clades diverged rapidly in the Cambrian—seems to be at odds with the molecular clock data, which suggest that clades diverged over a long period in the Proterozoic. Based on the new fossil discoveries, Andrew Smith (1999) suggests that

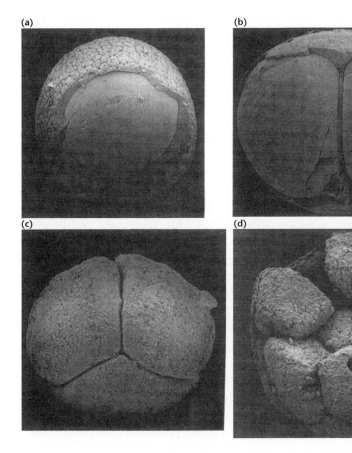

Figure 15.12 Precambrian embryos? These microfossils from the Doushantuo formation may be (a) zygotes and (b–d) cleavage-stage embryos of a Proterozoic bilateral animal, possibly an arthropod or a flatworm. Notice that an egg envelope surrounds each embryo, that the cells occur in multiples of two, and that they are arranged geometrically in a pattern similar to the embryos of some living arthropods and flatworms. Scale bar represents 200 μm (a, d), 150 μm (b), or 240 μm (c). From Xiao et al. (1998). (Photos from Shuhia Xiao, Harvard University)

the lineages leading to the living Bilateria diverged from each other over a prolonged period in the Proterozoic, but that the morphological traits that characterize the living bilateral phyla arose in an explosive frenzy of evolutionary innovation during the early Cambrian. His view is that between the gradual divergence of lineages and the appearance of large, complex forms in the Cambrian, many of these lineages must have existed as small, larvalike organisms without leaving fossils.

According to this point of view, the Cambrian explosion is an explosion of animal morphologies, but not necessarily an explosion of lineages. The only difficulty with this hypothesis is that it fails to explain why so many lineages evolved dramatic changes in body size and body during the same brief period. We end our discussion of the Cambrian explosion by considering some mechanisms behind the rapid morphological diversification that occurred.

What Caused the Cambrian Explosion?

What really exploded during the Cambrian was not lineages, but morphologies and ways of making a living.

Although we have emphasized the variety of body plans, cell types, and developmental patterns that evolved, the Cambrian explosion was also ecological in nature. Appreciating this point is a key to understanding why the event occurred. Most of the Ediacaran animals are either sessile filter feeders or predators that floated high in the water column and fed on planktonic organisms. But the Burgess Shale fauna introduces a huge variety of benthic and pelagic predators, filter feeders, grazers, scavengers, and detritivores that burrowed, walked, floated, clung, and swam. The Cambrian explosion filled many of the niches present in shallow marine habitats. Also, the explosion was clearly powered, in part, by entirely new modes of locomotion. What events triggered this morphological and ecological radiation?

Rising oxygen concentrations in seawater, due to the increase of photosynthetic algae in the Proterozoic, was clearly a key to the origin of multicellularity (Valentine 1994). Increased availability of oxygen makes larger size and higher metabolic rates possible. Large size is a prerequisite for the evolution of tissues, and higher metabolic rates are required for powered movement. Both of these traits are recorded in the Ediacaran faunas.

The origin of hard parts, another hallmark of the radiation, is documented in other late Precambrian fossils. Why did shells evolve? Stefan Bengtson and Yue Zhao (1992) addressed this question by analyzing the first mineralized exoskeletons in the fossil record. These are found in a widespread, tube-forming, 1- to 2-cm-long species called *Cloudina hartmannae*. Bengtson and Zhao examined 524 *Cloudina* tubes excavated in Shaanxi Province, China, and found conspicuous rounded holes, 40 to 400 μm in diameter, in 17 of them (Figure 15.13). Some of the holes penetrate only the outer layers of the tubes, but most go all the way through. Bengtson and Zhao suggest that the holes were drilled by predators. There is, for example, a positive correlation between the size of the hole and the width of the tube at the hole, which implies that predators were selecting tubes by size, with larger predators attacking larger prey. This work supports an important hypothesis: that predation set up selection gradients favoring mineralized shells. Natural selection by predators is an important theme throughout much of later Phanerozoic evolution (Vermeij 1987).

Other ecological interactions may have contributed to the selective processes favoring early Cambrian evolution of large and complex animal bodies. These might include new types of food, such as diversifying phytoplankton groups, lead-

Figure 15.13 Evidence for predatorial borings in the first exoskeletons This is a fossilized skeleton of *Cloudina*, showing a hole made by predators. (Stefan Bengtson. Reproduced by permission from *Science*. © American Association for the Advancement of Science).

ing to selection favoring novel feeding methods and anatomy, or structures and behaviors that permit swimming (Conway Morris 1995). It is easy to imagine feedbacks in this process. For example, the entry of large filter-feeding Cambrian animals into the plankton would create a new niche for a new kind of predator that also swam in the plankton and fed on the new planktivores.

Predation pressure may have favored increased body size, new modes of locomotion, and mineralized skeletons, but a response to selection occurred only because genetic variability existed. The massive amount of morphological evolution we have just documented requires a significant increase in the amount of information contained in the genome. Specifically, there must have been important changes in the genes that control embryonic development. The study of evolutionary developmental biology has become a field of its own, which we discuss in detail in Chapter 17.

15.3 Macroevolutionary Patterns

Understanding the great innovations in evolution is fundamental. That is why we spent most of Chapter 14 and much of this one on subjects like the origin of DNA synthesis, endosymbiosis, and multicellularity. But documenting evolution's "greatest hits" is only part of historical biology's portfolio. Searching for broad patterns in the fossil record is an equally important research program. The microevolutionary processes of mutation, natural selection, drift, and gene flow have produced an almost overwhelming diversity of life forms. The composition of this biota has changed radically over vast reaches of time. What patterns can we discern to make sense of all this variation?

The literature on macroevolutionary patterns is enormous, and we can only touch on a few of the major results. We start by summarizing a pattern that we touched on briefly in earlier chapters: The rapid diversification of species in response to a morphological innovation or ecological opportunity.

Adaptive Radiations

An **adaptive radiation** occurs when a single ancestral species diversifies into a large number of descendant species that occupy a wide variety of ecological niches (Figure 15.14). We introduced two examples in earlier chapters: the Galápagos finches and Hawaiian *Drosophila*.

What factors trigger adaptive radiations? Why do only certain lineages diversify broadly and rapidly? The answers vary from time to time and clade to clade.

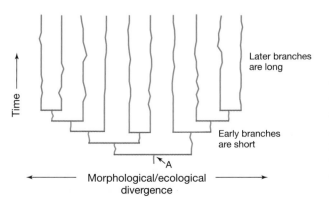

Later branches
are long

Early branches
are short

A

Time

Morphological/ecological
divergence

Figure 15.14 Adaptive radiation   This diagram shows the branching pattern produced by a hypothetical adaptive radiation. Time is plotted on the vertical axis; morphological or ecological differentiation on the horizontal axis. Note that the initial radiation is rapid, producing lineages with widely divergent forms. When phylogenetic trees are estimated for lineages that have undergone adaptive radiation, the early branches on the tree are often extremely short and later branches much longer. The pattern is "bushy" or "stemmy." The event that triggered the radiation occured at node A.

Ecological opportunity was a key factor in several of the examples we have reviewed. The ancestors of the Hawaiian *Drosophila* and Galápagos finches colonized a habitat that had few competitors and a wide variety of resources to exploit. These circumstances created conditions favorable to rapid diversification and speciation. But why these particular populations colonized the lake or island was largely a matter of blind luck. Similarly, in the last half of this chapter we will examine the aftermath of a mass extinction at the end of the Cretaceous period. Mammals diversified rapidly after the dinosaurs became extinct. The leading hypothesis for why they did so was that they lacked competitors, not that they had superior adaptations.

In contrast, many adaptive radiations are correlated with morphological innovations. The diversification of arthropods is a prime example. The variety of ecological niches occupied by insects, crustaceans, and spiders is remarkable and closely associated with modifications and elaborations of their jointed limbs.

Adaptive radiations are triggered by many different factors, ranging from the colonization of new habitats to the evolution of innovative new structures such as the flower.

Adaptive radiations among land plants have occurred at several different taxonomic levels. Two of the most notable ones were unique events, similar to the Cambrian explosion of animal diversity. The first was the radiation of terrestrial plants from aquatic ancestors in the early Devonian, about 400 Ma. During this period, early terrestrial plants evolved both the morphological features (such as leaves and vascular tissue) and the life histories (alternating gametophytic and sporophytic generations) that characterize their living descendants (Niklas et al. 1983; Bateman et al. 1998). The second was the Cretaceous explosion of flowering plants or angiosperms about 100 Ma. New phylogenies of angiosperms and their close relatives, the gymnosperms and gnetales, have produced a surprising result: A little-known shrub from the island of New Caledonia called *Amborella* (Figure 15.15) is the sister group to all other flowering plants (Qiu et al. 1999; Soltis et al. 1999). Identifying *Amborella* as the direct descendant of the angiosperm ancestor is an important clue to what that ancestor might have looked like and, by extension, what traits of that ancestor might have contributed to the early evolutionary success of flowering plants (Brown 1999).

Stasis

Perhaps the most prominent pattern in the history of life is that new morphospecies appear in the fossil record suddenly and then persist for millions of years without apparent change. In many cases, evolutionary innovations seem to arise at the same time as new species. As a result, morphological evolution seems to

Figure 15.15 Like the ancestral flower? The tropical shrub *Amborella* is the sister group to all of the other living flowering plants. If it has undergone less evolutionary change from the ancestral condition compared to other flowering plants, then *Amborella* may provide clues to the nature of the ancestral angiosperm and the conditions that led to the angiosperm adaptive radiation in the Cretaceous period. (Sandra Floyd, University of Colorado)

consist of long periods of stasis that are occasionally punctuated by speciation events that appear instantaneously in geological time. Gradual series of transitional forms do occur, but they are relatively rare.

Darwin (1859) was well aware of these observations and considered them a problem for his theory. Because his ideas were presented in opposition to the Theory of Special Creation, which predicts the instantaneous creation of new forms, Darwin repeatedly emphasized the gradual nature of evolution by natural selection. He attributed the sudden appearance of new taxa to the incompleteness of the fossil record and predicted that as specimen collections grew, the apparent gaps between fossil forms caused by stasis (and punctuated by sudden jumps) would be filled in by forms showing gradual transitions between morphospecies. For a century thereafter, most paleontologists followed his lead.

In 1972, however, Niles Eldredge and Stephen Jay Gould published an influential paper that accepted stasis as real and the sudden appearance of new forms as a predictable outcome of normal speciation processes. They called their proposal the theory of punctuated equilibrium. Subsequent versions of the theory were strongly criticized because they appeared to make some controversial claims about the relative importance of natural selection versus other kinds of evolutionary processes thought to operate during speciation events. The theory and its meaning were hotly debated for twenty years, but both proponents and critics now seem to agree that, at heart, the disagreements stemmed from the differences between observing patterns of speciation and morphological change on a biological time scale of years or decades (during which gradual change and natural selection should be important) versus a geological scale of millions of years (during which sudden changes in form might seem instantaneous; Figure 15.16) (see Eldredge and Gould 1972; Lande 1980; Gould 1982; Barton and Charlesworth 1984; Gould and Eldredge 1993; Eldredge et al. 1997).

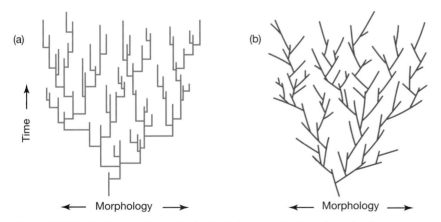

(a)

(b)

Time

← Morphology →   ← Morphology →

Figure 15.16   Patterns of morphological change: stasis and gradualism
When time is plotted against morphology, two extreme patterns are possible, along with many intermediate or mixed patterns. Diagram (a) shows the extreme called punctuated equilibrium: Morphology is static within species (lineages), and all morphological variation occurs at the time of a speciation (branching) event. Diagram (b) shows the other extreme, called phyletic gradualism: All morphological change occurs steadily and gradually and is completely unrelated to speciation events. Eldredge and Gould (1972) said that stasis dominates the history of species; Darwin (1859; see his Figure 1) said that the gradual pattern does.

Demonstrating Stasis

One benefit of this vigorous debate was that it spurred many paleontologists to ask whether stasis is in fact real: Do the data support the claim that stasis, punctuated by morphological change at speciation events, is the predominant feature of species histories through time?

Before looking at some tests of this hypothesis, we need to clarify several points. The first concerns time and scales of observation. Biologists and paleontologists often talk past one another when they talk about time. The fossil record rarely allows researchers to resolve events that occurred on time scales much less than a million years in length. When a researcher is studying the history of a lineage over 10 or 20 millions years, as paleontologists routinely do, a couple of hundred thousand years is trivial. Consequently, paleontologists say that speciation occurs "instantaneously" in geological time (see Gould 1982). But to a biologist measuring selection on an extant population, a long-term study lasts a decade or less. From this perspective, a hundred thousand years is not instantaneous—it is overwhelming. In the case of distinguishing stasis from gradual change in the fossil record, the finer the time scale, the better off we are.

The second point of clarification concerns rules for testing the pattern. Our goal is to follow changes in morphology in speciating clades through time and determine (1) whether change occurs in conjunction with speciation events or independently and (2) whether rapid change is followed by stasis or continued change. As critics of the theory have emphasized, a rigorous test for stasis vs. gradualism is exceptionally difficult. This is because the theory of punctuated equilibrium can become tautological if we are not careful. We define fossil species on the basis of morphology, so it might be trivial to observe a strong correlation between speciation and morphological change. To avoid circularity, an acceptable test requires that

- The phylogeny of the clade is known, so we can identify which species are ancestral and which descendant.
- Ancestral species survive long enough to co-occur with the new species in the fossil record.

The second criterion is critical. If it is not fulfilled, it is impossible to know whether the new morphospecies is indeed a product of a splitting event or whether it is the result of rapid evolution in the ancestral form without speciation taking place. This second possibility is called **phyletic transformation**, or **anagenesis**.

These are demanding criteria, especially when compounded with other difficult practical issues: the problem of misidentifying cryptic species in the fossil record, the need for analyzing change at the level of species, the requirement of sampling at frequent time intervals, and the necessity of sampling multiple localities to distinguish normal, within-species geographic variation from authentically different morphospecies.

Stasis and Speciation in Bryozoans

There are relatively few fossil series available anywhere in the world that meet these stringent requirements. One of them is a series of late Cenozoic fossils of the marine invertebrate phylum Bryozoa. In Chapter 12, we introduced experimental studies by Jeremy Jackson and Alan Cheetham (1990, 1994) on a subgroup called the cheilostome bryozoans. Their research showed that the criteria used to differentiate fossil morphospecies also work for genetically differentiated species alive today. This phylum is a good group for testing the idea of stasis for two reasons: We are confident that species designations can be made rigorously, and they are abundant in the fossil record of the past 100 million years.

Cheetham (1986) and Jackson and Cheetham (1994) performed a high-resolution analysis of speciation and morphologic change in cheilostomes from the Caribbean, starting in the Miocene and ending with living taxa. They began the study by defining 19 morphospecies in the genus *Stylopoma*, based on an analysis of 15 skeletal characters. Seven of these species still have living representatives, so they could check the validity of the morphospecies designations. Recall that Jackson and Cheetham did this by surveying allozymes in the living species and confirming that almost all morphospecies pairs were distinguished either by unique alleles or by significant differences in the frequencies of shared alleles. Then they estimated the phylogeny of the 19 morphospecies from differences in skeletal characters and scaled the tree so that the branch points and branch tips lined up with the dates of first and last appearance for fossil forms. They did a similar analysis for 19 living or extinct morphospecies in the genus *Metrarabdotos*.

The trees generated by the study are pictured in Figure 15.17. The phylogenies show an unequivocal pattern of stasis punctuated by rapid morphological change. The fact that ancestral and descendant species co-occur defends the idea that morphologic change was strongly associated with speciation events. This is an almost flawless example of stasis punctuated by evolutionary change at speciation.

What Is the Relative Frequency of Stasis and Gradualism?

Cheilostome bryozoans are not the only known example of stasis. Doug Erwin and Robert Anstey (1995a, b) reviewed a total of 58 studies conducted to test the theory of punctuated equilibrium. The analyses represent a wide variety of taxa and periods. Although the studies varied in their ability to meet the strict criteria we

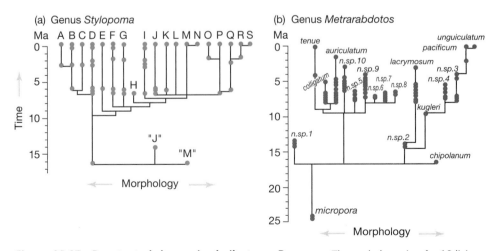

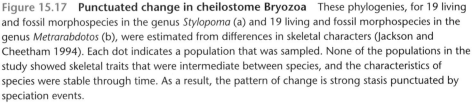

Figure 15.17 **Punctuated change in cheilostome Bryozoa** These phylogenies, for 19 living and fossil morphospecies in the genus *Stylopoma* (a) and 19 living and fossil morphospecies in the genus *Metrarabdotos* (b), were estimated from differences in skeletal characters (Jackson and Cheetham 1994). Each dot indicates a population that was sampled. None of the populations in the study showed skeletal traits that were intermediate between species, and the characteristics of species were stable through time. As a result, the pattern of change is strong stasis punctuated by speciation events.

listed, their sheer number may compensate somewhat. Erwin and Anstey's conclusion was that "Paleontological evidence overwhelmingly supports a view that speciation is sometimes gradual and sometimes punctuated, and that no one mode characterizes this very complicated process in the history of life." Further, Erwin and Anstey noted that a quarter of the studies reported a third pattern: gradualism and stasis.

Although partisans in the debate might be disappointed in this conclusion, it does suggest new avenues for research. For example, is it possible that different types of organisms exhibit different patterns of change? Hunter et al. (1988) suggested that a "tentative consensus" is emerging among researchers who have worked on the problem: that gradualist patterns tend to predominate in foraminifera, radiolarians, and other microscopic marine forms, while stasis occurs more often in macroscopic fossils such as marine arthropods, bivalves, corals, and bryozoans. If so, why? Research continues.

Why Does Stasis Occur?

One of Eldredge and Gould's prominent claims about the fossil record is that "stasis is data." That is, lack of change is a pattern that needs to be explained. Jackson and Cheetham's study of bryozoans showed that virtually no change occurred in these sessile invertebrates over millions of years. Why have such lineages remained unchanged for millions of years (at least, morphologically)? One approach to answering this question has focused on the so-called living fossils. These are species or clades showing little or no measurable morphological change over extended periods (Figure 15.18). Horseshoe crabs are a spectacular example. The extant species, in the genus *Limulus,* are virtually identical to fossil species in a different family that existed 150 Ma. While these horseshoe crab lineages stayed virtually unchanged, the entire radiation of birds, mammals, and flowering plants took place.

Have these species failed to change simply because they lack genetic variation? John Avise and colleagues (1994) answered this question by sequencing several genes in the mitochondrial DNA of horseshoe crabs and comparing the amount of genetic divergence they found to a previously published study of genetic distances in another arthropod clade: the king and hermit crabs (Cunningham et al. 1992). The result is striking: The horseshoe crabs show just as much genetic divergence as the king–hermit crab clade, even though far less morphological change has occurred (Figure 15.19). This is strong evidence that stasis is not from a lack of genetic variability.

What about some of the other possibilities? When Steve Stanley and Xiangning Yang (1987) looked at bivalve species that have shown remarkably little change over the past 15 million years, they discovered an interesting pattern. When they mapped change in 24 different shell characters over this interval, they found that even though most showed little net change within species, many had undergone large fluctuations, or what Stanley and Yang called "zigzag evolution" (Figure 15.20). These clam populations probably changed in response to environmental changes over time. But because these changes tended to fluctuate about a mean value, we perceive stasis as a result. This phenomenon has also been called **habitat tracking**, or **dynamic stasis**.

As a consequence of studies like these, the current view is that there is no single and general explanation for the low rates of morphological change that occur in particular lineages. Stasis is best tested and explained case by case (see Eldredge 1984; Stanley 1984; Thomson 1986).

Figure 15.18 "Living fossils" Here are just a few of numerous examples of contemporary and fossil species that are extremely similar. (a) Contemporary stromatolite-forming bacteria from Australia (left: Francois Gohier/Photo Researchers, Inc.) and 1800-Ma fossil forms (right: Biological Photo Service) from the Great Slave Lake area of Canada. (b) Leaves from a living gingko tree (left: Hugh Spencer/Photo Researchers, Inc.) and 40-Ma impression fossils made by gingko leaves (right: Sinclair Stammers/Science Photo Library/Photo Researchers, Inc.). We put the term "living fossils" in quotes because it is an oxymoron.

15.4 Extinction

Extinction is the ultimate fate of all species. Are there broad tendencies or patterns in the history of extinctions through time? Although research on extinction "is still at a reconnaissance level" (Raup 1994: 6758), several patterns are clear. Perhaps the most striking is that the global extinction rate has been anything but constant through time. Instead, there have been periods of particularly intense extinction. Consider a plot that David Raup (1991, 1994) constructed by calculating, for each one-million-year interval over the last 543 million years, the percent of taxa that went extinct in that interval (Figure 15.21). The histogram has a pronounced right skew, created by a few particularly large events. The most extreme of these intense periods are commonly referred to as **mass extinctions**. They represent intervals in which over 60% of the species that were alive went extinct in the span of a million years. Because of their speed and magnitude, they qualify as biological catastrophes.

How many mass extinctions have occurred during the Phanerozoic? To answer this question, M. J. Benton (1995) plotted the percent of families that died out during each stage in the fossil record of the past 510 million years. Figure 15.22 shows

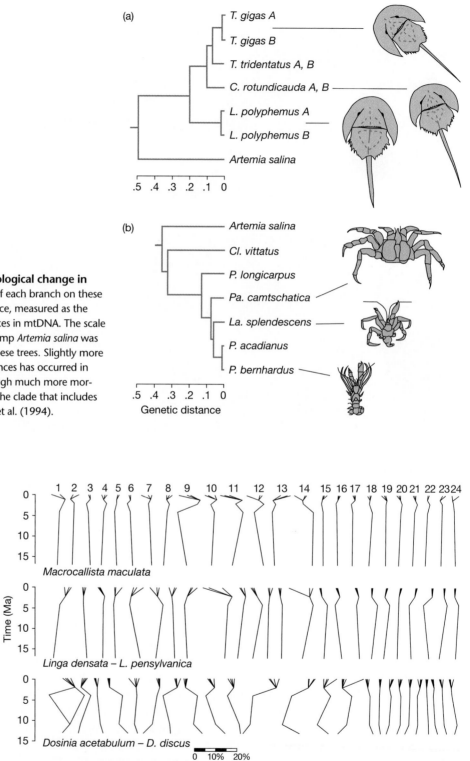

Figure 15.19 Genetic and morphological change in two arthropod clades The length of each branch on these phylogenies represents a genetic distance, measured as the percent difference in 16S rRNA sequences in mtDNA. The scale is the same for both trees. The fairy shrimp *Artemia salina* was used as the outgroup to root each of these trees. Slightly more genetic divergence in 16S rRNA sequences has occurred in the horseshoe crab clade (a), even though much more morphological divergence has occurred in the clade that includes hermit crabs and allies (b). From Avise et al. (1994).

Figure 15.20 Zigzag evolution in Pliocene bivalves results in stasis This diagram shows how 24 different morphological characters change through time in three lineages. The horizontal axis plots percent change in shell morphology between each time interval sampled in the study. From Stanley and Yang (1987).

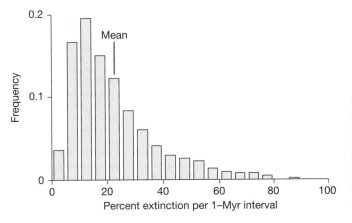

Figure 15.21   Distribution of extinction intensities during the Phanerozoic   David Raup (1994) broke the 543-million-year fossil record for the Phanerozoic into one-million-year intervals and calculated the percentage of species that went extinct during each such interval. Many of the larger events plotted in this figure were recognized early in the 19th century and were used to define boundaries for the eras, periods, and epochs that make up the geologic time scale.

the results of this analysis. The five prominent spikes that are circled on the graph are traditionally recognized as mass extinctions, and are referred to as the Big Five. On the geologic time scale, these events occurred at the terminal-Ordovician (ca. 440 Ma), late-Devonian (ca. 365 Ma), end-Permian (250 Ma), end-Triassic (ca. 215 Ma), and Cretaceous-Tertiary, or K–T (65 Ma). (Note that the Cretaceous is routinely symbolized with a K to distinguish it from other eras and periods that start with C.)

It is important to recognize, however, that the Big Five are responsible for perhaps 4% of all extinctions during the Phanerozoic. The other 96% of extinctions recorded in Figure 15.21 and 15.22 are referred to as **background extinctions**—meaning that they occurred at normal rates. What distinguishes mass extinctions from background extinctions? A mass extinction is global in extent, involves a broad range of organisms, and is rapid relative to the expected lifespan of taxa that are wiped out (Jablonski 1995). It is difficult to differentiate the two categories of extinction more precisely than this, however. As Raup's analysis makes clear, mass extinctions simply represent the tail of a continuous distribution of extinction events over time.

In this section we look at patterns that occur during times of background extinctions, delve into the causes of a particularly spectacular mass extinction, and consider an urgent question: Is a mass extinction event, caused by human beings, now underway?

The five largest extinction events in the Phanerozoic are known as the Big Five.

Background Extinction

Several interesting patterns have been resolved from data on background extinctions. First, within any radiation, the likelihood of subclades becoming extinct is constant and independent of how long the taxa have been in existence. Leigh Van

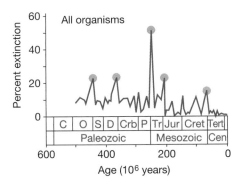

Figure 15.22   Patterns of extinctions of families through time   This diagram shows the distribution of mass extinctions events throughout the Phanerozoic. The Big Five extinctions are indicated by red dots. Redrawn with permission from Benton (1995). Copyright © 1995, American Association for the Advancement of Science.

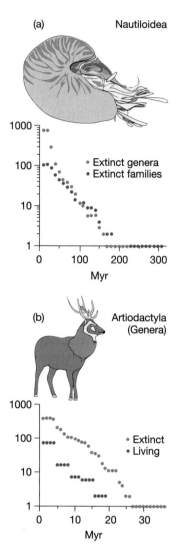

(a) Nautiloidea

(b) Artiodactyla (Genera)

Figure 15.23 Survivorship curves The first step in constructing these curves is to select a random sample of taxa from the fossil record of a particular clade—say, a family or class. The taxa included in the sample can come from any period. The logarithm of the number of genera or families in the clade that survive for different intervals is then plotted. The curves reproduced here are typical (Van Valen 1973): (a) is for genera and families of fossil marine invertebrates called nautiloids, and (b) is for genera in the deer family.

Valen (1973) discovered this when he plotted simple survivorship curves for a wide variety of fossil groups. Survivorship curves show the proportion of an original sample that survives for a particular amount of time. For fossil taxa, Van Valen plotted the number of species, genera, or families from an order or phylum of fossil animals that survived for different intervals. He put the number surviving on a logarithmic scale, so the slope of the curve at any point equaled the probability of becoming extinct at that time. Virtually every plot he constructed, from many different fossil groups and eras, produced a straight line. This means that the probability of subgroups becoming extinct was constant over the life span of the larger clade. The data in Figure 15.23 are typical. Note that the slopes of the lines vary from taxon to taxon, meaning that rates of extinction vary dramatically between lineages. In sum, during background times, extinction rates are constant within clades, but highly variable across clades.

Second, in marine organisms, extinction rates vary with how far larvae disperse after eggs are fertilized and begin development. David Jablonski (1986) came to this conclusion by studying extinction patterns in bivalve (clams and mussels) and gastropod (slugs and snails) species from the Gulf of Mexico and the Atlantic coastal plain region over the last 16 million years of the Cretaceous period. Jablonski found that marine invertebrate species with a planktonic larval stage survived longer, on average, than species whose young develop directly from the egg (Figure 15.24). In living species, planktonic larvae are carried on currents and often disperse long distances. This gives them greater colonizing ability, which might reduce the frequency of extinction. Populations with this life history also tend to have larger ranges. Indeed, Jablonski confirmed that geographic range also influences extinction rates: Species with large ranges survived longer than those with more limited ranges (Figure 15.25). Taxa found in small areas are less likely to survive sea-level changes, new predation, and new diseases, as well as other stresses that can lead to extinction.

Cretaceous–Tertiary: High-Impact Extinction

Why do mass extinctions occur? The answer is different for each of the Big Five. Here we examine the impact hypothesis for the extinction at the K–T boundary (Alvarez et al. 1980), which fired the imaginations of scientists and the lay public. The idea that a huge object hit the Earth and caused widespread extinctions, including the demise of the dinosaurs, provoked intense debate and research. In just 10 years after its publication, the hypothesis had inspired over 2000 scientific papers and innumerable accounts in the popular press (Glen 1990).

Evidence for the Impact Event

The discovery of anomalous concentrations of the element iridium in sediments that were laid down at the K–T boundary (Figure 15.26) was the first clue that an asteroid hit the Earth 65 Ma. Iridium is rare in the Earth's crust, but abundant in meteorites and other extra-terrestrial objects. Figure 15.27 shows a typical iridium spike found in strata that were laid down over the Cretaceous–Tertiary boundary. Glen (1990) counted 95 different sampling localities from all over the globe that had been found with iridium anomalies dating to the K–T boundary. On the basis of estimates for the amount of iridium needed to produce the anomalies and the density of iridium in typical meteorites, Alvarez et al. (1980) suggested that the asteroid was on the order of 10 km wide. It was, quite literally, the size of a mountain. It would also have been intensely hot, from friction with the atmosphere.

The discovery of two unusual minerals in K–T boundary layers provides additional support for the hypothesis. Shocked quartz particles (Figure 15.28) had been found only on the margins of well-documented meteorite impact craters until they were discovered at K–T boundary sites (Glen 1990). The other unusual structures are tiny glass particles called **microtektites** (Figure 15.29). Microtektites can have a variety of mineral compositions, depending on the source rock, but all originate as grains melted by the heat of an impact. If the melted particles are ejected from the crash site instead of being cooled in place, they are often teardrop or dumbbell shaped—a result of solidifying in flight.

The discovery of abundant shocked quartz and microtektites in K–T boundary layers from Haiti and other localities in the Caribbean helped investigators narrow the search for the crater (Florentin and Sen 1990; see also McKinnon

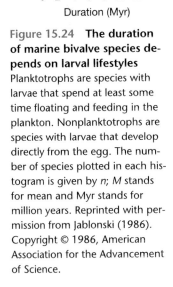

(a) Planktotrophs

$n = 50$
$M = 6$ Myr

(b) Nonplanktotrophs

$n = 50$
$M = 2$ Myr

Duration (Myr)

Figure 15.24 **The duration of marine bivalve species depends on larval lifestyles** Planktotrophs are species with larvae that spend at least some time floating and feeding in the plankton. Nonplanktotrophs are species with larvae that develop directly from the egg. The number of species plotted in each histogram is given by *n*; *M* stands for mean and Myr stands for million years. Reprinted with permission from Jablonski (1986). Copyright © 1986, American Association for the Advancement of Science.

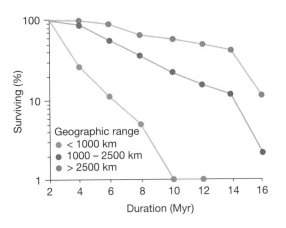

Figure 15.25 **Geographic range affects the survivorship of fossil bivalve and gastropod species** Jablonski (1986) broke the species in his study into three groups—those with broad, intermediate, and narrow geographic ranges along the Atlantic coast of North America—and created separate survivorship curves. The slopes of these curves give the extinction rate, as in Figure 15.22. Species with large ranges survived much longer in the fossil record than species with more restricted ranges. Reprinted with permission from Jablonski (1986). Copyright © 1986, American Association for the Advancement of Science.

Figure 15.26 **Clay lenses at the K–T boundary** This photograph shows a dark band of clay sandwiched between limestones laid down over the Cretaceous–Tertiary boundary. These rocks, which were formed in a deep-sea environment, are found today in Gubbio, Italy. The clay layer was laid down at the K–T boundary. Because limestone is made up of calcareous shells of marine invertebrates, the interruption of limestone formation and the deposition of clay represent the period immediately after the extinction. Few marine invertebrates were present in the ocean at this time. (Alessandro Montanari/Geological Observatory of Coldigioco, Frontale di Apiro, Italy)

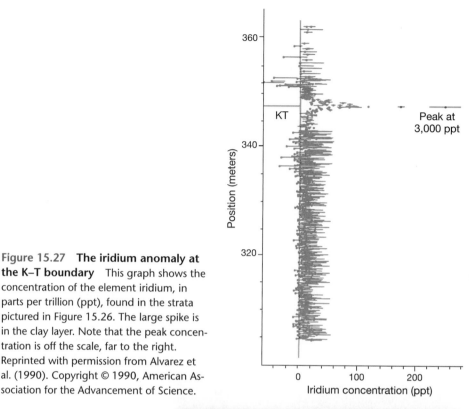

Figure 15.27   The iridium anomaly at the K–T boundary   This graph shows the concentration of the element iridium, in parts per trillion (ppt), found in the strata pictured in Figure 15.26. The large spike is in the clay layer. Note that the peak concentration is off the scale, far to the right. Reprinted with permission from Alvarez et al. (1990). Copyright © 1990, American Association for the Advancement of Science.

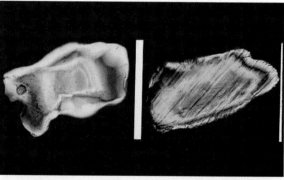

Figure 15.28   Shocked quartz   Small quartz grains (1–2 mm across) with parallel planes called lamellae are routinely found near meteorite strikes. The deformation is thought to be caused by the shock of impact. A shocked quartz grain is shown on the right, a normal grain on the left. (Glen A. Izett/U.S. Geological Survey, U.S. Department of the Interior)

Figure 15.29   Microtektites   Microtektites are spherical or teardrop-shaped particles of glass associated with impact sites. They are thought to originate as particles of rock melted by the heat of impact, after which they either cooled in place or splashed into the atmosphere. The tektites pictured here have been sectioned to show the interior. (Glen A. Izett/U.S. Geological Survey, U.S. Department of the Interior)

1992; Alvarez et al. 1995). Then, in the early 1990s, a series of papers on magnetic and gravitational anomalies confirmed the existence of a crater 180 kilometers in diameter, centered near a town called Chicxulub (cheek-soo-LOOB) in the northwest part of Mexico's Yucatán peninsula (Figure 15.30). Subsequent dating work confirmed that microtektites from the wall of the crater, recovered from cores drilled in the ocean floor, were 65.06 ± 0.18 million years old (Swisher et al. 1992). This is an almost exact match to dates for glasses ejected from the site and recovered in the Haitian K–T boundary layer.

The discovery of the crater was the long-sought smoking gun. It solidified a consensus among paleontologists, physicists, geologists, and astronomers that a large asteroid struck the Earth 65 Ma. The existence of the impact is no longer controversial; the consequences of the impact are.

Evidence for an asteroid impact at the end-Cretaceous is now overwhelming.

Killing Mechanisms

The 10-km asteroid that struck the ocean would have produced a series of events capable of affecting climate and atmospheric and oceanic chemistry all over the globe. The ocean floor near the impact site at Chicxulub consisted of carbonates, including large beds of anhydrite ($CaSO_4$), over a granitic basement layer. The distribution of shocked quartz and microtektites, far to the north and west of Chicxulub, confirms that a large quantity of this material was ejected from the site and that significant amounts were melted or vaporized by the heat generated at impact.

What consequences did the ejected material have? Vaporization of anhydrite and seawater would have contributed an enormous influx of SO_2 and water vapor to the atmosphere. The molecules would react to form H_2SO_4, with acid rain resulting. Sulfur dioxide is also a strong scatterer of solar radiation in the visible spectrum, which would lead to global cooling (McKinnon 1992). The cooling effect

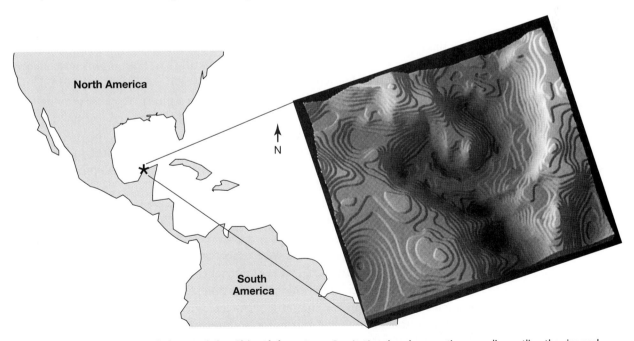

Figure 15.30   Location and shape of the Chicxulub crater   Gravitational and magnetic anomalies outline the rim and walls of a 180-km diameter crater, buried beneath sediments near Chicxulub on the Yucatán peninsula. The inset is a relief map of the impact site, showing the gravity field density (Schultz and D'Hondt 1996). (Peter H. Schultz, Brown University, Steven D'Hondt, University of Rhode Island Graduate School of Oceanography)

would have been enhanced by dust-sized carbonate, granitic, and other particles. These were ejected in quantities large enough to block incoming solar radiation. Further, soot deposits at numerous K–T boundary localities (for example, Wolbach et al. 1988; Heymann et al. 1994) suggest that widespread wildfires occurred, perhaps triggered by the fireball of hot gas and particulates expelled from the impact site. The ash and soot produced by the fires would have added to a widespread smog layer and the accentuation of global cooling. All these data indicate that the Earth became cold and dark soon after the impact.

A variety of models suggest that the force of the impact was also sufficient to trigger massive earthquakes, perhaps as large as magnitude 13 on the Richter scale, and to set off volcanoes. The second largest magma deposits of the Phanerozoic, the Deccan Traps of India, were contemporaneous with the K–T extinctions, although no causal connection has yet been established to directly tie this event to the impact. Widespread volcanism would have added sulfur dioxide, carbon dioxide, and ash to the atmosphere. The ash and sulfur dioxide would accentuate global cooling, while carbon dioxide would contribute to a longer-term greenhouse effect and warming.

Finally, the impact would have created an enormous tidal wave, or tsunami, in the Atlantic. If the asteroid was indeed 10 km wide, models suggest that the wave produced by the strike would have been as large as 4 km high. The mountain of rock made a mountain-sized splash. What evidence do we have that a tsunami of this size actually occurred? Joanne Bourgeois and colleagues (1988) discovered a huge sandstone deposit along the Brazos River in Texas, which has been mapped throughout northeast Mexico. It is 300 km long and several meters thick and is now interpreted by most geologists as a product of the rapid and massive deposition typical of tsunamis (see Kerr 1994). In the Haitian boundary layer, there is a thick jumble of coarse- and fine-grained particles sandwiched between the iridium-enriched clay layer above and extensive tektite deposits below. Florentin and Sen (1990) interpret this middle stratum as the product of tsunami-induced mixing and deposition. This occurred after the initial splash of microtektites, but before the fallout of iridium-enriched particulates from the atmosphere.

The asteroid strike would have affected the world's oceans in two ways. Globally, the primary productivity of phytoplankton would have been dramatically reduced by the cooling and darkening of the atmosphere. Locally, temperature regimes and chemical gradients in the Atlantic would have been disrupted by the largest tsunami ever recorded.

The asteroid impact had profound effects on both marine and terrestrial ecosystems.

These physical consequences of the asteroid impact are dramatic, and they undoubtedly led directly to the decimation of marine and terrestrial biotas in the days or months immediately following the blow. However, a large proportion of the end-Cretaceous extinctions must have been caused by ecological interactions between organisms and their traumatized environment. The decline of many groups was not instantaneous, but rather was drawn out over the 500,000 years following the impact. These extinctions were probably due to the disruption of ecological processes, biogeochemical cycles of nutrients, and interactions among species.

In sum, between acid rain, widespread wildfires, intense cooling, extensive darkness, an enormous tsunami, and subsequent ecological disruption, there is no shortage of plausible killing mechanisms for both terrestrial and marine environments. The question now becomes, Are there any patterns indicating which groups died out? If so, do these patterns tell us anything about which of the possible kill mechanisms were most important?

Extent and Selectivity of Extinctions

To date, our best estimate is that 60% to 80% of all species became extinct at the end-Cretaceous (Jablonski 1991). As in other mass extinctions, however, the losses were not distributed evenly across taxa. Among vertebrates, for example, amphibians, crocodilians, mammals, and turtles were little affected. Prominent terrestrial groups like the dinosaurs and pterosaurs (flying reptiles) were wiped out, and large-bodied marine reptiles like the ichthyosaurs, plesiosaurs, and mosasaurs disappeared. Only one order of birds, a shorebirdlike group, survived (Feduccia 1995; but see Cooper and Penny 1997), while insects escaped virtually unscathed. Among marine invertebrates, the ammonites and the rudists (a group of clams) were obliterated. Marine plankton became so scarce that micropaleontologists describe a "plankton line" at the K–T boundary. At some localities in North America, more than 35% of land plant species became extinct (Schultz and D'Hondt 1996). Pollen and spore deposits dated to just after the boundary show a prominent "fern spike" at these localities, implying that forest communities were dramatically reduced and replaced by widespread stands of ferns (Nichols et al. 1992; Sweet and Braman 1992).

What do these data imply regarding the kill mechanism? It has traditionally been argued that the K–T extinction was size selective, with large-bodied organisms suffering most. The logic was that an extended period of cold and dark after the impact would have affected large animals and plants the most, because of their higher nutritional requirements. But the analyses done to date have shown no difference in extinction rates between small- and large-bodied forms of marine bivalves and gastropods (Figure 15.31). Jablonski (1996) and others have pointed out that large crocodilians survived, while small-bodied and juvenile dinosaurs did not. The selectivity of the vertebrate extinctions is unresolved and still the focus of vigorous debate and research.

Although research on plant extinctions continues to gather momentum, an interesting pattern may be starting to emerge: Losses seem to be much more severe in North America. The reason may be that species there occupied the splash zone. Peter Shultz and Steven D'Hondt (1996) have proposed that the impact occurred at an oblique angle, with the asteroid approaching North America from the southeast and ejecting material and heat predominantly to the north and west.

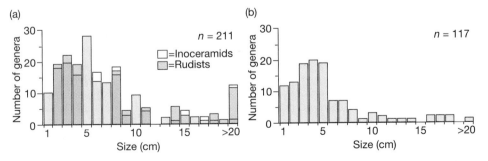

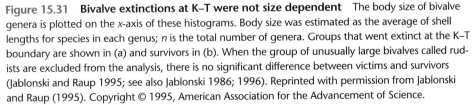

Figure 15.31   Bivalve extinctions at K–T were not size dependent   The body size of bivalve genera is plotted on the *x*-axis of these histograms. Body size was estimated as the average of shell lengths for species in each genus; *n* is the total number of genera. Groups that went extinct at the K–T boundary are shown in (a) and survivors in (b). When the group of unusually large bivalves called rudists are excluded from the analysis, there is no significant difference between victims and survivors (Jablonski and Raup 1995; see also Jablonski 1986; 1996). Reprinted with permission from Jablonski and Raup (1995). Copyright © 1995, American Association for the Advancement of Science.

We have made more progress in understanding the selectivity of the marine invertebrate extinctions, largely because the fossil record is so much more extensive (Jablonski 1986, 1991). For example, the probability that a bivalve genus survived the K–T boundary had nothing to do with whether it lived by burrowing or in exposed positions, whether it lived close to shore or far offshore, or whether it occupied tropical or polar latitudes (Raup and Jablonski 1993; Jablonski and Raup 1995). The outstanding pattern that has emerged in studies of bivalves and gastropods is that genera with wide geographic ranges were less susceptible to being eliminated than genera with narrower ranges (Figure 15.32). This is exactly the same result in earlier mass extinction events and for bivalves and gastropod species undergoing background extinction in the interval just before the bolide impact at the end-Cretaceous (Figure 15.25). Wide geographic ranges clearly buffer marine invertebrate clades against extinction. To date, this pattern represents the most robust result to emerge from the study of both background and mass extinctions.

Extinctions of marine invertebrates at K–T were selective: Genera with broad geographic ranges survived better.

15.5 Recent Extinctions: The Human Meteorite

Concern about widespread extinction is on the minds of people from all walks of life, from grade school children to heads of state. But despite celebrity examples like the dodo, passenger pigeon, and Carolina parakeet, is anything of special evolutionary significance going on now? That is, are we currently experiencing or contemplating an event anything like the Big Five in scale and speed? To answer this question, we examine data on extinctions that have occurred over the past 2000 years.

Polynesian Avifauna

David Steadman (1995) has amassed compelling evidence that an important extinction has just occurred among birds. Steadman estimates that 2,000 avian species have been extinguished over the past two millennia in the Pacific region alone, as a result of human colonization of islands. Because slightly over 9,000 species of birds exist today, Steadman's work means that the clade called Aves has recently lost almost 20% of its species.

The evidence for this claim comes from archaeological digs throughout the Pacific islands, conducted over the past 20 years by Steadman, Storrs Olson, Helen James, and colleagues. For example:

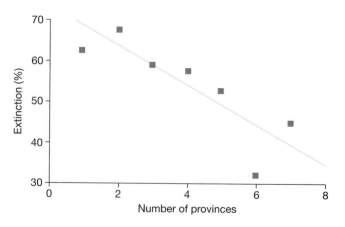

Figure 15.32 Marine bivalve genera with wide geographic ranges were less likely to become extinct at the K–T boundary On this graph, the percentage of bivalve genera that went extinct at the K–T boundary is plotted as a function of the number of biogeographic provinces occupied by those genera. (Biogeographic provinces are regions of the world that share similar floras and faunas.) There is a statistically significant trend for more localized genera to be wiped out during the mass extinction. For additional data on this pattern, see Jablonski (1986). Reprinted from Fig. 3 in D. Jablonski and D. M. Raup, *Background and Mass Extinctions: The Alternation of Evolutionary Regimes*, copyright © 1995, American Association for the Advancement of Science, 268: 389–391, with permission from the Association for the Advancement of Science.

- Sixty bird species endemic to the Hawaiian archipelago became extinct after the arrival of settlers there about 1,500 years ago. These species begin dropping out of the fossil record soon after the first fire pits, middens (trash piles), and tools appear in the digs.

- In New Zealand, 44 bird species became extinct after human colonization, but before historical times. The losses included 8 species of gigantic flightless birds called moas.

- On the island of 'Eua in the Kingdom of Tonga, only 6 of the 27 land birds represented in the prehuman fossil record are still extant (Figure 15.33).

- On each of the seven best-studied islands in central Polynesia, where research has been intense enough to recover and identify at least 300 bones and thus provide a reasonable sample of the fossil avifauna, at least 20 endemic species or populations were wiped out after human settlement.

The total losses are staggering. Extrapolating on the basis of data from the best-studied sites, Steadman suggests that a minimum of 10 species or populations have been lost on each of Oceania's 800 major islands. This estimate is probably conservative (see Balmford 1996). Consider, for example, the fossil record for the small, forest-dwelling, flightless birds called rails. One to four endemic species have been recovered as fossils from each of the 19 islands on which enough research has been done to uncover at least 50 land-bird bones. Extrapolating from this well-studied subset implies that there may have been 2,000 species of rail in the Pacific alone. But just 4 species are left in Oceania. As Steadman (1995: 1127) has written: "Only the bones remain as evidence of one of the most spectacular examples of avian speciation."

Several possible agents of human-caused extinction have been documented in the digs (Olson and James 1982; James et al. 1987; Steadman 1995). In Hawaii, for example, bird bones have been found that are split or charred by fire. These observations suggest direct predation by humans. The presence of pig, dog, and rat remains in the human-associated deposits indicates that these animals were imported by the colonists. Because most islands in the world lacked mammalian predators before the arrival of people, the introduction of rats and dogs was potentially devastating for ground-nesting birds and for the many species, like rails, that had evolved flightlessness. Habitat destruction was also a factor. Records from early European visitors to Hawaii indicate that slash-and-burn agriculture, or permanent irrigated fields for the cultivation of taro, had virtually wiped out the lowland forests of Hawaii before the 1700s.

As a control for the hypothesis that humans were the causative agent, Steadman offers the Galápagos archipelago. The Galápagos lacked any permanent human settlement until 1535 and had a tiny population until 1800. Archaeological and paleontological work has been extensive on the five largest islands in the group: Over 500,000 vertebrate bones have been excavated. The extensive fossil record documents the loss of only three populations in the 4,000 to 8,000 years preceding the arrival of humans. Over 20 taxa, however, have been eliminated since humans arrived.

Is a Mass Extinction Event Underway?

As impressive (and discouraging) as the data are for recent Polynesian birds, they do not approach the intensity or geographic scope of the Big Five, when between 50% and 90% of species were lost. What about more recent rates? The data in

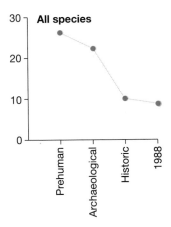

Figure 15.33   Extinction of forest birds on the island of 'Eua   This graph plots the number of species present on an island in the south Pacific at four different time intervals. "Archaeological" indicates data from cultures present on the island before contact with Europeans ("Historic"). Reprinted with permission from Steadman (1995). Copyright © 1995, American Association for the Advancement of Science.

Archaeological evidence suggests that humans were responsible for the extinction of at least 2000 bird species on islands in the Pacific.

Table 15.1 indicate only modest numbers and rates of extinctions documented since 1600. Is concern about an impending mass extinction overblown?

To answer this question, it is important to note that the majority of the losses represented in Table 15.1 occurred on islands. These island extinctions, in turn, usually resulted from the introduction of nonnative predators or competitors. This introduction process has probably peaked in intensity and should be less important in the future.

Instead, current concern is focused on a different agent of extinction: habitat loss due to expanding human populations. The current human population is about 6 billion and is growing at the rate of 1.6% per year. If this rate continues, world population will double in about 40 years (Cohen 1995). The point is that a new extinction agent has emerged and will grow in intensity, unless human population growth declines rapidly.

How Fast Are Species Disappearing?

Do we have any data to use in projecting the impact of human-induced habitat destruction as an agent of mass extinction? One approach (see Smith et al. 1993a) is to look at extinctions that have occurred in the very recent past. These events

Table 15.1   Extinction events in the past 400 years

Smith et al. (1993a), who compiled these data, point out that they should be interpreted with caution. The numbers reported here are most accurate for well-studied groups such as birds and mammals, but are almost certainly gross underestimates for poorly studied groups such as insects and plants. Note that the category called "threatened" lumps together species considered by the International Union for the Conservation of Nature (IUCN) as probably extinct, endangered (likely to go extinct if present trends continue), or vulnerable (likely to become endangered if present trends continue).

	No. of species certified extinct since 1600	No. of species listed as threatened	Approximate total of recorded, extant species	Percentage extinct	Percentage threatened
Animals					
Molluscs	191	354	10^5	0.2	0.4
Crustaceans	4	126	4×10^3	0.01	3
Insects	61	873	10^6	0.005	0.07
Vertebrates	229	2,212	4.7×10^4	0.5	5
Fish	29	452	2.4×10^4	0.1	2
Amphibians	2	59	3×10^3	0.07	2
Reptiles	23	167	6×10^3	0.4	3
Birds	116	1,029	9.5×10^3	1	11
Mammals	59	505	4.5×10^3	1	11
Total	485	3,565	1.4×10^6	0.04	0.3
Plants					
Gymnosperms	2	242	758	0.3	32
Dicotyledons	120	17,474	1.9×10^5	0.2	9
Monocotyledons	462	4,421	5.2×10^4	0.2	9
Palms	4	925	2820	0.1	33
Total	584	22,137	2.4×10^5	0.3	9

are much more likely to be due to habitat destruction than the extinctions recorded in Table 15.1. For example, from 1986 to 1990, 15 vertebrates were added to the list of species formally considered extinct. If this rate were sustained into the future, it would take just 7000 years to eliminate half of the 45,000 vertebrate species that have been named. If similar losses occurred among other taxonomic groups, this would clearly qualify as a mass extinction event comparable to the Big Five. The estimate is crude, however. Habitat destruction should eliminate particularly rare or localized species first. This fact suggests that the rate of 15 in four years is too high to be sustained, unless habitat destruction continues to accelerate.

Robert May and colleagues (1995) have reviewed three additional approaches used in estimating the current extinction rates:

- Multiply the number of species found per hectare in different environments by rates of habitat loss measured from satellite photos.
- Quantify the rate that well-known species are moving from threatened to endangered to extinct status in the lists maintained by conservation groups.
- Estimate the probability that all species currently listed as threatened or endangered will actually go extinct over the next 100 or 200 years.

All of these approaches to the estimation of current extinction rates suggest that extinctions are now occurring at 100 to 1000 times the normal, or background, rate of extinction (see also Smith et al. 1993a; 1993b; Pimm et al. 1995). If current rates of habitat destruction continue, the coming centuries or millennia will see a mass extinction on the same scale as the Big Five documented in the fossil record.

Where Is the Problem Most Acute?

Between 1600 and 1993, biologists observed the extinction of 486 animal species and 600 plant species (Smith et al. 1993b). Most of these extinctions occurred in North America, the Caribbean, Australasia, and the islands of the Pacific Ocean. Habitat destruction is now rampant worldwide (Figure 15.34), putting many more species at risk. For example, a 1996 report by the Nature Conservancy (Stein and Chipley 1996) listed roughly one-third of all American plant and animal species as threatened (Figure 15.35). The taxa at greatest risk in the United States are the freshwater mussels, crayfish, amphibians, and freshwater fish, including economically important species such as Pacific salmon. One hundred twenty-three of these freshwater animal species have become extinct in North America since 1900, a rate of about 4% per decade (Ricciardi and Rasmussen 1999).

This rate of loss of North American freshwater animals is comparable to the overall rate of extinctions of all species in tropical forests, although the problem in these forests is even more acute. There are two reasons for the importance of tropical forests:

- Tropical rain forests are extraordinarily rich in species. E. O. Wilson (1988) recounts that he once collected 43 species of ants belonging to 26 genera—numbers roughly equivalent to the entire ant fauna of the British Isles—from a single tree in a Peruvian rain forest. Peter Ashton identified 700 different species of trees—the same number found in all of North America—in just ten 1-hectare sample plots from a rain forest in Borneo. With the exception of

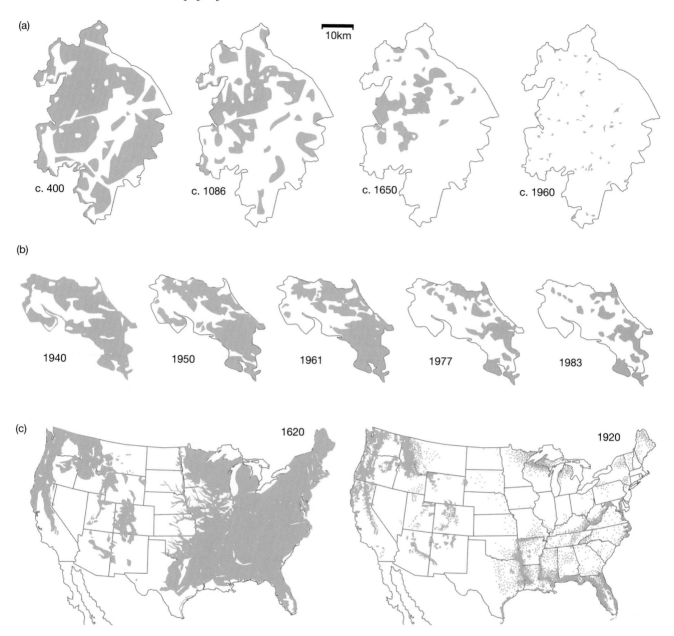

Figure 15.34 Deforestation in England, Costa Rica, and the United States Each set of maps shows the change over time in the fraction of land area covered by old-growth forests. (a) Warwickshire, England, 400 A.D. to 1960. (b) Costa Rica, 1940 to 1983 (c) United States, 1620 to 1920. From Dobson (1996). © 1998 by Scientific American Library. Used by permission of W. H. Freeman and Company. (c) Each dot on the U.S. map for 1920 represents 25,000 acres. From *Americans and Their Forests,* by M. Williams. Copyright © 1989 Cambridge University Press.

conifers, salamanders, and aphids, nearly every well-studied lineage on the tree of life shows a latitudinal gradient in diversity, with the largest number of species residing in the tropics. Why this pattern occurs is not clear (Ehrlich and Wilson 1991), but the results are striking. Tropical forests occupy less than 7% of the Earth's land area, but contain at least half of all plant and animal species.

- Tropical rain forests are presently under acute threat. Many nontropical habitats in the northern hemisphere, as well as most oceanic islands, have been under continuous occupation by high densities of humans for several hundred years. As a result, the flora and fauna of these nontropical habitats have already sus-

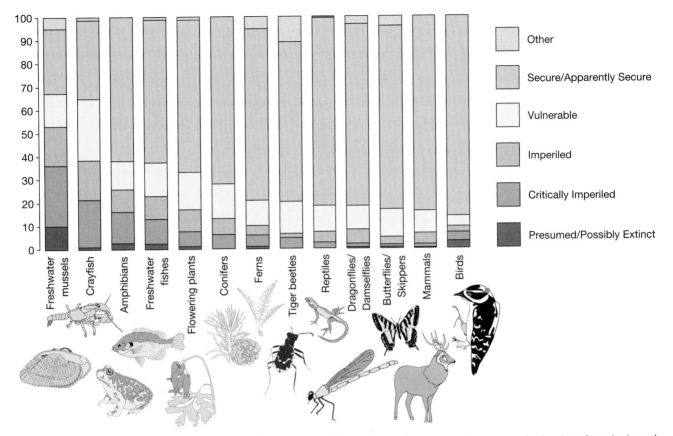

Figure 15.35 The status of species in the United States Each bar shows, for a group of organisms, the fraction of species in each of six status categories ranging from secure to presumed or possibly extinct. The Nature Conservancy report (Stein and Chipley 1996) considered a species critically imperiled if it had five or fewer populations or 1,000 or fewer individuals. Imperiled species had 6 to 20 populations, or 1,001 to 3,000 individuals. Vulnerable species were rare, but with larger populations. After Stein and Chipley (1996) and Dicke (1996). From *The New York Times*, January 2, 1996, B8. Copyright © 1996, The New York Times Co., Reprinted by permission.

tained numerous extinctions. Andrew Balmford (1996) suggests that the long history of dense human occupation, combined with extinctions caused by radical climate change during the ice ages of the Pliocene and Pleistocene, have put nontropical biomes through an "extinction filter." The plant and animal communities now living in these regions are expected to be relatively resilient in the face of continued human impact. In contrast, many areas of the tropics have been relatively unaffected by humans in recent history and were less affected by glaciation and sea-level changes in the Pleistocene. The tropics are now experiencing the highest rates of growth in human populations and the highest rates of habitat loss.

The threat to these forests is grave. According to the United Nations Food and Agriculture Organization (FAO), tropical forest loss between 1980 and 1990 totaled approximately 154,000 km^2 per year, about half of that in the Americas (Laurance 1990). An additional 56,000 km^2 per year was logged, but not completely destroyed. Because the total forest area of tropical Asia is comparatively small, the relative destruction of tropical forest on that continent was especially high during that same period: about 1.1% per year, compared to about 0.7% per year in Africa and the Americas (Laurance 1999).

The Brazilian Amazon is the largest continuous tropical forest in the world and is a special target of concern. David Skole and Compton Tucker (1993) used photographs from the Landsat satellite system to quantify rates of deforestation in the Brazilian Amazon (Figure 15.36). Their study indicates that an average of about 15,000 square kilometers of forest were lost each year during the interval between 1978 and 1988. Although this figure is much smaller than some earlier estimates of forest destruction rates (such as that from the FAO, which used indirect means to estimate forest destruction), it still represents an annual loss of forested area roughly equal in size to the state of Connecticut. Skole and Tucker also maintain

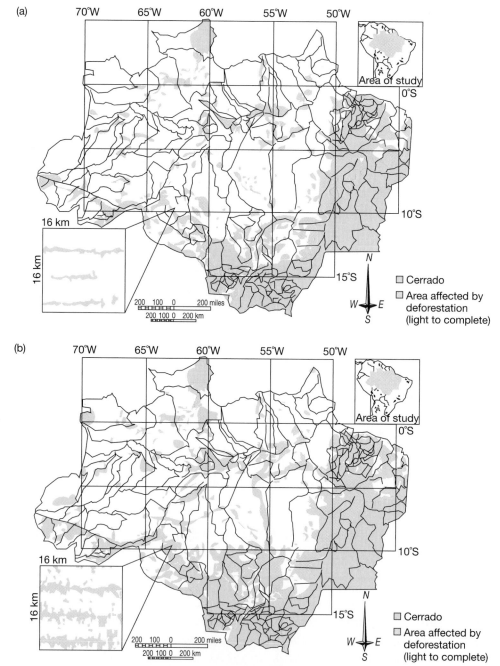

Figure 15.36   Deforestation in the Brazilian Amazon Skole and Tucker (1993) used satellite imagery to quantify the extent of forest, cerrado, and aquatic habitats in the Brazilian Amazon (cerrado is the term for tropical savanna, meaning grasslands with scattered trees). The top map was generated from photographs taken in 1978, the bottom map from images generated in 1988. Note that deforestation is most intense along the southern and eastern edges of the Amazon and along major transportation corridors, such as the Amazon River. Reprinted with permission from Skole and Tucker (1993). Copyright © 1993, American Association for the Advancement of Science.

that more than double this amount of land was adversely affected each year because of edge effects. Forested areas adjacent to clearings undergo dramatic changes: Light levels increase, soils dry, daily temperature fluctuations increase markedly, domestic livestock encroach, and hunting pressure by humans heightens. Skole and Tucker (1993: 1909) note that the "Implications for biological diversity are not encouraging."

Summary

The most efficient fossilization processes are compression, impression, casting, and permineralization. Because these events depend on the rapid burial of organic remains in water-saturated sediments, the fossil record is dominated by organisms with hard parts that lived in lowland or shallow-water marine environments. Thanks to new fossil finds and increasingly high-resolution dating techniques, the geological record of life on earth is steadily improving.

The Cambrian explosion was characterized by the sudden appearance of diverse animal body structures and the filling of ecological niches in shallow-water marine communities, although these events occurred independently in animal lineages that had diverged from each other earlier in the Proterozoic. The Cambrian explosion is just the most spectacular of a series of adaptive radiations that characterize the rise of morphological complexity and diversity through the Phanerozoic. Adaptive radiations can be triggered by ecological opportunities, key adaptations, or chance events such as colonizing a new habitat or surviving a mass extinction.

Prolonged stasis is another prominent pattern in evolutionary history. Stasis punctuated by sudden changes associated with speciation may be a common mode of evolution and implies that much of the morphological innovations that evolve in some taxonomic groups appear mainly in association with events that trigger speciation.

The eventual fate of both new taxa and new morphological traits is extinction. The five most intense extinctions are designated as mass extinctions and are commonly distinguished from background extinctions. The K–T extinction, for example, was caused by an asteroid that slammed into the Earth near Mexico's Yucatán peninsula. During both background and mass extinctions, geographically widespread species are less likely to go extinct.

During recent times, a prominent extinction has been documented in the loss of bird species on Polynesian islands. Although dramatic, this event was too local to qualify as a mass extinction. But current projections of species loss due to rapid habitat destruction indicate that a mass extinction may now be underway.

Questions

1. Terrestrial fossils from a particular time (say, 230 Ma) are patchily distributed. Instead of being evenly distributed over the continents in a continuous thin layer, they often occur in narrow strips or pockets a few miles wide. Why is this?

2. Most fossils of Mesozoic birds are from marine diving birds. Relatively few terrestrial species are known. Does this mean that most Mesozoic birds were, in fact, marine diving birds? Explain your reasoning.

3. The rise of angiosperm plants is widely thought to be due to competitive superiority over gymnosperms. However, it is not clear just what gave angiosperms the edge. Some of the hypotheses are that (1) angiosperms may have a better method of pollination, (2) angiosperms have a better method of water transport, and (3) angiosperms have more efficient leaves and can grow faster. Briefly outline experiments that would test these theories. (Remember that both angiosperms and gymnosperms are still alive today, including primitive angiosperms such as *Amborella*, and thus are available for experiments.)

4. Suppose that during the Pleistocene, a new species of lion arose gradually in western North America and then rapidly migrated to eastern North America. Now suppose you are a paleontologist whose only good fossil lions are from eastern North America. Would it appear to you that the new lion evolved gradually or suddenly? What further information would you need to correctly evaluate whether

Pleistocene lions evolved primarily via punctuated equilibrium vs. gradualism?

5. One of the (many) mysteries of the K–T extinction is the different fate of ammonites and nautiloids. These were mollusks with buoyant, chambered shells that lived in open-water habitats. Ammonites went extinct during the K–T extinction, but some nautiloids survived. The two groups had different reproductive strategies. Ammonites are thought to have produced many free-swimming young each year that fed near the ocean surface and grew rapidly. In contrast, a female nautilus produces just a few large eggs each year, each of which rests quietly in the depths for up to a year before hatching into a small, slow-growing nautilus. Based on these different reproductive strategies, suggest a possible hypothesis for why the nautiloids, but not the ammonites, might have been able to survive an asteroid impact. See the following reference for further discussion of the biology and history of ammonites and nautiloids:

Ward, P. 1992. *On Methuselah's Trail: Living Fossils and the Great Extinctions.* New York: W.H. Freeman & Co.

6. The Biscay coast of France and Spain has especially well-preserved K–T strata. Yet for most of the 20th century, no good ammonite specimens had been found near the K–T boundary, despite many years of study by two different teams. The paleontologists on these teams knew that ammonite fossils might not have been found simply due to limited searching time and limited personnel. Would this limited sampling make the extinction of ammonites appear to be earlier or later than it actually was? If ammonites had been more common, how would that affect the chance of discovering ammonite fossils and the resulting apparent time of extinction? How about if ammonites had been less common?

In the late 1980s there was a paleontology meeting near the Biscay Coast, and many of the paleontologists made a field trip to see the famous K–T boundary strata. The leader of one of the Biscay research teams took advantage of this opportunity by offering a prize for a good ammonite near the K–T boundary. The paleontologists fanned out and began combing the rocks, and one person discovered a lovely ammonite (and claimed the prize) after just a few hours of searching. What does this story illustrate about the effect of searching effort on apparent extinction times? Given the vagaries of fossil discovery, do you think mass extinctions will generally be reflected cleanly in the fossil record, or will mass extinctions appear to be spread out over time (even if they were actually simultaneous)? For more information, see

Raup, D. M. 1991. A kill curve for Phanerozoic marine species. *Paleobiology* 17: 37–48.

Marshall, C. R. and P. D. Ward. 1996. Sudden and gradual extinctions in the latest Cretaceous of western European Tethys. *Science* 274: 1360–1363.

7. In 1996, Gregory Retallack announced that he had found shocked quartz crystals dating from the same time as the end-Permian extinction. What is the implication of this finding? Other geologists point out that no one has found any evidence of elevated iridium from these strata (despite much searching). What is the significance of the lack of iridium?

8. Suppose you are talking to a friend about extinctions, and you mention that humans are known to have caused thousands of extinctions in the last few millenia. Your friend responds "So? Extinction is natural. Species have always gone extinct. So it's really not something we need to worry about." Is your friend correct that extinction is natural? Is the current rate of extinction typical? Is your friend correct that if extinctions are natural, then they are not a problem for the dominant life forms on Earth?

Exploring the Literature

9. The theory of punctuated equilibrium has been intensely controversial since it was proposed over 25 years ago. To begin a study of how scientific debates are conducted and how theories evolve over time, see

Eldredge, N., S. J. Gould, J. A. Coyne, and B. Charlesworth. 1997. On punctuated equilibria. *Science* 276: 338–341.

For a recent paleontological perspective on the theory and how to test it, see

Jackson, J. B. C., and A. H. Cheetham. 1999. Tempo and mode of speciation in the sea. *Trends in Ecology and Evolution* 14: 72–77.

10. Why are the tropics so rich in species? A number of hypotheses have been advanced over the years, but a robust result has been elusive. Check the following papers for recent work on this question:

Fjeldså, J. 1994. Geographical patterns for relict and young species of birds in Africa and South America and implications for conservation priorities. *Biodiversity and Conservation* 3: 207–226.

Gaston, K. J. 1996. Biodiversity—latitudinal gradients. *Progress in Physical Geography* 20: 466–476.

Gaston, K. J., and T. M. Blackburn. 1996. The tropics as a museum of biological diversity: An analysis of the New World avifauna. *Proceedings of the Royal Society of London,* Series B 263: 63–68.

Blackburn, T. M., and K. J. Gaston. 1997. The relationship between geographic area and the latitudinal gradient in species richness in New World birds. *Evolutionary Ecology* 11: 195–204.

Citations

Aguinaldo, A. M. A., J. M. Turbeville, L. S. Linford, M. C. Rivera, J. R. Garey, R. A. Raff, and J. A. Lake. 1997. Evidence for a clade of nematodes, arthropods, and other moulting animals. *Nature* 387: 489–493.

Alvarez, L. W., W. Alvarez, F. Asaro, and H. V. Michel. 1980. Extraterrestrial cause for the Cretaceous–Tertiary extinction. *Science* 208: 1095–1108.

Alvarez, W., F. Asaro, and A. Montanari. 1990. Iridium profile for 10 million years across the Cretaceous–Tertiary boundary at Gubbio (Italy). *Science* 250: 1700–1702.

Alvarez, W., P. Claeys, and S. W. Kieffer. 1995. Emplacement of Cretaceous–Tertiary boundary shocked quartz from Chicxulub crater. *Science* 269: 930–935.

Avise, J. C., W. S. Nelson, and H. Sugita. 1994. A speciational history of "living fossils:" Molecular evolutionary patterns in horseshoe crabs. *Evolution* 48: 1986–2001.

Ayala, F. J., A. Rzhetsky, and F. J. Ayala. 1998. Origin of the metazoan phyla: Molecular clocks confirm paleontological estimates. *Proceedings of the National Academy of Sciences, USA* 95: 606–611.

Balmford, A. 1996. Extinction filters and current resilience: The significance of past selection pressures for conservation biology. *Trends in Ecology and Evolution* 11: 193–196.

Barton, N. H., and B. Charlesworth. 1984. Genetic revolutions, founder effects, and speciation. *Annual Review of Ecology and Systematics* 15: 133–164.

Bateman, R. M., P. R. Crane, W. A. DiMichele, P. R. Kenrick, N. P. Rowe, T. Speck, and W. E. Stein. 1998. Early evolution of land plants: Phylogeny, physiology, and ecology of the primary terrestrial radiation. *Annual Review of Ecology and Systematics* 29: 263–292.

Bengtson, S. 1998. Animal embryos in deep time. *Nature* 391: 529–530.

Bengtson, S., and Y. Zhao. 1992. Predatorial borings in Late Precambrian mineralized exoskeletons. *Science* 257: 367–369.

Benton, M. J. 1995. Diversification and extinction in the history of life. *Science* 268: 52–58.

Bourgeois, J., T. A. Hansen, P. L. Wiberg, and E. G. Kauffman. 1988. A tsunami deposit at the Cretaceous–Tertiary boundary in Texas. *Science* 241: 567–570.

Briggs, D. E. G. 1991. Extraordinary fossils. *American Scientist* 79: 130–141.

Briggs, D. E. G., R. A. Fortey, and M. A. Wills. 1992. Morphological disparity in the Cambrian. *Science* 256: 1670–1673.

Brown, K. S. 1999. Deep Green rewrites evolutionary history of plants. *Science* 285: 990–991.

Budd, G. E. 1996. The morphology of *Opabinia regalis* and the reconstruction of the arthropod stem-group. *Lethaia* 29: 1–14.

Butterfield, N. J. 1990. A reassessment of the enigmatic Burgess Shale [British Columbia, Canada] fossil *Wiwaxia corrugata* (Matthew) and its relationship to the polychaete *Canadia spinosa* Walcott. *Paleobiology* 16: 287–303.

Cheetham, A. H. 1986. Tempo of evolution in a Neogene bryozoan: rates of morphologic change within and across species boundaries. *Paleobiology* 12: 190–202.

Cohen, J. E. 1995. Population growth and earth's human carrying capacity. *Science* 269: 341–346.

Conway Morris, S. 1989. Burgess shale faunas and the Cambrian explosion. *Science* 246: 339–346.

Conway Morris, S. 1993. The fossil record and the early evolution of the Metazoa. *Nature* 361: 219–225.

Conway Morris, S. 1995. Ecology in deep time. *Trends in Ecology and Evolution* 10: 290–294.

Conway Morris, S. 1998a. Early metazoan evolution: Reconciling paelontology and molecular biology. *American Zoologist* 38: 867–877.

Conway Morris, S. 1998b. *The Crucible of Creation.* Oxford: Oxford University Press.

Cooper, A., and D. Penny. 1997. Mass survival of birds across the Cretaceous–Tertiary boundary: Molecular evidence. *Science* 275: 1109–1113.

Cunningham, C. W., N. W. Blackstone, and L. W. Buss. 1992. Evolution of king crabs from hermit crab ancestors. *Nature* 355: 539–542.

Davidson, E. H., K. J. Peterson, and R. A. Cameron. 1995. Origin of bilaterian body plans: Evolution of developmental regulatory mechanisms. *Science* 270: 1319–1325.

Darwin, C. 1859. *The Origin of Species.* London: John Murray.

De Rosa, R., J. K. Grenier, T. Andreeva, C. E. Cook, A. Adoutte, M. Akam, S. B. Carroll, and G. Balavoine. 1999. *Hox* genes in brachiopods and priapulids and protostome evolution. *Nature* 399: 772–776.

Dicke, W. 1996. Numerous U.S. plant and freshwater species found in peril. *The New York Times,* January 2: B8.

Dobson, A. P. 1996. *Conservation and Biodiversity.* Scientific American Library, distributed by W. H. Freeman and Company, New York.

Ehrlich, P. R., and E. O. Wilson. 1991. Biodiversity studies: Science and policy. *Science* 253: 758–762.

Eldredge, N. 1984. Simpson's inverse: Bradytely and the phenomenon of living fossils. In N. Eldredge and S. M. Stanley, eds. *Living Fossils.* Berlin: Springer-Verlag, 272–277.

Eldredge, N., and S. J. Gould. 1972. Punctuated equilibria: An alternative to phyletic gradualism. In T. J. M. Schopf, ed. *Models in Paleobiology.* San Francisco: Freeman, Cooper & Company, 82–115.

Eldredge, N., S. J. Gould, J. A. Coyne, and B. Charlesworth. 1997. On punctuated equilibria. *Science* 276: 338–341.

Erwin, D. H. 1989a. Molecular clocks, molecular phylogenies and the origin of phyla. *Lethaia* 22: 251–257.

Erwin, D. H. 1989b. Regional paleoecology of Permian gastropod genera, southwestern United States and the end-Permian mass extinction. *Palaios* 4: 424–438.

Erwin, D. H. 1996. The geologic history of diversity. In R. C. Szaro and D. W. Johnston, eds. *Biodiversity in Managed Landscapes.* New York: Oxford University Press, 3–16.

Erwin, D. H., and R. L. Anstey. 1995a. Introduction. In D. H. Erwin and R. L. Anstey, eds. *New Approaches to Speciation in the Fossil Record.* New York: Columbia University Press, 1–8.

Erwin, D. H., and R. L. Anstey. 1995b. Speciation in the fossil record. In D. H. Erwin and R. L. Anstey, eds. *New Approaches to Speciation in the Fossil Record.* New York: Columbia University Press, 11–38.

Erwin, D., J. Valentine, and D. Jablonski. 1997. The origin of animal body plans. *American Scientist* 85: 126–137.

Feduccia, A. 1995. Explosive evolution in tertiary birds and mammals. *Science* 267: 637–638.

Florentin, J-M.R. M., and G. Sen. 1990. Impacts, tsunamis, and the Haitian Cretaceous–Tertiary boundary layer. *Science* 252: 1690–1693.

Glen, W. 1990. What killed the dinosaurs? *American Scientist* 78: 354–370.

Gould, S. J. 1982. The meaning of punctuated equilibrium and its role in validating a hierarchical approach to macroevolution. In R. Milkman, ed. *Perspectives on Evolution.* Sunderland, MA: Sinauer, 83–104.

Gould, S. J. 1989. *Wonderful Life.* New York: W. W. Norton.

Gould, S. J., and N. Eldredge. 1993. Punctuated equilibrium comes of age. *Nature* 366: 223–227.

Harland, W. B., R. L. Armstrong, A. V. Cox, L. E. Craig, A. G. Smith, and D. G. Smith. 1989. *A Geologic Time Scale.* Cambridge: Cambridge University Press.

Heymann, D., L. P. F. Chibante, R. R. Brooks, W. S. Wolbach, and R. E. Smalley. 1994. Fullerenes in the Cretaceous–Tertiary boundary layer. *Science* 265: 645–647.

Hunter, R. S. T., A. J. Arnold, and W. C. Parker. 1988. Evolution and homeomorphy in the development of the Paleocene *Planorotalites pseudomenardii* and the Miocene *Globorotalia (Globorotalia) maragritae* lineages. *Micropaleontology* 34: 181–192.

Irving, E. 1977. Drift of the major continental blocks since the Devonian. *Nature* 270: 304–309.

Jablonski, D. 1986. Background and mass extinctions: The alternation of evolutionary regimes. *Science* 231: 129–329.

Jablonski, D. 1991. Extinctions: A paleontological perspective. *Science* 253: 754–757.

Jablonski, D. 1995. Extinctions in the fossil record. In J. H. Lawton and R. M. May, eds., *Extinction Rates.* Oxford: Oxford University Press, 25–44.

Jablonski, D. 1996. Body size and macroevolution. In D. Jablonski, D. H. Erwin, and J. H. Lipps, eds. *Evolutionary Paleobiology.* Chicago: University of Chicago Press, 256–289.

Jablonski, D., and D. M. Raup. 1995. Selectivity of end-Cretaceous marine bivalve extinctions. *Science* 268: 389–391.

Jackson, J. B. C., and A. H. Cheetham. 1990. Evolutionary significance of morphospecies: A test with cheilostome bryozoa. *Science* 248: 579–583.

Jackson, J. B. C., and A. H. Cheetham. 1994. Phylogeny reconstruction and the tempo of speciation in the cheilostome Bryozoa. *Paleobiology* 20: 407–423.

James, H. F., T. W. Stafford, Jr., D. W. Steadman, S. L. Olson, P. S. Martin, A. J. T. Jull, and P. C. McCoy. 1987. Radiocarbon dates on bones of extinct birds from Hawaii. *Proceedings of the National Academy of Sciences, USA* 84: 2350–2354.

Janvier, P. 1999. Catching the first fish. *Nature* 402: 21–22.

Kerr, R. A. 1994. Testing an ancient impact's punch. *Science* 263: 1371–1372.

Lande, R. 1980. Genetic variation and phenotypic evolution during allopatric speciation. *American Naturalist* 116: 463–479.

Laurance, W. F. 1999. Reflections on the tropical deforestation crisis. *Biological Conservation* 91: 109–117.

Lee, M. S. Y. 1999. Molecular clock calibrations and metazoan divergence dates. *Journal of Molecular Evolution* 49: 385–391.

Li, H., E. L. Taylor, and T. N. Taylor. 1996. Permian vessel elements. *Science* 271: 188–189.

May, R. M., J. H. Lawton, and N. E. Stork. 1995. Assessing extinction rates. In J. H. Lawton and R. M. May, eds. *Extinction Rates* Oxford: Oxford University Press, 1–24.

McKinnon, W. B. 1992. Killer acid at the K–T boundary. *Nature* 357: 15–16.

Nichols, D. J., J. L. Brown, M. A. Attrep, Jr., and C. J. Orth. 1992. A new Cretaceous–Tertiary boundary locality in the western Powder River basin, Wyoming: Biological and geological implications. *Cretaceous Research* 13: 3–30.

Niklas, K. J. 1994. One giant step for life. *Natural History* 6: 22–25.

Niklas, K. J., B. H. Tiffney, and A. H. Knoll. 1983. Patterns in vascular land plant diversification. *Nature* 303: 614–616.

Olson, S. L., and H. F. James. 1982. Prodromus of the fossil avifauna of the Hawaiian islands. *Smithsonian Contributions in Zoology* 365: 1–59.

Pimm, S. L., G. J. Russell, J. L. Gittleman, and T. M. Brooks. 1995. The future of biodiversity. *Science* 269: 347–350.

Qiu, Y.-L., J. Lee, F. Bernasconi-Quadroni, D. E. Soltis, P. S. Soltis, M. Zanis, E. A. Zimmer, Z. Chen, V. Savolainen, and M. W. Chase. 1999. The earliest angiosperms: evidence from mitochondrial, plastid and nuclear genes. *Nature* 402: 404–407.

Raup, D. M. 1991. A kill curve for Phanerozoic marine species. *Paleobiology* 17: 37–48.

Raup, D. M. 1994. The role of extinction in evolution. *Proceedings of the National Academy of Sciences, USA* 91: 6758–6763.

Raup, D. M., and D. Jablonski. 1993. Geography of end-Cretaceous marine bivalve extinctions. *Science* 260: 971–973.

Raup, D. M., and J. J. Sepkoski, Jr. 1982. Mass extinction in the marine fossil record. *Science* 215: 1501–1503.

Ricciardi, A., and J. B. Rasmussen. 1999. Extinction rates of North American freshwater fauna. *Conservation Biology* 13: 1220–1222.

Runnegar, B. 1982. A molecular-clock date for the origin of the animal phyla. *Lethaia* 15: 199–205.

Schultz, P. H., and S. D'Hondt. 1996. Cretaceous–Tertiary (Chicxulub) impact angle and its consequences. *Geology* 24: 963–967.

Shu, D.-G., H.-L. Luo, S. Conway Morris, X.-L. Zhang, S.-X. Hu, L. Chen, J. Han, M. Zhu, Y. Li, and L.-Z. Chen. 1999. Lower Cambrian vertebrates from south China. *Nature* 402: 42–46.

Skole, D., and C. Tucker. 1993. Tropical deforestation and habitat fragmentation in the Amazon: Satellite data from 1978 to 1988. *Science* 260: 1905–1920.

Smith, F. D. M., R. M. May, R. Pellew, T. H. Johnson, and K. S. Walter. 1993a. Estimating extinction rates. *Nature* 364: 494–496.

Smith, F. D. M., R. M. May, R. Pellew, T. H. Johnson, and K. R. Walter. 1993b. How much do we know about the current extinction rate? *Trends in Ecology and Evolution* 8: 375–378.

Smith, A. B. 1999. Dating the origin of metazoan body plans. *Evolution and Development* 1: 138–142.

Soltis, P. S., D. E. Soltis, and M. W. Chase. 1999. Angiosperm phylogeny inferred from multiple genes as a tool for comparative biology. *Nature* 402: 402–404.

Stanley, S. M. 1984. Does bradytely exist? In N. Eldredge and S. M. Stanley, eds. *Living Fossils* Berlin: Springer-Verlag, 278–280.

Stanley, S. M., and X. Yang. 1987. Approximate evolutionary stasis for bivalve morphology over millions of years: A multivariate, multilineage study. *Paleobiology* 13: 113–139.

Stanley, S. M., and X. Yang. 1994. A double mass extinction at the end of the Paleozoic era. *Science* 266: 1340–1344.

Steadman, D. W. 1995. Prehistoric extinctions of Pacific island birds: Biodiversity meets zooarchaeology. *Science* 267: 1123–1131.

Stein, B. A., and R. M. Chipley, eds. 1996. *Priorities for Conservation: 1996 Annual Report Card for U.S. Plant and Animal Species.* Arlington, VA: The Nature Conservancy.

Sweet, A. R., and D. R. Braman. 1992. The K–T boundary and contiguous strata in western Canada: Interactions between paleoenvironments and palynological assemblages. *Cretaceous Research* 13: 31–79.

Swisher, C. C. III, J. M. Grajales-Nishimura, A. Montanari, S. V. Margolis, P. Claeys, W. Alvarez, P. Renne, E. Cedillo-Pardo, F. J-M. R. Maurrasse, G. H. Curtis, J. Smit, and M. O. McWilliams. 1992. Coeval ^{40}Ar/^{39}Ar ages of 65.0 million years ago from Chicxulub crater melt rock and Cretaceous–Tertiary boundary tektites. *Science* 257: 954–958.

Taylor, D. W., and L. J. Hickey. 1990. An Aptian plant with attached leaves and flowers: Implications for angiosperm origin. *Science* 247: 702–704.

Taylor, T. N., and E. L. Taylor. 1993. *The Biology and Evolution of Fossil Plants.* Englewood Cliffs, NJ: Prentice Hall.

Thomson, K. S. 1986. Living fossils. *Paleobiology* 12: 495–498.

Valentine, J. W. 1994. The Cambrian explosion. In S. Bengtson, ed. *Early Life on Earth,* Nobel Symposium No. 84. New York: Columbia University Press, 401–411.

Van Valen, L. 1973. A new evolutionary law. *Evolutionary Theory* 1: 1–30.

Vermeij, G. J. 1987. *Evolution and Escalation*. Princeton, NJ: Princeton University Press.

Waggoner, B. 1998. Interpreting the earliest metazoan fossils: What can we learn? *American Zoologist* 38: 975–982.

Weiguo, S. 1994. Early multicellular fossils. In S. Bengtson, ed. *Early Life on Earth,* Nobel Symposium No. 84. New York: Columbia University Press, 358–369.

Wilson, E. O. 1988. The current state of biodiversity. In E. O. Wilson, ed. *Biodiversity* Washington DC: National Academy Press, 3–18.

Wolbach, W. S., I. Gilmour, E. Anders, C. J. Orth, and R. R. Brooks. 1988. Global fire at the Cretaceous–Tertiary boundary. *Nature* 334: 665–669.

Wray, G. A., J. S. Levinton, and L. H. Shapiro. 1996. Molecular evidence for deep Precambrian divergences among metazoan phyla. *Science* 274: 568–573.

Xiao, S., Y. Zhang, and A. H. Knoll. 1998. Three-dimensional preservation of algae and animal embryos in a Neoproterozoic phosphorite. *Nature* 391: 553–558

Zimmer, C. 1999. Fossils give glimpse of old mother lamprey. *Science* 286: 1064–1065.

CHAPTER 16

Human Evolution

Australopithecus boisei (specimen KNM-ER 406, left) and *Homo ergaster* (specimen KNM-ER 3733, right) both lived in what is now Koobi Fora, Kenya, about 1.7 million years ago. (© 1985 David Brill, courtesy of National Museums of Kenya; from Johanson, Edgar, and Brill 1996)

THE FIRST PRINTING OF *ON THE ORIGIN OF SPECIES* SOLD OUT ON NOVEMBER 24, 1859, the very day it was published. Among the profound implications that made the book such a sensation was what it told its readers about themselves. Although Darwin saw this as clearly as anyone, his only explicit treatment of human evolution was a single paragraph in the last chapter, in which his strongest claim was that "Light will be thrown on the origin of man and his history" (Darwin 1859, p. 488).

Not until 12 years later did Darwin reveal the depth and breadth of his thinking about humans. In 1871 he published a two-volume work, *The Descent of Man, and Selection in Relation to Sex*. In the introduction, Darwin explained his initial reticence on the subject of human evolution: "During many years I collected notes on the origin or descent of man, without any intention of publishing on the subject, but rather with the determination not to publish, as I thought that I should thus only add to the prejudices against my views" (Darwin 1871, p. 1).

Darwin's apprehensions were well founded: The human implications of evolutionary biology have been, and remain, the focus of heated controversy. In 1925, Tennessee schoolteacher John T. Scopes was convicted of violating a new state law prohibiting the teaching of evolution (see Chapter 3). The Scopes case was popularly known as the Monkey Trial, indicating that for many observers the central issue at stake was the origin of the human species. In 1995, the Alabama state board of education ruled that all textbooks discussing evolution must carry a disclaimer that admonishes readers to consider evolution as theory, not fact. National Public Radio reporter Debbie Elliot interviewed members of the school board about their decision (National Public Radio 1995). Again, the origin of our species was a key issue. Board member Stephanie Bell, for example, told Elliot, "Most people do not believe that we evolved from apes."

In this chapter, we explore research on the evolutionary history of our species. We start by reviewing attempts to determine the evolutionary relationships among humans and the extant apes. Then we consider the fossil evidence bearing on the course of human evolution following the split between our lineage and the lineage of our closest living relatives. Next, we look at fossil and molecular evidence on the emergence of *Homo sapiens.* Finally, we consider the evolutionary origins of some of our species' defining characteristics, including tool use and language. Our exploration illustrates that the subject of human evolution generates controversies within the scientific community that, while different in focus, are as heated as those it generates among the lay public.

16.1 Relationships Among Humans and the Extant Apes

Humans (*Homo sapiens*) belong to the primate taxon Catarrhini, which includes the Old World monkeys, such as the baboons and macaques, and the Hominoidea (Figure 16.1). The Hominoidea, also referred to as the hominoids, include the gibbons (*Hylobates*) of southeast Asia and the Hominidae. The Hominidae (Wilson and Reeder 1993), or hominids, include the orangutan (*Pongo pygmaeus*), also of southeast Asia, and three African species: the gorilla (*Gorilla gorilla*), the common chimpanzee (*Pan troglodytes*), and the bonobo, or pygmy chimpanzee (*Pan paniscus*). Gibbons, orangutans, gorillas, and chimpanzees are referred to informally as apes. Orangutans, gorillas, and chimpanzees are referred to as great apes.

Humans Belong to the Same Clade as the Apes

Scientists universally agree that humans belong with the apes in the Hominoidea. Humans share with the apes numerous derived characteristics (synapomorphies). These evolutionary innovations distinguish the hominoids from the rest of the Catarrhini and indicate that the hominoids are descended from a common ancestor (see Chapter 13). The shared derived traits of the hominoids include relatively large brains, the absence of a tail, a more erect posture, greater flexibility of the hips and ankles, increased flexibility of the wrist and thumb, and changes in the structure and use of the arm and shoulder (Andrews 1992; see also Groves 1986; Andrews and Martin 1987; Begun et al. 1997). In addition to this morphological evidence, the molecular analyses described later in this chapter also unequivocally demonstrate that humans are hominoids.

Humans Belong to the Same Clade as the African Great Apes

Figure 16.1 includes a reconstruction of the phylogenetic relationships among the hominoids. This reconstruction places humans with the great apes in the Hominidae. Furthermore, it places humans with the African great apes in the Homininae. The

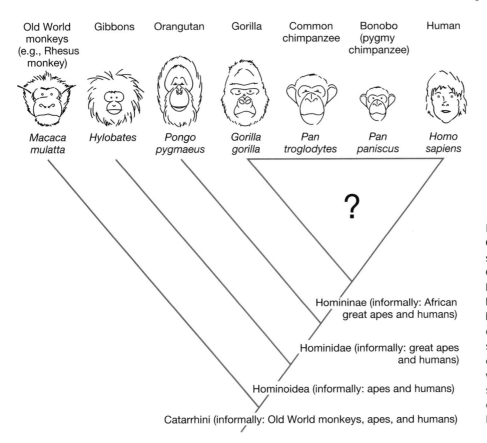

Figure 16.1 Phylogeny of the Catarrhini This evolutionary tree shows the relationships among the Old World monkeys, represented by a Rhesus monkey, and the apes and humans. Among the apes, the gibbons branch off first, followed by the orangutan. The evolutionary relationships among the gorilla, the two chimpanzees, and humans (triangle with question mark) have been the subject of considerable dispute. The classification used here follows that in Pough et al. (1996).

reconstruction was first proposed by Thomas Henry Huxley (Huxley 1863). Huxley's proposal raised dispute, but in recent years, as more data have been collected and analyzed, scientists in all fields have accepted the tree in Figure 16.1. Cladistic analyses of morphology support the tree. Humans and the African great apes share a number of derived traits that distinguish them from the rest of the apes. These synapomorphies include elongated skulls, enlarged brow ridges, shortened but stout canine teeth, changes in the front of the upper jaw (premaxilla), fusion of certain bones in the wrist, enlarged ovaries and mammary glands, changes in muscular anatomy, and reduced hairiness (Ward and Kimbel 1983; Groves 1986; Andrews and Martin 1987; Andrews 1992; Begun et al. 1997).

Molecular analyses concur. They have indicated a close relationship between humans and the African great apes since the beginnings of modern molecular systematics. George H. F. Nuttall (1904) used rabbit antibodies to human whole serum to show that chimpanzee serum reacted most like that of humans, followed by gorillas, orangutans, and monkeys. By the 1960s, more refined methods using specific blood proteins produced similar results (Goodman 1962; Sarich and Wilson 1967). Vincent Sarich and Allan Wilson took purified serum albumin, a blood protein, from humans, and injected it into rabbits. After giving the rabbits time to make antibodies against the human albumin protein, Sarich and Wilson took blood serum from the rabbits (which contained rabbit antihuman antibodies). The researchers mixed the rabbit serum with purified serum albumin from a variety of apes and Old World monkeys. Like Nuttall, Sarich and Wilson used the strength of the immune reaction between the rabbit antihuman antibodies and the pri-

Morphological and molecular analyses demonstrate that humans are closely related to gorillas and chimpanzees.

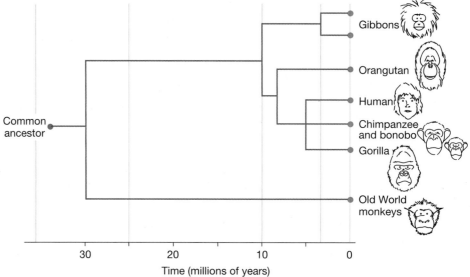

Figure 16.2 Sarich and Wilson's phylogeny of the Hominoidea The time line on the bottom is in millions of years before the present. Reprinted with permission from Sarich and Wilson (1967). Copyright © 1967, American Association for the Advancement of Science.

mate albumins as a measure of similarity among the albumins they tested, and they assumed that the similarity of two species' serum albumin proteins reflects the species' evolutionary kinship. The resulting phylogeny shows that humans are closely related to gorillas and the two chimpanzees (Figure 16.2).

Sarich and Wilson put a time line on their phylogeny by assuming that (1) serum albumin evolves at a constant rate and (2) the split between the apes and the Old World monkeys occurred 30 million years ago. (The latter assumption is based on the fossil record.) The time line suggests that humans and the African great apes shared a common ancestor about 5 million years ago, which is much more recently than scientists had previously suspected (see Lowenstein and Zihlman 1988). We will review other, more recent molecular phylogenies shortly; all are consistent with Sarich and Wilson's conclusions in showing a close kinship between humans and the African great apes.

The phylogenies in Figure 16.1 and Figure 16.2 show that humans, gorillas, and the two chimpanzees are closest relatives, but they do not resolve the evolutionary relationships among these four species. The true phylogeny for humans, gorillas, and the two chimpanzees could be any one of the four trees shown in Figure 16.3. It is probably safe to say that more scientists have invested more effort in attempting to determine which of these trees is correct than has been invested in any other species-level problem in the history of systematics.

Humans, Gorillas, Chimpanzees, and Bonobos: Morphological Evidence

Paleontologists have attempted to solve the human/African ape puzzle by performing cladistic analyses of morphology. These researchers point out several features shared by gorillas and the two chimpanzees, but lacking in humans. These traits mainly include skeletal features associated with knucklewalking (Andrews and Martin 1987). Knucklewalking is a derived trait in the African great apes. Gorillas and both species of chimpanzee are knucklewalkers, whereas humans are not. Considering knucklewalking in isolation, the simplest explanation for its distribution is that the human lineage diverged first from the lineage that would later pro-

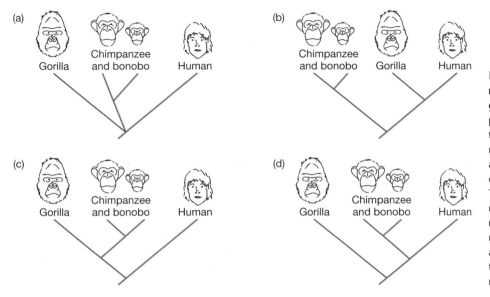

Figure 16.3 Possible phylogenies of humans and the African great apes The figure shows four possible resolutions of the evolutionary relationships among humans and the African great apes. All assume that the two species of chimpanzee are closest relatives. The true tree could have (a) a genuine simultaneous three-way split (trichotomy), (b) gorillas and humans as closest relatives, (c) gorillas and chimpanzees as closest relatives, or (d) chimpanzees and humans as closest relatives.

duce the gorilla and the two chimpanzees (Figure 16.3c). This scenario requires only one appearance of knucklewalking, in a common ancestor of the gorilla and chimpanzees, and no losses. However, the scenario does require that several traits shared only by humans and the two chimpanzees be interpreted either as ancestral traits that were lost in gorillas or as convergent traits that evolved independently in humans and the chimpanzees. These traits include features of the teeth, skull, and limbs, delayed sexual maturity, and prominent labia minora in females and a pendulous scrotum in males (Groves 1986; Begun 1992).

Cladistic reconstructions of evolutionary trees depend on the accurate identification of which traits are ancestral and which are derived (see Chapter 13). David R. Begun classified characteristics of the skulls of the great apes by including in his analysis an extinct European ape called *Dryopithecus,* which is known only from fossils about 10 million years old (Begun 1992; see also Begun 1995). *Dryopithecus* shares several cranial traits with gorillas that are absent in the two chimpanzees and humans. These traits might previously have been classified as uniquely derived in gorillas, but given their presence in *Dryopithecus,* the traits now appear to be ancestral. This, in turn, means that some of the traits thought to be ancestral or convergent in humans and chimpanzees now appear to be derived. When Begun reconstructed the ape evolutionary tree with the new classification of traits, he concluded that humans and chimpanzees are closest relatives (Figure 16.3d). This implies either that (1) the most recent common ancestor of humans, gorillas, and the chimpanzees was a knucklewalker, and that knucklewalking was subsequently lost in the human lineage or (2) that kucklewalking evolved independently in gorillas and the chimpanzees. It also implies that a number of characters of the teeth, skull, and limbs, as well as the delayed sexual maturity and shared genital anatomy of humans and chimpanzees, need have evolved only once.

A number of researchers have not been convinced by Begun's reasoning, arguing that some of the skull features that Begun believes are shared derived traits in humans and chimpanzees may be ancestral or convergent and that knucklewalking may not be so readily evolved or lost as Begun's phylogeny requires (Dean and Delson 1992; Andrews 1992). However, more recent analyses of much expanded

Recent morphological analyses suggest that humans and chimpanzees are closest relatives, but this conclusion is controversial.

data sets seem to confirm the close relationship among chimpanzees and humans (Shoshani et al. 1996; Begun et al. 1997).

Humans, Gorillas, Chimpanzees, and Bonobos: Molecular Evidence

Various molecular analyses indicate that humans and chimpanzees are closest relatives . . .

Molecular biologists have attempted to resolve the human/African great ape phylogeny by analyzing DNA sequences. These efforts are producing a consensus that humans and the chimpanzees are closest relatives (Figure 16.3d). The evidence is still controversial, however. Figure 16.4 shows three estimates of the phylogeny based on different parts of the genome. The tree in Figure 16.4a is based on the mitochondrial genome, which is inherited only along the maternal line. The tree in Figure 16.4b is based on a gene on the Y chromosome, which is inherited only along the paternal line. The tree in Figure 16.4c is based on autosomal nuclear genes, which are inherited both maternally and paternally. All three phylogenies strongly support the version of the human/African great ape phylogeny in which humans and the chimpanzees are closest relatives (Figure 16.3d).

Jonathan Marks (1994) objects to this interpretation of the molecular data. Marks emphasizes that phylogenies such as those in Figure 16.4 are actually gene trees, not species trees. Marks points out that if an ancestral species was genetically variable for the gene under study, then the gene tree estimated from sequence data may not be the same as the true species tree. Figure 16.5 illustrates the reasoning. If different descendant species lose different ancestral alleles, and if we sequence only one allele from each descendant species, then we can end up

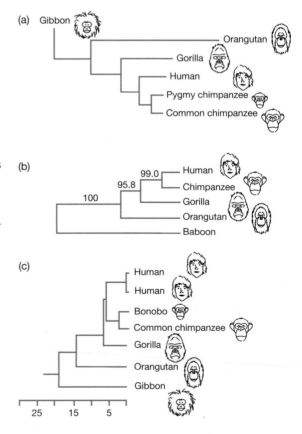

Figure 16.4   Three molecular phylogenies of humans and the African great apes (a) A phylogeny based on mitochondrial DNA. Satoshi Horai and colleagues (1992) analyzed variation in the sequence of a 4700-base-pair-long stretch of the mitochondrial genome, including genes for 11 transfer RNAs and 6 proteins. They estimated the phylogeny using the maximum likelihood method (see Chapter 13). Statistically, this tree, in which humans and chimpanzees are closest relatives, is about 10^{23} times as likely as the alternative tree in which chimpanzees and gorillas are closest relatives. Horai and colleagues estimate that the gorillas branched off the tree 7.7 $\pm$ 0.7 million years ago, and that humans and chimpanzees diverged 4.7 $\pm$ 0.5 million years ago. From Horai et al. (1992). (b) A phylogeny based on a Y-linked gene. Heui-Soo Kim and Osamu Takenaka (1996) analyzed variation in the sequence of the gene for testis-specific protein Y, a protein expressed only in the testes. Kim and Takenaka estimated the phylogeny by the maximum parsimony method (see Chapter 13). The tree has three nodes; the numbers on the branches indicate the percentage of bootstrap replicates in which the most parsimonious tree contained each node. These results provide strong statistical support for the hypothesis that humans and chimpanzees are closest relatives. From Kim and Takenaka (1996). (c) A phylogeny based on autosomal genes. Morris Goodman and colleagues (1994) analyzed sequence variation in noncoding regions of the β-globin gene cluster. They estimated the phylogeny by the maximum parsimony method (see Chapter 13). They also performed a maximum-likelihood analysis and found strong statistical support for the hypothesis that humans and chimpanzees are closest relatives. The scale bar below the phylogeny shows the estimated times for the branch points, in millions of years ago. From Goodman et al. (1994).

reconstructing only a portion of the original gene tree. This portion may imply a different branching pattern than that of the true species tree. Marks notes that his own study of chromosome morphology (1993; but see Borowik 1995), as well as Djian and Green's study of the involucrin gene (1989), examined variation within as well as between species. Both studies suggest that gorillas and the chimpanzees are closest relatives (Figure 16.3c). In light of the difference in conclusions between these studies and studies such as those shown in Figure 16.4, Marks asserts that the human/African great ape phylogeny is still an unresolved trichotomy (Figure 16.3a).

Maryellen Ruvolo and colleagues (1994) address Marks's objection by reconstructing a mitochondrial gene tree using several samples from each ape species (Figure 16.6). If the gene tree were different from the species tree, then different haplotypes from a single species might fail to cluster together. For example, the human haplotype *Hsa 1* might be more closely related to one of the chimpanzee haplotypes than to the other human haplotypes. In fact, all the haplotypes cluster

... but the phylogenies of genes and the phylogenies of species are not necessarily the same.

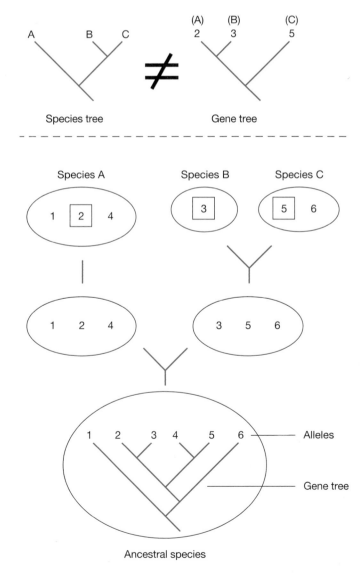

Figure 16.5   Gene trees versus species trees   Read the figure from the bottom to the top. The oval at the bottom represents an ancestral species containing six alleles for a gene. The alleles are related to each other as shown by the gene phylogeny inside the oval: Each allele was derived from its ancestors via a series of mutations. Moving up the figure, we see that a speciation event produces two sister species from the ancestral species. By selection or drift, one species loses alleles 3, 5, and 6, while the other species loses alleles 1, 2, and 4. Moving further up the figure, we find that another speciation event occurs, followed by the loss of alleles in its descendant species. We now have species A, containing alleles 1, 2, and 4; species B, containing allele 3; and species C, containing alleles 5 and 6. Now imagine that we sequence a single allele from each species (boxes) and reconstruct a phylogeny. In the true species tree, species B and C are closest relatives, but in the gene tree, alleles 2 and 3 are closest relatives. The gene tree and the species tree at the top of the figure show different branching patterns. Note that if we had sampled more extensively from each species and found all six alleles, we would have realized that the gene tree was a misleading guide for estimating the species phylogeny, because the alleles from species A, for example, would not cluster together. After Ruvolo (1994).

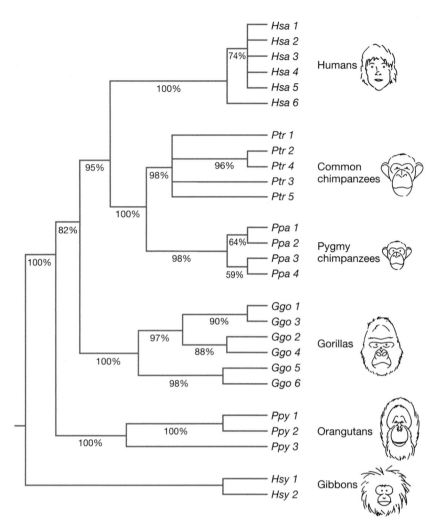

Figure 16.6   Phylogeny of mitochondrial cytochrome oxidase II alleles in humans and the African great apes   Ruvolo and colleagues estimated this tree using the maximum parsimony method; the numbers on the branches indicate the percentage of times each branch was present in the most parsimonious trees from 1000 bootstrap replicates (see Chapter 13). Note that the branch leading to the split between humans and chimpanzees has a bootstrap value of 95%. This represents strong support for the hypothesis that humans and chimpanzees are closest relatives. From Ruvolo et al. (1994).

A more thorough molecular analysis supports the hypothesis that humans and chimpanzees are closest relatives . . .

together by species, suggesting that the gene tree is a reliable estimate of the species tree and that humans and the chimpanzees are indeed closest relatives.

Ruvolo et al.'s test is not definitive, however. Had the mitochondrial sequences failed to cluster by species, it would have proved that the gene tree and the species tree are different. But failure to prove that the gene and species trees are different does not prove that they are the same. Look back at Figure 16.5 and consider the following scenario: Imagine that after the second speciation event, species A, B, and C all go through bottlenecks in which the number of individuals in each species is small. Now imagine that genetic drift eliminates alleles 1 and 4 in species A, and allele 6 in species C. Genetic diversity has been reduced to a single allele in each species. Now imagine that all three species recover from their bottlenecks and accumulate new genetic diversity. In species A, this diversity accumulates as a result of mutations in allele 2; in species B, the diversity accumulates as a result of mutations in allele 3; and in species C, the diversity accumulates as a result of mutations in allele 5. Now we reconstruct the phylogeny. No matter how many alleles we include from each species, the alleles will cluster by species, and the gene tree and species tree will be different.

A way around this problem is to reconstruct the phylogeny using several independent genes, and see if the independent phylogenies agree. Ruvolo (1995, 1997) reviews and tallies the independent data sets of DNA sequences that are informative about the human/African great ape phylogeny. She points out that if Marks is correct, and there is a real three-way split of the branches leading to humans, gorillas, and the two chimpanzees, then there should be equal numbers of independent data sets supporting a human–chimpanzee pairing, a human–gorilla pairing, and a chimpanzee–gorilla pairing. Ruvolo counts all mitochondrial DNA studies as a single data set, because all mitochondrial genes are linked and are thus not independent of each other. Likewise, any groups of nuclear genes that are near each other on the same chromosome count as a single data set, because the genes are linked. Ruvolo counts 14 independent data sets: 11 show humans and the chimpanzees as closest relatives, 2 show gorillas and the chimpanzees as closest relatives, and 1 shows humans and gorillas as closest relatives. Ruvolo calculates that under Marks's hypothesis of a true trichotomy, this distribution of results has a probability of only 0.002. Ruvolo concludes that the molecular phylogeny data reject the trichotomous tree and favor the tree in which humans and the chimpanzees are closest relatives (Figure 16.3d).

... as do combined analyses of several molecular data sets.

Most recently, Yoko Satta and colleagues have completed the most extensive survey of genetic data sets to date bearing on the question of great ape and human relations (Satta et al. 2000). These researchers examined 45 loci with a total of nearly 47,000 base pairs and reached the conclusion that the majority supported the chimpanzee–human pairing. Most significantly, Satta and colleagues seek to explain the reasons certain loci, and even certain sequences within specific loci, suggest alternative relationships. They point out that the percentage of loci that are incompatible with the hypothesis that chimpanzees and humans are closest relatives (40%, split more or less evenly between the human–gorilla and chimpanzee–gorilla hypotheses) can be explained in terms of the expected probability of random sorting of alleles, given certain estimates of population size and generation time. Incompatibility within loci results from the predictable effects of recombination and genetic drift (Sakka et al. 2000). In retrospect, as these researchers point out, it is not surprising that genetic data are somewhat ambiguous on this question, given the complexities of the molecular biology of the genome.

[For more of the controversy concerning the molecular evidence, see the exchange between Ruvolo (1994; 1995) and Green and Djian (1995), Marks (1994; 1995), and Rogers and Comuzzie (1995). See also papers by Deinard and Kidd (1999) and Rogers (1994) on the current state of the trichotomy theory.]

16.2 The Recent Ancestry of Humans

According to the evidence presented in Figures 16.4a and (c) humans and chimpanzees last shared an ancestor about 5 million years ago. This has interesting implications for our understanding of the nature of that last common ancestor. It is probable that at least some of the range of behaviors that are shared by chimpanzees and humans today were inherited from our last common ancestor. If this is the case, then that ancestor, in addition to being a knuckle-walker, would have had a broad, fruit-based diet and lived in a range of different habitats. It may well have used tools to obtain and process food, and it may well have hunted, as do living chimpanzees and humans. Finally, it may have engaged in complex social relations that included strategic alliances, deception, coalitionary killing (warfare), and cannibalism, as well as more positive behaviors like sharing, teaching, and compassion. Among the primates, all of these traits are most strongly developed in living chimpanzees and humans, and if they are inherited from our common ancestors, this has intriguing implications for understanding the biological basis of these behaviors. [For more on this subject, see recent reviews by Wrangham (1999), Boesch and Tomasello (1998) from the primatological perspective, and Begun (1994) from the paleontological perspective.]

What has been the pattern of evolution leading from our last common ancestor with the chimps to ourselves? Has our history involved only the steady transformation of a single lineage, finally culminating in *Homo sapiens,* or have there been repeated splits and extinctions in our recent evolutionary tree?

The Fossil Evidence

The fossil record includes a diversity of hominids that lived after the human lineage and the chimpanzee lineage separated.

Fossils provide the only data available for answering these questions. The fossil database for hominids is frustratingly sparse, but steadily improving (see Tattersall 1995; Johanson et al. 1996; Tattersall 1997). Illustrations of some of the key specimens appear in Figures 16.7 through 16.10.

Paleoanthropologists disagree about the most appropriate names for many of these specimens. In this text, we use the names used by Johanson et al. (1996) in the belief that they will be the names most familiar to readers. In several cases we note alternative names. Likewise, paleoanthropologists disagree about the number of species represented by the specimens in the figures (see Tattersall 1986, 1992). For example, the specimens of *Homo habilis* and *Homo rudolfensis* in Figures 16.9b and (c) are both from Koobi Fora, Kenya, and are both about 1.9 million years old. Some researchers consider them to be variants of the same species (*Homo habilis*), whereas others consider them different species. As with the names, we have followed the classification used by Johanson et al. (1996). The time ranges noted in the figures are also those given in Johanson et al. (1996); they differ somewhat from the estimates of other researchers, including those given by Strait et al. (1997) and used in Figure 16.11.

Figure 16.7 shows examples of the gracile australopithecines and *Ardipithecus.* The species depicted in Figure 16.7a and (b), *Australopithecus africanus* and *Australopithecus afarensis,* had skulls with small braincases (400 to just over 500 cm^3) and relatively large, projecting faces (Johanson et al. 1996). The females grew to heights of about 1.1 meters (3′7″), whereas the males were some 1.4 to 1.5 meters tall (4′7″ to 4′11″). Both species walked on two legs. Evidence for their erect posture comes from many bones of the skeleton, including the hips, knees, feet, limb proportions, and vertebral column, all of which are anatomically modified to permit upright posture and the support of the body mass on two rather than four feet. Other evidence for bipedal locomotion appears in the photo on page 21: fossilized footprints at Laetoli, Tanzania, of a pair of *A. afarensis* that walked side by side through fresh ash from the Sadiman volcano 3.6 million years ago (Stern and Susman 1983; White and Suwa 1987).

The two species depicted in Figures 16.7c and (d), *Australopithecus anamensis* and *Ardipithecus ramidus,* are less well known. The structure and size of a tibia from *A. anamensis* indicates that its owner was a biped somewhat larger than *A. afarensis* (Leakey et al. 1995). *Ardipithecus ramidus* is the least derived of these species, with teeth in many ways intermediate between chimpanzees and humans. It has skeletal features that suggest it was a biped, but its discoverers are withholding final

(a) Name: ***Australopithecus africanus***
    Specimen: Sts 5
    Age: 2.5 million years
    Found by: Robert Broom and
        John T. Robinson
    Location: Sterkfontein, South Africa
    Color photo: Johanson et al.
        (1996) pages 3; 135

    Species Time Range: ~2.4–2.8 Ma

(b) Name: ***Australopithecus afarensis***
    Also known as: *Praeanthropus africanus*
    Specimen: Reconstruction from fragments
    Color photo of same species:
        Johanson et al. (1996) page 129

    Species Time Range: ~3.0–3.9 Ma

(c) Name: ***Australopithecus anamensis***
    Specimen: KNM-KP 29281
    Age: 4.1 million years
    Found by: Peter Nzube
    Location: Kanapoi, Kenya
    Color photo:
        Johanson et al. (1996) page 123

    Species Time Range: ~3.9–4.2 Ma

(d) Name: ***Ardipithecus ramidus***
    Originally named as: *Australopithecus
        ramidus*
    Specimen: ARA-VP-1/128
    Age: 4.4 million years
    Found by: T. Assebework
    Location: Aramis, Ethiopia
    Color photo of same species:
        Johanson et al. (1996) page 116

    Species Time Range: ~4.4 Ma

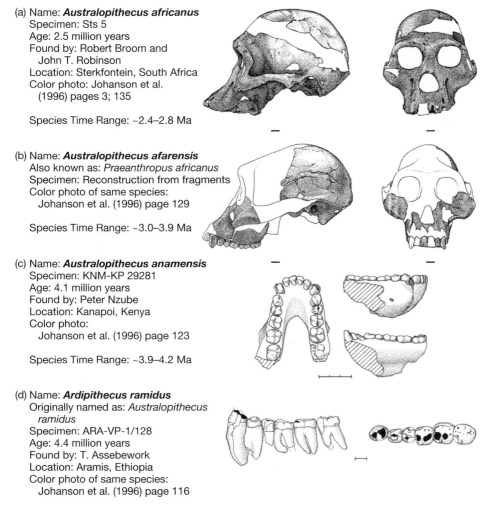

Figure 16.7   Gracile australopithecines and *Ardipithecus* (a) By Don McGranaghan, page 70 in Tattersall (1995). Scale bar = 1 cm. (b) By Don McGranaghan, page 146 in Tattersall (1995). Scale bar = 1 cm. (c) By Laszlo Meszoly, after Figure 1a in Leakey et al. (1995). Scale unit: 1 cm. (d) By Laszlo Meszoly, after Figure 3b in White et al. (1994); see also White et al. (1995). Scale bar = 1 cm.

judgment until they can complete a more thorough analysis (White et al. 1994; Johanson et al. 1996).

Figure 16.8 shows examples of the robust australopithecines. Like the gracile australopithecines, these species had relatively small braincases (in most cases between those of the gracile australopithecines and early *Homo* in relative size) and very large faces. Unlike the gracile australopithecines, they had enormous cheek teeth, robust jaws, and massive jaw muscles, sometimes anchored to a bony crest running along the centerline on the top of the skull (Johanson et al. 1996). These adaptations for powerful chewing have given one of the species, *A. boisei,* the nickname "nutcracker man." The robust australopithecines were about the same size as the gracile forms, and all were bipeds.

Figure 16.9 shows examples of early members of the genus *Homo,* all from Africa. Compared to the australopithecines, *Homo* species have larger braincases and relatively smaller faces (Johanson et al. 1996). Braincase volume in the genus *Homo* ranges from around 600 cm^3 in *H. habilis* to some 2000 cm^3 in examples of *H. sapiens.* In present-day humans, braincase volume averages about 1200 cm^3. The teeth and jaws of *Homo* species are smaller than those of the australopithecines. Members

(a) Name: ***Australopithecus robustus***
 Also known as: *Paranthropus robustus*
 Specimen: SK 48
 Age: 1.5–2.0 million years
 Found by: Fourie
 Location: Swartkrans, South Africa
 Color photo: Johanson et al.
  (1996) pages 108; 150

 Species Time Range: ~1.0–2.0 Ma

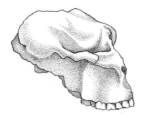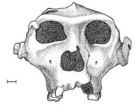

(b) Name: ***Australopithecus boisei***
 Also known as: *Paranthropus boisei*
 Specimen: KNM-ER 406
 Age: 1.7 million years
 Found by: Richard Leakey and H. Mutua
 Location: Koobi Fora, Kenya
 Color photo: Johanson et al.
  (1996) pages 54; 159; 160

 Species Time Range: ~1.4–2.3 Ma

Figure 16.8 Robust australopithecines
(a) By Laszlo Meszoly, after pages 108 and 150 in Johanson et al. (1996). Scale bar = 1 cm. (b) By Don McGranaghan, page 131 in Tattersall (1995). Scale bar = 1 cm. (c) By Don McGranaghan, page 195 in Tattersall (1995). Scale bar = 1 cm.

(c) Name: ***Australopithecus aethiopicus***
 Also known as: *Paranthropus aethiopicus*
 Specimen: KNM-WT 17000 (Black Skull)
 Age: 2.5 million years
 Found by: Alan C. Walker
 Location: Lake Turkana, Kenya
 Color photo: Johanson et al.
  (1996) pages 153; 154

 Species Time Range: ~1.9–2.7 Ma

(a) Name: ***Homo ergaster***
 Also known as: (African) *Homo erectus*
 Specimen: KNM-ER 3733
 Age: 1.75 million years
 Found by: Bernard Ngeneo
 Location: Koobi Fora, Kenya
 Color photo: Johanson et al.
  (1996) pages 180; 181

 Species Time Range: ~1.5–1.8 Ma

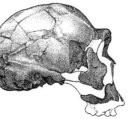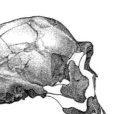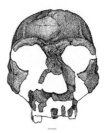

(b) Name: ***Homo habilis***
 Specimen: KNM-ER 1813
 Age: 1.9 million years
 Found by: Kamoya Kimeu
 Location: Koobi Fora, Kenya
 Color photo: Johanson et al.
  (1996) pages 6; 175

 Species Time Range: ~1.6–1.9 Ma

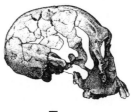

Figure 16.9 Early humans (a) By Don McGranaghan, page 138 in Tattersall (1995). Scale bar = 1 cm. (b) By Don McGranaghan, page 134 in Tattersall (1995). Scale bar = 1 cm. (c) By Don McGranaghan, page 133 in Tattersall (1995). Scale bar = 1 cm.

(c) Name: ***Homo rudolfensis***
 Also known as: *Homo habilis*
 Specimen: KNM-ER 1470
 Age: 1.8-1.9 million years
 Found by: Bernard Ngeneo
 Location: Koobi Fora, Kenya
 Color photo: Johanson et al.
  (1996) pages 178; 179

 Species Time Range: ~1.8–2.4 Ma

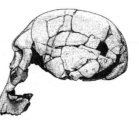

of the genus *Homo* tend to be taller than the australopithecines, with longer legs, although the earliest members of our genus overlap in tooth and body size with australopithecines. The difference in size between females and males within species tends to be less pronounced in *Homo* than in *Australopithecus*.

Figure 16.10 shows examples of more recent members of the genus *Homo* from Africa, Europe, and Java. Modern *H. sapiens,* like Cro-Magnon I, whose skull appears in Figure 16.10a, differ from earlier forms in a variety of traits (Johanson et al. 1996). Modern humans have very large braincases (Cro-Magnon I's is over 1600 cm^3, substantially larger than the present-day average). Associated with their large braincases, modern humans have high, steep foreheads. They also have relatively short, flat, vertical faces and prominent noses. Cro-Magnon I was a man who died in middle age about 30,000 years ago. His skeleton was found in a single prepared grave along with those of two other adult men, an adult woman, and an infant. The group had been buried with an assortment of animal bones, jewelry, and stone tools.

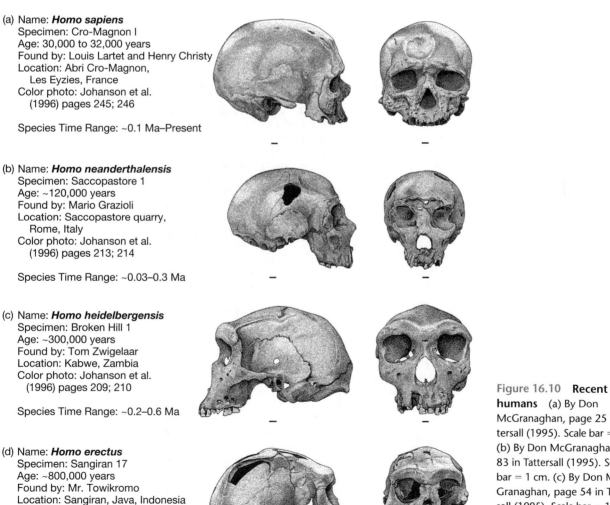

(a) Name: ***Homo sapiens***
 Specimen: Cro-Magnon I
 Age: 30,000 to 32,000 years
 Found by: Louis Lartet and Henry Christy
 Location: Abri Cro-Magnon,
  Les Eyzies, France
 Color photo: Johanson et al.
  (1996) pages 245; 246

 Species Time Range: ~0.1 Ma–Present

(b) Name: ***Homo neanderthalensis***
 Specimen: Saccopastore 1
 Age: ~120,000 years
 Found by: Mario Grazioli
 Location: Saccopastore quarry,
  Rome, Italy
 Color photo: Johanson et al.
  (1996) pages 213; 214

 Species Time Range: ~0.03–0.3 Ma

(c) Name: ***Homo heidelbergensis***
 Specimen: Broken Hill 1
 Age: ~300,000 years
 Found by: Tom Zwigelaar
 Location: Kabwe, Zambia
 Color photo: Johanson et al.
  (1996) pages 209; 210

 Species Time Range: ~0.2–0.6 Ma

(d) Name: ***Homo erectus***
 Specimen: Sangiran 17
 Age: ~800,000 years
 Found by: Mr. Towikromo
 Location: Sangiran, Java, Indonesia
 Color photo: Johanson et al.
  (1996) pages 192; 193

 Species Time Range: ~0.4–1.2 Ma

Figure 16.10 **Recent humans** (a) By Don McGranaghan, page 25 in Tattersall (1995). Scale bar = 1 cm. (b) By Don McGranaghan, page 83 in Tattersall (1995). Scale bar = 1 cm. (c) By Don McGranaghan, page 54 in Tattersall (1995). Scale bar = 1 cm. (d) By Don McGranaghan, page 172 in Tattersall (1995). Scale bar = 1 cm.

Interpreting the Fossil Evidence

Paleoanthropologists using fossils to reconstruct evolutionary history employ a two-step process (Strait et al. 1997). First the researchers use a cladistic analysis to estimate the evolutionary relationships among the various fossil species. Then they make educated guesses about which fossil species represent ancestors that lived at the branch points of the cladogram and which fossil species represent extinct side branches.

The results of one such study, by David S. Strait and colleagues (1997), appear in Figure 16.11. All of the species included, with the exception of *H. sapiens,* occur exclusively, or almost exclusively, in Africa. The first of our recent relatives to leave Africa were members of the genus *Homo.* They were probably descended from *H. ergaster.* We consider *H. ergaster* and its relationship to *H. sapiens* in the next section.

Figure 16.11 shows Strait et al.'s cladogram (a) and two hypotheses about what the cladogram tells us concerning the phylogenetic relationships among the various species [(b and c)]. The cladogram is based on a variety of skull and tooth characters. Note that in the cladogram, the lengths of the branches are meaningless; the only information encoded in the cladogram is in the order of branching.

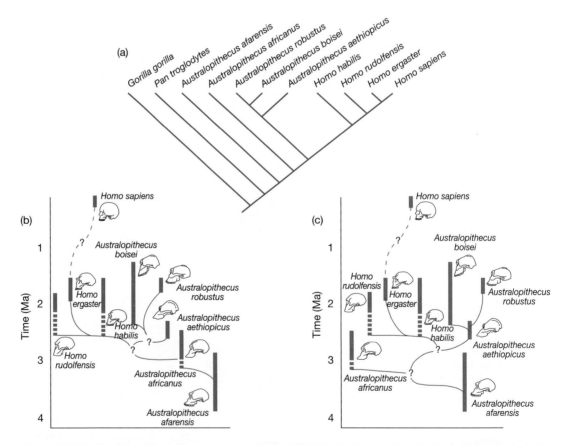

Figure 16.11   Phylogenies of *Homo sapiens* and its recent ancestors and extinct relatives   (a) A cladogram of three extant hominids (the gorilla, the common chimpanzee, and the modern human), and several extinct hominids known only from fossils. (b) and (c) Two hypotheses about the ancestor–descendant relationships implied by the cladogram in (a). In (b) and (c), the heavy green vertical bars indicate the known range of times over which each species existed, whereas the heavy green dashes represent the suspected range of times over which each species existed. The transition represented by the dashed blue line and question mark is the subject of debate. We discuss the evidence on this transition in Section 16.3.

The hypothesized phylogenies differ in the actual lengths ascribed to the branches in the cladogram. For example, in Figure 16.11b, the branch leading to *Australopithecus aethiopicus* is relatively long, so that *A. aethiopicus* is a sister species to *A. boisei* and *A. robustus*. In contrast, in Figure 16.11c, the length of the branch leading to *A. aethiopicus* is zero, so that *A. aethiopicus* is the common ancestor of *A. boisei* and *A. robustus*.

The hypothesized phylogenies in Figures 16.11b and (c) are testable, because they predict the existence of specific fossils that have not yet been discovered. For example, the phylogeny in Figure 16.11b predicts that paleontologists will one day find a new species of *Australopithecus*, between 2.6 and 3 million years old, that is the common ancestor of the three currently known robust australopithecines (*A. aethiopicus, A. robustus,* and *A. boisei*). This ancestral robust australopithecine should have all of the shared derived traits that define the robust group, but few, if any, of the unique traits that distinguish the three known species. The phylogeny in Figure 16.11c, in contrast, predicts that no such generalized robust australopithecine will be found, because the phylogeny asserts that the immediate ancestor of the group is already known.

Both of the hypothesized phylogenies in Figure 16.11b and (c) suggest that early *H. habilis* was the ancestor of the genus *Homo* and thus one of our own direct ancestors. Another possibility is that there existed an additional, as yet undiscovered, species of *Homo* that is the common ancestor to *H. rudolfensis, H. ergaster,* and *H. habilis*. In November 1994, a team of paleontologists lead by William H. Kimbel, Robert C. Walter, and Donald C. Johanson found a 2.3-million-year-old fossil jaw, together with a collection of simple stone tools, at Hadar, Ethiopia. The jaw may prove to be that of the heretofore undiscovered common ancestor of all members of the genus *Homo* (Gibbons 1996; Kimbel et al. 1996).

As the two different evolutionary trees offered by Strait and colleagues suggest, paleontologists have not reached a consensus on the correct phylogeny for *Homo, Australopithecus,* and their kin. Bernard Wood (1992) and Henry M. McHenry (1994), for example, propose phylogenies that are somewhat different from those suggested by Strait and colleagues (1997). Researchers disagree about which characters represent shared derived traits and which represent ancestral features. Two recently discovered fossil species may prove helpful in determining trait polarities. *Australopithecus anamensis* (Figure 16.7c), discovered in Kenya by Meave G. Leakey and colleagues (1995), lived 3.9 to 4.2 million years ago. *Ardipithecus ramidus* (Figure 16.7d), discovered in Ethiopia by Tim D. White and colleagues (1994; 1995), lived 4.4 million years ago. So far these two species are incompletely known; it may not be possible to confidently infer their precise relationships to the species included in Figure 16.11 until paleontologists find more complete skulls (Strait et al. 1997).

The phylogenetic relationships among the species of Australopithecus and Homo have not been definitively established.

Some Answers

Regardless of differences in the details, all recent estimates of the phylogeny of *Homo, Australopithecus,* and their relatives give the same general answers to the questions we posed at the beginning of this section. The pattern of evolution leading from our common ancestor with the chimpanzees to ourselves has been anything but simple. Frequent speciation produced a diversity of species. Throughout most of the last 4 million years, multiple species, as many as five at a time, have coexisted in Africa (see Tattersall 2000). For example, specimen KNM-ER 406 (Figure

16.8b) and specimen KNM-ER 3733 (Figure 16.9a) clearly represent different species (see the photo on page 549). Both were found at Koobi Fora, Kenya, in sediments of nearly the same age. We *Homo sapiens* are the lone survivors of an otherwise extinct radiation of bipedal African hominids.

16.3 The Origin of the Species *Homo sapiens*

Figure 16.11 represents the origin of our own species with only a dashed line and a question mark. The figure does not include three of the four recent *Homo* species shown in Figure 16.10. The question mark and the omissions reflect the considerable uncertainty over the origin of *Homo sapiens*.

Controversies over the Origin of Modern Humans

Paleoanthropologists are split on the taxonomic status of *H. ergaster* (Figure 16.9a) and *H. erectus* (Figure 16.10d). Some researchers consider these two forms to be regional variants of a single species (*H. erectus*), whereas others consider *H. erectus* to be a distinct Asian species descended from the African species *H. ergaster*. Likewise, some researchers consider *H. neanderthalensis* (Figure 16.10b) and *H. heidelbergensis* (Figure 16.10c) to be regional variants of transitional forms between *H. erectus* and modern *H. sapiens*. Others consider them to be distinct species, with *H. heidelbergensis* descended from *H. ergaster* and *H. neanderthalensis* descended from *H. heidelbergensis* (see Tattersall 1997). Most recently, a new species, *Homo antecessor*, has been suggested to be the common ancestor of both Neandertals and modern humans (Bermúdez de Castro et al. 1997; Arsuaga et al. 1999). Paleoanthropologists generally agree that modern humans are the descendants of some or all of the populations in the *H. ergaster/erectus* group. However, how and where the transition from *H. ergaster/erectus* to *H. sapiens* took place is a matter of debate.

The oldest examples of *H. ergaster/erectus* appear in the fossil record nearly simultaneously at Koobi Fora in Africa, at Dmanisi in the Caucasus region of Eastern Europe, at Longgupo Cave in China, and at Sangiran and Mojokerto in Java—all 1.6 to 1.9 million years ago (Gibbons 1994; Swisher et al. 1994; Gabunia and Vekua 1995; Huang Wanpo et al. 1995; Wood and Turner 1995; Gabunia et al. 2000). Because its immediate ancestors and closest relatives appeared to be confined to Africa, most paleontologists had assumed that *H. erectus* evolved in Africa and then moved to Asia. The fossils at Longgupo Cave, China, however, are similar enough to African *H. habilis* and *H. ergaster* to suggest that *H. erectus* may have evolved in Asia from earlier migrants (Huang Wanpo et al. 1995). Either way, prior to 2 million years ago the ancestors of our species within the genus *Homo* almost certainly lived in Africa.

The origin of modern Homo sapiens is controversial.

Anatomically modern *H. sapiens* first appear in the fossil record about 100,000 years ago in Africa and Israel, and somewhat later throughout Europe and Asia (Stringer 1988; Valladas et al. 1988; Aiello 1993). The range of hypotheses concerning the evolutionary transition from *H. ergaster/erectus* to *H. sapiens* is illustrated in Figure 16.12. At one extreme, the African replacement (or out-of-Africa) model (Figure 16.12a) posits that *H. sapiens* evolved in Africa, then migrated to Europe and Asia, replacing *H. erectus* and *H. neanderthalensis* without interbreeding. At the other extreme, the candelabra model (Figure 16.12d) holds that *H. sapiens* evolved independently in Europe, Africa, and Asia, without gene flow between regions. Be-

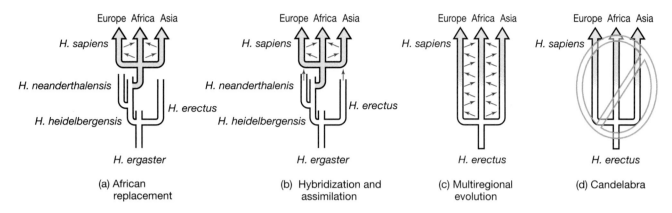

Figure 16.12 Hypotheses concerning the transition from *Homo ergaster/erectus* to *Homo sapiens*   The white portions of the phylogenies represent various archaic forms of *Homo,* including *H. ergaster, H. erectus, H. heidelbergensis,* and *H. neanderthalensis.* The colored portions represent modern *H. sapiens.* The small blue arrows represent gene flow. Note that specimens identified as *H. heidelbergensis* have been found in Europe and Africa, and specimens identified as *H. neanderthalensis* have been found in Europe and the Middle East. (a) The African replacement model. According to this model, modern *H. sapiens* evolved in Africa and then migrated to Europe and Asia. *H. sapiens* replaced the local forms without hybridization. No genes from these earlier forms persist in modern human populations. (b) The hybridization and assimilation model. According to this model, modern *H. sapiens* evolved in Africa and then migrated to Europe and Asia. *H. sapiens* largely replaced the local populations, but there was hybridization between the newcomers and the established residents. As a result, some genes from the archaic local populations were assimilated and persist in modern human populations. (c) The multiregional evolution model. According to this model, *H. sapiens* evolved concurrently in Europe, Africa, and Asia, with sufficient gene flow among populations to maintain their continuity as a single species. Gene pools of all present-day human populations are derived from a mixture of distant and local archaic populations. (d) The candelabra model: *H. sapiens* evolved independently in Europe, Africa, and Asia, without gene flow among populations. All genes in present-day European and Asian populations are derived from local archaic populations. The model names and characterizations we use are based on those used by Aiello (1993), Ayala et al. (1994), and Tattersall (1997). Not all authors would agree with our characterizations and names. Frayer et al. (1993), for example, apparently consider both models (b) and (c) to be variations of the multiregional evolution model they favor.

tween the extremes are hypotheses that postulate different combinations of migration, gene flow, and local evolutionary transition from *H. ergaster/erectus* to *H. sapiens.* These intermediates are the hybridization and assimilation model (Figure 16.12b), and the multiregional-origin model (Figure 16.12c).

At stake in the debate over these models is the nature and antiquity of the present-day geographic races of humans. If the African replacement model is correct, then present–day racial variation is the result of recent geographic differentiation that occurred within the last 100,000 to 200,000 years, after anatomically modern *H. sapiens* emerged from Africa. If one of the intermediate models is correct, then present-day racial variation represents some mixture of recent and ancient geographic differentiation. If the candelabra model is correct, then present-day racial variation derives from geographic differentiation among *H. ergaster/erectus* populations and may be as much as 1.5 to 2 million years old.

The candelabra model has been widely and thoroughly rejected by scientists in all fields (see Frayer et al. 1993; Ayala et al. 1994). It is flatly implausible that the same single descendant species, *H. sapiens,* could emerge in parallel in three different regions with no gene flow to maintain its continuity (see Chapters 6 and 12). However, this rejection of the candelabra model is the beginning and end of consensus in the field. Arguments over the remaining three models are based on archaeological and paleontological evidence and on genetic analyses. In much of the discussion that follows, we focus on distinguishing between the remaining extremes (the African replacement model of Figure 16.12a versus the multiregional evolution model of Figure 16.12c), but it is useful to keep in mind that these two models fall at the ends of a continuum of possibilities.

African Replacement versus Multiregional Evolution: Archaeological and Paleontological Evidence

David Frayer and colleagues (1993) use archaeological and paleontological data to argue against the African replacement model (Figure 16.12a) and in favor of some alternative [Figure 16.12b or (c)]. The researchers note that the African replacement model holds that long-established populations of one or more tool-using, hunter–gatherer species (*H. erectus* and other archaic forms of *Homo*) were supplanted wholesale throughout Europe and Asia by populations of another tool-using, hunter–gatherer species (modern *H. sapiens* emerging from Africa). It is hard to imagine how this could have happened, except by direct competition between the invaders and the established residents. It is implausible that modern *H. sapiens* could have been such a relentlessly superior competitor without a substantial technological advantage in the form of better tools or weapons. Thus, Frayer and colleagues conclude, the African replacement model predicts that the archaeological record will show evidence of abrupt changes in the level of technology in Europe and Asia as modern *H. sapiens* replaced archaic *Homo*. In fact, the researchers say, there is no evidence of any such abrupt technological changes.

Frayer and colleagues (1993) also argue that the African replacement model predicts that fossils of *Homo* populations in any given non-African region should show distinct changes in morphology when modern *H. sapiens* migrating from Africa replaced local archaic *Homo*. To refute this prediction, Frayer and colleagues point to distinctive traits of regional populations that have persisted from the distant past to the present. One-million-year-old fossils of *H. erectus* from Java, for example, have a straighter, more prominent browridge than their contemporaries elsewhere in the world. This strong browridge remains a distinctive feature of present-day Australian aborigines, whose ancestors may have arrived by boat from Java up to 60,000 years ago. Likewise, many present-day Asians have shovel-shaped upper front teeth, a trait that characterizes virtually all fossil specimens of Asian *H. erectus* and *H. sapiens*. (For other examples of the continuity of distinctive regional traits, see Thorne and Wolpoff 1981; Li Tianyuan and Etler 1992; and Frayer et al. 1993.) On these and other grounds, Frayer and colleagues reject the African replacement model.

Diane Waddle (1994) and Daniel Lieberman (1995) use statistical and cladistic approaches, respectively, to evaluate predictions from the African replacement model (Figure 16.12a) and the multiregional evolution model (Figure 16.12c). While Lieberman stresses that his data set of only 12 characters is too small to produce reliable inferences, both he and Waddle tentatively conclude that all modern humans are more closely related to archaic forms from Africa (Figure 16.13a) than regional groups are to local archaic forms (Figure 16.13b). If Lieberman and Waddle are correct, then the examples described by Frayer and colleagues of apparent long-term continuity of regionally distinctive traits must be the result of convergent evolution in *H. erectus* and *H. sapiens*.

The most recent analyses of Neandertal and pre-Neandertal fossils indicate that the story of modern human origins may be very complex indeed. Bermúdez de Castro and colleagues recently described a newly discovered set of human fossils

Morphological analyses, though not definitive, suggest that modern humans evolved in Africa, then replaced archaic humans elsewhere.

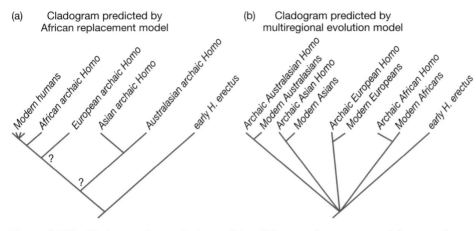

Figure 16.13 Phylogenetic predictions of the African replacement model versus the multiregional evolution model (a) The African replacement model predicts that all modern humans will be more closely related to each other than any is to any archaic species and that, among the archaic species, those from Africa will be the most closely related to modern humans (b) In contrast, the multiregional evolution model predicts that the archaic and modern humans in each region will be each other's closest relatives. From Lieberman (1995).

and associated artifacts and animal remains from the Gran Dolina section of the Atapuerca locality in Spain (Carbonell et al. 1999 and papers cited therein). The researchers have attributed the human fossils from this locality, which is dated to somewhere between 780,000 and 980,000 years ago, to a new species, *Homo antecessor*. Their analysis concludes that the Gran Dolina specimens share features of both modern humans and Neandertals, and may be the common ancestor of both. At the opposite end of the time range for Neandertals, Trinkaus and colleagues recently found evidence of mixed Neandertal and modern human charactersistics in two different specimens, one from France and the other from Portugal, both roughly 30,000 years old. Trinkaus and colleagues have interpreted these fossils as possible evidence of hybridization between Neandertals and modern humans (Trinkaus et al. 1998; Duarte et al. 1999; see, though, Tattersall and Schwartz 1999 for a dissenting opinion). *Homo antecessor* and possible Neandertal–modern human hybrids, all from Western Europe, may yet breathe new life into the multiregional model, at least insofar as it relates to Europe.

African Replacement versus Multiregional Evolution: Molecular Evidence

In principle, we could take Lieberman's cladistic approach and use it with DNA sequence data. If we had sequences of genes from both modern and archaic humans in all regions, we could estimate their phylogeny and see whether it most closely matches the tree predicted by the African replacement model or the tree predicted by the multiregional evolution model.

A team led by Svante Pääbo recovered a sequence of mitochondrial DNA from the skeleton of a *Homo neanderthalensis* that lived in Germany some 30,000 to 100,000 years ago (Krings et al. 1997, 1999). In the most recent analysis, the researchers compared the Neandertal sequence with 663 modern human mtDNA sequences, 7 chimpanzee sequences, and 2 bonobo sequences. Placing all the sequences in a phylogeny, the researchers found that modern humans from Europe,

Africa, Asia, America, Australia, and Oceania are all more closely related to each other than any of them is to the archaic European. In fact, the Neandertal sequence is no more similar to Europeans than to any other modern human and has, on average, more than three times as many differences from modern humans as are found among modern humans. Furthermore, using a divergence time of 4 to 5 million years for chimpanzees and humans, which is conservative, Krings and colleagues estimate the divergence of Neandertals from modern humans to have occurred between 317,000 and 741,000 years ago. These results are consistent with the African replacement model (Figure 16.12a). This conclusion is not definitive, however. It is based on a single archaic individual from only one geographic region, and it leaves open the possibility that modern Europeans inherited nuclear genes, if not mitochondrial genes, from archaic Europeans (see Nordborg 1998). Unfortunately, retrieving DNA from fossils is difficult, and may prove impossible for bones older than 100,000 years.

Working with DNA sequences of present-day humans only, researchers find it more difficult to design tests that distinguish the African replacement model from the multiregional evolution model [see, for example, the exchange between Sarah Tishkoff and colleagues (1996a) and Milford Wolpoff (1996)]. The trouble is that from a genetic perspective, the two models are identical in most respects. Both describe a species originating in Africa, spreading throughout Europe and Asia, and then differentiating into regionally distinct populations that nonetheless remain connected by gene flow [Figures 16.12a and (c)]. The only difference is that under the African origin model this process began 200,000 years ago or less, whereas under the multiregional evolution model it began some 1.8 million years ago. This means that any genetic patterns that might allow us to distinguish between the two models will involve quantitative differences rather than qualitative differences. Table 16.1 lists four criteria that molecular geneticists have used in efforts to distinguish between the African replacement versus multiregional evolution models. We will refer to the table in the paragraphs that follow.

S. Blair Hedges and colleagues (1992) attempt to distinguish between the African replacement model and the multiregional evolution model by analyzing sequence data for mitochondrial DNA. The data they used, collected by Linda Vigilant and colleagues (1991), are for sequences of noncoding mitochondrial DNA from 189 people from a diversity of geographic regions. Hedges and colleagues used a neighbor-joining technique to estimate the evolutionary tree linking the 189 mitochondrial sequences. Recall that mitochondrial DNA carries a record of direct maternal ancestry. By direct maternal ancestry we mean a person's mother, mother's mother, and so on. (A person's mother's father is part of the person's indirect maternal ancestry). Hedges et al.'s mitochondrial evolutionary tree traces the direct maternal ancestries of the 189 people back to a point at which they all converge to a single woman. Because the deepest branches in the mitochondrial phylogeny involve splits within African lineages, the tree suggests that this woman lived in Africa (Figure 16.14). (Note that the woman, often called by the misleading name Mitochondrial Eve, is not, in her generation, the sole female ancestor of the 189 present-day people. The 189 present-day people undoubtedly have a large number of indirect female ancestors who were contemporaries of their common direct maternal ancestor and from whom they inherited many of their nuclear genes. See Ayala et al. 1994; Ayala 1995.)

Table 16.1   Genetic predictions distinguishing the African replacement versus multiregional evolution models

Each of the criteria in the first column is a category of data we might use to distinguish between the African replacement model and the multiregional evolution model. The next two columns predict the patterns in each type of data under each model. The last column gives reasons why the distinctions implied by the predictions are not definitive. See text for more details.

Predictions

Criteria	African replacement	Multiregional evolution	Caveat
1. Location of ancestor of neutral alleles	Mostly Africa	Random	African origin of *H. ergaster/erectus* may bias the location of alleles toward African even under multiregional evolution.
2. Divergence time of African vs. non-African populations	200,000 years or less	1 million years or more	Gene flow among regional populations can reduce the apparent age of population divergence under multiregional evolution.
3. Genetic diversity	Genetic diversity is greater in Africa.	Genetic diversity is roughly equal in all regions.	African origin of *H. ergaster/erectus* and gene flow or selection may lead to greater diversity in Africa, even under multiregional evolution.
4. Sets of neutral alleles	Alleles present in Europe and Asia are subsets of those in Africa	Each region has some unique alleles; no region's alleles are a subset of another region's.	African origin of *H. ergaster/erectus* may mean that alleles present in Europe and Asia are subsets of those in Africa, even under multiregional evolution.

Hedges and colleagues note that their mitochodrial phylogeny is poorly supported statistically. Nonetheless, the conclusion suggested by the phylogeny may be correct (see Zischler et al. 1995). If we assume for the sake of argument that the common ancestor of all present-day human mitochondrial DNAs did, in fact, live in Africa, does that help us decide, by criterion 1 of Table 16.1, between the African replacement model and the multiregional evolution model? No, because both models include a common ancestry for all present-day humans that traces back to Africa. The key question is, When did the common ancestor of all human mitochondrial DNAs live?

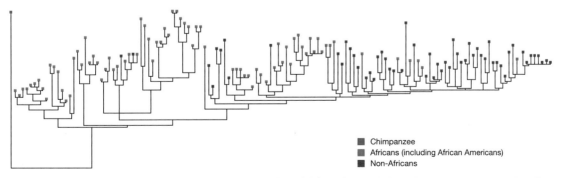

■ Chimpanzee
■ Africans (including African Americans)
■ Non-Africans

Figure 16.14   An evolutionary tree of human mitochondrial DNAs   Each branch tip represents a single individual. Red branch tips represent African individuals, blue branch tips represent non-Africans. The tree was rooted by using a chimpanzee sequence as the outgroup. Reprinted with permission from Hedges et al. (1992). Copyright © 1992, American Association for the Advancement of Science.

Linda Vigilant and colleagues (1991) estimate that the common ancestor of all present-day mitochondrial DNAs lived between 166,000 and 249,000 years ago. The researchers arrive at this figure by (1) assuming that mutations accumulate in mitochondrial DNAs at a constant rate, (2) comparing human sequences with chimpanzee sequences, and (3) using accepted estimates for the divergence time between humans and chimpanzees to calibrate the molecular clock (see Chapter 13). Ruvolo et al. (1993) use similar data and reasoning to estimate that the common ancestor of all human mitochondrial DNAs lived 129,000 to 536,000 years ago. The most precise, and possibly the most accurate, estimate, because it is based on sequences of entire mitochondrial genomes, is that of Horai et al. (1995): 125,000 to 161,000 years ago. Cavalli-Sforza (1997), using an alternative model, proposes a nearly identical date of 146,000 years ago. (Compare these dates with the range suggested by Krings et al. 1999 from Neandertal mtDNA.)

At first glance, these dates, which are consistent with the African replacement hypothesis, appear to refute the multiregional origin hypothesis by criterion 2 of Table 16.1. The dates suggest that non-African populations of humans diverged from African populations no more than a few hundred thousand years ago and certainly nothing like a million years ago. Examination of Figure 16.15, however,

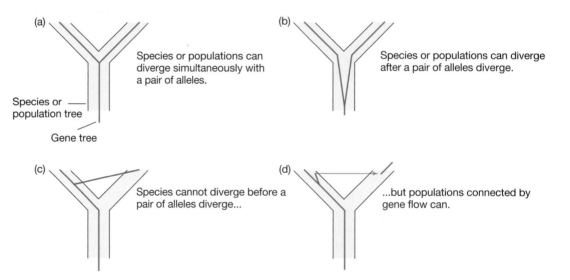

Figure 16.15 Divergence times of species trees, population trees, and gene trees These figures illustrate a hypothetical gene phylogeny within a species or population phylogeny. (a) In this scenario, a mutation creates a new allele to produce a split in the gene tree at the same time the species or population splits into two. One allele is then lost by drift or selection in each descendant species or population. The result is a gene tree that is exactly congruent with the species or population tree. If we use a molecular clock to estimate the divergence time for the species or populations, we will get the right answer. (b) Here, a mutation creates a new allele, producing a split in the gene tree. Some time later, the species or population splits into two. One allele is then lost by drift or selection in each descendant species or population. If we use a molecular clock for the gene tree to estimate the divergence time for the species or populations, the species or populations will appear to have diverged earlier than they actually did. (c) First, a species splits into two. Sometime later, a mutation creates a new allele, producing a split in the gene tree. Finally, one of the alleles moves from one species to the other. This last step is impossible; thus a molecular clock will not make a split between species appear more recent than it was. (d) A scenario like that in (c) is, however, possible for populations. First, a population splits into two. Some time later, a mutation creates a new allele, producing a split in the gene tree. Then a migrant carries the new allele to the other population (red arrow). Finally, the new allele is lost in one population, and the ancestral allele is lost in the other population. If we use a molecular clock for the gene tree to estimate the divergence time for the population tree, the populations will appear to have diverged more recently than they actually did.

shows that this is not necessarily the case. It is true that species cannot diverge any earlier than the divergence of any of their alleles, but populations connected by gene flow can. It is possible that the mitochondrial clock, which is effectively based on a single gene, makes the split between African and non–African populations look more recent than it actually was.

What we need to do is look at many loci at once and see whether, taken together, they tell the same story of a recent divergence between African and non–African populations. A. M. Bowcock and colleagues (1994) look at 30 nuclear microsatellite loci from people in each of 14 populations. Microsatellite loci are places in the genome where a short string of nucleotides, usually 2 to 5 bases long, is repeated in tandem. The number of repeats at any given locus is highly variable among individuals, meaning that each microsatellite locus has many alleles. Bowcock and colleagues calculate multilocus genetic distances among the 14 populations on the basis of the allele frequencies at each of the 30 microsatellite loci. They then use the genetic distances among populations to estimate the population phylogenetic tree (Figure 16.16).

On Bowcock et al.'s phylogenetic tree, geographically neighboring populations cluster together. Furthermore, the deepest branch point separates African from non–African populations. Analyzing the same data, D. B. Goldstein and colleagues (1995) estimate that the split between African and non–African populations occurred 75,000 to 287,000 years ago. This time range, which is consistent with the African replacement model, makes a more persuasive case than the mitochondrial clock date that the multiregional evolution model can be rejected by criterion 2 of Table 16.1. It is still possible under the multiregional evolution model to argue that there was enough gene flow to make the population split look more shallow than it was. But if there was that much gene flow, then it becomes hard to explain how any regional differentiation of characters could be maintained for a million years or more (Nei 1995).

Finally, we consider a study by Sarah Tishkoff and colleagues (1996b). These researchers examined allelic variation at a locus on chromosome 12 that is the site of a short-tandem-repeat polymorphism. This is a region of noncoding DNA in

Molecular analyses, though not definitive, suggest that modern humans evolved in Africa, then replaced archaic humans elsewhere.

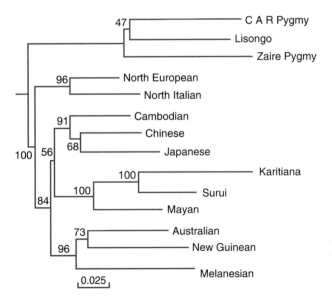

Figure 16.16 Phylogenetic tree for 14 human populations based on allele frequencies at 30 microsatellite loci   The number at a node indicates the percentage of times that the node was present in 100 bootstrap replicates (see Chapter 13). The deepest split in the tree was present in 100% of the bootstrap replicates, indicating strong statistical support for the conclusion that African (Central African Republic Pygmy, Lisongo, Zaire Pygmy) versus non-African is the most fundamental division in the population phylogeny. From Bowcock et al. (1994).

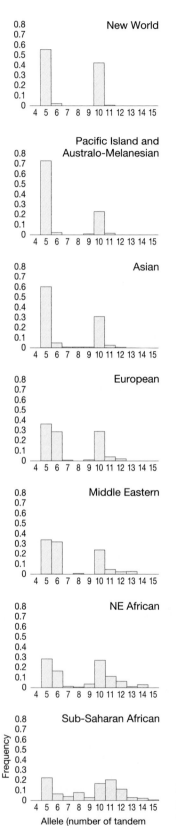

which the sequence TTTTC is repeated between 4 and 15 times, producing a total of 12 alleles. Tishkoff and colleagues determined the genotypes of more than 1600 people from seven different geographic regions. Figure 16.17 shows plots of the allele frequencies in each of the seven regions.

The African populations show much greater allelic diversity than non-African populations. This pattern is consistent with the African replacement model. If non-African populations were founded by small bands of people migrating out of Africa, then non-African populations should have reduced genetic diversity because of the founder effect (see Chapter 6).

Notice also that the graphs in Figure 16.17 are arranged by travel distance from sub-Saharan Africa, with the closest regions near the bottom and the most distant regions near the top. Moving up the figure from sub-Saharan Africa to northeast Africa, then to the Middle East, to Europe, and to Asia and beyond, we see that each region shows a set of alleles that is a subset of those present in the region below. Again, this is consistent with the African replacement model. It is what we would expect if each more-distant region were founded by a small band of people picking up from where their ancestors had settled and moving on. Not only is the pattern of allelic diversity consistent with the African replacement model, but it also tends to refute the multiregional evolution model by criteria 3 and 4 of Table 16.1. This refutation is not definitive, however, because the multiregional evolution model postulates the same pattern of migration and settlement, just earlier in time.

Tishkoff and colleagues can estimate when the founders of the non-African populations left Africa. They use a method based on linkage disequilibrium between the short-tandem-repeat locus and a second locus nearby (see Box 16.1). Tishkoff and colleagues estimate that the founders of the non-African populations left Africa not more than 102,000 to 450,000 years ago. These dates are consistent with the African replacement model and tend to refute the multiregional evolution model by criterion 2 of Table 16.1.

The balance of evidence we have reviewed appears to favor the African replacement model for the origin of *H. sapiens*. None of the tests are definitive, so some form of intermediate model (Figure 16.12b) cannot be ruled out. But taken together, the genetic data and at least some of the morphological data suggest that (1) all present-day people are descended from African ancestors and (2) all present-day non-African people are descended from *H. sapiens* ancestors who left Africa within the last few hundred thousand years. Present-day differences among races must have arisen since then.

16.4 The Evolution of Uniquely Human Traits

Humans have a number of traits that are unique among extant primates: We walk bipedally, we have very large brains, we manufacture and use complex tools, and we use language. We discussed bipedal locomotion and brain size briefly in Section 16.2. Here we consider evidence on the origin of tools and language.

Figure 16.17 Genetic diversity at a single locus among the people of seven geographic regions Each plot shows, for the people of a particular geographic region, the frequencies of the various alleles (numbered 4–15) at a short-tandem-repeat locus on chromosome 12. Plotted from tables in Tishkoff et al. (1996b).

BOX 16.1   Using linkage disequilibrium to date the divergence between African and non–African populations

Near the short-tandem-repeat locus on chromosome 12 is another locus, also noncoding, with a nucleotide sequence known as an Alu element. This Alu locus has two alleles: the ancestral, or *Alu(+)* allele; and a derived, *Alu(−)*, allele with a 256–base-pair deletion. Gorillas and chimpanzees lack the *Alu(−)* allele, so it probably arose in the human lineage after the split between the human lineage and the chimpanzee lineage.

The deletion mutation that created the *Alu(−)* allele probably occurred only once, in Africa, most likely in a chromosome that carried the six-repeat allele at the short-tandem-repeat locus (Tishkoff et al. 1996b). Upon its appearance in the population, the *Alu(−)* allele was in linkage disequilibrium with the short-tandem-repeat allele (see Chapter 7). The only kind of *Alu(−)* chromosome 12 in the population had the haplotype short-tandem-repeat-6-*Alu(−)*.

Recall from Chapter 7 that sexual reproduction in a population reduces linkage disequilibrium. After the *Alu(−)* allele appeared in Africa, mutations at the short-tandem-repeat locus and genetic recombination between it and the Alu locus created a great variety of other genotypic combinations on chromosome 12 (Figure 16.18,

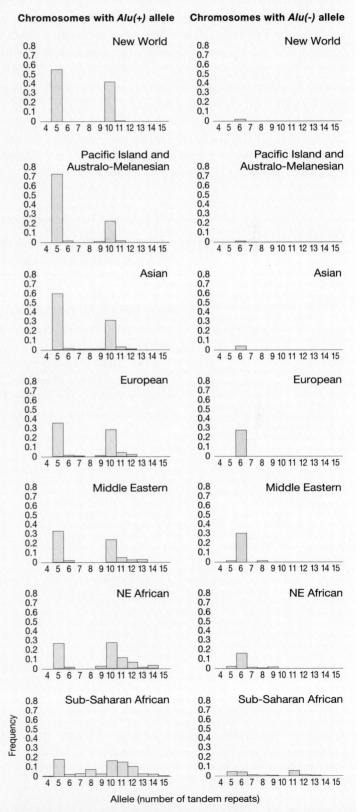

Figure 16.18   Genetic diversity at two linked loci in the people of seven geographic regions   These graphs represent the same populations and the same alleles represented in Figure 16.17. Near the short-tandem-repeat locus is an Alu element polymorphic for a deletion. There are two alleles at this locus: *Alu(+)* and *Alu(−)*. The left column shows allele frequencies at the short-tandem-repeat locus among chromosomes that carry the *Alu(+)* allele. The right column shows allele frequencies at the short-tandem-repeat locus among chromosomes that carry the *Alu(−)* allele. If a population is at linkage equilibrium for the short-tandem-repeat locus and the Alu locus, then the shape of the distributions of alleles in the two columns (but not necessarily the height of the distributions) will be the same. In sub-Saharan Africa, the distribution of alleles is roughly the same for *Alu(+)* and *Alu(−)* chromosomes. This pattern indicates that the short-tandem-repeat-locus and the Alu locus are near linkage equilibrium in this population. In the people of other regions, the distribution of alleles is dramatically different in *Alu(+)* versus *Alu(−)* chromosomes, indicating that the short-tandem-repeat locus and the Alu locus are in linkage disequilibrium. Plotted from tables in Tishkoff et al. (1996b).

BOX 16.1   Continued

African populations). The sub-Saharan African population is near linkage equilibrium at the two loci.

When the first migrants left Africa, it appears that they carried with them at appreciable frequency only three 2-locus genotypes: 5-*Alu(+)*; 10-*Alu(+)*; and 6-*Alu(−)*. In other words, genetic drift in the form of the founder effect put the migrant populations in linkage disequilibrium. The time since the migrants' departure from Africa has not been sufficient for mutation and recombination to create new two-locus genotypes and replenish the haplotype diversity to the lev-

els seen in Africa. In other words, the non-African populations are still in linkage disequilibrium, a population-genetic legacy from their ancient ancestors who left Africa. Using estimates of the rates of mutation and recombination, both of which affect the rate at which populations approach linkage equilibrium (see Chapter 7), Tishkoff and colleagues estimate that the founders of the non-African populations left Africa not more than 102,000 to 450,000 years ago (see also Pritchard and Feldman 1996; Risch et al. 1996).

Which of Our Ancestors Made and Used Stone Tools?

Chimpanzees make and use simple tools. They strip stems and twigs of leaves and use the resulting tools to fish termites out of termite mounds; they use rocks and sticks to hammer open nuts; they use leaves as umbrellas. Other animals use tools as well. One species of Darwin's finch, the woodpecker finch (*Camarhynchus pallidus*), uses cactus spines to extract insects from bark. So making and using tools is not, in itself, unique to humans. What is unique to humans is making and using complex tools.

The earliest uniquely complex tools that appear in the archaeological record are sharp-edged stone flakes and handheld chopping tools (Figure 16.19). A stone knapper making such tools began by selecting an appropriate cobble from a riverbed, preferably one of fine-grained volcanic rock (Schick and Toth, 1993). He

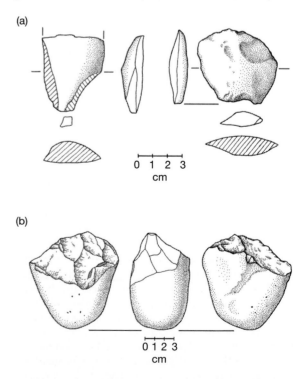

Figure 16.19   Oldowan stone tools from Hadar, Ethiopia   These 2.3-million-year-old stone tools are among the oldest known. (a) Two sharp-edged flakes. Each is shown from various sides. (b) A handheld chopper, shown from three sides. After Kimbel et al. (1996).

or she then struck the cobble with a second rock to chip off flakes. The flakes themselves were usable as cutting tools. Chipping numerous flakes off a cobble in an appropriate pattern produced a chopper. Tools of this style are said to belong to the Oldowan industrial complex, because they were first discovered at Olduvai Gorge, Tanzania. Archaeologists have learned firsthand that making Oldowan-style stone tools requires skill and experience.

The oldest known Oldowan tools are from Gona, Ethiopia. Based on the ages of the strata just above and below the sediment layer that contains the tools, Sileshi Semaw and colleagues (1997) established that the tools are 2.5 to 2.6 million years old. Who were the stone knappers who made them?

An obvious answer is, some early member of the genus *Homo.* The trouble with this answer is that we have no definitive evidence that any species of *Homo* had appeared by 2.5 million years ago. The oldest reliably dated Homo fossil is a 2.3-million-year-old upper jaw (maxilla) from Hadar, Ethiopia (Section 16.2; Gibbons 1996; Kimbel et al. 1996). Which species this fossil represents is unclear; it could be *H. habilis* (Figure 16.9b), *H. rudolfensis* (Figure 16.9c), or some heretofore unknown species.

The earliest known stone tools predate the earliest known Homo specimens.

Circumstantial evidence certainly suggests that the Hadar fossil may represent the same species that made the 2.5-million-year-old Gona tools. Hadar is geographically close to Gona, 2.3 million years ago is geologically close to 2.5 million years ago, and the Hadar fossil was found near 34 Oldowan tools. It is entirely possible that 2.5-million-year-old *Homo* fossils will eventually be found at Gona and that these early members of *Homo* were the Gona stone knappers. On the other hand, as Bernard Wood (1997) points out, circumstantial evidence is not proof. If other hominids were present at the same time and place, then they are suspects too. This is true even at Hadar, where the 2.3-million-year-old *Homo* jaw was found near Oldowan tools.

Wood notes that robust australopithecines existed with early *Homo* in the same part of Africa over approximately the same time span as the Oldowan industrial complex. While there is no good circumstantial evidence to indicate that robust australopithecines may have been responsible for Oldowan tools, some indirect evidence of their tool-using capabilities comes from their anatomy. Randall L. Susman (1994) makes an argument based on the anatomy of opposable thumbs. He starts by comparing the bones and muscles of the thumb in humans versus chimpanzees (Figure 16.20). Humans have three muscles that chimpanzees lack. Associated with these extra muscles, humans have thicker metacarpals with broader heads (Figure 16.21a). These differences in thumb anatomy make the human hand more adept at precision grasping than the chimpanzee hand. Susman argues that the modified anatomy of the human thumb evolved in response to selection pressures associated with the manufacture and use of complex tools.

Susman then compares the relative thickness of the thumb metacarpals in humans and chimpanzees with that in a variety of hominid fossils (Figure 16.21b). *H. neanderthalensis, H. erectus,* and *A. robustus* resemble *H. sapiens* in having thumb metacarpals with broad heads for their length. *A. afarensis,* a gracile australopithecine that disappears from the fossil record before Oldowan tools appear, is like the chimpanzees in having thumb metacarpals with narrow heads for their length. Susman asserts that we can use the thumb metacarpals to diagnose whether extinct hominids were makers and users of stone tools. Susman concludes that both *H. erectus* and *A. robustus* were toolmakers. Susman's argument has been controversial

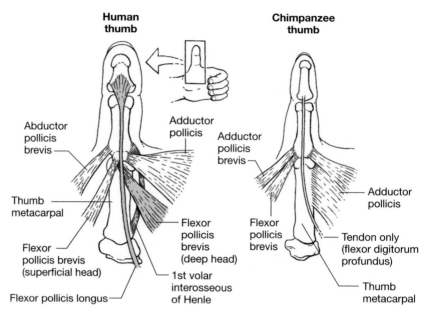

Figure 16.20 Human versus chimpanzee thumbs The three muscles in a human thumb that chimpanzees lack are highlighted with color. Reprinted with permission from Susman (1994). Copyright © 1994, American Association for the Advancement of Science.

(McGrew et al. 1995). More recently, Hamrick, et al. (1998) have argued that several australopithecine species possessed powerful grasping thumbs, including *A. africanus* (the appropriate bones are not known for *A. afarensis*), and that they may have used unmodified stones as tools.

None of the evidence we have discussed establishes for certain whether the Oldowan tools at Gona were made by a species of *Homo* or a robust australo-

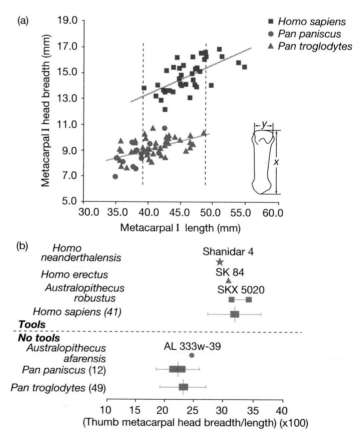

Figure 16.21 Thumb metacarpal bones in a variety of hominids (a) This graph plots the breadth of the thumb metacarpal head (the end of the bone pointing out along the thumb) against the length of the thumb metacarpal for a sample of humans, bonobos, and common chimpanzees. Longer metacarpals have broader heads, but human metacarpals have broader heads for their size than chimpanzee metacarpals. In other words, the ratio of metacarpal head breadth to length is greater in humans. (b) This plot shows the range of ratios (metacarpal head breadth to length) for several hominid fossils (the labels on the data points are the specimen numbers) and samples of present-day humans, bonobos, and common chimpanzees. For *Australopithecus robustus,* the ratio is reported as an estimated range, because the bone in question is not quite complete. Species above the dashed line are associated (at least temporally) with manufactured stone tools in the archaeological or fossil record or the present. Species below the dashed line are not. Reprinted with permission from Susman (1994) with some changes in the species names. Copyright © 1994, American Association for the Advancement of Science.

pithecine. Instead, the evidence argues that they could have been made by either or both. If we accept Susman's conclusion that the robust australopithecines were toolmakers, but the gracile australopithecines were not, and if we accept one of the phylogenies of Strait et al. (Figure 16.11), then we must make one of two inferences: (1) The manufacture and use of complex stone tools originated in an undiscovered common ancestor of *Homo* and the robust australopithecines, or (2) The manufacture and use of complex stone tools originated independently in at least two hominid lineages.

In the absence of definitive proof in the form of a fossil hand grasping a stone tool, we may never know the answer to these questions. Currently, however, most paleoanthropologists favor the view that early *Homo* is responsible for most if not all Oldowan tools. Whenever stone tools are found in association with fossil humans, *Homo* is always there, but australopithecines are often absent. (This includes Susman's fossils from Swartkrans, South Africa, which also contains *Homo erectus*.) And *Homo habilis,* if it is present at Hadar at 2.3 million years ago, comes very close to matching the time span of the Oldowan industrial complex. Three different species of robust australopithecine span the time from 2.5 to 1.0 million years ago, and there are no Oldowan tools after about 1.5 million years ago.

Some morphological analyses suggest that both early Homo species and the robust australopithecines could have been toolmakers.

Most paleoanthropologists, however, still believe that most, if not all, Oldowan stone tools are the handiwork of Homo.

Which of Our Ancestors Had Language?

If the history of hominid tool use is murky, the history of hominid language is even murkier. Like tool use, language is a behavior. Because behaviors do not fossilize, we have no direct evidence of their history. We are left to examine circumstantial evidence in the archaeological and fossil record. Before the invention of writing, language left even less circumstantial evidence than tool use.

Language is a complex adaptation located in the neural circuitry of the brain. The vocabulary and particular grammatical rules of any given language are transmitted culturally, but the capacity for language and a fundamental grammar are, in present-day humans, both innate and universal (see Pinker 1994). Among the evidence for this assertion is the observation that communities of deaf children, if isolated from native signers, invent their own signed languages from scratch. By the end of two generations of transmission to young children within the new deaf culture, these new sign languages develop all the hallmarks of genuine language. They have a standardized vocabulary and grammar, and their fluent users can efficiently communicate the full range of human ideas and emotions. Each of these new sign languages is unique, but all reflect the same universal grammar that linguists have identified in spoken languages.

Many of the brain's language circuits are concentrated in an area called the perisylvian cortex, usually in the brain's left hemisphere (see Pinker 1994). These language circuits include Broca's area and Wernicke's area. Homologous structures exist in the brains of monkeys (Galaburda and Pandya 1982). The monkey homologues of Wernicke's area function in the recognition of sounds, including monkey calls. The monkey homologues of Broca's area function in controlling the muscles of the face, tongue, mouth, and larynx. However, neither of these structures plays a role in the production of monkey vocal calls. Instead, vocal calls are generated by circuits in the brain stem and limbic system. These same structures control nonlinguistic vocalizations in humans, such as laughing, sobbing, and shouting in pain. Thus, the human language organ appears to be a derived modification of neural circuits common to all primates. But among extant species the nature of this modification—its specialization for linguistic communication—appears to be unique to humans. The implication is that the language organ, as such, evolved after our lineage split from the lineage of chimpanzees and bonobos.

Spoken language also relies on derived modifications of the larynx that are unique to humans. In modern human newborns and in other mammals, the larynx is high enough in the throat that it can rise to form a seal with the back opening of the nasal cavity (see Pinker 1994). This allows air to bypass the mouth and throat on its way from the nose to the lungs and prevents the infant from choking on accidentally inhaled food or water. When human babies are about three months old, the larynx descends to a lower position in the throat. This clears more space in which the tongue can move, changes the shape of a pair of resonating chambers, and makes it possible for humans to articulate a much greater diversity of vowel sounds than, for example, chimpanzees.

How far back in our evolutionary lineage can we trace the existence of language, and what evidence can we use? William Noble and Iain Davidson (1991) assert that the only reliable source of evidence is the archaeological record. In their view, the hallmark of language is the use of arbitrary symbols, standardized within a culture, to represent objects and ideas. To find language, then, we should look for such symbols in the archaeological record. The first unequivocally arbitrary symbols occur in cave paintings in Germany and France that are about 32,000 years old. Even Noble and Davidson cannot quite accept that language is as recent an innovation as that. They note that *Homo sapiens* had colonized Australia by 40,000 years ago (as early as 60,000, according to some) and confess that they cannot imagine how a group of people could build boats and cross the open ocean without language to help them conceive and coordinate the expedition. However, Noble and Davidson hold the line at about 40,000 years. This would imply that *H. sapiens* is the only species ever to use language.

Other support for the view that language is a recent innovation of *H. sapiens* came from anatomical studies of the skulls of *H. neanderthalensis*. These studies were attempts to reconstruct the Neandertal larynx. Analyses of the shape of the base of the skull and the position of its muscle attachment sites were used to argue that Neandertals had an undescended larynx that would have limited their ability to articulate vowels. This, it was argued, would have prevented them from developing language.

Early studies suggested that Neandertals could not talk . . .

B. Arensburg and colleagues (1989; 1990) refuted these arguments with a rare paleontological find. A 60,000-year-old Neandertal skeleton from Israel included an intact hyoid bone (Figure 16.22). The hyoid bone is located in the larynx and serves as the anchor for throat muscles that, in humans, are important in speaking. Arensburg et al. made a detailed analysis of the hyoid bone and compared it to the hyoids of chimpanzees and present-day humans. They found the Neandertal hyoid to be dramatically different from that of chimpanzees and virtually identical to that of present-day humans. Based on this hyoid bone, Arensburg and colleagues suggest that Neandertals had a descended larynx. Given that a descended larynx entails an increased risk of choking, it is hard to imagine why it would evolve unless it carried a substantial benefit. An obvious candidate for such a benefit is an improved ability to speak.

. . . but this conclusion was contradicted by the discovery of a key Neandertal bone.

If we accept the proposition that Neandertals could talk, can we trace language back any further in our lineage? Given that the language organ is in the brain, what

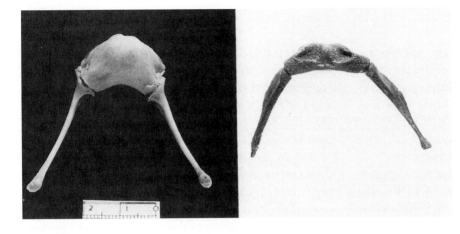

Figure 16.22   Hyoid bones from *Homo neanderthalensis* and a common chimpanzee   The Neandertal hyoid is on the right. The scale bar, which applies to both hyoids, is in cm. Reproduced by permission from Arensburg, Schepartz, Tillier, Vandermeersch, and Rak. 1990, "A reappraisal of the anatomical basis for speech in middle paleolithic hominids," *American Journal of Physical Anthropology* 83:137–146. © American Association of Physical Anthropologists.

can we glean from the fossil record about the brains of our ancestors? David Pilbeam and Stephen Jay Gould (1974) showed many years ago that even after taking body size into account, there has been something dramatically different about our brains since the first emergence of our genus. Figure 16.23 plots brain size as a function of body size in extant great apes, australopithecines, and three species of *Homo*. Inspection of the figure indicates that not only do *Homo* species have larger brains for their size, but brain size increases much more steeply with body size in *Homo* than in its hominid relatives. Nearly all the more recent analyses (for example, McHenry 1992; Kappelman 1996) benefiting from a much more complete fossil record of human brain and body size evolution, have essentially confirmed this result. Like a descended larynx, large brains come at a cost. They require a great deal of energy to maintain, and they generate considerable heat. What benefits could have compensated for these costs? Two possibilities are an increased capacity to make and use tools and an increased capacity for language.

Phillip Tobias (1987) examined casts of the insides of the braincases of specimens of *H. habilis*. In addition to their sheer size, these endocasts revealed the existence of derived structural traits unique to our genus. Among them are clear enlargments of Broca's and Wernicke's areas. By his own account, this discovery converted Tobias from a skeptic to an advocate of the hypothesis that language first emerged, at least in rudimentary form, in *H. habilis*. We do not know for certain, and perhaps never will, but language may be as much as 2 million years old.

An analysis of the skulls of Homo habilis suggest that they may have had at least rudimentary language.

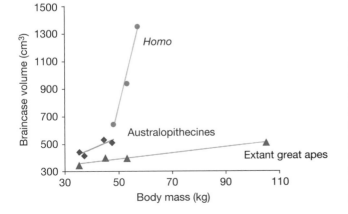

Figure 16.23   Brain size versus body size in a variety of hominoids   Data points represent species averages, with best-fit lines. In all three groups, species with larger brains have larger bodies. Australopithecines have larger brains for their size than extant great apes. *Homo* species have larger brains for their size, as well as a dramatically different relationship between brain size and body size. The extant great apes are the bonobo, common chimpanzee, orangutan, and gorilla; the australopithecines are *A. afarensis, A. africanus, A. boisei,* and *A. robustus;* the *Homo* species are *H. habilis, H. ergaster/erectus,* and *H. sapiens.* The data are from Tobias (1987) and Pilbeam and Gould (1974). See also McHenry (1994). After Pilbeam and Gould (1974).

Summary

As Darwin predicted, an evolutionary perspective throws light on the origin and nature of humans. Humans are relatives of the great apes. Although the evidence is conflicting, morphological and molecular studies suggest that our closest living relatives are the chimpanzees. Our most recent common ancestor with the chimpanzees lived about 5 million years ago.

Following its split from the chimpanzee lineage, our own lineage gave rise to several species of bipedal African hominids. Fossils provide strong evidence for the coexistence of at least two of these species and perhaps as many as five. We are the sole survivors of this evolutionary radiation.

The first members of genus *Homo* left Africa nearly 2 million years ago. Whether these populations ultimately contributed genes to present-day populations of humans is the subject of debate. No definitive tests have been performed, but the balance of the evidence suggests that all present-day non-African populations are descended from a more recent wave of emigrants that left Africa within the last 200,000 years. This implies that the present-day geographic variation among populations of humans is of relatively recent origin.

Among the derived traits unique to our species are the manufacture and use of complex tools and the capacity for language. Because behavior does not fossilize, researchers have to rely on circumstantial evidence to reconstruct the history of these traits. Tool use appeared at least 2.5 million years ago. It is most likely to have arisen in an early species of *Homo*, although robust australopithecines may also have used stone tools. The evidence on language is more tenuous, but it suggests that language may have emerged nearly as early as tool use.

Questions

1. Explain why Ruvolo et al. (1995) thought it was important to look at several nuclear genes, and not just the mtDNA genes, to study the relationships of humans and the great apes.

2. Suppose humans and chimpanzees are indeed each other's closest relatives. Is it then accurate to say that humans evolved from chimpanzees? How about that chimpanzees evolved from humans?

3. In a study of the phylogeny of Old World monkeys (Hayasaka et al. 1996), the three individual rhesus macaques that were studied did not form a monophyletic group. Instead, the mtDNA of one of the rhesus macaques was more similar to the mtDNA of Japanese and Taiwanese macaques (which are different species) than it was to the other rhesus macaques. How might this have happened? (There are at least two possibilities.) How does this relate to Jonathan Marks's objection to the conclusions many researchers draw from the human–ape molecular phylogenies (Figures 16.4 and 16.6)?

4. What's in a name? Jared Diamond (1992), who believes that humans and chimpanzees are closest relatives (Figure 16.3d), suggests that if we follow the naming traditions of cladistic taxonomy, then humans, chimpanzees, and bonobos should all be considered members of a single genus. Diamond proposes calling these species, respectively, *Homo sapiens, Homo troglodytes,* and *Homo paniscus.* Jonathan Marks objects to Diamond's taxonomic reasoning. Marks advocates a human–great-ape phylogeny in which either gorillas and chimpanzees are closest relatives (Figure 16.3c; Marks 1993) or there is an unresolved trichotomy (Figure 16.4a; Marks 1994, 1995). Concerning the nature of humans and apes, Marks (1994) asserts that

   > Popular works tell us that we are not merely genetically apes, but that we are literally apes (e.g., Diamond 1992). Sometimes there is profundity in absurdity, but I don't think this is one of those times. It merely reflects the paraphyletic nature of the category "apes"—humans are apes, but only in the same sense that pigeons are reptiles and horses are fish. . . . Focusing on the genetic relations obscures biologically significant patterns of phenotypic divergence.

   Do you think humans, chimpanzees, and bonobos should all be classified as members of the same genus? Is there more at stake in the disagreement between Diamond and Marks than just Latin names? If so, what?

5. Jared Diamond finds ethical dilemmas in the close kinship between humans and chimpanzees:

   > It's considered acceptable to exhibit caged apes in zoos, but it's not acceptable to do the same with humans. I wonder how the public will feel when the identifying label on the chimp cage in the zoo reads "*Homo troglodytes*" (Diamond 1992, p. 29).

Diamond finds the use of chimpanzees in medical research even more problematic. The scientific justification for the use of chimpanzees is that chimpanzee physiology is extremely similar to human physiology, so chimpanzees are the best substitute for human subjects. Diamond notes that jails are a very rough analogue to zoos, in the sense that they represent conditions under which we do consider it acceptable to keep people in cages without their consent (if not to display them). But there is no human analogue to research on chimpanzees: There are no conditions under which we consider it acceptable to do medical experiments on humans without their consent. Is it ethically justified to keep animals in zoos? To use animals in medical research? Does the phylogenetic relationship between ourselves and the animals in question matter? If so, how and why?

6. Look at the two different phylogenies illustrated in Figures 16.11b and (c). Suppose you are a hominid paleontologist who wants to figure out which of these trees (if either) is correct. What strata would you choose to search for hominid fossils? If you are lucky enough to find a hominid skull, what features of the skull could tell you which tree is correct? If you find no fossils at all, can you conclude anything about the two trees?

7. One of the most heated aspects of human racial politics is the contention that human races are genetically distinct. How do the African replacement versus multiregional evolution models address this issue? That is, which model predicts that human races are more genetically different from each other?

8. Several genetic studies show that African versus non-African is the most fundamental division in the phylogeny of present-day human populations. Some people might conclude from this data that modern African people are in some sense "primitive." What is the logical flaw in this thinking?

9. Jared Diamond and others have pointed out that different ethnic groups within Africa are more diverse than are the ethnic groups on all other continents taken together. What does this imply about the common U.S. practice of categorizing people into "African," "Caucasian," "Asian," and "Native American"?

10. Work by C. Swisher and colleagues (1996) indicates that *Homo erectus* may have persisted in Java until 53,000 years ago at Sambungmachan and until 27,000 years ago at Ngandong. If correct, these dates imply that *H. erectus* and *H. sapiens* coexisted in Java. Will this finding help settle the debate over the out-of-Africa model versus the multiregional origin model? Why or why not?

11. For the sake of argument, adopt the proposition implied by Wood (1997) that early species of *Homo* did not participate in the production of Oldowan stone tools. What other puzzles must we now solve? Note that after the invention of Oldowan tools, the next advance in toolmaking is marked in the archaeological record by the appearance of Acheulean tools, which are substantially more sophisticated than Oldowan tools (Johanson et al. 1996). Acheulean tools appear about 1.4 million years ago and persist until less than 200,000 years ago.

12. Derek Bickerton (1995) and Charles Catania (1995) object to the suggestion that *Homo habilis* had language. Bickerton writes, "If *H. habilis* already had all the necessary ingredients for language, what happened during the next $2\frac{1}{2}$ million years?" And Catania writes, "I . . . [deduce] that our hominid ancestors should have taken over the world 950,000 years ago if a million years ago they were really like us in their language competence. But they did not, so they were not; if they had been, those years would have been historic instead of prehistoric." How do you think Phillip Tobias would respond to Bickerton and Catania? Who do you think is right?

Exploring the Literature

13. Read Susman's (1994) paper on opposable thumbs and tool use. What weaknesses can you identify in Susman's argument? What additional data would you like to see? If you were Susman, how would you respond to these critiques? Read McGrew et al. (1995) to see if the critiques and responses therein are similar to your own.

McGrew, W. C., M. W. Hamrick, S. E. Inouye, J. C. Ohman, M. Slanina, G. Baker, R. P. Mensforth, and R. L. Susman. 1995. Thumbs, tools, and early humans. *Science* 268: 586–589.

Susman, R. L. 1994. Fossil evidence for early hominid tool use. *Science* 265: 1570–1573.

For an additional approach to the problem of tool use, see

Susman, R. L. 1998. Hand function and tool behavior in early hominids. *Journal of Human Evolution* 35: 23–46.

14. There were a couple of hints in this chapter that present-day humans have smaller bodies and brains than some of our recent ancestors. For documentation that this is indeed the case, see

Gibbons, A. 1997. Bone sizes trace the decline of man (and woman). *Science* 276: 896–897.

Ruff, C.B., E. Trinkaus, and T. W. Holliday. 1997. Body mass and encephalization in Pleistocene *Homo. Nature* 387: 173–176.

15. We noted that bipedal locomotion appears early in the *Australopithecus/Homo* lineage, but we did not discuss the adaptive significance of this trait. Researchers have offered a variety of hypotheses, some more intuitively plausible than others. For a start on this literature, see

Chaplin, G., N. G. Jablonski, and N. T. Cable. 1994. Physiology, thermoregulation, and bipedalism. *Journal of Human Evolution* 27: 497–510.

Hunt, E. D. 1994. The evolution of human bipedality: Ecology and functional morphology. *Journal of Human Evolution* 26: 183–199.

Jablonski, N. G., and G. Chaplin. 1993. Origin of habitual terrestrial bipedalism in the ancestor of the Hominidae. *Journal of Human Evolution* 24: 259–280.

Wheeler, P. E. 1994. The foraging times of bipedal and quadrupedal hominids in open equatorial environments (a reply to Chaplin, Jablonski & Cable, 1994). *Journal of Human Evolution* 27: 511–517.

Wheeler, P. E. 1994. The thermoregulatory advantages of heat storage and shade-seeking behavior to hominids foraging in equatorial savannah environments. *Journal of Human Evolution* 26: 339–350.

16. For another kind of evidence on the origin of language, see

Kay, R. F., M. Cartmill, and M. Balow. 1998. The hypoglossal canal and the origin of human vocal behavior. *Proceedings of the National Academy of Sciences, USA* 95: 5417–5419.

Citations

Aiello, L. C. 1993. The fossil evidence for modern human origins in Africa: A revised review. *American Anthropologist* 95: 73–96.

Andrews, P. 1992. Evolution and environment in the Hominoidea. *Nature* 360: 641–646.

Andrews, P., and L. Martin. 1987. Cladistic relationships of extant and fossil hominoids. *Journal of Human Evolution* 16: 101–118.

Arensburg, B., L. A. Schepartz, A. M. Tillier, B. Vandermeersch, and Y. Rak. 1990. A reappraisal of the anatomical basis for speech in Middle Paleolithic hominids. *American Journal of Physical Anthropology* 83: 137–146.

Arensburg, B., A. M. Tillier, B. Vandermeersch, H. Duday, L. A. Schepartz, and Y. Rak. 1989. A middle paleolithic human hyoid bone. *Nature* 338: 758–760.

Arsuaga, J. L., I. Martínez, C. Lorenzo, A. Gracia, A. Muñoz, O. Alonso, and J. Gallego 1999. The human cranial remains from Gran Dolina lower-Pleistocene site (Sierra de Atapuerca, Spain). *Journal of Human Evolution* 37: 431–457.

Ayala, F. J. 1995. The myth of Eve: Molecular biology and human origins. *Science* 270: 1930–1936.

Ayala, F. J., A. Escalante, C. O'hUigin, and J. Klein. 1994. Molecular genetics of speciation and human origins. *Proceedings of the National Academy of Sciences, USA* 91: 6787–6794.

Begun, D. R. 1992. Miocene fossil hominids and the chimp–human clade. *Science* 257: 1929–1933.

Begun, D. R. 1994 Relations among the great apes and humans: New interpretations based on the fossil great ape *Dryopithecus*. *Yearbook of Physical Anthropology* 37: 11–63.

Begun, D. R. 1995. Late-Miocene European orangutans, gorillas, humans, or none of the above? *Journal of Human Evolution* 29: 169–180.

Begun, D. R., C. V. Ward, and M. D. Rose. 1997. Events in hominoid evolution. In D. R. Begun, C. V. Ward, and M. D. Rose, eds.: *Function, Phylogeny and Fossils: Miocene Hominoid Evolution and Adaptations*. New York: Plenum Publishing Company, 389–415.

Begun, D. R., C. V. Ward, and M. D. Rose, eds. 1997. *Function, Phylogeny, and Fossils: Miocene Hominid Evolution and Adaptations*. New York: Plenum.

Bermúdez de Castro, J. M., J. L. Arsuaga, E. Carbonell, A. Rosas, I. Martínez, and M. Mosquera. 1997. A hominid from the lower Pleistocene of Atapuerca, Spain: possible ancestor to Neandertals and modern humans. *Science* 276: 1392–1395.

Beynon, A. D., M. C. Dean, and D. J. Reid. 1991. On thick and thin enamel in hominoids. *American Journal of Physical Anthropology* 86: 295–309.

Bickerton, Derek. 1995. Finding the true place of *Homo habilis* in language evolution. Open peer commentary in Wilkins and Wakefield 1995.

Boesch, C., and M. Tomasello. 1998 Chimpanzee and human cultures. *Current Anthropology* 39: 591–614.

Borowik, O. A. 1995 Coding chromosomal data for phylogenetic analysis: Phylogenetic resolution of the *Pan-Homo-Gorilla* trichotomy. *Systematic Biology* 44: 563–570.

Bowcock, A. M., A. Ruiz-Linares, J. Tomfohrde, E. Minch, J. R. Kidd, and L. L. Cavalli-Sforza. 1994. High resolution of human evolutionary trees with polymorphic microsatellites. *Nature* 368: 455–457.

Carbonell, E., J. M., Bermúdez de Castro, and J. L. Arsuaga. 1999. Preface: Special issue on Gran Dolina site: TD6 Aurora Stratum (Burgos, Spain). *Journal of Human Evolution* 37: 309–311.

Catania, A. C. 1995. Single words, multiple words, and the functions of language. Open peer commentary in Wilkins and Wakefield 1995.

Cavalli-Sforza, L. L. 1997. Genes, peoples and languages. *Proceedings of the National Academy of Sciences, USA* 94: 7719–7724.

Darwin, C. 1859. *On the Origin of Species by Means of Natural Selection, Or the Preservation of the Favoured Races in the Struggle for Life.* (London: John Murray).

Darwin, C. 1871. *The Descent of Man, and Selection in Relation to Sex.* London: John Murray.

Dean, D., and E. Delson. 1992. Second gorilla or third chimp? *Nature* 359: 676–677.

Deinard A., and K. Kidd. 1999. Evolution of a *HOX-B6* intergenic region within the great apes and humans. *Journal of Human Evolution* 36: 687–703.

Diamond, J. 1992. *The Third Chimpanzee.* New York: Harper Collins.

Djian, P., and H. Green. 1989. Vectorial expansion of the involucrin gene and the relatedness of the hominoids. *Proceedings of the National Academy of Sciences, USA* 86: 8447–8451.

Duarte, C., J. Maurício, et al. 1999. The early upper Paleolithic human skeleton from the Abrigo do Lagar Velho (Portugal) and modern human emergence in Iberia. *Proceedings of the National Academy of Sciences, USA* 96: 7604–7609.

Frayer, D. W., M. H. Wolpoff, A. G. Thorne, F. H. Smith, and G. G. Pope. 1993. Theories of modern human origins: The paleontological test. *American Anthropologist* 95: 14–50.

Gabunia, L., and A. Vekua. 1995. A Plio-Pleistocene hominid from Dmanisi, East Georgia, Caucasus. *Nature* 373: 509–512.

Gabunia, L., A. Vekua, et al. 2000. Earliest Pleistocene hominid cranial remains from Dmanisi, Republic of Georgia: Taxonomy, geological setting, and age. *Science* 288: 1019–1025.

Galaburda, A. M., and D. N. Pandya. 1982. Role of architectonics and connections in the study of primate brain evolution. In E. Armstrong and D. Falk, eds. *Primate Brain Evolution.* New York: Plenum.

Gibbons, A. 1994. Rewriting—and redating—prehistory. *Science* 263: 1087–1088.

Gibbons, A. 1996. A rare glimpse of an early human face. *Science* 274: 1298.

Goldstein, D. B., A. Ruiz Linares, L. L. Cavalli-Sforza, and M. W. Feldman. 1995. Genetic absolute dating based on microsatellites and the origin of modern humans. *Proceedings of the National Academy of Sciences, USA* 92: 6723–6727.

Goodman, M. 1962. Evolution of the immunologic species specificity of human serum proteins. *Human Biology* 34: 104–150.

Goodman, M., W. J. Bailey, K. Hayasaka, M. J. Stanhope, J. Slightom, and J. Czelusniak. 1994. Molecular evidence on primate phylogeny from DNA sequences. *American Journal of Physical Anthropology* 94: 3–24.

Green, H., and P. Djian. 1995. The involucrin gene and Hominoid relationships. *American Journal of Physical Anthropology* 98: 213–216.

Groves, C. P. 1986. Systematics of the great apes. In D. R. Swindler and J. Erwin, eds. *Comparative Primate Biology,* Volume 1: Systematics, Evolution, and Anatomy. New York: Alan R. Liss, Inc., 187–217.

Hamrick M. W., S. E. Churchill, D. Schmitt, and W. L. Hylander. 1998. EMG of the human flexor pollicis longus muscle: Implications for the evolution of Hominid tool use. *Journal of Human Evolution* 34: 123–136.

Hayasaka, K., K. Fujii, and S. Horai. 1996. Molecular phylogeny of macaques: Implications of nucleotide sequences from an 896-base-pair region of mitochondrial DNA. *Molecular Biology and Evolution* 13: 1044–1053.

Hedges, S. B., S. Kumar, K. Tamura, and M. Stoneking. 1992. Human origins and analysis of mitochondrial DNA sequences. *Science* 255: 737–739.

Horai, S., Y. Satta, K. Hayasaka, R. Kondo, T. Inoue, T. Ishida, S. Hayashi, and N. Takahata. 1992. Man's place in the Hominoidea revealed by mitochondrial DNA genealogy. *Journal of Molecular Evolution* 35: 32–43.

Horai, S., K. Hayasaka, R. Kondo, K. Tsugane, and N. Takahata. 1995. Recent African origin of modern humans revealed by complete sequences of hominoid mitochondrial DNAs. *Proceedings of the National Academy of Sciences, USA* 92: 532–536.

Huang Wanpo, R. G. Ciochon, R. Yúmin, F. Larick, H. Qiren, C. Schwarcz, J. de Vos Yonge, and W. Rink. 1995. Early *Homo* and associated artefacts from Asia. *Nature* 378: 275–278.

Huxley, T. H. 1863. *Evidence as to Man's Place in Nature.* New York: D. Appelton and Company.

Johanson, D. C., B. Edgar, and D. Brill. 1996. *From Lucy to Language.* (New York: Simon & Schuster Editions).

Kappelman, J. 1996. The evolution of body mass and relative brain size in fossil hominids. *Journal of Human Evolution* 30: 243–276.

Kim, H.-S., and O. Takenaka. 1996. A comparison of TSPY genes from Y-chromosomal DNA of the great apes and humans: Sequence, evolution, and phylogeny. *American Journal of Physical Anthropology* 100: 301–309.

Kimbel, W. H., R. C. Walter, D. C. Johanson, K. E. Reed, J. L. Aronson, Z. Assefa, C. W. Marean, G. G. Eck, R. Robe, E. Hovers, Y. Rak, C. Vondra, T. Yemane, D. York, Y. Chen, N. M. Evensen, and P. E. Smith. 1996. Late Pliocene *Homo* and Oldowan tools from the Hadar Formation (Kada Hadar member), Ethiopia. *Journal of Human Evolution* 31: 549–561.

Krings, M., A. Stone, R. W. Schmitz, H. Krainitzki, M. Stoneking, and S. Pääbo. 1997. Neandertal DNA sequences and the origin of modern humans. *Cell* 90: 19–30.

Krings M., H. Geisert, R. W. Schmitz, H. Krainitzki, and S. Pääbo. 1999 DNA sequence of the mitochondrial hypervariable region II from the Neandertal type specimen. *Proceedings of the National Academy of Sciences, USA* 96: 5581–5585.

Leakey, M. G., C. S. Feibel, I. McDougall, and A. Walker. 1995. New 4-million-year-old hominid species from Kanapoi and Allia Bay, Kenya. *Nature* 376: 565–571.

Lieberman, D. E. 1995. Testing hypotheses about recent human evolution from skulls: Integrating morphology, function, development, and phylogeny. *Current Anthropology* 36: 159–197.

Li Tianyuan and D. A. Etler. 1992. New middle pleistocene hominid crania from Yunxian in China. *Nature* 357: 404–407.

Lowenstein, J., and A. Zihlman. 1988. The invisible ape. *New Scientist* 3 (December): 56–59.

Marks, J. 1993. Hominoid heterochromatin: Terminal C-bands as a complex genetic trait linking chimpanzee and gorilla. *American Journal of Physical Anthropology* 90: 237–246.

Marks, J. 1994. Blood will tell (won't it?): A century of molecular discourse in anthropological systematics. *American Journal of Physical Anthropology* 94: 59–79.

Marks, J. 1995. Learning to live with a trichotomy. *American Journal of Physical Anthropology* 98: 211–213.

Martin, L. 1985. Significance of enamel thickness in hominoid evolution. *Nature* 314: 260–263.

McGrew, W. C., M. W. Hamrick, S. E. Inouye, J. C. Ohman, M. Slanina, G. Baker, R. P. Mensforth, and R. L. Susman. 1995. Thumbs, tools, and early humans. *Science* 268: 586–589.

McHenry, H. M. 1992. Body size and proportions in early hominids. *American Journal of Physical Anthropology* 87: 407–430.

McHenry, H. 1994. Tempo and mode in human evolution. *Proceedings of the National Academy of Sciences, USA* 91: 6780–6784.

National Public Radio. 1995. Evolution disclaimer to be placed in Alabama textbooks. Transcript 1747, segment 13.

Nei, M. 1995. Genetic support for the out-of-Africa theory of human evolution. *Proceedings of the National Academy of Sciences, USA* 92: 6720–6722.

Noble, W., and I. Davidson. 1991. The evolutionary emergence of modern human behavior: Language and its archaeology. *Man* 26: 223–253.

Nordborg, M. 1998. On the Probability of Neanderthal Ancestry. *American Journal of Human Genetics* 63: 1237–1240.

Nuttall, G. H. F. 1904. *Blood, Immunity, and Blood Relationship; a demonstration of certain blood-relationships amongst animals by means of the precipitin test for blood.* London: Cambridge University Press.

Pilbeam, D., and S. J. Gould. 1974. Size and scaling in human evolution. *Science* 186: 892–901.

Pinker, S. 1994. *The Language Instinct.* New York: Harper Collins.

Pough, F. H., J. B. Heiser, and W. N. McFarland. 1996. *Vertebrate Life* Upper Saddle River, NJ: Prentice Hall.

Pritchard, J. K., and M. W. Feldman. 1996. Genetic data and the African origin of humans. *Science* 274: 1548.

Rak, Y. 1983. *The Australopithecine Face.* New York: Academic Press.

Risch, N., K. K. Kidd, and S. A. Tishkoff. 1996. Genetic data and the African origin of humans—reply. *Science* 274: 1548–1549.

Rogers, J. 1994. Levels of genealogical hierarchy and the problem of hominoid phylogeny. *American Journal of Physical Anthropology* 94: 81–88.

Rogers, J., and A. G. Comuzzie. 1995. When is ancient polymorphism a potential problem for molecular phylogenetics? *American Journal of Physical Anthropology* 98: 216–218.

Ruvolo, M. 1994. Molecular evolutionary processes and conflicting gene trees: The Hominoid case. *American Journal of Physical Anthropology* 94: 89–113.

Ruvolo, M. 1995. Seeing the forest and the trees: Replies to Marks; Rogers and Commuzzie; Green and Djian. *American Journal of Physical Anthropology* 98: 218–232.

Ruvolo, M. 1997. Molecular phylogeny of the hominids: Inferences from multiple independent DNA sequence data sets. *Molecular Biology and Evolution* 14: 248–265.

Ruvolo, M., S. Zehr, M. von Dornum, D. Pan, B. Chang, and J. Lin. 1993. Mitochondrial COII sequences and modern human origins. *Molecular Biology and Evolution* 10: 1115–1135.

Ruvolo, M., D. Pan, S. Zehr, T. Goldberg, T. R. Disotell, and M. von Dornum. 1994. Gene trees and hominoid phylogeny. *Proceedings of the National Academy of Sciences, USA* 91: 8900–8904.

Sarich, V. M., and A. C. Wilson. 1967. Immunological time scale for Hominid evolution. *Science* 158: 1200–1203.

Satta, Y., J. Klein, and N. Takahata. 2000. DNA archives and our nearest relative: The trichotomy revisited. *Molecular Phylogenetic and Evolution* 14: 259–275.

Schick, K. D., and N. P. Toth NP. 1993. *Making Silent Stones Speak: Human Evolution and the Dawn of Technology.* New York: Simon & Schuster.

Semaw, S., P. Renne, J. W. K. Harris, C. S. Feibel, R. L. Bernor, N. Fesseha, and K. Mowbray. 1997. 2.5-million-year-old stone tools from Gona, Ethiopia. *Nature* 385: 333–336.

Shoshani, J., C. P. Groves, E. L. Simons, and G. F. Gunnel. 1996. Primate phylogeny: Morphological vs. molecular results. *Molecular Phylogenetics and Evolution* 5: 101–153.

Strait, D. S., F. E. Grine, and M. A. Moniz. 1997. A reappraisal of early hominid phylogeny. *Journal of Human Evolution* 32: 17–82.

Stern, J. T., and R. L. Susman. 1983. The locomotor anatomy of *Australopithecus afarensis*. *American Journal of Physical Anthropology* 60: 279–317.

Stringer, C. 1988. The dates of Eden. *Nature* 331: 565–566.

Susman, R. L. 1994. Fossil evidence for early hominid tool use. *Science* 265: 1570–1573.

Swisher, C. C., III, G. H. Curtis, et al. 1994. Age of the earliest known hominids in Java, Indonesia. *Science* 263: 1118–1121.

Swisher, C. C., III, W. J. Rink, S. C. Antón, H. P. Schwarcz, G. H. Curtis, A. Suprijo, and Widiasmoro. 1996. Latest *Homo erectus* of Java: Potential contemporaneity with *Homo sapiens* in Southeast Asia. *Science* 274: 1870–1874.

Tattersall, I. 1986. Species recognition in human paleontology. *Journal of Human Evolution* 15: 165–175.

Tattersall, I. 1992. Species concepts and species identification in human evolution. *Journal of Human Evolution* 22: 341–349.

Tattersall, I. 1995. *The Fossil Trail.* Oxford: Oxford University Press.

Tattersall, I. 1997. Out of Africa again . . . and again? *Scientific American* 276 (April): 60–67.

Tattersall, I. 2000. Once we were not alone. *Scientific American* 282 (January): 56–62.

Tattersall, I., and J. H. Schwartz. 1999. Hominids and hybrids: The place of Neanderthals in human evolution. *Proceedings of the National Academy of Sciences, USA* 96: 7117–7119.

Thorne, A. G., and M. H. Wolpoff. 1981. Regional continuity in Australasian pleistocene hominid evolution. *American Journal of Physical Anthropology* 55: 337–349.

Tishkoff, S. A., E. Dietzsch, W. Speed, A. . Pakstis, J. R. Kidd, K. Cheung, B. Bonné-Tamir, A. S. Santachiara-Benerecetti, P. Moral, M. Krings, S. Pääbo, E. Watson, N. Risch, T. Jenkins, and K. K. Kidd. 1996b. Global patterns of linkage disequilibrium at the CD4 locus and modern human origins. *Science* 271: 1380–1387.

Tishkoff, S. A., K. K. Kidd, and N. Risch. 1996a. Interpretations of multiregional evolution—reply. *Science* 274: 706–707.

Tobias, P. V. 1987. The brain of *Homo habilis*: A new level of organization in cerebral evolution. *Journal of Human Evolution* 16: 741–761.

Trinkaus, E., C. B. Ruff, S. E. Churchill, and B. Vandermeersch. 1998. Locomotion and body proportions of the Saint-Césaire 1 Châtelperronian Neandertal. *Proceedings of the National Academy of Sciences, USA.* 95: 5836–5840.

Valladas, H., J. L. Reyss, J. L. Joron, G. Valladas, O. Bar-Yosef, and B. Vandermeersch. 1988. Thermoluminescence dating of Mousterian "Proto-Cro-Magnon" remains from Israel and the origin of modern man. *Nature* 331: 614–616.

Vigilant, L., M. Stoneking, H. Harpending, K. Hawkes, and A. C. Wilson. 1991. African populations and the evolution of human mitochondrial DNA. *Science* 253: 1503–1507.

Waddle, D. M. 1994. Matrix correlation tests support a single origin for modern humans. *Nature* 368: 452–454.

Ward, S. C., and W. H. Kimbel. 1983. Subnasal alveolar morphology and the systematic position of *Sivapithecus*. *American Journal of Physical Anthropology* 61: 157–171.

White, T. D., and G. Suwa. 1987. Hominid footprints at Laetoli: Facts and interpretations. *American Journal of Physical Anthropology* 72: 485–514.

White, T. D., G. Suwa, and B. Asfaw. 1994. *Australopithecus ramidus*, a new species of early hominid from Aramis, Ethiopia. *Nature* 371: 306–312.

White, T. D., G. Suwa, and B. Asfaw. 1995. Corrigendum: *Australopithecus ramidus*, a new species of early hominid from Aramis, Ethiopia. *Nature* 375: 88.

Wilkins, W. K., and J. Wakefield. 1995. Brain evolution and neurolinguistic preconditions. *Behavioral and Brain Sciences* 18: 161–226.

Wilson, D. E., and D. M. Reeder, eds. 1993. *Mammal Species of the World,* 2nd ed. Washington: Smithsonian Institution Press.

Wolpoff, M. H. 1996. Interpretations of multiregional evolution. *Science* 274: 704–706.

Wood, B. 1992. Origin and evolution of the genus *Homo. Nature* 355: 783–790.

Wood, B. 1997. The oldest whodunnit in the world. *Nature* 385: 292–293.

Wood, B., and A. Turner. 1995. Out of Africa and into Asia. *Nature* 378: 239–240.

Wrangham, R. W. 1999. Evolution of coalitionary killing. *Yearbook of Physical Anthropology* 42: 1–30.

Zischler, H., H. Geisert, A. von Haeseler, and S. Pääbo. 1995. A nuclear "fossil" of the mitochondrial D-loop and the origin of modern humans. *Nature* 378: 489.

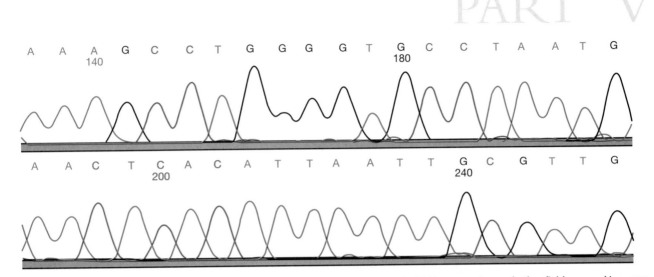

These plots are output from an automated DNA sequencer. Technological advances in DNA sequencing and other fields are making new data sets available for evolutionary analysis.

CURRENT RESEARCH—A SAMPLER

PRECEDING CHAPTERS HAVE EXPLORED THE MECHANISMS OF EVOLUTION, ADAPTATION, and the history of life. Part of this investigation involved developing a tool kit of knowledge, concepts, and techniques for asking and answering questions about organismal diversity and adaptation. In Part V, we use this tool kit to investigate questions from particularly active areas of research in evolutionary biology. The topics we cover represent but a small sample of the issues under investigation. We have chosen them in the hopes of highlighting research areas most relevant to readers' professional and personal lives.

We begin, in Chapter 17, with a look at recent experiments on the genetic basis of evolutionary innovations. Paleontologists and molecular geneticists are collaborating to explore the developmental basis of major changes observed in the fossil record, ranging from the origin of the flower to the diversification of animal eyes.

Chapter 18 considers a series of questions raised by the explosion of data on protein and nucleic acid sequences. Which of the four evolutionary forces are responsible for the differences observed in DNA sequences among species? Why are the genomes of so many species riddled with parasitic sequences?

Finally, Chapter 19 explores how evolutionary thinking is being applied to questions about human health. Does an understanding of how pathogens evolve help to direct vaccine development or patient care? Can evolutionary analysis produce medically useful insights into human physiology and behavior?

CHAPTER 17

Development and Evolution

This *Arabidopsis thaliana* individual has mutations that transform its flower into leaf-life structures. In 1790, the German poet and naturalist Johann Goethe hypothesized that flowers evolved from leaf-like structures. (Elliot M. Meyerowitz, California Institute of Technology)

MANY OF THE QUESTIONS ADDRESSED BY EVOLUTIONARY BIOLOGY (AND MANY of the examples in this book) concern the origin, maintenance, and diversification of the forms of multicellular organisms, from colonial algae to humans, plants, and worms. Until the middle 1980s, it was widely assumed that this diversity of forms had evolved through many large changes in the genes that control the development of a multicellular body from a single-celled zygote. Two discoveries helped to change this view. The first was the gradual revelation that there are often very few, small genetic differences between species with very different adult phenotypes. In Chapter 16, we discussed one example of this pattern: the extremely similar genomes of humans and other African great apes, particularly the chimpanzees. The second discovery was the fundamentally similar genetic and cellular mechanisms that underlie the development of embryos in species where adults are dissimilar (Raff and Kaufman 1983; Gerhart and Kirschner 1997). If the genomes and the developmental processes of very different organisms are largely the same, then what is the source of the tremendous morphological diversity that we see in nature and would like to understand? Developmental biologists and evolutionary biologists have merged their efforts recently in an attempt to answer this question.

587

It was not always so. Beginning in the late 19th century, embryologists focused their attention mainly on understanding the internal mechanics of animal and plant development. Their efforts were generally left out of the mainstream of evolutionary analysis, which, instead, focused on the unification of ecology, systematics, and population biology (see Gilbert 1997). However, the new tools of molecular genetics have allowed developmental biologists to dig deeper into the basis of morphological similarities and differences among species. It is now possible to identify the relatively small genetic changes responsible for the phenotypic variation seen among complex multicellular organisms. Using phylogenetic analyses, it is also possible to reconstruct the history of changes in these genes that have led to such striking phenotypic diversity. This research program, aimed at understanding the history of changes in genetic control of development, has sometimes been called "molecular paleontology" or "paleontology without fossils" (DiSilvestro 1997).

Researchers want to understand the genetic changes that were reponsible for the diversification of organisms.

17.1 Homeotic Genes, Pattern Formation, and Diversification

Complex multicellular organisms arise and become different from each other through the appearance of new features. Surprisingly, only a small proportion of these novel traits seem to require the evolution of new types of cells. One example of a new feature evolved through a novel cell type is the fast conduction of electical signals in the nervous systems of vertebrates. Rapid conduction is promoted by insulating the neurons of the peripheral nervous system with a lipid called myelin, which is secreted by a novel type of cell in the nervous system called the Schwann cell (see Gerhart and Kirschner 1997). The evolution of the Schwann cell allowed vertebrates to evolve large bodies with rapid information processing over long distances within the nervous system.

However, most new features of multicellular organisms arise when preexisting cell types appear at new locations or new times in the embryo. Multicellular organisms require a system for arranging cells in three-dimensional space and specifying the fate of those cells. Cells must be identified by their location—in relation to other cells and in relation to time—for cleavage and gastrulation to occur; for symmetry, segmentation, and body cavities to be organized properly; and for the correct differentiation of muscle, gut, and other tissues. Changes in the specification of cell fates is a major mechanism for the evolution of different organismal forms.

The genes that carry this information are called **homeotic loci**. The most important homeotic genes in animals are called the *HOM* loci in invertebrates and the *Hox* loci in vertebrates. (Often these homologous genes are generally referred to as *Hox* loci.) Determining how these gene complexes originated, changed through time, and interacted to produce morphological variation is currently one of the most active research topics in developmental and evolutionary biology.

Hox genes have been found in all major animal phyla, and share three key traits (Carroll 1995):

- They are organized in gene complexes. This means that related genes are found in close proximity on the chromosome, and suggests that the original genes were elaborated by duplication events. Every phylum or class that has been surveyed shows a unique pattern of gene duplication or loss compared to other groups (see Kenyon and Wang 1991; Balavoine and Telford 1995; Valentine et al. 1996; de Rosa et al. 1999.)

- There is a perfect correlation between the 3′–5′ order of genes along the chromosome and the anterior–posterior location of gene products in the embryo (Lewis 1978). Genes located at the 3′ end of the complex are expressed in the head region of the embryo, while genes located at the 5′ end are expressed toward the posterior of the embryo. Genes in the 3′ end of *Hox* complexes are also expressed earlier in development than genes located upstream. This phenomenon is called temporal and spatial colinearity (Krumlauf 1994), and is unique to *HOM* and *Hox* genes. Why it occurs is a complete mystery.

- Each locus within the complex contains a highly conserved 180-bp sequence called the homeobox. These bases code for a DNA binding motif. The discovery of the homeobox (McGinnis et al. 1984a, b) confirmed that *HOM* and *Hox* gene products are regulatory proteins that bind to DNA and control the transcription of other genes.

Homeotic loci code for regulatory proteins that control the transcription of other genes.

What do *Hox* loci regulate? The answer to this question came through experiments that produced fruit flies lacking specific *HOM* genes (Lewis 1978; Nüsslein-Volhard and Wieschaus 1980; Nüsslein-Volhard et al. 1987; McGinnis and Kuziora 1994). In *Drosophila,* homeotic genes are found in two main clusters called the *bithorax* and *Antennapedia* complexes (Figure 17.1). Mutations in *bithorax* complex genes tend to cause defects in the posterior half of embryos, while mutations in

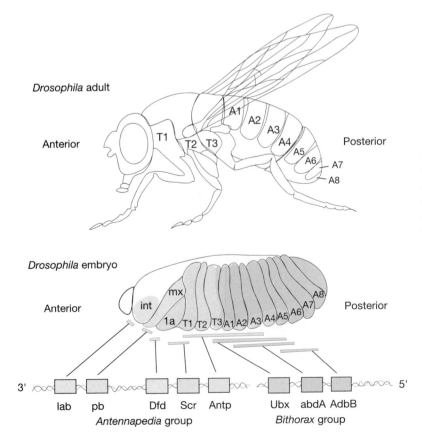

Figure 17.1 *Hox* genes in *Drosophila* The top diagram shows an adult fruit fly: T1–T3 and A1–A8 indicate the three thoracic and eight abdominal segments, respectively. Below the adult is a diagram of a *Drosophila* embryo in the same orientation (anterior to the left), showing the segments that will contribute to the head, thorax, and abdomen. The bottom part of the figure shows the relative locations of *antennapedia* group and *bithorax* group genes within the *HOM* cluster. Each *HOM* gene in the diagram is color coded to indicate the anterior or posterior portion of the body segments in which the gene affects segmental identity. (For example, *pb*-expression influences the identity of cells that make the proboscis or mouthparts of the fly; both are shaded greeen.) From Gerhart and Kirschner (1997). Copyright © 1997 Blackwell Science, Inc. Reprinted by permission.

Antennapedia genes affect the anterior. The key observation is that flies missing one or more of these gene products can produce segment-specific appendages in the wrong place. By mixing and matching mutations in the two complexes, researchers have been able to produce adult flies with legs growing from the head region, or with four pairs of wings instead of two pairs. (Figure 17.2). The cells that produced the misplaced antennae or wings acted as if they misunderstood their location. This means that, rather than specifying a particular structure, gene products from *Hox* loci demarcate relative positions in the embryo. That is, instead of signaling "make wing," a *Hox* protein indicates "this is thoracic segment 2." Other genes, presumably those regulated by the protein products of *Hox* loci, make the structure specified for each location. In sum, *Hox* genes regulate the fate of cells by specifying where they are in time (between the start of embryonic development and the achievement of the adult body plan) and space (especially along the anterior-to-posterior body axis).

The function of homeotic loci is to define where cells are within the embryo.

Changes in *HOM/Hox* Gene Number

The observations that the *Hox* genes share a conserved homeobox sequence and occur together in arrays on the chromosome suggest that the cluster has evolved through a series of gene duplication events. (We discussed the origin and evolution of such gene families in Chapter 4.) If so, then when did these loci originate? How did changes in the *Hox* gene cluster lead to phenotypic differences between major animal groups?

An early hypothesis to answer the first of these questions was that the *Hox* loci arose at the same time as the evolution of segmentation during the Cambrian explosion. This was inspired by the observation that homeotic genes tend to be expressed in a segment-specific manner in flies (Figure 17.1). But homologs to the

(a)

(b)

Figure 17.2   Homeotic mutants in *Drosophila*   (a) Mutations in the *HOM* complex loci *bithorax* (*bx*), *posterior bithorax* (*pbx*), and *anterobithorax* (*abx*) produce four-winged flies. The phenotype occurs because appendages that should appear on thoracic segment 2 (T2) also appear on T3. In effect, the mutations have changed the identity of T3 to equal that of T2. (Edward B. Lewis, California Institute of Technology) (b) Mutations in the *Antennapedia* (*antp*) locus can produce adult flies with legs growing from the head. For limb development, the identity of one head segment has been changed to that of a thoracic segment. (Dr. F. Rudolph Turner, Indiana University)

HOM/Hox genes have now been found in green plants, fungi, the roundworm *Caenorhabditis elegans,* and other nonsegmented animals, including groups such as jellyfish (see Kenyon and Wang 1991; Finnerty and Martindale 1997) and sponges (Kruse et al. 1994; Seimiya et al. 1994) that lack an anterior-to-posterior body axis altogether. These results push the origin of the *HOM/Hox* complex back in time, perhaps to the origin of multicellularity in the Precambrian. These results also show that the *Hox* genes must influence processes other than the specification of anterior-to-posterior cell fates. The answer to the first question, then, is that the oldest original *HOM/Hox* gene predates the diversification of the animals and the evolution of a differentiated body axis (Knoll and Carroll 1999).

Several research groups are focused on answering the second question. James Valentine and co-workers (1996; see also Erwin et al. 1997) have synthesized data on the origin and elaboration of homeotic genes by mapping the presence and absence of *Hox* loci onto the phylogeny of animals. Using a parsimony criterion, they inferred which genes in the *Hox* complex have been gained or lost at key branching points near the base of the tree. Their analysis (Figure 17.3) confirms that the loci called *labial (lab), proboscipedea (pb), Deformed (Dfd), Antennapedia (Antp),* and *Abdominal-B (Abd-B)* are ancestral to many other genes in the complex. Other loci within the cluster are probably descended by gene duplication from these original loci.

A more recent compilation by de Rosa et al. (1999) supports this conclusion. They show that each major clade of bilateral animals is characterized by a particular suite of *Hox* genes. Moreover, several specific changes in the *Hox* cluster are associated with particular animal clades. For example, the addition of the *Abdominal-B* locus is associated with the evolution of bilateral animals. Similarly, the entire *Hox* complex was duplicated several times in the lineage leading to the mouse and other vertebrates. These large duplication events may have played an important

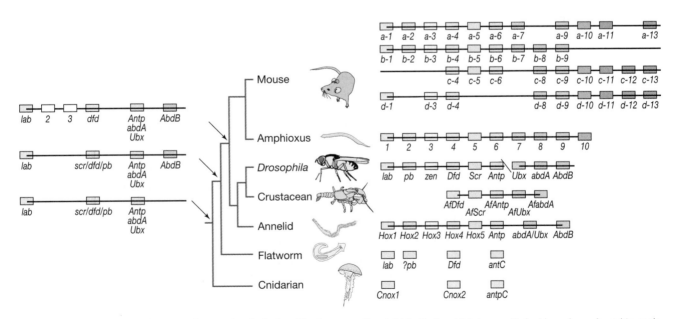

Figure 17.3   *Hox* genes from various animal phyla   The boxes on the right indicate which homeotic loci have been found in each of the taxa on the tree. Homologous loci are lined up vertically. Using these presence–absence data, Valentine et al. (1996) used parsimony to infer which loci were present at branch points early in the radiation. These loci are indicated at the left, with arrows pointing to the appropriate node. Note that amphioxus, a lancelet in the subphylum Cephalochordata, is the most "primitive" living chordate.

role in the divergence of vertebrates from other deuterostomes (sea urchins have just one *HOM* cluster compared to the four *Hox* clusters in mice). Thus, the answer to the second question is that some specific gene duplications are associated with the origin of distinctive animal clades.

However, as more and more diverse animal groups are examined for *Hox* genes, several puzzles have emerged. Sponges and cnidarians appear to have fewer *HOM* loci than the bilateral animals. At this level of complexity, the number of loci within the gene complex correlates roughly with the overall complexity of metazoan body plans (Knoll and Carroll 1999). This observation supports the hypothesis that the duplication and elaboration of genes in the *Hox* complex was a genetic innovation that helped make the Cambrian explosion possible.

To some degree, there is a positive correlation between the number of homeotic loci in an organism and its morphological complexity.

Among the bilaterally symmetric animals, though, there is not a strict correlation between morphological complexity and the number of *Hox* loci. Snails and slugs have 3–6 of these genes while tubeworms have 10; mice have 39 while zebrafish have 42. Further, when de Rosa and co-workers (1999) used a parsimony analysis to estimate how many *Hox* loci existed in the common ancestor of all bilateral animals, the answer was 10. Because sponges and cnidarians have just 3–4 of these genes, their result suggests that a dramatic jump in the number of *Hox* loci occurred sometime before the Cambrian explosion. When and where did this increase occur? Although the question is fundamental to understanding how genetic changes made the radiation of animals possible, it remains unanswered.

Changes in *HOM/Hox* Expression: Arthropod Segmentation

Because most of the living bilateral animals share a similar set of *Hox* genes, much of the post-Cambrian diversification of animal body plans must have involved changes in the timing or spatial location of *Hox* gene expression rather than changes in the number of loci in the cluster. One striking example of such changes in *Hox* expression is the evolution of segmental identity in arthropods.

Arthropods are far and away the most diverse animal phylum (Brusca and Brusca 1990). Over a million species have been described, and perhaps 50 times that many are still to be named. Members of the group are distinguished by having an exoskeleton (which necessitates growth by molting), a body organized into cephalic (head), thoracic, and abdominal regions (or tagma), pairs of appendages on each body segment, and an open circulatory system. Although estimating the phylogeny of arthropods is notoriously difficult and controversial (see Budd 1996), one recent classification identifies four subphyla:

- Chelicerates include scorpions, spiders, mites, horseshoe crabs, and a variety of extinct groups (Figure 17.4a). Prominent among the extinct forms are the eurypterids, or sea spiders, which occasionally reached three meters in length.
- Uniramians (Figure 17.4b) encompass the centipedes and millipedes (or myriapods) and insects.
- Trilobites (Figure 17.4c) were abundant in marine environments throughout the Paleozoic but became extinct at the end of the Permian.
- Crustaceans (Figure 17.4d) include the shrimp, copepods, barnacles, crabs, lobsters, crayfish, and pill bugs.

The onychophorans, or velvet worms, are probably the sister group to the true arthropods. They are burrowers restricted to the humid tropics. As Figure 17.4e shows, they have small, unbranched, saclike limbs called lobopods.

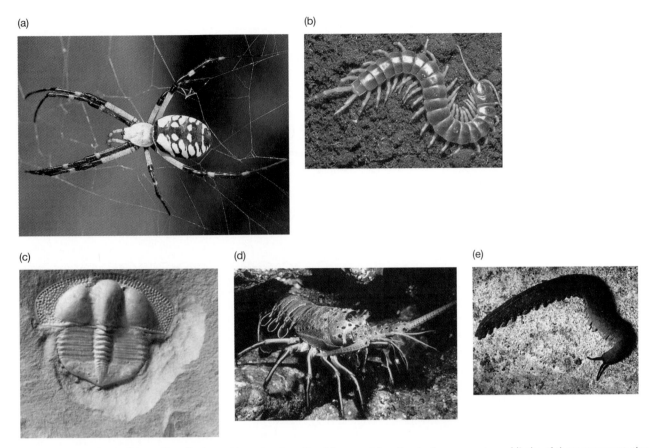

Figure 17.4 The arthropod radiation Note the diversity of form and function in the segments and limbs of these representative taxa. (a) Chelicerates. (Jeffrey Howe/Visuals Unlimited) (b) Uniramians. (Tom McHugh/Photo Researchers, Inc.) (c) Trilobites. (A. Graffham/Visuals Unlimited) (d) Crustaceans. (Marty Snyderman/Visuals Unlimited) (e) Onychophorans, the probable sister group to the four arthropod taxa. (R. Ashley/Visuals Unlimited)

The initial diversification of arthropods occurred during the Cambrian explosion, though many of the classes, orders, and families originated and diversified later. (For example, insects do not show up in the fossil record until early in the Devonian.) Much of the basic anatomical diversity of these four arthropod classes depends on the differentiation of segments from anterior to posterior. This differentiation is controlled by *Hox* gene expression, but all of these groups (including the onychophorans) have the same complement of 9 loci (Grenier et al. 1997). These genes are activated or repressed by a complex network of other gene products, and they influence the expression of a large number of other genes and developmental processes (Nagy 1998).

In short, morphological diversification of these groups occurred mainly by changes in the localized expression of these genes rather than by the addition of new *HOM* loci. These changes in gene expression were recently summarized Andrew Knoll and Sean Carroll (1999). Let's begin with onychophorans, which have a trunk made up of many segments. These segments are similar to each other, and each has a similar pair of simple lobe-shaped limbs. In these simplest arthropod relatives, *HOM* expression is limited to the loci called *Ubx* and *abd-A* in the most posterior part of the trunk (Figure 17.5). However, the arthropod groups are distinguished from each other by a complex set of differences in the anterior-to-posterior and

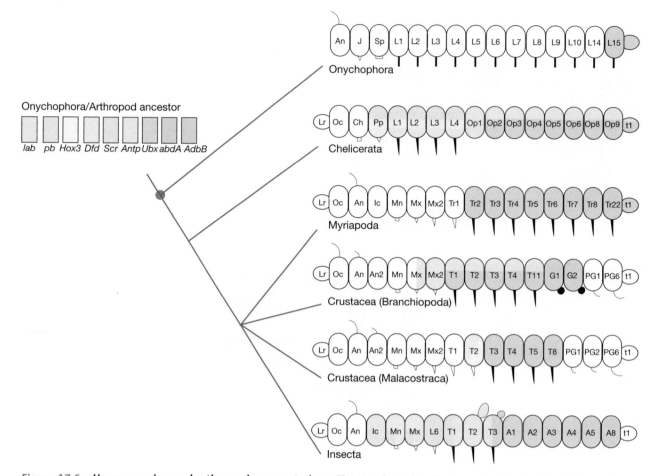

Onychophora/Arthropod ancestor

lab   pb   Hox3   Dfd   Scr   Antp Ubx abdA AdbB

Figure 17.5 *Hox* **expression and arthropod segmentation** The *Hox* cluster for all arthropods is similar (nine loci are color-coded on the left). The evolutionary tree shows the relationships between the four arthropod classes and the Onychophora. The diagrams on the right show where the various *Hox* loci are expressed in representatives of these taxa, including two crustaceans. (Note that appendages such as antennae, mouthparts, wings, and legs are shown for some segments.) The message of this figure is that differences in *Hox* gene expression distinguish the various arthropod segmentation patterns. Reprinted with permission from Knoll and Carroll (1999). Copyright © 1999, American Association for the Advancement of Science.

dorsal–ventral expression of all nine loci. Some patterns are similar among the groups:

1. *Ubx* and *abd-A* are expressed together in some posterior segments of the trunk, and
2. *abd-A* is expressed on the ventral side relative to *Ubx*.

All arthropods have the same suite of homeotic loci. Among species, however, the loci are expressed at different times and in different locations. The variation in homeotic gene expression correlates with variation in the arthropod body plan.

But other patterns distinguish the four classes from each other:

1. *Ubx/abd-A* expression does not occur in the last trunk segments of insects and crustaceans, and
2. In insects, the gene called *pb* is expressed in the ventral–anterior part of three head segments and helps to define the insect mouthparts.

Many other examples of group-specific differences are illustrated in Figure 17.5.

What is the main implication of this spatial variation in gene expression? Much of the evolutionary diversification of arthropods probably occurred as a result of changes in where *Hox* loci are expressed (Nagy 1998; Knoll and Carroll 1999).

17.2 The Genetics of Homology: Limbs

Analyzing homeotic loci is one basis for understanding the origins of evolutionary novelty. A second application of molecular genetics is understanding the underlying basis for homologies, especially among variant forms of a novel structure. One such structure is the limb. Among the vertebrates, limbs range from the wings of bats to the powerful legs of horses. Among the arthropods, limbs range from fly wings to crab claws. Are there any genetic similarities that link these diverse structures?

The Tetrapod Limb

The terrestrial vertebrates include the amphibians, crocodiles, birds, and mammals. The signature adaptation of this lineage, called the Tetrapoda, is the limb. This structure distinguished the early tetrapods from fish and allowed them to crawl about on land. Later, this innovation was modified into an enormous variety of shapes and sizes, with functions ranging from digging to flying. But in general, tetrapod limbs are variations on the same theme. From frogs to foxes, the number and arrangement of limb bones is similar (Figure 17.6).

What Structure Gave Rise to the Limb?

Establishing homologies between the tetrapod limb and ancestral forms is the key to answering this question. Phylogenetic analyses show that the sister group of the tetrapods is a lineage of lobe-finned fishes from the late Devonian (Ahlberg and Milner 1994; Ahlberg 1995). In particular, a group of fish called the Panderichthyidae share several synapomorphies with early tetrapods such as the early Devonian genus *Acanthostega*. The panderichthyids were large predators in shallow, freshwater habitats, and undoubtedly used their limbs to walk or pull themselves along much as modern lungfish do. As Figure 17.7 illustrates, there are numerous structural homologies between the fins of lobe-finned fish and the limbs of tetrapods. The phylogenetic and morphological analyses support the hypothesis that the tetrapod limb is derived from the fins of lobe-finned fish.

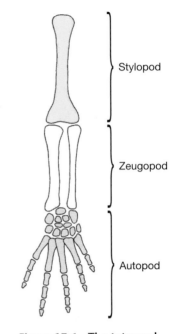

Figure 17.6 The tetrapod limb This diagram shows the basic elements of the tetrapod limb. There is a single proximal element, or stylopod (called the humerus in the forelimb and femur in the hindlimb) followed by an element with two bones (radius and ulna or tibia and fibula), a complex of small elements (carpus or wrist, tarsus or ankle), and the digits (fingers or toes).

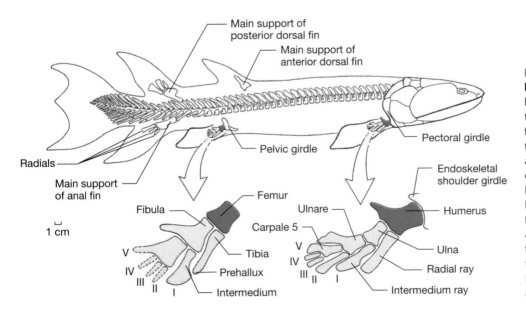

Figure 17.7 Fin bones in lobe-finned fish from the Devonian *Eusthenopteron*, the genus pictured here, had fins containing bones in the typical tetrapod limb arrangement. In the detailed drawing of the fins, dashed lines indicate elements that have not yet been found in fossilized specimens. These are inferred to have existed from the position and structure of adjoining elements. Modified from Carroll (1988) and Jarvik (1980).

What Developmental and Genetic Changes Were Responsible for the Origin of the Tetrapod Limb and Its Subsequent Modification?

To answer this question, we need to understand some developmental genetics. We start by investigating the mechanisms that create the common ground plan of the tetrapod limb, then consider how these mechanisms first appeared at the base of the tetrapod radiation.

Tetrapod limbs have a common ground plan. This results from a shared developmental program.

Evidence from developmental biology backs up our claim that tetrapod limbs share a common ground plan: Every tetrapod investigated to date shares the same basic features of limb development (Hinchcliffe 1989; Gilbert 1997). Specifically, all tetrapod limbs originate with a bud formed from mesodermal cells (Figure 17.8a). At the tip of this bud is a structure called the apical ectodermal ridge, or AER [Figure 17.8b and (c)]. Cells in the AER secrete a molecule that keeps the underlying cells in a growing and undifferentiated state. This population of cells is called the progress zone. Its descendants will eventually be instructed to form specific elements of the limb such as skeleton, muscle, and connective tissue. The progress zone grows outward and defines the long axis of the developing limb (Figure 17.8c).

At the base of the bud in Figure 17.8c is a group of cells called the zone of polarizing activity (ZPA). A molecule secreted from the ZPA diffuses into the surrounding tissue and establishes a gradient that supplies positional information to cells in the structure: The molecule is found at highest concentration near the ZPA and at lower concentration in the parts of the limb bud farthest from the ZPA. This concentration gradient is critical because the limb develops in a four-dimensional system consisting of time and the three spatial coordinates. The spatial dimensions are defined in relation to the main body orientation: anterior to posterior, dorsal to ventral, and proximal to distal (Figure 17.9). On your own forelimb, these axes run from the thumb to the last finger (a–p), from the back to the front of the hand (d–v), and from the arm to the fingertips (p–d), respectively. The concentrations of molecules diffusing from the AER and ZPA help set up a coordinate system and tell each cell where it is in the four-dimensional space.

These molecules are, in fact, now well characterized. Identifying them takes our recognition of homology among tetrapod limbs to the gene level. For exam-

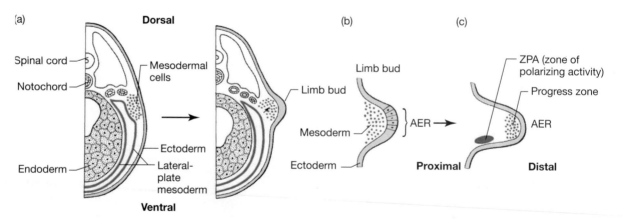

Figure 17.8 The developing limb bud (a) In tetrapods, the limb bud originates from cells that migrate from the lateral-plate mesoderm early in development. (b) These cells induce the formation of a structure called the apical ectodermal ridge (AER). (c) If the AER is removed experimentally, cells in the progress zone stop their rapid division and differentiate, so only part of the limb is formed.

ple, the products of a gene called fibroblast growth factor 2 (FGF-2) is found in the AER. Application of FGF-2, even when the AER is removed, leads to normal limb development in chicks (Cohn et al. 1995; Fallon et al. 1995). This is strong evidence that FGF-2 is the molecule that maintains the progress zone. Similarly, the product of a gene called *sonic hedgehog (shh)* localizes to the ZPA. Grafting *shh*-expressing cells onto limb buds produces an additional ZPA (Riddle et al. 1993; Tabin 1995). Finally, expression of a recently discovered gene called *Wnt7a* causes cells in developing limbs to form dorsal structures in mice (Parr and McMahon 1995; Yang and Niswander 1995). Relative to the axes of the limb, then,

1. a gradient of the *shh* gene product establishes the anterior–posterior axis of the limb,
2. *Fgf-2* establishes the proximal–distal axis, and
3. *Wnt7a* establishes the dorsal–ventral axis.

In sum, though we do not yet have a candidate clock molecule (to tell cells where they are in the temporal unfolding of limb development), the genes responsible for specifying the other three coordinates in the limb system are known (see Cameron et al. 1998; Tabin et al. 1999).

Equally important is the recent discovery that loci in the *Hox* family are critical to limb development. Earlier we noted how loci in the *Hox* complex are expressed along the embryo's anterior-to-posterior axis in the same order that they are found along the chromosome. This linear sequence of *Hox* genes is also expressed in the developing tetrapod limb—specifically in the progress zone (Duboule 1994). These data suggest that *Hox* loci tell cells where they are along the length of the limb, just as they tell cells where they are along the main body axis (Cohn et al. 1995; Tabin 1995).

This brief introduction to the developmental genetics of the vertebrate limb carries two messages. First, homologous genes and developmental pathways underlie the structural homology of amphibian, reptile, bird, and mammal limbs. Second, changes in the timing or level of expression in the pattern-forming genes we have reviewed—*Fgf-2, shh, Wnt,* or the *Hox* complex—could be responsible for adaptive changes in the limb. The evolution of longer limbs, for example, might be based on variation in *Fgf-2* gene expression and how long the progress zone is maintained.

Just this type of genetic connection has been made concerning the origin of a prominent part of the tetrapod limb: the hand and foot. The lobe-finned ancestors of the tetrapods, like extant fish, do not have feet or hands (see Figure 17.7); tetrapods do. A recent study by Paolo Sordino and co-workers (1995; see also Sordino and Duboule 1996) suggests a genetic mechanism for the change. Sordino et al. began by sequencing *shh* and members of the *Hox* gene family in zebrafish (*Danio rerio*). They labeled single-stranded copies of the genes with radioactive atoms, then hybridized the DNA with various stages of developing fins. As a result, cells with copies of the messenger RNAs for these genes became radioactively labeled. In this way, the researchers could determine when and where these genes are expressed.

The pattern of expression of *shh* and the various *Hox* loci, when contrasted with results for the same genes in a tetrapod (the mouse), is striking. Early in development, gene expression in tetrapods and fish is similar. *Hox* transcripts, for example, localize to the posterior margin of both the elongating fin and limb buds (Figure 17.10a, b). But in tetrapods, a switch occurs. Later in development, *Hox*

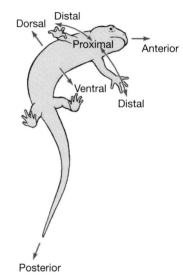

Figure 17.9 Developmental and morphological axes Dorsal = toward the back, ventral = toward the belly, anterior = toward the front, posterior = toward the rear, proximal = toward the main body axis, distal = away from the main body axis.

The shared developmental program observed in tetrapod limbs is produced by homologous genes.

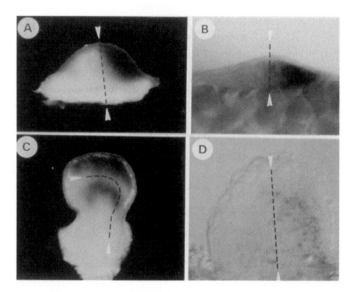

Figure 17.10   Expression of pattern-forming genes in fish and tetrapod limbs   (a) In early limb buds in mice, *Hoxd-11* expression (indicated by the dark splotch) is limited to the posterior half. (b) Early limb buds in zebrafish show the same pattern of *Hoxd-11* expression. (c) In later mouse embryos, a second wave of *Hoxd-11* expression is oriented anterior to posterior, in the most distal portion of the developing limb. (d) In contrast, in late fish embryos, *Hoxd-11* expression has decreased and is still restricted to the posterior margin. In each of the photographs, the dotted lines indicate Sordino et al.'s (1995) proposal for the axis of development. (Prof. Denis Duboule, Université de Genève, Geneva, Switzerland)

gene transcripts are found in the anterior and distal parts of the mouse limb bud. There is also a late expression of *shh* (Figure 17.10c). Why are these differences interesting? Because they do not happen in fish. There is no expression of *Hox* or *shh* in the late limb bud of zebrafish (Figure 17.10d); instead, development stops after formation of the tibia and fibula (a structure called the zeugopod in fish). The message of these data is that the late expression of *shh* and *Hox* genes is an evolutionary novelty and may have produced tissues that became the first hands and feet in the history of life. Hands and feet are not homologous to any part of a lobefin's fin. They are evolutionary add-ons.

Tetrapods may have gained hands and feet because of a change in the timing and location of homeotic gene expression.

This pattern of expression of *Hox* genes and *shh*—specifically the early-to-late change—also supports a recent hypothesis about the axis of development in the tetrapod limb. For years, morphologists had assumed that the axis of development, as specified by the progress zone and the sequence of bones formed, ran straight through the hand and foot; there were long debates about exactly which elements this straight axis ran through. (This is not trivial, because identifying the axis is important to hypotheses about patterns of digit reduction among tetrapods.) But in 1986, Neil Shubin and Pere Alberch proposed that the principal axis of limb development in tetrapods is not straight, as it obviously is in *Eusthenopteron* and extant fish, but bent (Figure 17.11). In a spectacular verification of the Shubin–Alberch hypothesis, Sordino et al.'s work suggests a mechanism for the bending: the switch in the pattern of expression of *shh* and *Hox*.

What is the regulatory change that produced the developmental switch, the change in the direction of the progress zone, and the resulting foot–hand structure? We do not yet know. Research is continuing.

The Arthropod Limb

As in tetrapods, the hallmark adaptation of arthropods, and the feature most responsible for the diversification of the group, is the limb. Arthropod appendages come in two basic forms: uniramous (one branch) and biramous (two branches). Chelicerates and Uniramians have unbranched appendages, while trilobite and crustacean limbs have two elements (Figure 17.12). In addition, several crustacean

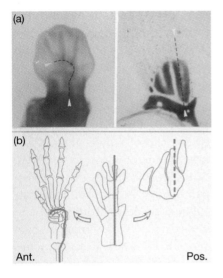

Figure 17.11 **The axis of development in the vertebrate limb** (a) These photographs show mouse (left) and fish (right) limbs at the stage of cartilage condensation. The proposed axis of development is indicated by dashed lines. (b) Adult limbs in mouse (left), *Eusthenopteron* (middle), and zebrafish (right), with the proposed axis of development indicated. (Prof. Denis Duboule, Université de Genève, Geneva, Switzerland)

groups have multiply branched or phyllopodous (leafy feet) appendages that are used for swimming. The position, form, and function of biramous, uniramous, and phyllopodous limbs can all vary enormously. Arthropod limbs may be located on the cephalic, thoracic, or abdominal segments and can be used for swimming, walking, breathing, fighting, or foraging. The ecological diversification of arthropods goes hand in hand with the elaboration of this morphological innovation.

Because their limbs are so diverse, arthropods are able to move in different ways and capture many different types of food.

What do we know about the genes involved in the evolution of arthropod appendages? Research on *Drosophila melanogaster* has established three major types of genetic control over limb formation in arthropods:

- The decision whether to make a limb depends on a gene called *wingless* (*wg*). Mutant flies that lack *wg* fail to make any limb primordia at all. *Wg* is expressed

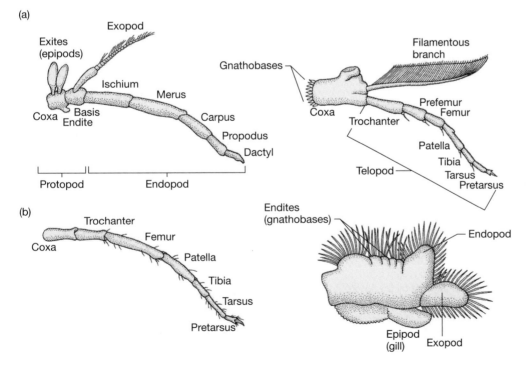

Figure 17.12 **Uniramous and biramous limbs of arthropods** This figure shows just a few of the many variations on the jointed limbs of arthropods. A major distinction among arthropod limbs is whether they are biramous (a) or uniramous (b). Classes within the phylum tend to have one type of limb or the other (except that some crustaceans have secondarily derived uniramous limbs). Modified from Brusca and Brusca (1990).

in the anterior section of limb primordia; the protein products of a locus called *engrailed (en)* show up in the posterior. These observations suggest that the two loci may be responsible for defining the anterior–posterior axis of the early limb.

- The decision to extend the limb primordium distally seems to hinge on the expression of a gene called *Distal-less (Dll)*. This is the first gene activated specifically in limb primordia.

- The decision on which type of limb will develop is controlled by homeotic *(HOM)* genes. This is not surprising, because the fate of a limb primordium should depend on its position in the embryo, and homeotics specify location.

What do these observations have to do with change in limb morphology through time, and the radiation of arthropods? There are hints, from studies on *Drosophila melanogaster,* that changes in these genes correlate with significant evolutionary events. Specifically, abnormal expression of *Dll* can lead to branched limbs in fruit flies, even though their appendages normally have just one element. This result suggests that changes in the regulation of the locus could have played a role in the evolution of uniramous, biramous, and phyllopodous appendages. Variation in when or where *Dll* is turned on or off might lead to new or modified limb outgrowths.

Grace Panganiban and colleagues (1995, 1997; see also Roush 1995) have extended our understanding of *Dll*'s role in the evolution of the arthropod limb. These researchers succeeded in making an antibody to the *Dll* protein product, and used it to stain early embryos of the common brine shrimp *Artemia franscicana* and the opossum shrimp *Mysidopsis bahia. Artemia* has phyllopodous appendages on its thorax that are used for swimming. *Mysidopsis* has biramous limbs: antennae on its head, walking legs on its thorax, and pleopods used for swimming on its abdomen. In *Artemia, Dll* is expressed in each of the outgrowths present in phyllopods (Figure 17.13). In *Mysidopsis,* the anti–*Dll* antibody localizes to each branch of the biramous appendages (Figure 17.14).

Dll seems to be regulated in different ways in different body regions of *Mysidopsis,* however. In the thorax, the two limb branches grow out from two independent clusters of cells, each of which expresses *Dll*. In the thorax, then, the expression of the gene varies spatially. But in the abdomen, the branches of pleopods form sequentially. In this case, the gene is turned on at two different times in the same location. This means that *Dll* is regulated temporally. In both

Variation in the timing or location of Dll *expression appears to correlate with variation in the shape of arthropod limbs.*

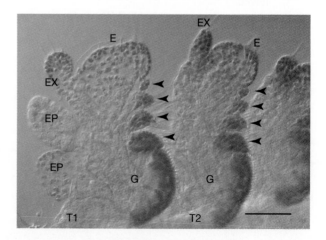

Figure 17.13 *Distal-less* **is expressed in each branch of crustacean phyllopodous limbs** The dark staining grains indicate the protein product of *Dll*. In both the first and second thoracic segments of the common brine shrimp (*Artemia franscicana*) pictured here, marked T1 and T2, all of the limb branches (indicated by the letters and arrowheads) contain *Dll* product. The scale bar represents 0.1 mm. Reproduced by permission from G. Panganiban et al., The development of crustacean limbs . . . , *Science.* © 1995 American Association for the Advancement of Science.

cases, there is a close correlation between when and where *Dll* is turned on or off and where limbs are produced. Just as it does in flies, *Dll* seems to signal "grow limb here" wherever it is expressed. Even more important, changes in when and where the gene is expressed appear to affect the branching of limbs.

We are still a long way, however, from understanding how changes in the expression of *Dll* and other genes affected the diversification of arthropod limbs. Two of the more urgent tasks now are to find out if *Distal-less* functions in similar ways in close relatives of the Arthropoda, like the groups called Onychophora, Tardigrada, and Pentastomida, and to expand the fossil record of these groups in the Vendian and Cambrian. For example, the earliest onychophorans in the Burgess Shale occur along with a diverse group of arthropods that have much more complex appendages (Gould 1989). The Tardigrada, or water bears, are tiny animals that live in moss or soil and lumber about on unbranched, clawed legs. The Pentastomida, or tongue worms, are a highly derived group of parasitic crustaceans that have tiny lobe-like legs with claws that allow them to cling to respiratory passages in their vertebrate hosts. We might predict that the earliest fossils in these forms, and in the ancestors to the arthropods, will have the same sort of unbranched, saclike legs.

Based on the research we have discussed so far, one could predict that *Distal-less* will be found in all three of these "worms with legs," but not in true worms such as annelids. In a spectacular set of experiments, Panganiban et al. (1997) both confirmed the first hypothesis and rejected the second. They used the same anti-*Dll* antibody to stain the body-wall outgrowths of a wide array of bilateral organisms. Their results are striking: *Dll* is expressed in the body wall and in the distal parts of the lobopods or limbs of the onychophoran *Peripatopsis capensis,* but also in the lateral body-wall outgrowths (called parapodia) of an annelid worm, *Chaetopterus variopedatus. Distal-less* is also expressed in the tube feet of a sea urchin and in the ampullae (attachment structures) of a tunicate. In all of these cases, *Dll* expression can be detected before these body-wall outgrowths are formed and, later, in the cells of the more distal parts of the outgrowth. Based on the type of structure that is formed, the timing of expression, and the location of gene expression, *Distal-less* seems to instruct cells to form an outgrowth with proximal–distal polarity in all of these groups. At genetic level, every type of "sticky-outey" (Nagy 1998) found in animals appears to be homologous (see Section 17.4).

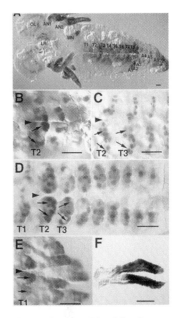

Figure 17.14 *Distal-less* **is expressed in both branches of biramous limbs in crustaceans** The dark staining grains in these photos indicate the protein product of *Dll.* The photos show how biramous limbs in the first and second thoracic segments (T1 and T2) of the opossum shrimp *Mysidopsis bahia* develop through time. The arrows highlight the way that *Dll* is localized in each branch of the biramous limb. The scale bars represent 0.1 mm. Reproduced by permission from G. Panganiban et al., The development of crustacean limbs . . . , *Science.* © 1995 American Association for the Advancement of Science.

17.3 Flowers

Our third example of "molecular paleontology" concerns the dominant land plants. The radiation of terrestrial life began in the Silurian period, when green algae first made the transition to land. There have been four major radiations of terrestrial plants since then: early groups called the Prilophyta and Rhyniophyta, ferns, early seed plants without flowers, and angiosperms. Each of these radiations was associated with one or more evolutionary innovations:

- Cuticle, a waxy covering over leaves, cut down on water loss through transpiration and allowed early land plants to tolerate drying. The evolution of guard cells and pores provided breaks in the cuticle and a mechanism for acquiring carbon dioxide from the atmosphere instead of from water.
- Vascular tissues permitted land plants to grow beyond small size, and set the stage for a radiation of ferns and other large Devonian plants (Niklas 1994).

These cells provide a mechanism for conducting water along the potential gradient created by uptake at the roots and loss at the leaves.

• Pollen and seeds are multicellular structures with a protective coat, and resist drying much better than sperm and spores. Pollen can be transported by wind while sperm are restricted to swimming; seeds can remain viable for years and contain a supply of nutrients. Gymnosperms (pines, junipers, and allies) were among the first plants with pollen and seeds, and radiated throughout the Triassic and Jurassic.

All of these features allowed plants to be less dependent on wet environments for growth and reproduction. The latest plant radiation, though, was not associated with an adaptation to dry conditions but with a new reproductive structure: the flower.

There are an estimated 200,000 to 300,000 living species of flowering plants, or Anthophyta. The angiosperms are far and away the most species-rich plant group in the history of life. The angiosperm flower consists of four concentric whorls of repeated organs. From outermost to innermost, these are the sepals, petals, stamens, and carpels (Figure 17.15). The sepals are sometimes fused together to form a calyx that encloses the developing flower bud. The petals are sometimes fused together to form a corolla. Stamens are the male organs (producing pollen), and the carpels are the female organ (in which ovules form); carpels are sometimes fused together to form a single, central pistil. All of these organs are thought to have evolved from leaf-like branching structures—a hypothesis that can be traced back to the 18th century poet and naturalist Johann Goethe (Coen 1999).

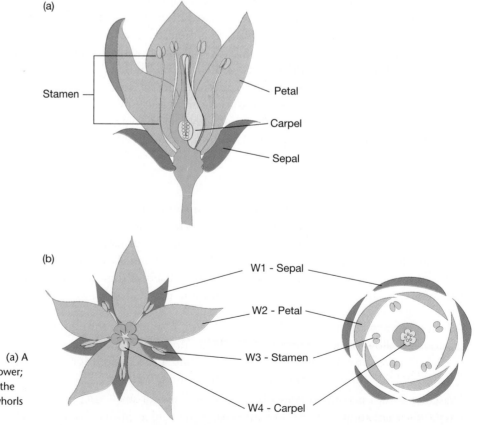

Figure 17.15 Parts of a flower (a) A long section through an idealized flower; (b) A top–down view showing how the four major organs are arranged in whorls (W1 through W4).

Floral Mutants

Understanding the genetic changes responsible for the origin of the flower is an active area of research. Not surprisingly, it turns out that flower morphology depends on the expression of homeotic genes. These loci are like the homeotic genes of animals, in two ways: They specify which organs appear in different locations, and they include a DNA-binding region called the *MADS* domain analogous to the homeodomain of *Hox* loci.

Like the discovery of the genetic basis of animal body plans and limbs, the discovery of plant homeotic genes began with the identification of homeotic mutant phenotypes with aberrant flower forms. And like the research on animal homeotic mutants, based mainly on mice and fruit flies, the research on plant homeotic mutants depended on two model organisms: the snapdragon *Antirrhinum majus* and the weedy mustard *Arabidopsis thaliana*. By exposing these plants to mutagens and examining thousands of individual plants for unusual flower phenotypes, researchers uncovered three main classes of homeotic mutations in which the normal arrangement of sepals, petals, stamens, and carpels is altered:

- Class A mutants, in which the outer whorls are replaced by sex organs, leaving (in order) carpels, stamens, stamens, and carpels (Figure 17.16a),
- Class B mutants, in which the middle two whorls are altered, leaving sepals, sepals, carpels, and carpels (Figure 17.16b), and
- Class C mutants, in which the inner two whorls are altered, leaving sepals, petals, petals, and sepals (Figure 17.16c). This is the opposite pattern to class A mutants, in which sex organs replace the outer vegetative organs.

In addition, combinations of these mutations in the same individual lead to the replacement of floral organs by leaf-like structures (Figure 17.17). This mutant appears to phenocopy the ancestral state of the flower (see Figure 17.18). These data strongly support Goethe's hypothesis that flowers are derived from whorls of leaves (reviewed in Weigel and Meyerowitz 1994). The results are also strongly reminiscent of homeotic mutants in *Drosophila* that replace one limb or segment type with another, or that prevent the differentiation of segments altogether.

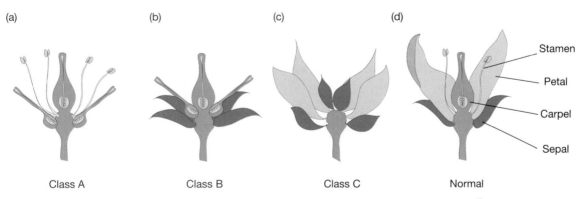

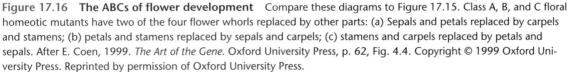

Figure 17.16 The ABCs of flower development Compare these diagrams to Figure 17.15. Class A, B, and C floral homeotic mutants have two of the four flower whorls replaced by other parts: (a) Sepals and petals replaced by carpels and stamens; (b) petals and stamens replaced by sepals and carpels; (c) stamens and carpels replaced by petals and sepals. After E. Coen, 1999. *The Art of the Gene.* Oxford University Press, p. 62, Fig. 4.4. Copyright © 1999 Oxford University Press. Reprinted by permission of Oxford University Press.

Figure 17.17 Null mutants in floral homeotic genes
(a) A wild-type flower of *Arabidopsis thaliana*. (Elliot M. Meyerowitz, California Institute of Technology) (b) In a triple mutant lacking the protein product of the *AP2, API,* and *AG* loci, sepals, petals, stamens, and ovules are replaced by leaf-like structures. (M.P. Running/Elliot M. Meyerowitz, California Institute of Technology)

Figure 17.18 The earliest flower (Leo J. Hickey, Yale University)

The expression of three loci specify the four structures found within a flower.

Floral Genes

Through careful genetic analysis, the floral homeotic mutants have been traced to a family of homeotic genes. Further, specific loci have now been assigned to each of the class A, B, and C mutants (Theissen and Saedler 1999). Three of these genes are *APETALA1* (or *AP1,* a class A gene), *APETALA3* (or *AP3,* class B), and *AGAMOUS* (or *AG,* class C). The AP1 protein is expressed early in the flower bud, and later is restricted to the two outer whorls. Failure to express *AP1* results in class A mutant phenotypes lacking sepals and petals. *AP3* is expressed later, and mainly in the middle whorls of the flower bud. Inactivation of *AP3* prevents normal development of petals plus stamens. *AG* also is expressed late, in the center of the flower bud. Inactivation of *AG* produces class C mutant phenotypes lacking sex organs.

The expression of these ABC homeotic loci depends on a master control gene called *LEAFY* (*LFY*) (Weigel and Meyerowitz 1993). The LFY protein activates *AP1, AP3,* and *AG* expression in the appropriate cells of the four whorls in the flower bud, beginning the specification of sepals, petals, stamens, and carpels.

Detlef Weigel and colleagues recently proposed a model for how all of these genes interact (Parcy et al. 1998). The model is based on several key observations. For example, *LFY* is expressed before the ABC genes, and it directly induces *AP1* expression in the middle whorls. However, *LFY* induces the class B and C homeotic genes through the action of other gene products. A gene called *UNUSUAL FLORAL ORGANS* (*UFO*) induces *AP3,* and an unknown factor X induces *AG* expression. As Figure 17.19 shows, the model suggests that *AG* expression (in the inner two whorls) directly inhibits *AP1* expression (limiting *AP1* to the outer whorls). This provides a simple mechanism for the differentiation of these two parts of the flower.

The network of gene interactions controlling flower development is complex, but a clearer picture is emerging as this book goes to press. The important conclusion is that the presence or absence of a combination of AP1, AG, and AP3 proteins in cells of the flower bud would provide enough coordinate information to instruct each cell to participate in the construction of a particular floral organ:

- AP1 without AG or AP3 induces sepals,
- AP1 with AP3 induces petals,
- AP3 with AG induces stamens, and
- AG without AP1 or AP3 induces carpels.

Like the control of animal pattern formation, this mechanism for control of flower formation could evolve through a small number of changes in the timing or lo-

(a)

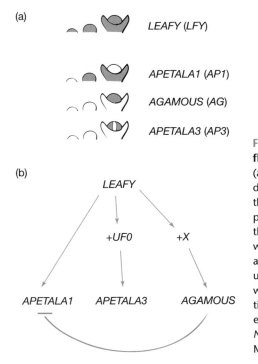

LEAFY (LFY)

APETALA1 (AP1)

AGAMOUS (AG)

APETALA3 (AP3)

(b)

LEAFY

+UF0   +X

APETALA1   APETALA3   AGAMOUS

Figure 17.19   A conceptual model of flower formation by homeotic genes
(a) The diagram shows three stages of flower development. The expression of *LEAFY (LFY)* throughout the flower bud (indicated in green) precedes the expression of the ABC genes in the outer (*AP1*), inner (*AG*), and middle (*AP3*) whorls, and induces flower formation. (b) *LFY* acts through intermediary genes (*UFO* and an unknown locus) to induce some homeotic loci, whose products themselves interact to differentiate the four whorls of the flower. From Parcy et al. (1998). Reprinted by permission from *Nature* 395: 561–568. © 1998, Macmillan Magazines Ltd.

cation of gene expression to produce relatively large changes in flower morphology (Baum 1998).

The story does not end with these genes. *LEAFY* induces the formation of a flower, so this gene is sufficient for flower development to occur. But is *LFY* expression necessary? In other words, are the loci we've been discussing general position-determining genes, similar to homeotics in animals, or are they flower-specific genes that originated at the same time angiosperms did? If we could identify such flower-specific genes, then plant molecular biologists would truly have accomplished some impressive molecular paleontology.

The answer now appears to be that *LFY* and its target genes are not *the* flower genes. Homologues of *LFY* (Mouradov et al. 1998), *AG* (Tandre et al. 1995), and other genes with *MADS*-boxes (Munster et al. 1997) have been discovered in pines, ferns, and other groups. In these species, the genes are involved in the formation of reproductive structures but not in the formation of flowers. This result alone strongly suggests that the genetic architecture for flower formation evolved well before the first appearance of the first flowering plant. A comparison of the amount of sequence divergence in these *MADS* boxes suggests that some of these genes may have first evolved their distinctive *MADS*-box sequences 450–500 Myr, around the time of the appearance of the first land plants (Purugganan 1997). Like the *HOM/Hox* genes of animals, the *MADS*-box genes of plants may have evolved early for some other function in the early land plants, and been co-opted later into the control of flower formation in the first angiosperms.

Finally, the analysis of flower homeotic polymorphisms may, like *HOM/Hox* polymorphisms, help us to understand the evolutionary forces that are currently shaping flower diversity. Some of this research is currently being done by Michael Purugganan and his colleagues. Investigators in this laboratory are pursuing two different approaches. The first involves the analysis of within-species variation in homeotic gene sequences of *Arabidopsis thaliana* (Purugganan and Suddith 1998,

1999). These studies are beginning to show how selection and genetic drift act on homeotic mutants. The second involves comparing homeotic sequences between closely related *Arabidopsis* species (Lawton-Rauh et al. 1999). These studies indicate that rates of sequence evolution vary widely among homeotic genes. Why would flower-determining genes evolve more rapidly in some species than others? It could be that pollinators select certain flower shapes created by variation in homeotic loci.

17.4. Deep Homology

The third section in this chapter pointed out that homologs of *LFY, AG,* and other flower-determining genes are found in plants that do not have flowers; Section 17.2 emphasized that *Distal-less* is found in animals that do not have limbs. It is important not to overlook just how remarkable these observations are. Before *Distal-less* expression was analyzed in a wide variety of animal groups, no one would have considered the insect leg and the tetrapod limb to be homologous structures. The leg of a grasshopper and a horse are so different in terms of their development in embryos and their adult morphology that evolutionary biologists have always maintained that they were independently derived. Arthropod limbs and tetrapod limbs have traditionally been viewed as structures that were not homologous. Yet based on molecular genetic studies, it is clear that all animal limbs—and perhaps all plant reproductive structures—are homologous on some level.

In some cases, homologous genes are involved in the formation of structures that did not appear to be homologous.

Some investigators have begun to call this level "deep homology." The idea is that at least a few of the genes responsible for prominent, but widely divergent, traits originated in a remote common ancestor. Among the animals, Douglas Erwin (1999) points out that deep homologies exist in a series of important structures and organs. For example, researchers have found strong similarities between genes in mice and fruit flies that are involved in the formation of eyes, hearts, dorsal or ventral nerve chords, and segmentation. Yet in adult mice and flies, these structures are very different.

The phenomenon of deep genetic homology is more than curiosity. If all animals share loci that are involved in the formation of eyes, hearts, nerve chords, and segments, it implies that their common ancestor also had these genes. In this way, understanding deep homology can help us reconstruct what that ancestor looked like, and perhaps how it lived—even though it is not present in the fossil record. In the case of bilaterally symmetric animals, deep homologies suggest that their common ancestor had the following traits (Arthur et al. 1999; Knoll and Carroll 1999):

- Serial repetition of some body parts without complex differentiation of segments,
- A simple type of appendage, but not a complex limb,
- Clusters of simple photoreceptors, but not an elaborate image-forming eye,
- A contractile blood vessel rather than a heart, and
- Condensations of nerve cell bodies into nerve cords, but not a brain.

In short, research into evolutionary aspects of development makes strong predictions for paleontologists to test. Results from the laboratory suggest what we should look for in the rocks. In the case of the bilateral animal ancestor, we know that the animal must have lived before the earliest part of the Cambrian. The deep homology underlying many organ systems in its descendants provides a reasonable sketch of its morphological features. Given these predictions for when the ancestor lived and what it looked like, paleontologists can test the hypothesis of deep homology. If it is correct, fossils resembling the bilateral animal ancestor should soon appear.

Summary

The diversity of animal and plant forms can largely be traced to evolutionary diversity in the number and interactions of a small number of genes that control pattern formation in the embryo. This control is administered by the actions of transcription factors encoded by loci in homeotic gene complexes. Changes in the number of genes in the homeotic complexes of animals may have been associated with the evolution of body-plan diversity and segmentation. Changes in the timing and spatial expression of loci in the *Hox* complexes and other genes were important in the diversity of segmented body plans and in the elaboration of the tetrapod and arthropod limbs.

Comparable evidence suggests that the diversification of angiosperms may also be associated with the origin of homeotic loci. Several homeotic genes that specify flower development and that control the identity of serially repeated flower parts have been identified in plants. The null mutants and interactions of these loci are analogous to those of *Hox* genes of animals. The genetic similarities that underlie otherwise dissimilar structures (such as arthropod and vertebrate limbs) suggests a kind of deep homology of these structures, and leads to a prediction about the kinds of structures that should be found in the common ancestor from which these genetic control networks were inherited.

Questions

1. What did the ancestor of the bilateral animals look like? Was it segmented? Did it have limbs, eyes, and a heart? In addition to the identification of homologous genes involved in the specification of these structures in arthropods, annelids, flatworms, vertebrates, echinoderms, and other distantly related groups, what other information from development, morphology, or the fossil record would you want in order to develop a more informed hypothesis about the common ancestor of the Bilateria?

2. Why do embryos tend to use diffusion gradients of at least 2 molecules to specify location of developing cells? Why isn't one diffusion gradient sufficient?

3. Many people have the impression that if foreign DNA is incorporated into the cells of an adult, the DNA will immediately begin to remodel the morphology of the host's body, to the point of growing new organs or new limbs—e.g., an adult human growing wings if given DNA from a bat. Why can't this happen? When are developmental genes normally active—and what size and shape is the body when they are active?

4. Should we expect plant homeotic genes that control flower formation to be organized in complexes on a chromosome in the same way that animal *Hox* genes are? Answering this question would require a hypothesis about the significance of spatial colinearity of *Hox* genes. Can you imagine such a hypothesis?

5. Suppose you are studying the development of monkeys, and you discover that monkeys have two (hypothetical) *Hox* genes that encode upper vs. lower trunk location. *Hox* gene *A* is expressed in the chest and shoulders, and *Hox* gene *B* in the abdomen and hips. Suppose you also discover that in monkeys, a gene you call "little" is active in the posterior limb buds of all limbs, and is necessary for forming little fingers and little toes (rather than thumbs and big toes).

   Draw a sketch of what do you think the limbs would look like if *Hox* genes *A* and *B* were switched with each other in a zygote. Would the resulting limbs have hands or feet, and where would the big toes, little toes, thumbs, or little fingers be? How about if *little* were knocked out? Explain your reasoning. (There are several possibilities.)

6. Now suppose you were hired by an eccentric millionaire who wanted to breed winged monkeys. Would it be sufficient to remove the (hypothetical) *Hox A* gene from the monkey genome and replace it with a bird's (hypothetical) *Hox A* gene? If you think not, describe what other genetic manipulations would have to be done, or what further information you would need, to breed monkeys that have wings. (If you think it will never be genetically possible to do this, explain why.)

7. As we have seen, vertebrates actually have a battery of 4 *Hox* gene clusters that are thought to have originated from duplication of one ancestral *Hox* gene cluster. Develop a hypothesis for what would happen in a mouse embryo if a gene from one *Hox* cluster were swapped with its sibling gene from another cluster. Assume that regulatory sequences are left in place and only the coding sequence of each gene is moved. Then look up the following paper and see if you were right.

   Greer, Joy M., John Puetz, Kirk R. Thomas, and Mario R. Capecchi. 2000. Maintenance of functional equivalence during paralogous *Hox* gene evolution. *Nature* 403: 661–665.

Exploring the Literature

8. The field of evolutionary developmental biology has grown dramatically in the last 10 years. For an introduction to this burgeoning literature, see Gerhart and Kirschner (1997), Gilbert (1997), or any of these other recent texts:

Coen, E. 1999. *The Art of Genes: How Organisms Make Themselves.* Oxford: Oxford University Press.

Gilbert, S. F., and A. M. Raunio. 1997. *Embryology: Constructing the Organism* Sunderland: Sinauer Associates.

Arthur, W. 1997. *The Origin of Animal Body Plans: A Study in Evolutionary Developmental Biology.* Cambridge: Cambridge University Press.

Hall, B. K. 1998. *Evolutionary Developmental Biology,* 2nd ed. London: Chapman and Hall.

Or read several editorial reviews of the field in the first few issues of the new journal *Development and Evolution.*

Malden, MA: Blackwell Science, which began publication in July 1999.

9. Devonian tetrapods had different numbers of digits on their fore- and hind limbs. Most later tetrapods have five digits on both limbs. To investigate the evolution of the pentadactyl limb, see

Coates, M. I., and J. A. Clack. 1990. Polydactyly in the earliest known tetrapod limbs. *Nature* 347: 66–69.

Gould, S. J. 1990. *Eight Little Piggies.* New York: W. W. Norton, Essay 4.

10. For a comprehensive introduction to "molecular paleontology", and the significance of this research for evolutionary analysis, read and see the citations in Knoll and Carroll (1999).

Citations

Ahlberg, P. E. 1995. *Elginerpeton panchei* and the earliest tetrapod clade. *Nature* 373: 420–425.

Ahlberg, P. E., and A. R. Milner. 1994. The origin and early diversification of tetrapods. *Nature* 368: 507–518.

Arthur, W., T. Jowett, and A. Panchen. 1999. Segments, limbs, homology, and co-option. *Evolution and Development* 1: 74–76.

Balavoine, G., and M. J. Telford. 1995. Identification of planarian homeobox sequences indicates the antiquity of most *Hox*/homeotic gene subclasses. *Proceedings of the National Academy of Sciences, USA* 92: 7227–7231.

Baum, D. A. 1998. The evolution of plant development. *Current Opinion in Plant Biology* 1: 79–86.

Brusca, R. C., and G. J. Brusca. 1990. *Invertebrates.* Sunderland, MA: Sinauer.

Budd, G. E. 1996. Progress and problems in arthropod phylogeny. *Trends in Ecology and Evolution* 11: 356–358.

Cameron, R. A., K. J. Peterson, and E. H. Davidson. 1998. Developmental gene regulation and the evolution of large animal body plans. *American Zoologist* 38: 609–620.

Carroll, R. L. 1988. *Vertebrate Paleontology and Evolution.* New York: W.H. Freeman.

Carroll, S. B. 1995. Homeotic genes and the evolution of the chordates. *Nature* 376: 479–485.

Coen, E. 1999. *The Art of the Gene.* Oxford: Oxford University Press.

Cohn, M. J., J. C. Izpisúa-Belmonte, H. Abud, J. K. Heath, and C. Tickle. 1995. Fibroblast growth factors induce additional limb development from the flank of chick embryos. *Cell* 80: 739–746.

De Rosa, R., J. K. Grenier, T. Andreeva, C. E. Cook, A. Adoutte, M. Akam, S. B. Carroll, and G. Balavoine. 1999. *Hox* genes in brachiopods and priapulids and protostome evolution. *Nature* 399: 772–776.

DiSilvestro, R. L. 1997. Out on a limb. *BioScience* 47: 729–731.

Duboule, D. 1994. How to make a limb. *Science* 266: 575–576.

Erwin, D. 1999. The origin of body plans. *American Zoologist* 39: 617–629.

Erwin, D., J. Valentine, and D. Jablonski. 1997. The origin of animal body plans. *American Scientist* 85: 126–137.

Fallon, J. F., A. Lopez, M. A. Ros, M. P. Savage, B. B. Olwin, and B. K. Simandi. 1995. *FGF-2*: Apical ectodermal ridge growth signal for chick limb development. *Science* 264: 104–106.

Finnerty, J. R., and M. Q. Martindale. 1997. Homeoboxes in sea anemones (Cnidaria; Anthozoa): A PCR-based survey *of Nematostella vectensis* and *Metridium senile. Biological Bulletin* 193: 62–76.

Finnerty, J. R., and M. Q. Martindale. 1999. Ancient origins of axial patterning genes: *Hox* genes and *ParaHox* genes in the Cnidaria. *Evolution and Development* 1: 16–23.

Gerhart, J., and M. Kirschner. 1997. *Cells, Embryos, and Evolution.* Malden, MA: Blackwell Science.

Gilbert, S. F. 1997. *Developmental Biology,* 5th ed. Sunderland, MA: Sinauer.

Gould, S. J. 1989. *Wonderful Life.* New York: W. W. Norton.

Grenier, J. K., T. L. Garber, R. Warren, P. M. Whitington, and S. B. Carroll. 1997. Evolution of the entire arthropod *Hox* gene set predated the origin and radiation of the onychophoran/arthropod clade. *Current Biology* 7: 547–553.

Hinchliffe, J. R. 1989. Reconstructing the archetype: Innovation and conservatism in the evolution and development of the pentadactyl limb. In D. B. Wake and G. Roth, eds., *Complex Organismal Functions: Integration and Evolution in Vertebrates.* Chichester, England: Wiley & Sons, 171–189.

Jarvik, E. 1980. *Basic Structure and Evolution of Vertebrates,* vol. 2. London: Academic Press.

Kenyon, C., and B. Wang. 1991. A cluster of Antennapedia-class homeobox genes in a nonsegmented animal. *Science* 253: 516–517.

Knoll, A. H., and S. B. Carroll. 1999. Early animal evolution: Emerging views from comparative biology and geology. *Science* 284: 2129–2137.

Krumlauf, R. 1994. *Hox* genes in vertebrate development. *Cell* 78: 191–201.

Kruse, M., A. Mikoc, H. Cetkovic, V. Gamulin, B. Rinkevich, I. M. Müller, and W. E. G. Müller. 1994. Molecular evidence for the presence of a developmental gene in the lowest animals: Identification of a homeobox-like gene in the marine sponge *Geodia cydonium. Mechanisms of Aging and Development* 77: 43–54.

Lawton-Rauh, A. L., E. S. Buckler, IV, and M. D. Purugganan. 1999. Patterns of molecular evolution among paralogous floral homeotic genes. *Molecular Biology and Evolution* 16: 1037–1045.

Lewis, E. B. 1978. A gene complex controlling segmentation in *Drosophila. Nature* 276: 565–570.

McGinnis, W., and M. Kuziora. 1994. The molecular architects of body design. *Scientific American* 270: 58–66.

McGinnis, W., M. S. Levine, E. Hafen, A. Kuroiwa, and W. J. Gehring. 1984a. A conserved DNA sequence in homoeotic genes of the *Drosophila* Antennapedia and bithorax complexes. *Nature* 308: 428–433.

McGinnis, W., R. L. Garber, J. Wirz, A. Kuroiwa, and W. J. Gehring. 1984b. A homologous protein-coding sequence in *Drosophila* homeotic genes and its conservation in other metazoans. *Cell* 37: 403–408.

Mouradov, A., T. Glassick, B. Hamdorf, L. Murphy, B. Fowler, S. Maria, and R. D. Teasdale. 1998. *NEEDLY,* a *Pinus radiata* ortholog of *FLORICAULA/LEAFY* genes, expressed in both reproductive and vegetative meristems. *Proceedings of the National Academy of Sciences USA* 95: 6537–6542.

Munster, T., J. Pahnke, A. DiRosa, J. T. Kim, W. Martin, H. Saedler, and G. Theissen. 1997. Floral homeotic genes were recruited from homologous *MADS*-box genes preexisting in the common ancestor of ferns and seed plants. *Proceedings of the National Academy of Sciences, USA* 94: 2415–2420.

Nagy, L. 1998. Changing patterns of gene regulation in the evolution of arthropod morphology. *American Naturalist* 38: 818–828.

Niklas, K. J. 1994. One giant step for life. *Natural History* 6: 22–25.

Nüsslein-Volhard, C., and E. Wieschaus. 1980. Mutations affecting segment number and polarity in *Drosophila. Nature* 287: 795–801.

Nüsslein-Volhard, C., H. G. Fröhnhofer, and R. Lehmann. 1987. Determination of anterioposterior polarity in *Drosophila. Science* 238: 1675–1681.

Panganiban, G., A. Sebring, L. Nagy, and S. Carroll. 1995. The development of crustacean limbs and the evolution of arthropods. *Science* 270: 1363–1366.

Panganiban, G., et al. 1997. The origin and evolution of animal appendages. *Proceedings of the National Academy of Sciences, USA* 94: 5162–5166.

Parcy, F., O. Nilsson, M. A. Busch, I. Lee, and D. Weigel. 1998. A genetic framework for floral patterning. *Nature* 395: 561–566.

Parr, B. A., and A. P. McMahon. 1995. Dorsalizing signal *Wnt-7a* required for normal polarity of D–V and A–P axes of mouse limb. *Nature* 374: 350–353.

Purugganan, M. D. 1997. The *MADS*-box floral homeotic gene lineages predate the origin of seed plants: Phylogenetic and molecular clock estimates. *Journal of Molecular Evolution* 45: 392–396.

Purugganan, M. D., and J. I. Suddith. 1998. Molecular population genetics of the *Arabidopsis CAULIFLOWER* regulatory gene: Nonneutral evolution and naturally occurring variation in floral homeotic function. *Proceedings of the National Academy of Sciences USA* 95: 8130–8134.

Purugganan, M. D., and J. I. Suddith. 1999. Molecular population genetics of floral homeotic loci: Departures from the equilibrium-neutral model at the *APETALA3* and *PISTILLATA* genes of *Arabidopsis thaliana. Genetics* 151: 839–848.

Raff, R. A., and T. C. Kaufmann. 1983. *Embryos, Genes, and Evolution.* New York: MacMillan.

Riddle, R. D., R. L. Johnson, E. Laufer, and C. Tabin. 1993. *Sonic hedgehog* mediates the polarizing activity of the ZPA. *Cell* 75: 1401–1416.

Roush, W. 1995. Gene ties arthropods together. *Science* 270: 1297.

Seimiya, M., H. Ishiguro, K. Miura, Y. Watanabe, and Y. Kurosawa. 1994. Homeobox-containing genes in the most primitive metazoa, the sponges. *European Journal of Biochemistry* 221: 219–225.

Shubin, N. H., and P. Alberch. 1986. A morphogenetic approach to the origin and basic organization of the tetrapod limb. In M. K. Hecht, B. Wallace, and G. T. Prance eds., *Evolutionary Biology,* vol. 20. New York: Plenum Press, 319–387.

Sordino, P., and D. Duboule. 1996. A molecular approach to the evolution of vertebrate paired appendages. *Trends in Ecology and Evolution* 11: 114–119.

Sordino, P., F. van der Hoeven, and D. Duboule. 1995. *Hox* gene expression in teleost fins and the origin of vertebrate digits. *Nature* 375: 678–681.

Tabin, C. 1995. The initiation of the limb bud: growth factors, *Hox* genes, and retinoids. *Cell* 80: 671–674.

Tabin, C. J., S. B. Carroll, and G. Panganiban. 1999. Out on a limb: Parallels in vertebrate and invertebrate limb patterning and the origin of appendages. *American Zoologist* 39: 650–663.

Tandre, K., V. A. Albert, A. Sundas, and P. Engstrom. 1995. Conifer homologs to genes that control floral development in angiosperms. *Plant Molecular Biology* 27: 69–78.

Theissen, G., and H. Saedler. 1999. The golden decade of molecular floral development (1990–1999): A cheerful obituary. *Developmental Genetics* 25: 181–193.

Valentine, J. W., D. H. Erwin, and D. Jablonski. 1996. Developmental evolution of metazoan body plans: The fossil evidence. *Developmental Biology* 173: 373–381.

Weigel, D., and E. M. Meyerowitz. 1993. Activation of floral homeotic genes in *Arabidopsis. Science* 261: 1723–1726.

Weigel, D., and E. M. Meyerowitz. 1994. The ABC's of floral homeotic genes. *Cell* 78: 203–209.

Yang, Y., and L. Niswander. 1995. Interaction between the signaling molecules *WNT7a* and *SHH* during vertebrate limb development: dorsal signals regulate anteroposterior patterning. *Cell* 80: 939–947.

CHAPTER 18

Molecular Evolution

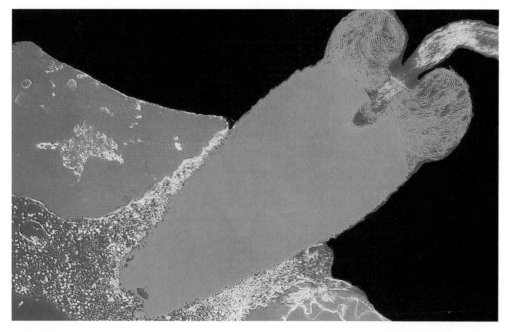

A sea urchin sperm penetrating the egg membrane. The proteins involved in the initial contact between sperm and egg evolve rapidly. (Dr. Everett Anderson/Science Photo Library/Photo Researchers, Inc.)

T HE ADVENT OF TECHNIQUES FOR STUDYING PROTEIN AND DNA SEQUENCES opened up a new arena for the study of evolution. Beginning with the first large-scale amino acid sequencing studies in the 1960s and continuing to the current explosion of data on the sequences of entire genomes, researchers have addressed a wide variety of questions about how proteins and nucleic acids change through time.

Several important issues in molecular evolution were introduced in earlier chapters. In Chapter 4, we explored the origin of genetic variation by analyzing how new alleles arise by point mutation and investigating how new loci are created by gene duplication, polyploidy, and other events. The observation of steady rates of change in certain proteins and genes, noted in Chapter 13, led to the development, testing, and implementation of the molecular-clock hypothesis.

The goal of this chapter is to explore two additional questions addressed by current research on molecular evolution. The first is how the evolutionary processes of drift and selection act at the molecular level. If we compare homologous protein or DNA sequences in two organisms and observe differences, how can we discern whether the changes resulted from drift or natural selection? Sections 18.1

and 18.2 are devoted to this topic. Then we turn to the larger scale issue of how whole genomes evolve. Section 18.3 explains that genomes are not cohesive units in which all loci cooperate to maximize the fitness of the individual. Most genomes are infected with parasitic sequences that detract from the fitness of the host. What are these parasites, and what limits their spread? The chapter's final section discusses the origin and history of the chloroplast and mitochondrial genomes. Once free-living bacteria were acquired by eukaryotes via endosymbiosis (see Chapter 14), what happened to their genes? Addressing these questions will reinforce the chapter's overall objective: introducing the data, theory, and analyses that biologists are using to study evolution at the molecular level.

18.1 The Amount and Rate of Sequence Change

The field of molecular evolution was launched in the mid-1960s, when biochemists succeeded in determining the amino acid sequences of hemoglobin, cytochrome *c*, and other particularly abundant and well-studied proteins found in humans and other vertebrates. These data sets provided the first opportunity for evolutionary biologists to compare the amount and rate of molecular change among species.

Early workers in the field made several striking observations about these data sets. Foremost among these were calculations made by Motoo Kimura (1968). Kimura showed that, if the number of sequence differences observed in the well-studied proteins of humans and horses were scaled for time, using divergence dates from the fossil record, and if these rates of molecular evolution were then extrapolated to all of the protein-coding loci in the genome, it meant that a mutation leading to an amino acid substitution had increased to fixation every two years, on average, as vertebrates diverged. This rate seemed far too high to be due to natural selection, given that most mutations should be deleterious. Beneficial mutations fixed by natural selection should be extremely rare.

A second observation, contributed by Emil Zuckerkandl and Linus Pauling (1965), was that the rate of amino acid sequence change in certain proteins appeared to be constant through time, or clocklike, during the diversification of vertebrates. This result also seemed to be inconsistent with the action of natural selection, which should be episodic in nature and correlated with changes in the environment rather than correlated with of time.

Early analyses of molecular evolution suggested that rates of change were high and were constant through time.

In short, early data on molecular evolution did not gibe with the expectation that natural selection was responsible for most evolutionary change. The results raised an important question: If natural selection does not explain evolution at the molecular level, then what process *is* responsible for the observation of rapid, clocklike change?

The Neutral Theory of Molecular Evolution

Kimura (1968, 1983) formulated the Neutral Theory of Molecular Evolution to explain the observed patterns of amino acid sequence divergence. This theory claims that the vast majority of base substitutions that become fixed in populations are neutral with respect to fitness and that genetic drift dominates evolution at the level of the DNA sequence. Kimura held that natural selection on beneficial mutations is largely inconsequential as an explanation for the differences among species observed at the molecular level.

Kimura modeled the evolution of neutral mutations as follows:

- If there are N individuals in a diploid population, then there are $2N$ copies of each gene present in that population.
- All of the $2N$ copies present in the current population are descendants of a single allele that existed at some time in the past. Conversely, of the $2N$ copies currently in existence, only one will become the ancestor of all copies present in the population at some point in the future.
- If all $2N$ copies of the gene are selectively equivalent, or neutral in their effect on the bearer's fitness, then each has an equal chance of becoming the allele that becomes fixed in the population. This chance is equal to $1/2N$.
- In every generation, mutation will introduce new neutral alleles to the population. If v is the mutation rate per gene per successful gamete, then $2Nv$ new mutants will be introduced into the population each generation.
- Based on the preceding statements, the rate at which new neutral mutants become fixed in the population is equal to $(2Nv)(1/2N)$, or simply v.

This derivation means that the rate of sequence evolution, if all of the alleles are neutral, is simply equal to the mutation rate. This is an elegant result. It is also astonishing, for two reasons:

The neutral theory models the fate of alleles whose selection coefficient is zero.

1. Positive natural selection is excluded. The theory's central claim is that the vast majority of base substitutions are neutral. The overall rate of evolution will be equal to the neutral mutation rate only when this proposition is true. Proponents of the neutral theory go further, however, by pointing out that even if a small proportion of nonneutral mutations occur in a population, they are likely to be deleterious and rapidly eliminated by natural selection. Thus, v will represent the maximum rate of evolutionary change measured.
2. The size of the population has no role. The models and experiments reviewed in Chapter 6 showed that genetic drift is far more effective at changing the frequency of alleles in small populations than in large ones. But Kimura's result shows that, for strictly neutral mutations, the rate of fixation of novel alleles due to drift does not depend on population size.

Although Kimura's theory appeared to explain why the amino acid sequences of hemoglobin, cytochrome *c,* and other proteins change steadily through time, the theory was inspired by fairly limited amounts of protein sequence data. How did the neutral theory hold up, once large volumes of DNA sequence data became available?

Patterns in DNA Sequence Divergence

During the late 1970s and 1980s researchers mined growing databases of DNA sequences to analyze the amounts and rates of change in different loci. To discern meaningful patterns in the data, it became routine to create categories defined by the type of sequence being considered. The most basic distinction was between coding and noncoding sequences. Coding sequences contain the instructions for tRNAs, rRNAs, or proteins; noncoding sequences include introns, regions that flank coding regions, regulatory sites, and the unusual loci called pseudogenes that were introduced in Chapter 4. What predictions does the neutral theory make about the rate and pattern of change in these different types of sequences? Have these predictions been verified or rejected?

Pseudogenes Establish a Canonical Rate of Neutral Evolution

Pseudogenes are functionless stretches of DNA that result from gene duplication events (see Chapter 4). Because they do not code for proteins, mutations in pseudogenes should be completely neutral with respect to fitness and should increase to fixation solely in response to genetic drift. For this reason, pseudogenes are considered a paradigm of neutral evolution (Li et al. 1981). As predicted by the neutral theory, the divergence rates recorded in pseudogenes—which should be equal to the neutral mutation rate ν—are among the highest observed among loci and sites in nuclear genomes (Li et al. 1981, Li and Graur 1991). This finding provided strong support for the neutral theory as an explanation for evolutionary change at the molecular level. It also quantified the rate of evolution due to drift. How do rates of change in other types of sequences compare to this standard, or canonical, rate of evolution caused by drift?

The evolution of pseudogenes conforms to the assumptions and predictions of the neutral theory.

Silent Sites Change Faster than Replacement Sites in Most Coding Loci

Chapter 4 pointed out that two basic types of point mutations occur in the coding regions of genes. To review, recall that bases in DNA are read in groups of three, called codons, and that only 20 different amino acids need to be specified by the 64 codons in the genetic code. As a result, base-pair substitutions may or may not lead to amino acid sequence changes. (To capture this point, biologists say that the genetic code is redundant.) DNA sequence changes that do not result in amino acid changes are called **silent-site (or synonymous) substitutions**; sequence changes that do result in an amino-acid change are called **replacement (or nonsynonymous) substitutions**.

Table 18.1 presents data on the rate of replacement versus silent substitution, based on comparisons of 28 homologous loci from humans and either mice or rats. The important point in these data is that, in every locus surveyed, the rate of silent changes is far higher than the rate of replacement changes.

In most coding sequences, substitution rates are higher at silent sites than at replacement sites.

This pattern accords with the neutral theory in important ways. Silent changes are not exposed to natural selection on protein function because they do not alter the amino acid sequence. As a result, silent substitutions should increase or decrease in frequency through time largely as a result of drift. Replacement substitutions, in contrast, change the amino-acid sequences of proteins. If most of these alterations are deleterious, then they should be eliminated by natural selection. (This type of natural selection is called **negative** or **purifying selection**, as opposed to positive selection on beneficial mutations.) Less frequently, replacement substitutions occur that have no effect on protein function and may be fixed by drift. Because these observations are consistent with the patterns predicted if drift dominates molecular evolution, they support the central tenet of the neutral theory.

Natural selection against deleterious mutations is called negative selection. Natural selection favoring beneficial mutations is called positive selection.

Variation Among Loci: Evidence for Functional Constraints

The data in Table 18.1 contain another important pattern. When homologous coding sequences from humans and rodents are compared, some loci are found to be nearly identical, while others have undergone rapid divergence. This result turns out to be typical. Rates of molecular evolution vary widely among loci.

Table 18.1 **Rates of nucleotide substitution vary among genes and among sites within genes**

These data report rates of replacement and silent substitutions in a series of protein-coding genes compared between humans and either mice or rats. The data are expressed as the average number of substitutions per site per billion years, plus or minus a statistical measure of uncertainty called the standard error.

Gene	L	Replacement rate ($\times 10^9$)	Silent rate ($\times 10^9$)
Histones			
Histone 3	135	0.00 ± 0.00	6.38 ± 1.19
Histone 4	101	0.00 ± 0.00	6.12 ± 1.32
Contractile system proteins			
Actin *a*	376	0.01 ± 0.01	3.68 ± 0.43
Actin *b*	349	0.03 ± 0.02	3.13 ± 0.39
Hormones, neuropeptides, and other active peptides			
Somatostatin-28	28	0.00 ± 0.00	3.97 ± 2.66
Insulin	51	0.13 ± 0.13	4.02 ± 2.29
Thyrotropin	118	0.33 ± 0.08	4.66 ± 1.12
Insulin-like growth factor II	179	0.52 ± 0.09	2.32 ± 0.40
Erythropoietin	191	0.72 ± 0.11	4.34 ± 0.65
Insulin C-peptide	35	0.91 ± 0.30	6.77 ± 3.49
Parathyroid hormone	90	0.94 ± 0.18	4.18 ± 0.98
Luteinizing hormone	141	1.02 ± 0.16	3.29 ± 0.60
Growth hormone	189	1.23 ± 0.15	4.95 ± 0.77
Urokinase-plasminogen activator	435	1.28 ± 0.10	3.92 ± 0.44
Interleukin I	265	1.42 ± 0.14	4.60 ± 0.65
Relaxin	54	2.51 ± 0.37	7.49 ± 6.10
Hemoglobins and myoglobin			
α-globin	141	0.55 ± 0.11	5.14 ± 0.90
Myoglobin	153	0.56 ± 0.10	4.44 ± 0.82
β-globin	144	0.80 ± 0.13	3.05 ± 0.56
Apolipoproteins			
E	283	0.98 ± 0.10	4.04 ± 0.53
A-I	243	1.57 ± 0.16	4.47 ± 0.66
A-IV	371	1.58 ± 0.12	4.15 ± 0.47
Immunoglobulins			
Ig V_H	100	1.07 ± 0.19	5.66 ± 1.36
Ig γ1	321	1.46 ± 0.13	5.11 ± 0.64
Ig κ	106	1.87 ± 0.26	5.90 ± 1.27
Interferons			
α1	166	1.41 ± 0.13	3.53 ± 0.61
β1	159	2.21 ± 0.24	5.88 ± 1.08
γ	136	2.79 ± 0.31	8.59 ± 2.56

Source: Li and Graur (1991)

The key to explaining this pattern lies in the following observation: Genes that are responsible for the most vital cellular functions appear to have the lowest rates of replacement substitutions. Histone proteins, for example, interact with DNA to form structures called nucleosomes. These protein–DNA complexes are a major feature of the chromatin fibers in eukaryotic cells. Changes in the amino acid sequences of histones disrupt the structural integrity of the nucleosome, with negative consequences for DNA transcription and synthesis. In contrast, genes that are less vital to the cell, and thus under less stringent functional constraints, show more rapid rates of replacement substitutions. When functional constraints are lower, a larger percentage of replacement substitutions are neutral with respect to fitness and may be fixed by drift.

The Nearly Neutral Model

Although the neutral theory appeared to account for several important patterns in DNA sequence data, data sets that indicated clocklike change in proteins presented a problem. The issue was that the neutral mutation rate v should vary among species as a function of generation time. Over any given time interval, more neutral mutations should occur in species with short generation times than in species with long generation times. Contrary to expectation, at least some protein sequence comparisons appeared to undergo clocklike change in absolute time—independent of differences in generation time among the species being compared.

To account for this observation, Tomoko Ohta and Motoo Kimura (1971; Ohta 1972, 1977) developed mathematical models exploring how drift and selection would effect mutations that are slightly deleterious, instead of being strictly neutral. Ohta's work showed that mutations are effectively neutral—meaning that they are fixed or eliminated by drift instead of selection—when $s \leq 1/(2N_e)$, where s is the selection coefficient introduced in Chapter 5 and N_e is the effective population size (meaning, the number of breeding adults).

How does this nearly neutral model explain the observation of molecular clocks in absolute time? As Lin Chao and David Carr (1993) have shown, there is a strong negative correlation between average population size in a species and its generation time. Species with short generation times tend to have large populations; species with long generation times tend to have small populations (Figure 18.1a). This is important because, according to Ohta's model, drift fixes a larger percentage of mutations in organisms with small population sizes. The upshot is that an increase in evolutionary rate due to the fixation of nearly neutral mutations in these small-population, long-generation species offsets the higher mutation rate in short-generation species and results in the molecular clock (Figure 18.1b). Consistent with this view, most studies have shown that replacement substitutions show relatively small differences in rates among mammalian lineages with different generation times (see Li and Tanimura 1987; Li et al. 1987). As predicted by the neutral theory, silent substitutions in mammals show much more pronounced generation-time effects.

Since its inception, however, the neutral and nearly neutral theories have been controversial (see Berry 1996; Ohta and Kreitman 1996). Discussion has focused on the claims by Kimura (1983) and King and Jukes (1969) that the number of beneficial mutations fixed by positive natural selection is inconsequential compared to the number of mutations that change in frequency under the influence of drift.

The nearly neutral model explains why, in some cases, rates of sequence change correlate with absolute time instead of generation time.

The neutralist controversy is a debate about the relative importance of selection and drift in explaining molecular evolution.

(a)

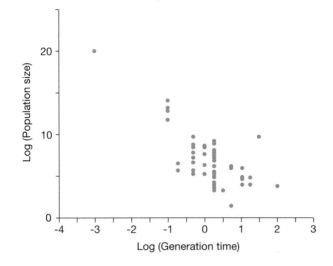

(b) Why are rates of replacement substitution constant in species with long and short generation time?

Note: Mutations are effectively neutral when $s \le \dfrac{1}{2N_e}$

Short generation time

- *Many* mutations on a per-year basis

- *Few* mutations are effectively neutral (because N_e is large)

Long generation time

- *Few* mutations on a per-year basis

- *Many* mutations are effectively neutral (because N_e is small)

Result: The differences between mutation rate and the frequency of nearly neutral mutations cancel out

Figure 18.1   Generation time, population size, and nearly neutral mutations
(a) This graph plots the logarithm of population size versus the logarithm of generation time. Statistical tests confirm that there is a strong negative correlation between the two variables. From L. Chao and D. E. Carr, 1993. The molecular clock and the relationship between population size and generation time. *Evolution* 47: 688–690, Fig. 1. Copyright © 1993 Evolution. Reprinted by permission of Evolution. (b) Differences in generation time and population size can lead to clocklike change in replacement substitutions that are nearly neutral with respect to fitness.

Is this claim accurate? How can we determine that natural selection has been responsible for changes observed at the molecular level?

18.2 Detecting Natural Selection on DNA Sequences

When researchers compare homologous DNA sequences among individuals and want to explain the differences they observe, they routinely use the neutral theory as a null hypothesis. The neutral theory specifies the rates and patterns of sequence change that occur in the absence of natural selection. If the changes that are actually observed are significantly different from the predictions made by the neutral theory, and if a researcher can defend the proposition that the sequences in question have functional significance for the organism, then there is convincing evidence that natural selection has caused molecular evolution.

Here we examine a few of the strategies that are being used to detect molecular evolution due to natural selection. We begin with studies of replacement changes, then explore evidence that many silent-site substitutions are also under selection.

Positive Selection on Replacement Substitutions

When mammalian cells are infected by a bacterium or a virus, they respond by displaying pieces of bacterial or viral protein on their surfaces. Immune system cells react by killing the infected cell. (This prevents the bacterium or virus inside the cell from replicating and slows the rate of infection.) The membrane proteins that display bacterial and viral proteins are encoded by a cluster of genes called the major histocompatibility complex, or MHC. The part of an MHC protein that binds to the foreign peptide is called the antigen recognition site (ARS). Austin Hughes and Masatoshi Nei (1988) set out to test the neutral theory by studying sequence changes in the ARS of MHC loci in humans and mice.

Hughes and Nei's specific goal was to compare the number of base substitutions that have occurred in silent versus replacement sites within the ARS. According to the neutral theory, silent-site substitutions should always be more common than replacement substitutions. This prediction follows from the proposition that most replacement substitutions are deleterious and thus eliminated by negative selection. More precisely, the rate of silent site substitutions should approximate ν and represent the highest rate of molecular evolution possible in a coding sequence. The data shown in Table 18.1 were consistent with this pattern. But is it true for the ARS of MHC genes in mice and humans?

The answer is no. When Hughes and Nei compared alleles from the MHC complexes of 12 different humans and counted the number of differences observed in silent versus replacement sites, they found significantly more replacement site than silent site changes. The same pattern occurred in the ARS of mouse MHC genes, although the differences were not as great. This pattern could only result if the replacement changes were selectively advantageous. The logic here is that positive selection causes replacement changes to spread through the population much more quickly than neutral alleles can spread by chance.

It is important to note, however, that Hughes and Nei found this pattern only in the ARS. Other exons within the MHC showed more silent than replacement changes or no difference. At sites other than the ARS, then, they could not rule out the null hypothesis that sequence change is dominated by drift.

Comparing Silent and Replacement Changes Within and Between Species

The work by Hughes and Nei provides a clear example of gene segments where neutral substitutions do not predominate. Subsequent to their study, many other loci have been found where replacement substitutions outnumber silent substitutions.

Even though the Hughes and Nei criterion for detecting positive selection has been useful, Paul Sharp (1997) points out that it is extremely conservative. Replacement substitutions will only outnumber silent substitutions when positive selection has been very strong. In a comparison of 363 homologous loci in mice and rats, for example, only one showed an excess of replacement over silent changes. But as Sharp notes (1997: 111), "it would be most surprising if this were the only one of these genes that had undergone adaptive changes during the divergence of the two species." Are more sensitive methods for detecting natural selection available?

Several different techniques are available for detecting when sequences have changed due to natural selection.

John McDonald and Martin Kreitman (1991) invented a test for natural selection that is in increasingly widespread use. The McDonald–Kreitman, or MK, test is based on an important corollary to the neutral theory's prediction that silent substitutions occur more rapidly than replacement substitutions. According to the neutral theory, the ratio of replacement to silent-site substitutions in any particular locus should be constant through time. Based on this proposition, McDonald and Kreitman predicted that the ratio of replacement to silent site substitutions in between-species comparisons should be the same as the ratio observed in within-species comparisons.

Their initial test of this prediction compared sequence data from the alcohol dehydrogenase (*Adh*) gene of 12 *Drosophila melanogaster,* 6 *D. simulans,* and 12 *D. yakuba* individuals. *Adh* was an interesting locus to study for two reasons: fruit flies feed on rotting fruit that may contain toxic concentrations of ethanol, and the alcohol dehydrogenase enzyme catalyzes the conversion of ethanol to a nontoxic product. Because of the enzyme's importance to these species, and because ethanol concentrations vary among food sources, it is reasonable to suspect that the locus is under strong selection when populations begin exploiting different fruits.

In an attempt to sample as much within-species variation as possible, the individuals chosen for the study were from geographically widespread locations. McDonald and Kreitman aligned the *Adh* sequences from each individual in the study and identified sites where a base differed from the most commonly observed nucleotide, or what is called the consensus sequence. The researchers counted differences as fixed if they were present in all individuals from a particular species, and as **polymorphisms**—or allele differences within species—if they were present in only some individuals from a particular species. Differences that were fixed in one species and polymorphic in another were counted as polymorphic.

McDonald and Kreitman found that 29% of the differences that were fixed between species were replacement substitutions. Within species, however, only 5% of the polymorphisms in the study represented replacements. Rather than being the same, these ratios show an almost sixfold, and statistically significant, difference ($P = 0.006$). This is strong evidence against the neutral model's prediction. McDonald and Kreitman's interpretation is that the replacement substitutions fixed between species are selectively advantageous. They suggest that these mutations occurred after *D. melanogaster, D. simulans,* and *D. yakuba* had diverged, and spread rapidly to fixation due to positive selection in the differing environments occupied by these species. Using the MK test, natural selection has now been detected in loci from plants and protists as well as animals (Escalante et al. 1998; Purugganen and Suddith 1998).

Which Loci are Under Strong Positive Selection?

Thanks to studies employing the Hughes and Nei analysis, the MK test, and other strategies, a few tentative generalizations are beginning to emerge concerning the types of loci where positive natural selection has been particularly strong. Replacement substitutions appear to be particularly abundant in recently duplicated genes that have attained new functions, in loci involved in sex determination or in species-specific interactions between sperm and egg at fertilization, in genes that code for certain enzymes or regulatory proteins, and in disease resistance loci (see Table 18.2).

As data accumulate from genome-sequencing projects in closely related species, such as humans and chimpanzees, the number and quality of comparative studies should virtually explode. Even before the era of genome sequencing began, however, it became clear that silent substitutions, as well as replacement changes, are subject to natural selection.

Selection on "silent" substitutions

The term silent substitution was coined to reflect two aspects of base substitutions at certain positions of codons: They do not result in a change in the amino acid sequence of the protein product, and they are not exposed to natural selection. The

Table 18.2 Studies that confirm positive selection on replacement substitutions

Although this list is by no means exhaustive, it underscores a general point: Evidence for positive selection is particularly strong in genes that code for proteins involved in fertilization, disease resistance, and feeding.

Gene	Species	Rationale	Reference
Lysin protein and receptor	Abalone	Strong selection on species-specific egg-recognition proteins on sperm	Swanson, W.J. and V.D. Vacquier. 1998. *Science* 281: 710–712.
Bindin protein	Sea urchins	Strong selection on species-specific egg-recognition proteins on sperm	Metz, E.C. and S. R. Palumbi. 1996. *Molecular Biology and Evolution* 13: 397–406.
Self-incompatibility loci	Tomato family plants	Strong selection for divergence among proteins involved in avoiding self-fertilization	Clark, A.G. and T.-H. Kao. 1991. *Proceedings of the National Academy of Sciences, USA* 88: 9823–9827.
Eosinophil cationic protein	Primates	Strong selection on a recently duplicated gene involved in disease resistance	Zhang, J., H.F. Rosenberg, and M. Nei. 1998. *Proceedings of the National Academy of Sciences, USA* 95: 3708–3713.
MHC Class II	Humans	Strong selection for divergence among antigen-recognition proteins	Hughes, A.L. and M. Nei. 1989. *Proceedings of the National Academy of Sciences, USA* 86: 958-962.
Immunoglobulins	Humans	Strong selection for divergence among loci that code for antibody proteins	Tanaka, T. and M. Nei. 1989. *Molecular Biology and Evolution* 6: 447–459.
Lysozyme	Hanuman langur	Strong selection on a digestive enzyme in a leaf-eating monkey	Messier, W. and C.-B. Smith. 1997. *Nature* 385: 151–154.

second proposition had to be discarded, however, in the face of data on phenomena known as codon bias, hitchhiking, and background selection. How can mutations that do not alter an amino acid sequence be affected by natural selection?

Codon Bias

Most of the 20 amino acids are encoded by more than one codon. We have emphasized that changes among redundant codons do not cause changes in the amino-acid sequences of proteins, and have implied that these silent-site changes are neutral with respect to fitness. If this were strictly true, we would expect that codon usage would be random, and that each codon in a suite of synonymous codons would be present in equal numbers throughout the genome of a particular organism. But early sequencing studies confirmed that codon usage is highly nonrandom (Table 18.3). This phenomenon is known as **codon bias**.

Several important patterns have emerged from studies of codon bias. In every organism studied to date, codon bias is strongest in highly expressed genes—such as those for the proteins found in ribosomes—and weak to nonexistent in rarely expressed genes. In addition, the suite of codons that are used most frequently correlates strongly with the most abundant species of tRNA in the cell (Figure 18.2).

The leading hypothesis to explain these observations is natural selection for translational efficiency (Sharp and Li 1986; Sharp et al. 1988; Akashi 1994). The logic here is that if a "silent" substitution in a highly expressed gene creates a codon that is rare in the pool of tRNAs, the mutation will be selected against. The selective agent is the speed and accuracy of translation. Speed and accuracy are especially important when the proteins encoded by particular genes are turning

Table 18.3 Codon bias

This table reports the relative frequencies of codons found in genes from three different species: the bacterium *E. coli*, baker's yeast (*Saccharomyces cerevisiae*), and the fruit fly *Drosophila melanogaster*. If each codon were used equally in each genome, the relative frequencies would all be 1. Deviations from 1.00 indicate codon bias. The amino acids listed are leucine, valine, isoleucine, phenylalanine, and methionine. "High" and "Low" differentiate data from highly transcribed versus rarely transcribed genes. In every case reported here, codon bias is more extreme in highly expressed genes.

Amino acid	Codon	*Escherichia coli* High	*Escherichia coli* Low	*Saccharomyces cerevisiae* High	*Saccharomyces cerevisiae* Low	*Drosophila melanogaster* High	*Drosophila melanogaster* Low
Leu	UUA	0.06	1.24	0.49	1.49	0.03	0.62
	UUG	0.07	0.87	5.34	1.48	0.69	1.05
	CUU	0.13	0.72	0.02	0.73	0.25	0.80
	CUC	0.17	0.65	0.00	0.51	0.72	0.90
	CUA	0.04	0.31	0.15	0.95	0.06	0.60
	CUG	5.54	2.20	0.02	0.84	4.25	2.04
Val	GUU	2.41	1.09	2.07	1.13	0.56	0.74
	GUC	0.08	0.99	1.91	0.76	1.59	0.93
	GUA	1.12	0.63	0.00	1.18	0.06	0.53
	GUG	0.40	1.29	0.02	0.93	1.79	1.80
Ile	AUU	0.48	1.38	1.26	1.29	0.74	1.27
	AUC	2.51	1.12	1.74	0.66	2.26	0.95
	AUA	0.01	0.50	0.00	1.05	0.00	0.78
Phe	UUU	0.34	1.33	0.19	1.38	0.12	0.86
	UUC	1.66	0.67	1.81	0.62	1.88	1.14
Met	AUG	1.00	1.00	1.00	1.00	1.00	1.00

Source: Sharp et al. (1988)

over rapidly and the corresponding genes must be transcribed continuously. It is reasonable, then, to observe the strongest codon bias in highly expressed genes.

Selection against certain synonymous substitutions represents a form of negative selection; it slows the rate of molecular evolution. As a result, codon bias may explain the observation that silent changes do not accumulate as quickly as base substitutions in pseudogenes. The general message here is that, in genomes where codon bias occurs, not all redundant substitutions are "silent" with respect to natural selection.

Hitchhiking and Background Selection

Another phenomenon that affects the rate and pattern of change at silent sites is referred to as **hitchhiking**, or a **selective sweep**. Hitchhiking can occur when strong positive selection acts on a particular amino acid substitution. As a favorable mutation increases in frequency, neutral or even slightly deleterious mutations closely

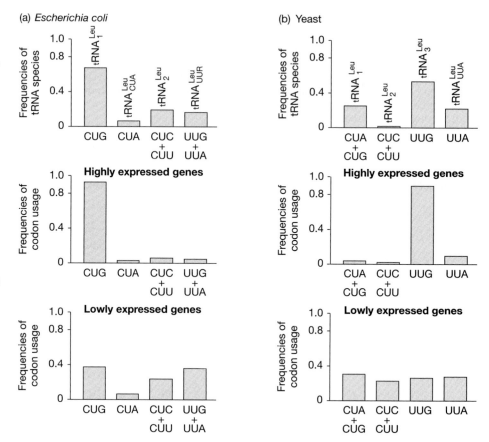

Figure 18.2 Codon bias correlates with the relative frequencies of tRNA species The bar chart in the top row of both (a) and (b) shows the frequencies of four different tRNA species that carry leucine in *E. coli* (a) and the yeast *Saccharomyces cerevisiae* (b). The bar charts in the middle and bottom rows report the frequency of the mRNA codons corresponding to each of these tRNA species in the same organisms. The mRNA codons were measured in two different classes of genes: those that are highly transcribed (middle) and those that are rarely transcribed (bottom). The data show that codon usage correlates strongly with tRNA availability in highly expressed genes, but not at all in rarely expressed genes. From Li and Graur (1991). Copyright © 1991 Sinauer Associates, Inc. Reprinted by permission of Sinauer Associates, Inc.

linked to the favored site will increase in frequency along with the beneficial locus. These linked mutations are swept along by selection and can actually "hitchhike" to fixation. Note that this process only occurs when recombination fails to break up the linkage between the hitchhiking sites and the site that is under selection.

Perhaps the best example of hitchhiking discovered to date was found on the fourth chromosome of fruit flies. The fourth chromosome in *Drosophila* is remarkable because no recombination occurs along its entire length. As a consequence, the entire chromosome represents a linkage group and is inherited like a single gene.

Andrew Berry and colleagues (1991) sequenced a 1.1-kb region of the fourth chromosome in 10 *Drosophila melanogaster* and 9 *D. simulans*. This chromosome region includes the introns and exons of a gene expressed in fly embryos, called *cubitus interruptus Dominant* (*ciD*). Berry et al. found a remarkable pattern in the sequence data: No differences were observed among the *D. melanogaster* individuals surveyed. The entire 1.1 kb of sequence was identical in the 10 individuals. Further, only one base difference was observed among the *D. simulans* in the study. This means that there was almost no polymorphism (variation within species) in this region. In contrast, a total of 54 substitutions were found when the same sequences were compared between the two species. A key observation here is that genes on other chromosomes surveyed in the same individuals showed normal amounts of polymorphism. These latter data serve as a control and confirm that the lack of variation observed in and around the *ciD* locus is not caused by an unusual sample of individuals. Rather, there is something unusual about the fourth chromosome in these flies.

Berry et al. suggest that selective sweeps recently eliminated all or most of the variation on the fourth chromosome in each of these two species. Their argument is that an advantageous mutation anywhere on the fourth chromosome would eliminate all within-species polymorphism as it increased to fixation. New variants, like the one substitution observed in the *D. simulans* sampled, will arise only through mutation.

In this way, selective sweeps create a "footprint" in the genome: a startling lack of polymorphism within linkage groups. Similar footprints have been found in other chromosomal regions where the frequency of recombination is low, including the ZFY locus of the human Y chromosome (Dorit et al. 1995) and a variety of loci in *D. melanogaster* and other fruit flies (for example, see Nurminsky et al. 1998).

Has hitchhiking produced all of these regions of reduced polymorphism? The answer is probably not. Another process, called background selection, can produce a similar pattern (Charlesworth et al. 1993). Background selection results from negative selection against deleterious mutations, rather than positive selection for advantageous mutations. Like hitchhiking, it occurs in regions of reduced recombination. The idea here is that selection against deleterious mutations removes closely linked neutral substitutions and produces a reduced level of polymorphism.

Background selection and hitchhiking are contrasting processes that lead to the same pattern.

Although the processes called hitchhiking and background selection are not mutually exclusive, their effects can be distinguished in at least some cases. Hitchhiking results in dramatic reductions in polymorphism as an occasional advantageous mutation quickly sweeps through a population. Background selection causes a slow, steady decrease in polymorphism as frequent deleterious mutations remove individuals from the population. The current consensus is that hitchhiking is probably responsible for the most dramatic instances of reduced polymorphism in linked regions—for example, where sequence variation is entirely eliminated—while background selection causes the less extreme cases.

18.3 Transposable Elements

The field of molecular evolution was launched by observations about rates and patterns of change in protein and DNA sequences. The message in these data—that many of the evolutionary changes observed at the molecular level are caused by drift, and not by positive natural selection—was astonishing to most biologists. Early data sets on the nature of the eukaryotic genome delivered an equally surprising conclusion: Genomes are not simple collections of sequences that code for proteins. Instead they contain a bestiary of sequence types. In humans, for example, only a small fraction of the DNA present codes for proteins, rRNAs, or tRNAs. The human genome, like the genomes of many other eukaryotes, is dominated by parasitic sequences that do not code for products used by the cell.

Two pioneering observations hinted at this conclusion and launched research on how genomes evolve. One observation was the C-value paradox. The data in Table 18.4 show that, in eukaryotes, the total amount of DNA found in a cell (also known as its C-value) does not correlate with the organism's degree of morphological complexity or its phylogenetic position. This finding suggests that much or most of the DNA in eukaryotes is functionless from the cell's viewpoint. A second important observation was Barbara McClintock's discovery of transposable genetic elements, or "jumping genes." While studying the inheritance of kernel

Table 18.4 *C* Values of eukaryotes

The total amount of DNA in a haploid genome is called the C value. Genome sizes show extreme variation within and between taxonomic groups. In flowering plants, for example, haploid genomes range by a factor of 2,500, from 50,000 to 125 billion kilobases (kb).

Species	C value (kb)
Navicola pelliculosa (diatom)	35,000
Drosophila melanogaster (fruit fly)	180,000
Paramecium aurelia (ciliate)	190,000
Gallus domesticus (chicken)	1,200,000
Erysiphe cichoracearum (fungus)	1,500,000
Cyprinus carpio (carp)	1,700,000
Lampreta planeri (lamprey)	1,900,000
Boa constrictor (snake)	2,100,000
Parascaris equorum (roundworm)	2,500,000
Carcarias obscurus (shark)	2,700,000
Rattus norvegicus (rat)	2,900,000
Xenopus laevis (toad)	3,100,000
Homo sapiens (human)	3,400,000
Nicotiana tabaccum (tobacco)	3,800,000
Paramecium caudatum (ciliate)	8,600,000
Schistocerca gregaria (locust)	9,300,000
Allium cepa (onion)	18,000,000
Coscinodiscus asteromphalus (diatom)	25,000,000
Lilium formosanum (lily)	36,000,000
Pinus resinosa (pine)	68,000,000
Amphiuma means (newt)	84,000,000
Protopterus aethiopicus (lungfish)	140,000,000
Ophioglossum petiolatum (fern)	160,000,000
Amoeba proteus (amoeba)	290,000,000
Amoeba dubia (amoeba)	670,000,000

Source: Li and Graur (1991)

color in corn, McClintock found loci that produced novel color patterns by moving to new locations in the genome.

Sequencing studies later revealed that much of the "extra" DNA responsible for the C-value paradox consists of these transposable genetic elements. We begin our look at how genomes evolve with some basic questions about these loci: What are transposable elements? Where did they come from? What effect do they have on the genomes that host them?

Transposable Elements Are Genomic Parasites

As the ensuing discussion will show, "transposable element" is really an umbrella term for genes with a diverse set of characteristics. The majority of transposable elements contain only the sequences required for transposition. All share the ability to move from one location to another in the genome, and most leave a copy of themselves behind when they move. In this way, transposition events lead to an increase in the number of transposable elements in the host genome.

If a transposable element disrupts an important coding sequence when it inserts into a new location in the genome, deleterious "knock-out" mutations result. In humans, for example, transposition events have resulted in tumor formation and cases of hemophilia (see Hutchison et al. 1989). The sheer bulk of transposable element DNA would also appear to have deleterious consequences for the host genome. The time, energy, and resources required to replicate a genome burdened by parasitic DNA could place a limit on growth rates, particularly in small, rapidly dividing organisms. Because they disrupt coding sequences and place an energetic burden on the cell, transposable elements are most accurately characterized as genomic parasites. Surveys done to date suggest that they exist in every organism.

Instead of cooperating to increase the fitness of the individual, some loci are parasitic.

What is the key to their success? The answer is that transposable elements can increase in a population in two ways, even when they reduce their host's fitness slightly. They can become fixed in small populations due to drift, or increase because transposition events result in more than one copy of the parasitic locus being present in many gametes. To grasp this second point more fully, consider that, if a transposition event reduces the survival and reproductive capacity of the host slightly, the extra copies of the transposable element in the gene pool can make up the deficit and result in the parasite's spread throughout the population. According to models developed by Brian Charlesworth and Charles Langley (1989), the transposable elements that replicate themselves most efficiently and with the least fitness cost to the host genome are favored by natural selection and tend to spread.

Another key to understanding transposable elements is to appreciate that their success depends on sex (Hickey 1992). In eukaryotes, outcrossing results in haploid genomes being mixed. This presents transposable elements with new targets for transposition and allows them to spread throughout a population. In prokaryotes and other groups where gene transfer is one-way and where most reproduction is by fission, transposable elements in the main chromosome tend to be eliminated by selection or drift. As a result, the transposable elements found in bacteria and archaea tend to reside on the circular extrachromosomal elements called plasmids.

Transposable genetic elements are grouped into two broad classes, based on whether they move via an RNA or DNA intermediate sequence. These Class I and Class II elements come in an enormous array of sizes, copy numbers, and structurally related families. The literature on mobile genetic elements is vast and growing rapidly (Voytas 1996). Although we can only touch on this body of knowledge here, understanding a few of the basic types of transposable elements is important. In sheer numbers, they represent some of the most successful genes in the history of life.

Class I Elements

Class I elements, also called **retrotransposons**, are the product of reverse transcription events. Although the molecular mechanism of transposition is not yet fully known, we do know that movement of Class I elements occurs through a ribonucleic acid (RNA) intermediate. Transposition is also replicative, meaning that the original copy of the sequence is intact after the event. The long interspersed

Transposable elements that spread via an RNA intermediate have to be reverse transcribed, by reverse transcriptase, before being inserted into a new location.

elements (LINEs) are retrotransposons that contain the coding sequence for reverse transcriptase and are thought to catalyze their own transposition. In mammals, LINEs are typically 6–7 kb in length (Hutchison et al. 1989; Wichman et al. 1992). The first human chromosome to be completely sequenced, number 22, contains over 14,000 LINEs. These elements represent greater than 13% of the total DNA on this chromosome (Dunham et al. 1999).

Another important category of retrotransposons is distinguished by the presence of long terminal repeats (LTRs). LTRs are one of the hallmarks of retroviral genomes. When retroviruses insert themselves into a host DNA to initiate an infection, LTRs mark the insertion point. In corn, 10 different families of retrotransposons with LTRs have been identified, each of which is found in 10 to 30,000 copies per haploid genome (SanMiguel et al. 1996). The complete genome of baker's yeast, *Saccharomyces cerevisiae,* has been sequenced and found to contain 52 complete LTR-containing sequences called Ty elements, and 264 naked LTRs that lack the coding regions of normal retrotransposons. These "empty" LTRs are interpreted as transposition footprints, meaning that they are sequences left behind when Ty elements were somehow excised from the genome (see Boeke 1989; Goffeau et al. 1996).

Retrotransposons and retroviruses may be closely related.

Where did retrotransposons come from? One hypothesis is that LINEs and the LTR-containing retrotransposons evolved from retroviruses. Retrotransposons resemble retroviruses that have lost the coding sequences required to make capsule proteins. This hypothesis proposes that retrotransposons have adopted a novel evolutionary strategy. In contrast to retroviruses, which replicate in their host cell, move on to infect new cells, and eventually infect new host individuals of the same generation, retrotransposons replicate by infecting the germline. Instead of being transmitted horizontally—meaning, from host to host in the same generation—they replicate by being transmitted vertically, to the next generation of hosts. Their transmission is much slower than that of conventional retroviruses, but retrotransposons also escape detection by the immune system.

A second type of Class I element is called a retrosequence. Retrosequences do not contain the coding sequence for reverse transcriptase, but amplify via RNA intermediates that are reverse transcribed and inserted into the genome. The short interspersed elements (SINEs) of mammals are among the best-studied examples. SINEs are grouped into several different families, each of which is distinguished by its sequence homology with a different functioning gene. The *Alu* family of sequences in primates, for example, is about 90% identical to the 7SL RNA gene, which is involved in transmembrane protein transport; other families of SINEs are homologous with various tRNA genes. SINEs are typically under 500 bp in length and lack the sequences necessary for translation of a transcribed RNA message. They can also be extremely abundant. Human chromosome number 22, for example, contains 20,188 *Alu* elements (Dunham et al. 1999). These SINEs account for 16.8% of the total DNA in this chromosome.

Retrosequences do not code for an RNA or protein that functions in the cell. They are closely related to functioning genes, however.

Where did SINEs and other retrosequences come from, and how are they replicated? In most SINE families it appears that only one or a very few master copies of the locus are actively transposing, and that the remainder represent inactive copies analogous to pseudogenes (see Shen et al. 1997). As with LINEs and other Class I elements, the mechanism of this transposition is not known. We do not know how transcription of the master gene locus is regulated, where the reverse transcriptase comes from, or how insertion of the resulting DNA copy proceeds.

We can, however, date the origin of some families. The *Alu* sequences, for example, have been found in every primate surveyed to date, but do not occur in rodents. This observation suggests that the *Alu* family entered the genome of a primate ancestor after these mammal lineages diverged.

Class II Elements

Class II transposable elements replicate via a DNA intermediate and are the dominant type of transposable genetic element in bacteria. Their transposition can be replicative, as in Class I elements, or conservative. In conservative transposition, the element is excised during the move so that copy number does not increase.

The first Class II elements ever described were the insertion sequences, or IS elements, discovered in the bacterium *Escherichia coli* (Table 18.5). When insertion sequences contain one or more coding sequences, they are called transposons. In addition to being inserted into the main bacterial chromosome, however, transposons are commonly inserted into plasmids. Plasmids replicate independently of the main chromosome and are readily transferred from one bacterial cell to another during conjugation.

Transposons encode a protein, called a transposase, that catalyzes transposition. In bacterial transposons, the coding region frequently also codes for a protein that confers resistance to an antibiotic (Table 18.6). As a result, plasmid-borne transposons have been responsible for the rapid evolution of drug resistance in disease-causing bacteria. Transposons that confer antibiotic resistance are the first example we have encountered of a transposable element that creates a fitness advantage for its host.

Class II transposable elements are also found in eukaryotes. The *Ac* and *Ds* elements of corn, discovered by Barbara McClintock in the 1950s, belong to this group. These Class II sequences code for a transposase as well as other proteins. The P elements found in *Drosophila melanogaster* are another example. A typical fly

The chromosomes of bacteria are relatively free of transposable elements. But transposable elements are frequently found in the extrachromosomal elements called plasmids.

Table 18.5   Insertion sequences found in the genome of *Escherichia Coli*

Some of the insertion sequences listed here are found on extra chromosomal genetic elements in *E. coli*. IS2 and IS3, for example, are located on F (or "fertility") plasmids. F plasmids are circular DNA molecules found in bacteria that can either replicate independently or integrate into the main chromosome and replicate with it. The presence of an F plasmid in a bacterial cell confers the ability to donate genetic material to another cell during conjugation.

Insertion sequence	Normal occurrence in *E. coli*
IS1	5–8 copies on chromosome
IS2	5 on chromosome; 1 on F
IS3	5 on chromosome; 2 on F
IS4	1 or 2 copies on chromosome
IS5	Unknown
γ-δ (TN1000)	1 or more on chromosome; 1 on F
pSC101 segment	On plasmid pSC101

Source: Griffiths et al. (1993)

Table 18.6	**Transposons can carry genes conferring antibiotic resistance**

This listing illustrates the diversity and number of plasmid-borne transposons that carry antibiotic-resistance genes.

Transposon	Confers resistance to
Tn1	Ampicillin
Tn2	Ampicillin
Tn3	Ampicillin
Tn4	Ampicillin, streptomycin, sulfanilamide
Tn5	Kanamycin
Tn6	Kanamycin
Tn7	Trimethoprim, streptomycin
Tn9	Chloramphenicol
Tn10	Tetracycline
Tn204	Chloramphenicol, fusidic acid
Tn402	Trimethoprim
Tn551	Erythromycin
Tn554	Erythromycin, spectinomycin
Tn732	Gentamicin, tobramycin
Tn903	Kanamycin
Tn917	Erythromycin
Tn1721	Tetracycline

Source: Griffiths et al. (1993)

genome contains 30–50 copies of these elements, which have insertion-sequence-like repeats at their ends and as many as three coding regions for protein products (Ajioka and Hartl 1989).

Evolutionary Impact of Transposable Elements

Research on transposable elements has delivered an important message: Genomes are not cohesive collections of loci that contribute to the fitness of the individual. Instead, they are riddled with parasites that transmit themselves at their host's expense. This realization raises some important questions. Do host genomes have mechanisms to counter transposable elements? Does natural selection limit their spread? Are transposable elements ever beneficial to their hosts?

Limiting the Spread of Transposable Elements

When a transposable element inserts into the coding region of a gene, the mutation that results should quickly be eliminated by natural selection. But Charlesworth and Langley (1989) proposed that natural selection might also remove transposable elements from a population in the absence of transposition events. Their hypothesis hinges on the idea that multiple copies of a transposable element on the same chromosome frequently leads to errors in meiosis. As Figure 18.3 shows, ectopic recombination can occur between transposable elements located at differ-

Normal crossing-over
and recombination

Ectopic crossing-over
and recombination

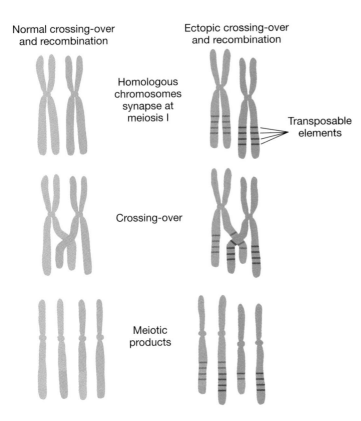

Homologous
chromosomes
synapse at
meiosis I

Transposable
elements

Crossing-over

Meiotic
products

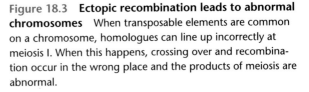

Figure 18.3 Ectopic recombination leads to abnormal chromosomes   When transposable elements are common on a chromosome, homologues can line up incorrectly at meiosis I. When this happens, crossing over and recombination occur in the wrong place and the products of meiosis are abnormal.

ent sites on homologous chromosomes. (Ectopic means "to occur in an abnormal location.") Half of the chromosomes that result will lack certain loci and be strongly selected against. The frequency of ectopic recombination, and thus the strength of selection, should correlate strongly with the frequency of transposable elements.

Paul Sniegowski and Charlesworth (1994) tested the ectopic recombination hypothesis by surveying the distribution of transposable elements in the *Drosophila melanogaster* genome. Their specific goal was to quantify the frequency of transposable elements inside the flipped chromosome segments, called inversions, that were introduced in Chapter 4. Recall that crossing over does not occur within inversions if one chromosome contains the inversion but its homologous chromosome does not. Because crossing over is infrequent inside inverted segments, ectopic recombination should be rare. As predicted by the Charlesworth–Langley hypothesis, transposable elements are much more common inside inversions than they are in homologous, noninverted chromosome segments, where crossing over occurs normally. This observation supports the hypothesis that ectopic recombination serves as a brake on the spread of transposable elements.

Do host genomes themselves have mechanisms to actively suppress transposition? Recent work by Rachel Waugh O'Neill and co-workers (1998) suggests that the answer may be yes. These biologists conducted a study inspired by the hypothesis that organisms add methyl ($-CH_3$) groups to their DNA as a way to thwart mobile genetic elements (Bester and Tycko 1996; Yoder et al. 1997). To look for an association between DNA methylation and the spread of transposable elements, Waugh O'Neill and associates analyzed the chromosomes found in the hybrid offspring of a mating between a tammar wallaby and a swamp wallaby. For unknown reasons, the DNA of the hybrid individual was virtually unmethylated.

The researchers also found that, in many of this individual's chromosomes, a retro-transposon called KERV-1 had virtually exploded in copy number (see Figure 18.4). According to Waugh O'Neill and colleagues, this correlation is strong support for the hypothesis that methylation protects host DNA from insertion by parasites.

Even if future work confirms that methylation protects DNA from parasitic sequences, a series of important questions remains. If methylation is effective, why are transposable elements present in such large numbers? Why are the numbers of parasites so variable among species? Why have so many different types of elements been able to take up residence in genomes?

It is not clear why eukaryotes are unable to eliminate parasitic sequences more efficiently.

To summarize, forty years of careful descriptive studies have produced important insights into the number, distribution, and types of transposable elements present in genomes. The natural history of these parasites is well known. Now, innovations like the host-defense and ectopic-recombination hypotheses are transforming research on transposable elements. Researchers are beginning to ask highly focused questions about the evolutionary dynamics of these elements. Recent experimental work has even shown that transposition events may occasionally have important beneficial effects for their hosts.

Positive Impacts of Transposable Elements

For decades, the only example of a positive fitness benefit provided by transposable elements was the antibiotic resistance conferred by some plasmid-borne transposons in bacteria. Recent work by John Moran and colleagues (1999) suggests that transposition events in eukaryotes may occasionally result in mutations that confer a fitness benefit. This conclusion is based on experiments with the LINE elements found in humans. To analyze how these loci move from one location to another, Moran and co-workers used recombinant DNA techniques to attach a marker gene to LINE-1 sequences. They introduced the engineered sequence into human cells growing in culture, allowed the parasitic sequences to insert themselves into the genome, and sequenced the LINE-1 elements that succeeded in transposing to new locations in the genome. In several cases, the sequence data showed

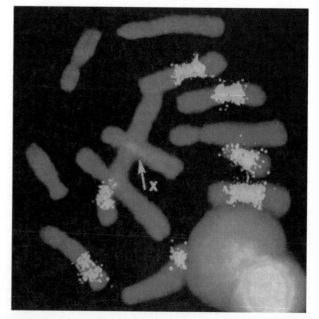

Figure 18.4 When DNA is unmethylated, transposable elements explode in number To produce this photograph, Waugh O'Neill and co-workers labeled single-stranded DNA from the KERV-1 transposable element with a fluorescent molecule, then allowed this probe DNA to hybridize to KERV-1 sequences in the chromosomes of a wallaby hybrid individual. The pink dots indicate the location of KERV-1 elements. (Note that they cluster near the centromeres of certain chromosomes.) No hybridization was observed in the parental wallaby species. From Waugh O'Neill et al. (1998).

that the mobile element had carried a chunk of host DNA along with it and the marker gene during transposition. In essence, the LINEs had duplicated segments of host DNA and moved them to new locations. Figure 18.5 illustrates how this happened. The diagram also shows that, if the transposed host DNA segment happens to contain an exon or regulatory sequence, the transposition event results in a novel gene. Moran and co-workers contend that these types of transposition events are important in the evolution of genomes and furnish a mechanism for the phenomenon called exon shuffling that was introduced in Chapter 14.

In a similar vein, work by Alka Agrawal and colleagues (1998) suggests that a key feature of the vertebrate immune system originated in a transposition event. As Figure 18.6a shows, the proteins that serve as antigen–recognition sites on the surface of immune system cells are encoded by three gene segments. As immune system cells develop in an embryo, a series of reactions take place that result in portions of the V (variable), D (diversity), and J (joining) gene segments being excised and recombined (see Figure 18.6b). These reactions are catalyzed by proteins called RAG1 and RAG2. What Agrawal et al. showed is that RAG1 and RAG2 can also catalyze the transposition of gene constructs that are unrelated to the V, J, and D regions. The experimental reaction they set up, diagrammed in Figure 18.6c, took place outside of a cell. Further, the reaction mechanism is identical to the chemical events that take place during movement by transposable elements. To make sense of this result, the researchers propose that the RAG proteins constitute a transposase homologous to those found in present-day transposable elements.

The implications of this work are remarkable. Agrawal and associates hypothesize that the V(J)D excision and rearrangement reactions observed in today's vertebrates are possible because of an insertion event by a transposable element several hundred million years ago. According to the hypothesis outlined in Figure 18.6d, a transposable element bearing *RAG1* and *RAG2* inserted into a membrane receptor gene early in the evolution of vertebrates. The transposase could catalyze the recombination of the resulting receptor gene segments, however. If so, then gene duplication events later in evolution could have expanded the membrane receptor locus and resulted in the extensive V, D, and J regions observed today.

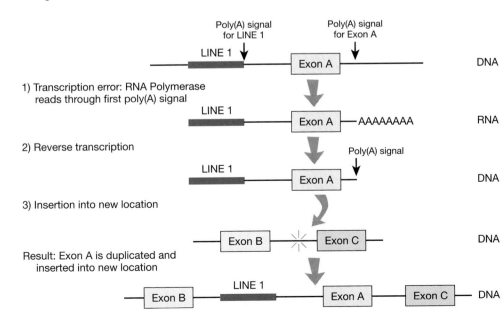

Figure 18.5 **"Exon shuffling" via transposition events** Based on the series of events diagrammed here, transposition by LINE elements can result in exons or regulatory sequences being moved to new locations in the genome. This phenomenon is known as exon shuffling and was introduced in Chapter 14. The experiment by Moran et al., described in the text, shows that each of the steps illustrated here can actually occur.

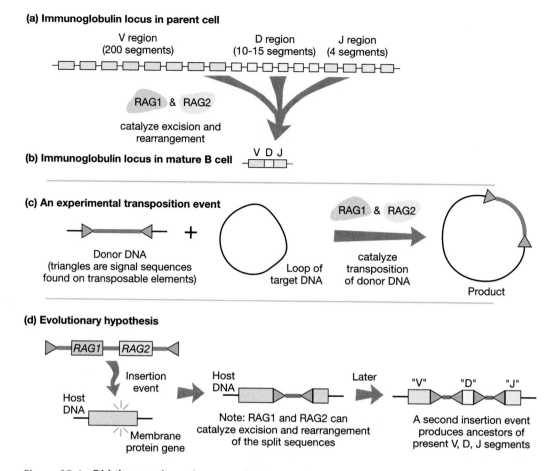

(a) Immunoglobulin locus in parent cell

V region
(200 segments)

D region
(10-15 segments)

J region
(4 segments)

RAG1 & RAG2

catalyze excision and
rearrangement

(b) Immunoglobulin locus in mature B cell

V D J

(c) An experimental transposition event

Donor DNA
(triangles are signal sequences
found on transposable elements)

+

Loop of
target DNA

RAG1 & RAG2

catalyze
transposition
of donor DNA

Product

(d) Evolutionary hypothesis

RAG1 RAG2

Insertion
event

Host
DNA

Membrane
protein gene

Host
DNA

Note: RAG1 and RAG2 can
catalyze excision and rearrangement
of the split sequences

Later

"V"   "D"   "J"

A second insertion event
produces ancestors of
present V, D, J segments

Figure 18.6 **Did the vertebrate immunoglobulin locus originate in a transposition event?** (a) As a human embryo develops, the cells that serve as precursors to immune system cells have immunoglobulin loci with many different V, D, and J regions. (b) As individual immune system cells mature, RAG1 and RAG2 catalyze reactions that result in a single V, D, and J segment being combined and the others being excised. (c) The experiment by Agrawal et al. showed that RAG1 and RAG2 can catalyze transposition events. (d) According to the evolutionary hypothesis outlined here, insertion events by transposable elements containing the *RAG1* and *RAG2* genes created the basic structure of the immunoglobulin locus. Gene duplication events could later produce the variety of V, D, and J segments observed today.

18.4 Organelle Genomes

Chloroplasts and mitochondria have their own DNA. They are dervied from bacteria that lived symbiotically in the cytoplasm of early eukaryotic cells.

Studies on transposable elements underscore the dynamic nature of the genome. Far from being static, stable entities, the composition and structure of genomes changes dramatically over time. This conclusion is reinforced by research on the history of the genomes present in eukaryotic chloroplasts and mitochondria. These organelles contain DNA that codes for functioning proteins, rRNAs, and tRNAs. The organization of these genomes differs markedly from nuclear DNA (nDNA), however—largely because both organelles originated as bacteria that took up residence in eukaryotic cells early in the history of life. As Chapter 14 showed, chloroplasts are derived from cyanobacteria and mitochondria from α–proteobacteria. Here, we describe the organization and gene content of organelle genomes and consider an important question about how they evolve.

Chloroplast DNA

In chloroplasts, the energy captured from photons is used to power the synthesis of sugars. Many of the proteins used in photosynthesis are encoded by chloroplast DNA (cpDNA). This is a circular double helix that lacks the histones and other chromosomal proteins found in nuclear DNA. Algal and plant cells typically contain many chloroplasts; each organelle normally has 20–80 copies of cpDNA.

Several cpDNAs have now been completely sequenced (see Wolfe et al. 1991). The 156-kb cpDNA genome of tobacco, for example, contains a total of 113 genes. The functions of these loci fall into two broad classes: sequences required for gene expression, and sequences required for photosynthesis. The loci involved in transcription and translation include RNA polymerase subunits, ribosomal proteins, a translation initiation factor, tRNAs, and rRNAs. The 29 genes involved in photosynthesis code for many of the proteins found in the thylakoid of the organelle, where photon capture takes place. Compared to nDNA, relatively few noncoding sequences are present.

The genome size and gene content of tobacco cpDNA appears to be typical of plants from other taxonomic groups (see Clegg et al. 1994), indicating that the information content of this genome has been conserved during evolution. Even the order of genes along the chromosome appears to be highly conserved, if not identical (Palmer 1987). Further, all cpDNA genomes examined to date use the same genetic code as nDNA.

The inheritance of chloroplast genomes is normally uniparental. In most flowering plants, all of the cpDNA present in embryos is derived from the mother; but in at least some gymnosperms, inheritance is paternal. Biparental inheritance has also been recorded in some flowering plants (Palmer 1987). Recombination between cpDNA copies is extremely rare, having been recorded only in the green alga *Chlamydomonas*. In most organisms, then, the entire cpDNA genome is inherited like a single allele.

Relative to plant nuclear genes, sequence evolution in cpDNA is slow. Compared to silent sites surveyed in a sample of nuclear genes, silent sites in a sample of cpDNA loci evolve about one-quarter as fast (Wolfe et al. 1987, 1989).

The gene content and size of the chloroplast genome is highly conserved. The rate of sequence change in cpDNA is low relative to the nuclear genome.

Structure and Evolution of mtDNA

In mitochondria, the energy stored in sugars is used to power the synthesis of adenosine triphosphate (ATP). Many of the proteins used in cellular respiration are encoded by mitochondrial DNA (mtDNA). Like cpDNA, mtDNA is usually a circular double helix that lacks chromosomal proteins. The number of mitochondria per cell varies from four in some unicellular fungi to thousands in the muscle cells of vertebrates; in vertebrates each organelle has 5–10 copies of mtDNA.

The mitochondrial DNAs of plants, fungi, and animals are very different from one another. As the observations summarized in Table 18.7 show, these genomes vary in characteristics ranging from structure, composition, and mode of inheritance to rate of evolution. Why these differences evolved as plants, fungi, and animals diverged from their common ancestor is not known (Gray et al. 1999).

The size, structure, and rate of sequence change in plant, fungal, and animal mtDNA varies enormously, but the information content is highly conserved.

Evolution of Organelle Genomes

Chapter 14 introduced evidence that genetic information has been transferred between species throughout the history of life. The largest horizontal-transfer events ever recorded occurred when eukaryotic cells acquired intact bacterial

Table 18.7 **Variation among mitochondrial DNAs**

	Plants	Fungi	Animals
Genome size	Extremely variable—ranges from 300 kb to 2,400 kb in melon family alone	Highly variable—ranges from 26.7 kb to 115 kb in filamentous ascomycetes alone	Relatively small, relatively little variation among species; many about 16 kb
Chromosome structure	Variable—circular or linear, sometimes several circular DNA molecules	Circular	All circular
Information content	Conserved: rRNAs, tRNAs, ribosomal proteins, and proteins involved in respiration	Conserved: rRNAs, tRNAs, ribosomal proteins, and proteins involved in respiration	Conserved: tRNAs, rRNAs, and proteins involved in respiration
Introns present?	No, but noncoding sequences are abundant	Yes	No—few noncoding sequences
Modifications to universal genetic code?	No	* AUA for methionine (not isoleucine) * CUN for threonine (not leucine) * UGA for tryptophan (not stop)	* AUA for methionine (not isoleucine) * UGA for tryptophan (not stop)
Mode of inheritance	Variable, but usually maternal	Either parent	Maternal
Recombination?	Frequent	Frequent	None to infrequent
Rate of sequence divergence relative to nDNA	Slow	May be similar—still under investigation	Fast

genomes via endosymbiosis. What happened to these genomes as endosymbiotic bacteria began evolving into mitochondria and chloroplasts?

Present-day organelle genomes have only a tiny fraction of the genetic information present in a bacterium (Gray 1992). This observation suggests that numerous loci have been lost completely or transferred to the nucleus. The genes that encode what may be the most abundant protein in nature, ribulose bisphosphate carboxylase (RuBPCase), are a vivid example. This enzyme catalyzes the fixation of CO_2 during the Calvin–Benson cycle, which is the central pathway in the light-independent reactions of photosynthesis. RuBPCase is made up of two subunits. The gene for the protein's small subunit is found in the nuclear genome, while the gene for the large subunit is part of the cpDNA genome (Gillham et al. 1985). The same type of evidence for gene transfers exists in mitochondria. The organelle's ribosomes, for example, are composed of rRNAs encoded by mtDNA and proteins encoded by nDNA.

Because the gene content of chloroplasts and mitochondria is highly conserved, most of these gene transfers probably took place very early in the evolution of endosymbiosis (Gillham et al. 1985; Clegg et al. 1994). Recent gene transfers have also been documented, however. For example, the gene *tufA,* which codes for a translation factor active only in the chloroplast, is found in the cpDNA of green algae but in the nuclear genome of the flowering plant *Arabidopsis.* This observation suggests that *tufA* was transferred to the nucleus after algae and higher plants diverged. But because copies of the gene exist in both genomes of some green algae, it is more likely that the gene was duplicated and transferred to the nucleus early in plant evolution and then subsequently lost from the cpDNA of some derived lineages, like flowering plants (Baldauf and Palmer 1990; Baldauf et al. 1990).

Evidence for an even more recent gene transfer involves the *cox*2 locus found in the mtDNA of plants. This gene codes for one of the large subunits of cytochrome oxidase, a key component of the respiratory chain. In most plants, this locus is part of the mitochondrial genome. Most legumes (members of the pea family) have a copy in both nuclear DNA and mtDNA, however, and in mung bean and cowpea, the only copy is located in the nucleus. In these species, the structure of the gene closely resembles the structure of an edited mRNA transcript. Because RNAs can be reverse transcribed to DNA, these facts suggest that gene transfer from the mitochondrion to the nucleus took place recently via the reverse transcription of an edited mRNA intermediate (Nugent and Palmer 1991; Covello and Gray 1992).

Research on interactions between the nuclear and organelle genomes continues. In the yeast *Saccharomyces cerevisiae,* the nuclear loci involved in the synthesis of mitochondria are scattered over many different chromosomes, yet are regulated in concert; how this regulation is accomplished is still an active area of research (see Grivell et al. 1993). Many other questions remain about the evolution of these cohabitating genomes. Why have some genes, but not all, been transferred from organelles to the nucleus? Is there a selective advantage to having certain genes located in each genome, or were the movements merely chance events? Is reverse transcription of mRNAs the usual mechanism of transfer, or are other processes involved?

Based on the data reviewed here, it is clear that organelle genomes are just as dynamic as nuclear genomes, if not more so. Explaining how and why particular loci move among genomes, and why organelle genomes vary so dramatically in gene content and organization, remain as important challenges.

Summary

A major goal of molecular evolutionary studies is to understand how the four evolutionary forces (mutation, migration, drift, and selection) produce the diversity of sequences observed today. The neutral theory is important to this effort because it specifies the rate of fixation for alleles that are neutral with respect to fitness and thus change in frequency solely due to drift. As a result, it provides a null model for testing whether positive selection or drift is responsible for the observed rates and patterns of molecular evolution. The nearly neutral model is an important extension to the theory because it specifies how drift and selection interact as a function of population size.

The rate of evolution due to drift can be quantified by studying pseudogenes. As predicted by the neutral theory, pseudogenes have the highest rates of change observed among coding and noncoding sequences. In the vast majority of coding sequences, silent substitutions accumulate faster than replacement substitutions. Among loci, the rate of molecular evolution varies as a function of a gene's importance to the cell. These observations suggest that molecular evolution is dominated by drift and selection against deleterious mutations. In important cases, however, researchers have been able to show that high rates of replacement substitutions have occurred as a product of selection on beneficial mutations.

Even though silent substitutions do not change the amino acid sequences of proteins, they can still increase or decrease in frequency in response to natural selection. Variation in the availability of tRNAs can cause selection for translation efficiency and lead to codon bias. Selection on advantageous mutations can sweep linked silent substitutions to fixation; selection on deleterious mutations can also lead to the elimination of linked silent substitutions.

Transposable genetic elements are prominent components of eukaryotic and prokaryotic genomes. These loci move by either an RNA or DNA intermediate, but they do not code for products that increase the fitness of the host organism. Because they can cause deleterious mutations when they move to new locations in the genome, they are considered parasitic.

Chloroplasts and mitochondria originated as endosymbiotic bacteria and represent the largest horizontal gene transfer events in evolutionary history. Although the gene order, gene organization, and rate of sequence change in chloroplast and mitochondrial DNA vary among species, the information content of these organelles is remarkably well conserved.

Questions

1. Why are the rates of silent-site substitutions reported in Table 18.1 higher than the rates of replacement substitutions? Why do pseudogenes have the highest rates of evolution?

2. By using the start codon AUG as a guidepost, researchers can determine whether substitutions in pseudogenes correspond to silent changes or replacement changes. In contrast to most other loci, the rate of silent and replacement changes is identical in pseudogenes. Explain this observation in light of the neutral theory of evolution.

3. Why did Kimura and Ohta develop the nearly neutral model of molecular evolution? According to the neutral and nearly neutral theories, how does population size affect the rate of molecular evolution?

4. When researchers compare a gene in closely related species, why is it logical to infer that positive natural selection has taken place if replacement substitutions outnumber silent substitutions?

5. What is codon bias? Why is the observation of nonrandom codon use evidence that certain codons might be favored by natural selection? If you were given a series of gene sequences from the human genome, how would you determine whether codon usage is random or nonrandom?

6. The complete gene sequences of both humans and chimpanzees will be available soon. Outline how you would analyze homologous genes in the two species to determine which of the observed sequence differences result from drift and which result from selection.

7. Recall that the fourth chromosome of *Drosophila melanogaster* does not recombine during meiosis. The lack of genetic polymorphism on this chromosome has been interpreted as the product of a selective sweep. If the fourth chromosome had normal rates of recombination, would you expect the level of polymorphism to be different? Why?

8. Consider the possible costs and benefits of transposable elements to their hosts. For a eukaryote, what are the costs of carrying retrotransposons? Are there any benefits to hosting these genetic elements? For a prokaryote, what are the costs or benefits of carrying plasmids?

9. Suggest a hypothesis to explain why the size of plant mitochondrial genomes is so variable. What prediction(s) does your hypothesis make? How would you go about testing these predictions?

Exploring the Literature

10. Steve Palumbi and co-workers have published data suggesting that in animals, the rate of sequence change varies as a function of metabolic rate as well as generation time. To learn how molecular evolution varies between endothermic (warm-blooded) and poikilothermic (cold-blooded) animals, see

Martin, A. P., G. J. P. Naylor, and S. R. Palumbi. 1992. Rates of mitochondrial DNA evolution in sharks are slow compared with mammals. *Nature* 357: 153–155

Martin, A. P. and S. R. Palumbi. 1993. Body size, metabolic rate, generation time, and the molecular clock. *Proceedings of the National Academy of Sciences, USA* 90: 4087–4091.

11. Convergent evolution occurs when organisms that are not closely related evolve similar adaptations to similar environments (see Chapter 2). To explore a spectacular example of convergent evolution at the molecular level, see

Chen, L., A. L. DeVries, and C.-H. Cheng. 1997. Evolution of antifreeze glycoprotein gene from a trypsinogen gene in Antarctic notothenioid fish. *Proceedings of the National Academy of Sciences, USA* 94: 3811–3816.

Chen, L., A. L. DeVries, and C.-H. Cheng. 1997. Convergent evolution of antifreeze glycoproteins in Antarctic notothenioid fish and Arctic cod. *Proceedings of the National Academy of Sciences, USA* 94: 3817–3822.

Citations

Agrawal, A., Q. M Eastman, and D. G. Schatz. 1998. Transposition mediated by RAG1 and RAG2 and its implications for the evolution of the immune system. *Nature* 394: 744–751.

Ajioka, J. W., and D. L. Hartl. 1989. Population dynamics of transposable elements. In D. E. Berg and M. M. Howe, eds. *Mobile DNA.* Washington, DC: American Society of Microbiology, 939–958.

Akashi, H. 1994. Synonymous codon usage in *Drosophila melanogaster:* natural selection and translational accuracy. *Genetics* 144: 927–935.

Baldauf, S. L., and J. D. Palmer. 1990. Evolutionary transfer of the chloroplast *tufA* gene to the nucleus. *Nature* 344: 262–265.

Baldauf, S. L., J. R. Manhart, and J. D. Palmer. 1990. Different fates of the chloroplast *tufA* gene following its transfer to the nucleus in green algae. *Proceedings of the National Academy of Sciences, USA* 87: 5317–5321.

Berry, A. 1996. Non-non-Darwinian evolution. *Evolution* 50: 462–466.

Berry, A., J. W. Ajioka, and M. Kreitman. 1991. Lack of polymorphism on the *Drosophila* fourth chromosome resulting from selection. *Genetics* 129: 1111–1117.

Bester, T. H. and B. Tycko. 1996. Creation of genomic methylation patterns. *Nature Genetics* 12: 363–367.

Boeke, J. D. 1989. Transposable elements in *Saccharomyces cerevisiae.* In D. E. Berg and M. M. Howe, eds. *Mobile DNA.* Washington, DC: American Society of Microbiology, 335–374.

Charlesworth, B., and C. H. Langley. 1989. The population genetics of *Drosophila* transposable elements. *Annual Review of Genetics* 23: 251–287.

Charlesworth, B., M. T. Morgan, and D. Charlesworth. 1993. The effects of deleterious mutations on neutral molecular variation. *Genetics* 134: 1289–1303.

Chao, L. and D. E. Carr. 1993. The molecular clock and the relationship between population size and generation time. *Evolution* 47: 688–690.

Clegg, M. T., B. S. Gaut, G. H. Learn, Jr., and B. R. Morton. 1994. Rates and patterns of chloroplast DNA evolution. *Proceedings of the National Academy of Sciences, USA* 91: 6795–6801.

Covello, P. S., and M. W. Gray. 1992. Silent mitochondrial and active nuclear genes for subunit 2 of cytochrome *c* oxidase (*cox2*) in soybean: Evidence for RNA-mediated gene transfer. *EMBO Journal* 11: 3815–3820.

Dorit, R. L., H. Akashi, and W. Gilbert. 1995. Absence of polymorphism at the *ZFY* locus on the human Y chromosome. *Science* 268: 1183–1185.

Dunham, I., N. Shimizu, B. A. Roe, S. Chissoe, et al. 1999. The DNA sequence of human chromosome 22. *Nature* 402: 489–495.

Escalante, A. A., A. A. Lal, and F. J. Ayala. 1998. Genetic polymorphism and natural selection in the malaria parasite *Plasmodium falciparum. Genetics* 149: 189–202.

Gillham, N. W., J. E. Boynton, and E. H. Harris. 1985. Evolution of plastid DNA. In T. Cavalier-Smith, ed. *The Evolution of Genome Size.* New York: John Wiley & Sons, 299–351.

Goffeau, A., B. G. Barrell, H. Bussey, R. W. Davis, B. Dujon, H. Feldmann, F. Galibert, J. D. Hoheisel, C. Jacq, M. Johnston, E. J. Louis, H. W. Mewes, Y. Murakami, P. Philippsen, H. Tettelin, and S. G. Oliver. 1996. Life with 6000 genes. *Science* 274:546–567.

Gray, M. W. 1992. The endosymbiont hypothesis revisited. In D. R. Wolstenholme and K. W. Jeon, eds. *Mitochondrial Genomes.* San Diego, CA: Academic Press, 233–357.

Gray, M. W., G. Burger, and B. F. Lang. 1999. Mitochondrial evolution. *Science* 283: 1476–1481.

Griffiths, A. J. F., J. H. Miller, D. T. Suzuki, R. C. Lewontin, and W. M. Gelbart. 1993. *An Introduction to Genetic Analysis.* New York: W. H. Freeman.

Grivell, L. A., J. H. de Winde, and W. Mulder. 1993. Global regulation of mitochondrial biogenesis in yeast. In P. Broda, S. G. Oliver, and P. F. G. Sims, eds. *The Eukaryotic Genome.* Cambridge University Press, 321–332.

Hickey, D. A. 1992. Evolutionary dynamics of tranposable elements in prokaryotes and eukaryotes. *Genetica* 86: 269–274.

Hughes, A. L., and M. Nei. 1988. Pattern of nucleotide substitution at major histocompatibility complex class I loci reveals overdominant selection. *Nature* 335: 167–170.

Hutchison, C. A. III, S. C. Hardies, D. D. Loeb, W. R. Shehee, and M. H. Edgell. 1989. LINEs and related retroposons: Long interspersed repeated sequences in the eucaryotic genome. In D. E. Berg and M. M. Howe, eds. *Mobile DNA.* Washington, DC: American Society of Microbiology, 593–617.

Kimura, M. 1968. Evolutionary rate at the molecular level. *Nature* 217: 624–626.

Kimura, M. 1983. The neutral theory of molecular evolution. In M. Nei and R. K. Koehn, eds. *Evolution of Genes and Proteins.* Sunderland, MA: Sinauer, 208–233.

King, J. L., and T. H. Jukes. 1969. Non-Darwinian evolution. *Science* 164: 788–798.

Li, W.-H., T. Gojobori, and M. Nei. 1981. Pseudogenes as a paradigm of neutral evolution. *Nature* 292: 237–239.

Li, W.-H. and M. Tanimura. 1987. The molecular clock runs more slowly in man than in apes and monkeys. *Nature* 326: 93–96.

Li, W.-H., M. Tanimura, and P. M. Sharp. 1987. An evaluation of the molecular clock hypothesis using mammalian DNA sequences. *Journal of Molecular Evolution* 25: 330–342.

Li, W.-H., and D. Graur. 1991. *Fundamentals of Molecular Evolution.* Sunderland, MA: Sinauer.

McDonald, J. H., and M. Kreitman. 1991. Adaptive protein evolution at the *Adh* locus in *Drosophila. Nature* 351: 652–654.

Moran, J. V., R. J. DeBerardinis, H. H. Kazazian, Jr. 1999. Exon shuffling by L1 retrotransposition. *Science* 283: 1530–1534.

Nugent, J. M., and J. D. Palmer. 1991. RNA-mediated transfer of the gene *coxII* from the mitochondrion to the nucleus during flowering plant evolution. *Cell* 66: 473–481.

Nurminsky, D. I., M. V. Nurminskaya, D. De Aguiar, and D. L. Hartl. 1998. Selective sweep of a newly evolved sperm-specific gene in *Drosophila. Nature* 396: 572–575.

Ohta, T., and M. Kimura. 1971. On the constancy of the evolutionary rate of cistrons. *Journal of Molecular Evolution* 1:18–25.

Ohta, T. 1972. Evolutionary rate of cistrons and DNA divergence. *Journal of Molecular Evolution* 1: 150–157.

Ohta, T. 1977. Extension to the neutral mutation random drift hypothesis. In M. Kimura, ed. *Molecular Evolution and Polymorphism.* Mishima, Japan: National Institute of Genetics, 148–176.

Ohta, T., and M. Kreitman. 1996. The neutralist-selectionist debate. *BioEssays* 18: 673–683.

Palmer, J. D. 1987. Chloroplast DNA evolution and biosystematic uses of chloroplast DNA variation. *American Naturalist* 130: S6–S29.

Purugganen, M. D. and J. I. Suddith. 1998. Molecular population genetics of the *Arabidopsis* CAULIFLOWER regulatory gene: Nonneutral evolution and naturally occurring variation in floral homeotic evolution. *Proceedings of the National Academy of Sciences USA* 95: 8130–8134.

SanMiguel, P., A. Tikhonov, Y.-K. Jin, N. Motchoulskaia, D. Zakharov, A. Melake-Berhan, P. S. Springer, K. J. Edwards, M. Lee, Z. Avramova, and J. L. Bennetzen. 1996. Nested retrotransposons in the intergenic regions of the maize genome. *Science* 274: 765–768.

Sharp, P. M. 1997. In search of molecular Darwinism. *Nature* 385: 111–112.

Sharp, P. M., and W.-H. Li. 1986. An evolutionary perspective on synonymous codon usage in unicellular organisms. *Journal of Molecular Evolution* 24: 28–38.

Sharp, P. M., E. Cowe, D. G. Higgins, D. Shields, K. H. Wolfe, and F. Wright. 1988. Codon usage patterns in *Escherichia coli, Bacillus subtilis, Saccharomyces cerevisiae, Schizosaccharomyces pombe, Drosophila melanogaster,* and *Homo sapiens:* A review of the considerable within-species diversity. *Nucleic Acids Research* 16: 8207–8211.

Shen, M. R., J. Brosius, and P. L. Deininger. 1997. BC1 RNA, the transcript from a master gene for ID element amplification, is able to prime its own reverse transcription. *Nucleic Acids Research* 25: 1641–1648.

Sniegowski, P. D. and B. Charlesworth. 1994. Transposable element numbers in cosmopolitan inversions from a natural population of *Drosophila melanogaster*. *Genetics* 137: 815–827.

Voytas, D. F. 1996. Retroelements in genome organization. *Science* 274: 737–738.

Waugh O'Neill, R. J., M. J. O'Neill, and J. A. Marshall Graves. 1998. Undermethylation associated with retroelement activation and chromosome remodelling in an interspecific mammalian hybrid. *Nature* 393: 68–72.

Wichman, H. A., R. A. Van Den Bussche, M. J. Hamilton, and R. J. Baker. 1992. Transposable elements and the evolution of genome organization in mammals. *Genetica* 86: 287–293.

Wolfe, K. H., W.-H. Li, and P. M. Sharp. 1987. Rates of nucleotide substitution vary greatly among plant mitochondrial, chloroplast, and nuclear DNAs. *Proceedings of the National Academy of Sciences, USA* 84: 9054–9058.

Wolfe, K. H., P. M. Sharp, and W.-H. Li. 1989. Rates of synonymous substitution in plant nuclear genes. *Journal of Molecular Evolution* 29: 208–211.

Wolfe, K. H., C. W. Morden, and J. D. Palmer. 1991. Ins and outs of plastid genome evolution. *Current Opinions in Genetics and Development* 1: 523–529.

Yoder, J. A., C. P. Walsh, and T. H. Bestor. 1997. Cytosine methylation and the ecology of intragenomic parasites. *Trends in Genetics* 13: 335–340.

Zuckerkandl, E. and L. Pauling. 1965. Evolutionary divergence and convergence in proteins. In V. Bryson and H. J. Vogel, eds. *Evolving Genes and Proteins*. New York: Academic Press, 97–165.

CHAPTER 19

Evolution and Human Health

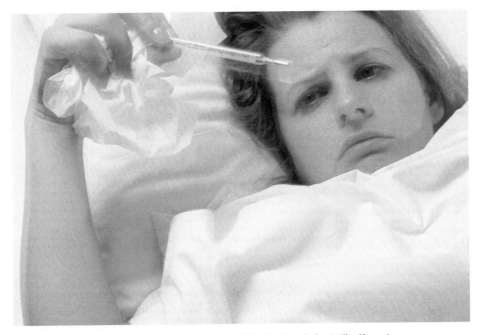

A human with a fever, caused by a viral or bacterial infection. (John Millar/Stone)

I N 1854 A CHOLERA EPIDEMIC STRUCK CENTRAL LONDON. THE DISEASE, WHICH causes severe diarrhea and dehydration, killed more than 500 people. In a famous act of medical detection, John Snow prepared a map of the affected neighborhood (see Tufte 1997). On it, he plotted the homes of the victims and the locations of the area's 11 water pumps (Figure 19.1). The fatalities clustered around the Broad Street pump, at the center of Snow's map. Sealing the case were the deaths of two women in distant neighborhoods, shortly after drinking water delivered by special arrangement from Broad Street. Although cholera's cause remained to be discovered, it was clearly associated with contaminated water.

In 1858, Louis Pasteur proposed that contagious diseases like cholera are caused by germs. Pasteur had been studying the fermentation of beer, wine, and milk, and had also been working to stop an epidemic of childbirth fever in a Paris maternity hospital. In a paper on lactic acid fermentation, Pasteur suggested that just as a particular microorganism was the cause of each kind of fermentation, so too might a particular microorganism be the cause of each infectious illness. Inspired by Pasteur, Robert Koch and others soon discovered the bacteria responsible for anthrax, wound infections, gonorrhea, typhoid fever, and tuberculosis. In 1883, Koch showed that cholera is caused by the bacterium *Vibrio cholerae*.

Figure 19.1 **John Snow's map of central London** The cholera deaths during the 1854 epidemic were concentrated around the Broad Street pump.

The germ theory of disease was arguably the most important breakthrough in the development of modern medicine. It laid the foundation not only for the identification of numerous pathogens, but also for the development of antiseptic surgery by Joseph Lister, the discovery of antibiotics by Alexander Fleming and others, and dramatic improvements in sanitation. The impact of sanitation and antibiotics on public health can be seen in Figure 19.2. The figure plots the death rate due to tuberculosis, in the United States, from 1900 through 1997. From 1900 to 1945, the tuberculosis death rate dropped from nearly 200 per 100,000 to about 40 per 100,000. This decline was largely due to improvements in sanitation, housing conditions, and nutrition. Then, in 1945, the death rate began falling more sharply still. The accelerated decline was due to the introduction of antibiotics, including streptomycin and isoniazid. By 1997, the tuberculosis death rate was fewer than 0.4 per 100,000, less than two-tenths of one percent what it was in 1900.

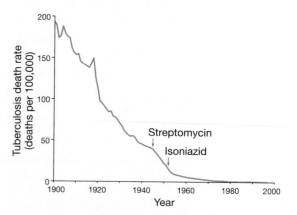

Figure 19.2 **Tuberculosis death rate as a function of time in the United States** Between 1900 and 1950, the TB death rate declined dramatically, largely as a result of improvements in sanitation and housing. The introduction of antibiotics at mid-century further hastened the decline.

Charles Darwin published *On the Origin of Species* in 1859, the year after Pasteur proposed the germ theory of disease. Evolutionary biology and modern medicine were born at the same time and have grown up in parallel. The relevance of evolutionary biology to medicine is deep and, in some ways, has only recently begun to be appreciated. George C. Williams and Randolph Nesse have been leaders in this field, which they call Darwinian medicine (Williams and Nesse 1991; Nesse and Williams 1994, 1998). Throughout this book we have highlighted medical applications of evolutionary analysis. In Chapter 1, for example, we discussed the evolution of HIV. In Chapters 4 through 7 we considered the impact of infectious diseases like AIDS, sickle-cell anemia, and cystic fibrosis on the evolution of human populations. In Chapter 11 we explored the evolution of senescence and menopause. Here we devote an entire chapter to medical applications we have not elsewhere had the opportunity to address.

The chapter is divided into two parts. In Sections 19.1 through 19.4, we consider medical consequences of the fact that populations evolve. Examples include the evolution of pathogen populations and the evolution of cell populations within individual patients. In Sections 19.5 through 19.7, we turn our attention from pathogen and cell populations to the human animal as it has been shaped by natural selection. We consider applications of the adaptationist program, introduced in Chapter 8, in understanding puzzling aspects of human physiology and behavior.

19.1 Evolving Pathogens: Evasion of the Host's Immune Response

The fundamental event in evolution is a change in the frequencies of various genotypes within a population. It is with this phenomenon that we begin our discussion of evolution and human health. There are two kinds of evolving populations important in medicine: populations of pathogens, and populations of human cells within individual patients (as in a cancer). We consider evolving pathogen populations first, and then evolving cell populations.

Pathogens and their hosts are, by definition, in conflict. The pathogens attempt to consume the host's tissues, converting them into more pathogens. The host attempts to kill the pathogens. When we become hosts, our bodies employ an impressive array of weapons against the invaders. Our immune systems are capable of recognizing billions of foreign proteins, of mounting an aggressive and multifaceted response, and of remembering the structure of the pathogen's proteins so that the response will be mobilized more quickly in the event of future invasions. The pathogens are formidable enemies, however. Many pathogens have large population sizes, short generation times, and high mutation rates. These traits mean that pathogen populations evolve quickly. Any mutation that enables its possessors to evade or withstand the host's immune response should be strongly selected, and should quickly increase in frequency. Walter Fitch and colleagues (1991) investigated whether selection imposed by the human immune system is responsible for detectable evolution in populations of influenza A viruses.

Conflicts among organisms are inevitable. In the conflict between a parasite and its host, the host's immune system selects for parasites that can evade detection.

Flu Virus Evolution

Influenza A is responsible for annual flu epidemics, and for occasional global pandemics, such as occurred in 1918, 1957, and 1968. Most of us think of the flu as merely an annoyance—worse than a cold, certainly, but not as bad as chicken pox. In fact, however, the flu can be deadly. In an ordinary flu season, the disease kills about 20,000 Americans. The 1918 pandemic flu was among the most devastating plagues in history. Within a period of months, it sickened some 20% of the world's population and killed between 20 and 100 million people (Kolata 1999).

Influenza A has a genome composed of eight RNA strands that encodes a total of 10 proteins (Figure 19.3), including polymerases, structural proteins, and coat proteins (Webster et al. 1992). The predominant coat protein is called hemagglutinin. Hemagglutinin initiates an infection by binding to sialic acid on the surface of a host cell (Laver et al. 1999). Hemagglutinin is also the primary protein recognized, attacked, and remembered by the host's immune system. In order to stay alive, any given strain of influenza A must find a steady supply of naive hosts who have never been exposed to its version of hemagglutinin. Fitch and colleagues focused on mutations that alter the amino acids in hemagglutinin's **antigenic sites**. Antigenic sites are the specific parts of a protein that the immune system recognizes and remembers. Fitch and colleagues hypothesized that flu strains with novel antigenic sites would enjoy a selective advantage.

To test their hypothesis, the researchers examined the hemagglutinin genes of influenza A viruses isolated from infected humans and stored in freezers between 1968 and 1987. Flu viruses evolve a million times faster than mammals, so the 20 years spanned by the frozen virus samples is equivalent to roughly four times the duration that separates humans from our common ancestor with the chimpanzees. In other words, the frozen flu samples constitute a fossil record—but one from which we can sequence genes.

From the sequences of the hemagglutinin genes, Fitch and colleagues estimated the phylogeny of the frozen flu samples. The result appears in Figure 19.4. Two patterns are immediately apparent. First, the flu strains accumulated nucleotide substitutions in their hemagglutinin genes at a steady rate, about 6.7×10^{-3} per nucleotide per year. Second, most of the flu samples represent extinct side branches on the evolutionary tree. The flu viruses of the late 1970s and early 1980s were not a diverse assembly of strains descended from different ancestors of the late 1960s. Instead they were close relatives, and all were descended from a single ancestor of the late 1960s. The progeny of the other late-sixties strains had all died out.

What allowed the surviving lineage to endure while the other lineages perished? According to the researchers' hypothesis, it was nucleotide substitutions resulting in amino-acid replacements in hemagglutinin's antigenic sites. From the

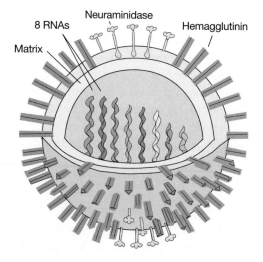

Figure 19.3 The influenza A virus The flu virus has two major surface proteins, hemagglutinin and neuraminidase. The viral genome, consisting of ten genes, is carried on eight separate pieces of RNA. Redrawn from Webster et al. (1992) by permission of the American Society for Microbiology.

(a)

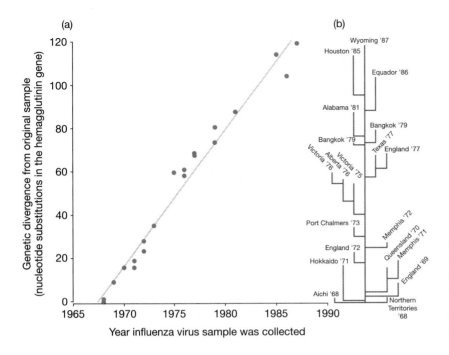

(b)

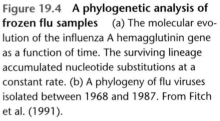

Figure 19.4   A phylogenetic analysis of frozen flu samples   (a) The molecular evolution of the influenza A hemagglutinin gene as a function of time. The surviving lineage accumulated nucleotide substitutions at a constant rate. (b) A phylogeny of flu viruses isolated between 1968 and 1987. From Fitch et al. (1991).

nucleotide sequences, the researchers inferred all of the amino-acid replacements that had occurred in the surviving lineage and in the extinct lineages. Then they noted whether each replacement had occurred in an antigenic site or a nonantigenic site. Fitch and colleagues predicted that, compared with the extinct lineages, the surviving lineage would have a higher fraction of its amino-acid replacements in its antigenic sites. The amino-acid replacements, 109 in total, were distributed as follows:

	Surviving lineage	Extinct lineages
In antigenic sites	33	31
In nonantigenic sites	10	35

Consistent with the researchers' prediction, more than three-quarters of the surviving lineage's replacements had occurred in regions of hemagglutinin recognized by the immune system, compared to fewer than half of the extinct lineages' replacements. This association between a lineage's fate and the location of its replacements is statistically significant ($P = 0.002$).

Robin Bush, Walter Fitch, and colleagues (1999) followed up on this result by examining nucleotide substitutions in a phylogeny of hemagglutinin genes from 357 influenza A strains isolated between 1985 and 1996. The researchers took as their null hypothesis the neutral theory of molecular evolution. Recall that under the neutral theory two processes dominate molecular evolution: (1) Mutations resulting in amino-acid replacements are typically deleterious and are eliminated by selection, and (2) mutations to synonymous codons are neutral and may become fixed in the population by genetic drift. According to the neutral theory, when we look at the nucleotide substitutions that have occurred on an evolutionary tree, silent substitutions should outnumber replacement substitutions. Of the 331 nucleotide substitutions that Bush, Fitch, and colleagues analyzed, 191 (58%) were silent and 140 (42%) were replacement substitutions. This result is consistent with the neutral theory.

However, the researchers also identified 18 codons in the hemagglutinin gene in which there had been significantly more replacement substitutions than silent

Phylogenetic analyses show that flu strains are more likely to survive if they have novel amino-acid sequences in proteins recognized by the host's immune system.

substitutions. The ratios in these 18 codons ranged from 4 replacement substitutions and 0 silent substitutions to 20 replacement substitutions and 1 silent substitution. An excess of replacement substitutions over silent substitutions is not consistent with the neutral theory. Bush, Fitch, and colleagues concluded that these 18 codons had been under positive selection for changes in the encoded amino acid. All 18 of the positively selected codons were for amino acids in antigenic sites of the hemagglutinin protein. It appears that the human immune system does, indeed, exert strong selection on flu virus hemagglutinin genes, and that virus populations evolve in response.

This result is potentially useful to the makers of flu vaccines. Flu vaccines work by exposing the patient's immune system to killed flu viruses. Even though the viruses are dead, the immune system recognizes the viral proteins as foreign, mounts a response against them, and remembers their structure. In the event of a later infection by live viruses, the immune system can respond immediately. It can respond immediately, that is, as long as the hemagglutinin on the live invaders is similar enough to the hemagglutinin on the dead viruses that were in the vaccine. The problem is that vaccines take months to prepare in large quantities. Vaccine makers must begin production well in advance of the flu season. That means their scientific advisors must try to predict which among recently circulating flu strains are most likely to be responsible for next season's epidemic, so that they know which strains to include in the vaccine.

Robin Bush, Catherine Bender, and colleagues (1999) devised a way to predict which of the currently circulating flu strains is most likely to have surviving descendents in the future. The survivor, they reasoned, is most likely to be the currently circulating strain with the most mutations in the 18 codons known to be under positive selection. On this basis, the researchers were able to accurately "predict," for 9 of 11 recent flu seasons, which of each season's strains would be the one to survive while the rest became extinct. Bush, Bender, and colleagues are careful to note that predicting which of this season's flu strains will be the ancestor of future lineages is not the same as predicting which, if any, of this season's strains will be responsible for next season's epidemic. Nonetheless, the predictive technique devised by Bush, Bender, and colleagues should be a valuable addition to the forecasting methods already in use.

The Origin of Pandemic Flu Strains

The fact that flu viruses with novel hemagglutinin genes appear to be at a selective advantage suggests a mechanism by which a strain could acquire the ability to cause a global pandemic. If a flu strain could somehow radically alter the structure of its hemagglutinin so that it was different from any hemagglutinin that had ever been seen by any human's immune system, then the strain could sweep the world and potentially infect everyone alive.

How could a flu strain radically alter the structure of its hemagglutinin? The organization of the influenza genome indicates a way (Figure 19.3). Recall that the flu genome has eight different RNA strands that encode a total of 10 different genes. If two flu strains simultaneously infect the same host cell, their genomes can recombine. That is, when new virions form, they may contain some RNA strands from Strain 1 and other RNA strands from Strain 2. Strain 1, for example, might produce offspring carrying Strain 2's hemagglutinin gene.

The phylogeny in Figure 19.5 provides evidence that flu strains do, in fact, swap genes. This phylogeny, produced by Owen Gorman and colleagues (1991), is based on nucleotide sequences of influenza nucleoprotein genes. Nucleoprotein is thought

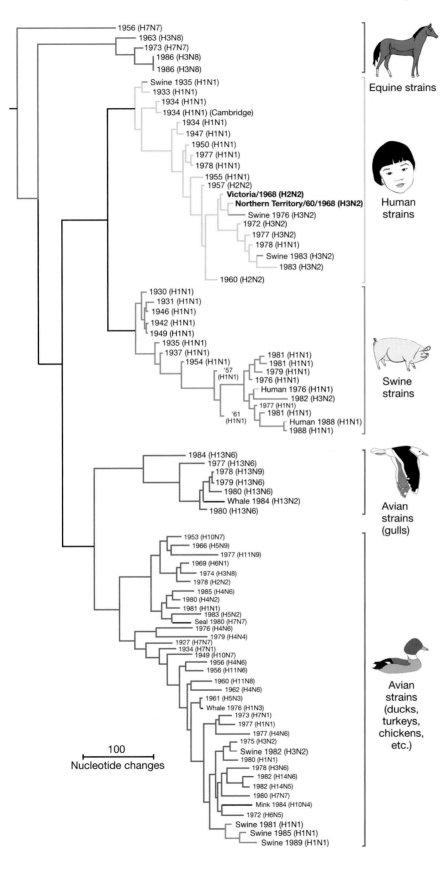

100
Nucleotide changes

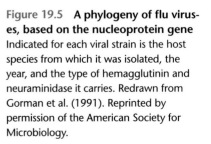

Figure 19.5   A phylogeny of flu viruses, based on the nucleoprotein gene
Indicated for each viral strain is the host species from which it was isolated, the year, and the type of hemagglutinin and neuraminidase it carries. Redrawn from Gorman et al. (1991). Reprinted by permission of the American Society for Microbiology.

to be the viral protein most responsible for host-specificity. That is, the structure of a strain's nucleoprotein both enables the strain to infect particular species of hosts, and tends to confine the strain to those species. Phylogenies based on the nucleoprotein gene should therefore be reliable indicators of strain history. Notice first that the nucleoprotein phylogeny has several distinct clades: one that infects mainly humans, another that infects mainly pigs, two that infect mainly birds, and so on.

Now look at the branch tips and their labels. These give the species from which each strain was isolated, the year of isolation, and the viral subtype. The subtype H3N2, for example, means "hemagglutinin-3, neuraminidase-2." Neuraminidase, like hemagglutinin, is a coat protein. The numbers refer to groups of hemagglutinins or neuraminidases, defined by the ability of host antibodies to recognize them. The most important point for our purposes is that each hemagglutinin group constitutes a clade. All H1s are more closely related to each other than to any H2 or H3 or H4. The same is true of the neuraminidases. Find the human strains Human/Victoria/1968 (H2N2) and Human/Northern Territory/60–1968 (H3N2); they are set in boldface type. These two strains have nucleoproteins that are very closely related, and neuraminidases that are closely related, but hemagglutinins that are distantly related. How is this possible? The simplest explanation is that it is possible because flu strains can trade genes. An examination of the phylogeny will reveal numerous additional examples.

Prior to the global pandemic of 1968, human flu viruses had never carried H3. This suggests that it was the acquisition of H3 from a nonhuman strain that allowed the 1968 flu to infect huge numbers of people worldwide. What was the source of the H3 gene? Figure 19.6 shows a phylogeny, from W. J. Bean and colleagues (1992),

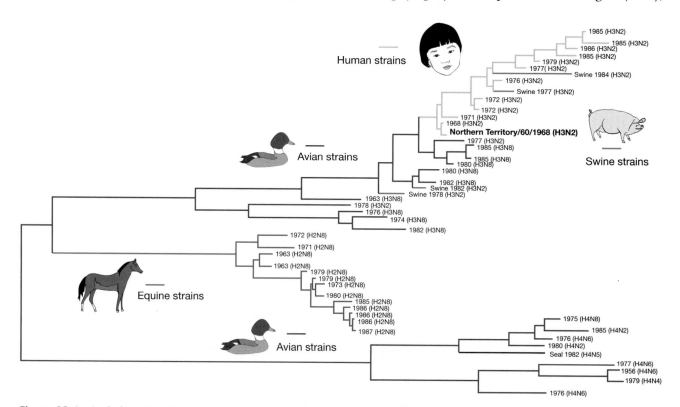

Figure 19.6 A phylogeny of flu virus hemagglutinin genes Indicated for each viral strain is the species from which it was isolated, the year, and the type of neuraminidase and hemagglutinin it carries. The 1968 human flu appears to have acquired its hemagglutinin gene from a bird flu strain. Redrawn from Bean et al. (1992). Reprinted by permission of the American Society for Microbiology.

of human and nonhuman H3 genes, with H4 genes as the outgroup. The human H3 genes branch from within the bird H3s, with Northern Territory/60–1968's H3 near the base of the human clade. Apparently, the 1968 human pandemic flu strain acquired its H3 gene from birds.

How could a human flu strain acquire an H3 gene from a bird strain? Look back at the nucleoprotein phylogeny (Figure 19.5). The nucleoprotein phylogeny reveals that human flu strains sometimes infect pigs (for example, Swine 1976, which was isolated in Hong Kong), bird strains sometimes infect pigs (Swine 1982, also from Hong Kong), and pig strains sometimes infect humans (Human 1976, from New Jersey). One popular hypothesis among flu researchers is that flu pandemics begin when human strains and bird strains simultaneously infect a pig, swap genes, and later move from pigs to people (Webster et al. 1992). An international team of researchers maintains constant surveillance of flu strains circulating in pigs, birds, and humans. Their goal is to spot new pandemic strains early enough to allow the production and distribution of large quantities of vaccine. The flu surveillance researchers keep an especially keen watch for recombinant strains, and for strains that that are moving from species to species.

Genetic recombination allows flu strains to acquire new coat proteins. The acquisition of a new coat protein appears to be one factor that enables a flu strain to cause a global epidemic.

19.2 Evolving Pathogens: Antibiotic Resistance

Antibiotics are chemicals that kill bacteria. For human patients, antibiotics are life-saving drugs. For populations of bacteria, however, antibiotics are powerful agents of selection. When applied to a population of bacteria, an antibiotic quickly sorts the resistant individuals (those that can tolerate the drug) from the susceptible ones (those that cannot). An evolutionary perspective suggests that antibiotics should be used judiciously; otherwise these miracle drugs may undermine their own effectiveness.

There are dozens of antibiotics, and dozens of molecular mechanisms whereby bacteria can become resistant (see Baquero and Blázquez 1997 for a review). Some of these mechanisms of resistance involve losses of function. Resistance to isoniazid in the tuberculosis pathogen *Mycobacterium tuberculosis* provides an example. Isoniazid poisons bacteria by interfering with the production of components of the cell wall (Rattan et al. 1998). Before it can do so, however, isoniazid must be converted by a bacterial enzyme into its biologically active form. The conversion is performed by the enzyme catalase/peroxidase, encoded by a gene called *KatG*. Mutations in the *KatG* gene that reduce or eliminate catalase/peroxidase activity render bacteria tolerant or immune to isoniazid's effects.

Other mechanisms of resistance involve gains of function. Numerous extrachromosomal elements of bacteria, such as plasmids and transposons, carry genes conferring resistance to one or more antibiotics. The plasmid *Tn3,* for example, found in *E. coli,* contains a gene called *bla*. This gene encodes an enzyme, β-lactamase, that breaks down the antibiotic ampicillin.

Evidence That Antibiotics Select for Resistant Bacteria

Evidence that antibiotics select in favor of resistant bacteria comes from a variety of studies. On the smallest scale are studies of bacterial evolution within individual patients. William Bishai and colleagues (1996) monitored an AIDS patient with tuberculosis. When they initially determined that the patient had tuberculosis, the researchers cultured bacteria from the patient's lungs and found them sensitive to a variety of antibiotics, including rifampin. They and other doctors treated the patient with rifampin in combination with several other drugs. The patient responded well to treatment; at one point, the researchers were unable to culture tuberculosis bacteria from his lungs. Soon, however, the patient relapsed and died. After his death, the researchers found that the tuberculosis bacteria in the patient's lungs had resurged. They screened these bacteria for resistance to antibiotics. The bacteria were still susceptible to most drugs, but they were resistant to rifampin. The researchers sequenced the *rpoB* gene from some of the resistant bacteria. In the gene, they found a point mutation known to confer rifampin resistance.

Did the rifampin-resistant strain of tuberculosis evolve in the patient's lungs, or had he been infected with a new strain that was already resistant when he got it? The researchers prepared genetic fingerprints of rifampin-sensitive bacteria from the patient's initial infection and rifampin-resistant bacteria from the patient's autopsy. Other than the *rpoB* point mutation, the genetic fingerprints of the two groups of bacteria were identical. The researchers examined over 100 other strains of bacteria from patients living in the same city at the same time. Only two of the strains had genetic fingerprints matching the strain that killed the patient, and neither of these was rifampin-resistant. The simplest explanation for these results is that the *rpoB* point mutation occurred in bacteria living in the patient's lungs, and ultimately rose to high frequency due to selection imposed by treatment with rifampin.

On a larger scale, researchers can compare the incidence of susceptible versus resistant bacterial strains among patients who are newly diagnosed, and thus have not been previously treated with antibiotics, versus patients who have relapsed after antibiotic treatment. If antibiotics select in favor of drug resistance, then a higher fraction of relapsed patients should harbor antibiotic-resistant bacteria. Alan Bloch and colleagues (1994) reported the results of a survey of tuberculosis patients conducted by the Centers for Disease Control. The results for bacterial susceptibility to isoniazid are as follows:

Several lines of evidence show that antibiotics select in favor of resistant bacteria, and that bacterial populations evolve rapidly in response.

	New cases	Relapsed cases
Number with resistant bacteria	243	41
Number with susceptible bacteria	2728	150
% Resistant	8.2%	21.5%

These numbers are consistent with the notion that populations of bacteria within patients evolve in response to treatment.

Finally, on the largest scale, researchers can examine the relationship over time between the fraction of patients with resistant bacteria and the society-wide level of antibiotic use. If antibiotics select in favor of resistance, then the level of resistance should track antibiotic consumption. D. J. Austin and colleagues (1999) plotted data on penicillin resistance among *Pneumococcus* bacteria in children in Iceland (Figure 19.7). In the late 1980s and the early 1990s, the fraction of children whose bacteria were resistant to penicillin rose dramatically. In response, Icelandic public health authorities waged a campaign to reduce the use of penicillin. Between 1992 and 1995, the per capita consumption of penicillin by children dropped by about 13%. The level of penicillin resistance peaked at just under 20% in 1993, then fell below 15% by 1996. Again, the data are consistent with

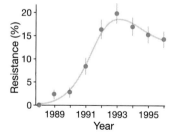

Figure 19.7   The frequency of penicillin resistance among *Pneumococcus* bacteria in Icelandic children as a function of time   When public health officials waged a public campaign to reduce the use of penicillin, the incidence of resistant strains declined. Redrawn from Austin et al. (1999). Copyright © 1999, National Academy of Sciences, USA.

the hypothesis that bacterial populations evolve in response to selection imposed by antibiotics.

Evaluating the Costs of Resistance to Bacteria

Presumably, penicillin resistance in Iceland fell with declining penicillin use because penicillin resistance is costly to bacteria. If resistance comes at a cost, then when antibiotics are absent, sensitive bacteria will have higher fitness.

Costs of resistance are generally assumed to be common. When antibiotic resistance is conferred by loss-of-function mutations, costs might be levied by the loss-of-function itself. When antibiotic resistance is conferred by gains of function, costs might be levied by the expense of maintaining new genes and their associated proteins.

Costs to bacteria associated with antibiotic resistance would be good news for public health. If an antibiotic begins to lose its effectiveness because too many bacteria are resistant, doctors and patients could simply make a collective agreement to suspend the use of the antibiotic until bacterial populations have evolved back to the point that they are again dominated by susceptible strains.

While resistance may, indeed, typically come at a cost, the cost may not always persist. Additional mutations elsewhere in the bacterial genome may compensate for the costs, making resistant bacteria equal in fitness to sensitive bacteria, even when antibiotics are absent. Stephanie Schrag and colleagues (1997) investigated the possibility that compensatory mutations could alleviate the cost of streptomycin resistance in *E. coli*. Their experimental design and results appear in Figure 19.8.

Antibiotic resistance is generally assumed to be costly to bacteria, but over the long term costs of resistance can be eliminated by natural selection.

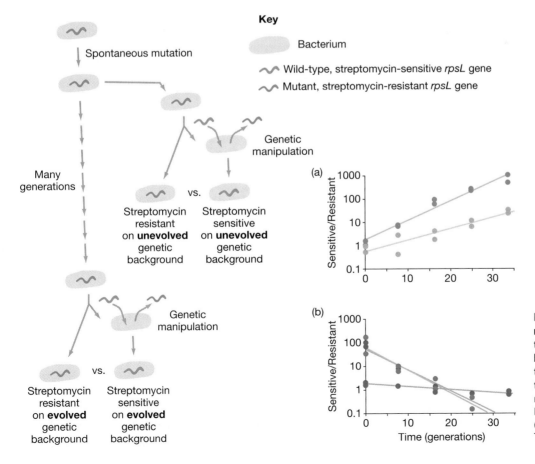

Figure 19.8   An assessment of the costs of antibiotic resistance to bacteria, over the short term and over the long term   See text for explanation. Scatter plots from Fig. 1 in Schrag et al. (1997). Copyright © 1997, The Royal Society.

Schrag and colleagues started with a population of streptomycin-sensitive *E. coli,* and screened for new streptomycin-resistant mutants. Streptomycin interferes with protein synthesis by binding to a ribosomal protein encoded by the *rpsL* gene. Point mutations in the *rpsL* gene render bacteria resistant to streptomycin. In one set of experiments, the researchers competed streptomycin-resistant strains against identical strains restored to sensitivity by replacement of the mutant version of *rpsL* with the wild-type version. If streptomycin resistance comes at a cost, at least in the short-term, then in mixed cultures the streptomycin sensitive bacteria should increase in frequency over time. This is exactly what happened (Figure 19.8, graph a).

In a second set of experiments, Schrag and colleagues let streptomycin-resistant strains evolve for many generations in the lab. Then the researchers again competed streptomycin-resistant strains against identical strains restored to sensitivity by genetic manipulation. If compensatory mutations had occurred and become fixed while the resistant strains were allowed to evolve, then in mixed cultures the streptomycin-sensitive strains should fail to increase in frequency over time. In fact, the result was even more dramatic: The streptomycin-sensitive strains decreased in frequency over time (Figure 19.8, graph b). Schrag and colleagues concluded that compensatory evolution had not only alleviated the cost of streptomycin resistance, it had created a multilocus genotype, or genetic background, on which the resistant allele of *rpsL* enjoyed a fitness advantage over the sensitive allele.

Judicious Use of Antibiotics

If Schrag et al.'s results are general, there is no guarantee that an antibiotic can be restored to medical effectiveness simply by withdrawing it from use until bacterial populations have evolved back to sensitivity. If we want to maintain an arsenal of potent antibiotics that we can use when patients' lives are at stake, it seems that our best chance is to try to avoid letting bacterial populations evolve resistance in the first place.

Bacteriologist Stuart Levy (1998) recommends guidelines for the limitation of antibiotic resistance. The guidelines are intended to prevent people from contracting bacterial infections in the first place, restrict unnecessary uses of antibiotics that may select for resistance in potentially pathogenic bacterial bystanders, and ensure that, when antibiotics *are* used, they exterminate the targeted bacterial population before resistance evolves. Among Levy's guidelines are these:

The best defense against antibiotic-resistant bacteria is to avoid letting bacterial populations evolve resistance in the first place.

- To avoid contracting foodborne bacteria, consumers should wash fruits and vegetables and avoid raw eggs and undercooked meat.
- Consumers should use antibacterial soaps and cleaners only when they are needed to prevent infection in patients with compromised immune systems.
- Patients should not request antibiotics for viral infections, such as colds or flu.
- When they do take antibiotics, patients should complete the full course of treatment. Patients should never save antibiotics prescribed for one infection and use them to treat another.
- To avoid spreading infections from patient to patient, doctors should wash their hands thoroughly between patients.
- Doctors should never prescribe unneeded antibiotics, even when patients request them.
- When they do prescribe antibiotics, doctors should use drugs that target the narrowest possible range of bacterial species.

- Doctors should isolate patients infected with bacteria resistant to several drugs to reduce the risk that such bacteria will spread.

19.3 Evolving Pathogens: Virulence

The final issue we will consider relating to the evolution of pathogen populations is **virulence**. Virulence is the harm done by a pathogen to the host during the course of an infection. Virulence varies dramatically among human pathogens. Some pathogens, like cholera and smallpox, are often lethal; others, like herpes viruses and cold viruses, may produce no symptoms at all. Evolutionary biologists investigating virulence seek to explain this diversity.

How Virulence Evolves

There are three general models to explain the evolution of virulence (Bull 1994, Ewald 1994, Levin 1996):

1. **The coincidental evolution hypothesis.** The virulence of many pathogens in humans may not be a target of selection itself, but rather an accidental by-product of selection on other traits. For example, tetanus is caused by a soil bacterium, *Claustridium tetanae*. When tetanus bacteria find themselves inside a human wound, they can grow and divide. They also produce a potent neuro-toxin, making tetanus infections highly lethal. However, tetanus bacteria ordinarily do not live in humans and are not transmitted by humans. The ability of these bacteria to produce tetanus toxin is probably the result of selection during their ordinary life in the soil, not selection inside human hosts.

2. **The short-sighted evolution hypothesis.** Pathogens may experience many generations of evolution by natural selection within an individual host before they have the opportunity to move to a new host. As a result, traits that enhance the within-host fitness of pathogen strains may rise to high frequency even if they are detrimental to transmission of the pathogen to new hosts. Poliovirus may provide an example. Ordinarily, polioviruses infect only cells that line the digestive tract, produce no symptoms, and are transmitted via feces. Occasionally, however, polio virions invade the cells of the nervous system. Acquiring the ability to invade the nervous system probably increases within-host fitness, because the virions that can do so have fewer intraspecific competitors. But the virions living in the nervous system are unlikely to ever be transmitted to a new host.

3. **The trade-off hypothesis.** Biologists traditionally believed that all pathogen populations would evolve toward ever-lower virulence. The reasoning was that damage to the host must ultimately be detrimental to the interests of the pathogens that live within it. This traditional view was naive. Recall our discussion of the evolution of aging in Chapter 11. There, we concluded that genes that hasten the death of their carriers can nonetheless rise in frequency if they confer a sufficient enhancement of early-life reproductive success. As with genes, so too with pathogens. A virulent strain may increase in frequency in the total pathogen population if, in the process of killing its hosts, it sufficiently increases its chances of being transmitted. Natural selection should favor pathogens that strike an optimal balance between the costs and benefits of harming their hosts.

Virulence, the harm done by a parasite to its host, is a trait that can evolve.

It is the trade-off hypothesis that we will focus on here. A key assumption of the trade-off hypothesis is that a pathogen cannot reproduce inside its host without doing the host some harm. Every offspring that the pathogen makes is constructed with energy and nutrients stolen from the host. In addition, the pathogen produces metabolic wastes that the host must detoxify and eliminate. These are the reasons the host mounts an immune response against the pathogen—an expensive endeavor that compounds the hosts' costs, but that may be better for the host than the alternative.

All else being equal, pathogens with higher within-host reproductive rates should be transmitted to new hosts at higher rates as well. But all else is equal only up to a point. Because reproducing faster within the host necessarily means harming the host more severely, it is possible for the pathogen to reproduce *too* fast. Reproducing too fast means debilitating the host so severely that the rate of transmission to new hosts is reduced.

Sharon Messenger, Ian Molineux, and James Bull (1999) tested the trade-off hypothesis by using *E. coli* as the host and a virus, bacteriophage f1, as the pathogen. Phage f1 produces lasting, nonlethal infections in *E. coli*. The phage invades a bacterium and lives inside it as a plasmid. It induces the machinery of the host cell to produce new phage copies that are secreted from the cell as phage chromosomes encased in protein filaments. Production of new phage copies slows the growth rate of the host bacterium to about one-third of normal. But when the host bacterium does divide, copies of phage f1 typically travel with both daughter cells. Thus phage f1 has two modes of transmission: It is transmitted vertically, from one host generation to the next, when the host cell divides, and it is transmitted horizontally, from one host to another, when secreted virions invade new hosts.

Messenger and colleagues maintained cultures of phage f1 in which they forced the viruses to alternate between the two modes of transmission. During the vertical-transmission phase, the researchers prevented secreted virions from infecting new bacterial cells. The only way phages could spread was via the reproduction of their hosts. During the horizontal-transmission phase, the researchers harvested secreted virions and introduced them to cultures of uninfected bacteria. Now the only way the phages could spread was via secretion.

The researchers maintained two sets of cultures for 24 days. For one set of cultures, they alternated one-day-long vertical-transmission phases with brief horizontal-transmission phases. For another set of cultures they alternated eight-day-long vertical-transmission phases with brief horizontal-transmission phases. At the end of 24 days, the researchers measured phage virulence and phage reproductive rate. They measured virulence as the growth rate of infected hosts, with lower host growth rates indicating more virulent viruses. They measured phage reproductive rate as the rate of virion secretion from infected hosts, with more rapid secretion indicating faster phage reproduction.

Messenger and colleagues made two predictions. First, they predicted that, across their cultures, they would find a correlation between phage virulence and phage reproduction rate. In other words, phages that induced their hosts to produce and secrete more phage copies would slow the growth of their hosts more severely. Second, they predicted that the cultures subjected to eight-day vertical-transmission phases would evolve lower reproductive rates and lower virulence than the cultures subjected to one-day vertical-transmission phases. Their reasoning here was that during the vertical-transmission phase natural selection should favor viral

According to the trade-off hypothesis, selection favors parasites that reproduce more quickly within their hosts—until the parasites begin to harm the hosts so severely that the probability of transmission begins to fall.

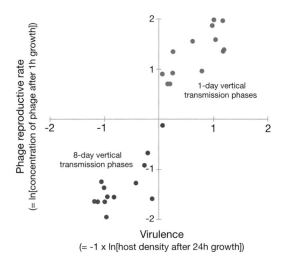

Figure 19.9 A trade-off between the virulence and within-host reproductive rate of a virus that infects *E. coli* When researchers gave the viruses more opportunities for horizontal transmission (red dots), the viruses evolved higher virulence and higher reproductive rates than viruses given fewer opportunities for horizontal transmission (blue dots). Redrawn from Fig. 3 in Messenger et al. (1999). Copyright © 1999, The Royal Society.

strains that allow their host bacteria to divide more quickly, whereas during the horizontal transmission phase selection should favor viral stains that induce their host bacteria to secrete more viral copies. The eight-day cultures experienced more selection to allow their hosts to divide and less selection to induce secretion, so they should evolve lower virulence.

The results appear in Figure 19.9. The figure is a scatterplot showing the reproductive rate and virulence of each of 13 one-day cultures and 13 eight-day cultures. First, note the strong correlation across both experiments between viral reproductive rate and virulence. As Messenger and colleagues predicted, the phage strains that slow their hosts' growth rate most dramatically are the strains that reproduce more quickly within their hosts. Second, note that the eight-day cultures had lower reproductive rates and lower virulence than the one-day cultures. As the researchers predicted, different patterns of selection favor different levels of virulence. These results are consistent with the trade-off hypothesis.

Virulence in Human Pathogens

Paul Ewald (1993, 1994) considered how the trade-off hypothesis might apply to human pathogens. He used the hypothesis to make predictions about how the details of disease transmission should select for different levels of virulence. In Chapter 1 we discussed Ewald's predictions about how rates of partner exchange will select on the virulence of HIV. Here we will discuss two more of Ewald's predictions, along with his tests of them using comparative data compiled from the literature.

Ewald's first prediction concerns the virulence of diseases like colds and flu, that are transmitted by direct contact between an infected person and an uninfected person, versus diseases like malaria, that are transmitted by insect vectors. Ewald reasoned that diseases transmitted by direct contact cannot afford to be virulent. If a host is so incapacitated by illness that he or she stays home in bed and avoids contact with uninfected individuals, the pathogen has no chance to be transmitted. Vectorborne diseases, on the other hand, can afford to be highly virulent. An insect vector can carry pathogens away from even a severely debilitated host, and might, in fact, be at less risk of being killed in the process. Ewald compiled data on the mortality rates of a variety of vectorborne and directly transmitted diseases. These data are consistent with Ewald's prediction (Figure 19.10). The vast

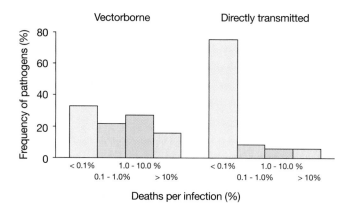

majority of directly transmitted diseases have mortality rates under 0.1%, whereas more than 60% of vectorborne diseases have mortality rates of 0.1% or higher.

Ewald's second prediction concerns bacteria that infect the digestive tract and cause diarrhea. These bacteria can typically be transmitted both directly from person to person and via contaminated water. Ewald reasoned that contaminated water can play the same role that insect vectors did in his first prediction. That is, when sewage enters the drinking water supply, even a severely incapacitated host can transmit bacteria to remote individuals over long distances. Ewald assembled data on roughly 1000 outbreaks of illness caused by nine different kinds of bacteria. Some of these bacteria have a stronger tendency to be spread by direct contact, others have a stronger tendency to be spread by contaminated water. For each kind of bacteria, Ewald calculated both the fraction of outbreaks attributed to contaminated water and the victim mortality rate. He predicted that diseases with a higher frequency of waterborne transmission would be more virulent. The data, plotted in Figure 19.11, are consistent with Ewald's prediction. The most virulent of the nine bacteria in the study is classical *Vibrio cholerae,* the pathogen responsible for the waterborne, and lethal, cholera outbreak in London during 1854.

The trade-off hypothesis for the evolution of virulence implies that human behavior can affect the severity of human diseases. For example, when people dump untreated sewage directly into rivers, or when health-care workers fail to wash their hands thoroughly between patients, they create conditions that may select for increased virulence in human pathogens. Conversely, when people keep their drinking water supplies pure, and when health-care workers avoid becoming inadvertent vectors, they create conditions that may select for reduced virulence in human pathogens.

Human diseases borne by insects or water tend to be more virulent than diseases transmitted by direct contact. When pathogens are carried by insects or water, even a severely debilitated host can still transmit the disease.

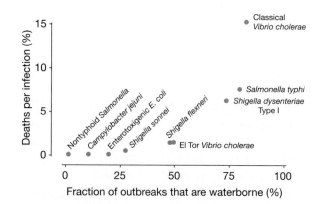

We now turn our attention from evolving populations of pathogens to another group of evolving populations important to human health: populations of humans cells inside individual humans. It may seem surprising to even propose that human cell populations can evolve, because we are used to thinking of all of the cells in a human body as being genetically identical. In fact, however, there are mechanisms that produce genetic diversity among somatic cells, and conditions under which somatic cell populations can evolve.

19.4 Tissues as Evolving Populations of Cells

All of the cells in an individual's body are descended from a common ancestor, the zygote. If, during the development of a tissue, a mutation occurs in a cell still capable of continued division, then we can think of the tissue as a population of reproducing cells with heritable genetic variation. If one of the genetic variants leads to increased cell survival or faster reproduction, then the tissue will evolve by natural selection, just like a population of free-living organisms.

A Patient's Spontaneous Recovery

Rochelle Hirschhorn and colleagues (1996) documented a case in which tissue evolution saved the life of a boy with a serious genetic disease. The disease is called adenosine deaminase deficiency. Adenosine deaminase (ADA), encoded by a locus on chromosome 20, is a housekeeping enzyme normally made in all cells of the body. ADA's job is to recycle purines. Cells lacking ADA accumulate two poisonous metabolites—deoxyadenosine and deoxyadenosine triphosphate. The cells in the body most susceptible to these poisons are the lymphocytes, including the T cells and B cells vital to the immune system (Youssoufian 1996). Individuals who inherit loss-of-function mutations in both copies of the ADA gene have no T cells, and have B cells that are nonfunctional or absent (Klug and Cummings 1997). In consequence, these individuals suffer from severe combined immunodeficiency. Without treatment, they usually die of opportunistic infections at an early age.

Both parents of the boy that Hirschhorn and colleagues studied carry, in heterozygous state, a recessive loss-of-function allele for ADA. One of the boy's older siblings inherited two loss-of-function alleles; made no ADA, and died of severe combined immunodeficiency at age two. The boy himself also inherited both his parents' mutant alleles, and during his first five years he suffered the recurrent bacterial and fungal infections characteristic of severe combined immunodeficiency. Between the ages of five and eight, however, the boy spontaneously and mysteriously recovered. He was 12 years old when Hirschhorn and colleagues published their paper, and he had been clinically healthy for four years.

With a careful genetic analysis of the mother, father, and son, Hirschhorn and colleagues were able to reconstruct a plausible explanation for the boy's recovery. Although the boys' parents are both carriers for ADA deficiency, they are carriers for different loss-of-function alleles. Hirschhorn and colleagues showed that the son's blood cells are a genetic mosaic (Box 19.1). The father's mutation is present in all of the boy's peripheral leukocytes (white blood cells) and lymphoid B cells. The mother's mutation is present in all of the boy's peripheral leukocytes, but absent in most of his B cells.

BOX 19.1 Genetic sleuthing solves a medical mystery

Figure 19.12 shows the nucleotide sequence of a short piece of the ADA gene. Both parents have point mutations substituting an A for a G. The father's point mutation is in an intron/exon splice site, resulting in the deletion of an entire exon from the processed ADA transcript. The mother's point mutation results in the substitution of one amino acid for another in the primary structure of the ADA enzyme. Both mutations are known, from other individuals who have them, to virtually destroy the enzymatic activity of ADA.

Notice that the wild-type, or normal, sequence contains a cutting site for the restriction enzyme *Hin*p1I

and a cutting site for the restriction enzyme *Bsr*I. The father's mutation eliminates the *Bsr*I cutting site, whereas the mother's mutation eliminates the *Hin*p1I cutting site. Hirschhorn and colleagues amplified a 254-bp-long sequence from the ADA genes in both parents and in different populations of the boy's blood cells. When digested with *Bsr*I, the wild-type 254-bp fragment yields a 182-bp fragment and a 72-bp fragment. Figure 19.13a illustrates a *Bsr*I restriction fragment analysis of the family. Look first at the mother's lane on the gel. Both her wild-type allele and her mutant allele have the *Bsr*I cutting site, so her gel shows only two bands: the 182-bp fragment and the 72-bp

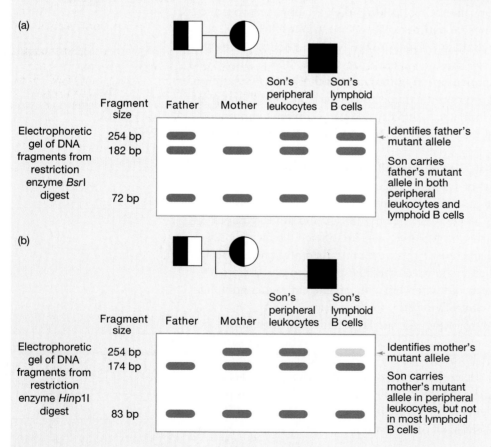

                    *Hin*p1I   *Bsr*I

Wild type: ...G C G C C A C C A G C C C A G T...

Father's mutation: ...G C G C C A C C A G C C C A **A** T...

Mother's mutation: ...G C **A** C C A C C A G C C C A G T...

Figure 19.12 **A short piece of the gene for adenosine deaminase** Point mutations are shown in red; the bars indicate sites recognized by the restriction enzymes *Hin*p1I and *Bsr*I. From Hirschhorn et al. (1996).

(a)

	Fragment size	Father	Mother	Son's peripheral leukocytes	Son's lymphoid B cells	

Electrophoretic gel of DNA fragments from restriction enzyme *Bsr*I digest

254 bp
182 bp

72 bp

Identifies father's mutant allele

Son carries father's mutant allele in both peripheral leukocytes and lymphoid B cells

(b)

	Fragment size	Father	Mother	Son's peripheral leukocytes	Son's lymphoid B cells	

Electrophoretic gel of DNA fragments from restriction enzyme *Hin*p1I digest

254 bp
174 bp

83 bp

Identifies mother's mutant allele

Son carries mother's mutant allele in peripheral leukocytes, but not in most lymphoid B cells

Figure 19.13 **Restriction fragment analysis of the genetics of ADA deficiency in a family** (a) A *Bsr*I restriction fragment assay for the father's mutant allele. The 254-bp band at the top of the gel is a marker for the presence of the father's mutation, which eliminates a cutting site recognized by *Bsr*I. (b) A *Hin*p1I restriction fragment assay for the mother's mutant allele. The 254-bp band at the top of the gel is a marker for the presence of the mother's mutation, which eliminates a cutting site recognized by *Hin*p1I. After Hirschhorn et al. (1996).

BOX 19.1   Continued

fragment. Now look at the father's lane. His wild-type allele has the *Bsr*I cutting site, so his lane shows the 180-bp and 72-bp bands. His mutant allele, however, lacks the *Bsr*I cutting site, so his lane also shows a 254-bp fragment. Finally, look at the boy's lanes. They show that the father's mutant allele is present in both the boy's peripheral leukocytes (white blood cells) and his lymphoid B cells.

Figure 19.13b illustrates a *Hin*p1I restriction fragment analysis. This time, the mother's mutant allele shows up as an uncut 254-bp fragment. The son's lanes show something unexpected: His mother's mutant allele is present in his peripheral leukocytes, but absent in most of his lymphoid B cells. We say most, because

there is a faint band from a 254-bp fragment in the boy's B-cell lane. Closer examination of 15 different B-cell lines revealed that all carry the father's mutant allele, but only two carry the mother's mutant allele.

Hirschhorn and colleagues discovered that the mother's mutant allele also contains a unique neutral marker. When they checked the boy's lymphoid B cell lines that appear to lack the mother's mutant allele, the researchers found that these cells in fact carry the mother's neutral marker. Hirschhorn and colleagues concluded that the ancestral cell to most of the boy's existing lymphoid B cells had sustained a lucky back mutation in the allele from the mother, thus spontaneously reverting to wild type.

How could this happen? Hirschhorn and colleagues found evidence that the cell ancestral to most of the boy's existing lymphoid B cells had sustained a lucky back mutation in the allele the boy inherited from his mother, thus spontaneously reverting to wild type, or normal (Box 19.1). Over time, the descendants of this reverted cell apparently became more and more abundant in the boy's B-cell population. Eventually, reverted B cells became abundant enough, and made and released enough ADA, that the boy's clinical symptoms of ADA deficiency vanished.

Hirschhorn and colleagues believe that the increase in frequency of reverted B cells in the boy's B-cell population happened by natural selection. It is also possible that the increase happened by drift. It is likely, however, that reverted cells are at a distinct selective advantage. Because they make their own supply of a crucial housekeeping enzyme, they should live longer than cells that have to pick up the enzyme after it has been released by the cells that make it.

The boy's story may have implications for the treatment of other individuals with ADA deficiency. The outlook for patients with ADA deficiency has improved in recent years. In 1987, researchers developed a form of injectable ADA that is an effective enzyme replacement treatment for many ADA patients (Hershfield et al. 1987). In the early 1990s, researchers began the first clinical experiments with somatic-cell gene therapy (Blaese et al. 1995, Bordignon et al. 1995). Somatic-cell gene therapy involves removing lymphocytes and/or bone marrow cells from the patient, inserting a functioning version of the ADA gene with its own promoter into their chromosomes, and returning the cells to the patient's body. In other words, gene therapy is an attempt to accomplish by design the reversion mutation that happened spontaneously in the boy studied by Hirschhorn and colleagues. The early trials have shown that the engineered cells can survive for years, and that they can grow and divide. In some early trials, gene therapy appears to have been responsible for dramatic improvements in the patients' clinical health.

As a precaution against the failure of gene therapy, researchers conducting gene-therapy trials have kept their patients on enzyme replacement therapy. Hirschhorn and colleagues suggest, however, that enzyme replacement may reduce the effectiveness

Populations of cells inside an individual's body may exhibit genetic variation and differential fitness, and thus may evolve.

of gene therapy. If enzyme replacement reduces the selective advantage the engineered cells have over ADA–deficient cells, it will slow the rate at which the patients' blood cell population evolves by natural selection. If further research supports this suggestion, then in the future doctors will have to balance the benefits of encouraging rapid selective fixation of the engineered cells against the risks of depriving patients of the insurance provided by continued enzyme replacement therapy.

Reconstructing the History of a Cancer

Another medical context in which it is productive to view tissues as evolving populations of cells is cancer (Shibata et al. 1996). A cancer starts with a cell that has accumulated mutations that free it from the normal controls on cell division. The cell divides, and its offspring divide, and so on, to produce a large population of descendants—that is, a tumor. The cells in some kinds of cancer have extremely high mutation rates, allowing tumors to accumulate measurable genetic diversity. If we assume

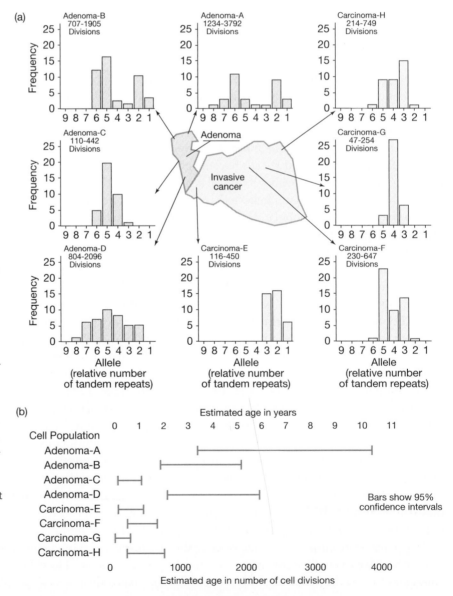

Figure 19.14 An evolutionary reconstruction of the history of a cancer within an individual patient (a) The diagram at the center is a map of a pair of neighboring tumors in a 43-year-old patient. At left is an adenoma, a tumor of glandular tissue that is usually benign. At right is a colorectal carcinoma, an invasive malignant tumor. The histograms that surround the map show the genetic diversity within samples of cells taken from the locations indicated by the arrows. Each histogram shows the frequencies of various alleles at a tandem repeat locus. Different alleles are identified by the number of tandem repeats they carry. Some cell samples, such as Adenoma-D, show high genetic diversity (many different alleles present). Other cell samples, such as Carcinoma-E, show low genetic diversity. From Shibata et al. (1996). (b) Estimates of the age of the different tumor subpopulations shown in (a). The higher genetic diversity of the samples from the adenoma indicates that the adenoma is older than the carcinoma. Redrawn from Shibata et al. (1996).

that the mutation rate per cell division is constant, and that the mutations are neutral, then the genetic diversity within a particular tumor is a measure of the tumor's age.

Occasionally, a cell may leave the tumor it was born in and migrate elsewhere in the body to initiate a new tumor. (This process is called metastasis.) This new tumor represents a new population of cells. Because it was founded by a single individual, this new population will have low genetic diversity. As it grows, however, the population will evolve. Like the population from which its founder came, the new tumor will accumulate genetic diversity as a result of mutation and genetic drift.

Darryl Shibata and colleagues (1996) use the amount of genetic variation in tumors to reconstruct the history of cancers within individual patients. Their approach is similar to that used by Sarah Tishkoff and colleagues to reconstruct the history of populations in the species *Homo sapiens* (Chapter 16). Non-African populations have lower genetic diversity than African populations. With other evidence, this pattern indicates that non-African populations were founded by emigrants from Africa.

Figure 19.14a shows the allelic diversity at a tandem repeat locus in subpopulations of two adjacent tumors in a 43-year-old patient. The patient has a type of cancer called a colorectal adenocarcinoma. One tumor, called an adenoma, is in glandular tissue; the other tumor, called a carcinoma, is in colorectal tissue. Three of the four adenoma subpopulations show substantially greater genetic diversity than the carcinoma subpopulations. Figure 19.14b shows estimates of the ages of the eight subpopulations, based on the estimated mutation rate for this kind of cancer and the assumption that each cancer cell divides once per day. It appears that the adenoma grew benignly for up to 10 years before it spawned a cell that migrated into the patient's colon to initiate the malignant carcinoma.

The analytical tools evolutionists use to reconstruct the history of populations can also be used to reconstruct the history of tissues and tumors within a patient's body.

19.5   The Adaptationist Program Applied to Humans

We now shift our focus from pathogen and cell populations that evolve within humans to the human animal itself. Our goal is to illustrate how researchers use the analytical tools of the adaptationist program to understand aspects of human physiology and behavior relevant to medicine and public health. Researchers following the adaptationist program identify traits that appear to be adaptive (see Chapter 8). On the assumption that these traits were produced by natural selection, the researchers formulate and test hypotheses about how the traits enhance fitness.

Section 19.5 considers complications that arise when applying the adaptationist program to a species that alters its own environment at a rate that outpaces evolution by natural selection. Section 19.6 uses fever to show how the adaptationist program can be applied to physiological puzzles. Finally, Section 19.7 explores parenting as an example of how the adaptationist program can be used to analyze human behavior. We should warn the reader now that few of the hypotheses we discuss have been adequately tested. Our coverage of the adaptationist approach to human health is partly a plea for more research.

The analytical tools evolutionists use to study form and function in other organisms can, with appropriate caution, also be used to study form and function in humans.

Adaptation to What Environment?

Before any attempt to apply the adaptationist program to humans, it is crucial to ask: To what environment are humans adapted? Until the advent of agriculture some 10,000 years ago, all humans lived as hunter–gatherers (Figure 19.15). Hunter–gatherers occupied, and still do occupy, a wide variety of habitats, ranging

Figure 19.15   **A Yagua hunter–gatherer from the Amazon rain forest of Peru**   Dressed for hunting, he is leaning on his blow gun. He uses the blow gun to shoot poisoned darts at game animals. (James Holland/Stock Boston)

from deserts to arctic tundra. But none of these environments resembles that of a modern urbanite. In other words, our lives are not like the lives of our ancestors.

Boyd Eaton and various colleagues have attempted to reconstruct some of the basic features of our ancestral Stone Age-lifestyle (Eaton et al. 1997; Eaton and Cordain 1997; Cordain et al. 1997). Their evidence comes from observations of present–day hunter–gatherers, archaeological remains, and analyses of uncultivated edible plants and wild game animals. Figure 19.16 shows their estimate of the energy sources in a typical hunter–gatherer diet, compared to the energy sources in a typical modern American diet. Hunter–gatherers eat more fruits and vegetables, leaner meat, fewer cereal grains, and fewer milk products. Hunter–gatherers also get considerably more exercise. Among the !Kung hunter–gatherers of the Kalahari Desert, for example, a typical individual walks 10 to 15 kilometers each day (6 to 9 miles), and uses energy at more than 1.5 times his or her resting metabolic rate. A sedentary modern American, in contrast, may walk effectively no distance each day, and uses energy at less than 1.2 times his or her resting metabolic rate.

Eaton points out that the impact of our novel lifestyle on health has not been assessed through rigorous controlled experiments. However, in observational studies, features of the modern lifestyle, such as a high-fat diet, are associated with a variety of diseases, including heart disease, strokes, and cancer. These conditions, rare in hunter–gatherers, are diseases of civilization (Nesse and Williams 1994).

There are a few cases in which researchers have shown that human populations have evolved in response to selection imposed by a modern lifestyle. Production of lactase provides an example. Lactase is the enzyme that enables us to digest milk sugar, or lactose. The only source of milk sugar in the diet of most mammals is mother's milk, so there would be no advantage, and probably some cost, in continuing to produce lactase after weaning. Indeed, individuals stop producing lactase

Figure 19.16   **A typical hunter–gatherer diet versus a typical modern American diet**   From Eaton and Cordain (1997). Copyright © 1997, S. Krager AG. Reprinted by permission of S. Karger AG.

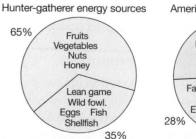

Hunter-gatherer energy sources

65% Fruits Vegetables Nuts Honey

Lean game Wild fowl. Eggs Fish Shellfish

35%

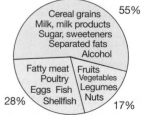

American energy sources

Cereal grains Milk, milk products Sugar, sweeteners Separated fats Alcohol   55%

Fatty meat Poultry Eggs Fish Shellfish   28%

Fruits Vegetables Legumes Nuts   17%

around the age of weaning in most mammal species, and in many populations of humans. However, for modern humans with lifelong access to cows' milk, continuing to produce lactase after weaning can be advantageous. In human populations with a long history of drinking fresh milk, and only in these populations, many individuals have a heritable ability to continue producing lactase as adults (Durham 1991).

In a milk-drinking culture, the ability to produce lactase is probably beneficial throughout life. Many diseases of civilization, however, strike only late in life. As a result, selection on genetic susceptibility to these diseases is probably weak (see Chapter 11). This consideration, combined with the fact that a modern lifestyle dates back at most a few hundred generations, implies that we should not expect that evolution by natural selection in human populations will have kept pace with our changing lifestyle. In other words, even the most sophisticated modern urbanites have bodies and brains that are largely designed for life in the Stone Age.

The pace of cultural change has been so much faster than the pace of evolution by natural selection that modern humans are largely still adapted to life in the Stone Age.

The recognition that the environment in which we live is different from the environment we are adapted to has at least two implications. It helps make sense of some otherwise puzzling features of our physiology, and it suggests ways to reduce some of the risks associated with modern life. Myopia provides an example of the former, and menstruation an example of the latter.

Myopia

In many populations the incidence of myopia, or nearsightedness, is 25% or more. Researchers have used twin studies to determine whether variation in vision has a genetic basis. For example, J. M. Teikari and colleagues (1991) assessed the similarities in vision between monozygotic versus dizygotic twins. As mentioned in Chapter 7, if a trait is heritable, then monozygotic twins should resemble each other more strongly than dizygotic twins. Teikari et al.'s data are as follows:

	Monozygotic Twins	Dizygotic Twins
Concordant pairs	36	19
Discordant pairs	18	36

Two-thirds of the monozygotic twin pairs were concordant—that is, both nearsighted or both normal—versus only one-third of the dizygotic twin pairs. These data suggest that nearsightedness is partially heritable.

But how could nearsightedness be even partially heritable? Many Americans are legally blind without their corrective lenses. These people would be at a serious disadvantage if forced to live as hunter–gatherers. Surely natural selection among hunter–gatherers would quickly eliminate alleles associated with myopia. And if natural selection eliminated alleles that caused myopia among our hunter–gatherer ancestors, they cannot have passed such alleles to us.

The solution to the puzzle of myopia is to recognize that modern humans live a lifestyle that is as novel in its visual demands as it is in its diet and activity levels. Hunter–gatherers do not spend their childhoods indoors reading under artificial light. Perhaps the alleles that predispose some of us to myopia actually cause myopia only in a modern environment.

The fact that human populations have not had time to adapt to the environment we live in can help explain some of our apparently maladaptive traits.

Evidence to evaluate this hypothesis comes from populations of people who have only recently adopted a modern lifestyle. Francis Young and colleagues (Young et al. 1969, Sorsby and Young 1970) went to Barrow, Alaska, to measure the incidence of myopia among Eskimos. The researchers chose this population because most of the families in it had moved to Barrow from isolated communities during and after World War II, drawn by the economic activity associated with a naval research laboratory, a radar station, and oil exploration. As the town of Barrow grew, a school system grew with it. Most of the children Young and colleagues examined attended formal, American-style schools and did a great deal of reading. Most of the adults over 35 had attended less formal, ungraded schools for a maximum of six years. Young et al.'s data on the incidence of myopia in younger versus older individuals are as follows:

Age	Number myopic	Number not myopic	Fraction Myopic
6–35	146	202	42%
36–88	8	152	5%

The children, who had been much more strongly exposed to a modern visual lifestyle, had a substantially higher incidence of myopia.

Young et al.'s study was observational, not experimental, and the amount of schooling was just one of many differences between the environments of the children versus the adults. Nonetheless, the data are consistent with a variety of studies on humans and animals indicating that the shape of the growing eye is molded by visual experience (see Norton and Wildsoet 1999 for a review). This body of human and animal research suggests that myopia is caused by a combination of genetic susceptibility and close-in visual work. The older Eskimos in Barrow had grown up in a visual environment more like that of our hunter–gatherer ancestors. The older adults had the same alleles they passed to their children and grandchildren, but they themselves were not myopic. In other words, myopia can be partially heritable because the alleles that predispose some modern humans to myopia do not cause myopia in a hunter–gatherer environment.

Menstrual Cycling

The monthly menstrual cycling experienced by most modern women is usually considered normal. However, a growing body of epidemiological evidence suggests that continuous menstrual cycling increases a woman's risk of breast cancer. A woman's risk of cancer is higher the earlier she begins to menstruate, the later she has her first child, and the less time she spends nursing (see, for example, Layde et al. 1989, Berkey et al. 1999). Menstrual cycling appears to elevate the risk of breast cancer because the combination of estrogen and progesterone present during the postovulatory phase of the cycle stimulates cell division in the lining of the milk ducts (Henderson et al. 1993). With more cell divisions come more opportunities for mutations that may create cancers. Given the high incidence of breast cancer among modern women—about one in eight in North America—it is worth knowing whether continuous menstrual cycling really is normal.

Beverly Strassmann (1999) spent two years observing menstrual cycling among the Dogon of Mali. The Dogon are a premodern people who use no contraceptives. Their culture is an easy one within which to study menstrual cycling, because custom dictates that the women sleep in special menstrual huts while they are menstruating. Strassmann confirmed that menstruating women do, in fact, sleep in the huts, and that women sleeping in the huts are menstruating, by regularly collecting urine samples from 93 women for two and one-half months and checking for metabolites of estrogen and progesterone. Strassmann then tracked visits to the huts by the Dogon women over a period of two years. She found that women between the ages of 20 and 35 spend little time cycling (Figure 19.17a). Instead, they are usually either pregnant or experiencing lactational amenorrhea—a suppression of cycling due to nursing. On any given day, about 25% of adult Dogon women are cycling, about 15% are pregnant, about 30% are in lactational amenorrhea, and about 30% are past menopause (Figure 19.17b). Strassmann estimates that the average Dogon woman has a total of about 100 menstrual cycles during her lifetime. This is less than one-third the number for a typical modern woman.

Strassmann's data on the Dogon suggest that women's bodies may not have been designed by natural selection to tolerate long periods of continuous menstrual cycling. If continuous menstrual cycling is not normal for women, then we can think of the high rates of breast cancer among modern women as another maladaptive consequence of life in a novel environment. Strassmann does not have data on the incidence of breast cancer among Dogon women, but she notes that among urban West African women, whose menstrual patterns parallel those of the Dogon, the breast cancer rate is about one-twelfth that among North American women.

Perhaps modern women should consider using hormonal treatments that maintain their bodies in a hormonal state more consistent with the state experienced by our ancestors. Oral contraceptives reduce the risk of endometrial and ovarian cancer among modern women who use them, but they do not reduce the risk of breast cancer (Henderson et al. 1993). D.V. Spicer and colleagues are developing an oral contraceptive that suppresses ovarian function, does not stimulate cell division in the breast, and contains sufficient concentrations of sex steroids to avoid adverse side effects such as accelerated osteoporosis (Spicer et al. 1991, Henderson et al. 1993). They predict that their new strategy will be equally as effective as present oral contraceptives at reducing the risk of ovarian cancer, about half as effective at reducing the risk of endometrial cancer, and much better at reducing the risk of breast cancer.

Keeping in mind that the environment modern humans live in may not be the environment to which we are adapted, we devote the remaining two sections of this chapter to examples showing how researchers use an adaptationist framework to develop and test hypotheses about medical physiology and hypotheses about fundamental aspects of human behavior.

The fact that human populations have not had time to adapt to the lifestyle we live may explain why we lack physiological defenses against many diseases of civilization.

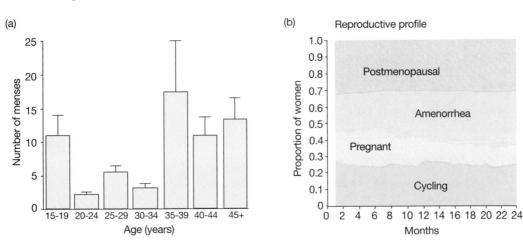

Figure 19.17 **Menstrual cycling in Dogon women** (a) Most Dogon women had relatively few menstrual cycles during a two-year period. (b) This is because, at any given time, a relatively high fraction of the Dogon women were pregnant or experiencing lactational amenorrhea. From Strassmann (1999). Reprinted by permission of Mary Ann Liebert, Inc.

19.6 Adaptation and Medical Physiology: Fever

Many people consider the symptoms that accompany illness to be a nuisance. A common response to the fever associated with a cold or flu, for example, is to take aspirin, acetaminophen, or ibuprofen. These drugs reduce the fever, but they do not combat the virus that is causing the cold or flu. Here we ask whether taking drugs to reduce fever is a good idea. To answer the question, we need to know why people run a fever when they are sick.

An evolutionary perspective suggests two interpretations of fever. One interpretation is that fever may represent manipulation of the host by the pathogen. Viruses or bacteria may release chemicals that cause the host to elevate its body temperature so as to increase the pathogen's growth rate or reproductive rate. If this hypothesis is correct, then reducing the fever would probably help the host combat the infection. The second interpretation is that fever may be an adaptive defense against the pathogen. The pathogen may grow and reproduce more slowly at higher temperatures, or the host's immune response may be more effective at higher temperatures. If this hypothesis is correct, then taking drugs that alleviate fever might be counterproductive to recovery.

Matthew Kluger has, for over 20 years, advocated the second hypothesis—that fever is an adaptive defense against disease. In 1974, Linda Vaughn, Harry Bernheim, and Kluger discovered that the desert iguana (*Dipsosaurus dorsalis*, Figure 19.18a) develops a behavioral fever in response to infection with a bacterium called *Aeromonas hydrophila*. Recall from Chapter 8 that iguanas, being ectotherms,

An evolutionary perspective can help medical researchers develop hypotheses about physiological function.

Figure 19.18 **Behavioral fever in the desert iguana (*Dipsosaurus dorsalis*)** (a) A desert iguana. (b) Linda Vaughn and colleagues injected saline into nine control iguanas. The gray and blue bars show the average upper and lower preferred temperatures, or set-points, (± standard error) for these lizards before and after the injection. There was no significant change in either set-point. Vaughn and colleagues then injected killed bacteria (*Aeromonas hydrophila*) into 10 experimental iguanas. The light and dark orange bars show the average set-points for these lizards before and after the injection. The increase in both set-points after injection with killed bacteria was statistically significant ($P < 0.001$). From Vaughn et al. (1974). Reprinted by permission from Nature. © 1974, Macmillan Magazines, Ltd. (c) This graph shows the fraction of experimentally infected lizards surviving at each temperature over time (numbers of lizards in parentheses). From Kluger, Ringler, and Anver (1975). Copyright © 1975, American Association for the Advancement of Science.

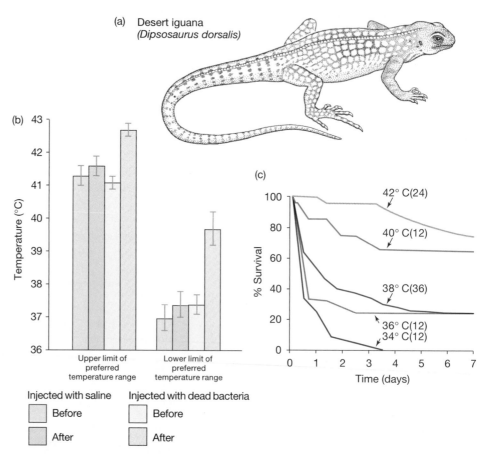

(a) Desert iguana (*Dipsosaurus dorsalis*)

use behavior instead of physiology to regulate their body temperatures. They move to hot spots to warm themselves and to cold spots to cool themselves. Vaughn et al. found that when they injected desert iguanas with dead bacteria, the lizards chose body temperatures about 2°C higher than they normally choose (Figure 19.18b).

Is behavioral fever an adaptive response to infection, or are the bacteria manipulating the iguanas? To distinguish between these hypotheses, Kluger, Daniel Ringler and Miriam Anver (1975) infected desert iguanas with live bacteria, then prevented the lizards from thermoregulating by keeping them in fixed-temperature incubators. Most of the iguanas kept at temperatures mimicking behavioral fever survived, whereas most of the iguanas kept at lower temperatures died (Figure 19.18c). This result suggests that behavioral fever is, in fact, adaptive for desert iguanas infected with *A. hydrophila*.

If fever is an adaptive defense against *A. hydrophila,* then it would probably be a bad idea for infected lizards to take aspirin. This is not as silly a statement as it sounds, at least in one sense: The researchers found that the aspirin-like drug sodium salicylate reduces behavioral fever in iguanas just as it reduces physiological fever in mammals. Apparently thermoregulation is controlled by similar neurological mechanisms in both groups of animals. Bernheim and Kluger (1976) infected 24 desert iguanas with bacteria, then gave half of the infected lizards sodium salicylate. The researchers allowed all the iguanas to behaviorally thermoregulate. All of the control iguanas developed behavioral fever, and all but one of them survived the infection. Five of the medicated lizards developed behavioral fever in spite of the medication, and all of them survived. The other seven medicated iguanas failed to develop behavioral fever, and all of them died.

Since the mid-1970s, researchers have documented behavioral fever in a wide variety of reptiles, amphibians, fishes, and invertebrates. In several animal studies researchers have shown that fever increases survival (see Kluger 1992 for a review). These results broadly support the hypothesis that fever is an adaptive response to infection.

Fever is much harder to study in endotherms than it is in ectotherms. Researchers cannot force endotherms to be at an arbitrary body temperature simply by putting them in an incubator. And as we will see shortly, drugs that reduce fever have effects on the immune system that are independent of fever.

In an attempt to disentangle the effects of the increased metabolic rate that accompanies a fever from the increase in body temperature per se, Manuel Banet used rats implanted with cooling devices and infected with *Salmonella enteritidis*. First, Banet (1979) implanted cooling devices in the brains of rats and used them to chill the hypothalamus. This induced the infected rats to develop extra high fevers, without dramatically elevating their metabolic rates. The extra-high-fever rats survived the bacterial infection at much lower rates than control rats. Second, Banet (1981a) cooled the spinal cords of infected rats. This induced the rats to greatly increase their metabolic rates, while preventing them from elevating their body temperatures. The high-metabolic-rate rats survived the infections at somewhat higher rates than normal-metabolic-rate rats. Finally, Banet (1981b) carefully monitored the body temperatures and metabolic rates of a group of infected rats, some of which had implants, but none of which were heated or cooled. Banet found that the rats that ran the highest fevers had the lowest rates of survival, but that the rats that showed the highest metabolic rates had the highest rates of survival. Together, Banet's results suggest that moderate fever is beneficial to infected rats, that the benefits of fever may be associated not so much with elevated temperature itself as with increased metabolic rate or other effects on the immune system, and that high fever in rats is deleterious to survival.

It is unclear how the results with iguanas and rats might apply to humans. Fewer clinical studies have been done on this question than one might expect (Kluger 1992; Green and Vermeulen 1994). We review three studies, none of which is conclusive.

Fever and Chickenpox

Timothy Doran and colleagues (1989) studied 68 children with chickenpox. After getting the informed consent of the children's parents, the researchers divided the children at random into two groups. The experimental group took acetaminophen, a common over-the-counter antifever medicine. The control group took a placebo (pills that looked like acetaminophen, but contained no medicine). The assignment of children to study groups was double blind: Neither the researchers nor the parents (nor the children) knew which children were in which group until after the study was over.

By most measures of the duration and severity of illness, there was no difference between the acetaminophen and placebo groups. Where the results hint at a difference, the children taking the placebo recovered faster. It took less time, on average, for all of the vesicles to scab over in the placebo children (5.6 ± 2.5 days) than in the acetaminophen children (6.7 ± 2.3 days). By itself, this result is statistically significant at $P = 0.048$; but given that several other measures failed to turn up any difference, it is not persuasive. (Remember that if we do 20 statistical tests, one of them is likely to be significant at $P < 0.05$ simply due to chance.) The children's itching seemed to subside faster in the placebo group (Figure 19.19), but this pattern is not statistically significant.

The simple interpretation is that the antifever medicine acetaminophen has little or no effect on the course of chickenpox, and therefore that fever is neither adaptive nor maladaptive as a defense against the disease. Kluger (1992) points out, however, that only slightly over half of the children in the study ran a fever (defined by Doran et al. as a temperature of 100.4°F or higher) and that the fraction of children who did run a fever was the same in the acetaminophen group (57%) and the placebo group (55%). Kluger's interpretation is that the acetaminophen children were not given enough medicine to reduce fever, and thus that the study did not test the hypothesis that fever is adaptive. Kluger's analysis illustrates that it

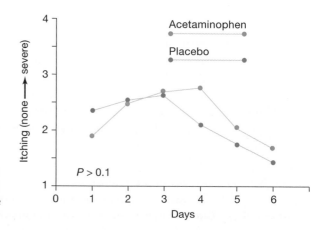

Figure 19.19   Does the antifever medicine acetaminophen have any effect on the course of chickenpox?   The graph shows the severity of itching as a function of time (as judged by the children's parents) in 37 children taking acetaminophen versus 31 children taking a placebo. The placebo group appears to have recovered faster, but the difference is not significant. From Doran et al. (1989).

is important to critically examine a study's methods and results before accepting the conclusions.

Fever and the Common Cold

Neil Graham and colleagues (1990) intentionally infected 56 consenting adult volunteers with rhinovirus type 2, one of the viruses that can cause the common cold. Assigned to groups double blind and at random, 14 subjects took a placebo. The rest took common over-the-counter antifever medications: 13 took ibuprofen, 15 took aspirin, and 14 took acetaminophen. The volunteers taking the placebo suffered less nose stuffiness (Figure 19.20a) and made more antibodies against the rhinovirus (Figure 19.20b) than did the volunteers taking antifever medicines. The reason for the reduced antibody response in volunteers taking medicine may be that the medicines prevented monocytes, a class of white blood cells, from

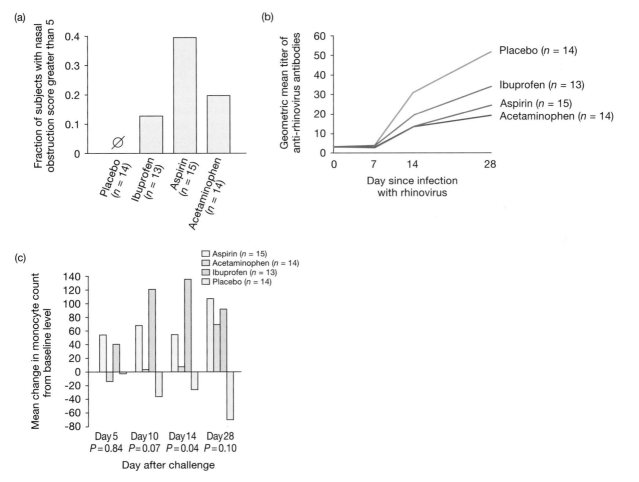

Figure 19.20 Do antifever medicines have any effect on the course of the common cold? (a) Volunteers taking a placebo had less-stuffy noses (fewer had "nasal obstruction scores" over 5) than volunteers taking one of three antifever medicines ($P = 0.02$). (b) Volunteers taking a placebo made more antibodies against the rhinovirus that caused their cold than did volunteers taking one of three antifever medicines. On day 28, the difference between the placebo group and the other three groups combined was significant at $P = 0.03$. (c) Monocytes are white blood cells that circulate to infected tissues, then leave the blood and differentiate into macrophages. Volunteers taking a placebo showed a drop in the concentration of monocytes in their blood over time (indicating that the cells had moved into the tissues), whereas volunteers taking one of three antifever medicines showed an increase in the concentration of monocytes. From Graham et al. (1990). Copyright © 1990, Journal of Infectious Diseases. Reprinted by permission of The University of Chicago Press.

moving from the blood to the infected tissues (Figure 19.20c). Once in the infected tissues, monocytes differentiate into macrophages, which help mount an immune response against the virus (Graham et al. 1990).

This time, the simple interpretation is that the antifever medications interfered with the immune response to the common cold, and therefore that fever is an adaptive defense against the disease. Kluger (1992) points out, however, that few of the subjects in the study ran a fever. Futhermore, the fraction of subjects taking the placebo who did run a fever (14%) was not significantly higher than the fraction of subjects taking medicine who ran a fever (7%). Kluger's interpretation is that few people infected with rhinovirus type 2 run a fever, and thus that the study did not test the hypothesis that fever is adaptive.

The study by Graham and colleagues did show, however, that antifever medicines interfered with the immune response to the virus. This demonstrates that antifever medicines have multiple physiological effects. As we noted above, this fact makes it extremely difficult to design studies on the adaptive significance of fever in mammals. Studies using traditional antifever medicines cannot separate fever from other aspects of the immune response.

Fever and Sepsis

Gordon Bernard and colleagues (1997) studied human patients with severe systemic bacterial infections, or sepsis. They randomly assigned 455 patients to receive either ibuprofen or a placebo, in addition to standard treatments for sepsis. The ibuprofen group had significantly lower body temperatures and metabolic rates. However, both groups had about the same mortality rate, roughly 40%. Because all of the patients in the study were already gravely ill when they were enrolled, and because virtually all of them received aggressive treatment with antibiotics and other medications, it is difficult to interpret the results with regard to our two evolutionary hypotheses about fever.

Fever and Medical Practice

Even if researchers find clear evidence that fever in humans is, indeed, an adaptive response to some infections, no responsible doctor (or evolutionary biologist) would suggest that it is always a bad idea to suppress fever. There are several reasons:

- It may be that fever is an adaptive response against some pathogens, but not against others. Some bacteria or viruses may grow and reproduce faster at fever temperatures than at normal temperatures. In other words, the adaptive-response and pathogen-manipulation-of-host hypotheses may be mutually exclusive for any particular pathogen, but they are not mutually exclusive across all pathogens.

- Even when fever is beneficial, it carries costs as well (Nesse and Williams 1994). In the case of mild illness and low fever, sometimes the benefits of antifever medicines in alleviating symptoms and allowing people to continue their normal activities outweigh the costs of a somewhat diminished immune response. In the case of serious illness and high fever, fever itself can deplete nutrient reserves and can even cause temporary or permanent tissue damage.

- There are circumstances in which fever may cause damage directly, unconnected with its role in infections. For example, experiments with animals and observational studies of humans suggest that fever following a stroke causes neurological damage and reduces the likelihood of survival (Azzimondi et al. 1995).

More research is needed on the adaptive significance of fever in humans, and on the costs and benefits of using various antifever medications.

19.7 Adaptation and Human Behavior: Parenting

In using the adaptationist approach to understand the behavior of humans, evolutionary psychologists adopt the view that the brain is an organ whose properties as a regulator of behavior have been shaped by natural selection. As a regulator of behavior, the brain is a flexible machine, not a computer slavishly converting input to output according to some fixed program. The human brain runs on a complex mix of conscious and unconscious perception, emotion, experience, and calculation, in pursuit of a variety of goals. But in the view of evolutionary psychologists,

> The ultimate objective of our conspicuously purposive physiology and psychology is not longevity or pleasure or self-actualization or health or wealth or peace of mind. It is fitness. Our appetites and ambitions and intellects and revulsions exist because of their historical contributions to this end. Our perceptions of self-interest have evolved as proximal tokens of expected gains and losses of fitness, "expected" being used here in its statistical sense of what would be anticipated on average from the cumulative evidence of the past (Daly and Wilson 1988a, page 10).

This adaptationist approach to human behavior requires caution. In its capacity as a regulator of behavior, the human brain is influenced by culture as well as by evolutionary history. Culture evolves by its own set of rules (see Box 19.2). Furthermore, culture can manifestly induce individuals to behave in ways contrary to the interests of their genetic fitness. The mass suicide of 39 members of the Heaven's Gate cult in March of 1997, for example, defies adaptationist explanation.

The influence of culture on human behavior means that studies on the behavior of people within a single society cannot disentangle the effects of culture from those of evolutionary history. To make a plausible claim that a psychological trait or pattern of behavior is a product of natural selection, evolutionary psychologists must show that the trait or behavior pattern is broadly cross-cultural. Cross-cultural diversity has fallen dramatically during the last century. All but the most remote and isolated traditional societies have been contacted, and western ideas and artifacts have spread virtually everywhere (see Diamond 1992). Some evolutionary biologists feel that it is no longer possible to conduct a genuine cross-cultural study. Others feel that cross-cultural studies are still worth pursuing, particularly when new findings are combined with information extracted from databases of earlier anthropological research.

Another caveat for the study of human behavior is one we have already discussed. The environments most humans live in today are strikingly different from the environments all humans lived in for most of our evolutionary history. From the time of the earliest members of the genus *Homo*, over 2 million years ago (see Chapter 16), until the advent of agriculture, at about 8000 B.C., all humans lived in small groups and made their living by hunting and gathering. The rapid pace of cultural change in the last ten thousand years has been largely too fast for genetic

BOX 19.2 Is cultural evolution Darwinian?

A treatment of the mechanisms of cultural evolution is beyond the scope of this chapter. In fact, the mechanisms of cultural evolution are probably beyond the scope of evolutionary biology altogether.

Richard Dawkins, in his 1976 book *The Selfish Gene,* suggested that we might develop a theory of cultural evolution by natural selection that works exactly like our theory of biological evolution. Central to this suggestion is the idea that natural selection is a generalizable process. Natural selection works on organisms because organisms have four key features: mutation, reproduction, inheritance, and differential reproductive success. In principle, natural selection should operate on any class of entities that have the same four properties.

Dawkins noted that elements of culture have these four properties, and thus should evolve by natural selection. A new word, song, idea, or style is analogous to a new allele created by mutation. The Shakers' austere and beautiful style of furniture design, for example, is an element of culture. A new piece of culture reproduces when other people adopt it and pass it on, as when a woodworker admires a Shaker table, then goes home to her shop and imitates the design. Some pieces of culture are more successful than others at getting themselves transmitted from person to person. For example, Shaker-style furniture has achieved much wider adoption than Shaker-style lifelong celibacy. Culture evolves as the relative frequencies of styles and ideas change.

Dawkins coined the term *meme* for the fundamental unit of cultural evolution. He saw the meme as analogous to the gene, the fundamental unit of biological evolution. Dawkins envisioned a detailed theory of population memetics that would be similar to the theory of population genetics we covered in Chapters 4 through 7. [For a more recent exposition on the potential explanatory power of this idea, see Dennett (1995).]

The trouble with Dawkins's suggestion, noted by Dawkins himself (see also the 1989 edition of his book), is that the effectiveness of natural selection as a mechanism of evolution depends not just on the property of inheritance, but also on the details of how inheritance works. This fact was first recognized by Fleeming Jenkin, one of Darwin's critics, in 1867. In Darwin's time, the prevailing model of inheritance involved the blending in the offspring of infinitely divisible particles

contributed by the parents. Jenkin pointed out that blending inheritance undermines evolution by natural selection because of the fate it implies for new variations. In a sexual population with blending inheritance, any new variation would quickly vanish, like a single drop of black paint dissolving into a bucket of white. Mendelian genetics rescues Darwin's theory, because Mendelian inheritance is particulate. Genes do not blend. A new recessive mutation can remain hidden in a population for generations. Eventually, the mutant allele may reach a high enough frequency that heterozygotes start to mate with each other, producing among their offspring a few homozygous recessives.

In its correct form, then, the generalizable theory of evolution by natural selection applies to any class of entities that has the properties of mutation, reproduction, *particulate* inheritance, and differential reproductive success. The crucial question for the theory of cultural evolution by natural selection is whether memes are transmitted by particulate or blending inheritance. As Allen Orr (1996) puts it, "Do street fashion and high fashion segregate like good genes, or do they first mix before replicating in magazines or storefronts?" Nobody knows. If memes are transmitted by blending inheritance, then natural selection is, at best, a weak mechanism of cultural evolution. We need other mechanisms to explain cultural evolution—perhaps mechanisms entirely different from those responsible for biological evolution.

Although biological evolution and cultural evolution may proceed by different mechanisms, this does not mean that either is irrelevant to the other. Cultural evolution can set the stage for biological evolution. Most humans, for example, stop producing the enzyme lactase in childhood, but the cultural practice of dairy farming led to the evolution of lifelong lactase production in many human populations (Durham 1991). Likewise, biological evolution can influence cultural evolution. For example, the division of the visible light spectrum into verbally distinguished colors follows cross-culturally universal patterns (Durham 1991). These patterns are determined by the way our eyes and brains encode visual information, indicating that the structure of our nervous systems has constrained cultural variation in color terminology. Cultural and biological evolution are distinct but interdependent.

evolution to keep up. As a result, it is of little use to ask why natural selection would have produced human behaviors, such as a willingness to ski down a mountainside at 75 miles per hour, that can only occur in a modern context. So long as we are careful to allow for our incomplete understanding of the hunter–gatherer lifestyle, however, it may make sense to ask why natural selection would have built into us a desire for the social rewards that we can attain, under the right circumstances, through dramatic demonstrations of superior athleticism and bravery. Scientists pursuing such an inquiry would formulate and test hypotheses about how a recipient of these social rewards, living in a hunter–gatherer society, might convert them into reproductive success.

Evolution and Parenthood

We now explore evolutionary psychology by considering aspects of parenthood. We begin with a prediction. On the assumption that the psychology of parenting has been shaped by natural selection, evolutionary psychologists predict that human adults should direct more of their parental caregiving to their own genetic offspring than to the genetic offspring of others. We would, in fact, make the same prediction about any organism that provides parental care. Parental care is expensive to the caregiver, and caregivers who reserve their efforts for their own genetic young should enjoy higher lifetime reproductive success than caregivers who are indiscriminate. The generality of this prediction gives us confidence that it is legitimate for human hunter–gatherers. And the prediction has hidden subtleties, as an animal example will show.

An evolutionary perspective can help researchers develop hypotheses about patterns of human behavior.

Reed buntings (*Emberiza shoeniclus*) are small ground-nesting birds in which both males and females provide parental care. Most nesting pairs are socially monogamous, meaning that each member of the pair tends no other nest than the one they tend together. Genetic testing by Andrew Dixon and colleagues (1994) revealed that there is more to the reed bunting mating system than meets the eye. The researchers found that 55% of all chicks were sired by males other than their mother's social mate, and that 86% of all nests included at least one such chick. Dixon and colleagues predicted that males, if they could tell what fraction of the chicks they had sired in any given nest, would adjust their parental effort accordingly. Dixon et al. tested this prediction by looking at the chick-feeding behavior of 13 pairs of buntings that raised two clutches of chicks in a single season. The males fed the chicks more often in the nest in which they had sired a higher proportion of the chicks (Figure 19.21a). The females, who were the genetic mothers of all the chicks in both nests, showed no such pattern (Figure 19.21b).

We have presented the reed bunting example because evolutionary biologists using Darwinism to understand human behavior are often accused of genetic determinism (see, for example, Lewontin 1980). Genetic determinism is the notion that fundamental characteristics of human societies are unchangeably programmed into our genes. Note, however, the sense in which genes do and do not determine the parental behavior of male reed buntings. A male's genotype does not specify a particular level of parental care that the male will provide at all times, no matter what. Instead, each male's genotype specifies a range of phenotypic plasticity in parental care (see Chapter 8). That is, the bird's brain has a mechanism that adjusts the effort the male expends in caring for a brood, based on cues that indicate his probable level of paternity in that brood. If a male's social or biological environment changes, he alters his level of parental care accordingly, as Figure 19.21a

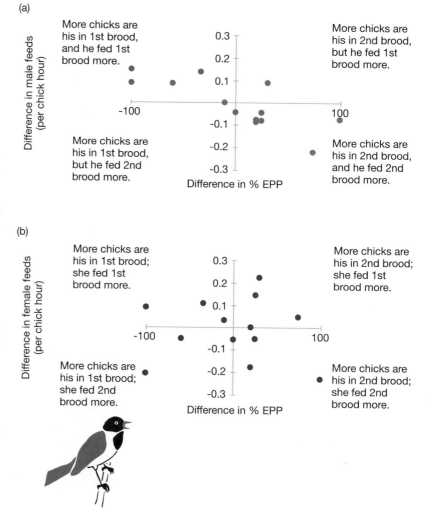

Figure 19.21 Male reed buntings adjust their parental effort depending on who they are feeding (a) Each dot represents a single male who raised two broods of chicks in a season. The horizontal axis represents the difference between the two broods in the percentage of extra-pair paternity (% EPP), or fraction of chicks sired by some other male. The vertical axis represents the difference between how frequently the male fed chicks in the first brood and how frequently he fed chicks in the second brood. Most males fed chicks more often in the nest in which they had sired a higher fraction of chicks. This association was significant at $P = 0.0064$. (b) Each dot represents a single female who raised two broods of chicks in a season. The horizontal axis represents the difference between the percentage of chicks in the first brood sired by extra-pair males and the percentage of chicks in the second brood sired by extra-pair males (% EPP). The vertical axis represents the difference between how frequently the female fed chicks in the first brood and how frequently she fed chicks in the second brood. Females showed no relationship between the number of feedings they provided and the relative number of chicks sired by extra-pair males. Redrawn from Dixon et al. (1994).

shows. The pattern of phenotypic plasticity in a trait is called the trait's **reaction norm**. Reaction norms for reed bunting parental care presumably vary from male to male—or at least did vary in ancestral populations. It is this genetic variation in reaction norms that provides the raw material for the evolution of parental behavior. The average reaction norm of today's reed buntings appears to be adaptive, and might be described as "reed bunting nature."

Evolutionary psychologists studying human behavior are likewise interested in phenotypic plasticity—that is, reaction norms. They recognize that human reaction norms allow wide latitude for social and environmental circumstances to modify human behavior, and they recognize that reaction norms vary from person to person. What evolutionary psychologists do is formulate and test hypotheses about average human reaction norms.

Are humans as discriminating as male reed buntings in adjusting their provision of parental care? The question is difficult to study directly, at least in modern western cultures, in which most of the interactions between parents and children take place in private. Other cultures, however, are more amenable to study.

Mark Flinn (1988) conducted an extensive and detailed observational study of the interactions between parents and offspring in a small village in rural Trinidad.

Flinn interviewed all the residents of the village to determine which people were genetically related and which people were living together. Then, once or twice a day for six months, Flinn walked a standard route through the village that took him within 20 meters of every house and public building. He started at a different, randomly determined point each day, so as not to regularly pass particular places at particular times. Each time he saw any one of the village's 342 residents, Flinn recorded what the person was doing, who he or she was with, and the nature of the interaction they were having. The houses and buildings in the village are all rather open, so Flinn was able to see much that went on indoors as well as out.

Fourteen of the village's 112 households included mothers that were the genetic mothers of all the resident children, and fathers that were the genetic fathers of some of the resident children and the stepfathers of others. These 14 families included 28 genetic offspring and 26 stepoffspring of the fathers. There can be no hidden differences between the genetic fathers and the stepfathers, because the genetic fathers and the stepfathers are the same men.

Flinn calculated the amount of time the fathers spent with their children, and the fraction of their interactions with their children that were agonistic. An agonistic interaction was one that "involved physical or verbal combat (e.g., spanking or arguing) or expressions of injury inflicted by another individual (e.g., screaming in pain or anguish or crying)" (Flinn 1988). Note that, overall, only 6% of the parent–offspring interactions Flinn saw were agonistic, and 94% of these involved only verbal exchanges. During his study, Flinn was not aware of any interactions between parents and children that would be considered physical child abuse. ("Screaming in pain or anguish" may sound like evidence of physical child abuse, but anyone who has spent time with a two-year-old knows that this is not necessarily so.) In other words, Flinn's research concerns parent–offspring interactions that most anyone would consider normal.

Flinn found that the 14 fathers with both genetic offspring and stepoffspring spent more of their time with their genetic offspring (Figure 19.22a). Furthermore,

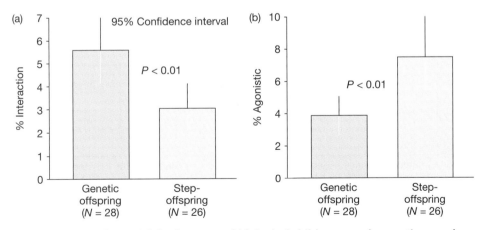

Figure 19.22 Fathers with both step- and biological children spend more time, and get along better, with their biological children (a) The bars show the fraction of their time (% of all observed interactions) that 14 fathers spent with their biological children versus their stepchildren. The 95% confidence interval is an estimate of the certainty that Flinn's estimated percentage is close to the truth. Roughly speaking, we can be 95% certain that the true number is within the 95% confidence interval. (b) The bars show the fraction of interactions between 14 fathers and their children that were agonistic (see text for definition). From Flinn (1988).

a smaller fraction of the father–genetic offspring interactions were agonistic (Figure 19.22b). These results are consistent with the prediction that parents discriminate among children on the basis of their genetic relationship with those children.

This is an observational study, however, and there is a potentially confounding variable. The pattern in Flinn's data could be explained by the late arrival of the stepfathers in the lives of their stepchildren. Men might feel less affection and concern for their stepchildren simply because they joined the family when the stepchildren were older, whereas the men were already in the family when all of their own genetic children were born.

Flinn's data set for this village, remarkably, includes 23 stepchildren who were born when their mothers and stepfathers were already living together, as well as 11 stepchildren born before their mothers and stepfathers moved in together. (This sample includes all the stepfathers in the village, not just those who also have genetic children living in the same house.) If a man's parental affection is simply a function of the fraction of the child's life during which the man has lived with the child, then the stepfathers who had lived with their stepchildren from the children's births should be more affectionate. In fact, the opposite appears to be true. The stepfathers in Flinn's study spent more of their time, and had a lower fraction of agonistic interactions, with stepchildren born before the stepfathers joined the family (Figure 19.23).

We noted earlier that studies within a single society offer no means of disentangling the influences of culture and evolutionary history. We could argue that the pattern of discrimination revealed in Flinn's study is simply a product of culture, and has nothing to do with our species' adaptive history. Evidence is accumulating, however, that parental discrimination between own and others' genetic offspring is a cross-cultural phenomenon. For example, Kim Hill and Hillard Kaplan (1988) studied the survival of biological children versus stepchildren in the Ache Indians, a traditional foraging culture in Paraguay. Hill and Kaplan found that 81% of children raised by both biological parents survived to their 15th birthday, whereas only 57%

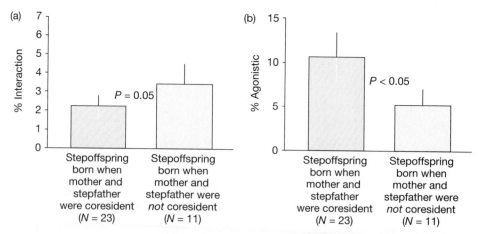

Figure 19.23 **Stepfathers spend more time with their stepchildren, and get along better with them, when the stepchildren are born before the stepfather joins the family**
(a) Fraction of time (% of all observed interactions) that stepfathers spent with their stepchildren, with the data separated by whether the stepfather was living in the house when the stepchild was born. (b) Fraction of interactions between stepfathers and their stepchildren that were agonistic (see text for definition). From Flinn (1988).

of children raised by one biological parent and one stepparent survived. Napoleon Chagnon (1992; see also 1988) studied the Yąnomamö Indians, a traditional hunting, gathering, and gardening culture in Venezuela and Brazil. The Yąnomamö are polygynous, which means that women have little trouble finding husbands, but men often have difficulty finding wives. Chagnon reports that men work harder to find wives for their biological sons than for their stepsons. Frank Marlowe (1999) studied Hadza hunter–gatherers in Tanzania. He found that compared to stepfathers, genetic fathers spend more time near their children, and play with them, talk with them, and nurture them more. Kermyt Anderson and colleagues (1999) studied modern American men living in Albuquerque, New Mexico. They found that men invest more toward the college education of their genetic children than that of their stepchildren.

Parental Discrimination and Children's Health

Discrimination by parents against stepchildren becomes a public-health issue when we consider its impact on the childrens' physiological state. Mark Flinn and Barry England (1995, 1997), working in another rural Caribbean village, this time in Dominica, gave chewing gum to children, and then asked the children to provide saliva samples. In the saliva samples, the researchers measured the concentration of cortisol. Cortisol is a hormone that animals produce when they are under stress. Over the short term, high levels of cortisol cause an animal to divert resources to immediate demands, for example by increasing metabolic rate and alertness and inhibiting growth and reproduction. Over the long term, chronically high levels of cortisol can inhibit the immune system, deplete energy stores, and induce social withdrawal.

Flinn and England found that among the children in the village they studied, individuals with relatively high concentrations of cortisol in their saliva were sick more often (Figure 19.24a). Not surprisingly, it was the stepchildren who had the highest concentrations of cortisol (Figure 19.24b) and higher frequencies of illness (Figure 19.24c). Ultimately, stepchildren had lower reproductive success during early adulthood (Figure 19.24d) and were more likely to leave town.

Martin Daly and Margo Wilson approached the public-health consequences of parental discrimination by analyzing case files of homicides in which parents killed their children (Daly and Wilson 1988a; see also Daly and Wilson 1988b; 1994a; 1994b). Daly and Wilson predicted that children would be killed at a higher rate by stepparents than by biological parents.

Data on murders of children in Canada dramatically confirm Daly and Wilson's prediction: Stepparents kill stepchildren at a much higher rate than biological parents kill biological children (Figure 19.25).

It is worth discussing this result a bit. In absolute numbers (that is, simply counting up homicides) more children are killed by biological parents than by stepparents (341 versus 67 in Daly and Wilson's study). But this is because only a small minority of children have stepparents. This is especially true for young children, the most common victims of parental homicide. In 1984, only 0.4% of Canadian children one to four years old lived with a stepparent. To adjust for the fact that few young children live with stepparents, Daly and Wilson reported the data in Figure 19.25 as rates: the number of homicides per million child–years that parents or stepparents and children spend living together. Epidemiologists often summarize the

Researchers investigating Darwinian predictions have discovered patterns of human behavior with profound consequences for public health.

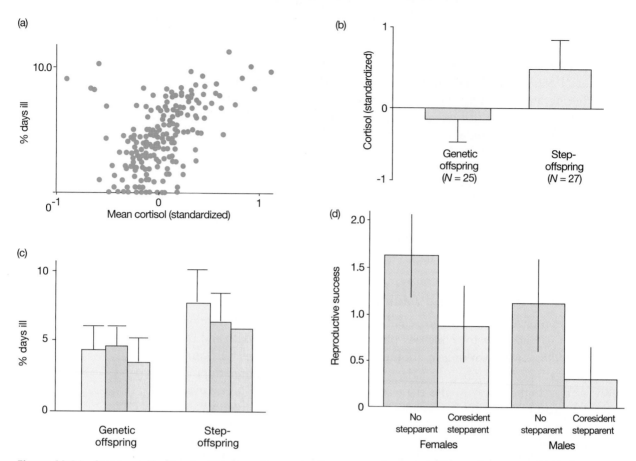

Figure 19.24 **Stress, cortisol levels, illness, and reproductive success for stepchildren versus genetic children**
(a) Children with higher levels of the stress hormone cortisol in their blood get sick more often. Negative numbers indicate lower than average cortisol concentrations; positive numbers indicate higher than average concentrations. (b) Stepoffspring have higher concentrations of cortisol in their blood than do biological children. (a, b) From Flinn and England (1995). Copyright © 1995, Current Anthropology. Reprinted by permission of The University of Chicago Press. (c) The difference in health between biological children and stepchildren is larger than the differences attributable to socioeconomic status (different colors indicate different levels of status). From Flinn and England (1997). Reprinted by permission of Wiley-LIss, Inc., a subsidiary of John Wiley & Sons, Inc. (d) Biological children have higher reproductive success during early adulthood (ages 18–28 for women, 20–30 for men) than do stepchildren. From Flinn (1988). Copyright © 1988, Elsevier Science. Reprinted with permission of Elsevier Science.

results of such a study by reporting a relative risk. Here, the relative risk of homicide in stepchildren versus biological children is the rate at which stepparents kill stepchildren divided by the rate at which biological parents kill biological children. For children zero to two years old, the relative risk of parental homicide for

Figure 19.25 **The risk to children of being killed by a biological parent versus a stepparent**
The graphs show, for biological parents (left) and stepparents (right), the rate at which parents killed children (number of homicides per million child-years that parents and children spent living in the same house). Children aged two or younger are killed by stepparents at a rate about 70 times higher than such children are killed by biological parents. The data are for homicides in Canada, 1974 through 1983. From Daly and Wilson (1988a; 1988b). Copyright © 1988, Aldine de Gruyter. Reprinted by permission of Aldine de Gruyter.

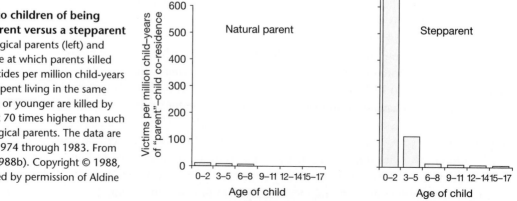

stepchildren versus biological children is about 70. This is an extraordinarily high relative risk. For comparison, the relative risk of lung cancer in smokers versus nonsmokers is about 11.

Daly and Wilson do not suggest that killing stepchildren, in and of itself, is or ever was adaptive for humans. Anyone who kills someone else's child, even in a traditional hunter–gatherer society, is likely to suffer social penalties that outweigh any potential benefits of eliminating an unwelcome demand for stepparental investment. Instead, what Daly and Wilson suggest is adaptive is the combination of two traits: (1) an intellectual and psychological apparatus that perceives a personal interest in the distinction between one's own and others' genetic offspring, and (2) the emotional motivation to turn this perception into active discrimination between the two kinds of children. Whenever such an apparatus exists, individuals will, rarely, commit errors of excess. These errors of excess become Daly and Wilson's data.

Daly and Wilson's data come from an observational study in which it was impossible to control, as Flinn (1988) was able to, for differences between biological parents and stepparents. Nonetheless, they provide an argument that research conducted within a Darwinian framework can yield insights useful to public-health workers and providers of social services.

Summary

Evolutionary biology has numerous applications in medicine. This chapter has considered two general ways in which evolutionary analysis improves our understanding of issues relating to human health. First, we used our knowledge of the mechanisms of evolution to study pathogens and tumors. Second, we used the methods of the adaptationist program to address questions about human physiology and behavior.

Pathogens and their hosts are locked in a perpetual evolutionary arms race. Our immune systems and the drugs we take impose strong selection on the viruses and bacteria that infect our tissues. Because pathogens have short generation times, large population sizes, and often high mutation rates, populations of viruses and bacteria evolve quickly. Phylogenetic analysis helps us reconstruct the history of pathogen evolution, and in the case of flu, understand some of the mechanisms that create pathogen strains capable of causing epidemics. Selection thinking also helps us predict when pathogen populations will become resistant to drugs, whether drug resistance will persist in pathogen populations if drug use is suspended, and what makes some diseases virulent and others benign.

Humans, like other organisms, are a product of evolution by natural selection. As a result, the tools of the adaptationist program can help us understand aspects of our own form and function. Selection thinking suggests, for example, that symptoms of disease, such as fever, may be adaptive facets of our immune response. And aspects of our behavior with significant public-health consequences, such as cross-culturally consistent patterns in the way we treat children, may be interpretable as psychological adaptations. It is also important to keep in mind that change in our environment in recent centuries has far outpaced the rate of adaptive evolution. Modern epidemics of myopia and breast cancer may be the result of our exposure to novel environments.

Recognition that evolutionary biology is a medical science has, in some respects, been slow in arriving. We expect that interactions between evolutionary biologists and medical researchers will become more frequent and more productive in the years to come.

Questions

1. Some biologists regard our bodies as small ecosystems that exert selective pressure for the evolution of invasive metastatic cancer. If this is true, why don't we all get cancer? (*Hint:* Consider the speed of evolution vs. human

lifespan.) However, these same biologists believe that humans have certain genes that have evolved specifically to prevent cancer. How is it possible to have both strong selection for cancer, and strong selection for anticancer genes? (*Hint:* Consider evolution within one person's body vs. evolution within a population of many people.)

2. We have seen how the genetic diversity within a tumor can be used to estimate the tumor's age (see Figure 19.14). This analysis depends on mutation rate per cell division being constant. If a cancerous tumor has evolved a high mutation rate, how will this bias the results? Do the genetic markers we use to estimate a tumor's age need to be selectively neutral? Why or why not?

3. Pathogens require a minimum population size of potential hosts. If the host population is too small, in a short time the entire population has either been killed by the pathogen, or has survived the initial infection and become immune. If this occurs, the pathogen dies out. What evolutionary changes in a pathogen might increase its ability to survive in a smaller population? For example, measles requires a host population of about 500,000 humans, while diphtheria can get by with only about 50,000 humans. Develop some hypotheses for why diphtheria can survive with just one-tenth the number of hosts. For example, how might these two diseases differ from each other in transmission rate, virulence, latency to infection, or mutation rate?

4. In the experiment illustrated in Figure 19.18b, why was it important to give one group of lizards a saline injection?

5. Suppose that your 2-year-old child is ill, and has a dangerously high fever of 106°F (about 41°C). You and your doctor decide to put your child on an antifever medication. Would you want your child's temperature to return all the way to normal, or to stabilize at a "mild fever" level (say, 101°F, or about 38.5°C)? If you feel that you don't have enough information to make this important decision, what further information would you want? (That is, what further studies should be done?)

6. Review the studies on fever that were presented in this chapter. Do you agree with Kluger that several of the studies did not really test the adaptive fever hypothesis? If so, can you design an experiment that will truly test the hypothesis? Is your experiment ethical?

7. The male reed buntings in Dixon et al.'s study (Figure 19.21) seem to be consciously aware of genetic relationships and "trying" to increase their reproductive success. Can evolution cause reed buntings (and other animals) to behave as if they are aware of the evolutionary consequences of their actions, without actually being aware of those consequences? Do you think the same could be true of some human behaviors?

8. Daly and Wilson's data on infanticide risks might be explained by stepfathers having, on average, more violent personalities than biological fathers. Could this "violent personality" explanation also apply to Flinn's data from the Trinidad village? Why or why not? Daly and Wilson's study involved general data about a large number of families, whereas Flinn's study involved detailed data on a small number of families. What are the advantages and disadvantages of each kind of study?

9. An evolutionary biologist once hypothesized that if evolution has affected human social behavior, then a mother's brothers should take a particular interest in her children—more so than the father's brothers, and perhaps even more so than the father himself. Why did he hypothesize this? (As it turns out, there are many cultures in which men do, in fact, direct parental care primarily toward their sisters' children.)

10. In 1999, a mysterious outbreak of human encephalitis occurred in the northeastern United States. The cause was tentatively identified as St. Louis encephalitis virus. At the same time, an unusual number of dead birds were noticed along the northeastern Atlantic coast. Figure 19.26 shows genetic relationships of three known encephalitis viruses (St. Louis, Japanese, and West Nile) with viruses isolated from the birds, from two human patients that died, from one dead horse, and from mosquitoes. (Data compiled from Anderson et al. 1999 and Lanciotti et al. 1999.)

   Were the birds, the horse, and the humans all suffering from the same disease?

   Do you think the outbreak was caused by St. Louis encephalitis virus?

   Does this cladogram suggest how the disease might be spread?

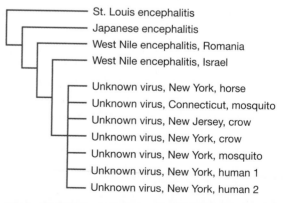

Figure 19.26 A phylogeny of encephalitis viruses isolated from various hosts in the northeastern United States during 1999 Based on data and analyses in Anderson et al. (1999) and Lanciotti et al. (1999).

Exploring the Literature

11. For another study on the effects of antifever medication on human illness, see

Sugimura, T., T. Fujimoto, H. Motoyama, T. Maruoka, and S. Korematu, et al. 1994. Risks of antipyretics in young children with fever due to infectious disease. *Acta Paediatrica Japonica* 36: 375–378.

How strong is the evidence in this paper that fever is an adaptive response to bacterial respiratory infection? Consider that acetaminophen affects aspects of the immune response other than fever ("Fever and the Common Cold" in Section 19.6). Also consider that Sugimura and colleagues conducted an observational study, not an experimental one. That is, the researchers did not randomly assign their subjects to acetaminophen versus placebo groups. Instead, the researchers asked parents to keep a diary of the number of doses of acetaminophen they gave their children.

12. In many countries, meat producers routinely feed antibiotics to livestock to promote growth. The reasons why antibiotics promote growth are unclear, but this use accounts for a large fraction of global antibiotic production. For strong circumstantial evidence that antibiotics fed to livestock select for resistant bacteria that later infect people, see

Wegener, H. C., et al. 1999. Use of antimicrobial growth promoters in food animals and *Enterococcus faecium* resistance to therapeutic antimicrobial drugs in Europe. *Emerging Infectious Diseases* 5: 329–325. Available from URL http://www.cdc.gov/ncidod/EID/eid.htm

13. A crucial question in deciding whether modern women should use hormonal treatments that suppress menstruation is whether menstruation itself is adaptive. The issue is controversial. See

Profet, M. 1993. Menstruation as a defense against pathogens transported by sperm. *Quarterly Review of Biology* 68: 335–381.

Strassmann, B. I. 1996. The evolution of endometrial cycles and menstruation. *Quarterly Review of Biology* 71: 181–220.

Citations

Anderson, J. F., T. G. Andreadis, et al. 1999. Isolation of West Nile virus from mosquitoes, crows, and a Cooper's hawk in Connecticut. *Science* 286: 2331–2333.

Anderson, K., H. Kaplan, and J. Lancaster. 1999. Paternal care by genetic fathers and stepfathers I: Reports from Albuquerque men. *Evolution and Human Behavior* 405–431.

Austin, D. J., K. G. Kristinsson, and R. M. Anderson. 1999. The relationship between the volume of antimicrobial consumption in human communities and the frequency of resistance. *Proceedings of the National Academy of Sciences, USA:* 96: 1152–1156.

Azzimondi, G., L. Bassein, F. Nonino, L. Fiorani, L. Vignatelli, et al. 1995. Fever in acute stroke worsens prognosis: A prospective study. *Stroke* 26: 2040–2043.

Banet, M. 1979. Fever and survival in the rat. *Pflügers Archiv* 381: 35–38.

Banet, M. 1981a. Fever and survival in the rat. The effect of enhancing the cold defense response. *Experientia* 37: 985–986.

Banet, M. 1981b. Fever and survival in the rat. Metabolic rate versus temperature response. *Experientia* 37: 1302–1304.

Baquero, F., and J. Blázquez. 1997. Evolution of antibiotic resistance. *Trends in Ecology and Evolution* 12: 482–487.

Bean, W. J., Schell, M., et al. 1992. Evolution of the H3 influenza virus hemagglutinin from human and nonhuman hosts. *Journal of Virology* 66: 1129–1138.

Berkey, C. S., A. L. Frazier et al. 1999. Adolescence and breast carcinoma risk. *Cancer* 85: 2400–2409.

Bernard, G. R., A. P. Wheeler, et al. 1997. The effects of ibuprofen on the physiology and survival of patients with sepsis. *New England Journal of Medicine* 336: 912–918.

Bernheim, H. A., and M. J. Kluger. 1976. Fever: Effect of drug-induced antipyresis on survival. *Science* 193: 237–239.

Bishai, W. R., N. M. H. Graham, et al. 1996. Rifampin-resistant tuberculosis in a patient receiving rifabutin prophylaxis. *New England Journal of Medicine* 334: 1573–1576.

Blaese, R. M., K. W. Culver, A. D. Miller, C. S. Carter, T. Fleisher, et al. 1995. T. lymphocyte-directed gene therapy for ADA-SCID: Initial trial results after 4 years. *Science* 270: 475–480.

Bloch, A. B., G. M. Cauthen, et al. 1994. Nationwide survey of drug-resistant tuberculosis in the United States. *Journal of the American Medical Association* 271: 665–671.

Bordignon, C., L. D. Notarangelo, N. Nobili, G. Ferrari, and G. Casorati, et al. 1995. Gene therapy in peripheral blood lymphocytes and bone marrow for ADA-immunodeficient patients. *Science* 270: 470–575.

Bull, J. J. 1994. Virulence. *Evolution* 48: 1423–1437.

Bush, R. M., C. A. Bender, et al. 1999. Predicting the evolution of human influenza A. *Science* 286: 1921–1925.

Bush, R. M., W. M. Fitch, et al. 1999. Positive selection on the H3 hemagglutinin gene of human influenza virus A. *Molecular Biology and Evolution* 16: 1457–1465.

Chagnon, N. A. 1988. Male manipulations of kinship classifications of female kin for reproductive advantage. In L. Betzig, M. Borgerhoff Mulder, and P. Turke, eds. *Human Reproductive Behavior: A Darwinian Perspective.* Cambridge: Cambridge University Press, 23–48.

Chagnon, N. A. 1992. *Yąnomamö.* Fort Worth: Harcourt Brace College Publishers.

Cordain, L., R. W. Gotshall, and S. B. Eaton. 1997. Evolutionary aspects of exercise. *World Review of Nutrition and Dietetics* 81: 49–60.

Daly, M., and M. Wilson. 1984. A sociobiological analysis of human infanticide. In G. Hausfater and S. B. Hrdy, eds. *Infanticide: Comparative and Evolutionary Perspectives.* New York: Aldine de Gruyter, 487–502.

Daly, M., and M. Wilson. 1988a. *Homicide.* New York: Aldine de Gruyter.

Daly, M., and M. Wilson. 1988b. Evolutionary social psychology and family homicide. *Science* 242: 519–524.

Daly, M., and M. Wilson. 1989. Evolution and family homicide, reply. *Science* 243: 463–464.

Daly, M., and M. I. Wilson. 1994a. Some differential attributes of lethal assaults on small children by stepfathers versus genetic fathers. *Ethology and Sociobiology* 15: 207–217.

Daly, M., and M. Wilson. 1994b. Stepparenthood and the evolved psychology of discriminative parental solicitude. In S. Parmigiani and F. S. vom Saal, eds. *Infanticide and Parental Care.* London: Harwood Academic Publishers, 121–133.

Dawkins, R. 1976, 1989. *The Selfish Gene,* 1st and 2nd ed. Oxford: Oxford University Press.

Dennett, D. C. 1995. *Darwin's Dangerous Idea.* New York: Simon and Schuster.

Diamond, J. 1992. *The Third Chimpanzee.* New York: Harper Collins.

Dixon, A., D. Ross, S. L. C. O'Malley, and T. Burke. 1994. Paternal investment inversely related to degree of extra-pair paternity in the reed bunting. *Nature* 371: 698–700.

Doran, T. F., C. De Angelis, R. Baumgardner, and E. D. Mellits. 1989. Acetaminophen: More harm than good for chickenpox? *Journal of Pediatrics* 114: 1045–1048.

Durham, W. H. 1991. *Coevolution: Genes, culture, and human diversity.* Stanford University Press, Stanford.

Eaton, S. B., and L. Cordain. 1997. Evolutionary aspects of diet: Old genes, new fuels. *World Review of Nutrition and Dietetics* 81: 26–37.

Eaton, S. B., S. B. Eaton III, and M. J. Konner. 1997. Paleolithic nutrition revisited: A twelve-year retrospective on its nature and implications. *European Journal of Clinical Nutrition* 51: 207–216.

Ewald, P. W. 1991. Waterborne transmission and the evolution of virulence among gastrointestinal bacteria. *Epidemiology and Infection* 106: 83–119.

Ewald, P. W. 1993. The evolution of virulence. *Scientific American* (April): 86–93.

Ewald, P. W. 1994. *Evolution of Infectious Disease.* Oxford: Oxford University Press.

Fitch, W. M., J. M. Leiter, et al. 1991. Positive Darwinian evolution in human influenza A viruses. *Proceedings of the National Academy of Sciences, USA* 88: 4270–4274.

Flinn, M. V. 1988. Step- and genetic-parent–offspring relationships in a Caribbean village. *Ethology and Sociobiology* 9: 335–369.

Flinn, M. V., and B. G. England. 1995. Childhood stress and family environment. *Current Anthropology* 36: 854–866

Flinn, M. V., and B. G. England. 1997. Social economics of childhood glucocorticoid stress response. *American Journal of Physical Anthropology* 102: 33–53.

Gorman, O. T., W. J. Bean, et al. 1991. Evolution of influenza A virus nucleoprotein genes: Implications for the origins of *H1N1* human and classical swine viruses. *Journal of Virology* 65: 3704–3714.

Graham, N. M. H., C. J. Burrell, R. M. Douglas, P. Debelle, and L. Davies. 1990. Adverse effects of aspirin, acetaminophen, and ibuprofen on immune function, viral shedding, and clinical status in rhinovirus-infected volunteers. *Journal of Infectious Diseases* 162: 1277–1282.

Green, M. H., and C. W. Vermeulen. 1994. Fever and the control of gram-negative bacteria. *Research in Microbiology* 145: 269–272.

Henderson, B. E., P. K. Ross, and M. C. Pike. 1993. Hormonal chemoprevention of cancer in women. *Science* 259: 633–638.

Hershfield, M. S., R. H. Buckley, M. L. Greenberg, A. L. Melton, L. Schiff, et al. 1987. Treatment of adenosine deaminase deficiency with polyethylene glycol-modified adenosine deaminas. *New England Journal of Medicine* 316: 589–596.

Hill, K., and H. Kaplan. 1988. Trade-offs in male and female reproductive strategies among the Ache, part 2. In L. Betzig, M. Borgerhoff Mulder, and P. Turke, eds. *Human Reproductive Behavior: A Darwinian Perspective.* Cambridge: Cambridge University Press, 291–305.

Hirschhorn, R., D. R. Yang, et al. 1996. Spontaneous in vivo reversion to normal of an inherited mutation in a patient with adenosine deaminase deficiency. *Nature Genetics* 13: 290–295.

Klug, W. S., and M. R. Cummings. 1997. *Concepts of Genetics,* 5th ed. Upper Saddle River, NJ: Prentice Hall.

Kluger, M. J., D. H. Ringler, and M. R. Anver. 1975. Fever and survival. *Science* 188: 166–168.

Kluger, M. J. 1992. Fever revisited. *Pediatrics* 90: 846–850.

Kolata, Gina. 1999. *Flu.* New York. Farrar, Straus, and Giroux.

Lanciotti, R. S., J. T. Roehrig, et al. 1999. Origin of West Nile virus responsible for an outbreak of encephalitis in the northeastern United States. *Science* 286:2333–2337.

Laver, W. G., N. Bischofberger, and R. G. Webster. 1999. Disarming flu viruses. *Scientific American* 280 (January): 78–87.

Layde P. M., L. A. Webster, et al. 1989. The independent associations of parity, age at first full term pregnancy, and duration of breastfeeding with the risk of breast cancer. *Journal of Clinical Epidemiology* 42: 963–73.

Levin, B. R. 1996. The evolution and maintenance of virulence in microparasites. *Emerging Infectious Diseases* 2: 93–102.

Levy, S. B. 1998. The challenge of antibiotic resistance. *Scientific American* 278 (March): 46–53.

Lewontin, R. C. 1980. Sociobiology: Another biological determinism. *International Journal of Health Services* 10: 347–363.

Marlowe, F. 1999. Showoffs or providers? The parenting effort of Hadza men. *Evolution and Human Behavior* 20: 391–404.

Messenger, S. L., I. J. Molineux, and J. J. Bull. 1999. Virulence evolution in a virus obeys a trade off. *Proceedings of the Royal Society, London,* Series B 266: 397–404.

Nesse, R. M., and G. C. Williams. 1994. *Why We Get Sick.* New York: Vintage Books.

Nesse, R. M., and G. C. Williams. 1998. Evolution and the origins of disease. *Scientific American* (November): 86–93.

Neutel, C. I., and H. Johansen. 1995. Measuring drug effectiveness by default: The case of Bendectin. *Canadian Journal of Public Health* 86: 66–70.

Norton, T. T., and C. Wildsoet. 1999. Toward controlling myopia progression? *Optometry and Vision Science* 76: 341–342.

Orr, H. A. 1996. Dennett's dangerous idea. *Evolution* 50: 467–472.

Rattan, A., A. Kalia, and N. Ahmad. 1998. Multidrug-resistant *Mycobacterium tuberculosis:* Molecular perspectives. *Emerging Infectious Diseases* 4: 195–209.

Schrag, S. J., V. Perrot, and B. R. Levin. 1997. Adaptation to the fitness costs of antibiotic resistance in *Escherichia coli. Proceedings of the Royal Society, London,* Series B 264: 1287–1291.

Shibata, D., W. Navidi, R. Salovaara, Z.-H. Li, and L. A. Aaltonen. 1996. Somatic microsatellite mutations as molecular tumor clocks. *Nature Medicine* 2: 676–681.

Sorsby, A., and F. A. Young. 1970. Transmission of refractive errors within Eskimo families. *American Journal of Optometry* 47: 244–249.

Spicer, D. V., D. Shoupe, and M. C. Pike. 1991. GnRH agonists as contraceptive agents: predicted significantly reduced risk of breast cancer. *Contraception* 44: 289–310.

Strassmann, B. I. 1999. Menstrual cycling and breast cancer: An evolutionary perspective. *Journal of Women's Health* 8: 193–202.

Teikari, J. M., J. O'Donnell, J. Kaprio, and M. Koskenvuo. 1991. Impact of heredity in myopia. *Human Heredity* 41: 151–156.

Tufte, E. R. 1997. *Visual Explanations.* Cheshire, CT: Graphics Press.

Vaughn, L. K., H. A. Bernheim, and M. Kluger. 1974. Fever in the lizard *Dipsosaurus dorsalis. Nature* 252: 473–474.

Webster, R. G., W. J. Bean, et al. 1992. Evolution and ecology of influenza A viruses. *Microbiological Reviews* 56: 152–179.

Williams, G. C., and R. M. Nesse. 1991. The dawn of Darwinian medicine. *The Quarterly Review of Biology* 66: 1–22.

Young, F. A., W. R. Baldwin, et al. 1969. The transmission of refractive errors with Eskimo families. *American Journal of Optometry* 46: 676–685.

Youssoufian, H. 1996. Natural gene therapy and the Darwinian legacy. *Nature Genetics* 13: 255–256.

Glossary

adaptation   A trait that increases the ability of an individual to survive or reproduce compared to individuals without the trait.

adaptive radiation   The divergence of a clade into populations adapted to many different ecological niches.

adaptive trait   A trait that increases the fitness of its bearer.

additive effect   The contribution an allele makes to the phenotype that is independent of the identity of the other alleles at the same or different loci.

additive genetic variation   Differences among individuals in a population that are due to the additive effects of genes.

agent of selection   Any factor that causes individuals with certain phenotypes to have, on average, higher fitness than individuals with other phenotypes.

alleles   Variant forms of a gene, or variant nucleotide sequences at a particular locus.

allopatric model   The hypothesis that speciation occurs when populations become geographically isolated and diverge because selection and drift act on them independently.

allopatry   Living in different geographic areas.

allozymes   Distinct forms of an enzyme, encoded by different alleles at the same locus.

altruism   Behavior which decreases the fitness of the actor and increases the fitness of the recipient.

ancestral   Describes a trait that was possessed by the common ancestor of the species on a branch of an evolutionary tree; used in contrast with derived.

antigenic site   A portion of a protein that is recognized by the immune system and initiates a response.

assortative mating   Occurs when individuals tend to mate with other individuals with the same genotype or phenotype.

back mutation   A mutation that reverses the effect of a previous mutation; typically a mutation that restores function after a loss-of-function mutation.

background extinction   Extinctions that occur during "normal" times, as opposed to during mass extinction events.

best-fit line   The line that most accurately represents the trend of the data in a scatterplot; typically, best-fit lines are calculated by least-squares linear regression.

blending inheritance   The hypothesis that heritable factors blend to produce a phenotype and are passed on to offspring in this blended form.

bootstrapping   In phylogeny reconstruction, a technique for estimating the strength of the evidence that a particular node in a tree exists. Bootstrap values ranges from 0% to 100%, with higher values indicating stronger support.

bottleneck   A large-scale but short-term reduction in population size followed by an increase in population size.

branch (of a phylogenetic tree)   Lines that indicate a specific population or taxonomic group through time.

catastrophism   In geology, the view that most or all landforms are the product of catastrophic events, such as the flood at the time of Noah described in the Bible. See uniformitarianism.

cenancestor   The last common ancestor of all extant organisms.

chromosome inversion   A region of DNA that has been flipped, so that the genes are in reverse order; results in lower rates of crossing-over and thus tighter linkage among loci within the inversion.

clade   The set of species descended from a particular common ancestor; synonymous with monophyletic group.

cladistics   A classification scheme based on the historical sequence of divergence events (phylogeny); also used to identify a method of inferring phylogenies based on the presence of shared derived characters (synapomorphies).

cladogram   An evolutionary tree reflecting the results of a cladistic analysis.

cline   A systematic change along a geographic transect in the frequency of a genotype or phenotype.

clone   An individual that is genetically identical to its parent, or a group of individuals that are genetically identical to each other.

codon   A set of three bases in DNA that specifies a particular amino acid–carrying tRNA.

codon bias   A nonrandom distribution of codons in a DNA sequence.

coefficient of inbreeding (F)   The probability that the alleles at any particular locus in the same individual are identical by descent from a common ancestor.

coefficient of linkage disequilibrium (D)   A calculated value that quantifies the degree to which genotypes at one locus are nonrandomly associated with genotypes at another locus.

coefficient of relatedness (r)   The probability that the alleles at any particular locus in two different individuals are identical by descent from a common ancestor.

coevolution   Occurs when interactions between species over time lead to reciprocal adaptation.

common garden experiment   An experiment in which individuals from different populations or treatments are reared together under identical conditions.

comparative method   A research program that compares traits and environments across taxa and looks for correlations that test hypotheses about adaptation.

confidence interval An indication of the statistical certainty of an estimate; if a study yielding an estimate is done repeatedly, and a 95% confidence interval is calculated for each estimate, the confidence interval will include the true value 95% of the time.

constraint Any factor that tends to slow the rate of adaptive evolution or prevent a population from evolving the optimal value of a trait.

control group A reference group that provides a basis for comparison; in an experiment, the control group is exposed to all conditions affecting the experimental group except one—the potential causative agent of interest.

convergent evolution Similarity between species that is caused by a similar, but evolutionarily independent, response to a common environmental problem.

cryptic species Species which are indistinguishable morphologically, but divergent in songs, calls, odor, or other traits.

Darwinian fitness The extent to which an individual contributes genes to future generations, or an individual's score on a measure of performance expected to correlate with genetic contribution to future generations (such as lifetime reproductive success).

derived Describes a trait that was not possessed by the common ancestor of the species on a branch of an evolutionary tree; an evolutionary novelty; used in contrast with ancestral.

differential success A difference between the average survival, fecundity, or number of matings achieved by individuals with certain phenotypes versus individuals with other phenotypes.

dioecious Describes a species in which male and female reproductive function occurs in separate individuals; usually used with plants.

direct fitness Fitness that is due to the production of offspring. See indirect fitness.

directional selection Occurs when individual fitness tends to increase or decrease with the value of phenotypic trait; can result in steady evolutionary change in the mean value of the trait in the population.

disruptive selection Occurs when individuals with more extreme values of a trait have higher fitness; can result in increased phenotypic variation in a population.

dominance genetic variation Differences among individuals in a population that are due to the nonadditive effects of genes, such as dominance; typically means the genetic variation left over after the additive genetic variation has been taken into account.

drift Synonym for genetic drift.

effective population size (N_e) The size of an ideal random mating population (with no selection, mutation, or migration) that would lose genetic variation via drift at the same rate as is observed in an actual population.

endosymbiosis theory The hypothesis that eukaryotic organelles such as mitochondria and chloroplasts originated when bacteria took up residence inside early eukaryotic cells.

environmental variation Differences among individuals in a population that are due to differences in the environments they have experienced.

epitope The specific part of a protein that is recognized by the immune system and initiates a response. Synonymous with antigenic site and antigenic determinant.

eugenics The study and practice of social control over the evolution of human populations; positive eugenics seeks to increase the frequency of desirable traits, whereas negative eugenics seeks to decrease the frequency of undesirable traits.

eusocial A social system characterized by overlapping generations, cooperative brood care, and specialized reproductive and non-reproductive castes.

evolution Originally defined as descent with modification, or change in the characteristics of populations over time. Currently defined as changes in allele frequencies over time.

evolutionarily stable strategy (ESS) In game theory, a strategy or set of strategies that cannot be invaded by a new, alternative strategy.

evolutionary arms race Occurs when an adaptation in one species (a parasite, for example) reduces the fitness of individuals in a second species (such as a host), thereby selecting in favor of counter-adaptations in the second species. These counter-adaptations, in turn, select in favor of new adaptations in the first species, and so on.

exon A nucleotide sequence that occurs between introns, and that remains in the messenger RNA after the introns have been spliced out.

extant Living today.

fecundity The number of gametes produced by an individual; usually used in reference to the number of eggs produced by a female.

fitness The extent to which an individual contributes genes to future generations, or an individual's score on a measure of performance expected to correlate with genetic contribution to future generations (such as lifetime reproductive success).

fixation The elimination from a population of all the alleles at a locus but one; the one remaining allele, now at a frequency of 1.0, is said to have achieved fixation, or to be fixed.

fossil Any trace of an organism that lived in the past.

fossil record The complete collection of fossils, located in many institutions around the world.

founder effect A change in allele frequencies that occurs after a founder event, due to genetic drift in the form of sampling error in drawing founders from the source population.

founder event The establishment of a new population, usually by a small number of individuals.

founder hypothesis The hypothesis that many speciation events begin when small populations colonize new geographic areas.

frequency The proportional representation of a phenotype, genotype, gamete, or allele in a population; if six out of ten individuals have brown eyes, the frequency of brown eyes is 60%, or 0.6.

frequency-dependent selection Occurs when an individual's fitness depends on the frequency of its phenotype in the population; typically occurs when a phenotype has higher fitness when it is rare, and lower fitness when it is common.

gamete pool The set of all copies of all gamete genotypes in a population that could potentially be contributed by the members of one generation to the members of the next generation.

gene duplication Generation of an extra copy of a locus, usually via unequal crossing-over.

gene family A group of loci related by common descent, and sharing identical or similar function.

gene flow   The movement of alleles from one population to another population, typically via the movement of individuals or via the transport of gametes by wind, water, or pollinators.

gene pool   The set of all copies of all alleles in a population that could potentially be contributed by the members of one generation to the members of the next generation.

genetic distance   A statistic that summarizes the number of genetic differences observed between populations or species.

genetic drift   Change in the frequencies of alleles in a population resulting from sampling error in drawing gametes from the gene pool to make zygotes, and from chance variation in the survival and/or reproductive success of individuals; results in non-adaptive evolution.

genetic load   Reduction in the mean fitness of a population due to the presence of deleterious alleles.

genetic recombination   The placement of allele copies into multilocus genotypes (on chromosomes or within gametes) that are different from the multilocus genotypes they belonged to in the previous generation; results from meiosis with crossing-over and sexual reproduction with outcrossing.

genetic variation   Differences among individuals in a population that are due to differences in genotype.

genotype-by-environment interaction   Differences in the effect of the environment on the phenotype displayed by different genotypes; for example, among people living in the same location some change their skin color with the seasons and others do not.

geologic column   A composite, older-to-younger sequence of rock formations that describes geological events at a particular locality.

geologic time scale   A sequence of eons, eras, periods, epochs, and stages that furnishes a chronology of Earth history.

h²   Symbol for the narrow-sense heritability (see heritability).

habitat tracking   Morphological evolution in response to short-term environmental change that in the long-term results in variation around a mean value. Also called dynamic stasis.

half-life   The time required for half of the atoms of a radioactive material, present at any time, to decay into a daughter isotope.

Hamilton's rule   An inequality that predicts when alleles for altruism should increase in frequency.

haplodiploidy   A reproductive system in which males are haploid and develop from unfertilized eggs, while females are diploid and develop from fertilized eggs.

haplotype   Genotype for a suite of linked loci on a chromosome; typically used for mitochondrial genotypes, because mitochondria are haploid and all loci are linked.

Hardy-Weinberg equilibrium   A situation in which allele and genotype frequencies in an ideal population do not change from one generation to the next, because the population experiences no selection, no mutation, no migration, no genetic drift, and random mating.

heritability   In the broad sense, that fraction of the total phenotypic variation in a population that is caused by genetic differences among individuals; in the narrow sense, that fraction of the total variation that is due to the additive effects of genes.

hermaphroditic   In general, describes a species in which male and female reproductive function occur in the same individual; with plants, describes a species with perfect flowers (that is, flowers with both male and female reproductive function)

heterozygosity   That fraction of the individuals in a population that are heterozygotes.

heterozygote inferiority (underdominance)   Describes a situation in which heterozygotes at a particular locus tend to have lower fitness than homozygotes.

heterozygote superiority (overdominance)   Describes a situation in which heterozygotes at a particular locus tend to have higher fitness than homozygotes.

histogram   A bar chart that represents the variation among individuals in a sample; each bar represents the number of individuals, or the frequency of individuals, with a particular value (or within a particular range of values) for the measurement in question.

hitchhiking   Change in the frequency of an allele due to selection on a closely linked locus. Also called a selective sweep.

homeotic loci   Genes whose products provide positional information in a multicellular embryo.

homology   Similarity between species that results from inheritance of traits from a common ancestor.

homoplasy   Similarity in the characters found in different species that is due to convergent evolution, parallelism, or reversal—not common descent.

hybrid zone   A geographic region where differentiated populations interbreed.

identical by descent   Describes alleles, within a single individual or different individuals, that have been inherited from the same ancestral copy of the allele.

inbreeding   Mating among kin.

inbreeding depression   Reduced fitness in individuals or populations resulting from kin matings; often due to the decrease in heterozygosity associated with kin matings, either because heterozygotes are superior or because homozygotes for deleterious alleles become more common.

inclusive fitness   An individual's total fitness; the sum of its indirect fitness, due to reproduction by relatives made possible by its actions, and direct fitness, due to its own reproduction.

independence (statistical)   Lack of association among data points, such that the value of a data point does not affect the value of any other data point.

indirect fitness   Fitness that is due to increased reproduction by relatives made possible by the focal individual's actions. See direct fitness.

inheritance of acquired characters   The hypothesis that phenotypic changes in the parental generation can be passed on, intact, to the next generation.

interaction   In genetics, occurs when the effect of an allele on the phenotype depends on the other alleles present at the same or different loci; in statistics, occurs when the effect of a treatment depends on the value of other treatments.

intersexual selection   Differential mating success among individuals of one sex due to interactions with members of the other sex; for example, variation in mating success among males due to female choosiness.

intrasexual selection   Differential mating success among individuals of one sex due to interactions with members of the same sex; for example, differences in mating success among males due to male-male competition over access to females.

intron (intervening sequence)   A noncoding stretch of DNA nucleotides that occurs between the coding regions of a gene and must be spliced out after transcription to produce a functional messenger RNA.

iteroparous   Describes a species or population in which individuals experience more than one bout of reproduction over the course of typical lifetime; humans provide an example.

kin recognition   The ability to discern the degree of genetic relatedness of other individuals.

kin selection   Natural selection based on indirect fitness gains.

lateral gene transfer   Transfer of genetic material across species barriers.

life history   An individual's pattern of allocation, throughout life, of time and energy to various fundamental activities, such as growth, repair of cell and tissue damage, and reproduction.

law of succession   The observation that fossil types are succeeded, in the same geographic area, by similar fossil or living species.

lineage   A group of ancestral and descendant populations or species that are descended from a common ancestor. Synonymous with clade.

linkage   The tendency for alleles at different loci on a chromosome to be inherited together. Also called genetic linkage.

linkage (dis)equilibrium   If, within a population, genotypes at one locus are randomly distributed with respect to genotypes at another locus, then the population is in linkage equilibrium for the two loci; otherwise, the population is in linkage disequilibrium.

loss-of-function mutation   A mutation that incapacitates a gene, so that no functional product is produced; also called a forward, knock-out, or null mutation.

macroevolution   Large evolutionary change, usually in morphology; typically refers to the evolution of differences among populations that would warrant their placement in different genera or higher-level taxa.

mass extinction   A large-scale, sudden extinction event that is geographically and taxonomically widespread.

maternal effect   Variation among individuals due to variation in non-genetic influences exerted by their mothers; for example, chicks whose mothers feed them more may grow to larger sizes, and thus be able to feed their own chicks more, even when size is not heritable.

maximum likelihood   In phylogeny inference, a method for choosing a preferred tree among many possible trees. In using maximum likelihood, a researcher asks how likely a particular tree is given a particular data set and a specific model of character change.

Mendelian gene   A locus whose alleles obey Mendel's laws of segregation and independent assortment.

microevolution   Changes in gene frequencies and trait distributions that occur within populations and species.

microtektites   Tiny glass particles created when minerals are melted by the heat generated in a meteorite or asteroid impact.

midoffspring value   The mean phenotype of the offspring within a family.

midparent value   The mean phenotype of an individual's two parents.

migration   In evolution, the movement of alleles from one population to another, typically via the movement of individuals or via the transport of gametes by wind, water, or pollinators.

Modern Synthesis   The broad-based effort, accomplished during the 1930's and 1940's, to unite Mendelian genetics with the theory of evolution by natural selection; also called the Evolutionary Synthesis.

molecular clock   The hypothesis that base substitutions accumulate in populations in a clock-like fashion; that is, as a linear function of time.

monoecious   Typically used for plants, to describe either: (1) a species in which male and female reproductive function occur in the same individual; or (2) a species in which separate male and female flowers are present on the same individual (see also hermaphroditic).

monophyletic group   The set of species (or populations) descended from a common ancestor.

morphology   Structural form, or physical phenotype; also the study of structural form.

morphospecies   Populations that are designated as separate species based on morphological differences.

mutation-selection balance   Describes an equilibrium in the frequency of an allele that occurs because new copies of the allele are created by mutation at exactly the same rate that old copies of the allele are eliminated by natural selection.

mutualism   An interaction between two individuals, typically of different species, in which both individuals benefit.

natural selection   A difference, on average, between the survival or fecundity of individuals with certain phenoypes compared to individuals with other phenotypes.

negative selection   Selection against deleterious mutations. Also called purifying selection.

neutral (mutation)   A mutation that has no effect on the fitness of the bearer.

neutral evolution (neutral theory)   A theory that models the rate of fixation of alleles with no effect on fitness; also associated with the claim that the vast majority of observed base substitutions are neutral with respect to fitness.

node   A point on an evolutionary tree at which a branch splits into two or more sub-branches.

null hypothesis   The predicted outcome, under the simplest possible assumptions, of an experiment or observation; in a test of whether populations are different, the null hypothesis is typically that they are not different, and that apparent differences are due to chance.

null model   The set of simple and explicit assumptions that allows a researcher to state a null hypothesis.

outbreeding   Mating among unrelated individuals.

outgroup   A taxonomic group that diverged prior to the rest of the taxa in a phylogenetic analysis.

overdominance   Describes a situation in which heterozygotes at a particular locus tend to have higher fitness than homozygotes.

P value   An estimate of the statistical support for a claim about a pattern in data, with smaller values indicating stronger support; an estimate of the probability that apparent violations of the null hypothesis are due to chance (see statistically significant).

paleontology   The study of fossil organisms.

paraphyletic group   A set of species that includes a common ancestor and some, but not all, of its descendants.

parental investment   Expenditure of time and energy on the provision, protection, and care of an offspring; more specifically, investment by a parent that increases the fitness of a particular offspring, and reduces the fitness the parent can gain by investing in other offspring.

parsimony   A criterion for selecting among alternative patterns or explanations based on minimizing the total amount of change or complexity.

parthenogenesis   A reproductive mode in which offspring develop from unfertilized eggs.

phenetics   A classification scheme based on grouping populations according to their similarities.

phenotypic plasticity   Variation, under environmental influence, in the phenotype associated with a genotype.

phenotypic variation   The total variation among the individuals in a population.

phyletic transformation   The evolution of a new morphospecies by the gradual transformation of an ancestral species, without a speciation or splitting event taking place. Also called anagenesis.

phylogeny   The evolutionary history of a group.

phylogenetic tree   A diagram (typically an estimate) of the relationships of ancestry and descent among a group of species or populations; in paleontological studies the ancestors may be known from fossils, whereas in studies of extant species the ancestors may be hypothetical constructs. Also called an evolutionary tree

point mutation   Alteration of a single base in a DNA sequence.

polyandry   A mating system in which at least some females mate with more than one male.

polygyny   A mating system in which at least some males mate with more than one female.

polymorphism   The existence within a population of more than one variant for a phenotypic trait, or of more than one allele.

polyphyletic group   A set of species that are grouped by similarity, but not descended from a common ancestor.

polyploid   Having more than two haploid sets of chromosomes.

polytomy   A node, or branch point, on a phylogeny with more than two descendent lineages emerging.

population   For sexual species, a group of interbreeding individuals and their offspring; for asexual species, a group of individuals living in the same area.

population genetics   The branch of evolutionary biology responsible for investigating processes that cause changes in allele and genotype frequencies in populations.

positive selection   Selection in favor of advantageous mutations.

postzygotic isolation   Reproductive isolation between populations caused by dysfunctional development or sterility in hybrid forms.

preadaptation   A trait that changes due to natural selection and acquires a new function.

prezygotic isolation   Reproductive isolation between populations caused by differences in mate choice or timing of breeding, so that no hybrid zygotes are formed.

primordial form   The first organism; the first entity capable of: (1) replicating itself through the directed chemical transformation of its environment, and (2) evolving by natural selection.

proximate causation   Explanations for how, in terms of physiological or molecular mechanisms, traits function.

pseudogene   DNA sequences that are homologous to functioning genes, but are not transcribed.

qualitative trait   A trait for which phenotypes fall into discrete categories (such as affected versus unaffected with cystic fibrosis).

quantitative trait   A trait for which phenotypes do not fall into discrete categories, but instead show continuous variation among individuals; a trait determined by the combined influence of the environment and many loci of small effect. See qualitative trait.

radiometric dating   Techniques for assigning absolute ages to rock samples, based on the ratio of parent to daughter radioactive isotopes present.

reaction norm   The pattern of phenotypic plasticity exhibited by a genotype.

reciprocal altruism   An exchange of fitness benefits, separated in time, between two individuals.

recombination rate (r)   The frequency, during meiosis, of crossing over between two linked loci; ranges from 0 to 0.5.

reinforcement   Natural selection that results in assortative mating in recently diverged populations in secondary contact; also known as reproductive character displacement.

relative dating   Techniques for assigning relative ages to rock strata, based on assumptions about the relationships between newer and older rocks.

relative fitness   The fitness of an individual, phenotype, or genotype compared to others in the population; can be calculated by dividing the individual's fitness by either (1) the mean fitness of the individuals in the population, or (2) the highest individual fitness found in the population; method (1) must be used when calculating the selection gradient.

replacement substitution   A DNA substitution that changes the amino acid or RNA sequence specified by a gene. Also called a nonsynonymous substitution.

reproductive success (RS)   The number of viable, fertile offspring produced by an individual.

response to selection (R)   In quantitative genetics, the difference between the mean phenotype of the offspring of the selected individuals in a population and the mean phenotype of the offspring of all the individuals.

retrotransposons   Transposable elements that move via an RNA intermediate and contain the coding sequence for reverse transcriptase; closely related to retroviruses.

retrovirus   An RNA virus whose genome is reverse transcribed, to DNA, by reverse transcriptase.

reversal   An event that results in the reversion of a derived trait to the ancestral form.

ribozyme   An RNA molecule that has the ability to catalyze a chemical reaction.

root   The location on a phylogeny of the common ancestor of a clade.

selection   Synonym for natural selection.

selection coefficient   A variable used in population genetics to represent the difference in fitness between one genotype and another.

686 Glossary

selection differential (S) A measure of the strength of selection used in quantitative genetics; equal to the difference between the mean phenotype of the selected individuals (for example, those that survive to reproduce) and the mean phenotype of the entire population.

selection gradient A measure of the strength of selection used in quantitative genetics; for selection on a single trait it is equal to the slope of the best-fit line in a scatterplot showing relative fitness as a function of phenotype.

selectionist theory The viewpoint that natural selection is responsible for a significant percentage of substitution events observed at the molecular level.

selfishness An interaction between individuals that results in a fitness gain for one individual and a fitness loss for the other.

semelparous Describes a species or population in which individuals experience only one bout of reproduction over the course of typical lifetime; salmon provide an example.

senescence A decline with age in reproductive performance, physiological function, or probability of survival.

sexual dimorphism A difference between the phenotypes of females versus males within a species.

sexual selection A difference, among members of the same sex, between the average mating success of individuals with a particular phenotype versus individuals with other phenotypes.

significant In scientific discussions, typically a synonym for statistically significant.

silent substitution A DNA substitution that does not change the amino acid or RNA sequence specified by the gene. Also called a synonymous substitution.

sister species The species that diverged from the same ancestral node on a phylogenetic tree.

species Groups of interbreeding populations that are evolutionarily independent of other populations.

spite Behavior which decreases the fitness of both the actor and the recipient.

stabilizing selection Occurs when individuals with intermediate values of a trait have higher fitness; can result in reduced phenotypic variation in a population, and can prevent evolution in the mean value of the trait.

standard deviation A measure of the variation among the numbers in a list; equal to the square root of the variance (see variance).

standard error The likely size of the error, due to chance effects, in an estimated value, such as the average phenotype for a population.

statistically significant Describes a claim for which there is a degree of evidence in the data; by convention, a result is considered statistically significant if the probability is less than or equal to 0.05 that the observed violation of the null hypothesis is due to chance effects.

substitution Fixation of a new mutation. This entails the replacement of an existing allele.

sympatric Living in the same geographic area.

synapomorphy A shared, derived character; in a phylogenetic analysis, synapomorphies are used to define clades and distinguish them from outgroups.

systematics A scientific field devoted to the classification of organisms.

taxon Any named group of organisms, (the plural form is taxa).

Theory of Evolution by Natural Selection The hypothesis that descent with modification is caused in large part by the action of natural selection.

tip (of a phylogenetic tree) The ends of the branches on a phylogenetic tree, which represent extinct or living taxa.

trade-off An inescapable compromise between one trait and another.

transition In DNA, a mutation that substitutes a purine for a purine, or a pyrimidine for a pyrimidine.

transitional form A species that exhibits traits common to ancestral and derived groups, especially when the groups are sharply differentiated.

transposable elements Any DNA sequence capable of transmitting itself or a copy of itself to a new location in the genome.

transposons Transposable elements that move via a DNA intermediate, and contain insertion sequences along with a transposase enzyme and possibly other coding sequences.

transversion In DNA, a mutation that substitutes a purine for a pyrimidine, or a pyrimidine for a purine.

ultimate causation Explanations for why, in terms of fitness benefits, traits evolved.

underdominance Describes a situation in which heterozygotes at a particular locus tend to have lower fitness than homozygotes.

unequal cross-over A crossing-over event between mispaired DNA strands that results in the duplication of sequences in some daughter strands and deletions in others.

uniformitarianism The assumption (sometimes called a "law") that processes identical to those at work today are responsible for events that occurred in the past; first articulated by James Hutton, the founder of modern geology.

variance A measure of the variation among the numbers in a list; to calculate the variance of a list of numbers, first square the difference between each number and the mean of the list, then take the sum of the squared differences and divide it by the number of items in the list. (For technical reasons, when researchers calculate the variance for a sample of individuals, they usually divide the sum of the squared differences by the sample size minus one).

vestigial traits (or structures) Rudimentary traits that are homologous to fully functional traits in closely related species.

vicariance Splitting of a population's former range into two or more isolated patches.

virulence The damage inflicted by a pathogen on its host; occurs because the pathogen extracts energy and nutrients from the host, and because the pathogen produces toxic metabolic wastes.

wild type A phenotype or allele common in nature.

Index

Note: Pages in **bold** locate figures; pages in *italic* locate tables.